Universitext

For other titles in this series, go to
www.springer.com/series/223

Haim Brezis

Functional Analysis, Sobolev Spaces and Partial Differential Equations

 Springer

Haim Brezis
Distinguished Professor
Department of Mathematics
Rutgers University
Piscataway, NJ 08854
USA
brezis@math.rutgers.edu

and

Professeur émérite, Université Pierre et Marie Curie (Paris 6)

and

Visiting Distinguished Professor at the Technion

ISBN 978-0-387-70913-0   ISBN 978-0-387-70914-7
DOI 10.1007/978-0-387-70914-7
Springer New York Dordrecht Heidelberg London

Library of Congress Control Number: 2010938382

Mathematics Subject Classification (2010): 35Rxx, 46Sxx, 47Sxx

Springer is part of Springer Science+Business Media (www.springer.com)

*To Felix Browder, a mentor and close friend,
who taught me to enjoy PDEs through the
eyes of a functional analyst*

Preface

This book has its roots in a course I taught for many years at the University of Paris. It is intended for students who have a good background in real analysis (as expounded, for instance, in the textbooks of G. B. Folland [2], A. W. Knapp [1], and H. L. Royden [1]). I conceived a program mixing elements from two distinct "worlds": functional analysis (FA) and partial differential equations (PDEs). The first part deals with abstract results in FA and operator theory. The second part concerns the study of spaces of functions (of one or more real variables) having specific differentiability properties: the celebrated Sobolev spaces, which lie at the heart of the modern theory of PDEs. I show how the abstract results from FA can be applied to solve PDEs. The Sobolev spaces occur in a wide range of questions, in both pure and applied mathematics. They appear in linear and nonlinear PDEs that arise, for example, in differential geometry, harmonic analysis, engineering, mechanics, and physics. They belong to the toolbox of any graduate student in analysis.

Unfortunately, FA and PDEs are often taught in separate courses, even though they are intimately connected. Many questions tackled in FA originated in PDEs (for a historical perspective, see, e.g., J. Dieudonné [1] and H. Brezis–F. Browder [1]). There is an abundance of books (even voluminous treatises) devoted to FA. There are also numerous textbooks dealing with PDEs. However, a synthetic presentation intended for graduate students is rare. and I have tried to fill this gap. Students who are often fascinated by the most abstract constructions in mathematics are usually attracted by the elegance of FA. On the other hand, they are repelled by the never-ending PDE formulas with their countless subscripts. I have attempted to present a "smooth" transition from FA to PDEs by analyzing first the simple case of one-dimensional PDEs (i.e., ODEs—ordinary differential equations), which looks much more manageable to the beginner. In this approach, I expound techniques that are possibly too sophisticated for ODEs, but which later become the cornerstones of the PDE theory. This layout makes it much easier for students to tackle elaborate higher-dimensional PDEs afterward.

A previous version of this book, originally published in 1983 in French and followed by numerous translations, became very popular worldwide, and was adopted as a textbook in many European universities. A deficiency of the French text was the lack of exercises. The present book contains a wealth of problems. I plan to add even more in future editions. I have also outlined some recent developments, especially in the direction of nonlinear PDEs.

Brief user's guide

1. Statements or paragraphs preceded by the bullet symbol • **are extremely important**, and it is essential to grasp them well in order to understand what comes afterward.
2. Results marked by the star symbol ⋆ **can be skipped** by the beginner; they are of interest only to advanced readers.
3. In each chapter I have labeled propositions, theorems, and corollaries in a continuous manner (e.g., Proposition 3.6 is followed by Theorem 3.7, Corollary 3.8, etc.). Only the remarks and the lemmas are numbered separately.
4. In order to simplify the presentation I assume that all vector spaces are over $\mathbb{R}$. Most of the results remain valid for vector spaces over $\mathbb{C}$. I have added in Chapter 11 a short section describing similarities and differences.
5. Many chapters are followed by numerous exercises. Partial solutions are presented at the end of the book. More elaborate problems are proposed in a separate section called "Problems" followed by "Partial Solutions of the Problems." The problems usually require knowledge of material coming from various chapters. I have indicated at the beginning of each problem which chapters are involved. Some exercises and problems expound results stated without details or without proofs in the body of the chapter.

Acknowledgments

During the preparation of this book I received much encouragement from two dear friends and former colleagues: Ph. Ciarlet and H. Berestycki. I am very grateful to G. Tronel, M. Comte, Th. Gallouet, S. Guerre-Delabrière, O. Kavian, S. Kichenassamy, and the late Th. Lachand-Robert, who shared their "field experience" in dealing with students. S. Antman, D. Kinderlehrer, and Y. Li explained to me the background and "taste" of American students. C. Jones kindly communicated to me an English translation that he had prepared for his personal use of some chapters of the original French book. I owe thanks to A. Ponce, H.-M. Nguyen, H. Castro, and H. Wang, who checked carefully parts of the book. I was blessed with two extraordinary assistants who typed most of this book at Rutgers: Barbara Miller, who is retired, and now Barbara Mastrian. I do not have enough words of praise and gratitude for their constant dedication and their professional help. They always found attractive solutions to the challenging intricacies of PDE formulas. Without their enthusiasm and patience this book would never have been finished. It has been a great pleasure, as

ever, to work with Ann Kostant at Springer on this project. I have had many oppor-
tunities in the past to appreciate her long-standing commitment to the mathematical
community.

The author is partially supported by NSF Grant DMS-0802958.

Haim Brezis
Rutgers University
March 2010

Contents

Chapter 1
The Hahn–Banach Theorems. Introduction to the Theory of Conjugate Convex Functions

1.1 The Analytic Form of the Hahn–Banach Theorem: Extension of Linear Functionals

Let E be a vector space over $\mathbb{R}$. We recall that a *functional* is a function defined on E, or on some subspace of E, *with values in* $\mathbb{R}$. The main result of this section concerns the extension of a linear functional defined on a linear subspace of E by a linear functional defined on all of E.

Theorem 1.1 (Helly, Hahn–Banach analytic form). *Let* $p : E \to \mathbb{R}$ *be a function satisfying*[1]

(1) $\qquad\qquad p(\lambda x) = \lambda p(x) \qquad \forall x \in E \quad and \quad \forall \lambda > 0,$

(2) $\qquad\qquad p(x + y) \le p(x) + p(y) \quad \forall x, y \in E.$

Let $G \subset E$ *be a linear subspace and let* $g : G \to \mathbb{R}$ *be a linear functional such that*

(3) $\qquad\qquad\qquad g(x) \le p(x) \quad \forall x \in G.$

Under these assumptions, there exists a linear functional f *defined on all of* E *that extends* g, *i.e.,* $g(x) = f(x) \,\forall x \in G$, *and such that*

(4) $\qquad\qquad\qquad f(x) \le p(x) \quad \forall x \in E.$

The proof of Theorem 1.1 depends on Zorn's lemma, which is a celebrated and very useful property of ordered sets. Before stating Zorn's lemma we must clarify some notions. Let P be a set with a (partial) order relation $\le$. We say that a subset $Q \subset P$ is *totally ordered* if for any pair (a, b) in Q either $a \le b$ or $b \le a$ (or both!). Let $Q \subset P$ be a subset of P; we say that $c \in P$ is an *upper bound* for Q if $a \le c$ for every $a \in Q$. We say that $m \in P$ is a *maximal* element of P if there is *no* element

---

[1] A function p satisfying (1) and (2) is sometimes called a *Minkowski functional*.

H. Brezis, *Functional Analysis, Sobolev Spaces and Partial Differential Equations*, DOI 10.1007/978-0-387-70914-7_1, © Springer Science+Business Media, LLC 2011

$x \in P$ such that $m \leq x$, except for $x = m$. Note that a maximal element of P need not be an upper bound for P.

We say that P is *inductive* if every totally ordered subset Q in P has an upper bound.

• **Lemma 1.1 (Zorn).** *Every nonempty ordered set that is inductive has a maximal element.*

Zorn's lemma follows from the axiom of choice, but we shall not discuss its derivation here; see, e.g., J. Dugundji [1], N. Dunford–J. T. Schwartz [1] (Volume 1, Theorem 1.2.7), E. Hewitt–K. Stromberg [1], S. Lang [1], and A. Knapp [1].

Remark 1. Zorn's lemma has many important applications in analysis. It is a *basic tool* in proving some *seemingly innocent existence statements* such as "every vector space has a basis" (see Exercise 1.5) and "on any vector space there are nontrivial linear functionals." Most analysts do not know how to prove Zorn's lemma; but it is quite essential for an analyst to understand the statement of Zorn's lemma and to be able to use it properly!

Proof of Lemma 1.2. Consider the set

$$
P = \left\{ h : D(h) \subset E \to \mathbb{R} \, \middle| \, \begin{array}{l} D(h) \text{ is a linear subspace of } E, \\ h \text{ is linear, } G \subset D(h), \\ h \text{ extends } g, \text{ and } h(x) \leq p(x) \quad \forall x \in D(h) \end{array} \right\}.
$$

On P we define the order relation

$$
(h_1 \leq h_2) \Leftrightarrow (D(h_1) \subset D(h_2) \text{ and } h_2 \text{ extends } h_1).
$$

It is clear that P is nonempty, since $g \in P$. We claim that P is *inductive*. Indeed, let $Q \subset P$ be a totally ordered subset; we write Q as $Q = (h_i)_{i \in I}$ and we set

$$
D(h) = \bigcup_{i \in I} D(h_i), \quad h(x) = h_i(x) \quad \text{if } x \in D(h_i) \text{ for some } i.
$$

It is easy to see that the definition of h makes sense, that $h \in P$, and that h is an upper bound for Q. We may therefore apply Zorn's lemma, and so we have a maximal element f in P. We claim that $D(f) = E$, which completes the proof of Theorem 1.1.

Suppose, by contradiction, that $D(f) \neq E$. Let $x_0 \notin D(f)$; set $D(h) = D(f) + \mathbb{R}x_0$, and for every $x \in D(f)$, set $h(x + tx_0) = f(x) + t\alpha \, (t \in \mathbb{R})$, where the constant $\alpha \in \mathbb{R}$ will be chosen in such a way that $h \in P$. We must ensure that

$$
f(x) + t\alpha \leq p(x + tx_0) \quad \forall x \in D(f) \quad \text{and} \quad \forall t \in \mathbb{R}.
$$

In view of (1) it suffices to check that

$$\begin{cases} f(x) + \alpha \leq p(x + x_0) & \forall x \in D(f), \\ f(x) - \alpha \leq p(x - x_0) & \forall x \in D(f). \end{cases}$$

In other words, we must find some α satisfying

$$\sup_{y \in D(f)} \{f(y) - p(y - x_0)\} \leq \alpha \leq \inf_{x \in D(f)} \{p(x + x_0) - f(x)\}.$$

Such an α exists, since

$$f(y) - p(y - x_0) \leq p(x + x_0) - f(x) \quad \forall x \in D(f), \quad \forall y \in D(f);$$

indeed, it follows from (2) that

$$f(x) + f(y) \leq p(x + y) \leq p(x + x_0) + p(y - x_0).$$

We conclude that $f \leq h$; but this is impossible, since f is maximal and $h \neq f$.

We now describe some simple applications of Theorem 1.1 to the case in which E is a *normed vector space* (n.v.s.) with norm $\| \ \|$.

Notation. We denote by $E^\star$ the *dual space* of E, that is, the space of all *continuous linear functionals on E*; the (dual) *norm on $E^\star$* is defined by

$$(5) \qquad \|f\|_{E^\star} = \sup_{\substack{\|x\| \leq 1 \\ x \in E}} |f(x)| = \sup_{\substack{\|x\| \leq 1 \\ x \in E}} f(x).$$

When there is no confusion we shall also write $\|f\|$ instead of $\|f\|_{E^\star}$.

Given $f \in E^\star$ and $x \in E$ we shall often write $\langle f, x \rangle$ instead of $f(x)$; we say that $\langle \ , \ \rangle$ is the *scalar product for the duality $E^\star, E$*.

It is well known that $E^\star$ is a Banach space, i.e., $E^\star$ is complete (even if E is not); this follows from the fact that $\mathbb{R}$ is complete.

• **Corollary 1.2.** *Let $G \subset E$ be a linear subspace. If $g : G \to \mathbb{R}$ is a continuous linear functional, then there exists $f \in E^\star$ that extends g and such that*

$$\|f\|_{E^\star} = \sup_{\substack{x \in G \\ \|x\| \leq 1}} |g(x)| = \|g\|_{G^\star}.$$

Proof. Use Theorem 1.1 with $p(x) = \|g\|_{G^\star} \|x\|$. ∎

• **Corollary 1.3.** *For every $x_0 \in E$ there exists $f_0 \in E^\star$ such that*

$$\|f_0\| = \|x_0\| \ and \ \langle f_0, x_0 \rangle = \|x_0\|^2.$$

Proof. Use Corollary 1.2 with $G = \mathbb{R}x_0$ and $g(tx_0) = t\|x_0\|^2$, so that $\|g\|_{G^\star} = \|x_0\|$. ∎

Remark 2. The element f_0 given by Corollary 1.3 is in general not unique (try to construct an example or see Exercise 1.2). However, if $E^\star$ is strictly convex[2]—for example if E is a Hilbert space (see Chapter 5) or if $E = L^p(\Omega)$ with $1 < p < \infty$ (see Chapter 4)—then f_0 is unique. In general, we set, for every $x_0 \in E$,

$$F(x_0) = \left\{ f_0 \in E^\star; \; \|f_0\| = \|x_0\| \text{ and } \langle f_0, x_0 \rangle = \|x_0\|^2 \right\}.$$

The (multivalued) map $x_0 \mapsto F(x_0)$ is called the *duality map* from E into $E^\star$; some of its properties are described in Exercises 1.1, 1.2, and 3.28 and Problem 13.

• **Corollary 1.4.** *For every $x \in E$ we have*

$$(6) \qquad\qquad \|x\| = \sup_{\substack{f \in E^\star \\ \|f\| \leq 1}} |\langle f, x \rangle| = \max_{\substack{f \in E^\star \\ \|f\| \leq 1}} |\langle f, x \rangle|.$$

Proof. We may always assume that $x \neq 0$. It is clear that

$$\sup_{\substack{f \in E^\star \\ \|f\| \leq 1}} |\langle f, x \rangle| \leq \|x\|.$$

On the other hand, we know from Corollary 1.3 that there is some $f_0 \in E^\star$ such that $\|f_0\| = \|x\|$ and $\langle f_0, x \rangle = \|x\|^2$. Set $f_1 = f_0/\|x\|$, so that $\|f_1\| = 1$ and $\langle f_1, x \rangle = \|x\|$.

Remark 3. Formula (5)—which is a *definition*—should not be confused with formula (6), which is a *statement*. In general, the "sup" in (5) is *not achieved*; see, e.g., Exercise 1.3. However, the "sup" in (5) is achieved if E is a reflexive Banach space (see Chapter 3); a deep result due to R. C. James asserts the converse: if E is a Banach space such that for every $f \in E^\star$ the sup in (5) is achieved, then E is reflexive; see, e.g., J. Diestel [1, Chapter 1] or R. Holmes [1].

1.2 The Geometric Forms of the Hahn–Banach Theorem: Separation of Convex Sets

We start with some preliminary facts about hyperplanes. In the following, E denotes an n.v.s.

Definition. An affine *hyperplane* is a subset H of E of the form

$$H = \{x \in E \; ; \; f(x) = \alpha\},$$

where f is a linear functional[3] that does not vanish identically and $\alpha \in \mathbb{R}$ is a given constant. We write $H = [f = \alpha]$ and say that $f = \alpha$ is the equation of H.

---

[2] A normed space is said to be *strictly convex* if $\|tx + (1-t)y\| < 1, \forall t \in (0,1), \forall x, y$ with $\|x\| = \|y\| = 1$ and $x \neq y$; see Exercise 1.26.

[3] We do not assume that f is continuous (in every infinite-dimensional normed space there exist discontinuous linear functionals; see Exercise 1.5).

Proposition 1.5. *The hyperplane $H = [f = \alpha]$ is closed if and only if f is continuous.*

Proof. It is clear that if f is continuous then H is closed. Conversely, let us assume that H is closed. The complement H^c of H is open and nonempty (since f does not vanish identically). Let $x_0 \in H^c$, so that $f(x_0) \neq \alpha$, for example, $f(x_0) < \alpha$.

Fix $r > 0$ such that $B(x_0, r) \subset H^c$, where

$$B(x_0, r) = \{x \in E \,;\, \|x - x_0\| < r\}.$$

We claim that

(7) $f(x) < \alpha \quad \forall x \in B(x_0, r).$

Indeed, suppose by contradiction that $f(x_1) > \alpha$ for some $x_1 \in B(x_0, r)$. The segment

$$\{x_t = (1 - t)x_0 + tx_1 \,;\, t \in [0, 1]\}$$

is contained in $B(x_0, r)$ and thus $f(x_t) \neq \alpha$, $\forall t \in [0, 1]$; on the other hand, $f(x_t) = \alpha$ for some $t \in [0, 1]$, namely $t = \frac{f(x_1) - \alpha}{f(x_1) - f(x_0)}$, a contradiction, and thus (7) is proved. It follows from (7) that

$$f(x_0 + rz) < \alpha \quad \forall z \in B(0, 1).$$

Consequently, f is continuous and $\|f\| \leq \frac{1}{r}(\alpha - f(x_0))$.

Definition. Let A and B be two subsets of E. We say that the hyperplane $H = [f = \alpha]$ *separates A and B* if

$$f(x) \leq \alpha \quad \forall x \in A \quad \text{and} \quad f(x) \geq \alpha \quad \forall x \in B.$$

We say that H *strictly separates* A and B if there exists some $\varepsilon > 0$ such that

$$f(x) \leq \alpha - \varepsilon \quad \forall x \in A \text{ and } f(x) \geq \alpha + \varepsilon \quad \forall x \in B.$$

Geometrically, the separation means that A lies in one of the half-spaces determined by H, and B lies in the other; see Figure 1.

Finally, we recall that a subset $A \subset E$ is *convex* if

$$tx + (1 - t)y \in A \quad \forall x, y \in A, \quad \forall t \in [0, 1].$$

• **Theorem 1.6 (Hahn–Banach, first geometric form).** *Let $A \subset E$ and $B \subset E$ be two nonempty convex subsets such that $A \cap B = \emptyset$. Assume that one of them is open. Then there exists a closed hyperplane that separates A and B.*

The proof of Theorem 1.6 relies on the following two lemmas.

Lemma 1.2. *Let $C \subset E$ be an open convex set with $0 \in C$. For every $x \in E$ set*

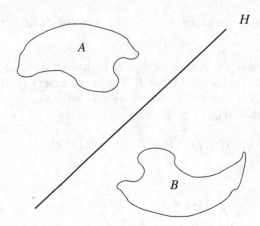

Fig. 1

$$(8) \qquad\qquad p(x) = \inf\{\alpha > 0; \, \alpha^{-1}x \in C\}$$

(p is called the gauge of C or the Minkowski functional of C).

   Then p satisfies (1), (2), *and the following properties:*

(9)   *there is a constant M such that* $0 \le p(x) \le M\|x\| \quad \forall x \in E$,

(10) $C = \{x \in E \,;\, p(x) < 1\}$.

Proof of Lemma 1.2. It is obvious that (1) holds.

Proof of (9). Let $r > 0$ be such that $B(0, r) \subset C$; we clearly have

$$p(x) \le \frac{1}{r}\|x\| \quad \forall x \in E.$$

Proof of (10). First, suppose that $x \in C$; since C is open, it follows that $(1+\varepsilon)x \in C$ for $\varepsilon > 0$ small enough and therefore $p(x) \le \frac{1}{1+\varepsilon} < 1$. Conversely, if $p(x) < 1$ there exists $\alpha \in (0, 1)$ such that $\alpha^{-1}x \in C$, and thus $x = \alpha(\alpha^{-1}x) + (1 - \alpha)0 \in C$.

Proof of (2). Let $x, y \in E$ and let $\varepsilon > 0$. Using (1) and (10) we obtain that $\frac{x}{p(x)+\varepsilon} \in C$ and $\frac{y}{p(y)+\varepsilon} \in C$. Thus $\frac{tx}{p(x)+\varepsilon} + \frac{(1-t)y}{p(y)+\varepsilon} \in C$ for all $t \in [0, 1]$. Choosing the value $t = \frac{p(x)+\varepsilon}{p(x)+p(y)+2\varepsilon}$, we find that $\frac{x+y}{p(x)+p(y)+2\varepsilon} \in C$. Using (1) and (10) once more, we are led to $p(x + y) < p(x) + p(y) + 2\varepsilon, \; \forall \varepsilon > 0$.

Lemma 1.3. *Let $C \subset E$ be a nonempty open convex set and let $x_0 \in E$ with $x_0 \notin C$. Then there exists $f \in E^\star$ such that $f(x) < f(x_0) \quad \forall x \in C$. In particular, the hyperplane $[f = f(x_0)]$ separates $\{x_0\}$ and C.*

Proof of Lemma 1.3. After a translation we may always assume that $0 \in C$. We may thus introduce the gauge p of C (see Lemma 1.2). Consider the linear subspace $G = \mathbb{R}x_0$ and the linear functional $g : G \to \mathbb{R}$ defined by

$$g(tx_0) = t, \quad t \in \mathbb{R}.$$

It is clear that

$$g(x) \leq p(x) \quad \forall x \in G$$

(consider the two cases $t > 0$ and $t \leq 0$). It follows from Theorem 1.1 that there exists a linear functional f on E that extends g and satisfies

$$f(x) \leq p(x) \quad \forall x \in E.$$

In particular, we have $f(x_0) = 1$ and that f is continuous by (9). We deduce from (10) that $f(x) < 1$ for every $x \in C$.

Proof of Theorem 1.6. Set $C = A - B$, so that C is convex (check!), C is open (since $C = \bigcup_{y \in B}(A - y)$), and $0 \notin C$ (because $A \cap B = \emptyset$). By Lemma 1.3 there is some $f \in E^*$ such that

$$f(z) < 0 \quad \forall z \in C,$$

that is,

$$f(x) < f(y) \quad \forall x \in A, \quad \forall y \in B.$$

Fix a constant α satisfying

$$\sup_{x \in A} f(x) \leq \alpha \leq \inf_{y \in B} f(y).$$

Clearly, the hyperplane $[f = \alpha]$ separates A and B.

• **Theorem 1.7 (Hahn–Banach, second geometric form).** *Let $A \subset E$ and $B \subset E$ be two nonempty convex subsets such that $A \cap B = \emptyset$. Assume that A is closed and B is compact. Then there exists a closed hyperplane that strictly separates A and B.*

Proof. Set $C = A - B$, so that C is convex, closed (check!), and $0 \notin C$. Hence, there is some $r > 0$ such that $B(0, r) \cap C = \emptyset$. By Theorem 1.6 there is a closed hyperplane that separates $B(0, r)$ and C. Therefore, there is some $f \in E^*$, $f \not\equiv 0$, such that

$$f(x - y) \leq f(rz) \quad \forall x \in A, \quad \forall y \in B, \quad \forall z \in B(0, 1).$$

It follows that $f(x - y) \leq -r\|f\| \quad \forall x \in A, \forall y \in B$. Letting $\varepsilon = \frac{1}{2}r\|f\| > 0$, we obtain

$$f(x) + \varepsilon \leq f(y) - \varepsilon \quad \forall x \in A, \quad \forall y \in B.$$

Choosing α such that

$$\sup_{x \in A} f(x) + \varepsilon \leq \alpha \leq \inf_{y \in B} f(y) - \varepsilon,$$

we see that the hyperplane $[f = \alpha]$ strictly separates A and B.

Remark 4. Assume that $A \subset E$ and $B \subset E$ are two nonempty convex sets such that $A \cap B = \emptyset$. If we make *no further assumption*, it is in general *impossible* to separate

A and B by a closed hyperplane. One can even construct such an example in which A and B are both closed (see Exercise 1.14). However, if E is *finite-dimensional* one can *always* separate any two nonempty convex sets A and B such that $A \cap B = \emptyset$ (no further assumption is required!); see Exercise 1.9.

We conclude this section with a very useful fact:

• **Corollary 1.8.** *Let $F \subset E$ be a linear subspace such that $\overline{F} \neq E$. Then there exists some $f \in E^\star$, $f \not\equiv 0$, such that*

$$\langle f, x \rangle = 0 \quad \forall x \in F.$$

Proof. Let $x_0 \in E$ with $x_0 \notin \overline{F}$. Using Theorem 1.7 with $A = \overline{F}$ and $B = \{x_0\}$, we find a closed hyperplane $[f = \alpha]$ that strictly separates $\overline{F}$ and $\{x_0\}$. Thus, we have

$$\langle f, x \rangle < \alpha < \langle f, x_0 \rangle \quad \forall x \in F.$$

It follows that $\langle f, x \rangle = 0 \quad \forall x \in F$, since $\lambda \langle f, x \rangle < \alpha$ for every $\lambda \in \mathbb{R}$.

• *Remark* 5. Corollary 1.8 is used very often in proving that a linear subspace $F \subset E$ is dense. It suffices to show that *every continuous linear functional on E that vanishes on F must vanish everywhere on E.*

1.3 The Bidual $E^{\star\star}$. Orthogonality Relations

Let E be an n.v.s. and let $E^\star$ be the dual space with norm

$$\|f\|_{E^\star} = \sup_{\substack{x \in E \\ \|x\| \leq 1}} |\langle f, x \rangle|.$$

The bidual $E^{\star\star}$ is the dual of $E^\star$ with norm

$$\|\xi\|_{E^{\star\star}} = \sup_{\substack{f \in E^\star \\ \|f\| \leq 1}} |\langle \xi, f \rangle| \quad (\xi \in E^{\star\star}).$$

There is a *canonical injection* $J : E \to E^{\star\star}$ defined as follows: given $x \in E$, the map $f \mapsto \langle f, x \rangle$ is a continuous linear functional on $E^\star$; thus it is an element of $E^{\star\star}$, which we denote by Jx.[4] We have

$$\langle Jx, f \rangle_{E^{\star\star}, E^\star} = \langle f, x \rangle_{E^\star, E} \quad \forall x \in E, \quad \forall f \in E^\star.$$

It is clear that J is linear and that J is an *isometry*, that is, $\|Jx\|_{E^{\star\star}} = \|x\|_E$; indeed, we have

---

[4] J should not be confused with the duality map $F : E \to E^\star$ defined in Remark 2.

$$\|Jx\|_{E^{**}} = \sup_{\substack{f \in E^* \\ \|f\| \leq 1}} |\langle Jx, f \rangle| = \sup_{\substack{f \in E^* \\ \|f\| \leq 1}} |\langle f, x \rangle| = \|x\|$$

(by Corollary 1.4).

It may happen that J is *not surjective* from E onto E^{**} (see Chapters 3 and 4). However, it is convenient to *identify E with a subspace of E^{**} using J*. If J turns out to be surjective then one says that E is reflexive, and E^{**} is identified with E (see Chapter 3).

Notation. If $M \subset E$ is a linear subspace we set

$$M^{\perp} = \{f \in E^*; \ \langle f, x \rangle = 0 \quad \forall x \in M\}.$$

If $N \subset E^*$ is a linear subspace we set

$$N^{\perp} = \{x \in E \ ; \langle f, x \rangle = 0 \quad \forall f \in N\}.$$

Note that—by definition—$N^{\perp}$ is a subset of E *rather than* E^{**}. It is clear that $M^{\perp}$ (resp. $N^{\perp}$) is a closed linear subspace of E^* (resp. E). We say that $M^{\perp}$ (resp. $N^{\perp}$) is the space orthogonal to M (resp. N).

Proposition 1.9. *Let $M \subset E$ be a linear subspace. Then*

$$(M^{\perp})^{\perp} = \overline{M}.$$

Let $N \subset E^*$ be a linear subspace. Then

$$(N^{\perp})^{\perp} \supset \overline{N}.$$

Proof. It is clear that $M \subset (M^{\perp})^{\perp}$, and since $(M^{\perp})^{\perp}$ is closed we have $\overline{M} \subset (M^{\perp})^{\perp}$. Conversely, let us show that $(M^{\perp})^{\perp} \subset \overline{M}$. Suppose by contradiction that there is some $x_0 \in (M^{\perp})^{\perp}$ such that $x_0 \notin \overline{M}$. By Theorem 1.7 there is a closed hyperplane that strictly separates $\{x_0\}$ and $\overline{M}$. Thus, there are some $f \in E^*$ and some $\alpha \in \mathbb{R}$ such that

$$\langle f, x \rangle < \alpha < \langle f, x_0 \rangle \quad \forall x \in M.$$

Since M is a linear space it follows that $\langle f, x \rangle = 0 \quad \forall x \in M$ and also $\langle f, x_0 \rangle > 0$. Therefore $f \in M^{\perp}$ and consequently $\langle f, x_0 \rangle = 0$, a contradiction.

It is also clear that $N \subset (N^{\perp})^{\perp}$ and thus $\overline{N} \subset (N^{\perp})^{\perp}$.

Remark 6. It may happen that $(N^{\perp})^{\perp}$ is strictly bigger than $\overline{N}$ (see Exercise 1.16). It is, however, instructive to "try" to prove that $(N^{\perp})^{\perp} = \overline{N}$ and see where the argument breaks down. Suppose $f_0 \in E^*$ is such that $f_0 \in (N^{\perp})^{\perp}$ and $f_0 \notin \overline{N}$. Applying Hahn–Banach in E^*, we may strictly separate $\{f_0\}$ and $\overline{N}$. Thus, there is some $\xi \in E^{**}$ such that $\langle \xi, f_0 \rangle > 0$. But we cannot derive a contradiction, since

$\xi \notin N^{\perp}$—unless we happen to know (by chance!) that $\xi \in E$, or more precisely that $\xi = Jx_0$ for some $x_0 \in E$. In particular, if E is reflexive, it is indeed true that $(N^{\perp})^{\perp} = \overline{N}$. In the general case one can show that $(N^{\perp})^{\perp}$ coincides with the closure of N in the weak* topology $\sigma(E^{\star}, E)$ (see Chapter 3).

1.4 A Quick Introduction to the Theory of Conjugate Convex Functions

We start with some basic facts about lower semicontinuous functions and convex functions. In this section we consider functions φ defined on a set E with values in $(-\infty, +\infty]$, so that φ can take the value $+\infty$ (but $-\infty$ is excluded). We denote by $D(\varphi)$ the domain of φ, that is,

$$D(\varphi) = \{x \in E ; \; \varphi(x) < +\infty\}.$$

Notation. The *epigraph* of φ is the set[5]

$$\text{epi } \varphi = \{[x, \lambda] \in E \times \mathbb{R}; \; \varphi(x) \leq \lambda\}.$$

We assume now that E is a *topological space*. We recall the following.

Definition. A function $\varphi : E \to (-\infty, +\infty]$ is said to be *lower semicontinuous* (l.s.c.) if for every $\lambda \in \mathbb{R}$ the set

$$[\varphi \leq \lambda] = \{x \in E; \; \varphi(x) \leq \lambda\}$$

is closed.

Here are some well-known elementary facts about l.s.c. functions (see, e.g., G. Choquet, [1], J. Dixmier [1], J. R. Munkres [1], H. L. Royden [1]):

1. If φ is l.s.c., then epi φ is closed in $E \times \mathbb{R}$; and conversely.
2. If φ is l.s.c., then for every $x \in E$ and for every $\varepsilon > 0$ there is some neighborhood V of x such that
$$\varphi(y) \geq \varphi(x) - \varepsilon \quad \forall y \in V;$$
and conversely.
In particular, if φ is l.s.c., then for every sequence (x_n) in E such that $x_n \to x$, we have
$$\liminf_{n \to \infty} \varphi(x_n) \geq \varphi(x)$$
and conversely if E is a metric space.
3. If φ_1 and φ_2 are l.s.c., then $\varphi_1 + \varphi_2$ is l.s.c.

---

[5] We insist on the fact that $\mathbb{R} = (-\infty, \infty)$, so that λ does not take the value ∞.

4. If $(\varphi_i)_{i \in I}$ is a family of l.s.c. functions then their *superior envelope* is also *l.s.c.*, that is, the function φ defined by

$$\varphi(x) = \sup_{i \in I} \varphi_i(x)$$

is l.s.c.
5. If E is *compact* and φ is l.s.c., then $\inf_E \varphi$ is achieved.

(If E is a compact metric space one can argue with minimizing sequences. For a general topological compact space consider the sets $[\varphi \leq \lambda]$ for appropriate values of λ.)

We now assume that E is a *vector space*. Recall the following definition.

Definition. A function $\varphi : E \to (-\infty, +\infty]$ is said to be *convex* if

$$\boxed{\varphi(tx + (1 - t)y) \leq t\varphi(x) + (1 - t)\varphi(y) \quad \forall x, y \in E, \quad \forall t \in (0, 1).}$$

We shall use some elementary properties of convex functions:

1. If φ is a convex function, then epi φ is a convex set in $E \times \mathbb{R}$; and conversely.
2. If φ is a convex function, then for every $\lambda \in \mathbb{R}$ the set $[\varphi \leq \lambda]$ is convex; but the converse is *not* true.
3. If φ_1 and φ_2 are convex, then $\varphi_1 + \varphi_2$ is convex.
4. If $(\varphi_i)_{i \in I}$ is a family of convex functions, then the superior envelope, $\sup_i \varphi_i$, is convex.

We assume hereinafter that E is an n.v.s.

Definition. Let $\varphi : E \to (-\infty, +\infty]$ be a function such that $\varphi \not\equiv +\infty$ (i.e., $D(\varphi) \neq \emptyset$). We define the *conjugate function* $\varphi^* : E^* \to (-\infty, +\infty]$ to be[6]

$$\boxed{\varphi^*(f) = \sup_{x \in E} \{\langle f, x \rangle - \varphi(x)\} \quad (f \in E^*).}$$

Note that φ^* is convex and l.s.c. on E^*. Indeed, for each fixed $x \in E$ the function $f \mapsto \langle f, x \rangle - \varphi(x)$ is convex and continuous (and thus l.s.c.) on E^*. It follows that the superior envelope of these functions (as x runs through E) is convex and l.s.c.

Remark 7. Clearly we have the inequality

(11) $$\langle f, x \rangle \leq \varphi(x) + \varphi^*(f) \quad \forall x \in E, \quad \forall f \in E^*,$$

which is sometimes called *Young's inequality*. Of course, this fact is obvious with our definition of φ^*! The classical form of Young's inequality (see the proof of Theorem 4.6 in Chapter 4) asserts that

---

[6] φ^* is sometimes called the Legendre transform of φ.

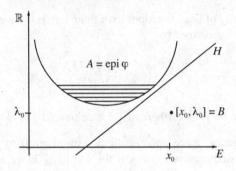

Fig. 2

$$(12) \qquad ab \leq \frac{1}{p}a^p + \frac{1}{p'}b^{p'} \quad \forall a, b \geq 0$$

with $1 < p < \infty$ and $\frac{1}{p} + \frac{1}{p'} = 1$. Inequality (12) becomes a special case of (11) with $E = E^\star = \mathbb{R}$ and $\varphi(t) = \frac{1}{p}|t|^p$, $\varphi^\star(s) = \frac{1}{p'}|s|^{p'}$ (see Exercise 1.18, question (h)).

Proposition 1.10. *Assume that $\varphi : E \to (-\infty, +\infty]$ is convex l.s.c. and $\varphi \not\equiv +\infty$. Then $\varphi^\star \not\equiv +\infty$, and in particular, φ is bounded below by an affine continuous function.*

Proof. Let $x_0 \in D(\varphi)$ and let $\lambda_0 < \varphi(x_0)$. We apply Theorem 1.7 (Hahn–Banach, second geometric form) in the space $E \times \mathbb{R}$ with $A = \operatorname{epi} \varphi$ and $B = \{[x_0, \lambda_0]\}$. So, there exists a closed hyperplane $H = [\Phi = \alpha]$ in $E \times \mathbb{R}$ that strictly separates A and B; see Figure 2. Note that the function $x \in E \mapsto \Phi([x, 0])$ is a continuous linear functional on E, and thus $\Phi([x, 0]) = \langle f, x \rangle$ for some $f \in E^\star$. Letting $k = \Phi([0, 1])$, we have

$$\Phi([x, \lambda]) = \langle f, x \rangle + k\lambda \quad \forall [x, \lambda] \in E \times \mathbb{R}.$$

Writing that $\Phi > \alpha$ on A and $\Phi < \alpha$ on B, we obtain

$$\langle f, x \rangle + k\lambda > \alpha, \quad \forall [x, \lambda] \in \operatorname{epi} \varphi,$$

and

$$\langle f, x_0 \rangle + k\lambda_0 < \alpha.$$

In particular, we have

$$(13) \qquad \langle f, x \rangle + k\varphi(x) > \alpha \quad \forall x \in D(\varphi)$$

and thus

$$\langle f, x_0 \rangle + k\varphi(x_0) > \alpha > \langle f, x_0 \rangle + k\lambda_0.$$

It follows that $k > 0$. By (13) we have

$$\left\langle -\frac{1}{k}f, x \right\rangle - \varphi(x) < -\frac{\alpha}{k} \quad \forall x \in D(\varphi)$$

and therefore $\varphi^\star(-\frac{1}{k}f) < +\infty$.

If we iterate the operation $\star$, we obtain a function $\varphi^{\star\star}$ defined on $E^{\star\star}$. Instead, we choose to restrict $\varphi^{\star\star}$ to E, that is, we define

$$\varphi^{\star\star}(x) = \sup_{f \in E^\star} \{\langle f, x \rangle - \varphi^\star(f)\} \quad (x \in E).$$

• **Theorem 1.11 (Fenchel–Moreau).** *Assume that $\varphi : E \to (-\infty, +\infty]$ is convex, l.s.c., and $\varphi \not\equiv +\infty$. Then $\varphi^{\star\star} = \varphi$.*

Proof. We proceed in two steps:

Step 1: We assume in addition that $\varphi \geq 0$ and we claim that $\varphi^{\star\star} = \varphi$.

First, it is obvious that $\varphi^{\star\star} \leq \varphi$, since $\langle f, x \rangle - \varphi^\star(f) \leq \varphi(x) \ \forall x \in E$ and $\forall f \in E^\star$. In order to prove that $\varphi^{\star\star} = \varphi$ we argue by contradiction, and we assume that $\varphi^{\star\star}(x_0) < \varphi(x_0)$ for some $x_0 \in E$. We could possibly have $\varphi(x_0) = +\infty$, but $\varphi^{\star\star}(x_0)$ is always finite. We apply Theorem 1.7 (Hahn–Banach, second geometric form) in the space $E \times \mathbb{R}$ with $A = \text{epi} \ \varphi$ and $B = [x_0, \varphi^{\star\star}(x_0)]$. So, there exist, as in the proof of Proposition 1.10, $f \in E^\star$, $k \in \mathbb{R}$, and $\alpha \in \mathbb{R}$ such that

(14) $$\langle f, x \rangle + k\lambda > \alpha \quad \forall[x, \lambda] \in \text{epi} \ \varphi,$$

(15) $$\langle f, x_0 \rangle + k\varphi^{\star\star}(x_0) < \alpha.$$

It follows that $k \geq 0$ (fix some $x \in D(\varphi)$ and let $\lambda \to +\infty$ in (14)). [Here we cannot assert, as in the proof of Proposition 1.10, that $k > 0$; we could possibly have $k = 0$, which would correspond to a "vertical" hyperplane H in $E \times \mathbb{R}$.]

Let $\varepsilon > 0$; since $\varphi \geq 0$, we have by (14),

$$\langle f, x \rangle + (k+\varepsilon)\varphi(x) \geq \alpha \quad \forall x \in D(\varphi).$$

Therefore

$$\varphi^\star\left(-\frac{f}{k+\varepsilon}\right) \leq -\frac{\alpha}{k+\varepsilon}.$$

It follows from the definition of $\varphi^{\star\star}(x_0)$ that

$$\varphi^{\star\star}(x_0) \geq \left\langle -\frac{f}{k+\varepsilon}, x_0 \right\rangle - \varphi^\star\left(-\frac{f}{k+\varepsilon}\right) \geq \left\langle -\frac{f}{k+\varepsilon}, x_0 \right\rangle + \frac{\alpha}{k+\varepsilon}.$$

Thus we have

$$\langle f, x_0 \rangle + (k+\varepsilon)\varphi^{\star\star}(x_0) \geq \alpha \quad \forall \varepsilon > 0,$$

which contradicts (15).

Step 2: The general case.

Fix some $f_0 \in D(\varphi^\star)$ $(D(\varphi^\star) \neq \emptyset$ by Proposition 1.10) and define

$$\overline{\varphi}(x) = \varphi(x) - \langle f_0, x \rangle + \varphi^\star(f_0),$$

so that $\overline{\varphi}$ is convex l.s.c., $\overline{\varphi} \not\equiv +\infty$, and $\overline{\varphi} \geq 0$. We know from Step 1 that $(\overline{\varphi})^{\star\star} = \overline{\varphi}$. Let us now compute $(\overline{\varphi})^\star$ and $(\overline{\varphi})^{\star\star}$. We have

$$(\overline{\varphi})^\star(f) = \varphi^\star(f + f_0) - \varphi^\star(f_0)$$

and

$$(\overline{\varphi})^{\star\star}(x) = \varphi^{\star\star}(x) - \langle f_0, x \rangle + \varphi^\star(f_0).$$

Writing that $(\overline{\varphi})^{\star\star} = \overline{\varphi}$, we obtain $\varphi^{\star\star} = \varphi$.

Let us examine some examples.

Example 1. Consider $\varphi(x) = \|x\|$. It is easy to check that

$$\varphi^\star(f) = \begin{cases} 0 & \text{if } \|f\| \leq 1, \\ +\infty & \text{if } \|f\| > 1. \end{cases}$$

It follows that

$$\varphi^{\star\star}(x) = \sup_{\substack{f \in E^\star \\ \|f\| \leq 1}} \langle f, x \rangle.$$

Writing the equality

$$\varphi^{\star\star} = \varphi,$$

we obtain again part of Corollary 1.4.

Example 2. Given a nonempty set $K \subset E$, we set

$$I_K(x) = \begin{cases} 0 & \text{if } x \in K, \\ +\infty & \text{if } x \notin K. \end{cases}$$

The function I_K is called the *indicator function* of K (and should not be confused with the characteristic function, χ_K, of K, which is 1 on K and 0 outside K). Note that I_K is a convex function iff K is a convex set, and I_K is l.s.c. iff K is closed. The conjugate function $(I_K)^\star$ is called the *supporting function* of K.

It is easy to see that if $K = M$ is a linear subspace then $(I_M)^\star = I_{M^\perp}$ and $(I_M)^{\star\star} = I_{(M^\perp)^\perp}$. Assuming that M is a closed linear space and writing that $(I_M)^{\star\star} = I_M$, we obtain $(M^\perp)^\perp = M$. In some sense, Theorem 1.11 can be viewed as a counterpart of Proposition 1.9.

We conclude this chapter with another useful property of conjugate functions.

★ **Theorem 1.12 (Fenchel–Rockafellar).** *Let $\varphi, \psi : E \to (-\infty, +\infty]$ be two convex functions. Assume that there is some $x_0 \in D(\varphi) \cap D(\psi)$ such that φ is continuous at x_0. Then*

$$\inf_{x \in E} \{\varphi(x) + \psi(x)\} = \sup_{f \in E^\star} \{-\varphi^\star(-f) - \psi^\star(f)\}$$

$$= \max_{f \in E^\star} \{-\varphi^\star(-f) - \psi^\star(f)\} = -\min_{f \in E^\star} \{\varphi^\star(-f) + \psi^\star(f)\}.$$

The proof of Theorem 1.12 relies on the following lemma.

Lemma 1.4. *Let $C \subset E$ be a convex set, then* Int C *is convex.*[7] *If, in addition,* Int $C \neq \emptyset$, *then*

$$\overline{C} = \overline{\text{Int } C}.$$

For the proof of Lemma 1.4, see, e.g., Exercise 1.7.

Proof of Theorem 1.12. Set

$$a = \inf_{x \in E} \{\varphi(x) + \psi(x)\},$$

$$b = \sup_{f \in E^\star} \{-\varphi^\star(-f) - \psi^\star(f)\}.$$

It is clear that $b \leq a$. If $a = -\infty$, the conclusion of Theorem 1.12 is obvious. Thus we may assume hereinafter that $a \in \mathbb{R}$. Let $C = \text{epi } \varphi$, so that Int $C \neq \emptyset$ (since φ is continuous at x_0). We apply Theorem 1.6 (Hahn–Banach, first geometric form) with $A = \text{Int } C$ and

$$B = \{[x, \lambda] \in E \times \mathbb{R}; \ \lambda \leq a - \psi(x)\}.$$

Then A and B are nonempty convex sets. Moreover, $A \cap B = \emptyset$; indeed, if $[x, \lambda] \in A$, then $\lambda > \varphi(x)$, and on the other hand, $\varphi(x) \geq a - \psi(x)$ (by definition of a), so that $[x, \lambda] \notin B$.

Hence there exists a closed hyperplane H that separates A and B. It follows that H also separates $\overline{A}$ and B. But we know from Lemma 1.4 that $\overline{A} = \overline{C}$. Therefore, there exist $f \in E^\star$, $k \in \mathbb{R}$, and $\alpha \in \mathbb{R}$ such that the hyperplane $H = [\Phi = \alpha]$ in $E \times \mathbb{R}$ separates C and B, where

$$\Phi([x, \lambda]) = \langle f, x \rangle + k\lambda \quad \forall [x, \lambda] \in E \times \mathbb{R}.$$

Thus we have

(16) $\langle f, x \rangle + k\lambda \geq \alpha \quad \forall [x, \lambda] \in C,$

(17) $\langle f, x \rangle + k\lambda \leq \alpha \quad \forall [x, \lambda] \in B.$

---

[7] As usual, Int C denotes the interior of C.

Choosing $x = x_0$ and letting $\lambda \to +\infty$ in (16), we see that $k \geq 0$. We claim that

(18) $$k > 0.$$

Assume by contradiction that $k = 0$; it follows that $\|f\| \neq 0$ (since $\Phi \neq 0$). By (16) and (17) we have

$$\langle f, x \rangle \geq \alpha \quad \forall x \in D(\varphi),$$
$$\langle f, x \rangle \leq \alpha \quad \forall x \in D(\psi).$$

But $B(x_0, \varepsilon_0) \subset D(\varphi)$ for some $\varepsilon_0 > 0$ (small enough), and thus

$$\langle f, x_0 + \varepsilon_0 z \rangle \geq \alpha \quad \forall z \in B(0, 1),$$

which implies that $\langle f, x_0 \rangle \geq \alpha + \varepsilon_0 \|f\|$. On the other hand, we have $\langle f, x_0 \rangle \leq \alpha$, since $x_0 \in D(\psi)$; therefore we obtain $\|f\| = 0$, which is a contradiction and completes the proof of (18).

From (16) and (17) we obtain

$$\varphi^\star\left(-\frac{f}{k}\right) \leq -\frac{\alpha}{k}$$

and

$$\psi^\star\left(\frac{f}{k}\right) \leq \frac{\alpha}{k} - a,$$

so that

$$-\varphi^\star\left(-\frac{f}{k}\right) - \psi^\star\left(\frac{f}{k}\right) \geq a.$$

On the other hand, from the definition of b, we have

$$-\varphi^\star\left(-\frac{f}{k}\right) - \psi^\star\left(\frac{f}{k}\right) \leq b.$$

We conclude that

$$a = b = -\varphi^\star\left(-\frac{f}{k}\right) - \psi^\star\left(\frac{f}{k}\right).$$

Example 3. Let K be a nonempty convex set. We claim that for every $x_0 \in E$ we have

(19) $$\mathrm{dist}(x_0, K) = \inf_{x \in K} \|x - x_0\| = \max_{\substack{f \in E^\star \\ \|f\| \leq 1}} \{\langle f, x_0 \rangle - I_K^\star(f)\}.$$

Indeed, we have

$$\inf_{x \in K} \|x - x_0\| = \inf_{x \in E} \{\varphi(x) + \psi(x)\},$$

with $\varphi(x) = \|x - x_0\|$ and $\psi(x) = I_K(x)$. Applying Theorem 1.12, we obtain (19). In the special case that $K = M$ is a linear subspace, we obtain the relation

$$\text{dist}(x_0, M) = \inf_{x \in M} \|x - x_0\| = \max_{\substack{f \in M^\perp \\ \|f\| \le 1}} \langle f, x_0 \rangle.$$

Remark 8. Relation (19) may provide us with some useful information in the case that $\inf_{x \in K} \|x - x_0\|$ is not achieved (see, e.g., Exercise 1.17). The theory of minimal surfaces provides an interesting setting in which the *primal problem* (i.e., $\inf_{x \in E}\{\varphi(x) + \psi(x)\}$) need not have a solution, while the *dual problem* (i.e., $\max_{f \in E^*}\{-\varphi^*(-f) - \psi^*(f)\}$) has a solution; see I. Ekeland–R. Temam [1].

Example 4. Let $\varphi : E \to \mathbb{R}$ be convex and continuous and let $M \subset E$ be a linear subspace. Then we have

$$\inf_{x \in M} \varphi(x) = -\min_{f \in M^\perp} \varphi^*(f).$$

It suffices to apply Theorem 1.12 with $\psi = I_M$.

Comments on Chapter 1

1. Generalizations and variants of the Hahn–Banach theorems.

The first geometric form of the Hahn–Banach theorem (Theorem 1.6) is still valid in general topological vector spaces. The second geometric form (Theorem 1.7) holds in *locally convex spaces*—such spaces play an important role, for example, in the *theory of distributions* (see, e.g., L. Schwartz [1] and F. Treves [1]). Interested readers may consult, e.g., N. Bourbaki [1], J. Kelley-I. Namioka [1], G. Choquet [2] (Volume 2), A. Taylor–D. Lay [1], and A. Knapp [2].

2. Applications of the Hahn–Banach theorems.

The Hahn–Banach theorems have a *wide* and *diversified* range of applications. Here are two examples:

(a) The Krein–Milman theorem.

The second geometric form of the Hahn–Banach theorem is a basic ingredient in the proof of the Krein–Milman theorem. Before stating this result we need some definitions. Let E be an n.v.s. and let A be a subset of E. The *convex hull* of A, denoted by conv A, is the smallest convex set containing A. Clearly, conv A consists of all *finite* convex combinations of elements in A, i.e.,

$$\text{conv } A = \left\{ \sum_{i \in I} t_i a_i;\ I \text{ finite},\ a_i \in A\ \forall i,\ t_i \ge 0\ \forall i,\ \text{and}\ \sum_{i \in I} t_i = 1 \right\}.$$

The *closed convex hull* of A, denoted by $\overline{\text{conv}}A$, is the closure of conv A. Given a convex set $K \subset E$ we say that a point $x \in K$ is *extremal* if x cannot be written as a convex combination of two points $x_0, x_1 \in K$, i.e., $x \ne (1 - t)x_0 + tx_1$ with $t \in (0, 1)$, and $x_0 \ne x_1$.

• **Theorem 1.13 (Krein–Milman).** *Let $K \subset E$ be a compact convex set. Then K coincides with the closed convex hull of its extremal points.*

The Krein–Milman theorem has itself numerous applications and extensions (such as Choquet's integral representation theorem, Bochner's theorem, Bernstein's theorem, etc.). On this vast subject, see, e.g., N. Bourbaki [1], G. Choquet [2] (Volume 2), R. Phelps [1], C. Dellacherie-P. A. Meyer [1] (Chapter 10), N. Dunford–J. T. Schwartz [1] (Volume 1), W. Rudin [1], R. Larsen [1], J. Kelley–I. Namioka [1], R. Edwards [1]. An interesting application to PDEs, due to Y. Pinchover, is presented in S. Agmon [2]. For a proof of the Krein–Milman theorem, see Problem 1.

(b) In the theory of partial differential equations.

Let us mention, for example, that the existence of a *fundamental solution* for a general differential operator $P(D)$ with constant coefficients (the Malgrange–Ehrenpreis theorem) relies on the analytic form of Hahn–Banach; see, e.g., L. Hörmander [1], [2], K. Yosida [1], W. Rudin [1], F. Treves [2], M. Reed-B. Simon [1] (Volume 2). In the same spirit, let us mention also the proof of the existence of the Green's function for the Laplacian by the method of P. Lax; see P. Lax [1] (Section 9.5) and P. Garabedian [1]. The proof of the existence of a solution $u \in L^\infty(\Omega)$ for the equation div $u = f$ in $\Omega \subset \mathbb{R}^N$, given any $f \in L^N(\Omega)$, relies on Hahn–Banach (see J. Bourgain–H. Brezis [1], [2]). Surprisingly, the u obtained via Hahn–Banach depends *nonlinearly* on f. In fact, there exists no bounded linear operator from L^N into L^∞ giving u in terms of f. This shows that the use of Zorn's lemma (and the underlying axiom of choice) in the proof of Hahn–Banach can be delicate and may destroy the linear character of the problem. Sometimes there is no way to circumvent this obstruction.

3. Convex functions.

Convex analysis and duality principles are topics which have considerably expanded and have become increasingly popular in recent years; see, e.g., J. J. Moreau [1], R. T. Rockafellar [1], [2], I. Ekeland–R. Temam [1], I. Ekeland–T. Turnbull [1], F. Clarke [1], J. P. Aubin–I. Ekeland [1], J. B. Hiriart–Urutty–C. Lemaréchal [1]. Among the applications let us mention the following:

(a) *Game theory, economics, optimization, convex programming*; see J. P. Aubin [1], [2], [3], J. P. Aubin–I. Ekeland [1], S. Karlin [1], A. Balakrishnan [1], V. Barbu–I. Precupanu [1], J. Franklin [1], J. Stoer–C. Witzgall [1].

(b) *Mechanics*; see J. J. Moreau [2], P. Germain [1], [2], G. Duvaut–J. L. Lions [1], R. Temam–G. Strang [1] and the comments by P. Germain following this paper, H. D. Bui [1] and the numerous references therein. Note also the use of (nonconvex) duality by J. F. Toland [1], [2], [3] (for the study of rotating chains), by A. Damlamian [1] (for a problem arising in plasma physics), and by G. Auchmuty [1].

(c) The theory of *monotone operators and nonlinear semigroups*; see H. Brezis [1], F. Browder [1], V. Barbu [1], and R. Phelps [2].

(d) Variational problems involving *periodic solutions of Hamiltonian systems and nonlinear vibrating strings*; see the recent works of F. Clarke, I. Ekeland,

J. M. Lasry, H. Brezis, J. M. Coron, L. Nirenberg (we refer, e.g., to F. Clarke–I. Ekeland [1], H. Brezis–J. M. Coron–L. Nirenberg [1], H. Brezis [2], J. P. Aubin–I. Ekeland [1], I. Ekeland [1], and their bibliographies).

(e) The theory of *large deviations in probability*; see, e.g., R. Azencott et al. [1], D. W. Stroock [1].

(f) The theory of *partial differential equations and complex analysis*; see L. Hörmander [3].

4. Extensions of bounded linear operators.

Let E and F be two Banach spaces and let $G \subset E$ be a closed subspace. Let $S : G \to F$ be a bounded linear operator. One may ask whether it is possible to extend S by a bounded linear operator $T : E \to F$. Note that Corollary 1.2 settles this question only when $F = \mathbb{R}$. In general, the answer is negative (even if E and F are reflexive spaces; see Exercise 1.27), except in some special cases; for example, the following:

(a) If dim $F < \infty$. One may choose a basis in F and apply Corollary 1.2 to each component of S.

(b) If G admits a topological complement (see Section 2.4). This is true in particular if dim $G < \infty$ or codim $G < \infty$ or if E is a Hilbert space.

One may also ask the question whether there is an extension T *with the same norm*, i.e., $\|T\|_{\mathcal{L}(E,F)} = \|S\|_{\mathcal{L}(G,F)}$. The answer is yes *only* in some *exceptional* cases; see L. Nachbin [1], J. Kelley [1], and Exercise 5.15.

Exercises for Chapter 1

$\boxed{1.1}$ *Properties of the duality map.*

Let E be an n.v.s. The duality map F is defined for every $x \in E$ by

$$F(x) = \{f \in E^\star;\ \|f\| = \|x\| \text{ and } \langle f, x \rangle = \|x\|^2\}.$$

1. Prove that

$$F(x) = \{f \in E^\star;\ \|f\| \leq \|x\| \text{ and } \langle f, x \rangle = \|x\|^2\}$$

and deduce that $F(x)$ is nonempty, closed, and convex.

2. Prove that if $E^\star$ is strictly convex, then $F(x)$ contains a single point.

3. Prove that

$$F(x) = \left\{ f \in E^\star;\ \frac{1}{2}\|y\|^2 - \frac{1}{2}\|x\|^2 \geq \langle f, y - x \rangle \quad \forall y \in E \right\}.$$

4. Deduce that

$$\langle F(x) - F(y), x - y \rangle \geq 0 \quad \forall x, y \in E,$$

and more precisely that

$$\langle f - g, x - y \rangle \geq 0 \quad \forall x, y \in E, \quad \forall f \in F(x), \quad \forall g \in F(y).$$

Show that, in fact,

$$\langle f - g, x - y \rangle \geq (\|x\| - \|y\|)^2 \quad \forall x, y \in E, \quad \forall f \in F(x), \quad \forall g \in F(y).$$

5. Assume again that $E^\star$ is strictly convex and let $x, y \in E$ be such that

$$\langle F(x) - F(y), x - y \rangle = 0.$$

Show that $Fx = Fy$.

1.2 Let E be a vector space of dimension n and let $(e_i)_{1 \leq i \leq n}$ be a basis of E. Given $x \in E$, write $x = \sum_{i=1}^{n} x_i e_i$ with $x_i \in \mathbb{R}$; given $f \in E^\star$, set $f_i = \langle f, e_i \rangle$.

1. Consider on E the norm

$$\|x\|_1 = \sum_{i=1}^{n} |x_i|.$$

(a) Compute explicitly, in terms of the f_i's, the dual norm $\|f\|_{E^\star}$ of $f \in E^\star$.
(b) Determine explicitly the set $F(x)$ (duality map) for every $x \in E$.

2. Same questions but where E is provided with the norm

$$\|x\|_\infty = \max_{1 \leq i \leq n} |x_i|.$$

3. Same questions but where E is provided with the norm

$$\|x\|_2 = \left(\sum_{i=1}^{n} |x_i|^2 \right)^{1/2},$$

and more generally with the norm

$$\|x\|_p = \left(\sum_{i=1}^{n} |x_i|^p \right)^{1/p}, \quad \text{where } p \in (1, \infty).$$

1.3 Let $E = \{u \in C([0, 1]; \mathbb{R}); u(0) = 0\}$ with its usual norm

$$\|u\| = \max_{t \in [0,1]} |u(t)|.$$

Consider the linear functional

$$f : u \in E \mapsto f(u) = \int_0^1 u(t)dt.$$

1. Show that $f \in E^\star$ and compute $\|f\|_{E^\star}$.
2. Can one find some $u \in E$ such that $\|u\| = 1$ and $f(u) = \|f\|_{E^\star}$?

1.4 Consider the space $E = c_0$ (sequences tending to zero) with its usual norm (see Section 11.3). For every element $u = (u_1, u_2, u_3, \dots)$ in E define

$$f(u) = \sum_{n=1}^{\infty} \frac{1}{2^n} u_n.$$

1. Check that f is a continuous linear functional on E and compute $\|f\|_{E^\star}$.
2. Can one find some $u \in E$ such that $\|u\| = 1$ and $f(u) = \|f\|_{E^\star}$?

1.5 Let E be an infinite-dimensional n.v.s.

1. Prove (using Zorn's lemma) that there exists an algebraic basis $(e_i)_{i \in I}$ in E such that $\|e_i\| = 1 \ \forall i \in I$.
   Recall that an algebraic basis (or Hamel basis) is a subset $(e_i)_{i \in I}$ in E such that every $x \in E$ may be written uniquely as

$$x = \sum_{i \in J} x_i e_i \text{ with } J \subset I, \ J \text{ finite.}$$

2. Construct a linear functional $f : E \to \mathbb{R}$ that is not continuous.
3. Assuming in addition that E is a Banach space, prove that I is not countable.
   [**Hint:** Use Baire category theorem (Theorem 2.1).]

1.6 Let E be an n.v.s. and let $H \subset E$ be a hyperplane. Let $V \subset E$ be an affine subspace containing H.

1. Prove that either $V = H$ or $V = E$.
2. Deduce that H is either closed or dense in E.

1.7 Let E be an n.v.s. and let $C \subset E$ be convex.

1. Prove that $\overline{C}$ and Int C are convex.
2. Given $x \in C$ and $y \in$ Int C, show that $tx + (1 - t)y \in$ Int $C \ \forall t \in (0, 1)$.
3. Deduce that $\overline{C} = \overline{\text{Int } C}$ whenever Int $C \neq \emptyset$.

1.8 Let E be an n.v.s. with norm $\| \ \|$. Let $C \subset E$ be an open convex set such that $0 \in C$. Let p denote the gauge of C (see Lemma 1.2).

1. Assuming C is symmetric (i.e., $-C = C$) and C is bounded, prove that p is a norm which is equivalent to $\| \ \|$.

2. Let $E = C([0, 1]; \mathbb{R})$ with its usual norm

$$\|u\| = \max_{t \in [0,1]} |u(t)|.$$

Let

$$C = \left\{ u \in E; \ \int_0^1 |u(t)|^2 dt < 1 \right\}.$$

Check that C is convex and symmetric and that $0 \in C$. Is C bounded in E? Compute the gauge p of C and show that p is a norm on E. Is p equivalent to $\| \ \|$?

1.9 *Hahn–Banach in finite-dimensional spaces.*

Let E be a finite-dimensional normed space. Let $C \subset E$ be a nonempty convex set such that $0 \notin C$. We claim that there always exists some hyperplane that separates C and $\{0\}$.

[Note that every hyperplane is closed (why?). The main point in this exercise is that no additional assumption on C is required.]

1. Let $(x_n)_{n \geq 1}$ be a countable subset of C that is dense in C (why does it exist?). For every n let

$$C_n = \text{conv}\{x_1, x_2, \dots, x_n\} = \left\{ x = \sum_{i=1}^n t_i x_i; \ t_i \geq 0 \ \forall i \text{ and } \sum_{i=1}^n t_i = 1 \right\}.$$

Check that C_n is compact and that $\bigcup_{n=1}^\infty C_n$ is dense in C.
2. Prove that there is some $f_n \in E^\star$ such that

$$\|f_n\| = 1 \text{ and } \langle f_n, x \rangle \geq 0 \quad \forall x \in C_n.$$

3. Deduce that there is some $f \in E^\star$ such that

$$\|f\| = 1 \text{ and } \langle f, x \rangle \geq 0 \quad \forall x \in C.$$

Conclude.
4. Let $A, B \subset E$ be nonempty disjoint convex sets. Prove that there exists some hyperplane H that separates A and B.

1.10 Let E be an n.v.s. and let I be any set of indices. Fix a subset $(x_i)_{i \in I}$ in E and a subset $(\alpha_i)_{i \in I}$ in $\mathbb{R}$. Show that the following properties are equivalent:

(A)   There exists some $f \in E^\star$ such that $\langle f, x_i \rangle = \alpha_i \quad \forall i \in I$.

(B) $\begin{cases} \text{There exists a constant } M \geq 0 \text{ such that for each finite subset} \\ J \subset I \text{ and for every choice of real numbers } (\beta_i)_{i \in J}, \text{ we have} \\ \left| \sum_{i \in J} \beta_i \alpha_i \right| \leq M \left\| \sum_{i \in J} \beta_i x_i \right\|. \end{cases}$

Note that in the proof of (B) $\Rightarrow$ (A) one may find some $f \in E^*$ with $\|f\|_{E^*} \leq M$.
[**Hint:** Try first to define f on the linear space spanned by the $(x_i)_{i \in I}$.]

1.11 Let E be an n.v.s. and let $M > 0$. Fix n elements $(f_1)_{1 \leq i \leq n}$ in E^* and n real numbers $(\alpha_i)_{1 \leq i \leq n}$. Prove that the following properties are equivalent:

(A)
$$\begin{cases} \forall \varepsilon > 0 \;\; \exists x_\varepsilon \in E \text{ such that} \\ \|x_\varepsilon\| \leq M + \varepsilon \text{ and } \langle f_i, x_\varepsilon \rangle = \alpha_i \quad \forall i = 1, 2, \dots, n. \end{cases}$$

(B)
$$\left| \sum_{i=1}^{n} \beta_i \alpha_i \right| \leq M \left\| \sum_{i=1}^{n} \beta_i f_i \right\| \quad \forall \beta_1, \beta_2, \dots, \beta_n \in \mathbb{R}.$$

[**Hint:** For the proof of (B) $\Rightarrow$ (A) consider first the case in which the f_i's are linearly independent and imitate the proof of Lemma 3.3.]
Compare Exercises 1.10, 1.11 and Lemma 3.3.

1.12 Let E be a vector space. Fix n linear functionals $(f_i)_{1 \leq i \leq n}$ on E and n real numbers $(\alpha_i)_{1 \leq i \leq n}$. Prove that the following properties are equivalent:

(A)   There exists some $x \in E$ such that $f_i(x) = \alpha_i \quad \forall i = 1, 2, \dots, n$.

(B)
$$\begin{cases} \text{For any choice of real numbers } \beta_1, \beta_2, \dots, \beta_n \text{ such that} \\ \sum_{i=1}^{n} \beta_i f_i = 0, \text{ one also has } \sum_{i=1}^{n} \beta_i \alpha_i = 0. \end{cases}$$

1.13 Let $E = \mathbb{R}^n$ and let

$$P = \{ x \in \mathbb{R}^n; \; x_i \geq 0 \quad \forall i = 1, 2, \dots, n \}.$$

Let M be a linear subspace of E such that $M \cap P = \{0\}$. Prove that there is some hyperplane H in E such that

$$M \subset H \text{ and } H \cap P = \{0\}.$$

[**Hint:** Show first that $M^\perp \cap \text{Int } P \neq \emptyset$.]

1.14 Let $E = \ell^1$ (see Section 11.3) and consider the two sets

$$X = \{ x = (x_n)_{n \geq 1} \in E; \; x_{2n} = 0 \;\, \forall n \geq 1 \}$$

and

$$Y = \left\{ y = (y_n)_{n \geq 1} \in E; \; y_{2n} = \frac{1}{2^n} y_{2n-1} \;\, \forall n \geq 1 \right\}.$$

1. Check that X and Y are closed linear spaces and that $\overline{X + Y} = E$.
2. Let $c \in E$ be defined by

$$\begin{cases} c_{2n-1} = 0 & \forall n \geq 1, \\ c_{2n} = \frac{1}{2^n} & \forall n \geq 1. \end{cases}$$

Check that $c \notin X + Y$.

3. Set $Z = X - c$ and check that $Y \cap Z = \emptyset$. Does there exist a closed hyperplane in E that separates Y and Z?
   Compare with Theorem 1.7 and Exercise 1.9.
4. Same questions in $E = \ell^p$, $1 < p < \infty$, and in $E = c_0$.

1.15 Let E be an n.v.s. and let $C \subset E$ be a convex set such that $0 \in C$. Set

(A) $C^\star = \{ f \in E^\star ;\ \langle f, x \rangle \leq 1 \quad \forall x \in C \},$
(B) $C^{\star\star} = \{ x \in E ;\ \langle f, x \rangle \leq 1 \quad \forall f \in C^\star \}.$

1. Prove that $C^{\star\star} = \overline{C}$.
2. What is $C^\star$ if C is a linear space?

1.16 Let $E = \ell^1$, so that $E^\star = \ell^\infty$ (see Section 11.3). Consider $N = c_0$ as a closed subspace of $E^\star$.
   Determine

$$N^\perp = \{ x \in E ;\ \langle f, x \rangle = 0 \quad \forall f \in N \}$$

and

$$N^{\perp\perp} = \{ f \in E^\star ;\ \langle f, x \rangle = 0 \quad \forall x \in N^\perp \}.$$

Check that $N^{\perp\perp} \neq N$.

1.17 Let E be an n.v.s. and let $f \in E^\star$ with $f \neq 0$. Let M be the hyperplane $[f = 0]$.

1. Determine $M^\perp$.
2. Prove that for every $x \in E$, $\operatorname{dist}(x, M) = \inf_{y \in M} \| x - y \| = \frac{|\langle f, x \rangle|}{\| f \|}$.
   [Find a direct method or use Example 3 in Section 1.4.]
3. Assume now that $E = \{ u \in C([0, 1]; \mathbb{R}) ;\ u(0) = 0 \}$ and that

$$\langle f, u \rangle = \int_0^1 u(t) dt, \quad u \in E.$$

Prove that $\operatorname{dist}(u, M) = | \int_0^1 u(t) dt |\ \forall u \in E$.
Show that $\inf_{v \in M} \| u - v \|$ is never achieved for any $u \in E \setminus M$.

1.18 Check that the functions $\varphi : \mathbb{R} \to (-\infty, +\infty]$ defined below are convex l.s.c. and determine the conjugate functions $\varphi^\star$. Draw their graphs and mark their epigraphs.

(a) $\varphi(x) = ax + b,$ where $a, b \in \mathbb{R}.$

(b) $\varphi(x) = e^x.$

(c) $\varphi(x) = \begin{cases} 0 & \text{if } |x| \leq 1, \\ +\infty & \text{if } |x| > 1. \end{cases}$

(d) $\varphi(x) = \begin{cases} 0 & \text{if } x = 0, \\ +\infty & \text{if } x \neq 0. \end{cases}$

(e) $\varphi(x) = \begin{cases} -\log x & \text{if } x > 0, \\ +\infty & \text{if } x \leq 0. \end{cases}$

(f) $\varphi(x) = \begin{cases} -(1 - x^2)^{1/2} & \text{if } |x| \leq 1, \\ +\infty & \text{if } |x| > 1. \end{cases}$

(g) $\varphi(x) = \begin{cases} \frac{1}{2}|x|^2 & \text{if } |x| \leq 1, \\ |x| - \frac{1}{2} & \text{if } |x| > 1. \end{cases}$

(h) $\varphi(x) = \dfrac{1}{p}|x|^p,$ where $1 < p < \infty.$

(i) $\varphi(x) = x^+ = \max\{x, 0\}.$

(j) $\varphi(x) = \begin{cases} \frac{1}{p}x^p & \text{if } x \geq 0, \text{ where } 1 < p < +\infty, \\ +\infty & \text{if } x < 0. \end{cases}$

(k) $\varphi(x) = \begin{cases} -\frac{1}{p}x^p & \text{if } x \geq 0, \text{ where } 0 < p < 1, \\ +\infty & \text{if } x < 0. \end{cases}$

(l) $\varphi(x) = \dfrac{1}{p}[(|x| - 1)^+]^p,$ where $1 < p < \infty.$

$\boxed{1.19}$ Let E be an n.v.s.

1. Let $\varphi, \psi : E \to (-\infty, +\infty]$ be two functions such that $\varphi \leq \psi$. Prove that $\psi^* \leq \varphi^*$.
2. Let $F : \mathbb{R} \to (-\infty, +\infty]$ be a convex l.s.c. function such that $F(0) = 0$ and $F(t) \geq 0 \; \forall t \in \mathbb{R}$. Set $\varphi(x) = F(\|x\|)$.
   Prove that φ is convex l.s.c. and that $\varphi^*(f) = F^*(\|f\|) \; \forall f \in E^*$.

$\boxed{1.20}$ Let $E = \ell^p$ with $1 \leq p < \infty$ (see Section 11.3). Check that the functions $\varphi : E \to (-\infty, +\infty]$ defined below are convex l.s.c. and determine φ^*. For $x = (x_1, x_2, \ldots, x_n, \ldots)$ set

(a) $\varphi(x) = \begin{cases} \sum_{k=1}^{+\infty} k|x_k|^2 & \text{if } \sum_{k=1}^{\infty} k|x_k|^2 < +\infty, \\ +\infty & \text{otherwise.} \end{cases}$

(b) $\varphi(x) = \displaystyle\sum_{k=2}^{+\infty} |x_k|^k.$ (Check that $\varphi(x) < \infty$ for every $x \in E$.)

(c) $\varphi(x) = \begin{cases} \sum\limits_{k=1}^{+\infty} |x_k| & \text{if } \sum\limits_{k=1}^{\infty} |x_k| < +\infty, \\ +\infty & \text{otherwise.} \end{cases}$

$\boxed{1.21}$ Let $E = E^\star = \mathbb{R}^2$ and let

$$C = \{[x_1, x_2]; \; x_1 \geq 0, \; x_2 \geq 0\}.$$

On E define the function

$$\varphi(x) = \begin{cases} -\sqrt{x_1 x_2} & \text{if } x \in C, \\ +\infty & \text{if } x \notin C. \end{cases}$$

1. Prove that φ is convex l.s.c. on E.
2. Determine $\varphi^\star$.
3. Consider the set $D = \{[x_1, x_2]; \; x_1 = 0\}$ and the function $\psi = I_D$. Compute the value of the expressions

$$\inf_{x \in E} \{\varphi(x) + \psi(x)\} \quad \text{and} \quad \sup_{f \in E^\star} \{-\varphi^\star(-f) - \psi^\star(f)\}.$$

4. Compare with the conclusion of Theorem 1.12 and explain the difference.

$\boxed{1.22}$ Let E be an n.v.s. and let $A \subset E$ be a closed nonempty set. Let

$$\varphi(x) = \text{dist}(x, A) = \inf_{a \in A} \|x - a\|.$$

1. Check that $|\varphi(x) - \varphi(y)| \leq \|x - y\| \; \forall x, y \in E$.
2. Assuming that A is convex, prove that φ is convex.
3. Conversely, assuming that φ is convex, prove that A is convex.
4. Prove that $\varphi^\star = (I_A)^\star + I_{B_{E^\star}}$ for every A not necessarily convex.

$\boxed{1.23}$ *Inf-convolution.*

Let E be an n.v.s. Given two functions $\varphi, \psi : E \to (-\infty, +\infty]$, one defines the *inf-convolution* of φ and ψ as follows: for every $x \in E$, let

$$(\varphi \nabla \psi)(x) = \inf_{y \in E} \{\varphi(x - y) + \psi(y)\}.$$

Note the following:

(i) $(\varphi \nabla \psi)(x)$ may take the values $\pm\infty$,
(ii) $(\varphi \nabla \psi)(x) < +\infty$ iff $x \in D(\varphi) + D(\psi)$.

1. Assuming that $D(\varphi^\star) \cap D(\psi^\star) \neq \emptyset$, prove that $(\varphi \nabla \psi)$ does not take the value $-\infty$ and that

$$(\varphi \nabla \psi)^* = \varphi^* + \psi^*.$$

2. Assuming that $D(\varphi) \cap D(\psi) \neq \emptyset$, prove that

$$(\varphi + \psi)^* \leq (\varphi^* \nabla \psi^*) \text{ on } E^*.$$

3. Assume that φ and ψ are convex and there exists $x_0 \in D(\varphi) \cap D(\psi)$ such that φ is continuous at x_0. Prove that

$$(\varphi + \psi)^* = (\varphi^* \nabla \psi^*) \text{ on } E^*.$$

4. Assume that φ and ψ are convex and l.s.c., and that $D(\varphi) \cap D(\psi) \neq \emptyset$. Prove that

$$(\varphi^* \nabla \psi^*)^* = (\varphi + \psi) \text{ on } E.$$

Given a function $\varphi : E \to (-\infty, +\infty]$, set

$$\text{epist } \varphi = \{[x, \lambda] \in E \times \mathbb{R}; \ \varphi(x) < \lambda\}.$$

5. Check that φ is convex iff epist φ is a convex subset of $E \times \mathbb{R}$.
6. Let $\varphi, \psi : E \to (-\infty, +\infty]$ be functions such that $D(\varphi^*) \cap D(\psi^*) \neq \emptyset$. Prove that

$$\text{epist}(\varphi \nabla \psi) = (\text{epist } \varphi) + (\text{epist } \psi).$$

7. Deduce that if $\varphi, \psi : E \to (-\infty, +\infty]$ are convex functions such that $D(\varphi^*) \cap D(\psi^*) \neq \emptyset$, then $(\varphi \nabla \psi)$ is a convex function.

$\boxed{1.24}$ *Regularization by inf-convolution.*

Let E be an n.v.s. and let $\varphi : E \to (-\infty, +\infty]$ be a convex l.s.c. function such that $\varphi \not\equiv +\infty$. Our aim is to construct a sequence of functions (φ_n) such that we have the following:

(i)   For every n, $\varphi_n : E \to (-\infty, +\infty)$ is convex and continuous.
(ii) For every x, the sequence $(\varphi_n(x))_n$ is nondecreasing and converges to $\varphi(x)$.

For this purpose, let

$$\varphi_n(x) = \inf_{y \in E} \{n\|x - y\| + \varphi(y)\}.$$

1. Prove that there is some N, large enough, such that for $n \geq N$, $\varphi_n(x)$ is finite for all $x \in E$. From now on, one chooses $n \geq N$.
2. Prove that φ_n is convex (see Exercise 1.23) and that

$$|\varphi_n(x_1) - \varphi_n(x_2)| \leq n\|x_1 - x_2\| \quad \forall x_1, x_2 \in E.$$

3. Determine $(\varphi_n)^*$.
4. Check that $\varphi_n(x) \leq \varphi(x) \ \forall x \in E, \forall n$. Prove that for every $x \in E$, the sequence $(\varphi_n(x))_n$ is nondecreasing.

5. Given $x \in D(\varphi)$, choose $y_n \in E$ such that

$$\varphi_n(x) \leq n\|x - y_n\| + \varphi(y_n) \leq \varphi_n(x) + \frac{1}{n}.$$

Prove that $\lim_{n \to \infty} y_n = x$ and deduce that $\lim_{n \to \infty} \varphi_n(x) = \varphi(x)$.

6. For $x \notin D(\varphi)$, prove that $\lim_{n \to \infty} \varphi_n(x) = +\infty$.
   [**Hint:** Argue by contradiction.]

$\boxed{1.25}$ *A semiscalar product.*

Let E be an n.v.s.

1. Let $\varphi : E \to (-\infty, +\infty)$ be convex. Given $x, y \in E$, consider the function

$$h(t) = \frac{\varphi(x + ty) - \varphi(x)}{t}, \quad t > 0.$$

Check that h is nondecreasing on $(0, +\infty)$ and deduce that

$$\lim_{t \downarrow 0} h(t) = \inf_{t > 0} h(t) \text{ exists in } [-\infty, +\infty).$$

Define the semiscalar product $[x, y]$ by

$$[x, y] = \inf_{t > 0} \frac{1}{2t}[\|x + ty\|^2 - \|x\|^2].$$

2. Prove that $|[x, y]| \leq \|x\|\|y\| \quad \forall x, y \in E$.
3. Prove that

$$[x, \lambda x + \mu y] = \lambda\|x\|^2 + \mu[x, y] \quad \forall x, y \in E, \quad \forall \lambda \in \mathbb{R}, \quad \forall \mu \geq 0$$

and

$$[\lambda x, \mu y] = \lambda \mu[x, y] \quad \forall x, y \in E, \quad \forall \lambda \geq 0, \quad \forall \mu \geq 0.$$

4. Prove that for every $x \in E$, the function $y \mapsto [x, y]$ is convex. Prove that the function $G(x, y) = -[x, y]$ is l.s.c. on $E \times E$.
5. Prove that

$$[x, y] = \max_{f \in F(x)} \langle f, y \rangle \quad \forall x, y \in E,$$

where F denotes the duality map (see Remark 2 following Corollary 1.3 and Exercise 1.1).

[**Hint:** Set $\alpha = [x, y]$ and apply Theorem 1.12 to the functions φ and ψ defined as follows:

$$\varphi(z) = \frac{1}{2}\|x + z\|^2 - \frac{1}{2}\|x\|^2, \quad z \in E,$$

and

$$\psi(z) = \begin{cases} -t\alpha & \text{when } z = ty \text{ and } t \geq 0, \\ +\infty & \text{otherwise.}\end{cases}]$$

6. Determine explicitly $[x, y]$, where $E = \mathbb{R}^n$ with the norm $\|x\|_p$, $1 \le p \le \infty$ (see Section 11.3).

[**Hint:** Use the results of Exercise 1.2.]

1.26 *Strictly convex norms and functions.*

Let E be an n.v.s. One says that the *norm* $\| \ \|$ is *strictly convex* (or that the *space* E is *strictly convex*) if

$$\|tx + (1 - t)y\| < 1, \quad \forall x, y \in E \text{ with } x \ne y, \ \|x\| = \|y\| = 1, \quad \forall t \in (0, 1).$$

One says that a *function* $\varphi : E \to (-\infty, +\infty]$ is *strictly convex* if

$$\varphi(tx + (1 - t)y) < t\varphi(x) + (1 - t)\varphi(y) \quad \forall x, y \in E \text{ with } x \ne y, \quad \forall t \in (0, 1).$$

1. Prove that the *norm* $\| \ \|$ is strictly convex iff the *function* $\varphi(x) = \|x\|^2$ is strictly convex.
2. Same question with $\varphi(x) = \|x\|^p$ and $1 < p < \infty$.

1.27 Let E and F be two Banach spaces and let $G \subset E$ be a closed subspace. Let $T : G \to F$ be a continuous linear map. The aim is to show that sometimes, T cannot be extended by a continuous linear map $\widetilde{T} : E \to F$. For this purpose, let E be a Banach space and let $G \subset E$ be a closed subspace that admits no complement (see Remark 8 in Chapter 2). Let $F = G$ and $T = I$ (the identity map). Prove that T cannot be extended.

[**Hint:** Argue by contradiction.]

Compare with the conclusion of Corollary 1.2.

Chapter 2
The Uniform Boundedness Principle and the Closed Graph Theorem

2.1 The Baire Category Theorem

The following classical result plays an essential role in the proofs of Chapter 2.

• **Theorem 2.1 (Baire).** *Let X be a complete metric space and let $(X_n)_{n\geq 1}$ be a sequence of closed subsets in X. Assume that*

$$\text{Int } X_n = \emptyset \quad \text{for every } n \geq 1.$$

Then

$$\text{Int}\left(\bigcup_{n=1}^{\infty} X_n\right) = \emptyset.$$

Remark 1. The Baire category theorem is often used in the following form. Let X be a nonempty complete metric space. Let $(X_n)_{n\geq 1}$ be a sequence of closed subsets such that

$$\bigcup_{n=1}^{\infty} X_n = X.$$

Then there exists some n_0 such that $\text{Int } X_{n_0} \neq \emptyset$.

Proof. Set $O_n = X_n^c$, so that O_n is open and dense in X for every $n \geq 1$. Our aim is to prove that $G = \bigcap_{n=1}^{\infty} O_n$ is dense in X. Let ω be a nonempty open set in X; we shall prove that $\omega \cap G \neq \emptyset$.

As usual, set

$$B(x,r) = \{y \in X; \ d(y,x) < r\}.$$

Pick any $x_0 \in \omega$ and $r_0 > 0$ such that

$$\overline{B(x_0, r_0)} \subset \omega.$$

Then, choose $x_1 \in B(x_0, r_0) \cap O_1$ and $r_1 > 0$ such that

H. Brezis, *Functional Analysis, Sobolev Spaces and Partial Differential Equations*, DOI 10.1007/978-0-387-70914-7_2, © Springer Science+Business Media, LLC 2011

$$\begin{cases} \overline{B(x_1, r_1)} \subset B(x_0, r_0) \cap O_1, \\ 0 < r_1 < \frac{r_0}{2}, \end{cases}$$

which is always possible since O_1 is open and dense. By induction one constructs two sequences (x_n) and (r_n) such that

$$\begin{cases} \overline{B(x_{n+1}, r_{n+1})} \subset B(x_n, r_n) \cap O_{n+1}, \quad \forall n \geq 0, \\ 0 < r_{n+1} < \frac{r_n}{2}. \end{cases}$$

It follows that (x_n) is a Cauchy sequence; let $x_n \to \ell$.

Since $x_{n+p} \in B(x_n, r_n)$ for every $n \geq 0$ and for every $p \geq 0$, we obtain at the limit (as $p \to \infty$),

$$\ell \in \overline{B(x_n, r_n)}, \quad \forall n \geq 0.$$

In particular, $\ell \in \omega \cap G$.

2.2 The Uniform Boundedness Principle

Notation. Let E and F be two n.v.s. We denote by $\mathcal{L}(E, F)$ the space of *continuous* (= bounded) *linear* operators from E into F equipped with the norm

$$\|T\|_{\mathscr{L}(E,F)} = \sup_{\substack{x \in E \\ \|x\| \leq 1}} \|Tx\|.$$

As usual, one writes $\mathscr{L}(E)$ instead of $\mathscr{L}(E, E)$.

• **Theorem 2.2 (Banach–Steinhaus, uniform boundedness principle).** *Let E and F be two Banach spaces and let $(T_i)_{i \in I}$ be a family (not necessarily countable) of continuous linear operators from E into F. Assume that*

(1) $$\sup_{i \in I} \|T_i x\| < \infty \quad \forall x \in E.$$

Then

(2) $$\sup_{i \in I} \|T_i\|_{\mathscr{L}(E,F)} < \infty.$$

In other words, there exists a constant c such that

$$\|T_i x\| \leq c\|x\| \quad \forall x \in E, \quad \forall i \in I.$$

Remark 2. The conclusion of Theorem 2.2 is quite remarkable and surprising. From *pointwise estimates* one derives a *global* (uniform) *estimate*.

Proof. For every $n \geq 1$, let

$$X_n = \{x \in E; \quad \forall i \in I, \quad \|T_i x\| \leq n\},$$

so that X_n is closed, and by (1) we have

$$\bigcup_{n=1}^{\infty} X_n = E.$$

It follows from the Baire category theorem that $\text{Int}(X_{n_0}) \neq \emptyset$ for some $n_0 \geq 1$. Pick $x_0 \in E$ and $r > 0$ such that $B(x_0, r) \subset X_{n_0}$. We have

$$\|T_i(x_0 + rz)\| \leq n_0 \quad \forall i \in I, \quad \forall z \in B(0, 1).$$

This leads to

$$r \|T_i\|_{\mathcal{L}(E,F)} \leq n_0 + \|T_i x_0\|,$$

which implies (2).

Remark 3. Recall that in general, a *pointwise limit* of continuous maps need *not* be continuous. The linearity assumption plays an essential role in Theorem 2.2. Note, however, that in the setting of Theorem 2.2 it does *not* follow that $\|T_n - T\|_{\mathcal{L}(E,F)} \to 0$.

Here are a few direct consequences of the uniform boundedness principle.

Corollary 2.3. *Let E and F be two Banach spaces. Let (T_n) be a sequence of continuous linear operators from E into F such that for every $x \in E$, $T_n x$ converges (as $n \to \infty$) to a limit denoted by Tx. Then we have*

(a) $\sup_n \|T_n\|_{\mathcal{L}(E,F)} < \infty$,
(b) $T \in \mathcal{L}(E, F)$,
(c) $\|T\|_{\mathcal{L}(E,F)} \leq \liminf_{n \to \infty} \|T_n\|_{\mathcal{L}(E,F)}$.

Proof. (a) follows directly from Theorem 2.2, and thus there exists a constant c such that

$$\|T_n x\| \leq c \|x\| \quad \forall n, \quad \forall x \in E.$$

At the limit we find

$$\|Tx\| \leq c \|x\| \quad \forall x \in E.$$

Since T is clearly linear, we obtain (b).

Finally, we have

$$\|T_n x\| \leq \|T_n\|_{\mathcal{L}(E,F)} \|x\| \quad \forall x \in E,$$

and (c) follows directly.

• **Corollary 2.4.** *Let G be a Banach space and let B be a subset of G. Assume that*

(3)   *for every $f \in G^\star$ the set $f(B) = \{\langle f, x \rangle; \ x \in B\}$ is bounded (in $\mathbb{R}$).*

Then

(4)   *B is bounded.*

Proof. We shall use Theorem 2.2 with $E = G^\star$, $F = \mathbb{R}$, and $I = B$. For every $b \in B$, set

$$T_b(f) = \langle f, b \rangle, \quad f \in E = G^\star,$$

so that by (3),

$$\sup_{b \in B} |T_b(f)| < \infty \quad \forall f \in E.$$

It follows from Theorem 2.2 that there exists a constant c such that

$$|\langle f, b \rangle| \le c \|f\| \quad \forall f \in G^\star \quad \forall b \in B.$$

Therefore we find (using Corollary 1.4) that

$$\|b\| \le c \quad \forall b \in B.$$

Remark 4. Corollary 2.4 says that in order to prove that a set B is bounded it suffices to "look" at B through the bounded linear functionals. This is a familiar procedure in finite-dimensional spaces, where the linear functionals are the components with respect to some basis. In some sense, Corollary 2.4 replaces, in infinite-dimensional spaces, the use of components. Sometimes, one expresses the conclusion of Corollary 2.4 by saying that "weakly bounded" $\Longleftrightarrow$ "strongly bounded" (see Chapter 3).

Next we have a statement dual to Corollary 2.4:

Corollary 2.5. *Let G be a Banach space and let $B^\star$ be a subset of $G^\star$. Assume that*

(5)   *for every $x \in G$ the set $\langle B^\star, x \rangle = \{\langle f, x \rangle; f \in B^\star\}$ is bounded (in $\mathbb{R}$).*

Then

(6) $B^\star$ *is bounded.*

Proof. Use Theorem 2.2 with $E = G$, $F = \mathbb{R}$, and $I = B^\star$. For every $b \in B^\star$ set

$$T_b(x) = \langle b, x \rangle \quad (x \in G = E).$$

We find that there exists a constant c such that

$$|\langle b, x \rangle| \le c \|x\| \quad \forall b \in B^\star, \quad \forall x \in G.$$

We conclude (from the definition of a dual norm) that

$$\|b\| \le c \quad \forall b \in B^\star.$$

2.3 The Open Mapping Theorem and the Closed Graph Theorem

Here are two basic results due to Banach.

• **Theorem 2.6 (open mapping theorem).** *Let E and F be two Banach spaces and let T be a continuous linear operator from E into F that is **surjective** (= onto). Then there exists a constant $c > 0$ such that*

$$(7) \qquad\qquad T(B_E(0, 1)) \supset B_F(0, c).$$

Remark 5. Property (7) implies that the image under T of any open set in E is an open set in F (which justifies the name given to this theorem!). Indeed, let us suppose U is open in E and let us prove that $T(U)$ is open. Fix any point $y_0 \in T(U)$, so that $y_0 = Tx_0$ for some $x_0 \in U$. Let $r > 0$ be such that $B(x_0, r) \subset U$, i.e., $x_0 + B(0, r) \subset U$. It follows that

$$y_0 + T(B(0, r)) \subset T(U).$$

Using (7) we obtain
$$T(B(0, r)) \supset B(0, rc)$$

and therefore
$$B(y_0, rc) \subset T(U).$$

Some important consequences of Theorem 2.6 are the following.

• **Corollary 2.7.** *Let E and F be two Banach spaces and let T be a continuous linear operator from E into F that is **bijective**, i.e., injective (= one-to-one) and surjective. Then T^{-1} is also continuous (from F into E).*

Proof of Corollary 2.7. Property (7) and the assumption that T is injective imply that if $x \in E$ is chosen so that $\|Tx\| < c$, then $\|x\| < 1$. By homogeneity, we find that

$$\|x\| \le \frac{1}{c}\|Tx\| \quad \forall x \in E$$

and therefore T^{-1} is continuous.

Corollary 2.8. *Let E be a vector space provided with two norms, $\|\ \|_1$ and $\|\ \|_2$. Assume that E is a Banach space for **both** norms and that there exists a constant $C \ge 0$ such that*
$$\|x\|_2 \le C\|x\|_1 \quad \forall x \in E.$$

*Then the two norms are **equivalent**, i.e., there is a constant $c > 0$ such that*

$$\|x\|_1 \le c\|x\|_2 \quad \forall x \in E.$$

Proof of Corollary 2.8. Apply Corollary 2.7 with

$$E = (E, \|\ \|_1), \ \ F = (E, \|\ \|_2), \ \text{and } T = I.$$

Proof of Theorem 2.6. We split the argument into two steps:

Step 1. Assume that T is a linear surjective operator from E onto F. Then there exists a constant $c > 0$ such that

(8) $$\overline{T(B(0,1))} \supset B(0,2c).$$

Proof. Set $X_n = n\overline{T(B(0,1))}$. Since T is surjective, we have $\bigcup_{n=1}^{\infty} X_n = F$, and by the Baire category theorem there exists some n_0 such that $\mathrm{Int}(X_{n_0}) \neq \emptyset$. It follows that

$$\mathrm{Int}\,[\overline{T(B(0,1))}] \neq \emptyset.$$

Pick $c > 0$ and $y_0 \in F$ such that

(9) $$B(y_0, 4c) \subset \overline{T(B(0,1))}.$$

In particular, $y_0 \in \overline{T(B(0,1))}$, and by symmetry,

(10) $$-y_0 \in \overline{T(B(0,1))}.$$

Adding (9) and (10) leads to

$$B(0,4c) \subset \overline{T(B(0,1))} + \overline{T(B(0,1))}.$$

On the other hand, since $\overline{T(B(0,1))}$ is convex, we have

$$\overline{T(B(0,1))} + \overline{T(B(0,1))} = 2\overline{T(B(0,1))},$$

and (8) follows.

Step 2. Assume T is a continuous linear operator from E into F that satisfies (8). Then we have

(11) $$T(B(0,1)) \supset B(0,c).$$

Proof. Choose any $y \in F$ with $\|y\| < c$. The aim is to find some $x \in E$ such that

$$\|x\| < 1 \quad \text{and} \quad Tx = y.$$

By (8) we know that

(12) $$\forall \varepsilon > 0 \quad \exists z \in E \text{ with } \|z\| < \frac{1}{2} \text{ and } \|y - Tz\| < \varepsilon.$$

Choosing $\varepsilon = c/2$, we find some $z_1 \in E$ such that

$$\|z_1\| < \frac{1}{2} \quad \text{and} \quad \|y - Tz_1\| < \frac{c}{2}.$$

By the same construction applied to $y - Tz_1$ (instead of y) with $\varepsilon = c/4$ we find some $z_2 \in E$ such that

$$\|z_2\| < \frac{1}{4} \text{ and } \|(y - Tz_1) - Tz_2\| < \frac{c}{4}.$$

Proceeding similarly, by induction we obtain a sequence (z_n) such that

$$\|z_n\| < \frac{1}{2^n} \quad \text{and} \quad \|y - T(z_1 + z_2 + \cdots + z_n)\| < \frac{c}{2^n} \quad \forall n.$$

It follows that the sequence $x_n = z_1 + z_2 + \cdots + z_n$ is a Cauchy sequence. Let $x_n \to x$ with, clearly, $\|x\| < 1$ and $y = Tx$ (since T is continuous).

• **Theorem 2.9 (closed graph theorem).** *Let E and F be two Banach spaces. Let T be a linear operator from E into F. Assume that the graph of T, $G(T)$, is closed in $E \times F$. Then T is continuous.*

Remark 6. The converse is obviously true, since the graph of any continuous map (linear or not) is closed.

Proof of Theorem 2.9. Consider, on E, the two norms

$$\|x\|_1 = \|x\|_E + \|Tx\|_F \quad \text{and} \quad \|x\|_2 = \|x\|_E$$

(the norm $\| \ \|_1$ is called the *graph norm*).

It is easy to check, using the assumption that $G(T)$ is closed, that E is a Banach space for the norm $\| \ \|_1$. On the other hand, E is also a Banach space for the norm $\| \ \|_2$ and $\| \ \|_2 \leq \| \ \|_1$. It follows from Corollary 2.8 that the two norms are equivalent and thus there exists a constant $c > 0$ such that $\|x\|_1 \leq c\|x\|_2$. We conclude that $\|Tx\|_F \leq c\|x\|_E$.

★ 2.4 Complementary Subspaces. Right and Left Invertibility of Linear Operators

We start with some geometric properties of closed subspaces in a Banach space that follow from the open mapping theorem.

★ **Theorem 2.10.** *Let E be a Banach space. Assume that G and L are two closed linear subspaces such that $G + L$ is closed. Then there exists a constant $C \geq 0$ such that*

(13) $$\begin{cases} \text{every } z \in G + L \text{ admits a decomposition of the form} \\ z = x + y \text{ with } x \in G, y \in L, \|x\| \leq C\|z\| \text{ and } \|y\| \leq C\|z\|. \end{cases}$$

Proof. Consider the product space $G \times L$ with its norm

$$\| [x, y] \| = \|x\| + \|y\|$$

and the space $G + L$ provided with the norm of E.

The mapping $T : G \times L \to G + L$ defined by $T[x, y] = x + y$ is continuous, linear, and surjective. By the open mapping theorem there exists a constant $c > 0$ such that every $z \in G + L$ with $\|z\| < c$ can be written as $z = x + y$ with $x \in G$, $y \in L$, and $\|x\| + \|y\| < 1$. By homogeneity every $z \in G + L$ can be written as

$$z = x + y \quad \text{with } x \in G, \, y \in L, \text{ and } \|x\| + \|y\| \le (1/c)\|z\|.$$

⋆ **Corollary 2.11.** *Under the same assumptions as in Theorem 2.10, there exists a constant C such that*

(14) $$\text{dist}(x, G \cap L) \le C\{\text{dist}(x, G) + \text{dist}(x, L)\} \quad \forall x \in E.$$

Proof. Given $x \in E$ and $\varepsilon > 0$, there exist $a \in G$ and $b \in L$ such that

$$\|x - a\| \le \text{dist}(x, G) + \varepsilon, \quad \|x - b\| \le \text{dist}(x, L) + \varepsilon.$$

Property (13) applied to $z = a - b$ says that there exist $a' \in G$ and $b' \in L$ such that

$$a - b = a' + b', \quad \|a'\| \le C\|a - b\|, \quad \|b'\| \le C\|a - b\|.$$

It follows that $a - a' \in G \cap L$ and

$$\begin{aligned}
\text{dist}(x, G \cap L) &\le \|x - (a - a')\| \le \|x - a\| + \|a'\| \\
&\le \|x - a\| + C\|a - b\| \le \|x - a\| + C(\|x - a\| + \|x - b\|) \\
&\le (1 + C)\,\text{dist}(x, G) + \text{dist}(x, L) + (1 + 2C)\varepsilon.
\end{aligned}$$

Finally, we obtain (14) by letting $\varepsilon \to 0$.

Remark 7. The converse of Corollary 2.11 is also true: If G and L are two closed linear subspaces such that (14) holds, then $G + L$ is closed (see Exercise 2.16).

Definition. Let $G \subset E$ be a *closed* subspace of a Banach space E. A subspace $L \subset E$ is said to be a *topological complement* or simply a *complement* of G if

(i) L is *closed*,
(ii) $G \cap L = \{0\}$ and $G + L = E$.

We shall also say that G and L are *complementary* subspaces of E. If this holds, then every $z \in E$ may be uniquely written as $z = x + y$ with $x \in G$ and $y \in L$. It follows from Theorem 2.10 that the *projection operators* $z \mapsto x$ and $z \mapsto y$ are *continuous* linear operators. (That property could also serve as a definition of complementary subspaces.)

Examples

1. Every *finite-dimensional* subspace G admits a complement. Indeed, let e_1, $e_2, \dots, e_n$ be a basis of G. Every $x \in G$ may be written as $x = \sum_{i=1}^{n} x_i e_i$. Set $\varphi_i(x) = x_i$. Using Hahn–Banach (analytic form)—or more precisely Corollary 1.2—each φ_i can be extended by a continuous linear functional $\tilde{\varphi}_i$ defined on E. It is easy to check that $L = \cap_{i=1}^{n}(\tilde{\varphi}_i)^{-1}(0)$ is a complement of G.

2. Every *closed* subspace G of *finite codimension* admits a complement. It suffices to choose any finite-dimensional space L such that $G \cap L = \{0\}$ and $G + L = E$ (L is closed since it is finite-dimensional).

Here is a typical example of this kind of situation. Let $N \subset E^*$ be a subspace of dimension p. Then

$$G = \{x \in E; \ \langle f, x \rangle = 0 \quad \forall f \in N\} = N^{\perp}$$

is closed and of codimension p. Indeed, let $f_1, f_2, \dots, f_p$ be a basis of N. Then there exist $e_1, e_2, \dots, e_p \in E$ such that

$$\langle f_i, e_j \rangle = \delta_{ij} \quad \forall i, j = 1, 2, \dots, p.$$

[Consider the map $\Phi : E \to \mathbb{R}^p$ defined by

$$(15) \qquad\qquad \Phi(x) = (\langle f_1, x \rangle, \langle f_2, x \rangle, \dots, \langle f_p, x \rangle)$$

and note that Φ is surjective; otherwise, there would exist—by Hahn–Banach (second geometric form)—some $\alpha = (\alpha_1, \alpha_2, \dots, \alpha_p) \neq 0$ such that

$$\alpha \cdot \Phi(x) = \left\langle \sum_{i=1}^{p} \alpha_i f_i, x \right\rangle = 0 \quad \forall x \in E,$$

which is absurd].

It is easy to check that the vectors $(e_i)_{1 \le i \le p}$ are linearly independent and that the space generated by the e_i's is a complement of G. Another proof of the fact that the codimension of $N^{\perp}$ equals the dimension of N is presented in Chapter 11 (Proposition 11.11).

3. In a Hilbert space every closed subspace admits a complement (see Section 5.2).

Remark 8. It is important to know that some closed subspaces (even in reflexive Banach spaces) have *no* complement. In fact, a remarkable result of J. Lindenstrauss and L. Tzafriri [1] asserts that in *every* Banach space that is not isomorphic to a Hilbert space, there exist closed subspaces *without* any complement.

Definition. Let $T \in \mathcal{L}(E, F)$. A *right inverse* of T is an operator $S \in \mathcal{L}(F, E)$ such that $T \circ S = I_F$. A *left inverse* of T is an operator $S \in \mathcal{L}(F, E)$ such that $S \circ T = I_E$.

Our next results provide necessary and sufficient conditions for the existence of such inverses.

★ **Theorem 2.12.** *Assume that $T \in \mathcal{L}(E, F)$ is* **surjective**. *The following properties are equivalent:*

(i) *T admits a right inverse.*
(ii) *$N(T) = T^{-1}(0)$ admits a complement in E.*

Proof.
(i) $\Rightarrow$ (ii). Let S be a right inverse of T. It is easy to see (please check) that $R(S) = S(F)$ is a complement of $N(T)$ in E.

(ii) $\Rightarrow$ (i). Let L be a complement of $N(T)$. Let P be the (continuous) projection operator from E onto L. Given $f \in F$, we denote by x any solution of the equation $Tx = f$. Set $Sf = Px$ and note that S is independent of the choice of x. It is easy to check that $S \in \mathcal{L}(F, E)$ and that $T \circ S = I_F$.

Remark 9. In view of Remark 8 and Theorem 2.12, it is easy to construct surjective operators T without a right inverse. Indeed, let $G \subset E$ be a closed subspace without complement, let $F = E/G$, and let T be the canonical projection from E onto F (for the definition and properties of the quotient space, see Section 11.2).

⋆ **Theorem 2.13.** *Assume that $T \in \mathcal{L}(E, F)$ is **injective**. The following properties are equivalent:*

(i) *T admits a left inverse.*
(ii) *$R(T) = T(E)$ is closed and admits a complement in F.*

Proof.
    (i) $\Rightarrow$ (ii). It is easy to check that $R(T)$ is closed and that $N(S)$ is a complement of $R(T)$ [write $f = TSf + (f - TSf)$].
    (ii) $\Rightarrow$ (i). Let P be a continuous projection operator from F onto $R(T)$. Let $f \in F$; since $Pf \in R(T)$, there exists a unique $x \in E$ such that $Tx = Pf$. Set $Sf = x$. It is clear that $S \circ T = I_E$; moreover, S is continuous by Corollary 2.7.

⋆ 2.5 Orthogonality Revisited

There are some simple formulas giving the orthogonal expression of a sum or of an intersection.

Proposition 2.14. *Let G and L be two closed subspaces in E. Then*

(16)
$$\boxed{G \cap L = (G^{\perp} + L^{\perp})^{\perp},}$$

(17)
$$\boxed{G^{\perp} \cap L^{\perp} = (G + L)^{\perp}.}$$

Proof of (16). It is clear that $G \cap L \subset (G^{\perp} + L^{\perp})^{\perp}$; indeed, if $x \in G \cap L$ and $f \in G^{\perp} + L^{\perp}$ then $\langle f, x \rangle = 0$. Conversely, we have $G^{\perp} \subset G^{\perp} + L^{\perp}$ and thus $(G^{\perp} + L^{\perp})^{\perp} \subset G^{\perp \perp} = G$ (note that if $N_1 \subset N_2$ then $N_2^{\perp} \subset N_1^{\perp}$); similarly $(G^{\perp} + L^{\perp})^{\perp} \subset L$. Therefore $(G^{\perp} + L^{\perp})^{\perp} \subset G \cap L$.

Proof of (17). Use the same argument as for the proof of (16).

Corollary 2.15. *Let G and L be two closed subspaces in E. Then*

(18)
$$(G \cap L)^{\perp} \supset \overline{G^{\perp} + L^{\perp}},$$

(19)
$$(G^{\perp} \cap L^{\perp})^{\perp} = \overline{G + L}.$$

Proof. Use Propositions 1.9 and 2.14.

Here is a deeper result.

★ **Theorem 2.16.** *Let G and L be two closed subspaces in a Banach space E. The following properties are equivalent:*

(a) *$G + L$ is closed in E,*
(b) *$G^\perp + L^\perp$ is closed in $E^\star$,*
(c) *$G + L = (G^\perp \cap L^\perp)^\perp$,*
(d) *$G^\perp + L^\perp = (G \cap L)^\perp$.*

Proof. (a) $\Longleftrightarrow$ (c) follows from (19). (d) $\Longrightarrow$ (b) is obvious.
    We are left with the implications (a) $\Rightarrow$ (d) and (b) $\Rightarrow$ (a).

    (a) $\Longrightarrow$ (d). In view of (18) it suffices to prove that $(G \cap L)^\perp \subset G^\perp + L^\perp$. Given $f \in (G \cap L)^\perp$, consider the functional $\varphi : G + L \to \mathbb{R}$ defined as follows. For every $x \in G + L$ write $x = a + b$ with $a \in G$ and $b \in L$. Set

$$\varphi(x) = \langle f, a \rangle.$$

Clearly, φ is independent of the decomposition of x, and φ is linear. On the other hand, by Theorem 2.10 we may choose a decomposition of x in such a way that $\|a\| \leq C \|x\|$, and thus

$$|\varphi(x)| \leq C \|x\| \quad \forall x \in G + L.$$

Extend φ by a continuous linear functional $\tilde{\varphi}$ defined on all of E (see Corollary 1.2). So, we have

$$f = (f - \tilde{\varphi}) + \tilde{\varphi} \quad \text{with} \quad f - \tilde{\varphi} \in G^\perp \quad \text{and} \quad \tilde{\varphi} \in L^\perp.$$

    (b) $\Longrightarrow$ (a). We know by Corollary 2.11 that there exists a constant C such that

$$(20) \qquad \operatorname{dist}(f, G^\perp \cap L^\perp) \leq C\{\operatorname{dist}(f, G^\perp) + \operatorname{dist}(f, L^\perp)\} \quad \forall f \in E^\star.$$

On the other hand, we have

$$(21) \qquad \operatorname{dist}(f, G^\perp) = \sup_{\substack{x \in G \\ \|x\| \leq 1}} \langle f, x \rangle \quad \forall f \in E^\star.$$

[Use Theorem 1.12 with $\varphi(x) = I_{B_E}(x) - \langle f, x \rangle$ and $\psi(x) = I_G(x)$, where

$$B_E = \{x \in E;\ \|x\| \leq 1\}.]$$

Similarly, we have

(22) $$\operatorname{dist}(f, L^{\perp}) = \sup_{\substack{x \in L \\ \|x\| \le 1}} \langle f, x \rangle \quad \forall f \in E^{\star}$$

and also (by (17))

(23) $$\operatorname{dist}(f, G^{\perp} \cap L^{\perp}) = \operatorname{dist}(f, (G + L)^{\perp}) = \sup_{\substack{x \in \overline{G+L} \\ \|x\| \le 1}} \langle f, x \rangle \quad \forall f \in E^{\star}.$$

Combining (20), (21), (22), and (23) we obtain

(24) $$\sup_{\substack{x \in \overline{G+L} \\ \|x\| \le 1}} \langle f, x \rangle \le C \left\{ \sup_{\substack{x \in G \\ \|x\| \le 1}} \langle f, x \rangle + \sup_{\substack{x \in L \\ \|x\| \le 1}} \langle f, x \rangle \right\} \quad \forall f \in E^{\star}.$$

It follows from (24) that

(25) $$B_G + G_L \supset \frac{1}{C} B_{\overline{G+L}}.$$

Indeed, suppose by contradiction that there existed some $x_0 \in \overline{G + L}$ with $\|x_0\| \le 1/C$ and $x_0 \notin \overline{B_G + B_L}$. Then there would be a closed hyperplane in E strictly separating $\{x_0\}$ and $\overline{B_G + B_L}$. Thus, there would exist some $f_0 \in E^{\star}$ and some $\alpha \in \mathbb{R}$ such that

$$\langle f_0, x \rangle < \alpha < \langle f_0, x_0 \rangle \quad \forall x \in B_G + B_L.$$

Therefore, we would have

$$\sup_{\substack{x \in G \\ \|x\| \le 1}} \langle f_0, x \rangle + \sup_{\substack{x \in L \\ \|x\| \le 1}} \langle f_0, x \rangle \le \alpha < \langle f_0, x_0 \rangle,$$

which contradicts (24), and (25) is proved.

Finally, consider the space $X = G \times L$ with the norm

$$\| [x, y] \| = \max\{\|x\|, \|y\|\}$$

and the space $Y = \overline{G + L}$ with the norm of E. The map $T : X \to Y$ defined by $T([x, y]) = x + y$ is linear and continuous. From (25) we know that

$$\overline{T(B_X)} \supset \frac{1}{C} B_Y.$$

Using Step 2 from the proof of Theorem 2.6 (open mapping theorem) we conclude that

$$T(B_X) \supset \frac{1}{2C} B_Y.$$

It follows that T is surjective from X onto Y, i.e., $G + L = \overline{G + L}$.

2.6 An Introduction to Unbounded Linear Operators. Definition of the Adjoint

Definition. Let E and F be two Banach spaces. An *unbounded linear operator* from E into F is a linear map $A : D(A) \subset E \to F$ defined on a linear subspace $D(A) \subset E$ with values in F. The set $D(A)$ is called the *domain* of A.

One says that A *is bounded* (or *continuous*) if $D(A) = E$ and if there is a constant $c \geq 0$ such that

$$\|Au\| \leq c\|u\| \quad \forall u \in E.$$

The norm of a bounded operator is defined by

$$\|A\|_{\mathscr{L}(E,F)} = \operatorname*{Sup}_{u \neq 0} \frac{\|Au\|}{\|u\|}.$$

Remark 10. It may of course happen that an unbounded linear operator turns out to be bounded. This terminology is slightly inconsistent, but it is commonly used and does not lead to any confusion.

Here are some important definitions and further notation:

$$\text{Graph of } A = G(A) = \{[u, Au]; \, u \in D(A)\} \subset E \times F,$$

$$\text{Range of } A = R(A) = \{Au; \, u \in D(A)\} \subset F,$$

$$\text{Kernel of } A = N(A) = \{u \in D(A); \, Au = 0\} \subset E.$$

A map A is said to be *closed* if $G(A)$ is closed in $E \times F$.

• *Remark* 11. In order to prove that an operator A is closed, one proceeds in general as follows. Take a sequence (u_n) in $D(A)$ such that $u_n \to u$ in E and $Au_n \to f$ in F. Then check two facts:

(a) $u \in D(A)$,
(b) $f = Au$.

Note that it does *not* suffice to consider sequences (u_n) such that $u_n \to 0$ in E and $Au_n \to f$ in F (and to prove that $f = 0$).

Remark 12. If A is closed, then $N(A)$ is closed; however, $R(A)$ need not be closed.

Remark 13. *In practice, most* unbounded operators are *closed* and are *densely defined*, i.e., $D(A)$ is dense in E.

Definition of the adjoint $A^\star$. Let $A : D(A) \subset E \to F$ be an unbounded linear operator that is *densely defined*. We shall introduce an unbounded operator $A^\star : D(A^\star) \subset F^\star \to E^\star$ as follows. First, one defines its domain:

$$D(A^\star) = \{v \in F^\star; \, \exists c \geq 0 \text{ such that } |\langle v, Au \rangle| \leq c\|u\| \quad \forall u \in D(A)\}.$$

It is clear that $D(A^\star)$ is a linear subspace of $F^\star$. We shall now define $A^\star v$. Given $v \in D(A^\star)$, consider the map $g : D(A) \to \mathbb{R}$ defined by

$$g(u) = \langle v, Au \rangle \quad \forall u \in D(A).$$

We have

$$|g(u)| \leq c\|u\| \quad \forall u \in D(A).$$

By Hahn–Banach (analytic form; see Theorem 1.1) there exists a linear map $f : E \to \mathbb{R}$ that extends g and such that

$$|f(u)| \leq c\|u\| \quad \forall u \in E.$$

It follows that $f \in E^\star$. Note that the extension of g is *unique*, since $D(A)$ is *dense* in E.

Set

$$A^\star v = f.$$

The unbounded linear operator $A^\star \colon D(A^\star) \subset F^\star \to E^\star$ is called the *adjoint* of A. In brief, the fundamental relation between A and $A^\star$ is given by

$$\boxed{\langle v, Au \rangle_{F^\star, F} = \langle A^\star v, u \rangle_{E^\star, E} \quad \forall u \in D(A), \quad \forall v \in D(A^\star).}$$

Remark 14. It is not necessary to invoke Hahn–Banach to extend g. It suffices to use the classical *extension by continuity*, which applies since $D(A)$ is dense, g is uniformly continuous on $D(A)$, and $\mathbb{R}$ is complete (see, e.g., H. L. Royden [1] (Proposition 11 in Chapter 7) or J. Dugundji [1] (Theorem 5.2 in Chapter XIV).

⋆ *Remark* 15. It may happen that $D(A^\star)$ is not dense in $F^\star$ (even if A is closed); but this is a rather pathological situation (see Exercise 2.22). It is always true that if A is closed then $D(A^\star)$ is dense in $F^\star$ for the weak⋆ topology $\sigma(F^\star, F)$ defined in Chapter 3 (see Problem 9). In particular, if F is reflexive, then $D(A^\star)$ is dense in $F^\star$ for the usual (norm) topology (see Theorem 3.24).

Remark 16. If A is a bounded operator then $A^\star$ is also a bounded operator (from $F^\star$ into $E^\star$) and, moreover,

$$\boxed{\|A^\star\|_{\mathscr{L}(F^\star, E^\star)} = \|A\|_{\mathscr{L}(E,F)}.}$$

Indeed, it is clear that $D(A^\star) = F^\star$. From the basic relation, we have

$$|\langle A^\star v, u \rangle| \leq \|A\| \, \|u\| \, \|v\| \quad \forall u \in E, \quad \forall v \in F^\star,$$

which implies that $\|A^\star v\| \leq \|A\| \, \|v\|$ and thus $\|A^\star\| \leq \|A\|$.

We also have

$$|\langle v, Au \rangle| \leq \|A^\star\| \, \|u\| \, \|v\| \quad \forall u \in E, \quad \forall v \in F^\star,$$

which implies (by Corollary 1.4) that $\|Au\| \leq \|A^\star\| \, \|u\|$ and thus $\|A\| \leq \|A^\star\|$.

Proposition 2.17. *Let $A : D(A) \subset E \to F$ be a densely defined unbounded linear operator. Then $A^\star$ is closed, i.e., $G(A^\star)$ is closed in $F^\star \times E^\star$.*

Proof. Let $v_n \in D(A^\star)$ be such that $v_n \to v$ in $F^\star$ and $A^\star v_n \to f$ in $E^\star$. One has to check that (a) $v \in D(A^\star)$ and (b) $A^\star v = f$.
   We have

$$\langle v_n, Au \rangle = \langle A^\star v_n, u \rangle \quad \forall u \in D(A).$$

At the limit we obtain

$$\langle v, Au \rangle = \langle f, u \rangle \quad \forall u \in D(A).$$

Therefore $v \in D(A^\star)$ (since $|\langle v, Au \rangle| \leq \|f\| \, \|u\| \ \forall u \in D(A)$) and $A^\star v = f$.

   The graphs of A and $A^\star$ are related by a very simple orthogonality relation: Consider the isomorphism $I : F^\star \times E^\star \to E^\star \times F^\star$ defined by

$$I([v, f]) = [-f, v].$$

Let $A : D(A) \subset E \to F$ be a densely defined unbounded linear operator. Then

$$\boxed{I[G(A^\star)] = G(A)^\perp.}$$

Indeed, let $[v, f] \in F^\star \times E^\star$, then

$$[v, f] \in G(A^\star) \iff \langle f, u \rangle = \langle v, Au \rangle \quad \forall u \in D(A)$$
$$\iff -\langle f, u \rangle + \langle v, Au \rangle = 0 \quad \forall u \in D(A)$$
$$\iff [-f, v] \in G(A)^\perp.$$

   Here are some standard orthogonality relations between ranges and kernels:

Corollary 2.18. *Let $A : D(A) \subset E \to F$ be an unbounded linear operator that is densely defined and closed. Then*

(i)
$$N(A) = R(A^\star)^\perp,$$

(ii)
$$N(A^\star) = R(A)^\perp,$$

(iii)
$$N(A)^\perp \supset \overline{R(A^\star)},$$

(iv)
$$N(A^\star)^\perp = \overline{R(A)}.$$

Proof. Note that (iii) and (iv) follow directly from (i) and (ii) combined with Proposition 1.9. There is a simple and direct proof of (i) and (ii) (see Exercise 2.18). However, it is instructive to relate these facts to Proposition 2.14 by the following device. Consider the space $X = E \times F$, so that $X^\star = E^\star \times F^\star$, and the subspaces of X

$$G = G(A) \quad \text{and} \quad L = E \times \{0\}.$$

It is very easy to check that

(26) $$N(A) \times \{0\} = G \cap L,$$

(27) $$E \times R(A) = G + L,$$

(28) $$\{0\} \times N(A^\star) = G^\perp \cap L^\perp,$$

(29) $$R(A^\star) \times F^\star = G^\perp + L^\perp.$$

Proof of (i). By (29) we have

$$R(A^\star)^\perp \times \{0\} = (G^\perp + L^\perp)^\perp = G \cap L \quad \text{(by (16))}$$
$$= N(A) \times \{0\} \quad \text{(by (26))}.$$

Proof of (ii). By (27) we have

$$\{0\} \times R(A)^\perp = (G + L)^\perp = G^\perp \cap L^\perp \quad \text{(by (17))}$$
$$= \{0\} \times N(A^\star) \quad \text{(by (28))}.$$

Remark 17. It may happen, even if A is a bounded linear operator, that $N(A)^\perp \neq \overline{R(A^\star)}$ (see Exercise 2.23). However, it is always true that $N(A)^\perp$ is the closure of $R(A^\star)$ for the weak* topology $\sigma(E^\star, E)$ (see Problem 9). In particular, if E is reflexive then $N(A)^\perp = \overline{R(A^\star)}$.

⋆ 2.7 A Characterization of Operators with Closed Range. A Characterization of Surjective Operators

The main result concerning operators with closed range is the following.

⋆ Theorem 2.19. *Let* $A : D(A) \subset E \to F$ *be an unbounded linear operator that is densely defined and closed. The following properties are equivalent:*

  (i) $R(A)$ *is closed,*
  (ii) $R(A^\star)$ *is closed,*
  (iii) $R(A) = N(A^\star)^\perp,$
  (iv) $R(A^\star) = N(A)^\perp.$

Proof. With the same notation as in the proof of Corollary 2.18, we have

  (i) $\Leftrightarrow G + L$ is closed in X (see (27)),
  (ii) $\Leftrightarrow G^\perp + L^\perp$ is closed in $X^\star$ (see (29)),
  (iii) $\Leftrightarrow G + L = (G^\perp \cap L^\perp)^\perp$ (see (27) and (28)),
  (iv) $\Leftrightarrow (G \cap L)^\perp = G^\perp + L^\perp$ (see (26) and (29)).

The conclusion then follows from Theorem 2.16.

Remark 18. Let $A : D(A) \subset E \to F$ be a closed unbounded linear operator. Then $R(A)$ is closed if and only if there exists a constant C such that

$$\text{dist}(u, N(A)) \le C\|Au\| \quad \forall u \in D(A);$$

see Exercise 2.14.

The next result provides a useful characterization of *surjective* operators.

$\star$ **Theorem 2.20.** *Let* $A : D(A) \subset E \to F$ *be an unbounded linear operator that is densely defined and closed. The following properties are equivalent:*

(a) *A is surjective, i.e.,* $R(A) = F$,
(b) *there is a constant C such that*

$$\|v\| \le C\|A^\star v\| \quad \forall v \in D(A^\star),$$

(c) $N(A^\star) = \{0\}$ *and* $R(A^\star)$ *is closed.*

Remark 19. The implication (b) $\Rightarrow$ (a) is sometimes useful in practice to establish that an operator A is surjective. One proceeds as follows. Assuming that v satisfies $A^\star v = f$, one tries to prove that $\|v\| \le C\|f\|$ (with C independent of f). This is called the method of *a priori estimates*. One is not concerned with the question whether the equation $A^\star v = f$ admits a solution; one assumes that v is a priori given and one tries to estimate its norm.

Proof.
(a) $\Rightarrow$ (b). Set
$$B^\star = \{v \in D(A^\star); \|A^\star v\| \le 1\}.$$

By homogeneity it suffices to prove that $B^\star$ is bounded. For this purpose—in view of Corollary 2.5 (uniform boundedness principle)—we have only to show that *given* any $f_0 \in F$ the set $\langle B^\star, f_0 \rangle$ is bounded (in $\mathbb{R}$). Since A is surjective, there is some $u_0 \in D(A)$ such that $Au_0 = f_0$. For every $v \in B^\star$ we have

$$\langle v, f_0 \rangle = \langle v, Au_0 \rangle = \langle A^\star v, u_0 \rangle$$

and thus $|\langle v, f_0 \rangle| \le \|u_0\|$.
(b) $\Rightarrow$ (c). Suppose $f_n = A^\star v_n \to f$. Using (b) with $v_n - v_m$ we see that (v_n) is Cauchy, so that $v_n \to v$. Since $A^\star$ is closed (by Proposition 2.17), we conclude that $A^\star v = f$.
(c) $\Rightarrow$ (a). Since $R(A^\star)$ is closed, we infer from Theorem 2.19 that $R(A) = N(A^\star)^\perp = F$.

There is a "dual" statement.

$\star$ **Theorem 2.21.** *Let* $A : D(A) \subset F$ *be an unbounded linear operator that is densely defined and closed. The following properties are equivalent:*

(a) $A^\star$ *is surjective, i.e.,* $R(A^\star) = E^\star$,

(b) *there is a constant C such that*

$$\|u\| \le C\|Au\| \quad \forall u \in D(A),$$

(c) $N(A) = \{0\}$ *and* $R(A)$ *is closed.*

Proof. It is similar to the proof of Theorem 2.20 and we shall leave it as an exercise.

Remark 20. If one assumes that *either* dim $E < \infty$ *or* that dim $F < \infty$, then the following are equivalent:

$$A \text{ surjective} \Leftrightarrow A^\star \text{ injective},$$
$$A^\star \text{ surjective} \Leftrightarrow A \text{ injective},$$

which is indeed a classical result for linear operators in finite-dimensional spaces. The reason that these equivalences hold is that $R(A)$ and $R(A^\star)$ are finite-dimensional (and thus closed).

In the *general case* one has only the implications

$$A \text{ surjective} \Rightarrow A^\star \text{ injective},$$
$$A^\star \text{ surjective} \Rightarrow A \text{ injective}.$$

The converses fail, as may be seen from the following simple example. Let $E = F = \ell^2$; for every $x \in \ell^2$ write $x = (x_n)_{n\ge 1}$ and set $Ax = \left(\frac{1}{n}x_n\right)_{n\ge 1}$. It is easy to see that A is a bounded operator and that $A^\star = A$; $A^\star$ (resp. A) is injective but A (resp. $A^\star$) is *not* surjective; $R(A)$ (resp. $R(A^\star)$) is dense and not closed.

Comments on Chapter 2

1. One may write down *explicitly* some simple closed subspaces without complement. For example c_0 is a closed subspace of ℓ^∞ without complement; see, e.g., C. DeVito [1] (the notation c_0 and ℓ^∞ is explained in Section 11.3). There are other examples in W. Rudin [1] (a subspace of L^1), G. Köthe [1], and B. Beauzamy [1] (a subspace of ℓ^p, $p \ne 2$).

2. Most of the results in Chapter 2 extend to *Fréchet spaces* (locally convex spaces that are metrizable and complete). There are many possible extensions; see, e.g., H. Schaefer [1], J. Horváth [1], R. Edwards [1], F. Treves [1], [3], G. Köthe [1]. These extensions are motivated by the *theory of distributions* (see L. Schwartz [1]), in which many important spaces are *not* Banach spaces. For the applications to the theory of partial differential equations the reader may consult L. Hörmander [1] or F. Treves [1], [2], [3].

3. There are various extensions of the results of Section 2.5 in T. Kato [1].

Exercises for Chapter 2

2.1 *Continuity of convex functions.*
  Let E be a Banach space and let $\varphi : E \to (-\infty, +\infty]$ be a convex l.s.c. function. Assume $x_0 \in \mathrm{Int} D(\varphi)$.

1. Prove that there exist two constants $R > 0$ and M such that

$$\varphi(x) \leq M \quad \forall x \in E \text{ with } \|x - x_0\| \leq R.$$

[**Hint**: Given an appropriate $\rho > 0$, consider the sets

$$F_n = \{x \in E; \ \|x - x_0\| \leq \rho \text{ and } \varphi(x) \leq n\}.]$$

2. Prove that $\forall r < R, \exists L \geq 0$ such that

$$|\varphi(x_1) - \varphi(x_2)| \leq L\|x_1 - x_2\| \quad \forall x_1, x_2 \in E \text{ with } \|x_i - x_0\| \leq r, \ i = 1, 2.$$

  More precisely, one may choose $L = \frac{2[M - \varphi(x_0)]}{R - r}$.

2.2 Let E be a vector space and let $p : E \to \mathbb{R}$ be a function with the following three properties:

 (i) $p(x + y) \leq p(x) + p(y) \ \forall x, y \in E$,
 (ii) for each fixed $x \in E$ the function $\lambda \mapsto p(\lambda x)$ is continuous from $\mathbb{R}$ into $\mathbb{R}$,
 (iii) whenever a sequence (y_n) in E satisfies $p(y_n) \to 0$, then $p(\lambda y_n) \to 0$ for every $\lambda \in \mathbb{R}$.

  Assume that (x_n) is a sequence in E such that $p(x_n) \to 0$ and (α_n) is a bounded sequence in $\mathbb{R}$. Prove that $p(0) = 0$ and that $p(\alpha_n x_n) \to 0$.
  [**Hint**: Given $\varepsilon > 0$ consider the sets

$$F_n = \{\lambda \in \mathbb{R}; \ |p(\lambda x_k)| \leq \varepsilon, \quad \forall k \geq n\}.]$$

Deduce that if (x_n) is a sequence in E such that $p(x_n - x) \to 0$ for some $x \in E$, and (α_n) is a sequence in $\mathbb{R}$ such that $\alpha_n \to \alpha$, then $p(\alpha_n x_n) \to p(\alpha x)$.

2.3 Let E and F be two Banach spaces and let (T_n) be a sequence in $\mathcal{L}(E, F)$. Assume that for every $x \in E$, $T_n x$ converges as $n \to \infty$ to a limit denoted by Tx. Show that if $x_n \to x$ in E, then $T_n x_n \to Tx$ in F.

2.4 Let E and F be two Banach spaces and let $a : E \times F \to \mathbb{R}$ be a bilinear form satisfying:

 (i) for each fixed $x \in E$, the map $y \mapsto a(x, y)$ is continuous;
 (ii) for each fixed $y \in F$, the map $x \mapsto a(x, y)$ is continuous.

  Prove that there exists a constant $C \geq 0$ such that

$$|a(x, y)| \leq C\|x\| \ \|y\| \quad \forall x \in E, \quad \forall y \in F.$$

[**Hint**: Introduce a linear operator $T : E \to F^\star$ and prove that T is bounded with the help of Corollary 2.5.]

$\boxed{2.5}$ Let E be a Banach space and let ε_n be a sequence of positive numbers such that $\lim \varepsilon_n = 0$. Further, let (f_n) be a sequence in $E^\star$ satisfying the property

$$\begin{cases} \exists r > 0, \quad \forall x \in E \quad \text{with } \|x\| < r, \ \exists C(x) \in \mathbb{R} \quad \text{such that} \\ \langle f_n, x \rangle \leq \varepsilon_n \|f_n\| + C(x) \quad \forall n. \end{cases}$$

Prove that (f_n) is bounded.
  [**Hint**: Introduce $g_n = f_n/(1 + \varepsilon_n \|f_n\|)$.]

$\boxed{2.6}$ *Locally bounded nonlinear monotone operators.*
  Let E be Banach space and let $D(A)$ be any subset in E. A (nonlinear) map $A : D(A) \subset E \to E^\star$ is said to be *monotone* if it satisfies

$$\langle Ax - Ay, x - y \rangle \geq 0 \quad \forall x, y \in D(A).$$

1. Let $x_0 \in \text{Int}D(A)$. Prove that there exist two constants $R > 0$ and C such that

$$\|Ax\| \leq C \quad \forall x \in D(A) \text{ with } \|x - x_0\| < R.$$

[**Hint**: Argue by contradiction and construct a sequence (x_n) in $D(A)$ such that $x_n \to x_0$ and $\|Ax_n\| \to \infty$. Choose $r > 0$ such that $B(x_0, r) \subset D(A)$. Use the monotonicity of A at x_n and at $(x_0 + x)$ with $\|x\| < r$. Apply Exercise 2.5.]
2. Prove the same conclusion for a point $x_0 \in \text{Int}[\text{conv } D(A)]$.
3. Extend the conclusion of question 1 to the case of A *multivalued*, i.e., for every $x \in D(A)$, Ax is a nonempty subset of $E^\star$; the monotonicity is defined as follows:

$$\langle f - g, x - y \rangle \geq 0 \quad \forall x, y \in D(A), \quad \forall f \in Ax, \quad \forall g \in Ay.$$

$\boxed{2.7}$ Let $\alpha = (\alpha_n)$ be a given sequence of real numbers and let $1 \leq p \leq \infty$. Assume that $\sum |\alpha_n||x_n| < \infty$ for every element $x = (x_n)$ in ℓ^p (the space ℓ^p is defined in Section 11.3).
  Prove that $\alpha \in \ell^{p'}$.

$\boxed{2.8}$ Let E be a Banach space and let $T : E \to E^\star$ be a linear operator satisfying

$$\langle Tx, x \rangle \geq 0 \quad \forall x \in E.$$

Prove that T is a bounded operator.
  [Two methods are possible: (i) Use Exercise 2.6 or (ii) Apply the closed graph theorem.]

$\boxed{2.9}$ Let E be a Banach space and let $T : E \to E^\star$ be a linear operator satisfying

$$\langle Tx, y \rangle = \langle Ty, x \rangle \quad \forall x, y \in E.$$

Prove that T is a bounded operator.

2.10 Let E and F be two Banach spaces and let $T \in \mathcal{L}(E, F)$ be surjective.

1. Let M be any subset of E. Prove that $T(M)$ is closed in F iff $M + N(T)$ is closed in E.
2. Deduce that if M is a closed vector space in E and dim $N(T) < \infty$, then $T(M)$ is closed.

2.11 Let E be a Banach space, $F = \ell^1$, and let $T \in \mathcal{L}(E, F)$ be surjective. Prove that there exists $S \in \mathcal{L}(F, E)$ such that $T \circ S = I_F$, i.e., S has a right inverse of T.

[**Hint**: Do not apply Theorem 2.12; try to define S explicitly using the canonical basis of ℓ^1.]

2.12 Let E and F be two Banach spaces with norms $\| \ \|_E$ and $\| \ \|_F$. Let $T \in \mathcal{L}(E, F)$ be such that $R(T)$ is closed and dim $N(T) < \infty$. Let $| \ |$ denote another norm on E that is weaker than $\| \ \|_E$, i.e., $|x| \leq M \|x\|_E \ \forall x \in E$.

Prove that there exists a constant C such that

$$\|x\|_E \leq C(\|Tx\|_F + |x|) \quad \forall x \in E.$$

[**Hint**: Argue by contradiction.]

2.13 Let E and F be two Banach spaces. Prove that the set

$$\Omega = \{T \in \mathcal{L}(E, F); \ T \text{ admits a left inverse}\}$$

is open in $\mathcal{L}(E, F)$.

[**Hint**: Prove first that the set

$$\mathcal{O} = \{T \in \mathcal{L}(E, F); \ T \text{ is bijective}\}$$

is open in $\mathcal{L}(E, F)$.]

2.14 Let E and F be two Banach spaces

1. Let $T \in \mathcal{L}(E, F)$. Prove that $R(T)$ is closed iff there exists a constant C such that

$$\text{dist}(x, N(T)) \leq C \|Tx\| \quad \forall x \in E.$$

[**Hint**: Use the quotient space $E/N(T)$; see Section 11.2.]
2. Let $A : D(A) \subset E \to F$ be a closed unbounded operator.
   Prove that $R(A)$ is closed iff there exists a constant C such that

$$\text{dist}(u, N(A)) \leq C \|Au\| \quad \forall u \in D(A).$$

[**Hint**: Consider the operator $T : E_0 \to F$, where $E_0 = D(A)$ with the graph norm and $T = A$.]

2.15 Let E_1, E_2, and F be three Banach spaces. Let $T_1 \in \mathcal{L}(E_1, F)$ and let $T_2 \in \mathcal{L}(E_2, F)$ be such that

$$R(T_1) \cap R(T_2) = \{0\} \quad \text{and} \quad R(T_1) + R(T_2) = F.$$

Prove that $R(T_1)$ and $R(T_2)$ are closed.

[**Hint**: Apply Exercise 2.10 to the map $T : E_1 \times E_2 \to F$ defined by

$$T(x_1, x_2) = T_1 x_1 + T_2 x_2.]$$

2.16 Let E be a Banach space. Let G and L be two closed subspaces of E. Assume that there exists a constant C such that

$$\text{dist}(x, G \cap L) \leq C \, \text{dist}(x, L), \quad \forall x \in G.$$

Prove that $G + L$ is closed.

2.17 Let $E = C([0, 1])$ with its usual norm. Consider the operator $A : D(A) \subset E \to E$ defined by

$$D(A) = C^1([0, 1]) \quad \text{and} \quad Au = u' = \frac{du}{dt}.$$

1. Check that $\overline{D(A)} = E$.
2. Is A closed?
3. Consider the operator $B : D(B) \subset E \to E$ defined by

$$D(B) = C^2([0, 1]) \quad \text{and} \quad Bu = u' = \frac{du}{dt}.$$

Is B closed?

2.18 Let E and F be two Banach spaces and let $A : D(A) \subset E \to F$ be a densely defined unbounded operator.

1. Prove that $N(A^\star) = R(A)^\perp$ and $N(A) \subset R(A^\star)^\perp$.
2. Assuming that A is also closed prove that $N(A) = R(A^\star)^\perp$.
   [Try to find direct arguments and do not rely on the proof of Corollary 2.18. For question 2 argue by contradiction: suppose there is some $u \in R(A^\star)^\perp$ such that $[u, 0] \notin G(A)$ and apply Hahn–Banach.]

2.19 Let E be a Banach space and let $A : D(A) \subset E \to E^\star$ be a densely defined unbounded operator.

1. Assume that there exists a constant C such that

(1) $$\langle Au, u \rangle \geq -C\|Au\|^2 \quad \forall u \in D(A).$$

Prove that $N(A) \subset N(A^\star)$.

2. Conversely, assume that $N(A) \subset N(A^\star)$. Also, assume that A is closed and $R(A)$ is closed. Prove that there exists a constant C such that (1) holds.

$\boxed{2.20}$ Let E and F be two Banach spaces. Let $T \in \mathcal{L}(E, F)$ and let $A : D(A) \subset E \to F$ be an unbounded operator that is densely defined and closed. Consider the operator $B : D(B) \subset E \to F$ defined by

$$D(B) = D(A), \quad B = A + T.$$

1. Prove that B is closed.
2. Prove that $D(B^\star) = D(A^\star)$ and $B^\star = A^\star + T^\star$.

$\boxed{2.21}$ Let E be an infinite-dimensional Banach space. Fix an element $a \in E, a \neq 0$, and a discontinuous linear functional $f : E \to \mathbb{R}$ (such functionals exist; see Exercise 1.5). Consider the operator $A : E \to E$ defined by

$$D(A) = E, \quad Ax = x - f(x)a.$$

1. Determine $N(A)$ and $R(A)$.
2. Is A closed?
3. Determine $A^\star$ (define $D(A^\star)$ carefully).
4. Determine $N(A^\star)$ and $R(A^\star)$.
5. Compare $N(A)$ with $R(A^\star)^\perp$ as well as $N(A^\star)$ with $R(A)^\perp$.
6. Compare with the results of Exercise 2.18.

$\boxed{2.22}$ The purpose of this exercise is to construct an unbounded operator $A : D(A) \subset E \to E$ that is densely defined, closed, and such that $\overline{D(A^\star)} \neq E^\star$.
   Let $E = \ell^1$, so that $E^\star = \ell^\infty$. Consider the operator $A : D(A) \subset E \to E$ defined by

$$D(A) = \left\{ u = (u_n) \in \ell^1; \ (nu_n) \in \ell^1 \right\} \text{ and } Au = (nu_n).$$

1. Check that A is densely defined and closed.
2. Determine $D(A^\star)$, $A^\star$, and $\overline{D(A^\star)}$.

$\boxed{2.23}$ Let $E = \ell^1$, so that $E^\star = \ell^\infty$. Consider the operator $T \in \mathcal{L}(E, E)$ defined by

$$Tu = \left(\frac{1}{n} u_n \right)_{n \geq 1} \quad \text{for every } u = (u_n)_{n \geq 1} \text{ in } \ell^1.$$

Determine $N(T)$, $N(T)^\perp$, $T^\star$, $R(T^\star)$, and $\overline{R(T^\star)}$.
   Compare with Corollary 2.18.

2.24 Let E, F, and G be three Banach spaces. Let $A : D(A) \subset E \to F$ be a densely defined unbounded operator. Let $T \in \mathcal{L}(F, G)$ and consider the operator $B : D(B) \subset E \to G$ defined by $D(B) = D(A)$ and $B = T \circ A$.

1. Determine $B^\star$.
2. Prove (by an example) that B need *not* be closed even if A is closed.

2.25 Let E, F, and G be three Banach spaces.

1. Let $T \in \mathcal{L}(E, F)$ and $S \in \mathcal{L}(F, G)$. Prove that

$$(S \circ T)^\star = T^\star \circ S^\star.$$

2. Assume that $T \in \mathcal{L}(E, F)$ is bijective. Prove that $T^\star$ is bijective and that $(T^\star)^{-1} = (T^{-1})^\star$.

2.26 Let E and F be two Banach spaces and let $T \in \mathcal{L}(E, F)$. Let $\psi : F \to (-\infty, +\infty]$ be a convex function. Assume that there exists some element in $R(T)$ where ψ is finite and continuous.
Set
$$\varphi(x) = \psi(Tx), \quad x \in E.$$
Prove that for every $f \in F^\star$

$$\varphi^\star(T^\star f) = \inf_{g \in N(T^\star)} \psi^\star(f - g) = \min_{g \in N(T^\star)} \psi^\star(f - g).$$

2.27 Le E, F be two Banach spaces and let $T \in \mathcal{L}(E, F)$. Assume that $R(T)$ has finite codimension, i.e., there exists a finite-dimensional subspace X of F such that $X + R(T) = F$ and $X \cap R(T) = \{0\}$.

Prove that $R(T)$ is closed.

Chapter 3
Weak Topologies. Reflexive Spaces. Separable Spaces. Uniform Convexity

3.1 The Coarsest Topology for Which a Collection of Maps Becomes Continuous

We begin this chapter by recalling a well-known concept in topology. Suppose X is a set (without any structure) and $(Y_i)_{i \in I}$ is a collection of *topological spaces*. We are given a collection of maps $(\varphi_i)_{i \in I}$ such that for every $i \in I$, φ_i maps X into Y_i and we consider the following:

Problem 1. Construct a topology on X that makes all the maps $(\varphi_i)_{i \in I}$ continuous. If possible, find a topology $\mathcal{T}$ that is the *most economical* in the sense that it has the *fewest open sets*.

Note that if we equip X with the discrete topology (i.e., every subset of X is open), then every map φ_i is continuous; of course, this topology is far from being the "cheapest"; in fact, it is the most expensive one! As we shall see, there is always a (unique) "cheapest" topology $\mathcal{T}$ on X for which every map φ_i is continuous. It is called the *coarsest* or *weakest* topology (or sometimes the initial topology) associated to the collection $(\varphi_i)_{i \in I}$.

If $\omega_i \subset Y_i$ is any open set, then $\varphi_i^{-1}(\omega_i)$ is *necessarily* an open set in $\mathcal{T}$. As ω_i runs through the family of open sets of Y_i and i runs through I we obtain a family of subsets of X, each of which *must* be open in the topology $\mathcal{T}$. Let us denote this family by $(U_\lambda)_{\lambda \in \Lambda}$. Of course, this family need not be a topology. Therefore, we are led to the following:

Problem 2. Given a set X and a family $(U_\lambda)_{\lambda \in \Lambda}$ of subsets in X, construct the cheapest topology $\mathcal{T}$ on X in which U_λ is open for all $\lambda \in \Lambda$.

In other words, we must find the cheapest family $\mathcal{F}$ of subsets of X that *is stable*[1] by $\cap_{\text{finite}}$ and $\cup_{\text{arbitrary}}$ and with the property that $U_\lambda \in \mathcal{F}$ for every $\lambda \in \Lambda$. The construction goes as follows. First, consider finite intersections of sets in $(U_\lambda)_{\lambda \in \Lambda}$, i.e., $\cap_{\lambda \in \Gamma} U_\lambda$ where $\Gamma \subset \Lambda$ is finite. In this way we obtain a new family, called Φ, of

---

[1] Meaning that a finite intersection of sets in $\mathcal{F}$ and an arbitrary union of sets in $\mathcal{F}$ both belong to $\mathcal{F}$.

subsets of X which includes $(U_\lambda)_{\lambda \in \Lambda}$ and which is stable under $\cap_{\text{finite}}$. However, it need not be stable under $\cup_{\text{arbitrary}}$. Therefore, we consider next the family $\mathscr{F}$ obtained by forming arbitrary unions of elements from Φ. It is clear that $\mathscr{F}$ is stable under $\cup_{\text{arbitrary}}$. It is not clear whether $\mathscr{F}$ is stable under $\cap_{\text{finite}}$; but indeed we have the following result:

Lemma 3.1. *The family $\mathscr{F}$ is stable under $\cap_{\text{finite}}$.*

The proof of Lemma 3.1—a delightful exercise in set theory—is left to the reader; see e.g., G. Folland [2]. It is now obvious that the above construction gives the cheapest topology with the required property.

Remark 1. *One cannot reverse the order of operations in the construction of $\mathscr{F}$.* It would have been equally natural to start with $\cup_{\text{arbitrary}}$ and then to take $\cap_{\text{finite}}$. The outcome is a family that is stable under $\cap_{\text{finite}}$; but it is not stable under $\cup_{\text{arbitrary}}$. One would have to consider once more $\cup_{\text{arbitrary}}$ and the process then stabilizes.

To summarize this discussion we find that the open sets of the topology $\mathscr{T}$ are obtained by considering first $\cap_{\text{finite}}$ of sets of the form $\varphi_i^{-1}(\omega_i)$ and then $\cup_{\text{arbitrary}}$. It follows that for every $x \in X$, we obtain a basis of neighborhoods of x for the topology $\mathscr{T}$ by considering sets of the form $\cap_{\text{finite}} \varphi_i^{-1}(V_i)$, where V_i is a neighborhood of $\varphi_i(x)$ in Y_i. Recall that in a topological space, a *basis of neighborhoods* of a point x is a family of neighborhoods of x, such that every neighborhood of x contains a neighborhood from the basis.

In what follows we equip X with the topology $\mathscr{T}$ that is the weakest topology associated to the collection $(\varphi_i)_{i \in I}$. Here are two simple properties of the topology $\mathscr{T}$.

• **Proposition 3.1.** *Let (x_n) be a sequence in X. Then $x_n \to x$ (in $\mathscr{T}$) if and only if $\varphi_i(x_n) \to \varphi_i(x)$ for every $i \in I$.*

Proof. If $x_n \to x$, then $\varphi_i(x_n) \to \varphi_i(x)$ for each i, since each φ_i is continuous for $\mathscr{T}$. Conversely, let U be a neighborhood of x. From the preceding discussion, we may always assume that U has the form $U = \cap_{i \in J} \varphi_i^{-1}(V_i)$ with $J \subset I$ finite. For each $i \in J$ there is some integer N_i such that $\varphi_i(x_n) \in V_i$ for $n \geq N_i$. It follows that $x_n \in U$ for $n \geq N = \max_{i \in J} N_i$.

• **Proposition 3.2.** *Let Z be a topological space and let ψ be a map from Z into X. Then ψ is continuous if and only if $\varphi_i \circ \psi$ is continuous from Z into Y_i for every $i \in I$.*

Proof. If ψ is continuous then $\varphi_i \circ \psi$ is also continuous for every $i \in I$. Conversely, we have to prove that $\psi^{-1}(U)$ is open (in Z) for every open set U (in X). But we know that U has the form $U = \cup_{\text{arbitrary}} \cap_{\text{finite}} \varphi_i^{-1}(\omega_i)$, where ω_i is open in Y_i. Therefore

$$\psi^{-1}(U) = \underset{\text{arbitrary}}{\cup} \underset{\text{finite}}{\cap} \psi^{-1}[\varphi_i^{-1}(\omega_i)] = \underset{\text{arbitrary}}{\cup} \underset{\text{finite}}{\cap} (\varphi_i \circ \psi)^{-1}(\omega_i),$$

which is open in Z since every map $\varphi_i \circ \psi$ is continuous.

3.2 Definition and Elementary Properties of the Weak Topology $\sigma(E, E^\star)$

Let E be a Banach space and let $f \in E^\star$. We denote by $\varphi_f : E \to \mathbb{R}$ the linear functional $\varphi_f(x) = \langle f, x \rangle$. As f runs through $E^\star$ we obtain a collection $(\varphi_f)_{f \in E^\star}$ of maps from E into $\mathbb{R}$. We now ignore the usual topology on E (associated to $\| \ \|$) and define a new topology on the set E as follows:

Definition. The *weak topology* $\sigma(E, E^\star)$ on E is the coarsest topology associated to the collection $(\varphi_f)_{f \in E^\star}$ (in the sense of Section 3.1 with $X = E$, $Y_i = \mathbb{R}$, for each i, and $I = E^\star$).

Note that every map φ_f is continuous for the usual topology and therefore *the weak topology is weaker than the usual topology.*

Proposition 3.3. *The weak topology $\sigma(E, E^\star)$ is Hausdorff.*

Proof. Given $x_1, x_2 \in E$ with $x_1 \neq x_2$ we have to find two open sets O_1 and O_2 for the weak topology $\sigma(E, E^\star)$ such that $x_1 \in O_1$, $x_2 \in O_2$, and $O_1 \cap O_2 = \emptyset$. By Hahn–Banach (second geometric form) there exists a closed hyperplane strictly separating $\{x_1\}$ and $\{x_2\}$. Thus, there exist some $f \in E^\star$ and some $\alpha \in \mathbb{R}$ such that

$$\langle f, x_1 \rangle < \alpha < \langle f, x_2 \rangle.$$

Set

$$O_1 = \{x \in E; \ \langle f, x \rangle < \alpha\} = \varphi_f^{-1}\left((-\infty, \alpha)\right),$$
$$O_2 = \{x \in E; \ \langle f, x \rangle > \alpha\} = \varphi_f^{-1}\left((\alpha, +\infty)\right).$$

Clearly, O_1 and O_2 are open for $\sigma(E, E^\star)$ and they satisfy the required properties.

• Proposition 3.4. *Let $x_0 \in E$; given $\varepsilon > 0$ and a **finite** set $\{f_1, f_2, \ldots, f_k\}$ in $E^\star$ consider*

$$V = V(f_1, f_2, \ldots, f_k; \ \varepsilon) = \{x \in E; \ |\langle f_i, x - x_0 \rangle| < \varepsilon \ \forall i = 1, 2, \ldots, k\}.$$

*Then V is a neighborhood of x_0 for the topology $\sigma(E, E^\star)$. Moreover, we obtain a **basis of neighborhoods** of x_0 for $\sigma(E, E^\star)$ by varying ε, k, and the f_i's in $E^\star$.*

Proof. Clearly $V = \cap_{i=1}^{k} \varphi_{f_i}^{-1}((a_i - \varepsilon, a_i + \varepsilon))$, with $a_i = \langle f_i, x_0 \rangle$, is open for the topology $\sigma(E, E^\star)$ and contains x_0. Conversely, let U be a neighborhood of x_0 for $\sigma(E, E^\star)$. From the discussion in Section 3.1 we know that there exists an open set W containing x_0, $W \subset U$, of the form $W = \cap_{\text{finite}} \varphi_{f_i}^{-1}(\omega_i)$, where ω_i is a neighborhood (in $\mathbb{R}$) of $a_i = \langle f_i, x_0 \rangle$. Hence there exists $\varepsilon > 0$ such that $(a_i - \varepsilon, a_i + \varepsilon) \subset \omega_i$ for every i. It follows that $x_0 \in V \subset W \subset U$.

Notation. If a sequence (x_n) in E converges to x in the weak topology $\sigma(E, E^\star)$ we shall write

$$\boxed{x_n \rightharpoonup x.}$$

To avoid any confusion we shall sometimes say, "$x_n \rightharpoonup x$ weakly in $\sigma(E, E^\star)$." In order to be totally clear we shall sometimes emphasize strong convergence by saying, "$x_n \to x$ strongly," meaning that $\|x_n - x\| \to 0$.

● **Proposition 3.5.** *Let (x_n) be a sequence in E. Then*

(i) *$[x_n \rightharpoonup x$ weakly in $\sigma(E, E^\star)] \Leftrightarrow [\langle f, x_n \rangle \to \langle f, x \rangle \ \forall f \in E^\star]$.*
(ii) *If $x_n \to x$ strongly, then $x_n \rightharpoonup x$ weakly in $\sigma(E, E^\star)$.*
(iii) *If $x_n \rightharpoonup x$ weakly in $\sigma(E, E^\star)$, then $(\|x_n\|)$ is bounded and $\|x\| \le \liminf \|x_n\|$.*
(iv) *If $x_n \rightharpoonup x$ weakly in $\sigma(E, E^\star)$ and if $f_n \to f$ strongly in $E^\star$ (i.e., $\|f_n - f\|_{E^\star} \to$ 0), then $\langle f_n, x_n \rangle \to \langle f, x \rangle$.*

Proof.

(i) follows from Proposition 3.1 and the definition of the weak topology $\sigma(E, E^\star)$.
(ii) follows from (i), since $|\langle f, x_n \rangle - \langle f, x \rangle| \le \|f\| \, \|x_n - x\|$; it is also clear from the fact that the weak topology is weaker than the strong topology.
(iii) follows from the uniform boundedness principle (see Corollary 2.4), since for every $f \in E^\star$ the set $(\langle f, x_n \rangle)_n$ is bounded. Passing to the limit in the inequality

$$|\langle f, x_n \rangle| \le \|f\| \, \|x_n\|,$$

we obtain

$$|\langle f, x \rangle| \le \|f\| \liminf \|x_n\|,$$

which implies (by Corollary 1.4) that

$$\|x\| = \sup_{\|f\| \le 1} |\langle f, x \rangle| \le \liminf \|x_n\|.$$

(iv) follows from the inequality

$$|\langle f_n, x_n \rangle - \langle f, x \rangle| \le |\langle f_n - f, x_n \rangle| + |\langle f, x_n - x \rangle| \le \|f_n - f\| \, \|x_n\| + |\langle f, x_n - x \rangle|,$$

combined with (i) and (iii).

● **Proposition 3.6.** *When E is finite-dimensional, the weak topology $\sigma(E, E^\star)$ and the usual topology are the same. In particular, a sequence (x_n) converges weakly if and only if it converges strongly.*

Proof. Since the weak topology has *always* fewer open sets than the strong topology, it suffices to check that every strongly open set is weakly open. Let $x_0 \in E$ and let U be a neighborhood of x_0 in the strong topology. We have to find a neighborhood V of x_0 in the weak topology $\sigma(E, E^\star)$ such that $V \subset U$. In other words, we have to find $f_1, f_2, \ldots, f_k$ in $E^\star$ and $\varepsilon > 0$ such that

$$V = \{x \in E; \ |\langle f_i, x - x_0 \rangle| < \varepsilon \ \ \forall i = 1, 2, \ldots, k\} \subset U.$$

Fix $r > 0$ such that $B(x_0, r) \subset U$. Pick a basis $e_1, e_2, \ldots, e_k$ in E such that $\|e_i\| = 1$, $\forall i$. Every $x \in E$ admits a decomposition $x = \sum_{i=1}^{k} x_i e_i$, and the maps $x \mapsto x_i$ are continuous linear functionals on E denoted by f_i. We have

$$\|x - x_0\| \leq \sum_{i=1}^{k} |\langle f_i, x - x_0 \rangle| < k\varepsilon$$

for every $x \in V$. Choosing $\varepsilon = r/k$, we obtain $V \subset U$.

Remark 2. Open (resp. closed) sets in the weak topology $\sigma(E, E^\star)$ are *always* open (resp. closed) in the strong topology. In *any infinite-dimensional space* the weak topology is *strictly coarser* than the strong topology; i.e., there exist open (resp. closed) sets in the strong topology that are *not* open (resp. closed) in the weak topology. Here are two examples:

Example 1. The unit *sphere* $S = \{x \in E; \; \|x\| = 1\}$, with E infinite-dimensional, is *never* closed in the weak topology $\sigma(E, E^\star)$. More precisely, we have

(1) $$\overline{S}^{\sigma(E,E^\star)} = B_E,$$

where $\overline{S}^{\sigma(E,E^\star)}$ denotes the closure of S in the topology $\sigma(E, E^\star)$ and B_E (already defined in Chapter 2) denotes the closed unit ball in E,

$$B_E = \{x \in E; \; \|x\| \leq 1\}.$$

First let us check that every $x_0 \in E$ with $\|x_0\| < 1$ belongs to $\overline{S}^{\sigma(E,E^\star)}$. Indeed, let V be a neighborhood of x_0 in $\sigma(E, E^\star)$. We have to prove that $V \cap S \neq \emptyset$. In view of Proposition 3.4 we may always assume that V has the form

$$V = \{x \in E; \; |\langle f_i, x - x_0 \rangle| < \varepsilon \; \forall i = 1, 2, \ldots, k\}$$

with $\varepsilon > 0$ and $f_1, f_2, \ldots, f_k \in E^\star$. Fix $y_0 \in E$, $y_0 \neq 0$, such that

$$\langle f_i, y_0 \rangle = 0 \qquad \forall i = 1, 2, \ldots, k.$$

[Such a y_0 exists; otherwise, the map $\varphi : E \to \mathbb{R}^k$ defined by $\varphi(x) = (\langle f_i, x \rangle)_{1 \leq i \leq k}$ would be injective and φ would be an isomorphism from E onto $\varphi(E)$, and thus $\dim E \leq k$, which contradicts the assumption that E is infinite-dimensional.][2] The function $g(t) = \|x_0 + t y_0\|$ is continuous on $[0, \infty)$ with $g(0) < 1$ and $\lim_{t \to +\infty} g(t) = +\infty$. Hence there exists some $t_0 > 0$ such that $\|x_0 + t_0 y_0\| = 1$. It follows that $x_0 + t_0 y_0 \in V \cap S$, and thus we have established that

$$S \subset B_E \subset \overline{S}^{\sigma(E,E^\star)}.$$

---

[2] The geometric interpretation of this construction is the following. When E is *infinite-dimensional*, every neighborhood V of x_0 in the topology $\sigma(E, E^\star)$ contains a line passing through x_0, even a "huge" affine space passing through x_0.

In order to complete the proof of (1) it suffices to know that B_E is closed in the topology $\sigma(E, E^\star)$. But we have

$$B_E = \bigcap_{\substack{f \in E^\star \\ \|f\| \le 1}} \{x \in E; \ |\langle f, x \rangle| \le 1\},$$

which is an intersection of weakly closed sets.

Example 2. The unit ball $U = \{x \in E; \ \|x\| < 1\}$, with E infinite-dimensional, is *never* open in the weak topology $\sigma(E, E^\star)$. Suppose, by contradiction, that U is weakly open. Then its complement $U^c = \{x \in E; \ \|x\| \ge 1\}$ is weakly closed. It follows that $S = B_E \cap U^c$ is also weakly closed; this contradicts Example 1.

$\star$ *Remark* 3. In infinite-dimensional spaces the weak topology is *never metrizable*, i.e., there is no metric (and a fortiori no norm) on E that induces on E the weak topology $\sigma(E, E^\star)$; see Exercise 3.8. However, as we shall see later (Theorem 3.29), if $E^\star$ is *separable* one can define a norm on E that induces on *bounded sets* of E the weak topology $\sigma(E, E^\star)$.

$\star$ *Remark* 4. *Usually*, in infinite-dimensional spaces, there exist sequences that converge weakly and do not converge strongly. For example, if $E^\star$ is *separable* or if E is *reflexive* one can construct a sequence (x_n) in E such that $\|x_n\| = 1$ and $x_n \rightharpoonup 0$ weakly (see Exercise 3.22). However, there are infinite-dimensional spaces with the property that *every* weakly convergent sequence is strongly convergent. For example, ℓ^1 has that unusual property (see Problem 8). Such spaces are quite "rare" and somewhat "*pathological*." This strange fact does not contradict Remark 2, which asserts that in infinite-dimensional spaces, the weak topology and the strong topology are *always* distinct: the weak topology is *strictly coarser* than the strong topology. Keep in mind that two *metric* (or metrizable) spaces with the same convergent sequences have identical topologies; however, if two *topological* spaces have the same convergent sequences they need *not* have identical topologies.

3.3 Weak Topology, Convex Sets, and Linear Operators

Every weakly closed set is strongly closed and the converse is false in infinite-dimensional spaces (see Remark 2). However, it is very useful to know that for *convex* sets, weakly closed = strongly closed:

$\bullet$ **Theorem 3.7.** *Let C be a convex subset of E. Then C is closed in the weak topology $\sigma(E, E^\star)$ if and only if it is closed in the strong topology.*

Proof. Assume that C is closed in the strong topology and let us prove that C is closed in the weak topology. We shall check that the complement C^c of C is open in the weak topology. To this end, let $x_0 \notin C$. By Hahn–Banach there exists a closed hyperplane strictly separating $\{x_0\}$ and C. Thus, there exist some $f \in E^\star$ and some $\alpha \in \mathbb{R}$ such that

$$\langle f, x_0 \rangle < \alpha < \langle f, y \rangle \quad \forall y \in C.$$

Set

$$V = \{x \in E; \ \langle f, x \rangle < \alpha\};$$

so that $x_0 \in V$, $V \cap C = \emptyset$ (i.e., $V \subset C^c$) and V is open in the weak topology.

Corollary 3.8 (Mazur). *Assume (x_n) converges* **weakly** *to x. Then there exists a sequence (y_n) made up of convex combinations of the x_n's that converges* **strongly** *to x.*

Proof. Let $C = \text{conv}(\cup_{p=1}^\infty \{x_p\})$ denote the convex hull of the x_n's. Since x belongs to the weak closure of $\cup_{p=1}^\infty \{x_p\}$ it belongs a fortiori to the weak closure of C. By Theorem 3.7, $x \in \overline{C}$, the strong closure of C, and the conclusion follows.

Remark 5. There are some variants of Corollary 3.8 (see Exercises 3.4 and 5.24). Also, note that the proof of Theorem 3.7 shows that every closed convex set C coincides with the intersection of all the closed half-spaces containing C.

• **Corollary 3.9.** *Assume that $\varphi : E \to (-\infty + \infty]$ is convex and l.s.c. in the strong topology. Then φ is l.s.c. in the weak topology $\sigma(E, E^\star)$.*

Proof. For every $\lambda \in \mathbb{R}$ the set

$$A = \{x \in E; \ \varphi(x) \leq \lambda\}$$

is convex and strongly closed. By Theorem 3.7 it is weakly closed and thus φ is weakly l.s.c.

• *Remark 6.* It may be rather difficult in practice to prove that a function is l.s.c. in the weak topology. Corollary 3.9 is often used as follows:

$$\boxed{\varphi \text{ convex and strongly continuous} \Rightarrow \varphi \text{ weakly l.s.c.}}$$

For example, the function $\varphi(x) = \|x\|$ is convex and strongly continuous; thus it is weakly l.s.c. In particular, if $x_n \rightharpoonup x$ weakly, it follows that $\|x\| \leq \liminf \|x_n\|$ (see also Proposition 3.5).

Theorem 3.10. *Let E and F be two Banach spaces and let T be a linear operator from E into F. Assume that T is continuous in the strong topologies. Then T is continuous from E weak $\sigma(E, E^\star)$ into F weak $\sigma(F, F^\star)$ and conversely.*

Proof. In view of Proposition 3.2 it suffices to check that for every $f \in F^\star$ the map $x \mapsto \langle f, Tx \rangle$ is continuous from E weak $\sigma(E, E^\star)$ into $\mathbb{R}$. But the map $x \mapsto \langle f, Tx \rangle$ is a continuous linear functional on E. Therefore, it is also continuous in the weak topology $\sigma(E, E^\star)$.

Conversely, suppose that T is continuous from E weak into F weak. Then $G(T)$ is closed in $E \times F$ equipped with the product topology $\sigma(E, E^\star) \times \sigma(F, F^\star)$, which is clearly the same as $\sigma(E \times F, (E \times F)^\star)$. It follows that $G(T)$ is strongly closed (any weakly closed set is strongly closed). We conclude with the help of the closed graph theorem (Theorem 2.9) that T is continuous from E strong into F strong.

Remark 7. The argument above shows more: that if a linear operator T is continuous from E strong into F weak then T is continuous from E strong into F strong. As a consequence, for linear operators, the following continuity properties are all the same: $S \to S, W \to W, S \to W$ (S = strong, W = weak). On the other hand, very *few* linear operators are continuous $W \to S$; this happens if and only if T is continuous $S \to S$ and, *moreover*, dim $R(T) < \infty$ (see Exercise 6.7).

Also, note that in general, *nonlinear* maps that are continuous from E strong into F strong are *not* continuous from E weak into F weak (see, e.g., Exercise 4.20). This is a *major source of difficulties in nonlinear problems*.

3.4 The Weak* Topology $\sigma(E^\star, E)$

So far, we have two topologies on $E^\star$:

(a) the usual (strong) topology associated to the norm of $E^\star$,
(b) the weak topology $\sigma(E^\star, E^{\star\star})$, obtained by performing on $E^\star$ the construction of Section 3.3.

We are now going to define a *third topology* on $E^\star$ called the weak* topology and denoted by $\sigma(E^\star, E)$ (the $\star$ is here to remind us that this topology is defined only on dual spaces). For every $x \in E$ consider the linear functional $\varphi_x : E^\star \to \mathbb{R}$ defined by $f \mapsto \varphi_x(f) = \langle f, x \rangle$. As x runs through E we obtain a collection $(\varphi_x)_{x \in E}$ of maps from $E^\star$ into $\mathbb{R}$.

Definition. The *weak* topology*, $\sigma(E^\star, E)$, is the coarsest topology on $E^\star$ associated to the collection $(\varphi_x)_{x \in E}$ (in the sense of Section 3.1 with $X = E^\star$, $Y_i = \mathbb{R}$, for all i, and $I = E$).

Since $E \subset E^{\star\star}$, it is clear that the topology $\sigma(E^\star, E)$ is coarser than the topology $\sigma(E^\star, E^{\star\star})$; i.e., the topology $\sigma(E^\star, E)$ has fewer open sets (resp. closed sets) than the topology $\sigma(E^\star, E^{\star\star})$, which in turn has fewer open sets (resp. closed sets) than the strong topology.

Remark 8. The reader probably wonders why there is such hysteria over weak topologies! The reason is the following: *a coarser topology has more compact sets*. For example, the closed unit ball $B_{E^\star}$ in $E^\star$, which is *never compact in the strong topology* (unless dim $E < \infty$; see Theorem 6.5), is *always compact in the weak* topology* (see Theorem 3.16). Knowing the basic role of compact sets—for example, in *existence* mechanisms such as minimization—it is easy to understand the importance of the weak* topology.

Proposition 3.11. *The weak* topology is Hausdorff.*

Proof. Given $f_1, f_2 \in E^*$ with $f_1 \neq f_2$ there exists some $x \in E$ such that $\langle f_1, x \rangle \neq \langle f_2, x \rangle$ (this does not use Hahn–Banach, but just the fact that $f_1 \neq f_2$). Assume for example that $\langle f_1, x \rangle < \langle f_2, x \rangle$ and choose α such that

$$\langle f_1, x \rangle < \alpha < \langle f_2, x \rangle.$$

Set

$$O_1 = \{f \in E^*; \ \langle f, x \rangle < \alpha\} = \varphi_x^{-1}((-\infty, \alpha)),$$
$$O_2 = \{f \in E^*; \ \langle f, x \rangle > \alpha\} = \varphi_x^{-1}((\alpha, +\infty)).$$

Then O_1 and O_2 are open sets in $\sigma(E^*, E)$ such that $f_1 \in O_1$, $f_2 \in O_2$, and $O_1 \cap O_2 = \emptyset$.

Proposition 3.12. *Let $f_0 \in E^*$; given a **finite** set $\{x_1, x_2, \ldots, x_k\}$ in E and $\varepsilon > 0$, consider*

$$V = V(x_1, x_2, \ldots, x_k; \varepsilon) = \left\{ f \in E^*; |\langle f - f_0, x_i \rangle| < \varepsilon \ \ \forall i = 1, 2, \ldots, k \right\}.$$

*Then V is a neighborhood of f_0 for the topology $\sigma(E^*, E)$. Moreover, we obtain a **basis of neighborhoods** of f_0 for $\sigma(E^*, E)$ by varying ε, k, and the x_i's in E.*

Proof. Same as the proof of Proposition 3.4.

Notation. If a sequence (f_n) in E^* converges to f in the weak* topology we shall write

$$\boxed{f_n \overset{*}{\rightharpoonup} f.}$$

To avoid any confusion we shall sometimes emphasize "$f_n \overset{*}{\rightharpoonup} f$ in $\sigma(E^*, E)$," "$f_n \rightharpoonup f$ in $\sigma(E^*, E^{**})$," and "$f_n \to f$ strongly."

• **Proposition 3.13.** *Let (f_n) be a sequence in E^*. Then*

(i) $[f_n \overset{*}{\rightharpoonup} f$ *in* $\sigma(E^*, E)] \Leftrightarrow [\langle f_n, x \rangle \to \langle f, x \rangle, \ \forall x \in E]$.

(ii) *If* $f_n \to f$ *strongly, then* $f_n \rightharpoonup f$ *in* $\sigma(E^*, E^{**})$.

    If $f_n \rightharpoonup f$ *in* $\sigma(E^*, E^{**})$, *then* $f_n \overset{*}{\rightharpoonup} f$ *in* $\sigma(E^*, E)$.

(iii) *If* $f_n \overset{*}{\rightharpoonup} f$ *in* $\sigma(E^*, E)$ *then* $(\|f_n\|)$ *is bounded and* $\|f\| \leq \liminf \|f_n\|$.

(iv) *If* $f_n \overset{*}{\rightharpoonup} f$ *in* $\sigma(E^*, E)$ *and if* $x_n \to x$ *strongly in* E, *then* $\langle f_n, x_n \rangle \to \langle f, x \rangle$.

Proof. Copy the proof of Proposition 3.5.

Remark 9. Assume $f_n \overset{*}{\rightharpoonup} f$ in $\sigma(E^*, E)$ (or even $f_n \rightharpoonup f$ in $\sigma(E^*, E^{**})$) and $x_n \rightharpoonup x$ in $\sigma(E, E^*)$. One *cannot* conclude, in general, that $\langle f_n, x_n \rangle \to \langle f, x \rangle$ (it is very easy to construct an example in Hilbert spaces).

Remark 10. When E is a finite-dimensional space the three topologies (strong, weak, weak*) on E^* coincide. Indeed, the canonical injection $J : E \to E^{**}$ (see Section 1.3) is surjective (since dim $E = $ dim E^{**}) and therefore $\sigma(E^*, E) = \sigma(E^*, E^{**})$.

★ **Proposition 3.14.** *Let $\varphi : E^* \to \mathbb{R}$ be a linear functional that is continuous for the weak* topology. Then there exists some $x_0 \in E$ such that*

$$\varphi(f) = \langle f, x_0 \rangle \quad \forall f \in E^*.$$

The proof relies on the following useful algebraic lemma:

Lemma 3.2. *Let X be a vector space and let $\varphi, \varphi_1, \varphi_2, \ldots, \varphi_k$ be $(k + 1)$ linear functionals on X such that*

$$(2) \qquad\qquad [\varphi_i(v) = 0 \quad \forall i = 1, 2, \ldots, k] \Rightarrow [\varphi(v) = 0].$$

Then there exist constants $\lambda_1, \lambda_2, \ldots, \lambda_k \in \mathbb{R}$ such that $\varphi = \sum_{i=1}^{k} \lambda_i \varphi_i$.

Proof of Lemma 3.2. Consider the map $F : X \to \mathbb{R}^{k+1}$ defined by

$$F(u) = [\varphi(u), \varphi_1(u), \varphi_2(u), \ldots, \varphi_k(u)].$$

It follows from assumption (2) that $a = [1, 0, 0, \ldots, 0]$ does not belong to $R(F)$. Thus, one can strictly separate $\{a\}$ and $R(F)$ by some hyperplane in $\mathbb{R}^{k+1}$; i.e., there exist constants $\lambda, \lambda_1, \lambda_2, \ldots, \lambda_k$ and α such that

$$\lambda < \alpha < \lambda \varphi(u) + \sum_{i=1}^{k} \lambda_i \varphi_i(u) \quad \forall u \in X.$$

It follows that

$$\lambda \varphi(u) + \sum_{i=1}^{k} \lambda_i \varphi_i(u) = 0 \quad \forall u \in X$$

and also $\lambda < 0$ (so that $\lambda \neq 0$).

Proof of Proposition 3.14. Since φ is continuous for the weak* topology, there exists a neighborhood V of 0 for $\sigma(E^*, E)$ such that

$$|\varphi(f)| < 1 \quad \forall f \in V.$$

We may always assume that

$$V = \{f \in E^*; |\langle f, x_i \rangle| < \varepsilon \quad \forall i = 1, 2, \ldots, k\}$$

with $x_i \in E$ and $\varepsilon > 0$. In particular,

$$[\langle f, x_i \rangle = 0 \quad \forall i = 1, 2, \ldots, k] \Rightarrow [\varphi(f) = 0].$$

It follows from Lemma 3.2 that

$$\varphi(f) = \sum_{i=1}^{k} \lambda_i \langle f, x_i \rangle = \left\langle f, \sum_{i=1}^{k} \lambda_i x_i \right\rangle \quad \forall f \in E^*.$$

⋆ **Corollary 3.15.** *Assume that H is a hyperplane in E^* that is closed in $\sigma(E^*, E)$. Then H has the form*

$$H = \{f \in E^*; \langle f, x_0 \rangle = \alpha\}$$

for some $x_0 \in E$, $x_0 \neq 0$, and some $\alpha \in \mathbb{R}$.

Proof. H may be written as

$$H = \{f \in E^*; \varphi(f) = \alpha\},$$

where φ is a linear functional on E^*, $\varphi \not\equiv 0$. Let $f_0 \notin H$ and let V be a neighborhood of f_0 for the topology $\sigma(E^*, E)$ such that $V \subset H^c$. We may assume that

$$V = \{f \in E^*; |\langle f - f_0, x_i \rangle| < \varepsilon \quad \forall i = 1, 2, \ldots, k\}.$$

Since V is convex we find that either

(3) $$\varphi(f) < \alpha \quad \forall f \in V$$

or

(3') $$\varphi(f) > \alpha \quad \forall f \in V.$$

Assuming, for example, that (3) holds, we obtain

$$\varphi(g) < \alpha - \varphi(f_0) \quad \forall g \in W = V - f_0,$$

and since $-W = W$ we are led to

(4) $$|\varphi(g)| \leq |\alpha - \varphi(f_0)| \quad \forall g \in W.$$

It follows from (4) that φ is continuous at 0 for the topology $\sigma(E^*, E)$ (since W is a neighborhood of 0). Applying Proposition 3.14, we conclude that there is some $x_0 \in E$ such that

$$\varphi(f) = \langle f, x_0 \rangle \quad \forall f \in E^*.$$

Remark 11. Assume that the canonical injection $J : E \to E^{**}$ is *not* surjective. Then the topology $\sigma(E^*, E)$ is strictly *coarser* than the topology $\sigma(E^*, E^{**})$. For example, let $\xi \in E^{**}$ with $\xi \notin J(E)$. Then the set

$$H = \{f \in E^*; \langle \xi, f \rangle = 0\}$$

is closed in $\sigma(E^*, E^{**})$ but—in view of Corollary 3.15—it is not closed in $\sigma(E^*, E)$. We also learn from this example that *convex* sets that are closed in the strong topology need *not* be closed in the weak* topology. There are *two types of closed convex* sets in E^*:

(a) the convex sets that are strongly closed (= closed in the topology $\sigma(E^*, E^{**})$ by Theorem 3.7),
(b) the convex sets that are closed in $\sigma(E^*, E)$.

● **Theorem 3.16 (Banach–Alaoglu–Bourbaki).** *The closed unit ball*

$$B_{E^*} = \{f \in E^*; \ \|f\| \leq 1\}$$

is compact in the weak topology $\sigma(E^*, E)$.*

Remark 12. The *compactness* of B_{E^*} is *the most essential property* of the weak* topology; see also Remark 8.

Proof. Consider the Cartesian product $Y = \mathbb{R}^E$, which consists of all maps from E into $\mathbb{R}$; we denote elements of Y by $\omega = (\omega_x)_{x \in E}$ with $\omega_x \in \mathbb{R}$. The space Y is equipped with the standard *product topology* (see, e.g., H. L. Royden [1], I. R. Munkres [1], A. Knapp [1], or J. Dixmier [1]), i.e., the coarsest topology on Y associated to the collection of maps $\omega \mapsto \omega_x$ (as x runs through E), which is, of course, the same as the topology of pointwise convergence (see, e.g., J. R. Munkres [1]). In what follows E^* is systematically equipped with the weak* topology $\sigma(E^*, E)$. Since E^* consists of special maps from E into $\mathbb{R}$ (i.e., continuous linear maps), we may consider E^* as a subset of Y. More precisely, let $\Phi : E^* \to Y$ be the canonical injection from E^* into Y, so that $\Phi(f) = (\omega_x)_{x \in E}$ with $\omega_x = \langle f, x \rangle$. Clearly, Φ is continuous from E^* into Y (use Proposition 3.2 and note that for every fixed $x \in E$ the map $f \in E^* \mapsto (\Phi(f))_x = \langle f, x \rangle$ is continuous). The inverse map Φ^{-1} is also continuous from $\Phi(E^*)$ equipped with the Y topology) into E^*: indeed, using Proposition 3.2 once more, it suffices to check that for every fixed $x \in E$ the map $\omega \mapsto \langle \Phi^{-1}(\omega), x \rangle$ is continuous on $\Phi(E^*)$, which is obvious since $\langle \Phi^{-1}(\omega), x \rangle = \omega_x$ (note that $\omega = \Phi(f)$ for some $f \in E^*$ and $\langle \Phi^{-1}(\omega), x \rangle = \langle f, x \rangle = \omega_x$). In other words, Φ is a *homeomorphism* from E^* onto $\Phi(E^*)$. On the other hand, it is clear that $\Phi(B_{E^*}) = K$, where K is defined by

$$K = \left\{ \omega \in Y \ \middle| \ \begin{array}{l} |\omega_x| \leq \|x\|, \ \omega_{x+y} = \omega_x + \omega_y \\ \text{and } \omega_{\lambda x} = \lambda \omega_x \ \ \forall \lambda \in \mathbb{R}, \ \ \forall x, y \in E \end{array} \right\}.$$

In order to complete the proof of Theorem 3.16 it suffices to check that K is a compact subset of Y. Write K as $K = K_1 \cap K_2$, where

$$K_1 = \{\omega \in Y; \ |\omega_x| \leq \|x\| \ \ \forall x \in E\}$$

and

$$K_2 = \left\{\omega \in Y; \omega_{x+y} = \omega_x + \omega_y \text{ and } \omega_{\lambda x} = \lambda \omega_x \ \ \forall \lambda \in \mathbb{R}, \ \ \forall x, y \in E\right\}.$$

The set K_1 may also be written as a product of compact intervals

$$K_1 = \prod_{x \in E} [-\|x\|, +\|x\|].$$

Let us recall that (arbitrary) *products of compact spaces are compact*—a deep theorem due to Tychonoff; see, e.g., H. L. Royden [1], G. B. Folland [2], J. R. Munkres [1], A. Knapp [1], or J. Dixmier [1]. Therefore K_1 is compact. On the other hand, K_2 is closed in Y; indeed, for each *fixed* $\lambda \in \mathbb{R}$, $x, y \in E$ the sets

$$A_{x,y} = \{\omega \in Y; \omega_{x+y} - \omega_x - \omega_y = 0\},$$
$$B_{\lambda,x} = \{\omega \in Y; \omega_{\lambda x} - \lambda \omega_x = 0\},$$

are closed in Y (since the maps $\omega \mapsto \omega_{x+y} - \omega_x - \omega_y$ and $\omega \mapsto \omega_{\lambda x} - \lambda \omega_x$ are continuous on Y) and we may write K_2 as

$$K_2 = \left[\bigcap_{x,y \in E} A_{x,y} \right] \cap \left[\bigcap_{\substack{x \in E \\ \lambda \in \mathbb{R}}} B_{\lambda,x} \right].$$

Finally, K is compact since it is the intersection of a compact set (K_1) and a closed set (K_2).

3.5 Reflexive Spaces

Definition. Let E be a Banach space and let $J : E \to E^{**}$ be the canonical injection from E into E^{**} (see Section 1.3). The space E is said to be *reflexive* if J is surjective, i.e., $J(E) = E^{**}$.

When E is reflexive, E^{**} is usually identified with E.

Remark 13. Many important spaces in analysis are reflexive. Clearly, *finite-dimensional* spaces are reflexive (since dim $E = $ dim $E^* = $ dim E^{**}). As we shall see in Chapter 4 (see also Chapter 11), L^p (and ℓ^p) spaces are reflexive for $1 < p < \infty$. In Chapter 5 we shall see that Hilbert spaces are reflexive. However, equally important spaces in analysis are *not* reflexive; for example:

- L^1 and L^∞ (and ℓ^1, ℓ^∞) are not reflexive (see Chapters 4 and 11);
- $C(K)$, the space of continuous functions on an infinite compact metric space K, is not reflexive (see Exercise 3.25).

$\star$ *Remark* 14. It is essential to use J in the above definition. R. C. James [1] has constructed a striking example of a nonreflexive space with the property that there exists a surjective isometry from E onto E^{**}.

Our next result describes a basic property of reflexive spaces:

• **Theorem 3.17 (Kakutani).** *Let E be a Banach space. Then E is reflexive if and only if*

$$B_E = \{x \in E; \ \|x\| \le 1\}$$

is compact in the weak topology $\sigma(E, E^\star)$.

Proof. Assume first that E is reflexive, so that $J(B_E) = B_{E^{\star\star}}$. We already know (by Theorem 3.16) that $B_{E^{\star\star}}$ is compact in the topology $\sigma(E^{\star\star}, E^\star)$. Therefore, it suffices to check that J^{-1} is continuous from $E^{\star\star}$ equipped with $\sigma(E^{\star\star}, E^\star)$ with values in E equipped with $\sigma(E, E^\star)$. In view of Proposition 3.2, we have only to prove that for every fixed $f \in E^\star$ the map $\xi \mapsto \langle f, J^{-1}\xi \rangle$ is continuous on $E^{\star\star}$ equipped with $\sigma(E^{\star\star}, E^\star)$. But $\langle f, J^{-1}\xi \rangle = \langle \xi, f \rangle$, and the map $\xi \mapsto \langle \xi, f \rangle$ is indeed continuous on $E^{\star\star}$ for the topology $\sigma(E^{\star\star}, E^\star)$. Hence we have proved that B_E is compact in $\sigma(E, E^\star)$.

The converse is more delicate and relies on the following two lemmas:

Lemma 3.3 (Helly). *Let E be a Banach space. Let $f_1, f_2, \ldots, f_k$ be given in $E^\star$ and let $\gamma_1, \gamma_2, \ldots, \gamma_k$ be given in $\mathbb{R}$. The following properties are equivalent:*

(i) $\forall \varepsilon > 0 \ \exists x_\varepsilon \in E$ *such that* $\|x_\varepsilon\| \le 1$ *and*

$$|\langle f_i, x_\varepsilon \rangle - \gamma_i| < \varepsilon \quad \forall i = 1, 2, \ldots, k,$$

(ii) $|\sum_{i=1}^{k} \beta_i \gamma_i| \le \|\sum_{i=1}^{k} \beta_i f_i\| \quad \forall \beta_1, \beta_2, \ldots, \beta_k \in \mathbb{R}$.

Proof. (i) $\Rightarrow$ (ii). Fix $\beta_1, \beta_2, \ldots, \beta_k$ in $\mathbb{R}$ and let $S = \sum_{i=1}^{k} |\beta_i|$. It follows from (i) that

$$\left| \sum_{i=1}^{k} \beta_i \langle f_i, x_\varepsilon \rangle - \sum_{i=1}^{k} \beta_i \gamma_i \right| \le \varepsilon S$$

and therefore

$$\left| \sum_{i=1}^{k} \beta_i \gamma_i \right| \le \left\| \sum_{i=1}^{k} \beta_i f_i \right\| \|x_\varepsilon\| + \varepsilon S \le \left\| \sum_{i=1}^{k} \beta_i f_i \right\| + \varepsilon S.$$

Since this holds for every $\varepsilon > 0$, we obtain (ii).

(ii) $\Rightarrow$ (i). Set $\gamma = (\gamma_1, \gamma_2, \ldots, \gamma_k) \in \mathbb{R}^k$ and consider the map $\varphi : E \to \mathbb{R}^k$ defined by

$$\varphi(x) = (\langle f_1, x \rangle, \ldots, \langle f_k, x \rangle).$$

Property (i) says precisely that $\gamma \in \overline{\varphi(B_E)}$. Suppose, by contradiction, that (i) fails, so that $\gamma \notin \overline{\varphi(B_E)}$. Hence $\{\gamma\}$ and $\overline{\varphi(B_E)}$ may be strictly separated in $\mathbb{R}^k$ by some hyperplane; i.e., there exists some $\beta = (\beta_1, \beta_2, \ldots, \beta_k) \in \mathbb{R}^k$ and some $\alpha \in \mathbb{R}$ such that

$$\beta \cdot \varphi(x) < \alpha < \beta \cdot \gamma \quad \forall x \in B_E.$$

It follows that

$$\left\langle \sum_{i=1}^{k} \beta_i f_i, x \right\rangle < \alpha < \sum_{i=1}^{k} \beta_i \gamma_i \quad \forall x \in B_E,$$

and therefore

$$\left\|\sum_{i=1}^{k}\beta_i f_i\right\| \le \alpha < \sum_{i=1}^{k}\beta_i \gamma_i,$$

which contradicts (ii).

Lemma 3.4 (Goldstine). *Let E be any Banach space. Then $J(B_E)$ is dense in $B_{E^{**}}$ with respect to the topology $\sigma(E^{**}, E^{*})$, and consequently $J(E)$ is dense in E^{**} in the topology $\sigma(E^{**}, E^{*})$.*

Proof. Let $\xi \in B_{E^{**}}$ and let V be a neighborhood of ξ for the topology $\sigma(E^{**}, E^{*})$. We must prove that $V \cap J(B_E) \ne \emptyset$. As usual, we may assume that V is of the form

$$V = \left\{\eta \in E^{**};\ |\langle \eta - \xi, f_i\rangle| < \varepsilon \quad \forall i = 1, 2, \ldots, k\right\}$$

for some given elements $f_1, f_2, \ldots, f_k$ in E^{*} and some $\varepsilon > 0$. We have to find some $x \in B_E$ such that $J(x) \in V$, i.e.,

$$|\langle f_i, x\rangle - \langle \xi, f_i\rangle| < \varepsilon \quad \forall i = 1, 2, \ldots, k.$$

Set $\gamma_i = \langle \xi, f_i\rangle$. In view of Lemma 3.3 it suffices to check that

$$\left|\sum_{i=1}^{k}\beta_i \gamma_i\right| \le \left\|\sum_{i=1}^{k}\beta_i f_i\right\|,$$

which is clear since $\sum_{i=1}^{k}\beta_i \gamma_i = \left\langle \xi, \sum_{i=1}^{k}\beta_i f_i\right\rangle$ and $\|\xi\| \le 1$.

Remark 15. Note that $J(B_E)$ is *closed* in $B_{E^{**}}$ in the strong topology. Indeed, if $\xi_n = J(x_n) \to \xi$ we see that (x_n) is a Cauchy sequence in B_E (since J is an isometry) and therefore $x_n \to x$, so that $\xi = Jx$. It follows that $J(B_E)$ is not dense in $B_{E^{**}}$ in the strong topology, unless $J(B_E) = B_{E^{**}}$, i.e., E is reflexive.

Remark 16. See Problem 9 for an alternative proof of Lemma 3.4 (based on a variant of Hahn–Banach in E^{**}).

Proof of Theorem 3.17, concluded. The canonical injection $J : E \to E^{**}$ is always continuous from $\sigma(E, E^{*})$ into $\sigma(E^{**}, E^{*})$, since for every fixed $f \in E^{*}$ the map $x \mapsto \langle Jx, f\rangle = \langle f, x\rangle$ is continuous with respect to $\sigma(E, E^{*})$. Assuming that B_E is compact in the topology $\sigma(E, E^{*})$, we deduce that $J(B_E)$ is compact—and thus closed—in E^{**} with respect to the topology $\sigma(E^{**}, E^{*})$. On the other hand, by Lemma 3.4, $J(B_E)$ is dense in $B_{E^{**}}$ for the same topology. It follows that $J(B_E) = B_{E^{**}}$ and thus $J(E) = E^{**}$.

In connection with the compactness properties of reflexive spaces we also have the following two results:

• **Theorem 3.18.** *Assume that E is a reflexive Banach space and let (x_n) be a bounded sequence in E. Then there exists a subsequence (x_{n_k}) that converges in the weak topology $\sigma(E, E^{*})$.*

The converse is also true, namely the following.

★ **Theorem 3.19 (Eberlein–Šmulian).** *Assume that E is a Banach space such that every bounded sequence in E admits a weakly convergent subsequence (in* $\sigma(E, E^\star)$*). Then E is reflexive.*

The proof of Theorem 3.18 requires a little excursion through separable spaces and will be given in Section 3.6. The proof of Theorem 3.19 is rather delicate and is omitted; see, e.g., R. Holmes [1], K. Yosida [1], N. Dunford–J. T. Schwartz [1], J. Diestel [2], or Problem 10.

Remark 17. In order to clarify the connection between Theorems 3.17, 3.18, and 3.19 it is useful to recall the following facts:

(i) If X is a *metric* space, then

$$[X \text{ is compact}] \Leftrightarrow [\text{every sequence in } X \text{ admits a convergent subsequence}].$$

(ii) There exist *compact topological* spaces X and some sequences in X *without* any convergent subsequence. A typical example is $X = B_{E^\star}$, which is compact in the topology $\sigma(E^\star, E)$; when $E = \ell^\infty$ it is easy to construct a sequence in X *without* any convergent subsequence (see Exercise 3.18).

(iii) If X is a *topological* space with the property that every sequence admits a convergent subsequence, then X need *not* be compact.

Here are some further properties of reflexive spaces.

● **Proposition 3.20.** *Assume that E is a reflexive Banach space and let $M \subset E$ be a closed linear subspace of E. Then M is reflexive.*

Proof. The space M—equipped with the norm of E—has a priori two distinct weak topologies:

(a) the topology induced by $\sigma(E, E^\star)$,
(b) its own weak topology $\sigma(M, M^\star)$.

In fact, these two topologies are the same (since, by Hahn–Banach, every continuous linear functional on M is the restriction to M of a continuous linear functional on E). In view of Theorem 3.17, we have to check that B_M is compact in the topology $\sigma(M, M^\star)$ or equivalently in the topology $\sigma(E, E^\star)$. However, B_E is compact in the topology $\sigma(E, E^\star)$ and M is closed in the topology $\sigma(E, E^\star)$ (by Theorem 3.7). Therefore B_M is compact in the topology $\sigma(E, E^\star)$.

Corollary 3.21. *A Banach space E is reflexive if and only if its dual space $E^\star$ is reflexive.*

Proof. E reflexive $\Rightarrow$ $E^\star$ reflexive. The idea of the proof is simple, since, roughly speaking, we have that $E^{\star\star} = E \Rightarrow E^{\star\star\star} = E^\star$. More precisely, let J be the canonical isomorphism from E into $E^{\star\star}$. Let $\varphi \in E^{\star\star\star}$ be given. The map $x \mapsto \langle \varphi, Jx \rangle$ is a continuous linear functional on E. Call it $f \in E^\star$, so that

$$\langle \varphi, Jx \rangle = \langle f, x \rangle \quad \forall x \in E.$$

But we also have

$$\langle \varphi, Jx \rangle = \langle Jx, f \rangle \quad \forall x \in E.$$

Since J is surjective, we infer that

$$\langle \varphi, \xi \rangle = \langle \xi, f \rangle \quad \forall \xi \in E^{**},$$

which means precisely that the canonical injection from E^* into E^{***} is surjective.

E^* *reflexive* $\Rightarrow$ E *reflexive*. From the step above we already know that E^{**} is reflexive. Since $J(E)$ is a closed subspace of E^{**} in the strong topology, we conclude (by Proposition 3.20) that $J(E)$ is reflexive. Therefore, E is reflexive.[3]

● **Corollary 3.22.** *Let E be a reflexive Banach space. Let $K \subset E$ be a bounded, closed, and convex subset of E. Then K is compact in the topology $\sigma(E, E^*)$.*

Proof. K is closed for the topology $\sigma(E, E^*)$ (by Theorem 3.7). On the other hand, there exists a constant m such that $K \subset m B_E$, and $m B_E$ is compact in $\sigma(E, E^*)$ (by Theorem 3.17).

● **Corollary 3.23.** *Let E be a reflexive Banach space and let $A \subset E$ be a nonempty, closed, convex subset of E. Let $\varphi : A \to (-\infty, +\infty]$ be a convex l.s.c. function such that $\varphi \not\equiv +\infty$ and*

$$(5) \qquad \lim_{\substack{x \in A \\ \|x\| \to \infty}} \varphi(x) = +\infty \quad \text{(no assumption if A is bounded).}$$

Then φ achieves its minimum on A, i.e., there exists some $x_0 \in A$ such that

$$\varphi(x_0) = \min_A \varphi.$$

Proof. Fix any $a \in A$ such that $\varphi(a) < +\infty$ and consider the set

$$\tilde{A} = \{x \in A; \ \varphi(x) \le \varphi(a)\}.$$

Then $\tilde{A}$ is closed, convex, and bounded (by (5)) and thus it is *compact* in the topology $\sigma(E, E^*)$ (by Corollary 3.22). On the other hand, φ is also l.s.c. in the topology $\sigma(E, E^*)$ (by Corollary 3.9). It follows that φ achieves its minimum on $\tilde{A}$ (see property 5 following the definition of l.s.c. in Chapter 1), i.e., there exists $x_0 \in \tilde{A}$ such that

$$\varphi(x_0) \le \varphi(x) \quad \forall x \in \tilde{A}.$$

If $x \in A \backslash \tilde{A}$, we have $\varphi(x_0) \le \varphi(a) < \varphi(x)$; therefore

$$\varphi(x_0) \le \varphi(x) \quad \forall x \in A.$$

---

[3] It is clear that if E and F are Banach spaces, and T is a linear surjective isometry from E onto F, then E is reflexive iff F is reflexive. Of course, there is no contradiction with Remark 14!

Remark 18. Corollary 3.23 is the main reason why *reflexive spaces* and *convex functions* are so important in many problems occurring in the *calculus of variations* and in *optimization*.

Theorem 3.24. *Let E and F be two reflexive Banach spaces. Let $A : D(A) \subset E \to F$ be an unbounded linear operator that is densely defined and closed. Then $D(A^\star)$ is dense in $F^\star$. Thus $A^{\star\star}$ is well defined ($A^{\star\star} : D(A^{\star\star}) \subset E^{\star\star} \to F^{\star\star}$) and it may also be viewed as an unbounded operator from E into F. Then we have*

$$\boxed{A^{\star\star} = A.}$$

Proof.

1. $D(A^\star)$ *is dense* in $F^\star$. Let φ be a continuous linear functional on $F^\star$ that vanishes on $D(A^\star)$. In view of Corollary 1.8 it suffices to prove that $\varphi \equiv 0$ on $F^\star$. Since F is reflexive, $\varphi \in F$ and we have

$$(6) \qquad\qquad \langle w, \varphi \rangle = 0 \quad \forall w \in D(A^\star).$$

If $\varphi \neq 0$ then $[0, \varphi] \notin G(A)$ in $E \times F$. Thus, one may strictly separate $[0, \varphi]$ and $G(A)$ by a closed hyperplane in $E \times F$; i.e., there exist some $[f, v] \in E^\star \times F^\star$ and some $\alpha \in \mathbb{R}$ such that

$$\langle f, u \rangle + \langle v, Au \rangle < \alpha < \langle v, \varphi \rangle \quad \forall u \in D(A).$$

It follows that

$$\langle f, u \rangle + \langle v, Au \rangle = 0 \quad \forall u \in D(A)$$

and

$$\langle v, \varphi \rangle \neq 0.$$

Thus $v \in D(A^\star)$, and we are led to a contradiction by choosing $w = v$ in (6).

2. $A^{\star\star} = A$. We recall (see Section 2.6) that

$$I[G(A^\star)] = G(A)^\perp$$

and

$$I[G(A^{\star\star})] = G(A^\star)^\perp.$$

It follows that

$$G(A^{\star\star}) = G(A)^{\perp\perp} = G(A),$$

since A is closed.

3.6 Separable Spaces

Definition. We say that a metric space E is *separable* if there exists a subset $D \subset E$ that is countable and dense.

Many important spaces in analysis are separable. Clearly, *finite-dimensional* spaces are separable. As we shall see in Chapter 4 (see also Chapter 11), L^p (and ℓ^p) spaces are separable for $1 \le p < \infty$. Also $C(K)$, the space of continuous functions on a compact metric space K, is separable (see Problem 24). However, L^∞ and ℓ^∞ are *not* separable (see Chapters 4 and 11).

Proposition 3.25. *Let E be a separable metric space and let $F \subset E$ be any subset. Then F is also separable.*

Proof. Let (u_n) be a countable dense subset of E. Let (r_m) be any sequence of positive numbers such that $r_m \to 0$. Choose any point $a_{m,n} \in B(u_n, r_m) \cap F$ whenever this set is nonempty. The set $(a_{m,n})$ is countable and dense in F.

Theorem 3.26. *Let E be a Banach space such that $E^\star$ is separable. Then E is separable.*

Remark 19. The converse is not true. As we shall see in Chapter 4, $E = L^1$ is separable but its dual space $E^\star = L^\infty$ is *not* separable.

Proof. Let $(f_n)_{n \ge 1}$ be countable and dense in $E^\star$. Since

$$\|f_n\| = \sup_{\substack{x \in E \\ \|x\| \le 1}} \langle f_n, x \rangle,$$

we can find some $x_n \in E$ such that

$$\|x_n\| = 1 \text{ and } \langle f_n, x_n \rangle \ge \frac{1}{2} \|f_n\|.$$

Let us denote by L_0 the vector space over $\mathbb{Q}$ generated by the $(x_n)_{n \ge 1}$; i.e., L_0 consists of all *finite* linear combinations with coefficients in $\mathbb{Q}$ of the elements $(x_n)_{n \ge 1}$. We claim that L_0 is *countable*. Indeed, for every integer n, let Λ_n be the vector space over $\mathbb{Q}$ generated by the $(x_k)_{1 \le k \le n}$. Clearly, Λ_n is countable and, moreover, $L_0 = \bigcup_{n \ge 1} \Lambda_n$.

Let L denote the vector space over $\mathbb{R}$ generated by the $(x_n)_{n \ge 1}$. Of course, L_0 is a dense subset of L. We claim that L is a dense subspace of E—and this will conclude the proof (L_0 will be a dense countable subset of E). Let $f \in E^\star$ be a continuous linear functional that vanishes on L; in view of Corollary 1.8 we have to prove that $f = 0$. Given any $\varepsilon > 0$, there is some integer N such that $\|f - f_N\| < \varepsilon$. We have

$$\frac{1}{2} \|f_N\| \le \langle f_N, x_N \rangle = \langle f_N - f, x_N \rangle < \varepsilon$$

(since $\langle f, x_N \rangle = 0$). It follows that $\|f\| \le \|f - f_N\| + \|f_N\| < 3\varepsilon$. Thus $f = 0$.

Corollary 3.27. *Let E be a Banach space. Then*

$$[E \text{ reflexive and separable}] \Leftrightarrow [E^\star \text{ reflexive and separable}].$$

Proof. We already know (Corollary 3.21 and Theorem 3.26) that

$$[E^* \text{ reflexive and separable}] \Rightarrow [E \text{ reflexive and separable}].$$

Conversely, if E is reflexive and separable, so is $E^{**} = J(E)$; thus E^* is reflexive and separable.

Separability properties are closely related to the *metrizability* of the weak topologies. Let us recall that a topological space X is said to be *metrizable* if there is a metric on X that induces the topology of X.

Theorem 3.28. *Let E be a separable Banach space. Then B_{E^*} is metrizable in the weak* topology $\sigma(E^*, E)$.*
Conversely, if B_{E^*} is metrizable in $\sigma(E^*, E)$, then E is separable.

There is a "dual" statement.

Theorem 3.29. *Let E be a Banach space such that E^* is separable. Then B_E is metrizable in the weak topology $\sigma(E, E^*)$.*
Conversely, if B_E is metrizable in $\sigma(E, E^*)$, then E^* is separable.

Proof of Theorem 3.28. Let $(x_n)_{n \geq 1}$ be a dense countable subset of B_E. For every $f \in E^*$ set

$$[f] = \sum_{n=1}^{\infty} \frac{1}{2^n} |\langle f, x_n \rangle|.$$

Clearly, $[\ ]$ is a norm on E^* and $[f] \leq \|f\|$. Let $d(f, g) = [f - g]$ be the corresponding metric. We shall prove that the topology induced by d on B_{E^*} is the same as the topology $\sigma(E^*, E)$ restricted to B_{E^*}.
(a) Let $f_0 \in B_{E^*}$ and let V be a neighborhood of f_0 for $\sigma(E^*, E)$. We have to find some $r > 0$ such that

$$U = \{f \in B_{E^*}; \ d(f, f_0) < r\} \subset V.$$

As usual, we may assume that V has the form

$$V = \{f \in B_{E^*}; \ |\langle f - f_0, y_i \rangle| < \varepsilon \quad \forall i = 1, 2, \ldots, k\}$$

with $\varepsilon > 0$ and $y_1, y_2, \ldots, y_k \in E$. Without loss of generality we may assume that $\|y_i\| \leq 1$ for every $i = 1, 2, \ldots, k$. For every i there is some integer n_i such that

$$\|y_i - x_{n_i}\| < \varepsilon/4$$

(since the set $(x_n)_{n \geq 1}$ is dense in B_E).
Choose $r > 0$ small enough that

$$2^{n_i} r < \varepsilon/2 \quad \forall i = 1, 2, \ldots, k.$$

We claim that for such r, $U \subset V$. Indeed, if $d(f, f_0) < r$, we have

$$\frac{1}{2^{n_i}}|\langle f - f_0, x_{n_i}\rangle| < r \quad \forall i = 1, 2, \ldots, k$$

and therefore, $\forall i = 1, 2, \ldots, k$,

$$|\langle f - f_0, y_i\rangle| = |\langle f - f_0, y_i - x_{n_i}\rangle + \langle f - f_0, x_{n_i}\rangle| < \frac{\varepsilon}{2} + \frac{\varepsilon}{2}.$$

It follows that $f \in V$.

(b) Let $f_0 \in B_{E^*}$. Given $r > 0$, we have to find some neighborhood V of f_0 for $\sigma(E^*, E)$ such that

$$V \subset U = \{f \in B_{E^*}; \ d(f, f_0) < r\}.$$

We shall choose V to be

$$V = \{f \in B_{E^*}; \ \langle f - f_0, x_i\rangle| < \varepsilon \quad \forall i = 1, 2, \ldots, k\}$$

with ε and k to be determined in such a way that $V \subset U$. For $f \in V$ we have

$$d(f, f_0) = \sum_{n=1}^{k} \frac{1}{2^n}|\langle f - f_0, x_n\rangle| + \sum_{n=k+1}^{\infty} \frac{1}{2^n}|\langle f - f_0, x_n\rangle|$$

$$< \varepsilon + 2 \sum_{n=k+1}^{\infty} \frac{1}{2^n} = \varepsilon + \frac{1}{2^{k-1}}.$$

Thus, it suffices to take $\varepsilon = \frac{r}{2}$ and k large enough that $\frac{1}{2^{k-1}} < \frac{r}{2}$.

★*Conversely*, suppose B_{E^*} is metrizable in $\sigma(E^*, E)$ and let us prove that E is separable. Set

$$U_n = \{f \in B_{E^*}; \ d(f, 0) < 1/n\}$$

and let V_n be a neighborhood of 0 in $\sigma(E^*, E)$ such that $V_n \subset U_n$. We may assume that V_n has the form

$$V_n = \{f \in B_{E^*}; \ |\langle f, x\rangle| < \varepsilon_n \quad \forall x \in \Phi_n\}$$

with $\varepsilon_n > 0$ and Φ_n is a finite subset of E. Set

$$D = \bigcup_{n=1}^{\infty} \Phi_n,$$

so that D is countable.

We claim that the vector space generated by D is dense in E (which implies that E is separable). Indeed, suppose $f \in E^*$ is such that $\langle f, x\rangle = 0 \ \forall x \in D$. It follows that $f \in V_n \ \forall n$ and therefore $f \in U_n \ \forall n$, so that $f = 0$.

Proof of Theorem 3.29. The proof of the implication

$$[E^\star \text{ separable}] \Rightarrow [B_E \text{ is metrizable in } \sigma(E, E^\star)]$$

is exactly the same as above—just change the roles of E and $E^\star$. The proof of the converse is more delicate (find where the proof above breaks down); we refer to N. Dunford–J. T. Schwartz [1] or Exercise 3.24.

Remark 20. One should emphasize again (see Remark 3) that in infinite-dimensional spaces the weak topology $\sigma(E, E^\star)$ (resp. weak* topology $\sigma(E^\star, E)$) on *all* of E (resp. $E^\star$) is *not* metrizable; see Exercise 3.8. In particular, the topology induced by the norm [ ] on all of $E^\star$ does not coincide with the weak* topology.

Corollary 3.30. *Let E be a separable Banach space and let (f_n) be a bounded sequence in $E^\star$. Then there exists a subsequence (f_{n_k}) that converges in the weak* topology $\sigma(E^\star, E)$.*

Proof. Without loss of generality we may assume that $\| f_n \| \leq 1$ for all n. The set $B_{E^\star}$ is compact and metrizable for the topology $\sigma(E^\star, E)$ (by Theorems 3.16 and 3.28). The conclusion follows.

We may now return to the proof of Theorem 3.18:

Proof of Theorem 3.18. Let M_0 be the vector space generated by the x_n's and let $M = \overline{M_0}$. Clearly, M is separable (see the proof of Theorem 3.26). Moreover, M is reflexive (by Proposition 3.20). It follows that B_M is compact and metrizable in the weak topology $\sigma(M, M^\star)$, since $M^\star$ is separable (we use here Corollary 3.27 and Theorem 3.29). We may thus find a subsequence (x_{n_k}) that converges weakly $\sigma(M, M^\star)$, and hence (x_{n_k}) converges also weakly $\sigma(E, E^\star)$ (as in the proof of Proposition 3.20).

3.7 Uniformly Convex Spaces

Definition. A Banach space is said to be *uniformly convex* if

$$\forall \varepsilon > 0 \ \exists \delta > 0 \text{ such that}$$

$$\left[x, y \in E, \|x\| \leq 1, \|y\| \leq 1 \text{ and } \|x - y\| > \varepsilon \right] \Rightarrow \left[\left\| \frac{x + y}{2} \right\| < 1 - \delta \right].$$

The uniform convexity is a *geometric* property of the unit ball: if we slide a rule of length $\varepsilon > 0$ in the unit ball, then its midpoint must stay within a ball of radius $(1 - \delta)$ for some $\delta > 0$. In particular, the unit *sphere* must be "round" and cannot include any line segment.

Example 1. Let $E = \mathbb{R}^2$. The norm $\|x\|_2 = \left[|x_1|^2 + |x_2|^2 \right]^{1/2}$ is uniformly convex, while the norm $\|x\|_1 = |x_1| + |x_2|$ and the norm $\|x\|_\infty = \max(|x_1|, |x_2|)$ are *not* uniformly convex. This can be easily seen by staring at the unit balls, as shown in Figure 3.

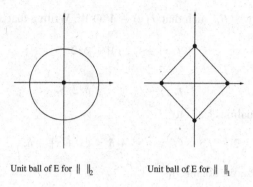

Unit ball of E for $\|\ \|_2$ Unit ball of E for $\|\ \|_1$

Fig. 3

Example 2. As we shall see in Chapters 4 and 5, the L^p spaces are uniformly convex for $1 < p < \infty$ and Hilbert spaces are also uniformly convex.

● **Theorem 3.31 (Milman–Pettis).** *Every uniformly convex Banach space is reflexive.*

Remark 21. Uniform convexity is a *geometric* property of the norm; an equivalent norm need *not* be uniformly convex. On the other hand, reflexivity is a *topological* property: a reflexive space remains reflexive for an equivalent norm. It is a striking feature of Theorem 3.31 that a geometric property implies a topological property. Uniform convexity is often used as a tool to prove reflexivity; but it is not the ultimate tool—there are some weird reflexive spaces that admit no uniformly convex equivalent norm!

Proof. Let $\xi \in E^{**}$ with $\|\xi\| = 1$. We have to show that $\xi \in J(B_E)$. Since $J(B_E)$ is closed in E^{**} in the strong topology, it suffices to prove that

$$(7) \qquad\qquad \forall \varepsilon > 0 \quad \exists x \in B_E \text{ such that } \|\xi - J(x)\| \le \varepsilon.$$

Fix $\varepsilon > 0$ and let $\delta > 0$ be the modulus of uniform convexity. Choose some $f \in E^*$ such that $\|f\| = 1$ and

$$(8) \qquad\qquad \langle \xi, f \rangle > 1 - (\delta/2)$$

(which is possible, since $\|\xi\| = 1$). Set

$$V = \{\eta \in E^{**};\ |\langle \eta - \xi, f \rangle| < \delta/2\},$$

so that V is a neighborhood of ξ in the topology $\sigma(E^{**}, E^*)$. Since $J(B_E)$ is dense in $B_{E^{**}}$ with respect to $\sigma(E^{**}, E^*)$ (Lemma 3.4), we know that $V \cap J(B_E) \ne \emptyset$ and thus there is some $x \in B_E$ such that $J(x) \in V$. We claim that this x satisfies (7).

Suppose, by contradiction, that $\|\xi - Jx\| > \varepsilon$, i.e., $\xi \in (Jx + \varepsilon B_{E^{**}})^c = W$. The set W is also a neighborhood of ξ in the topology $\sigma(E^{**}, E^*)$ (since $B_{E^{**}}$ is closed in $\sigma(E^{**}, E^*)$). Using Lemma 3.4 once more, we know that $V \cap W \cap J(B_E) \ne \phi$, i.e., there exists some $y \in B_E$ such that $J(y) \in V \cap W$. Writing that $J(x), J(y) \in V$, we obtain

$$|\langle f, x \rangle - \langle \xi, f \rangle| < \delta/2$$

and

$$|\langle f, y \rangle - \langle \xi, f \rangle| < \delta/2.$$

Adding these inequalities leads to

$$2\langle \xi, f \rangle < \langle f, x + y \rangle + \delta \leq \|x + y\| + \delta.$$

Combining with (8), we obtain

$$\left\| \frac{x + y}{2} \right\| > 1 - \delta.$$

It follows (by uniform convexity) that $\|x - y\| \leq \varepsilon$; this is absurd, since $J(y) \in W$ (i.e., $\|x - y\| > \varepsilon$).

We conclude with a useful property of uniformly convex spaces.

Proposition 3.32. *Assume that E is a uniformly convex Banach space. Let (x_n) be a sequence in E such that $x_n \rightharpoonup x$ weakly $\sigma(E, E^\star)$ and*

$$\limsup \|x_n\| \leq \|x\|.$$

Then $x_n \to x$ strongly.

Proof. We may always assume that $x \neq 0$ (otherwise the conclusion is obvious). Set

$$\lambda_n = \max(\|x_n\|, \|x\|), \quad y_n = \lambda_n^{-1} x_n, \text{ and } y = \|x\|^{-1} x,$$

so that $\lambda_n \to \|x\|$ and $y_n \rightharpoonup y$ weakly $\sigma(E, E^\star)$. It follows that

$$\|y\| \leq \liminf \|(y_n + y)/2\|$$

(see Proposition 3.5(iii)). On the other hand, $\|y\| = 1$ and $\|y_n\| \leq 1$, so that in fact, $\|(y_n + y)/2\| \to 1$. We deduce from the uniform convexity that $\|y_n - y\| \to 0$ and thus $x_n \to x$ strongly.

Comments on Chapter 3

1. The topologies $\sigma(E, E^\star)$, $\sigma(E^\star, E)$, etc., are locally convex topologies. As such, they enjoy all the properties of locally convex spaces; for example, Hahn–Banach (geometric form), Krein–Milman, etc., still hold; see, e.g., N. Bourbaki [1], A. Knapp [2], and also Problem 9.

2. Here is another remarkable property of the weak* topology that is worth mentioning.

★ **Theorem 3.33 (Banach–Dieudonné–Krein–Šmulian).** *Let E be a Banach space and let $C \subset E^*$ be convex. Assume that for every integer n the set $C \cap (nB_{E^*})$ is closed for the topology $\sigma(E^*, E)$. Then C is closed for the topology $\sigma(E^*, E)$.*

The proof may be found in, e.g., N. Bourbaki [1], R. Larsen [1], R. Holmes [1], N. Dunford–J. T. Schwartz [1], H. Schaefer [1], and Problem 11. The above references also include much material related to the Eberlein–Šmulian theorem (Theorem 3.19).

3. The theory of *vector spaces in duality*—which extends the duality $\langle E, E^* \rangle$—was very popular in the late forties and early fifties, especially in connection with the theory of distributions. One says that two vector spaces X and Y are in duality if there is a bilinear form $\langle\,,\,\rangle$ on $X \times Y$ that separates points (i.e., $\forall x \neq 0\ \exists y$ such that $\langle x, y \rangle \neq 0$ and $\forall y \neq 0\ \exists x$ such that $\langle x, y \rangle \neq 0$). Many topologies may be defined on X (or Y) such as the weak topology $\sigma(X, Y)$, Mackey's topology $\tau(X, Y)$, and the strong topology $\beta(X, Y)$. These topologies are of interest in spaces that are *not* Banach spaces, such as the spaces used in the theory of distributions. On this subject the reader may consult, e.g., N. Bourbaki [1], H. Schaefer [1], G. Köthe [1], F. Treves [1], J. Kelley–I. Namioka [1], R. Edwards [1], J. Horváth [1], etc.

4. The properties of separability, reflexivity, and uniform convexity are also related to the *differentiability* properties of the function $x \mapsto \|x\|$ (see, e.g., J. Diestel [1], B. Beauzamy [1], and Problem 13). The existence of equivalent norms with nice geometric properties has been extensively studied. For example, how does one know whether a Banach space admits an equivalent uniformly convex norm? how useful is this information? (such spaces are called superreflexive; see, e.g., J. Diestel [1] or B. Beauzamy [1]). The *geometry of Banach spaces* has flourished since the early sixties and has become an *active field* associated with the names A. Dvoretzky, A. Grothendieck, R. C. James, J. Lindenstrauss, V. Milman, L. Tzafriri (and their group in Israel), A. Pelczynski, P. Enflo, L. Schwartz (and his group including G. Pisier, B. Maurey, B. Beauzamy), W. B. Johnson, H. P. Rosenthal, J. Bourgain, D. Preiss, M. Talagrand, T. Gowers, and many others. On this subject the reader may consult the books of B. Beauzamy [1], J. Diestel [1], [2], J. Lindenstrauss–L. Tzafriri [2], L. Schwartz [2], R. Deville–G. Godefroy–V. Zizler [1], Y. Benyamini and J. Lindenstrauss [1], F. Albiac and N. Kalton [1], A. Pietsch [1], etc.

Exercises for Chapter 3

3.1 Let E be a Banach space and let $A \subset E$ be a subset that is compact in the weak topology $\sigma(E, E^*)$. Prove that A is bounded.

3.2 Let E be a Banach space and let (x_n) be a sequence such that $x_n \rightharpoonup x$ in the weak topology $\sigma(E, E^*)$. Set

$$\sigma_n = \frac{1}{n}(x_1 + x_2 + \cdots + x_n).$$

Prove that $\sigma_n \rightharpoonup x$ in the weak topology $\sigma(E, E^\star)$.

3.3 Let E be a Banach space. Let $A \subset E$ be a convex subset. Prove that the closure of A in the strong topology and that in the weak topology $\sigma(E, E^\star)$ are the same.

3.4 Let E be a Banach space and let (x_n) be a sequence in E such that $x_n \rightharpoonup x$ in the weak topology $\sigma(E, E^\star)$.

1. Prove that there exists a sequence (y_n) in E such that

   (a)
   $$y_n \in \operatorname{conv} \left(\bigcup_{i=n}^{\infty} \{x_i\} \right) \quad \forall n$$

   and

   (b)
   $$y_n \to x \quad \text{strongly.}$$

2. Prove that there exists a sequence (z_n) in E such that

   (a')
   $$z_n \in \operatorname{conv} \left(\bigcup_{i=1}^{n} \{x_i\} \right) \quad \forall n$$

   and

   (b')
   $$z_n \to x \quad \text{strongly.}$$

3.5 Let E be a Banach space and let $K \subset E$ be a subset of E that is compact in the strong topology. Let (x_n) be a sequence in K such that $x_n \rightharpoonup x$ weakly $\sigma(E, E^\star)$. Prove that $x_n \to x$ strongly.
   [**Hint**: Argue by contradiction.]

3.6 Let X be a topological space and let E be a Banach space. Let $u, v : X \to E$ be two continuous maps from X with values in E equipped with the weak topology $\sigma(E, E^\star)$.

1. Prove that the map $x \mapsto u(x) + v(x)$ is continuous from X into E equipped with $\sigma(E, E^\star)$.
2. Let $a : X \to \mathbb{R}$ be a continuous function. Prove that the map $x \mapsto a(x)u(x)$ is continuous from X into E equipped with $\sigma(E, E^\star)$.

3.7 Let E be a Banach space and let $A \subset E$ be a subset that is closed in the weak topology $\sigma(E, E^\star)$. Let $B \subset E$ be a subset that is compact in the weak topology $\sigma(E, E^\star)$.

1. Prove that $A + B$ is closed in $\sigma(E, E^*)$.
2. Assume, in addition, that A and B are convex, nonempty, and disjoint. Prove that there exists a closed hyperplane strictly separating A and B.

$\boxed{3.8}$ Let E be an infinite-dimensional Banach space. Our purpose is to show that E equipped with the weak topology is not metrizable. Suppose, by contradiction, that there is a metric $d(x, y)$ on E that induces on E the same topology as $\sigma(E, E^*)$.

1. For every integer $k \geq 1$ let V_k denote a neighborhood of 0 in the topology $\sigma(E, E^*)$, such that

$$V_k \subset \left\{ x \in E; \ d(x, 0) < \frac{1}{k} \right\}.$$

Prove that there exists a sequence (f_n) in E^* such that every $g \in E^*$ is a (finite) linear combination of the f_n's.
[**Hint:** Use Lemma 3.2.]
2. Deduce that E^* is finite-dimensional.
[**Hint:** Use the Baire category theorem as in Exercise 1.5.]
3. Conclude.
4. Prove by a similar method that E^* equipped with the weak* topology $\sigma(E^*, E)$ is not metrizable.

$\boxed{3.9}$ Let E be a Banach space; let $M \subset E$ be a linear subspace, and let $f_0 \in E^*$. Prove that there exists some $g_0 \in M^\perp$ such that

$$\inf_{g \in M^\perp} \| f_0 - g \| = \| f_0 - g_0 \|.$$

Two methods are suggested:

1. Use Theorem 1.12.
2. Use the weak* topology $\sigma(E^*, E)$.

$\boxed{3.10}$ Let E and F be two Banach spaces. Let $T \in \mathscr{L}(E, F)$, so that $T^* \in \mathscr{L}(F^*, E^*)$. Prove that T^* is continuous from F^* equipped with $\sigma(F^*, F)$ into E^* equipped with $\sigma(E^*, E)$.

$\boxed{3.11}$ Let E be a Banach space and let $A : E \to E^*$ be a monotone map defined on $D(A) = E$; see Exercise 2.6. Assume that for every $x, y \in E$ the map

$$t \in \mathbb{R} \mapsto \langle A(x + ty), y \rangle$$

is continuous at $t = 0$. Prove that A is continuous from E strong into E^* equipped with $\sigma(E^*, E)$.

$\boxed{3.12}$ Let E be a Banach space and let $x_0 \in E$. Let $\varphi : E \to (-\infty, +\infty]$ be a convex l.s.c. function with $\varphi \not\equiv +\infty$.

1. Show that the following properties are equivalent:

   (A) $\exists R, \exists M < +\infty$ such that $\varphi(x) \leq M, \quad \forall x \in E$ with $\|x - x_0\| \leq R$,

   (B) $\lim_{\substack{f \in E^\star \\ \|f\| \to \infty}} \{\varphi^\star(f) - \langle f, x_0 \rangle\} = +\infty.$

2. Assuming (A) or (B) prove that

   $$\inf_{f \in E^\star} \{\varphi^\star(f) - \langle f, x_0 \rangle\} \quad \text{is achieved.}$$

   [**Hint**: Use the weak* topology $\sigma(E^\star, E)$ or Theorem 1.12.]
   What is the value of this inf?

3.13 Let E be a Banach space. Let (x_n) be a sequence in E and let $x \in E$. Set

$$K_n = \overline{\text{conv}\left(\bigcup_{i=n}^{\infty}\{x_i\}\right)}.$$

1. Prove that if $x_n \rightharpoonup x$ weakly $\sigma(E, E^\star)$, then

   $$\bigcap_{n=1}^{\infty} K_n = \{x\}.$$

2. Assume that E is reflexive. Prove that if (x_n) is bounded and if $\bigcap_{n=1}^{\infty} K_n = \{x\}$, then $x_n \rightharpoonup x$ weakly $\sigma(E, E^\star)$.
3. Assume that E is finite-dimensional and $\bigcap_{n=1}^{\infty} K_n = \{x\}$. Prove that $x_n \to x$.
   [Note that we do not assume here that (x_n) is bounded.]
4. In $\ell^p, 1 < p < \infty$ (see Chapter 11), construct a sequence (x_n) such that $\bigcap_{n=1}^{\infty} K_n = \{x\}$, and (x_n) is *not* bounded.
   [I owe the results of questions 3 and 4 to Guy Amram and Daniel Baffet.]

3.14 Let E be a reflexive Banach space and let I be a set of indices. Consider a collection $(f_i)_{i \in I}$ in $E^\star$ and a collection $(\alpha_i)_{i \in I}$ in $\mathbb{R}$. Let $M > 0$.
   Show that the following properties are equivalent:

(A)   There exists some $x \in E$ with $\|x\| \leq M$ such that $\langle f_i, x \rangle = \alpha_i$
         for every $i \in I$.

(B)   One has $|\sum_{i \in J} \beta_i \alpha_i| \leq M \|\sum_{i \in J} \beta_i f_i\|$ for every collection $(\beta_i)_{i \in J}$
         in $\mathbb{R}$ with $J \subset I$, J finite.

Compare with Exercises 1.10, 1.11 and Lemma 3.3.

3.15 *Center of mass of a measure on a convex set.*

Let E be a reflexive Banach space and let $K \subset E$ be bounded, closed, and convex. In the following K is equipped with $\sigma(E, E^\star)$, so that K is compact. Let $F = C(K)$ with its usual norm. Fix some $\mu \in F^\star$ with $\|\mu\| = 1$ and assume that $\mu \geq 0$ in the sense that

$$\langle \mu, u \rangle \geq 0 \quad \forall u \in C(K), \quad u \geq 0 \text{ on } K.$$

Prove that there exists a unique element $x_0 \in K$ such that

(1) $$\langle \mu, f_{|K} \rangle = \langle f, x_0 \rangle \quad \forall f \in E^\star.$$

[**Hint**: Find first some $x_0 \in E$ satisfying (1), and then prove that $x_0 \in K$ with the help of Hahn–Banach.]

3.16 Let E be a Banach space.

1. Let (f_n) be a sequence in $(E^\star)$ such that for every $x \in E$, $\langle f_n, x \rangle$ converges to a limit. Prove that there exists some $f \in E^\star$ such that $f_n \overset{\star}{\rightharpoonup} f$ in $\sigma(E^\star, E)$.
2. Assume here that E is reflexive. Let (x_n) be a sequence in E such that for every $f \in E^\star$, $\langle f, x_n \rangle$ converges to a limit. Prove that there exists some $x \in E$ such that $x_n \rightharpoonup x$ in $\sigma(E, E^\star)$.
3. Construct an example in a nonreflexive space E where the conclusion of 2 fails.
   [**Hint**: Take $E = c_0$ (see Section 11.3) and $x_n = (1, 1, \ldots, \underset{(n)}{1}, 0, 0, \ldots)$.]

3.17

1. Let (x^n) be a sequence in ℓ^p with $1 \leq p \leq \infty$. Assuming $x^n \rightharpoonup x$ in $\sigma(\ell^p, \ell^{p'})$ prove that:

   (a) (x^n) is bounded in ℓ^p,
   (b) $x_i^n \xrightarrow[n\to\infty]{} x_i$ for every i, where $x^n = (x_1^n, x_2^n, \ldots, x_i^n, \ldots)$ and $x = (x_1, x_2, \ldots, x_i, \ldots)$.

2. Conversely, suppose (x^n) is a sequence in ℓ^p with $1 < p \leq \infty$. Assume that (a) and (b) hold (for some limit denoted by x_i). Prove that $x \in \ell^p$ and that $x^n \rightharpoonup x$ in $\sigma(\ell^p, \ell^{p'})$.

3.18 For every integer $n \geq 1$ let

$$e^n = (0, 0, \ldots, \underset{(n)}{1}, 0, \ldots).$$

1. Prove that $e^n \xrightarrow[n\to\infty]{} 0$ in ℓ^p weakly $\sigma(\ell^p, \ell^{p'})$ with $1 < p \leq \infty$.
2. Prove that there is no subsequence (e^{n_k}) that converges in ℓ^1 with respect to $\sigma(\ell^1, \ell^\infty)$.
3. Construct an example of a Banach space E and a sequence (f_n) in $E^\star$ such that $\|f_n\| = 1 \quad \forall n$ and such that (f_n) has no subsequence that converges in

$\sigma(E^\star, E)$. Is there a contradiction with the compactness of $B_{E^\star}$ in the topology $\sigma(E^\star, E)$?

[**Hint**: Take $E = \ell^\infty$.]

3.19 Let $E = \ell^p$ and $F = \ell^q$ with $1 < p < \infty$ and $1 < q < \infty$. Let $a : \mathbb{R} \to \mathbb{R}$ be a continuous function such that

$$|a(t)| \leq C|t|^{p/q} \quad \forall t \in \mathbb{R}.$$

Given

$$x = (x_1, x_2, \ldots, x_i, \ldots) \in \ell^p,$$

set

$$Ax = \big(a(x_1), a(x_2), \ldots, a(x_i), \ldots \big).$$

1. Prove that $Ax \in \ell^q$ and that the map $x \mapsto Ax$ is continuous from ℓ^p (strong) into ℓ^q (strong).
2. Prove that if (x^n) is a sequence in ℓ^p such that $x^n \rightharpoonup x$ in $\sigma(\ell^p, \ell^{p'})$ then $Ax^n \rightharpoonup Ax$ in $\sigma(\ell^q, \ell^{q'})$.
3. Deduce that A is continuous from B_E equipped with $\sigma(E, E^\star)$ into F equipped with $\sigma(F, F^\star)$.

3.20 Let E be a Banach space.

1. Prove that there exist a compact topological space K and an isometry from E into $C(K)$ equipped with its usual norm.
   [**Hint**: Take $K = B_{E^\star}$ equipped with $\sigma(E^\star, E)$.]
2. Assuming that E is separable, prove that there exists an isometry from E into ℓ^∞.

3.21 Let E be a separable Banach space and let (f_n) be a bounded sequence in $E^\star$. Prove directly—without using the metrizability of $E^\star$—that there exists a subsequence (f_{n_k}) that converges in $\sigma(E^\star, E)$.
   [**Hint**: Use a diagonal process.]

3.22 Let E be an infinite-dimensional Banach space satisfying *one* of the following assumptions:

(a) $E^\star$ is separable,
(b) E is reflexive.

Prove that there exists a sequence (x_n) in E such that

$$\|x_n\| = 1 \quad \forall n \quad \text{and} \quad x_n \rightharpoonup 0 \text{ weakly } \sigma(E, E^\star).$$

3.23 The proof of Theorem 2.16 becomes much easier if E is reflexive. Find, in particular, a simple proof of (b) $\Rightarrow$ (a).

3.24 The purpose of this exercise is to sketch part of the proof of Theorem 3.29, i.e., if E is a Banach space such that B_E is metrizable with respect to $\sigma(E, E^\star)$, then $E^\star$ is separable. Let $d(x, y)$ be a metric on B_E that induces on B_E the same topology as $\sigma(E, E^\star)$. Set

$$U_n = \left\{ x \in B_E; \ d(x, 0) < \frac{1}{n} \right\}.$$

Let V_n be a neighborhood of 0 for $\sigma(E, E^\star)$ such that $V_n \subset U_n$. We may assume that V_n has the form

$$V_n = \{ x \in E; \ |\langle f, x \rangle| < \varepsilon_n \quad \forall f \in \Phi_n \}$$

with $\varepsilon_n > 0$ and $\Phi_n \subset E^\star$ is some finite subset. Let $D = \cup_{n=1}^\infty \Phi_n$ and let F denote the vector space generated by D. We claim that F is dense in $E^\star$ with respect to the strong topology. Suppose, by contradiction, that $\overline{F} \neq E^\star$.

1. Prove that there exist some $\xi \in E^{\star\star}$ and some $f_0 \in E^\star$ such that

$$\langle \xi, f_0 \rangle > 1, \quad \langle \xi, f \rangle = 0 \quad \forall f \in F, \quad \text{and} \quad \|\xi\| = 1.$$

2. Let

$$W = \left\{ x \in B_E; \ |\langle f_0, x \rangle| < \frac{1}{2} \right\}.$$

Prove that there is some integer $n_0 \geq 1$ such that $V_{n_0} \subset W$.

3. Prove that there exists $x_1 \in B_E$ such that

$$\begin{cases} |\langle f, x_1 \rangle - \langle \xi, f \rangle| < \varepsilon_{n_0} \quad \forall f \in \Phi_{n_0}, \\ |\langle f_0, x_1 \rangle - \langle \xi, f_0 \rangle| < \dfrac{1}{2}. \end{cases}$$

4. Deduce that $x_1 \in V_{n_0}$ and that $\langle f_0, x_1 \rangle > \frac{1}{2}$.
5. Conclude.

3.25 Let K be a compact metric space that is not finite. Prove that $C(K)$ is not reflexive.

[**Hint**: Let (a_n) be a sequence in K such that $a_n \to a$ and $a_n \neq a$ $\forall n$. Consider the linear functional $f(u) = \sum_{n=1}^\infty \frac{1}{2^n} u(a_n)$, $u \in C(K)$, and proceed as in Exercises 1.3 and 1.4.]

3.26 Let F be a separable Banach space and let (a_n) be a dense subset of B_F. Consider the linear operator $T : \ell^1 \to F$ defined by

$$Tx = \sum_{i=1}^\infty x_i a_i \quad \text{with } x = (x_1, x_2, \ldots, x_n, \ldots) \in \ell^1.$$

1. Prove that T is bounded and surjective.

In what follows we assume, in addition, that F is infinite-dimensional and that F^* is separable.

2. Prove that T has no right inverse.
   [**Hint**: Use the results of Exercise 3.22 and Problem 8.]
3. Deduce that $N(T)$ has no complement in ℓ^1.
4. Determine E^*.

3.27 Let E be a separable Banach space with norm $\| \ \|$. The dual norm on E^* is also denoted by $\| \ \|$. The purpose of this exercise is to construct an equivalent norm on E that is strictly convex and whose dual norm is also strictly convex.

Let $(a_n) \subset B_E$ be a dense subset of B_E with respect to the strong topology. Let $(b_n) \subset B_{E^*}$ be a countable subset of B_{E^*} that is dense in B_{E^*} for the weak* topology $\sigma(E^*, E)$. Why does such a set exist?

Given $f \in E^*$, set

$$\|f\|_1 = \left\{ \|f\|^2 + \sum_{n=1}^{\infty} \frac{1}{2^n} |\langle f, a_n \rangle|^2 \right\}^{1/2}.$$

1. Prove that $\| \ \|_1$ is a norm equivalent to $\| \ \|$.
2. Prove that $\| \ \|_1$ is strictly convex.
   [**Hint**: Use Exercise 1.26.]

Given $x \in E$, set

$$\|x\|_2 = \left\{ \|x\|_1^2 + \sum_{n=1}^{\infty} \frac{1}{2^n} |\langle b_n, x \rangle|^2 \right\}^{1/2},$$

where $\|x\|_1 = \sup_{\|f\|_1 \leq 1} \langle f, x \rangle$.

3. Prove that $\| \ \|_2$ is a strictly convex norm that is equivalent to $\| \ \|$.
4. Prove that the dual norm of $\| \ \|_2$ is also strictly convex.
   [**Hint**: Use the result of Exercise 1.23, question 3.]
5. Find another approach based on the results of Problem 4.

3.28 Let E be a uniformly convex Banach space. Let F denote the (multivalued) duality map from E into E^*, see Remark 2 following Corollary 1.3 and also Exercise 1.1.

Prove that for every $f \in E^*$ there exists a unique $x \in E$ such that $f \in Fx$.

3.29 Let E be a uniformly convex Banach space.

1. Prove that $\forall M > 0, \forall \varepsilon > 0, \exists \delta > 0$ such that

$$\left\| \frac{x+y}{2} \right\|^2 \leq \frac{1}{2}\|x\|^2 + \frac{1}{2}\|y\|^2 - \delta$$

$$\forall x, y \in E \quad \text{with} \quad \|x\| \leq M, \|y\| \leq M \quad \text{and} \quad \|x - y\| > \varepsilon.$$

[**Hint**: Argue by contradiction.]

2. Same question when $\| \ \|^2$ is replaced by $\| \ \|^p$ with $1 < p < \infty$.

$\boxed{3.30}$ Let E be a Banach space with norm $\| \ \|$. Assume that there exists on E an equivalent norm, denoted by $| \ |$, that is uniformly convex.

Prove that given any $k > 1$, there exists a uniformly convex norm $||| \ |||$ on E such that

$$\|x\| \le |||x||| \le k\|x\| \quad \forall x \in E.$$

[**Hint**: Set $|||x|||^2 = \|x\|^2 + \alpha|x|^2$ with $\alpha > 0$ small enough and use Exercise 3.29.]

Example: $E = \mathbb{R}^n$.

$\boxed{3.31}$ Let E be a uniformly convex Banach space.

1. Prove that

$$\forall \varepsilon > 0, \quad \forall \alpha \in \left(0, \frac{1}{2}\right), \quad \exists \delta > 0 \quad \text{such that}$$

$$\|tx + (1-t)y\| \le 1 - \delta$$

$\forall t \in [\alpha, 1-\alpha], \quad \forall x, y \in E \quad$ with $\|x\| \le 1, \|y\| \le 1$ and $\|x - y\| \ge \varepsilon$.

[**Hint**: If $\alpha \le t \le \frac{1}{2}$ write $tx + (1-t)y = \frac{1}{2}(y + z)$.]

2. Deduce that E is strictly convex.

$\boxed{3.32}$ *Projection on a closed convex set in a uniformly convex Banach space.*

Let E be a uniformly convex Banach space and $C \subset E$ a nonempty closed convex set.

1. Prove that for every $x \in E$,

$$\inf_{y \in C} \|x - y\|$$

is achieved by some unique point in C, denoted by $P_C x$.

2. Prove that every minimizing sequence (y_n) in C converges strongly to $P_C x$.

3. Prove that the map $x \mapsto P_C x$ is continuous from E strong into E strong.

4. More precisely, prove that P_C is uniformly continuous on bounded subsets of E.
   [**Hint**: Use Exercise 3.29.]

Let $\varphi : E \to (-\infty, +\infty]$ be a convex l.s.c. function, $\varphi \not\equiv +\infty$.

5. Prove that for every $x \in E$ and every integer $n \ge 1$,

$$\inf_{y \in E} \left\{ n\|x - y\|^2 + \varphi(y) \right\}$$

is achieved at some unique point, denoted by y_n.

6. Prove that $y_n \xrightarrow[n \to \infty]{} P_C x$, where $C = \overline{D(\varphi)}$.

Chapter 4
L^p Spaces

Let $(\Omega, \mathcal{M}, \mu)$ denote a *measure space*, i.e., Ω is a set and

(i) $\mathcal{M}$ is a σ-*algebra* in Ω, i.e., $\mathcal{M}$ is a collection of subsets of Ω such that:

    (a) $\emptyset \in \mathcal{M}$,
    (b) $A \in \mathcal{M} \Rightarrow A^c \in \mathcal{M}$,
    (c) $\bigcup_{n=1}^{\infty} A_n \in \mathcal{M}$ whenever $A_n \in \mathcal{M}$ $\forall n$,

(ii) μ is a *measure*, i.e., $\mu : \mathcal{M} \to [0, \infty]$ satisfies

    (a) $\mu(\emptyset) = 0$,

    (b) $\begin{cases} \mu\left(\bigcup_{n=1}^{\infty} A_n\right) = \sum_{n=1}^{\infty} \mu(A_n) \text{ whenever } (A_n) \text{ is a disjoint} \\ \text{countable family of members of } \mathcal{M}. \end{cases}$

    The members of $\mathcal{M}$ are called the *measurable sets*. Sometimes we shall write $|A|$ instead of $\mu(A)$. We shall also assume—even though this is not essential—that

(iii) Ω is σ-*finite*, i.e., there exists a countable family (Ω_n) in $\mathcal{M}$ such that $\Omega = \bigcup_{n=1}^{\infty} \Omega_n$ and $\mu(\Omega_n) < \infty$ $\forall n$.

    The sets $E \in \mathcal{M}$ with the property that $\mu(E) = 0$ are called the *null sets*. We say that a property holds a.e. (or for almost all $x \in \Omega$) if it holds everywhere on Ω except on a null set.

    We assume that the reader is familiar with the notions of *measurable functions* and *integrable functions* $f : \Omega \to \mathbb{R}$; see, e.g., H. L. Royden [1], G. B. Folland [2], A. Knapp [1], D. L. Cohn [1], A. Friedman [3], W. Rudin [2], P. Halmos [1], E. Hewitt–K. Stromberg [1], R. Wheeden–A. Zygmund [1], J. Neveu [1], P. Malliavin [1], A. J. Weir [1], A. Kolmogorov–S. Fomin [1], I. Fonseca–G. Leoni [1]. We denote by $L^1(\Omega, \mu)$, or simply $L^1(\Omega)$ (or just L^1), the space of integrable functions from Ω into $\mathbb{R}$.

    We shall often write $\int f$ instead of $\int_{\Omega} f \, d\mu$, and we shall also use the notation

H. Brezis, *Functional Analysis, Sobolev Spaces and Partial Differential Equations*,
DOI 10.1007/978-0-387-70914-7_4, © Springer Science+Business Media, LLC 2011

$$\|f\|_{L^1} = \|f\|_1 = \int_\Omega |f| d\mu = \int |f|.$$

As usual, we identify two functions that coincide a.e. We recall the following basic facts.

4.1 Some Results about Integration That Everyone Must Know

• **Theorem 4.1 (monotone convergence theorem, Beppo Levi).** *Let (f_n) be a sequence of functions in L^1 that satisfy*

(a) $f_1 \le f_2 \le \cdots \le f_n \le f_{n+1} \le \cdots$ *a.e. on Ω,*
(b) $\sup_n \int f_n < \infty.$

Then $f_n(x)$ converges a.e. on Ω to a finite limit, which we denote by $f(x)$; the function f belongs to L^1 and $\|f_n - f\|_1 \to 0$.

• **Theorem 4.2 (dominated convergence theorem, Lebesgue).** *Let (f_n) be a sequence of functions in L^1 that satisfy*

(a) $f_n(x) \to f(x)$ *a.e. on Ω,*
(b) *there is a function $g \in L^1$ such that for all n, $|f_n(x)| \le g(x)$ a.e. on Ω.*

Then $f \in L^1$ and $\|f_n - f\|_1 \to 0$.

Lemma 4.1 (Fatou's lemma). *Let (f_n) be a sequence of functions in L^1 that satisfy*

(a) *for all n, $f_n \ge 0$ a.e.*
(b) $\sup_n \int f_n < \infty.$

For almost all $x \in \Omega$ we set $f(x) = \liminf_{n\to\infty} f_n(x) \le +\infty$. Then $f \in L^1$ and

$$\int f \le \liminf_{n\to\infty} \int f_n.$$

A basic example is the case in which $\Omega = \mathbb{R}^N$, $\mathcal{M}$ consists of the Lebesgue measurable sets, and μ is the Lebesgue measure on $\mathbb{R}^N$.

Notation. We denote by $C_c(\mathbb{R}^N)$ the space of all continuous functions on $\mathbb{R}^N$ with *compact support*, i.e.,

$$C_c(\mathbb{R}^N) = \{f \in C(\mathbb{R}^N); f(x) = 0 \quad \forall x \in \mathbb{R}^N \setminus K, \text{ where } K \text{ is compact}\}.$$

Theorem 4.3 (density). *The space $C_c(\mathbb{R}^N)$ is dense in $L^1(\mathbb{R}^N)$; i.e.,*

$$\forall f \in L^1(\mathbb{R}^N) \ \forall \varepsilon > 0 \ \exists f_1 \in C_c(\mathbb{R}^N) \text{ such that } \|f - f_1\|_1 \le \varepsilon.$$

Let $(\Omega_1, \mathcal{M}_1, \mu_1)$ and $(\Omega_2, \mathcal{M}_2, \mu_2)$ be two measure spaces that are σ-finite. One can define in a standard way the structure of measure space $(\Omega, \mathcal{M}, \mu)$ on the Cartesian product $\Omega = \Omega_1 \times \Omega_2$.

Theorem 4.4 (Tonelli). *Let* $F(x, y) : \Omega_1 \times \Omega_2 \to \mathbb{R}$ *be a measurable function satisfying*

(a) $\displaystyle\int_{\Omega_2} |F(x, y)| d\mu_2 < \infty$ *for* a.e. $x \in \Omega_1$

and

(b) $\displaystyle\int_{\Omega_1} d\mu_1 \int_{\Omega_2} |F(x, y)| d\mu_2 < \infty.$

$\quad$ *Then* $F \in L^1(\Omega_1 \times \Omega_2).$

Theorem 4.5 (Fubini). *Assume that* $F \in L^1(\Omega_1 \times \Omega_2)$. *Then for* a.e. $x \in \Omega_1$, $F(x, y) \in L_y^1(\Omega_2)$ *and* $\int_{\Omega_2} F(x, y) d\mu_2 \in L_x^1(\Omega_1)$. *Similarly, for* a.e. $y \in \Omega_2$, $F(x, y) \in L_x^1(\Omega_1)$ *and* $\int_{\Omega_1} F(x, y) d\mu_1 \in L_y^1(\Omega_2)$.
$\quad$ *Moreover, one has*

$$\int_{\Omega_1} d\mu_1 \int_{\Omega_2} F(x, y) d\mu_2 = \int_{\Omega_2} d\mu_2 \int_{\Omega_1} F(x, y) d\mu_1 = \iint_{\Omega_1 \times \Omega_2} F(x, y) d\mu_1 d\mu_2.$$

4.2 Definition and Elementary Properties of L^p Spaces

Definition. Let $p \in \mathbb{R}$ with $1 < p < \infty$; we set

$$L^p(\Omega) = \left\{ f : \Omega \to \mathbb{R}; \; f \text{ is measurable and } |f|^p \in L^1(\Omega) \right\}$$

with

$$\|f\|_{L^p} = \|f\|_p = \left[\int_{\Omega} |f(x)|^p d\mu \right]^{1/p}.$$

We shall check later on that $\| \; \|_p$ is a norm.

Definition. We set

$$L^\infty(\Omega) = \left\{ f : \Omega \to \mathbb{R} \;\middle|\; \begin{array}{l} f \text{ is measurable and there is a constant } C \\ \text{such that } |f(x)| \le C \text{ a.e. on } \Omega \end{array} \right\}$$

with

$$\|f\|_{L^\infty} = \|f\|_\infty = \inf\{C; \; |f(x)| \le C \text{ a.e. on } \Omega\}.$$

The following remark implies that $\| \; \|_\infty$ is a norm:

Remark 1. If $f \in L^\infty$ then we have

$$|f(x)| \le \|f\|_\infty \quad \text{a.e. on } \Omega.$$

Indeed, there exists a sequence C_n such that $C_n \to \|f\|_\infty$ and for each n, $|f(x)| \le C_n$ a.e. on Ω. Therefore $|f(x)| \le C_n$ for all $x \in \Omega \backslash E_n$, with $|E_n| = 0$. We set

$E = \cup_{n=1}^\infty E_n$, so that $|E| = 0$ and

$$|f(x)| \leq C_n \quad \forall n, \quad \forall x \in \Omega \backslash E;$$

it follows that $|f(x)| \leq \|f\|_\infty \quad \forall x \in \Omega \backslash E$.

Notation. Let $1 \leq p \leq \infty$; we denote by p' the *conjugate exponent*,

$$\boxed{\frac{1}{p} + \frac{1}{p'} = 1.}$$

• **Theorem 4.6 (Hölder's inequality).** *Assume that $f \in L^p$ and $g \in L^{p'}$ with $1 \leq p \leq \infty$. Then $fg \in L^1$ and*

(1) $$\boxed{\int |fg| \leq \|f\|_p \, \|g\|_{p'}.}$$

Proof. The conclusion is obvious if $p = 1$ or $p = \infty$; therefore we assume that $1 < p < \infty$. We recall *Young's inequality*:[1]

(2) $$\boxed{ab \leq \frac{1}{p}a^p + \frac{1}{p'}b^{p'} \quad \forall a \geq 0, \quad \forall b \geq 0.}$$

Inequality (2) is a straightforward consequence of the concavity of the function log on $(0, \infty)$:

$$\log\left(\frac{1}{p}a^p + \frac{1}{p'}b^{p'}\right) \geq \frac{1}{p}\log a^p + \frac{1}{p'}\log b^{p'} = \log ab.$$

We have
$$|f(x)g(x)| \leq \frac{1}{p}|f(x)|^p + \frac{1}{p'}|g(x)|^{p'} \text{ a.e. } x \in \Omega.$$

It follows that $fg \in L^1$ and

(3) $$\int |fg| \leq \frac{1}{p}\|f\|_p^p + \frac{1}{p'}\|g\|_{p'}^{p'}.$$

Replacing f by $\lambda f (\lambda > 0)$ in (3), yields

(4) $$\int |fg| \leq \frac{\lambda^{p-1}}{p}\|f\|_p^p + \frac{1}{\lambda p'}\|g\|_{p'}^{p'}.$$

Choosing $\lambda = \|f\|_p^{-1}\|g\|_p^{p'/p}$ (so as to minimize the right-hand side in (4)), we obtain (1).

---

[1] It is sometimes convenient to use the form $ab \leq \varepsilon a^p + C_\varepsilon b^{p'}$ with $C_\varepsilon = \varepsilon^{-1/(p-1)}$.

Remark 2. It is useful to keep in mind the following extension of Hölder's inequality: Assume that $f_1, f_2, \ldots, f_k$ are functions such that

$$f_i \in L^{p_i}, \ 1 \leq i \leq k \text{ with } \frac{1}{p} = \frac{1}{p_1} + \frac{1}{p_2} + \cdots + \frac{1}{p_k} \leq 1.$$

Then the product $f = f_1 f_2 \cdots f_k$ belongs to L^p and

$$\|f\|_p \leq \|f_1\|_{p_1} \|f_2\|_{p_2} \cdots \|f_k\|_{p_k}.$$

In particular, if $f \in L^p \cap L^q$ with $1 \leq p \leq q \leq \infty$, then $f \in L^r$ for all $r, p \leq r \leq q$, and the following "*interpolation inequality*" holds:

$$\|f\|_r \leq \|f\|_p^\alpha \|f\|_q^{1-\alpha}, \ \text{where } \frac{1}{r} = \frac{\alpha}{p} + \frac{1-\alpha}{q}, \ 0 \leq \alpha \leq 1;$$

see Exercise 4.4.

Theorem 4.7. *L^p is a vector space and $\| \ \|_p$ is a norm for any p, $1 \leq p \leq \infty$.*

Proof. The cases $p = 1$ and $p = \infty$ are clear. Therefore we assume $1 < p < \infty$ and let $f, g \in L^p$. We have

$$|f(x) + g(x)|^p \leq (|f(x)| + |g(x)|)^p \leq 2^p (|f(x)|^p + |g(x)|^p).$$

Consequently, $f + g \in L^p$. On the other hand,

$$\|f + g\|_p^p = \int |f + g|^{p-1} |f + g| \leq \int |f + g|^{p-1} |f| + \int |f + g|^{p-1} |g|.$$

But $|f + g|^{p-1} \in L^{p'}$, and by Hölder's inequality we obtain

$$\|f + g\|_p^p \leq \|f + g\|_p^{p-1} (\|f\|_p + \|g\|_p),$$

i.e., $\|f + g\|_p \leq \|f\|_p + \|g\|_p$.

• Theorem 4.8 (Fischer–Riesz). *L^p is a Banach space for any p, $1 \leq p \leq \infty$.*

Proof. We distinguish the cases $p = \infty$ and $1 \leq p < \infty$.

Case 1: $p = \infty$. Let (f_n) be a Cauchy sequence is L^∞. Given an integer $k \geq 1$ there is an integer N_k such that $\| f_m - f_n \|_\infty \leq \frac{1}{k}$ for $m, n \geq N_k$. Hence there is a null set E_k such that

(5) $$|f_m(x) - f_n(x)| \leq \frac{1}{k} \quad \forall x \in \Omega \backslash E_k, \quad \forall m, n \geq N_k.$$

Then we let $E = \bigcup_k E_k$—so that E is a null set—and we see that for all $x \in \Omega \backslash E$, the sequence $f_n(x)$ is Cauchy (in $\mathbb{R}$). Thus $f_n(x) \to f(x)$ for all $x \in \Omega \backslash E$. Passing to the limit in (5) as $m \to \infty$ we obtain

$$|f(x) - f_n(x)| \leq \frac{1}{k} \quad \text{for all } x \in \Omega\backslash E, \quad \forall n \geq N_k.$$

We conclude that $f \in L^\infty$ and $\|f - f_n\|_\infty \leq \frac{1}{k}$ $\forall n \geq N_k$; therefore $f_n \to f$ in L^∞.

Case 2: $1 \leq p < \infty$. Let (f_n) be a Cauchy sequence in L^p. In order to conclude, it suffices to show that a subsequence converges in L^p.

We extract a subsequence (f_{n_k}) such that

$$\|f_{n_{k+1}} - f_{n_k}\|_p \leq \frac{1}{2^k} \quad \forall k \geq 1.$$

[One proceeds as follows: choose n_1 such that $\|f_m - f_n\|_p \leq \frac{1}{2}$ $\forall m, n \geq n_1$; then choose $n_2 \geq n_1$ such that $\|f_m - f_n\|_p \leq \frac{1}{2^2}$ $\forall m, n \geq n_2$ etc.] We claim that f_{n_k} converges in L^p. In order to simplify the notation we write f_k instead of f_{n_k}, so that we have

$$(6) \qquad \|f_{k+1} - f_k\|_p \leq \frac{1}{2^k} \quad \forall k \geq 1.$$

Let

$$g_n(x) = \sum_{k=1}^{n} |f_{k+1}(x) - f_k(x)|,$$

so that

$$\|g_n\|_p \leq 1.$$

As a consequence of the monotone convergence theorem, $g_n(x)$ tends to a finite limit, say $g(x)$, a.e. on Ω, with $g \in L^p$. On the other hand, for $m \geq n \geq 2$ we have

$$|f_m(x) - f_n(x)| \leq |f_m(x) - f_{m-1}(x)| + \cdots + |f_{n+1}(x) - f_n(x)| \leq g(x) - g_{n-1}(x).$$

It follows that a.e. on Ω, $f_n(x)$ is Cauchy and converges to a finite limit, say $f(x)$. We have a.e. on Ω,

$$(7) \qquad |f(x) - f_n(x)| \leq g(x) \quad \text{for } n \geq 2,$$

and in particular $f \in L^p$. Finally, we conclude by dominated convergence that $\|f_n - f\|_p \to 0$, since $|f_n(x) - f(x)|^p \to 0$ a.e. and also $|f_n - f|^p \leq g^p \in L^1$.

Theorem 4.9. *Let (f_n) be a sequence in L^p and let $f \in L^p$ be such that $\|f_n - f\|_p \to 0$.*

Then, there exist a subsequence (f_{n_k}) and a function $h \in L^p$ such that

(a) $f_{n_k}(x) \to f(x)$ a.e. on Ω,
(b) $|f_{n_k}(x)| \leq h(x)$ $\forall k$, a.e. on Ω.

Proof. The conclusion is obvious when $p = \infty$. Thus we assume $1 \leq p < \infty$. Since (f_n) is a Cauchy sequence we may go back to the proof of Theorem 4.8 and consider a subsequence (f_{n_k})—denoted by (f_k)—satisfying (6), such that $f_k(x)$ tends a.e. to a limit[2] $f^\star(x)$ with $f^\star \in L^p$. Moreover, by (7), we have $|f^\star(x) - f_k(x)| \le g(x)$ $\forall k$, a.e. on Ω with $g \in L^p$. By dominated convergence we know that $f_k \to f^\star$ in L^p and thus $f = f^\star$ a.e. In addition, we also have $|f_k(x)| \le |f^\star(x)| + g(x)$, and the conclusion follows.

4.3 Reflexivity. Separability. Dual of L^p

We shall consider separately the following three cases:

(A) $1 < p < \infty$,
(B) $p = 1$,
(C) $p = \infty$.

A. Study of $L^p(\Omega)$ for $1 < p < \infty$.
   This case is the most "favorable": L^p is reflexive, separable, and the dual of L^p is $L^{p'}$.

• **Theorem 4.10.** L^p *is reflexive for any* p, $1 < p < \infty$.

The proof consists of three steps:
Step 1 (Clarkson's first inequality). Let $2 \le p < \infty$. We claim that

$$(8) \qquad \left\| \frac{f+g}{2} \right\|_p^p + \left\| \frac{f-g}{2} \right\|_p^p \le \frac{1}{2}(\|f\|_p^p + \|g\|_p^p) \quad \forall f, g \in L^p.$$

Proof of (8). Clearly, it suffices to show that

$$\left| \frac{a+b}{2} \right|^p + \left| \frac{a-b}{2} \right|^p \le \frac{1}{2}(|a|^p + |b|^p) \quad \forall a, b \in \mathbb{R}.$$

First we note that
$$\alpha^p + \beta^p \le (\alpha^2 + \beta^2)^{p/2} \quad \forall \alpha, \beta \ge 0$$

(by homogeneity, assume $\beta = 1$ and observe that the function

$$(x^2 + 1)^{p/2} - x^p - 1$$

increases on $[0, \infty)$). Choosing $\alpha = |\frac{a+b}{2}|$ and $\beta = |\frac{a-b}{2}|$, we obtain

$$\left| \frac{a+b}{2} \right|^p + \left| \frac{a-b}{2} \right|^p \le \left(\left| \frac{a+b}{2} \right|^2 + \left| \frac{a-b}{2} \right|^2 \right)^{p/2} = \left(\frac{a^2}{2} + \frac{b^2}{2} \right)^{p/2} \le \frac{1}{2}(|a|^p + |b|^p)$$

---

[2] A priori one should distinguish f and $f^\star$: by assumption $f_n \to f$ in L^p, and on the other hand, $f_{n_k}(x) \to f^\star(x)$ a.e.

(the last inequality follows from the convexity of the function $x \mapsto |x|^{p/2}$ since $p \geq 2$).

Step 2: L^p is uniformly convex, and thus reflexive for $2 \leq p < \infty$. Indeed, let $\varepsilon > 0$ and let $f, g \in L^p$ with $\|f\|_p \leq 1$, $\|g\|_p \leq 1$, and $\|f - g\|_p > \varepsilon$. We deduce from (8) that

$$\left\| \frac{f+g}{2} \right\|_p^p < 1 - \left(\frac{\varepsilon}{2} \right)^p$$

and thus $\|\frac{f+g}{2}\|_p < 1 - \delta$ with $\delta = 1 - [1 - (\frac{\varepsilon}{2})^p]^{1/p} > 0$. Therefore, L^p is uniformly convex and thus reflexive by Theorem 3.31.

Step 3: L^p is reflexive for $1 < p \leq 2$.

Proof. Let $1 < p < \infty$. Consider the operator $T : L^p \to (L^{p'})^*$ defined as follows: Let $u \in L^p$ be fixed; the mapping $f \in L^{p'} \mapsto \int uf$ is a continuous linear functional on $L^{p'}$ and thus it defines an element, say Tu, in $(L^{p'})^*$ such that

$$\langle Tu, f \rangle = \int uf \quad \forall f \in L^{p'}.$$

We claim that

(9) $$\|Tu\|_{(L^{p'})^*} = \|u\|_p \quad \forall u \in L^p.$$

Indeed, by Hölder's inequality, we have

$$|\langle Tu, f \rangle| \leq \|u\|_p \|f\|_{p'} \quad \forall f \in L^{p'}$$

and therefore $\|Tu\|_{(L^{p'})^*} \leq \|u\|_p$.
On the other hand, set

$$f_0(x) = |u(x)|^{p-2} u(x) \quad (f_0(x) = 0 \text{ if } u(x) = 0).$$

Clearly we have

$$f_0 \in L^{p'}, \quad \|f_0\|_{p'} = \|u\|_p^{p-1} \quad \text{and} \quad \langle Tu, f_0 \rangle = \|u\|_p^p;$$

thus

(10) $$\|Tu\|_{(L^{p'})^*} \geq \frac{\langle Tu, f_0 \rangle}{\|f_0\|_{p'}} = \|u\|_p.$$

Hence, we have shown that T is an isometry from L^p into $(L^{p'})^*$, which implies that $T(L^p)$ is a closed subspace of $(L^{p'})^*$ (because L^p is a Banach space).

Assume now $1 < p \leq 2$. Since $L^{p'}$ is reflexive (by Step 2), it follows that $(L^{p'})^*$

is also reflexive (Corollary 3.21). We conclude, by Proposition 3.20, that $T(L^p)$ is reflexive, and as a consequence, L^p is also reflexive.

Remark 3. In fact, L^p is also *uniformly convex for* $1 < p \leq 2$. This is a consequence of *Clarkson's second inequality*, which holds for $1 < p \leq 2$:

$$\left\| \frac{f+g}{2} \right\|_p^{p'} + \left\| \frac{f-g}{2} \right\|_p^{p'} \leq \left(\frac{1}{2} \|f\|_p^p + \frac{1}{2} \|g\|_p^p \right)^{1/(p-1)} \quad \forall f, g \in L^p.$$

This inequality is trickier to prove than Clarkson's first inequality (see, e.g., Problem 20 or E. Hewitt–K. Stromberg [1]). Clearly, it implies that L^p is uniformly convex when $1 < p \leq 2$; for another approach, see also C. Morawetz [1] (Exercise 4.12) or J. Diestel [1].

• **Theorem 4.11 (Riesz representation theorem).** *Let* $1 < p < \infty$ *and let* $\phi \in (L^p)^\star$. *Then there exists a unique function* $u \in L^{p'}$ *such that*

$$\langle \phi, f \rangle = \int uf \quad \forall f \in L^p.$$

Moreover,

$$\|u\|_{p'} = \|\phi\|_{(L^p)^\star}.$$

Remark 4. Theorem 4.11 is very important. It says that every continuous linear functional on L^p with $1 < p < \infty$ can be represented "concretely" as an integral. The mapping $\phi \mapsto u$, which is a linear surjective isometry, allows us to identify the "abstract" space $(L^p)^\star$ with $L^{p'}$.

In what follows, we shall systematically make the identification

$$\boxed{(L^p)^\star = L^{p'}.}$$

Proof. We consider the operator $T : L^{p'} \to (L^p)^\star$ defined by $\langle Tu, f \rangle = \int uf$ $\forall u \in L^{p'}$, $\forall f \in L^p$. The argument used in the proof of Theorem 4.10 (Step 3) shows that

$$\|Tu\|_{(L^p)^\star} = \|u\|_{p'} \quad \forall u \in L^{p'}.$$

We claim that T is surjective. Indeed, let $E = T(L^{p'})$. Since E is a closed subspace, it suffices to prove that E is dense in $(L^p)^\star$. Let $h \in (L^p)^{\star\star}$ satisfy $\langle h, Tu \rangle = 0$ $\forall u \in L^{p'}$. Since L^p is reflexive, $h \in L^p$, and satisfies $\int uh = 0 \ \forall u \in L^{p'}$. Choosing $u = |h|^{p-2}h$, we see that $h = 0$.

Theorem 4.12. *The space* $C_c(\mathbb{R}^N)$ *is dense in* $L^p(\mathbb{R}^N)$ *for any* p, $1 \leq p < \infty$.

Before proving Theorem 4.12, we introduce some notation.

Notation. The *truncation operation* $T_n : \mathbb{R} \to \mathbb{R}$ is defined by

$$T_n r = \begin{cases} r & \text{if } |r| \leq n, \\ \dfrac{nr}{|r|} & \text{if } |r| > n. \end{cases}$$

Given a set $E \subset \Omega$, we define the *characteristic function*[3] χ_E to be

$$\chi_E(x) = \begin{cases} 1 & \text{if } x \in E, \\ 0 & \text{if } x \in \Omega \backslash E. \end{cases}$$

Proof. First, we claim that given $f \in L^p(\mathbb{R}^N)$ and $\varepsilon > 0$ there exist a function $g \in L^\infty(\mathbb{R}^N)$ and a compact set K in $\mathbb{R}^N$ such that $g = 0$ outside K and

(11) $$\|f - g\|_p < \varepsilon.$$

Indeed, let χ_n be the characteristic function of $B(0, n)$ and let $f_n = \chi_n T_n f$. By dominated convergence we see that $\|f_n - f\|_p \to 0$ and thus we may choose $g = f_n$ with n large enough. Next, given $\delta > 0$ there exists (by Theorem 4.3) a function $g_1 \in C_c(\mathbb{R}^N)$ such that

$$\|g - g_1\|_1 < \delta.$$

We may always assume that $\|g_1\|_\infty \leq \|g\|_\infty$; otherwise, we replace g_1 by $T_n g_1$ with $n = \|g\|_\infty$. Finally, we have

$$\|g - g_1\|_p \leq \|g - g_1\|_1^{1/p} \|g - g_1\|_\infty^{1-(1/p)} \leq \delta^{1/p} (2\|g\|_\infty)^{1-(1/p)}.$$

We conclude by choosing $\delta > 0$ small enough that

$$\delta^{1/p}(2\|g\|_\infty)^{1-(1/p)} < \varepsilon.$$

Definition. The measure space Ω is called *separable* if there is a countable family (E_n) of members of $\mathcal{M}$ such that the σ-algebra generated by (E_n) coincides with $\mathcal{M}$ (i.e., $\mathcal{M}$ is the smallest σ-algebra containing all the E_n's).

Example. The measure space $\Omega = \mathbb{R}^N$ is separable. Indeed, we may choose for (E_n) any countable family of open sets such that every open set in $\mathbb{R}^N$ can be written as a union of E_n's. More generally, if Ω is a *separable metric space* and $\mathcal{M}$ consists of the Borel sets (i.e., $\mathcal{M}$ is the σ-algebra generated by the open sets in Ω), then Ω is a *separable measure space*.

Theorem 4.13. *Assume that Ω is a separable measure space. Then $L^p(\Omega)$ is separable for any p, $1 \leq p < \infty$.*

We shall consider only the case $\Omega = \mathbb{R}^N$, since the general case is somewhat tricky. Note that as a consequence, $L^p(\Omega)$ is also separable for any measurable set $\Omega \subset \mathbb{R}^N$. Indeed, there is a canonical isometry from $L^p(\Omega)$ into $L^p(\mathbb{R}^N)$ (the

---

[3] Not to be confused with the *indicator function* I_E introduced in Chapter 1.

extension by 0 outside Ω); therefore $L^p(\Omega)$ may be identified with a subspace of $L^p(\mathbb{R}^N)$ and hence $L^p(\Omega)$ is separable (by Proposition 3.25).

Proof of Theorem 4.13 *when* $\Omega = \mathbb{R}^N$. Let $\mathcal{R}$ denote the countable family of sets in $\mathbb{R}^N$ of the form $R = \prod_{k=1}^{N}(a_k, b_k)$ with $a_k, b_k \in \mathbb{Q}$. Let $\mathcal{E}$ denote the vector space over $\mathbb{Q}$ generated by the functions $(\chi_R)_{R \in \mathcal{R}}$, that is, $\mathcal{E}$ consists of finite linear combinations with rational coefficients of functions χ_R, so that $\mathcal{E}$ is countable.

We claim that $\mathcal{E}$ is dense in $L^p(\mathbb{R}^N)$. Indeed, given $f \in L^p(\mathbb{R}^N)$ and $\varepsilon > 0$, there exists some $f_1 \in C_c(\mathbb{R}^N)$ such that $\|f - f_1\|_p < \varepsilon$. Let $R \in \mathcal{R}$ be any cube containing supp f_1 (the support of f_1). Given $\delta > 0$ it is easy to construct a function $f_2 \in \mathcal{E}$ such that $\|f_1 - f_2\|_\infty < \delta$ and f_2 vanishes outside R: it suffices to split R into small cubes of $\mathcal{R}$ where the oscillation (i.e., $\sup - \inf$) of f_1 is less than δ. Therefore we have $\|f_1 - f_2\|_p \leq \|f_1 - f_2\|_\infty |R|^{1/p} < \delta |R|^{1/p}$. We conclude that $\|f - f_2\|_p < 2\varepsilon$, provided $\delta > 0$ is chosen so that $\delta |R|^{1/p} < \varepsilon$.

B. Study of $L^1(\Omega)$.
We start with a description of the dual space of $L^1(\Omega)$.

● **Theorem 4.14 (Riesz representation theorem).** *Let* $\phi \in (L^1)^\star$. *Then there exists a unique function* $u \in L^\infty$ *such that*

$$\langle \phi, f \rangle = \int uf \quad \forall f \in L^1.$$

Moreover,

$$\|u\|_\infty = \|\phi\|_{(L^1)^\star}.$$

● *Remark* 5. Theorem 4.14 asserts that every continuous linear functional on L^1 can be represented "concretely" as an integral. The mapping $\phi \mapsto u$, which is a linear surjective isometry, allows us to identify the "abstract" space $(L^1)^\star$ with L^∞. *In what follows, we shall systematically make the identification*

$$\boxed{(L^1)^\star = L^\infty.}$$

Proof. Let (Ω_n) be a sequence of measurable sets in Ω such that $\Omega = \cup_{n=1}^{\infty} \Omega_n$ and $|\Omega_n| < \infty$ $\forall n$. Set $\chi_n = \chi_{\Omega_n}$.

The *uniqueness* of u is obvious. Indeed, suppose $u \in L^\infty$ satisfies

$$\int uf = 0 \quad \forall f \in L^1.$$

Choosing $f = \chi_n$ sign u (throughout this book, we use the convention that sign $0 = 0$), we see that $u = 0$ a.e. on Ω_n and thus $u = 0$ a.e. on Ω.

We now prove the *existence* of u. First, we construct a function $\theta \in L^2(\Omega)$ such that

$$\theta(x) \geq \varepsilon_n > 0 \quad \forall x \in \Omega_n.$$

It is clear that such a function θ exists. Indeed, we define θ to be α_1 on Ω_1, α_2 on $\Omega_2 \backslash \Omega_1, \ldots, \alpha_n$ on $\Omega_n \backslash \Omega_{n-1}$, etc., and we adjust the constants $\alpha_n > 0$ in such a way that $\theta \in L^2$.

The mapping $f \in L^2(\Omega) \mapsto \langle \phi, \theta f \rangle$ is a continuous linear functional on $L^2(\Omega)$. By Theorem 4.11 (applied with $p = 2$) there exists a function $v \in L^2(\Omega)$ such that

$$(12) \qquad \langle \phi, \theta f \rangle = \int vf \quad \forall f \in L^2(\Omega).$$

Set $u(x) = v(x)/\theta(x)$. Clearly, u is well defined since $\theta > 0$ on Ω; moreover, u is measurable and $u\chi_n \in L^2(\Omega)$. We claim that u has all the required properties. We have

$$(13) \qquad \langle \phi, \chi_n g \rangle = \int u\chi_n g \quad \forall g \in L^\infty(\Omega) \quad \forall n.$$

Indeed, it suffices to choose $f = \chi_n g/\theta$ in (12) (note that $f \in L^2(\Omega)$ since f is bounded on Ω_n and $f = 0$ outside Ω_n).

Next, we claim that $u \in L^\infty(\Omega)$ and that

$$(14) \qquad \|u\|_\infty \leq \|\phi\|_{(L^1)^\star}.$$

Fix any constant $C > \|\phi\|_{(L^1)^\star}$ and set

$$A = \{x \in \Omega; \ |u(x)| > C\}.$$

Let us verify that A is a null set. Indeed, by choosing $g = \chi_A \operatorname{sign} u$ in (13) we obtain

$$\int_{A \cap \Omega_n} |u| \leq \|\phi\|_{(L^1)^\star} |A \cap \Omega_n|$$

and therefore

$$C|A \cap \Omega_n| \leq \|\phi\|_{(L^1)^\star} |A \cap \Omega_n|.$$

It follows that $|A \cap \Omega_n| = 0 \ \forall n$, and thus A is a null set. This concludes the proof of (14).

Finally, we claim that

$$(15) \qquad \langle \phi, h \rangle = \int uh \quad \forall h \in L^1(\Omega).$$

Indeed, it suffices to choose $g = T_n h$ (truncation of h) in (13) and to observe that $\chi_n T_n h \to h$ in $L^1(\Omega)$.

In order to complete the proof of Theorem 4.14 it remains only to check that $\|u\|_\infty = \|\phi\|_{(L^1)^\star}$. We have, by (15),

$$|\langle \phi, h \rangle| \leq \|u\|_\infty \|h\|_1 \quad \forall h \in L^1(\Omega),$$

and therefore $\|\phi\|_{(L^1)^\star} \leq \|u\|_\infty$. We conclude with the help of (14).

• *Remark* 6. The space $L^1(\Omega)$ is *never reflexive* except in the trivial case where Ω consists of a finite number of atoms—and then $L^1(\Omega)$ is finite-dimensional. Indeed suppose, by contradiction, that $L^1(\Omega)$ is reflexive and consider two cases:

(i) $\forall \varepsilon > 0 \; \exists \omega \subset \Omega$ measurable with $0 < \mu(\omega) < \varepsilon$.
(ii) $\exists \varepsilon > 0$ such that $\mu(\omega) \geq \varepsilon$ for every measurable set $\omega \subset \Omega$ with $\mu(\omega) > 0$.

In Case (i) there is a *decreasing* sequence (ω_n) of measurable sets such that $\mu(\omega_n) > 0 \; \forall n$ and $\mu(\omega_n) \to 0$ [choose first any sequence (ω'_k) such that $0 < \mu(\omega'_k) < 1/2^k$ and then set $\omega_n = \bigcup_{k=n}^\infty \omega'_k$].

Let $\chi_n = \chi_{\omega_n}$ and define $u_n = \chi_n/\|\chi_n\|_1$. Since $\|u_n\|_1 = 1$ there is a subsequence—still denoted by u_n—and some $u \in L^1$ such that $u_n \rightharpoonup u$ in the weak topology $\sigma(L^1, L^\infty)$ (by Theorem 3.18), i.e.,

$$(16) \qquad \int u_n \phi \to \int u\phi \quad \forall \phi \in L^\infty.$$

On the other hand, for fixed j, and $n > j$ we have $\int u_n \chi_j = 1$. At the limit, as $n \to \infty$, we obtain $\int u\chi_j = 1 \; \forall j$. Finally, we note (by dominated convergence) that $\int u\chi_j \to 0$ as $j \to \infty$—a contradiction.

In Case (ii) the space Ω is purely atomic and consists of a countable union of distinct atoms (a_n) (unless there is only a finite number of atoms!). In that case $L^1(\Omega)$ is isomorphic to ℓ^1 and it suffices to prove that ℓ^1 is not reflexive. Consider the canonical basis:

$$e_n = (0, 0, \ldots, \underset{(n)}{1}, 0, 0 \ldots).$$

Assuming ℓ^1 is reflexive, there exist a subsequence (e_{n_k}) and some $x \in \ell^1$ such that $e_{n_k} \rightharpoonup x$ in the weak topology $\sigma(\ell^1, \ell^\infty)$, i.e.,

$$\langle \varphi, e_{n_k} \rangle \underset{k\to\infty}{\longrightarrow} \langle \varphi, x \rangle \quad \forall \varphi \in \ell^\infty.$$

Choosing

$$\varphi = \varphi_j = (0, 0, \ldots, \underset{(j)}{1}, 1, 1, \ldots)$$

we find that $\langle \varphi_j, x \rangle = 1 \; \forall j$. On the other hand $\langle \varphi_j, x \rangle \to 0$ as $j \to \infty$ (since $x \in \ell^1$)—a contradiction.

C. Study of L^∞.

We already know (Theorem 4.14) that $L^\infty = (L^1)^\star$. Being a dual space, L^∞ enjoys some nice properties. In particular, we have the following:

(i) The closed unit ball B_{L^∞} is compact in the weak* topology $\sigma(L^\infty, L^1)$ (by Theorem 3.16).
(ii) If Ω is a measurable subset in $\mathbb{R}^N$ and (f_n) is a bounded sequence in $L^\infty(\Omega)$, there exists a subsequence (f_{n_k}) and some $f \in L^\infty(\Omega)$ such that $f_{n_k} \rightharpoonup f$ in

the weak* topology $\sigma(L^\infty, L^1)$ (this is a consequence of Corollary 3.30 and Theorem 4.13).

However $L^\infty(\Omega)$ is *not reflexive*, except in the trivial case where Ω consists of a finite number of atoms; otherwise $L^1(\Omega)$ would be reflexive (by Corollary 3.21) and we know that L^1 is not reflexive (Remark 6). As a consequence, it follows that the *dual space* $(L^\infty)^\star$ of L^∞ contains L^1 (since $L^\infty = (L^1)^\star$) and $(L^\infty)^\star$ is *strictly bigger* than L^1. In other words, there are continuous linear functionals ϕ on L^∞ which *cannot* be represented as

$$\langle \phi, f \rangle = \int uf \quad \forall f \in L^\infty \text{ and some } u \in L^1.$$

In fact, let us describe a "concrete" example of such a functional. Let $\phi_0 : C_c(\mathbb{R}^N) \to \mathbb{R}$ be defined by

$$\phi_0(f) = f(0) \text{ for } f \in C_c(\mathbb{R}^N).$$

Clearly ϕ_0 is a continuous linear functional on $C_c(\mathbb{R}^N)$ for the $\| \ \|_\infty$ norm. By Hahn–Banach, we may extend ϕ_0 into a continuous linear functional ϕ on $L^\infty(\mathbb{R}^N)$ and we have

$$(17) \qquad \langle \phi, f \rangle = f(0) \quad \forall f \in C_c(\mathbb{R}^N).$$

Let us verify that there exists *no* function $u \in L^1(\mathbb{R}^N)$ such that

$$(18) \qquad \langle \phi, f \rangle = \int uf \quad \forall f \in L^\infty(\mathbb{R}^N).$$

Assume, by contradiction, that such a function u exists. We deduce from (17) and (18) that

$$\int uf = 0 \quad \forall f \in C_c(\mathbb{R}^N) \text{ and } f(0) = 0.$$

Applying Corollary 4.24 (with $\Omega = \mathbb{R}^N \backslash \{0\}$) we see that $u = 0$ a.e. on $\mathbb{R}^N \backslash \{0\}$ and thus $u = 0$ a.e. on $\mathbb{R}^N$. We conclude (by (18)) that

$$\langle \phi, f \rangle = 0 \quad \forall f \in L^\infty(\mathbb{R}^N),$$

which contradicts (17).

$\star$ *Remark* 7. The dual space of L^∞ does not coincide with L^1 but we may still ask the question: what does $(L^\infty)^\star$ look like? For this purpose it is convenient to view $L^\infty(\Omega; \mathbb{C})$ as a commutative $C^\star$-algebra (see, e.g., W. Rudin [1]). By Gelfand's theorem $L^\infty(\Omega; \mathbb{C})$ is isomorphic and isometric to the space $C(K; \mathbb{C})$ of continuous complex-valued functions on some compact topological space K (K is the spectrum of the algebra L^∞; K is not metrizable except when Ω consists of a finite number of atoms). Therefore $(L^\infty(\Omega; \mathbb{C}))^\star$ may be identified with the space of complex-valued *Radon measures* on K and $L^\infty(\Omega; \mathbb{R})^\star$ may be identified with the space of real-valued Radon measures on K; for more details, see Comment 3 at the end of this chapter, W. Rudin [1] and K. Yosida [1] (p. 118).

Remark 8. The space $L^\infty(\Omega)$ is *not separable* except when Ω consists of a finite number of atoms. In order to prove this fact it is convenient to use the following.

Lemma 4.2. *Let E be a Banach space. Assume that there exists a family $(O_i)_{i \in I}$ such that*

(i) *for each $i \in I$, O_i is a nonempty open subset of E,*
(ii) *$O_i \cap O_j = \emptyset$ if $i \neq j$,*
(iii) *I is **uncountable**.*

    *Then E is **not** separable.*

Proof of Lemma 4.2. Suppose, by contradiction, that E is separable. Let $(u_n)_{n \in \mathbb{N}}$ denote a dense countable set in E. For each $i \in I$, the set $O_i \cap (u_n)_{n \in \mathbb{N}} \neq \emptyset$ and we may choose $n(i)$ such that $u_{n(i)} \in O_i$. The mapping $i \mapsto n(i)$ is injective; indeed, if $n(i) = n(j)$, then $u_{n(i)} = u_{n(j)} \in O_i \cap O_j$ and thus $i = j$. Therefore, I is countable—a contradiction.

    We now establish that $L^\infty(\Omega)$ is not separable. We claim that there is an uncountable family $(\omega_i)_{i \in I}$ of measurable sets in Ω which are all distinct, that is, the symmetric difference $\omega_i \triangle \omega_j$ has positive measure for $i \neq j$. We then conclude by applying Lemma 4.2 to the family $(O_i)_{i \in I}$ defined by

$$O_i = \{ f \in L^\infty(\Omega);\ \| f - \chi_{\omega_i} \|_\infty < 1/2 \}$$

(note that $\| \chi_\omega - \chi_{\omega'} \|_\infty = 1$ if ω and ω' are distinct). The existence of an uncountable family (ω_i) is clear when Ω is an open set in $\mathbb{R}^N$ since we may consider all the balls $B(x_0, r)$ with $x_0 \in \Omega$ and $r > 0$ small enough.

    When Ω is a general measure space we split Ω into its atomic part Ω_a and its nonatomic (= diffuse) part Ω_d; then we distinguish two cases:

(i) Ω_d is not a null set.
(ii) Ω_d is a null set.

    In Case (i), then for each real number t, $0 < t < \mu(\Omega_d)$, there is a measurable set ω with $\mu(\omega) = t$; see, e.g., P. Halmos [1], A. J. Weir [1], or J. Neveu [1]. In this way, we obtain an uncountable family of distinct measurable sets.

    In Case (ii) Ω consists of a countable union of distinct atoms (a_n) (unless Ω consists of a finite number of atoms). For any collection of integers, $A \subset \mathbb{N}$, we define $\omega_A = \bigcup_{n \in A} a_n$. Clearly, (ω_A) is an uncountable family of distinct measurable sets.

    The following table summarizes the main properties of the space $L^p(\Omega)$ when Ω is a measurable subset of $\mathbb{R}^N$:

	Reflexive	Separable	Dual space
L^p with $1 < p < \infty$	YES	YES	$L^{p'}$
L^1	NO	YES	L^∞
L^∞	NO	NO	Strictly bigger than L^1

4.4 Convolution and regularization

We first define the convolution product of a function $f \in L^1(\mathbb{R}^N)$ with a function $g \in L^p(\mathbb{R}^N)$.

• **Theorem 4.15 (Young).** *Let $f \in L^1(\mathbb{R}^N)$ and let $g \in L^p(\mathbb{R}^N)$ with $1 \le p \le \infty$. Then for a.e. $x \in \mathbb{R}^N$ the function $y \mapsto f(x-y)g(y)$ is integrable on $\mathbb{R}^N$ and we define*

$$(f \star g)(x) = \int_{\mathbb{R}^N} f(x-y)g(y)dy.$$

In addition $f \star g \in L^p(\mathbb{R}^N)$ and

$$\|f \star g\|_p \le \|f\|_1 \|g\|_p.$$

Proof. The conclusion is obvious when $p = \infty$. We consider two cases:

(i) $p = 1$,
(ii) $1 < p < \infty$.

   Case (i): $p = 1$. Set $F(x, y) = f(x-y)g(y)$.
For a.e. $y \in \mathbb{R}^N$ we have

$$\int_{\mathbb{R}^N} |F(x, y)|dx = |g(y)| \int_{\mathbb{R}^N} |f(x-y)|dx = |g(y)| \|f\|_1 < \infty$$

and, moreover,

$$\int_{\mathbb{R}^N} dy \int_{\mathbb{R}^N} |F(x, y)|dx = \|g\|_1 \|f\|_1 < \infty.$$

We deduce from Tonelli's theorem (Theorem 4.4) that $F \in L^1(\mathbb{R}^N \times \mathbb{R}^N)$. Applying Fubini's theorem (Theorem 4.5), we see that

$$\int_{\mathbb{R}^N} |F(x, y)|dy < \infty \text{ for a.e. } x \in \mathbb{R}^N$$

and, moreover,

$$\int_{\mathbb{R}^N} dx \int_{\mathbb{R}^N} |F(x, y)|dy = \int_{\mathbb{R}^N} dy \int_{\mathbb{R}^N} |F(x, y)|dx = \|f\|_1 \|g\|_1.$$

This is precisely the conclusion of Theorem 4.15 when $p = 1$.

   Case (ii): $1 < p < \infty$. By Case (i) we know that for a.e. *fixed* $x \in \mathbb{R}^N$ the function $y \mapsto |f(x-y)| \, |g(y)|^p$ is integrable on $\mathbb{R}^N$, that is,

$$|f(x-y)|^{1/p}|g(y)| \in L^p_y(\mathbb{R}^N).$$

Since $|f(x, y)|^{1/p'} \in L^{p'}_y(\mathbb{R}^N)$, we deduce from Hölder's inequality that

$$|f(x-y)||g(y)| = |f(x-y)|^{1/p'}|f(x-y)|^{1/p}|g(y)| \in L^1_y(\mathbb{R}^N)$$

and

$$\int_{\mathbb{R}^N} |f(x-y)||g(y)|dy \le \|f\|_1^{1/p'} \left(\int_{\mathbb{R}^N} |f(x-y)| \, |g(y)|^p dy \right)^{1/p},$$

that is,

$$|(f \star g)(x)|^p \le \|f\|_1^{p/p'} (|f| \star |g|^p)(x).$$

We conclude, by Case (i), that $f \star g \in L^p(\mathbb{R}^N)$ and

$$\|f \star g\|_p^p \le \|f\|_1^{p/p'} \|f\|_1 \|g\|_p^p,$$

that is,

$$\|f \star g\|_p \le \|f\|_1 \|g\|_p.$$

Notation. Given a function f on $\mathbb{R}^N$ we set $\check{f}(x) = f(-x)$.

Proposition 4.16. *Let* $f \in L^1(\mathbb{R}^N)$, $g \in L^p(\mathbb{R}^N)$ *and* $h \in L^{p'}(\mathbb{R}^N)$. *Then we have*

$$\int_{\mathbb{R}^N} (f \star g)h = \int_{\mathbb{R}^N} g(\check{f} \star h).$$

Proof. The function $F(x, y) = f(x-y)g(y)h(x)$ belongs to $L^1(\mathbb{R}^N \times \mathbb{R}^N)$ since

$$\int |h(x)|dx \int |f(x-y)| \, |g(y)|dy < \infty$$

by Theorem 4.15 and Hölder's inequality. Therefore we have

$$\int (f \star g)(x)h(x)dx = \int dx \int F(x, y)dy = \int dy \int F(x, y)dx$$
$$= \int g(y)(\check{f} \star h)(y)dy.$$

Support and convolution. The notion of support of a function f is standard: supp f is the complement of the biggest open set on which f vanishes; in other words supp f is the closure of the set $\{x; f(x) \ne 0\}$. This notion is not adequate when dealing with equivalence classes, such as the space L^p. We need a definition which is *intrinsic*, that is, supp f_1 and supp f_2 should be the same (or differ by a null set) if $f_1 = f_2$ a.e. The reader will easily admit that the usual notion does not make sense for $f = \chi_{\mathbb{Q}}$ on $\mathbb{R}$. In the following proposition we introduce the appropriate notion.

Proposition 4.17 (and definition of the support). *Let* $f : \mathbb{R}^N \to \mathbb{R}$ *be any function. Consider the family* $(\omega_i)_{i \in I}$ *of all open sets on* $\mathbb{R}^N$ *such that for each* $i \in I$, $f = 0$ *a.e. on* ω_i. *Set* $\omega = \bigcup_{i \in I} \omega_i$.
   Then $f = 0$ *a.e. on* ω.

By definition, supp f *is the complement of* ω *in* $\mathbb{R}^N$.

Remark 9.

(a) Assume $f_1 = f_2$ a.e. on $\mathbb{R}^N$; clearly we have supp $f_1 =$ supp f_2. Hence we may talk about supp f for a function $f \in L^p$—without saying what representative we pick in the equivalence class.

(b) If f is a continuous function on $\mathbb{R}^N$ it is easy to check that the new definition of supp f coincides with the usual definition.

Proof of Proposition 4.17. Since the set I need not be countable it is not clear that $f = 0$ a.e. on ω. However we may recover the *countable* case as follows. There is a *countable* family (O_n) of open sets in $\mathbb{R}^N$ such that *every* open set on $\mathbb{R}^N$ is the union of some O_n's. Write $\omega_i = \bigcup_{n \in A_i} O_n$ and $\omega = \bigcup_{n \in B} O_n$ where $B = \bigcup_{i \in I} A_i$. Since $f = 0$ a.e. on every set O_n with $n \in B$, we conclude that $f = 0$ a.e. on ω.

• **Proposition 4.18.** *Let* $f \in L^1(\mathbb{R}^N)$ *and* $g \in L^p(\mathbb{R}^N)$ *with* $1 \le p \le \infty$. *Then*

$$\boxed{\text{supp}(f \star g) \subset \overline{\text{supp} f + \text{supp} g}.}$$

Proof. Fix $x \in \mathbb{R}^N$ such that the function $y \mapsto f(x-y)g(y)$ is integrable (see Theorem 4.15). We have

$$(f \star g)(x) = \int f(x-y)g(y)dy = \int_{(x - \text{supp} f) \cap \text{supp} g} f(x-y)g(y)dy.$$

If $x \notin \text{supp} f + \text{supp} g$, then $(x - \text{supp} f) \cap \text{supp} g = \emptyset$ and so $(f \star g)(x) = 0$. Thus

$$(f \star g)(x) = 0 \quad \text{a.e. on } (\text{supp} f + \text{supp} g)^c.$$

In particular,

$$(f \star g)(x) = 0 \quad \text{a.e. on Int}[(\text{supp} f + \text{supp} g)^c]$$

and therefore

$$\text{supp}(f \star g) \subset \overline{\text{supp} f + \text{supp} g}.$$

• *Remark* 10. If *both* f and g have compact support, then $f \star g$ also has compact support. However, $f \star g$ need *not* have compact support if *only one* of them has compact support.

Definition. Let $\Omega \subset \mathbb{R}^N$ be open and let $1 \le p \le \infty$. We say that a function $f : \Omega \to \mathbb{R}$ belongs to $L^p_{\text{loc}}(\Omega)$ if $f \chi_K \in L^p(\Omega)$ for every compact set K contained in Ω.

Note that if $f \in L^p_{\text{loc}}(\Omega)$, then $f \in L^1_{\text{loc}}(\Omega)$.

Proposition 4.19. *Let* $f \in C_c(\mathbb{R}^N)$ *and* $g \in L^1_{\text{loc}}(\mathbb{R}^N)$. *Then* $(f \star g)(x)$ *is well defined for* *every* $x \in \mathbb{R}^N$, *and, moreover,* $(f \star g) \in C(\mathbb{R}^N)$.

Proof. Note that for *every* $x \in \mathbb{R}^N$ the function $y \mapsto f(x-y)g(y)$ is integrable on $\mathbb{R}^N$ and therefore $(f \star g)(x)$ is defined for *every* $x \in \mathbb{R}^N$.

Let $x_n \to x$ and let K be a fixed compact set in $\mathbb{R}^N$ such that $(x_n - \operatorname{supp} f) \subset K$ $\forall n$. Therefore, we have $f(x_n - y) = 0$ $\forall n$, $\forall y \notin K$. We deduce from the uniform continuity of f that

$$|f(x_n - y) - f(x-y)| \le \varepsilon_n \chi_K(y) \quad \forall n, \quad \forall y \in \mathbb{R}^N$$

with $\varepsilon_n \to 0$. We conclude that

$$|(f \star g)(x_n) - (f \star g)(x)| \le \varepsilon_n \int_K |g(y)| dy \longrightarrow 0.$$

Notation. Let $\Omega \subset \mathbb{R}^N$ be an open set.

$C(\Omega)$ is the space of continuous functions on Ω.

$C^k(\Omega)$ is the space of functions k times continuously differentiable on Ω ($k \ge 1$ is an integer).

$C^\infty(\Omega) = \cap_k C^k(\Omega)$.

$C_c(\Omega)$ is the space of continuous functions on Ω with compact support in Ω, i.e., which vanish outside some compact set $K \subset \Omega$.

$C_c^k(\Omega) = C^k(\Omega) \cap C_c(\Omega)$.

$C_c^\infty(\Omega) = C^\infty(\Omega) \cap C_c(\Omega)$,
(some authors write $\mathcal{D}(\Omega)$ or $C_0^\infty(\Omega)$ instead of $C_c^\infty(\Omega)$).

If $f \in C^1(\Omega)$, its gradient is defined by

$$\nabla f = \left(\frac{\partial f}{\partial x_1}, \frac{\partial f}{\partial x_2}, \ldots, \frac{\partial f}{\partial x_N} \right).$$

If $f \in C^k(\Omega)$ and $\alpha = (\alpha_1, \alpha_2, \ldots, \alpha_N)$ is a multi-index of length $|\alpha| = \alpha_1 + \alpha_2 + \cdots + \alpha_N$, less than k, we write

$$D^\alpha f = \frac{\partial^{\alpha_1}}{\partial x_1^{\alpha_1}} \frac{\partial^{\alpha_2}}{\partial x_2^{\alpha_2}} \cdots \frac{\partial^{\alpha_N}}{\partial x_N^{\alpha_N}} f.$$

• **Proposition 4.20.** *Let $f \in C_c^k(\mathbb{R}^N)(k \ge 1)$ and let $g \in L_{\text{loc}}^1(\mathbb{R}^N)$. Then $f \star g \in C^k(\mathbb{R}^N)$ and*

$$\boxed{D^\alpha(f \star g) = (D^\alpha f) \star g \quad \forall \alpha \text{ with } |\alpha| \le k.}$$

In particular, if $f \in C_c^\infty(\mathbb{R}^N)$ and $g \in L_{\text{loc}}^1(\mathbb{R}^N)$, then $f \star g \in C^\infty(\mathbb{R}^N)$.

Proof. By induction it suffices to consider the case $k = 1$. Given $x \in \mathbb{R}^N$ we claim that $f \star g$ is differentiable at x and that

$$\nabla(f \star g)(x) = (\nabla f) \star g(x).$$

Let $h \in \mathbb{R}^N$ with $|h| < 1$. We have, for all $y \in \mathbb{R}^N$,

$$|f(x + h - y) - f(x - y) - h \cdot \nabla f(x - y)|$$

$$= \left| \int_0^1 [h \cdot \nabla f(x + sh - y) - h \cdot \nabla f(x - y)] ds \right| \le |h| \varepsilon(|h|)$$

with $\varepsilon(|h|) \to 0$ as $|h| \to 0$ (since ∇f is uniformly continuous on $\mathbb{R}^N$).

Let K be a fixed compact set in $\mathbb{R}^N$ large enough that $x + B(0, 1) - \operatorname{supp} f \subset K$. We have

$$f(x + h - y) - f(x - y) - h \cdot \nabla f(x - y) = 0 \quad \forall y \notin K, \quad \forall h \in B(0, 1)$$

and therefore

$$|f(x+h-y)-f(x-y)-h \cdot \nabla f(x-y)| \le |h| \varepsilon(|h|) \chi_K(y) \, \forall y \in \mathbb{R}^N, \; \forall h \in B(0, 1).$$

We conclude that for $h \in B(0, 1)$,

$$|(f \star g)(x + h) - (f \star g)(x) - h \cdot (\nabla f \star g)(x)| \le |h| \varepsilon(|h|) \int_K |g(y)| dy.$$

It follows that $f \star g$ is differentiable at x and $\nabla(f \star g)(x) = (\nabla f) \star g(x)$.

Mollifiers

Definition. A sequence of *mollifiers* $(\rho_n)_{n \ge 1}$ is any sequence of functions on $\mathbb{R}^N$ such that

$$\rho_n \in C_c^\infty(\mathbb{R}^N), \quad \operatorname{supp} \rho_n \subset \overline{B(0, 1/n)}, \quad \int \rho_n = 1, \rho_n \ge 0 \text{ on } \mathbb{R}^N.$$

In what follows *we shall systematically use the notation* (ρ_n) *to denote a sequence of mollifiers*.

It is easy to generate a sequence of mollifiers starting with a *single* function $\rho \in C_c^\infty(\mathbb{R}^N)$ such that $\operatorname{supp} \rho \subset \overline{B(0, 1)}$, $\rho \ge 0$ on $\mathbb{R}^N$, and ρ does not vanish identically—for example the function

$$\rho(x) = \begin{cases} e^{1/(|x|^2 - 1)} & \text{if } |x| < 1, \\ 0 & \text{if } |x| > 1. \end{cases}$$

We obtain a sequence of mollifiers by letting $\rho_n(x) = C \, n^N \rho(nx)$ with $C = 1/\int \rho$.

Proposition 4.21. *Assume $f \in C(\mathbb{R}^N)$. Then $(\rho_n \star f) \xrightarrow[n \to \infty]{} f$ uniformly on compact sets of $\mathbb{R}^N$.*

Proof.[4] Let $K \subset \mathbb{R}^N$ be a fixed compact set. Given $\varepsilon > 0$ there exists $\delta > 0$ (depending on K and ε) such that

$$|f(x - y) - f(x)| < \varepsilon \quad \forall x \in K, \quad \forall y \in B(0, \delta).$$

We have, for $x \in \mathbb{R}^N$,

$$(\rho_n \star f)(x) - f(x) = \int [f(x - y) - f(x)]\rho_n(y)dy$$

$$= \int_{B(0,1/n)} [f(x - y) - f(x)]\rho_n(y)dy.$$

For $n > 1/\delta$ and $x \in K$ we obtain

$$|(\rho_n \star f)(x) - f(x)| \leq \varepsilon \int \rho_n = \varepsilon.$$

• **Theorem 4.22.** *Assume* $f \in L^p(\mathbb{R}^N)$ *with* $1 \leq p < \infty$. *Then* $(\rho_n \star f) \underset{n \to \infty}{\longrightarrow} f$ *in* $L^p(\mathbb{R}^N)$.

Proof. Given $\varepsilon > 0$, we fix a function $f_1 \in C_c(\mathbb{R}^N)$ such that $\|f - f_1\|_p < \varepsilon$ (see Theorem 4.12). By Proposition 4.21 we know that $(\rho_n \star f_1) \to f_1$ uniformly on every compact set of $\mathbb{R}^N$. On the other hand, we have (by Proposition 4.18) that

$$\text{supp}(\rho_n \star f_1) \subset \overline{B(0, 1/n)} + \text{supp } f_1 \subset \overline{B(0, 1)} + \text{supp } f_1,$$

which is a fixed compact set. It follows that

$$\|(\rho_n \star f_1) - f_1\|_p \underset{n \to \infty}{\longrightarrow} 0.$$

Finally, we write

$$(\rho_n \star f) - f = [\rho_n \star (f - f_1)] + [(\rho_n \star f_1) - f_1] + [f_1 - f]$$

and thus

$$\|(\rho_n \star f) - f\|_p \leq 2\|f - f_1\|_p + \|(\rho_n \star f_1) - f_1\|_p$$

(by Theorem 4.15).

We conclude that

$$\limsup_{n \to \infty} \|(\rho_n \star f) - f\|_p \leq 2\varepsilon \quad \forall \varepsilon > 0$$

and therefore $\lim_{n \to \infty} \|(\rho_n \star f) - f\|_p = 0$.

• **Corollary 4.23.** *Let* $\Omega \subset \mathbb{R}^N$ *be an open set. Then* $C_c^\infty(\Omega)$ *is dense in* $L^p(\Omega)$ *for any* $1 \leq p < \infty$.

---

[4] The technique of regularization by convolution was originally introduced by Leray and Friedrichs.

Proof. Given $f \in L^p(\Omega)$ we set

$$\bar{f}(x) = \begin{cases} f(x) & \text{if } x \in \Omega, \\ 0 & \text{if } x \in \mathbb{R}^N \setminus \Omega, \end{cases}$$

so that $\bar{f} \in L^p(\mathbb{R}^N)$.

Let (K_n) be a sequence of compact sets in $\mathbb{R}^N$ such that

$$\bigcup_{n=1}^{\infty} K_n = \Omega \quad \text{and } \operatorname{dist}(K_n, \Omega^c) \geq 2/n \quad \forall n.$$

[We may choose, for example, $K_n = \{x \in \mathbb{R}^N; |x| \leq n \text{ and } \operatorname{dist}(x, \Omega^c) \geq 2/n\}$.]
Set $g_n = \chi_{K_n} \bar{f}$ and $f_n = \rho_n \star g_n$, so that

$$\operatorname{supp} f_n \subset \overline{B(0, 1/n)} + K_n \subset \Omega.$$

It follows that $f_n \in C_c^{\infty}(\Omega)$. On the other hand, we have

$$\begin{aligned}
\|f_n - f\|_{L^p(\Omega)} &= \|f_n - \bar{f}\|_{L^p(\mathbb{R}^N)} \\
&\leq \|(\rho_n \star g_n) - (\rho_n \star \bar{f})\|_{L^p(\mathbb{R}^N)} + \|(\rho_n \star \bar{f}) - \bar{f}\|_{L^p(\mathbb{R}^N)} \\
&\leq \|g_n - \bar{f}\|_{L^p(\mathbb{R}^N)} + \|(\rho_n \star \bar{f}) - \bar{f}\|_{L^p(\mathbb{R}^N)}.
\end{aligned}$$

Finally, we note that $\|g_n - \bar{f}\|_{L^p(\mathbb{R}^N)} \to 0$ by dominated convergence and $\|(\rho_n \star \bar{f}) - \bar{f}\|_{L^p(\mathbb{R}^N)} \to 0$ by Theorem 4.22. We conclude that $\|f_n - f\|_{L^p(\Omega)} \to 0$.

Corollary 4.24. *Let $\Omega \subset \mathbb{R}^N$ be an open set and let $u \in L^1_{\text{loc}}(\Omega)$ be such that*

$$\int uf = 0 \quad \forall f \in C_c^{\infty}(\Omega).$$

Then $u = 0$ a.e. on Ω.

Proof. Let $g \in L^{\infty}(\mathbb{R}^N)$ be a function such that $\operatorname{supp} g$ is a compact set contained in Ω. Set $g_n = \rho_n \star g$, so that $g_n \in C_c^{\infty}(\Omega)$ provided n is large enough. Therefore we have

$$(19) \qquad \int u\, g_n = 0 \quad \forall n.$$

Since $g_n \to g$ in $L^1(\mathbb{R}^N)$ (by Theorem 4.22) there is a subsequence—still denoted by g_n—such that $g_n \to g$ a.e. on $\mathbb{R}^N$ (see Theorem 4.9). Moreover, we have $\|g_n\|_{L^{\infty}(\mathbb{R}^N)} \leq \|g\|_{L^{\infty}(\mathbb{R}^N)}$. Passing to the limit in (19) (by dominated convergence), we obtain

$$(20) \qquad \int u g = 0.$$

Let K be a compact set contained in Ω. We choose as function g the function

$$
g = \begin{cases} \text{sign } u & \text{on } K, \\ 0 & \text{on } \mathbb{R}^N \setminus K. \end{cases}
$$

We deduce from (20) that $\int_K |u| = 0$ and thus $u = 0$ a.e. on K. Since this holds for any compact $K \subset \Omega$, we conclude that $u = 0$ a.e. on Ω.

4.5 Criterion for Strong Compactness in L^p

It is important to be able to decide whether a family of functions in $L^p(\Omega)$ has compact closure in $L^p(\Omega)$ (for the strong topology). We recall that the Ascoli–Arzelà theorem answers the same question in $C(K)$, the space of continuous functions over a *compact metric* space K with values in $\mathbb{R}$.

• **Theorem 4.25 (Ascoli–Arzelà).** *Let K be a compact metric space and let $\mathcal{H}$ be a bounded subset of $C(K)$. Assume that $\mathcal{H}$ is uniformly equicontinuous, that is,*

(21) $\quad \forall \varepsilon > 0 \; \exists \delta > 0$ *such that* $d(x_1, x_2) < \delta \Rightarrow |f(x_1) - f(x_2)| < \varepsilon \quad \forall f \in \mathcal{H}.$

Then the closure of $\mathcal{H}$ in $C(K)$ is compact.

For the proof of the Ascoli–Arzelà theorem, see, e.g., W. Rudin [1], [2], A. Knapp [1], J. Dixmier [1], A. Friedman [3], G. Choquet [1], K. Yosida [1], H. L. Royden [1], J. R. Munkres [1], G. B. Folland [2], etc.

Notation (shift of function). We set $(\tau_h f)(x) = f(x + h)$, $x \in \mathbb{R}^N$, $h \in \mathbb{R}^N$.

The following theorem and its corollary are "L^p-versions" of the Ascoli–Arzelà theorem.

• **Theorem 4.26 (Kolmogorov–M. Riesz–Fréchet).** *Let $\mathcal{F}$ be a bounded set in $L^p(\mathbb{R}^N)$ with $1 \leq p < \infty$. Assume that*[5]

(22) $\quad\quad\quad \lim_{|h| \to 0} \|\tau_h f - f\|_p = 0 \quad$ *uniformly in $f \in \mathcal{F}$,*

i.e., $\forall \varepsilon > 0 \; \exists \delta > 0$ such that $\|\tau_h f - f\|_p < \varepsilon \; \forall f \in \mathcal{F}, \forall h \in \mathbb{R}^N$ with $|h| < \delta$.
Then the closure of $\mathcal{F}_{|\Omega}$ in $L^p(\Omega)$ is eompact for any measurable set $\Omega \subset \mathbb{R}^N$ with finite measure.

[Here $\mathcal{F}_{|\Omega}$ denotes the restrictions to Ω of the functions in $\mathcal{F}$.]

The proof consists of four steps:
Step 1: We claim that

(23) $\quad\quad\quad \|(\rho_n \star f) - f\|_{L^p(\mathbb{R}^N)} \leq \varepsilon \quad \forall f \in \mathcal{F}, \quad \forall n > 1/\delta.$

---

[5] Assumption (22) should be compared with (21). It is an "integral" equicontinuity assumption.

Indeed, we have

$$|(\rho_n \star f)(x) - f(x)| \leq \int |f(x-y) - f(x)| \rho_n(y) dy$$

$$\leq \left[\int |f(x-y) - f(x)|^p \rho_n(y) dy \right]^{1/p}$$

by Hölder's inequality.

Thus we obtain

$$\int |(\rho_n \star f)(x) - f(x)|^p dx \leq \int \int |f(x-y) - f(x)|^p \rho_n(y) dx \, dy$$

$$= \int_{B(0,1/n)} \rho_n(y) dy \int |f(x-y) - f(x)|^p dx \leq \varepsilon^p,$$

provided $1/n < \delta$.

Step 2: We claim that

(24) $$\left\| \rho_n \star f \right\|_{L^\infty(\mathbb{R}^N)} \leq C_n \left\| f \right\|_{L^p(\mathbb{R}^N)} \quad \forall f \in \mathcal{F}$$

and

(25) $$\begin{aligned} |(\rho_n \star f)(x_1) - (\rho_n \star f)(x_2)| &\leq C_n \|f\|_p |x_1 - x_2| \\ &\forall f \in \mathcal{F}, \quad \forall x_1, x_2 \in \mathbb{R}^N, \end{aligned}$$

where C_n depends only on n.

Inequality (24) follows from Hölder's inequality with $C_n = \|\rho_n\|_{p'}$. On the other hand, we have $\nabla(\rho_n \star f) = (\nabla \rho_n) \star f$ and therefore

$$\|\nabla(\rho_n \star f)\|_{L^\infty(\mathbb{R}^N)} \leq \|\nabla \rho_n\|_{L^{p'}(\mathbb{R}^N)} \|f\|_{L^p(\mathbb{R}^N)}.$$

Thus we obtain (25) with $C_n = \|\nabla \rho_n\|_{L^{p'}(\mathbb{R}^N)}$.

Step 3: Given $\varepsilon > 0$ and $\Omega \subset \mathbb{R}^N$ of finite measure, there is a bounded measurable subset ω of Ω such that

(26) $$\|f\|_{L^p(\Omega \setminus \omega)} < \varepsilon \quad \forall f \in \mathcal{F}.$$

Indeed, we write

$$\|f\|_{L^p(\Omega \setminus \omega)} \leq \|f - (\rho_n \star f)\|_{L^p(\mathbb{R}^N)} + \|\rho_n \star f\|_{L^p(\Omega \setminus \omega)}.$$

In view of (24) it suffices to choose ω such that $|\Omega \setminus \omega|$ is small enough.

Step 4: *Conclusion.* Since $L^p(\Omega)$ is complete, it suffices (see, e.g., A. Knapp [1] or J. R. Munkres [1], Section 7.3) to show that $\mathcal{F}_{|\Omega}$ is *totally bounded*, i.e., given any $\varepsilon > 0$ there is a finite covering of $\mathcal{F}_{|\Omega}$ by balls of radius ε. Given $\varepsilon > 0$ we *fix* a bounded measurable set ω such that (26) holds. Also we *fix* $n > 1/\delta$. The family $\mathcal{H} = (\rho_n \star \mathcal{F})_{|\bar{\omega}}$ satisfies all the assumptions of the Ascoli–Arzelà theorem (by Step 2). Therefore $\mathcal{H}$ has compact closure in $C(\bar{\omega})$; consequently $\mathcal{H}$ also has compact closure in $L^p(\omega)$. Hence we may cover $\mathcal{H}$ by a finite number of balls of radius ε in $L^p(\omega)$, say,

$$\mathcal{H} \subset \bigcup_i B(g_i, \varepsilon) \text{ with } g_i \in L^p(\omega).$$

Consider the functions $\bar{g}_i : \Omega \to \mathbb{R}$ defined by

$$\bar{g}_i = \begin{cases} g_i & \text{on } \omega, \\ 0 & \text{on } \Omega \backslash \omega, \end{cases}$$

and the balls $B(\bar{g}_i, 3\varepsilon)$ in $L^p(\Omega)$.

We claim that they cover $\mathcal{F}_{|\Omega}$. Indeed, given $f \in \mathcal{F}$ there is some i such that

$$\left\| (\rho_n \star f) - g_i \right\|_{L^p(\omega)} < \varepsilon.$$

Since

$$\left\| f - \bar{g}_i \right\|_{L^p(\Omega)}^p = \int_{\Omega \backslash \omega} |f|^p + \int_\omega |f - g_i|^p$$

we have, by (26),

$$\left\| f - \bar{g}_i \right\|_{L^p(\Omega)} \leq \varepsilon + \left\| f - g_i \right\|_{L^p(\omega)}$$
$$\leq \varepsilon + \left\| f - (\rho_n \star f) \right\|_{L^p(\mathbb{R}^N)} + \left\| (\rho_n \star f) - g_i \right\|_{L^p(\omega)} < 3\varepsilon.$$

We conclude that $\mathcal{F}_{|\Omega}$ has compact closure in $L^p(\Omega)$.

Remark 11. When trying to establish that a family $\mathcal{F}$ in $L^p(\Omega)$ has compact closure in $L^p(\Omega)$, with Ω bounded, it is usually convenient to extend the functions to all of $\mathbb{R}^N$, then apply Theorem 4.26 and consider the restrictions to Ω.

Remark 12. Under the assumptions of Theorem 4.26 we cannot conclude in general that $\mathcal{F}$ itself has compact closure in $L^p(\mathbb{R}^N)$ (construct an example, or see Exercise 4.33). An additional assumption is required; we describe it next:

Corollary 4.27. *Let $\mathcal{F}$ be a bounded set in $L^p(\mathbb{R}^N)$ with $1 \leq p < \infty$. Assume (22) and also*

$$(27) \qquad \begin{cases} \forall \varepsilon > 0 \; \exists \Omega \subset \mathbb{R}^N, \; \text{bounded, measurable such that} \\ \|f\|_{L^p(\mathbb{R}^n \backslash \Omega)} < \varepsilon \quad \forall f \in \mathcal{F}. \end{cases}$$

Then $\mathcal{F}$ has compact closure in $L^p(\mathbb{R}^N)$.

Proof. Given $\varepsilon > 0$ we fix $\Omega \subset \mathbb{R}^N$ bounded measurable such that (27) holds. By Theorem 4.26 we know that $\mathcal{F}_{|\Omega}$ has compact closure in $L^p(\Omega)$. Hence we may cover $\mathcal{F}_{|\Omega}$ with a finite number of balls of radius ε in $L^p(\Omega)$, say

$$\mathcal{F}_{|\Omega} \subset \bigcup_i B(g_i, \varepsilon) \quad \text{with } g_i \in L^p(\Omega).$$

Set

$$\bar{g}_i(x) = \begin{cases} g_i(x) & \text{in } \Omega, \\ 0 & \text{on } \mathbb{R}^N \backslash \Omega. \end{cases}$$

It is clear that $\mathcal{F}$ is covered by the balls $B(\bar{g}_i, 2\varepsilon)$ in $L^p(\mathbb{R}^N)$.

Remark 13. The converse of Corollary 4.27 is also true (see Exercise 4.34). Therefore we have a complete characterization of compact sets in $L^p(\mathbb{R}^N)$.

We conclude with a useful application of Theorem 4.26:

Corollary 4.28. *Let G be a fixed function in $L^1(\mathbb{R}^N)$ and let*

$$\mathcal{F} = G \star \mathcal{B},$$

where $\mathcal{B}$ is a bounded set in $L^p(\mathbb{R}^N)$ with $1 \leq p < \infty$. Then $\mathcal{F}_{|\Omega}$ has compact closure in $L^p(\Omega)$ for any measurable set Ω with finite measure.

Proof. Clearly $\mathcal{F}$ is bounded in $L^p(\mathbb{R}^N)$. On the other hand, if we write $f = G \star u$ with $u \in \mathcal{B}$ we have

$$\|\tau_h f - f\|_p = \|(\tau_h G - G) \star u\|_p \leq C\|\tau_h G - G\|_1,$$

and we conclude with the help of the following lemma:

Lemma 4.3. *Let $G \in L^q(\mathbb{R}^N)$ with $1 \leq q < \infty$.*
Then
$$\lim_{h \to 0} \|\tau_h G - G\|_q = 0.$$

Proof. Given $\varepsilon > 0$, there exists (by Theorem 4.12) a function $G_1 \in C_c(\mathbb{R}^N)$ such that $\|G - G_1\|_q < \varepsilon$.

We write

$$\|\tau_h G - G\|_q \leq \|\tau_h G - \tau_h G_1\|_q + \|\tau_h G_1 - G_1\|_q + \|G_1 - G\|_q$$
$$\leq 2\varepsilon + \|\tau_h G_1 - G_1\|_q.$$

Since $\lim_{h \to 0} \|\tau_h G_1 - G_1\|_q = 0$ we see that

$$\limsup_{h \to 0} \|\tau_h G - G\|_q \leq 2\varepsilon \quad \forall \varepsilon > 0.$$

Comments on Chapter 4

1. Egorov's theorem.

Some basic results of integration theory have been recalled in Section 4.1. One useful result that has not been mentioned is the following.

⋆ **Theorem 4.29 (Egorov).** *Assume that Ω is a measure space with finite measure. Let (f_n) be a sequence of measurable functions on Ω such that*

$$f_n(x) \to f(x) \text{ a.e. on } \Omega \text{ (with } |f(x)| < \infty \text{ a.e.).}$$

Then $\forall \varepsilon > 0$ $\exists A \subset \Omega$ measurable such that $|\Omega \backslash A| < \varepsilon$ and $f_n \to f$ uniformly on A.

For a proof, see Exercise 4.14, P. Halmos [1], G. B. Folland [2], E. Hewitt–K. Stromberg [1], R. Wheeden–A. Zygmund [1], K. Yosida [1], A. Friedman [3], etc.

2. Weakly compact sets in L^1.
Since L^1 is not reflexive, bounded sets of L^1 do not play an important role with respect to the weak topology $\sigma(L^1, L^\infty)$. The following result provides a useful characterization of weakly compact sets of L^1.

⋆ **Theorem 4.30 (Dunford–Pettis).** *Let $\mathcal{F}$ be a bounded set in $L^1(\Omega)$. Then $\mathcal{F}$ has compact closure in the weak topology $\sigma(L^1, L^\infty)$ if and only if $\mathcal{F}$ is equi-integrable, that is,*

(a)
$$\begin{cases} \forall \varepsilon > 0 \; \exists \delta > 0 \quad \text{such that} \\ \displaystyle\int_A |f| < \varepsilon \;\; \forall A \subset \Omega, \text{measurable with } |A| < \delta, \;\; \forall f \in \mathcal{F} \end{cases}$$

and

(b)
$$\begin{cases} \forall \varepsilon > 0 \; \exists \omega \subset \Omega, \; \text{measurable with } |\omega| < \infty \text{ such that} \\ \displaystyle\int_{\Omega \backslash \omega} |f| < \varepsilon \;\; \forall f \in \mathcal{F}. \end{cases}$$

For a proof and discussion of Theorem 4.30 see Problem 23 or N. Dunford–J. T. Schwartz [1], B. Beauzamy [1], J. Diestel [2], I. Fonseca–G. Leoni [1], and also J. Neveu [1], C. Dellacherie–P. A. Meyer [1] for the probabilistic aspects; see also Exercise 4.36.

3. Radon measures.
As we have just pointed out, bounded sets of L^1 enjoy no compactness properties. To overcome this lack of compactness it is sometimes very useful *to embed L^1 into a large space: the space of Radon measures.*

Assume, for example, that Ω is a bounded open set of $\mathbb{R}^N$ with the Lebesgue measure. Consider the space $E = C(\overline{\Omega})$ with its norm $\|u\| = \sup_{x \in \overline{\Omega}} |u(x)|$. Its dual space, denoted by $\mathcal{M}(\overline{\Omega})$, is called the space of *Radon measures* on $\overline{\Omega}$. The weak⋆ topology on $\mathcal{M}(\overline{\Omega})$ is sometimes called the "vague" topology.

We shall identify $L^1(\Omega)$ with a subspace of $\mathcal{M}(\overline{\Omega})$. For this purpose we introduce the mapping $L^1(\Omega) \to \mathcal{M}(\overline{\Omega})$ defined as follows. Given $f \in L^1(\Omega)$, the mapping $u \in C(\overline{\Omega}) \mapsto \int_\Omega fu\,dx$ is a continuous linear functional on $C(\overline{\Omega})$, which we denote Tf, so that

$$\langle Tf, u \rangle_{E^*, E} = \int_\Omega f \, u \, dx \quad \forall u \in E.$$

Clearly T is linear, and, moreover, T is an *isometry*, since

$$\|Tf\|_{\mathcal{M}(\overline{\Omega})} = \sup_{\substack{u \in E \\ \|u\| \leq 1}} \int_\Omega fu = \|f\|_1 \quad \text{(see Exercise 4.26)}.$$

Using T we may identify $L^1(\Omega)$ with a subspace of $\mathcal{M}(\overline{\Omega})$. Since $\mathcal{M}(\overline{\Omega})$ is the dual space of the separable space $C(\overline{\Omega})$, it has some compactness properties in the weak* topology. In particular, if (f_n) *is a bounded sequence in* $L^1(\Omega)$, *there exist a subsequence* (f_{n_k}) *and a Radon measure* μ *such that* $f_{n_k} \overset{\star}{\rightharpoonup} \mu$ *in the weak* topology* $\sigma(E^*, E)$, that is,

$$\int_\Omega f_{n_k} u \to \langle \mu, u \rangle \quad \forall u \in C(\overline{\Omega}).$$

For example, a sequence in L^1 can converge to a Dirac measure with respect to the weak* topology. Some futher properties of Radon measures are discussed in Problem 24.

The terminology "measure" is justified by the following result, which connects the above definition with the standard notion of measures in the set-theoretic sense:

Theorem 4.31 (Riesz representation theorem). *Let* μ *be a Radon measure on* $\overline{\Omega}$. *Then there is a unique signed Borel measure* v *on* $\overline{\Omega}$ *(that is, a measure defined on Borel sets of* $\overline{\Omega}$) *such that*

$$\langle \mu, u \rangle = \int_{\overline{\Omega}} u dv \quad \forall u \in C(\overline{\Omega}).$$

It is often convenient to replace the space $E = C(\overline{\Omega})$ by the subspace

$$E_0 = \{f \in C(\overline{\Omega}); f = 0 \text{ on the boundary of } \overline{\Omega}\}.$$

The dual of E_0 is denoted by $\mathcal{M}(\Omega)$ (as opposed to $\mathcal{M}(\overline{\Omega})$). The Riesz representation theorem remains valid with the additional condition that $|v|$(boundary of $\overline{\Omega}) = 0$.

On this vast and classical subject, see, e.g., H. L. Royden [1], W. Rudin [2], G. B. Folland [2], A. Knapp [1], P. Malliavin [1], P. Halmos [1], I. Fonseca–G. Leoni [1].

4. The Bochner integral of vector-valued functions.

Let Ω be a measure space and let E be a Banach space. The space $L^p(\Omega; E)$ consists of all functions f defined on Ω with values into E that are *measurable in some appropriate sense* and such that $\int_\Omega \|f(x)\|^p d\mu < \infty$ (with the usual modification when $p = \infty$). Most of the properties described in Sections 4.2 and 4.3 still hold under some additional assumptions on E. For example, if E is reflexive and $1 < p < \infty$, then $L^p(\Omega; E)$ is reflexive and its dual space is $L^{p'}(\Omega; E^*)$. For more details, see K. Yosida [1], D. L. Cohn [1], E. Hille [1], B. Beauzamy [1], L. Schwartz [3]. The space $L^p(\Omega; E)$ is very useful in the study of evolution equations when Ω is an interval in $\mathbb{R}$ (see Chapter 10).

5. Interpolation theory.
The most striking result, which began interpolation theory, is the following.

Theorem 4.32 (Schur, M. Riesz, Thorin). *Assume that Ω is a measure space with $|\Omega| < \infty$, and that $T : L^1(\Omega) \to L^1(\Omega)$ is a bounded linear operator with norm*

$$M_1 = \|T\|_{\mathcal{L}(L^1, L^1)}.$$

Assume, in addition, that $T : L^\infty(\Omega) \to L^\infty(\Omega)$ is a bounded linear operator with norm

$$M_\infty = \|T\|_{\mathcal{L}(L^\infty, L^\infty)}.$$

Then T is a bounded operator from $L^p(\Omega)$ into $L^p(\Omega)$ for all $1 < p < \infty$, and its norm M_p satisfies

$$M_p \leq M_1^{1/p} M_\infty^{1/p'}.$$

Interpolation theory was originally discovered by I. Schur, M. Riesz, G. O. Thorin, J. Marcinkiewicz, and A. Zygmund. Decisive contributions have been made by a number of authors including J.-L. Lions, J. Peetre, A. P. Calderon, E. Stein, and E. Gagliardo. It has become a *useful tool in harmonic analysis* (see, e.g., E. Stein–G. Weiss [1], E. Stein [1], C. Sadosky [1]) and in *partial differential equations* (see, e.g., J.-L. Lions–E. Magenes [1]). On these questions see also G. B. Folland [2], N. Dunford–J. T. Schwartz [1] (Volume 1 p. 520), J. Bergh–J. Löfström [1], M. Reed–B. Simon [1], (Volume 2, p. 27) and Problem 22.

6. Young's inequality.
The following is an extension of Theorem 4.15.

Theorem 4.33 (Young). *Assume $f \in L^p(\mathbb{R}^N)$ and $g \in L^q(\mathbb{R}^N)$ with $1 \leq p \leq \infty$, $1 \leq q \leq \infty$ and $\frac{1}{r} = \frac{1}{p} + \frac{1}{q} - 1 \geq 0$.*
Then $f \star g \in L^r(\mathbb{R}^N)$ and $\|f \star g\|_r \leq \|f\|_p \|g\|_q$.

For a proof see, e.g., Exercise 4.30.

7. The notion of convolution—extended to distributions (see L. Schwartz [1] or A. Knapp [2])—plays a fundamental role in the theory of partial differential equations. For example, the equation $P(D)u = f$ in $\mathbb{R}^N$, where $P(D)$ is any differential operator with constant coefficients, has a solution of the form $u = E \star f$, where E is the *fundamental solution* of $P(D)$ (theorem of Malgrange–Ehrenpreis; see also Comment 2b in Chapter 1). In particular, the equation $\Delta u = f$ in $\mathbb{R}^3$ has a solution of the form $u = E \star f$, where $E(x) = -(4\pi|x|)^{-1}$.

Exercises for Chapter 4

Except where otherwise stated, Ω denotes a σ-finite measure space.

4.1 Let $\alpha > 0$ and $\beta > 0$. Set

$$f(x) = \left\{1 + |x|^{\alpha}\right\}^{-1}\left\{1 + |\log|x||^{\beta}\right\}^{-1}, \quad x \in \mathbb{R}^N.$$

Under what conditions does f belong to $L^p(\mathbb{R}^N)$?

4.2 Assume $|\Omega| < \infty$ and let $1 \le p \le q \le \infty$. Prove that $L^q(\Omega) \subset L^p(\Omega)$ with continuous injection. More precisely, show that

$$\|f\|_p \le |\Omega|^{\frac{1}{p}-\frac{1}{q}}\|f\|_q \quad \forall f \in L^q(\Omega).$$

[**Hint:** Use Hölder's inequality.]

4.3

1. Let $f, g \in L^p(\Omega)$ with $1 \le p \le \infty$. Prove that

$$h(x) = \max\{f(x), g(x)\} \in L^p(\Omega).$$

2. Let (f_n) and (g_n) be two sequences in $L^p(\Omega)$ with $1 \le p \le \infty$ such that $f_n \to f$ in $L^p(\Omega)$ and $g_n \to g$ in $L^p(\Omega)$. Set $h_n = \max\{f_n, g_n\}$ and prove that $h_n \to h$ in $L^p(\Omega)$.
3. Let (f_n) be a sequence in $L^p(\Omega)$ with $1 \le p < \infty$ and let (g_n) be a bounded sequence in $L^\infty(\Omega)$. Assume $f_n \to f$ in $L^p(\Omega)$ and $g_n \to g$ a.e. Prove that $f_n g_n \to fg$ in $L^p(\Omega)$.

4.4

1. Let $f_1, f_2, \ldots, f_k$ be k functions such that $f_i \in L^{p_i}(\Omega)$ $\forall i$ with $1 \le p_i \le \infty$ and $\sum_{i=1}^k \frac{1}{p_i} \le 1$.
   Set

$$f(x) = \prod_{i=1}^k f_i(x).$$

   Prove that $f \in L^p(\Omega)$ with $\frac{1}{p} = \sum_{i=1}^k \frac{1}{p_i}$ and that

$$\|f\|_p \le \prod_{i=1}^k \|f_i\|_{p_i}.$$

[**Hint:** Start with $k = 2$ and proceed by induction.]
2. Deduce that if $f \in L^p(\Omega) \cap L^q(\Omega)$ with $1 \le p \le \infty$ and $1 \le q \le \infty$, then $f \in L^r(\Omega)$ for every r between p and q. More precisely, write

$$\frac{1}{r} = \frac{\alpha}{p} + \frac{1-\alpha}{q} \quad \text{with } \alpha \in [0, 1]$$

and prove that

$$\|f\|_r \le \|f\|_p^{\alpha} \|f\|_q^{1-\alpha}.$$

4.5 Let $1 \le p < \infty$ and $1 \le q \le \infty$.

1. Prove that $L^1(\Omega) \cap L^\infty(\Omega)$ is a dense subset of $L^p(\Omega)$.
2. Prove that the set

$$\{ f \in L^p(\Omega) \cap L^q(\Omega) \; ; \; \|f\|_q \le 1 \}$$

is closed in $L^p(\Omega)$.
3. Let (f_n) be a sequence in $L^p(\Omega) \cap L^q(\Omega)$ and let $f \in L^p(\Omega)$. Assume that

$$f_n \to f \text{ in } L^p(\Omega) \text{ and } \|f_n\|_q \le C.$$

Prove that $f \in L^r(\Omega)$ and that $f_n \to f$ in $L^r(\Omega)$ for every r between p and q, $r \ne q$.

4.6 Assume $|\Omega| < \infty$.

1. Let $f \in L^\infty(\Omega)$. Prove that $\lim_{p\to\infty} \|f\|_p = \|f\|_\infty$.
2. Let $f \in \cap_{1 \le p < \infty} L^p(\Omega)$ and assume that there is a constant C such that

$$\|f\|_p \le C \quad \forall\, 1 \le p < \infty.$$

Prove that $f \in L^\infty(\Omega)$.
3. Construct an example of a function $f \in \cap_{1 \le p < \infty} L^p(\Omega)$ such that $f \notin L^\infty(\Omega)$ with $\Omega = (0, 1)$.

4.7 Let $1 \le q \le p \le \infty$. Let $a(x)$ be a measurable function on Ω. Assume that $au \in L^q(\Omega)$ for every function $u \in L^p(\Omega)$.
Prove that $a \in L^r(\Omega)$ with

$$r = \begin{cases} \dfrac{pq}{p - q} & \text{if } p < \infty, \\ q & \text{if } p = \infty. \end{cases}$$

[**Hint:** Use the closed graph theorem.]

4.8 Let $X \subset L^1(\Omega)$ be a closed vector space in $L^1(\Omega)$. Assume that

$$X \subset \bigcup_{1 < q \le \infty} L^q(\Omega).$$

1. Prove that there exists some $p > 1$ such that $X \subset L^p(\Omega)$.

   [**Hint:** For every integer $n \geq 1$ consider the set

   $$X_n = \left\{ f \in X \cap L^{1+(1/n)}(\Omega) \, ; \, \|f\|_{1+(1/n)} \leq n \right\}.]$$

2. Prove that there is a constant C such that

   $$\|f\|_p \leq C\|f\|_1 \quad \forall f \in X.$$

4.9 *Jensen's inequality.*

Assume $|\Omega| < \infty$. Let $j : \mathbb{R} \to (-\infty, +\infty]$ be a convex l.s.c. function, $j \not\equiv +\infty$. Let $f \in L^1(\Omega)$ be such that $f(x) \in D(j)$ a.e. and $j(f) \in L^1(\Omega)$. Prove that

$$j\left(\frac{1}{|\Omega|} \int_\Omega f\right) \leq \frac{1}{|\Omega|} \int_\Omega j(f).$$

4.10 *Convex integrands.*

Assume $|\Omega| < \infty$. Let $1 \leq p < \infty$ and let $j : \mathbb{R} \to \mathbb{R}$ be a convex and continuous function. Consider the function $J : L^p(\Omega) \to (-\infty, +\infty]$ defined by

$$J(u) = \begin{cases} \int_\Omega j(u(x))dx & \text{if } j(u) \in L^1(\Omega), \\ +\infty & \text{if } j(u) \notin L^1(\Omega). \end{cases}$$

1. Prove that J is convex.
2. Prove that J is l.s.c.

   [**Hint:** Start with the case $j \geq 0$ and use Fatou's lemma.]
3. Prove that the conjugate function $J^\star : L^{p'}(\Omega) \to (-\infty, +\infty]$ is given by

   $$J^\star(f) = \begin{cases} \int_\Omega j^\star(f(x))dx & \text{if } j^\star(f) \in L^1(\Omega), \\ +\infty & \text{if } j^\star(f) \notin L^1(\Omega). \end{cases}$$

   [**Hint:** When $1 < p < \infty$ consider $J_n(u) = J(u) + \frac{1}{n}\int |u|^p$ and determine $J_n^\star$.]
4. Let ∂j (resp. ∂J) denote the subdifferential of j (resp. J) (see Problem 2). Let $u \in L^p(\Omega)$ and let $f \in L^{p'}(\Omega)$; prove that

   $$f \in \partial J(u) \iff f(x) \in \partial j(u(x)) \quad \text{a.e. on } \Omega.$$

4.11 *The spaces $L^\alpha(\Omega)$ with $0 < \alpha < 1$.*

Let $0 < \alpha < 1$. Set

$$L^\alpha(\Omega) = \left\{ u : \Omega \to \mathbb{R}; \quad u \text{ is measurable and } |u|^\alpha \in L^1(\Omega) \right\}$$

and

$$[u]_\alpha = \left(\int |u|^\alpha \right)^{1/\alpha}.$$

1. Check that L^α is a vector space but that $[\ ]_\alpha$ is not a norm. More precisely, prove that if $u, v \in L^\alpha(\Omega)$, $u \geq 0$ a.e. and $v \geq 0$ a.e., then

$$[u + v]_\alpha \geq [u]_\alpha + [v]_\alpha.$$

2. Prove that

$$[u + v]_\alpha^\alpha \leq [u]_\alpha^\alpha + [v]_\alpha^\alpha \quad \forall u, v \in L^\alpha(\Omega).$$

$\boxed{4.12}$ *L^p is uniformly convex for $1 < p \leq 2$ (by the method of C. Morawetz).*

1. Let $1 < p < \infty$. Prove that there is a constant C (depending only on p) such that

$$|a - b|^p \leq C(|a|^p + |b|^p)^{1-s} \left(|a|^p + |b|^p - 2 \left| \frac{a+b}{2} \right|^p \right)^s \quad \forall a, b \in \mathbb{R},$$

where $s = p/2$.
2. Deduce that $L^p(\Omega)$ is uniformly convex for $1 < p \leq 2$.

[**Hint:** Use question 1 and Hölder's inequality.]

$\boxed{4.13}$

1. Check that

$$\big| |a + b| - |a| - |b| \big| \leq 2|b| \quad \forall a, b \in \mathbb{R}.$$

2. Let (f_n) be a sequence in $L^1(\Omega)$ such that

   (i) $f_n(x) \to f(x)$ a.e.,
   (ii) (f_n) is bounded in $L^1(\Omega)$ i.e., $\|f_n\|_1 \leq M \quad \forall n$.

Prove that $f \in L^1(\Omega)$ and that

$$\lim_{n \to \infty} \int \{ |f_n| - |f_n - f| \} = \int |f|.$$

[**Hint:** Use question 1 with $a = f_n - f$ and $b = f$, and consider the sequence $\varphi_n = \big| |f_n| - |f_n - f| - |f| \big|.$]
3. Let (f_n) be a sequence in $L^1(\Omega)$ and let f be a function in $L^1(\Omega)$ such that

   (i) $f_n(x) \to f(x)$ a.e.,
   (ii) $\|f_n\|_1 \to \|f\|$.

Prove that $\|f_n - f\|_1 = 0$.

$\boxed{4.14}$ *The theorems of Egorov and Vitali.*

Assume $|\Omega| < \infty$. Let (f_n) be a sequence of measurable functions such that $f_n \to f$ a.e. (with $|f| < \infty$ a.e.).

1. Let $\alpha > 0$ be fixed. Prove that

$$\mathrm{meas}[|f_n - f| > \alpha]] \xrightarrow[n \to \infty]{} 0.$$

2. More precisely, let

$$S_n(\alpha) = \bigcup_{k \geq n} [|f_k - f| > \alpha].$$

Prove that $|S_n(\alpha)| \xrightarrow[n \to \infty]{} 0$.

3. (*Egorov*). Prove that

$$\begin{cases} \forall \delta > 0 \quad \exists A \subset \Omega \quad \text{measurable such that} \\ |A| < \delta \text{ and } f_n \to f \text{ uniformly on } \Omega \backslash A. \end{cases}$$

[**Hint:** Given an integer $m \geq 1$, prove with the help of question 2 that there exists $\Sigma_m \subset \Omega$, measurable, such that $|\Sigma_m| < \delta/2^m$ and there exists an integer N_m such that

$$|f_k(x) - f(x)| < \frac{1}{m} \quad \forall k \geq N_m, \quad \forall x \in \Omega \backslash \Sigma_m.]$$

4. (*Vitali*). Let (f_n) be a sequence in $L^p(\Omega)$ with $1 \leq p < \infty$. Assume that

   (i) $\forall \varepsilon > 0 \;\; \exists \delta > 0$ such that $\int_A |f_n|^p < \varepsilon \;\;\; \forall n$ and $\forall A \subset \Omega$ measurable with $|A| < \delta$.

   (ii) $f_n \to f$ a.e.

   Prove that $f \in L^p(\Omega)$ and that $f_n \to f$ in $L^p(\Omega)$.

4.15 *Let $\Omega = (0, 1)$.*

1. Consider the sequence (f_n) of functions defined by $f_n(x) = ne^{-nx}$. Prove that

   (i) $f_n \to 0$ a.e.
   (ii) f_n is bounded in $L^1(\Omega)$.
   (iii) $f_n \nrightarrow 0$ in $L^1(\Omega)$ strongly.
   (iv) $f_n \nrightarrow 0$ weakly $\sigma(L^1, L^\infty)$.

   More precisely, there is no subsequence that converges weakly $\sigma(L^1, L^\infty)$.

2. Let $1 < p < \infty$ and consider the sequence (g_n) of functions defined by $g_n(x) = n^{1/p} e^{-nx}$. Prove that

   (i) $g_n \to 0$ a.e.
   (ii) (g_n) is bounded in $L^p(\Omega)$.
   (iii) $g_n \nrightarrow 0$ in $L^p(\Omega)$ strongly.
   (iv) $g_n \rightharpoonup 0$ weakly $\sigma(L^p, L^{p'})$.

4.16 Let $1 < p < \infty$. Let (f_n) be a sequence in $L^p(\Omega)$ such that

(i) f_n is bounded in $L^p(\Omega)$.
(ii) $f_n \to f$ a.e. on Ω.

1. Prove that $f_n \rightharpoonup f$ weakly $\sigma(L^p, L^{p'})$.
   [**Hint:** First show that if $f_n \rightharpoonup \tilde{f}$ weakly $\sigma(L^p, L^{p'})$ and $f_n \to f$ a.e., then $f = \tilde{f}$ a.e. (use Exercise 3.4).]

2. Same conclusion if assumption (ii) is replaced by

   (ii') $\|f_n - f\|_1 \to 0$.

3. Assume now (i), (ii), and $|\Omega| < \infty$. Prove that $\|f_n - f\|_q \to 0$ for every q with $1 \leq q < p$.
   [**Hint:** Introduce the truncated functions $T_k f_n$ or alternatively use Egorov's theorem.]

4.17 *Brezis–Lieb's lemma.*
   Let $1 < p < \infty$.

1. Prove that there is a constant C (depending on p) such that

$$\left| |a + b|^p - |a|^p - |b|^p \right| \leq C \left(|a|^{p-1}|b| + |a|\,|b|^{p-1} \right) \quad \forall a, b \in \mathbb{R}.$$

2. Let (f_n) be a bounded sequence in $L^p(\Omega)$ such that $f_n \to f$ a.e. on Ω. Prove that $f \in L^p(\Omega)$ and that

$$\lim_{n \to \infty} \int_\Omega \left\{ |f_n|^p - |f_n - f|^p \right\} = \int_\Omega |f|^p.$$

   [**Hint:** Use question 1 with $a = f_n - f$ and $b = f$. Note that by Exercise 4.16, $|f_n - f| \rightharpoonup 0$ weakly in L^p and $|f_n - f|^{p-1} \rightharpoonup 0$ weakly in $L^{p'}$.]

3. Deduce that if (f_n) is a sequence in $L^p(\Omega)$ satisfying

   (i) $f_n(x) \to f(x)$ a.e.,
   (ii) $\|f_n\|_p \to \|f\|_p$,

   then $\|f_n - f\|_p \to 0$.
4. Find an alternative method for question 3.

4.18 *Rademacher's functions.*
   Let $1 \leq p \leq \infty$ and let $f \in L^p_{\text{loc}}(\mathbb{R})$. Assume that f is T-periodic, i.e., $f(x+T) = f(x)$ a.e. $x \in \mathbb{R}$.
   Set

$$\overline{f} = \frac{1}{T} \int_0^T f(t)dt.$$

Consider the sequence (u_n) in $L^p(0, 1)$ defined by

$$u_n(x) = f(nx), \quad x \in (0, 1).$$

1. Prove that $u_n \rightharpoonup \overline{f}$ in $L^p(0, 1)$ with respect to the topology $\sigma(L^p, L^{p'})$.
2. Determine $\lim_{n \to \infty} \|u_n - \overline{f}\|_p$.
3. Examine the following examples:

   (i) $u_n(x) = \sin nx$,
   (ii) $u_n(x) = f(nx)$ where f is 1-periodic and

$$f(x) = \begin{cases} \alpha & \text{for } x \in (0, 1/2), \\ \beta & \text{for } x \in (1/2, 1). \end{cases}$$

The functions of example (ii) are called *Rademacher's functions*.

4.19

1. Let (f_n) be a sequence in $L^p(\Omega)$ with $1 < p < \infty$ and let $f \in L^p(\Omega)$. Assume that

   (i) $f_n \rightharpoonup f$ weakly $\sigma(L^p, L^{p'})$,
   (ii) $\|f_n\|_p \to \|f\|_p$.

   Prove that $f_n \to f$ strongly in $L^p(\Omega)$.
2. Construct a sequence (f_n) in $L^1(0, 1)$, $f_n \geq 0$, such that:

   (i) $f_n \rightharpoonup f$ weakly $\sigma(L^1, L^\infty)$,
   (ii) $\|f_n\|_1 \to \|f\|_1$,
   (iii) $\|f_n - f\|_1 \nrightarrow 0$.

Compare with the results of Exercise 4.13 and with Proposition 3.32.

4.20 Assume $|\Omega| < \infty$. Let $1 \leq p < \infty$ and $1 \leq q < \infty$.
Let $a : \mathbb{R} \to \mathbb{R}$ be a continuous function such that

$$|a(t)| \leq C\{|t|^{p/q} + 1\} \quad \forall t \in \mathbb{R}.$$

Consider the (nonlinear) map $A : L^p(\Omega) \to L^q(\Omega)$ defined by

$$(Au)(x) = a(u(x)), \quad x \in \Omega.$$

1. Prove that A is continuous from $L^p(\Omega)$ strong into $L^q(\Omega)$ strong.
2. Take $\Omega = (0, 1)$ and assume that for every sequence (u_n) such that $u_n \rightharpoonup u$ weakly $\sigma(L^p, L^{p'})$ then $Au_n \rightharpoonup Au$ weakly $\sigma(L^q, L^{q'})$.
   Prove that a is an affine function.

   [**Hint:** Use Rademacher's functions; see Exercise 4.18.]

4.21 Given a function $u_0 : \mathbb{R} \to \mathbb{R}$, set $u_n(x) = u_0(x + n)$.

1. Assume $u_0 \in L^p(\mathbb{R})$ with $1 < p < \infty$. Prove that $u_n \rightharpoonup 0$ in $L^p(\mathbb{R})$ with respect to the weak topology $\sigma(L^p, L^{p'})$.

2. Assume $u_0 \in L^\infty(\mathbb{R})$ and that $u_0(x) \to 0$ as $|x| \to \infty$ in the following weak
   sense:

   for every $\delta > 0$ the set $[|u_0| > \delta]$ has finite measure.

   Prove that $u_n \overset{\star}{\rightharpoonup} 0$ in $L^\infty(\mathbb{R})$ weak* $\sigma(L^\infty, L^1)$.

3. Take $u_0 = \chi_{(0,1)}$.
   Prove that there exists no subsequence (u_{n_k}) that converges in $L^1(\mathbb{R})$ with respect
   to $\sigma(L^1, L^\infty)$.

4.22

1. Let (f_n) be a sequence in $L^p(\Omega)$ with $1 < p \leq \infty$ and let $f \in L^p(\Omega)$.
   Show that the following properties are equivalent:
   (A) $f_n \rightharpoonup f$ in $\sigma(L^p, L^{p'})$.
   (B) $\begin{cases} \|f_n\|_p \leq C \\ \text{and} \\ \int_E f_n \to \int_E f \;\; \forall E \subset \Omega, \; E \text{ measurable and } |E| < \infty. \end{cases}$
2. If $p = 1$ and $|\Omega| < \infty$ prove that (A) $\Leftrightarrow$ (B).
3. Assume $p = 1$ and $|\Omega| = \infty$. Prove that (A) $\Rightarrow$ (B).
   Construct an example showing that in general, (B) $\not\Rightarrow$ (A).
   [**Hint:** Use Exercise 4.21, question 3.]
4. Let (f_n) be a sequence in $L^1(\Omega)$ and let $f \in L^1(\Omega)$ with $|\Omega| = \infty$. Assume that

   (a) $f_n \geq 0 \;\; \forall n$ and $f \geq 0$ a.e. on Ω,
   (b) $\int_\Omega f_n \to \int_\Omega f$,
   (c) $\int_E f_n \to \int_E f \;\; \forall E \subset \Omega, \; E$ measurable and $|E| < \infty$.

   Prove that $f_n \rightharpoonup f$ in $L^1(\Omega)$ weakly $\sigma(L^1, L^\infty)$.
   [**Hint:** Show that $\int_F f_n \to \int_F f \;\; \forall F \subset \Omega, \; F$ measurable and $|F| \leq \infty$.]

4.23 Let $f : \Omega \to \mathbb{R}$ be a measurable function and let $1 \leq p \leq \infty$. The purpose
of this exercise is to show that the set

$$C = \big\{ u \in L^p(\Omega) ; \;\; u \geq f \quad \text{a.e.} \big\}$$

is closed in $L^p(\Omega)$ with respect to the topology $\sigma(L^p, L^{p'})$.

1. Assume first that $1 \leq p < \infty$. Prove that C is convex and closed in the strong
   L^p topology. Deduce that C is closed in $\sigma(L^p, L^{p'})$.
2. Taking $p = \infty$, prove that

$$C = \left\{ u \in L^\infty(\Omega) \,\middle|\, \begin{array}{l} \int u\varphi \geq \int f\varphi \quad \forall \varphi \in L^1(\Omega) \\ \text{with} \;\; f\varphi \in L^1(\Omega) \;\; \text{and} \;\; \varphi \geq 0 \;\; \text{a.e.} \end{array} \right\}.$$

[**Hint:** Assume first that $f \in L^\infty(\Omega)$; in the general case introduce the sets
$\omega_n = [|f| < n]$.]

3. Deduce that when $p = \infty$, C is closed in $\sigma(L^\infty, L^1)$.
4. Let $f_1, f_2 \in L^\infty(\Omega)$ with $f_1 \leq f_2$ a.e. Prove that the set

$$C = \{u \in L^\infty(\Omega) ; \quad f_1 \leq u \leq f_2 \quad \text{a.e.}\}$$

is compact in $L^\infty(\Omega)$ with respect to the topology $\sigma(L^\infty, L^1)$.

4.24 Let $u \in L^\infty(\mathbb{R}^N)$. Let (ρ_n) be a sequence of mollifiers. Let (ζ_n) be a sequence in $L^\infty(\mathbb{R}^N)$ such that

$$\|\zeta_n\|_\infty \leq 1 \quad \forall n \quad \text{and} \quad \zeta_n \to \zeta \text{ a.e. on } \mathbb{R}^N.$$

Set

$$v_n = \rho_n \star (\zeta_n u) \quad \text{and} \quad v = \zeta u.$$

1. Prove that $v_n \overset{\star}{\rightharpoonup} v$ in $L^\infty(\mathbb{R}^N)$ weak* $\sigma(L^\infty, L^1)$.
2. Prove that $\int_B |v_n - v| \to 0$ for every ball B.

4.25 *Regularization of functions in $L^\infty(\Omega)$.*
Let $\Omega \subset \mathbb{R}^N$ be open.

1. Let $u \in L^\infty(\Omega)$. Prove that there exists a sequence (u_n) in $C_c^\infty(\Omega)$ such that

   (a) $\|u_n\|_\infty \leq \|u\|_\infty \ \forall n$,
   (b) $u_n \to u$ a.e. on Ω,
   (c) $u_n \overset{\star}{\rightharpoonup} u$ in $L^\infty(\Omega)$ weak* $\sigma(L^\infty, L^1)$.

2. If $u \geq 0$ a.e. on Ω, show that one can also take

   (d) $u_n \geq 0$ on Ω $\forall n$.

3. Deduce that $C_c^\infty(\Omega)$ is dense in $L^\infty(\Omega)$ with respect to the topology $\sigma(L^\infty, L^1)$.

4.26 Let $\Omega \subset \mathbb{R}^N$ be open and let $f \in L^1_{\text{loc}}(\Omega)$.

1. Prove that $f \in L^1(\Omega)$ iff

$$A = \sup \left\{ \int f\varphi ; \varphi \in C_c(\Omega), \quad \|\varphi\|_\infty \leq 1 \right\} < \infty.$$

If $f \in L^1(\Omega)$ show that $A = \|f\|_1$.
2. Prove that $f^+ \in L^1(\Omega)$ iff

$$B = \sup \left\{ \int f\varphi ; \varphi \in C_c(\Omega), \quad \|\varphi\|_\infty \leq 1 \quad \text{and } \varphi \geq 0 \right\} < \infty.$$

If $f^+ \in L^1(\Omega)$ show that $B = \|f^+\|_1$.
3. Same questions when $C_c(\Omega)$ is replaced by $C_c^\infty(\Omega)$.

4. Deduce that

$$\left[\int f\varphi = 0 \quad \forall \varphi \in C_c^\infty(\Omega)\right] \Longrightarrow [f = 0 \quad \text{a.e.}]$$

and

$$\left[\int f\varphi \geq 0 \quad \forall \varphi \in C_c^\infty(\Omega), \varphi \geq 0\right] \Longrightarrow [f \geq 0 \quad \text{a.e.}].$$

$\boxed{4.27}$ Let $\Omega \subset \mathbb{R}^N$ be open. Let $u, v \in L_{\text{loc}}^1(\Omega)$ with $u \neq 0$ a.e. on a set of positive measure. Assume that

$$\left[\varphi \in C_c^\infty(\Omega) \text{ and } \int u\varphi > 0\right] \Longrightarrow \left[\int v\varphi \geq 0\right].$$

Prove that there exists a constant $\lambda \geq 0$ such that $v = \lambda u$.

$\boxed{4.28}$ Let $\rho \in L^1(\mathbb{R}^N)$ with $\int \rho = 1$. Set $\rho_n(x) = n^N \rho(nx)$. Let $f \in L^p(\mathbb{R}^N)$ with $1 \leq p < \infty$. Prove that $\rho_n \star f \to f$ in $L^p(\mathbb{R}^N)$.

$\boxed{4.29}$ Let $K \subset \mathbb{R}^N$ be a compact subset. Prove that there exists a sequence of functions (u_n) in $C_c^\infty(\mathbb{R}^N)$ such that

(a) $0 \leq u_n \leq 1$ on $\mathbb{R}^N$,
(b) $u_n = 1$ on K,
(c) $\text{supp } u_n \subset K + B(0, 1/n)$,
(d) $|D^\alpha u_n(x)| \leq C_\alpha n^{|\alpha|} \, \forall x \in \mathbb{R}^N$, $\forall$ multi-index α (where C_α depends only on α and not on n).

[**Hint:** Let χ_n be the characteristic function of $K + B(0, 1/2n)$; take $u_n = \rho_{2n} \star \chi_n$.]

$\boxed{4.30}$ *Young's inequality.*
  Let $1 \leq p \leq \infty$, $1 \leq q \leq \infty$ be such that $\frac{1}{p} + \frac{1}{q} \geq 1$.
  Set $\frac{1}{r} = \frac{1}{p} + \frac{1}{q} - 1$, so that $1 \leq r \leq \infty$.
  Let $f \in L^p(\mathbb{R}^N)$ and $g \in L^q(\mathbb{R}^N)$.

1. Prove that for a.e. $x \in \mathbb{R}^N$, the function $y \mapsto f(x - y) g(y)$ is integrable on $\mathbb{R}^N$.

   [**Hint:** Set $\alpha = p/q'$, $\beta = q/p'$ and write

   $$|f(x - y)g(y)| = |f(x - y)|^\alpha |g(y)|^\beta \left(|f(x - y)|^{1-\alpha}|g(y)|^{1-\beta}\right).]$$

2. Set

   $$(f \star g)(x) = \int_{\mathbb{R}^N} f(x - y)g(y)dy.$$

   Prove that $f \star g \in L^r(\mathbb{R}^N)$ and that $\|f \star g\|_r \leq \|f\|_p \|g\|_q$.
3. Assume here that $\frac{1}{p} + \frac{1}{q} = 1$. Prove that

   $$f \star g \in C(\mathbb{R}^N) \cap L^\infty(\mathbb{R}^N)$$

and, moreover, if $1 < p < \infty$ then $(f \star g)(x) \to 0$ as $|x| \to \infty$.

4.31 Let $f \in L^p(\mathbb{R}^N)$ with $1 \le p < \infty$. For every $r > 0$ set

$$f_r(x) = \frac{1}{|B(x,r)|} \int_{B(x,r)} f(y)dy, \ x \in \mathbb{R}^N.$$

1. Prove that $f_r \in L^p(\mathbb{R}^N) \cap C(\mathbb{R}^N)$ and that $f_r(x) \to 0$ as $|x| \to \infty$ (r being fixed).
2. Prove that $f_r \to f$ in $L^p(\mathbb{R}^N)$ as $r \to 0$.

[**Hint:** Write $f_r = \varphi_r \star f$ for some appropriate φ_r.]

4.32

1. Let $f, g \in L^1(\mathbb{R}^N)$ and let $h \in L^p(\mathbb{R}^N)$ with $1 \le p \le \infty$. Show that $f \star g = g \star f$ and $(f \star g) \star h = f \star (g \star h)$.
2. Let $f \in L^1(\mathbb{R}^N)$. Assume that $f \star \varphi = 0 \ \ \forall \varphi \in C_c^\infty(\mathbb{R}^N)$. Prove that $f = 0$ a.e. on $\mathbb{R}^N$. Same question for $f \in L^1_{\text{loc}}(\mathbb{R}^N)$.
3. Let $a \in L^1(\mathbb{R}^N)$ be a fixed function. Consider the operator $T_a : L^2(\mathbb{R}^N) \to L^2(\mathbb{R}^N)$ defined by

$$T_a(u) = a \star u.$$

Check that T_a is bounded and that $\|T_a\|_{\mathcal{L}(L^2)} \le \|a\|_{L^1(\mathbb{R}^N)}$. Compute $T_a \circ T_b$ and prove that $T_a \circ T_b = T_b \circ T_a \ \ \forall a, b \in L^1(\mathbb{R}^N)$. Determine $(T_a)^\star$, $T_a \circ (T_a)^\star$ and $(T_a)^\star \circ T_a$. Under what condition on a is $(T_a)^\star = T_a$?

4.33 Fix a function $\varphi \in C_c(\mathbb{R})$, $\varphi \not\equiv 0$, and consider the family of functions

$$\mathcal{F} = \bigcup_{n=1}^\infty \{\varphi_n\},$$

where $\varphi_n(x) = \varphi(x+n)$, $x \in \mathbb{R}$.

1. Assume $1 \le p < \infty$. Prove that $\forall \varepsilon > 0 \ \exists \delta > 0$ such that

$$\|\tau_h f - f\|_p < \varepsilon \ \ \forall f \in \mathcal{F} \text{ and } \forall h \in \mathbb{R} \text{ with } |h| < \delta.$$

2. Prove that $\mathcal{F}$ does *not* have compact closure in $L^p(\mathbb{R})$.

4.34 Let $1 \le p < \infty$ and let $\mathcal{F} \subset L^p(\mathbb{R}^N)$ be a compact subset of $L^p(\mathbb{R}^N)$.

1. Prove that $\mathcal{F}$ is bounded in $L^p(\mathbb{R}^N)$.
2. Prove that $\forall \varepsilon > 0 \ \ \exists \delta > 0$ such that

$$\|\tau_h f - f\|_p < \varepsilon \ \ \forall f \in \mathcal{F} \text{ and } \forall h \in \mathbb{R}^N \text{ with } |h| < \delta.$$

3. Prove that $\forall \varepsilon > 0 \ \ \exists \Omega \subset \mathbb{R}^N$ bounded, open, such that

$$\|f\|_{L^p(\mathbb{R}^N\setminus\Omega)} < \varepsilon \quad \forall f \in \mathcal{F}.$$

Compare with Corollary 4.27.

4.35 Fix a function $G \in L^p(\mathbb{R}^N)$ with $1 \le p < \infty$ and let $\mathcal{F} = G \star \mathcal{B}$, where $\mathcal{B}$ is a bounded set in $L^1(\mathbb{R}^N)$.

Prove that $\mathcal{F}_{|\Omega}$ has compact closure in $L^p(\Omega)$ for any measurable set $\Omega \subset \mathbb{R}^N$ with finite measure. Compare with Corollary 4.28.

4.36 *Equi-integrable families.*

A subset $\mathcal{F} \subset L^1(\Omega)$ is said to be *equi-integrable* if it satisfies the following properties:[6]

(a) $\mathcal{F}$ is bounded in $L^1(\Omega)$,

(b) $\begin{cases} \forall \varepsilon > 0 \quad \exists \delta > 0 \quad \text{such that } \int_E |f| < \varepsilon \\ \forall f \in \mathcal{F}, \quad \forall E \subset \Omega, \ E \text{ measurable and } |E| < \delta, \end{cases}$

(c) $\begin{cases} \forall \varepsilon > 0 \quad \exists \omega \subset \Omega \text{ measurable with } |\omega| < \infty \\ \text{such that } \int_{\Omega\setminus\omega} |f| < \varepsilon \quad \forall f \in \mathcal{F}. \end{cases}$

Let (Ω_n) be a nondecreasing sequence of measurable sets in Ω with $|\Omega_n| < \infty$ $\forall n$ and such that $\Omega = \bigcup_n \Omega_n$.

1. Prove that $\mathcal{F}$ is equi-integrable iff

(d) $$\lim_{t\to\infty} \sup_{f\in\mathcal{F}} \int_{[|f|>t]} |f| = 0$$

and

(e) $$\lim_{n\to\infty} \sup_{f\in\mathcal{F}} \int_{\Omega\setminus\Omega_n} |f| = 0.$$

2. Prove that if $\mathcal{F} \subset L^1(\Omega)$ is compact, then $\mathcal{F}$ is equi-integrable. Is the converse true?

4.37 Fix a function $f \in L^1(\mathbb{R})$ such that

$$\int_{-\infty}^{+\infty} f(t)dt = 0 \quad \text{and} \quad \int_0^{+\infty} f(t)dt > 0,$$

and let $u_n(x) = nf(nx)$ for $x \in I = (-1, +1)$.

1. Prove that

$$\lim_{n\to\infty} \int_I u_n(x)\varphi(x)dx = 0 \quad \forall \varphi \in C([-1, +1]).$$

---

[6] One can show that (a) follows from (b) and (c) if the measure space Ω is diffuse (i.e., Ω has no atoms). Consider for example $\Omega = \mathbb{R}^N$ with the Lebesgue measure.

2. Check that the sequence (u_n) is bounded in $L^1(I)$. Show that no subsequence of (u_n) is equi-integrable.
3. Prove that there exists no function $u \in L^1(I)$ such that

$$\lim_{k \to \infty} \int_I u_{n_k}(x)\varphi(x)dx = \int_I u(x)\varphi(x)dx \quad \forall \varphi \in L^\infty(I),$$

along some subsequence (u_{n_k}).
4. Compare with the Dunford–Pettis theorem (see question A3 in Problem 23).
5. Prove that there exists a subsequence (u_{n_k}) such that $u_{n_k}(x) \to 0$ a.e. on I as $k \to \infty$.

[**Hint:** Compute $\int_{[n^{-1/2} < |x| < 1]} |u_n(x)|dx$ and apply Theorem 4.9.]

4.38 Set $I = (0, 1)$ and consider the sequence (u_n) of functions in $L^1(I)$ defined by

$$u_n(x) = \begin{cases} n & \text{if } x \in \bigcup_{j=0}^{n-1} \left(\frac{j}{n}, \frac{j}{n} + \frac{1}{n^2}\right), \\ 0 & \text{otherwise.} \end{cases}$$

1. Check that $|\operatorname{supp} u_n| = \frac{1}{n}$ and $\|u_n\|_1 = 1$.
2. Prove that

$$\lim_{n \to +\infty} \int_I u_n(x)\varphi(x)dx = \int_I \varphi(x)dx \quad \forall \varphi \in C([0, 1]).$$

[**Hint:** Start with the case $\varphi \in C^1([0, 1])$.]

3. Show that no subsequence of (u_n) is equi-integrable.
4. Prove that there exists no function $u \in L^1(I)$ such that

$$\lim_{k \to \infty} \int_I u_{n_k}(x)\varphi(x)dx = \int_I u(x)\varphi(x)dx \quad \forall \varphi \in L^\infty(I),$$

along some subsequence (u_{n_k}).

[**Hint:** Use a further subsequence $(u_{n_k'})$ such that $\sum_k |\operatorname{supp} u_{n_k'}| < 1$.]

5. Prove that there exists a subsequence (u_{n_k}) such that $u_{n_k}(x) \to 0$ a.e. on I as $k \to \infty$.

Chapter 5
Hilbert Spaces

5.1 Definitions and Elementary Properties. Projection onto a Closed Convex Set

Definition. Let H be a vector space. A *scalar product* (u, v) is a bilinear form on $H \times H$ with values in $\mathbb{R}$ (i.e., a map from $H \times H$ to $\mathbb{R}$ that is linear in both variables) such that

$$(u, v) = (v, u) \quad \forall u, v \in H \quad \text{(symmetry)},$$
$$(u, u) \geq 0 \quad\quad \forall u \in H \quad\quad \text{(positive)},$$
$$(u, u) \neq 0 \quad\quad \forall u \neq 0 \quad\quad \text{(definite)}.$$

Let us recall that a scalar product satisfies the Cauchy–Schwarz inequality

$$|(u, v)| \leq (u, u)^{1/2}(v, v)^{1/2} \quad \forall u, v \in H.$$

[It is sometimes useful to keep in mind that the proof of the Cauchy–Schwarz inequality does not require the assumption $(u, u) \neq 0 \; \forall u \neq 0$.] It follows from the Cauchy–Schwarz inequality that the quantity

$$\boxed{|u| = (u, u)^{1/2}}$$

is a norm—we shall often denote by $|\;|$ (instead of $\|\;\|$) norms arising from scalar products. Indeed, we have

$$|u + v|^2 = (u + v, u + v) = |u|^2 + (u, v) + (v, u) + |v|^2 \leq |u|^2 + 2|u|\,|v| + |v|^2,$$

and thus $|u + v| \leq |u| + |v|$.

Let us recall the classical *parallelogram law*:

$$(1) \qquad \left|\frac{a+b}{2}\right|^2 + \left|\frac{a-b}{2}\right|^2 = \frac{1}{2}(|a|^2 + |b|^2) \quad \forall a, b \in H.$$

H. Brezis, *Functional Analysis, Sobolev Spaces and Partial Differential Equations*,
DOI 10.1007/978-0-387-70914-7_5, © Springer Science+Business Media, LLC 2011

Definition. A *Hilbert space* is a vector space H equipped with a scalar product such that H is *complete* for the norm $|\ |$.

In what follows, H *will always denote a Hilbert space.*

Basic example. $L^2(\Omega)$ equipped with the scalar product

$$(u, v) = \int_\Omega u(x)v(x)d\mu$$

is a Hilbert space. In particular, ℓ^2 is a Hilbert space. The Sobolev space H^1 studied in Chapters 8 and 9 is another example of a Hilbert space; it is "modeled" on $L^2(\Omega)$.

• **Proposition 5.1.** H *is uniformly convex, and thus it is reflexive.*

Proof. Let $\varepsilon > 0$ and $u, v \in H$ satisfy $|u| \leq 1, |v| \leq 1$, and $|u - v| > \varepsilon$. In view of the parallelogram law we have

$$\left|\frac{u+v}{2}\right|^2 < 1 - \frac{\varepsilon^2}{4} \text{ and thus } \left|\frac{u+v}{2}\right| < 1 - \delta \text{ with } \delta = 1 - \left(1 - \frac{\varepsilon^2}{4}\right)^{1/2} > 0.$$

• **Theorem 5.2 (projection onto a closed convex set).** *Let $K \subset H$ be a nonempty closed convex set. Then for every $f \in H$ there exists a unique element $u \in K$ such that*

$$(2) \qquad |f - u| = \min_{v \in K} |f - v| = \operatorname{dist}(f, K).$$

*Moreover, u is **characterized** by the property*

$$(3) \qquad u \in K \text{ and } (f - u, v - u) \leq 0 \quad \forall v \in K.$$

Notation. The above element u is called the *projection* of f onto K and is denoted by

$$\boxed{u = P_K f.}$$

Inequality (3) says that the scalar product of the vector $\overrightarrow{uf}$ with any vector $\overrightarrow{uv}$ ($v \in K$) is ≤ 0, i.e., the angle θ determined by these two vectors is $\geq \pi/2$; see Figure 4.

Proof. (a) *Existence.* We shall present two different proofs:

1. The function $\varphi(v) = |f - v|$ is convex, continuous and $\lim_{|v| \to \infty} \varphi(v) = +\infty$. It follows from Corollary 3.23 that φ achieves its minimum on K since H is *reflexive.*

2. The second proof does *not* rely on the theory of reflexive and uniformly convex spaces. It is a *direct* argument. Let (v_n) be a *minimizing sequence* for (2), i.e., $v_n \in K$ and

$$d_n = |f - v_n| \to d = \inf_{v \in K} |f - v|.$$

We claim that (v_n) is a *Cauchy sequence.* Indeed, the parallelogram law applied with $a = f - v_n$ and $b = f - v_m$ leads to

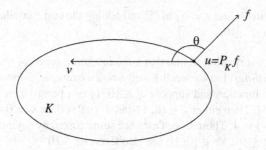

Fig. 4

$$\left| f - \frac{v_n + v_m}{2} \right|^2 + \left| \frac{v_n - v_m}{2} \right|^2 = \frac{1}{2}(d_n^2 + d_m^2).$$

But $\frac{v_n + v_m}{2} \in K$ and thus $\left| f - \frac{v_n + v_m}{2} \right| \geq d$. It follows that

$$\left| \frac{v_n - v_m}{2} \right|^2 \leq \frac{1}{2}(d_n^2 + d_m^2) - d^2 \text{ and } \lim_{m,n \to \infty} |v_n - v_m| = 0.$$

Therefore the sequence (v_n) converges to some limit $u \in K$ with $d = |f - u|$.

(b) *Equivalence of* (2) *and* (3).
Assume that $u \in K$ satisfies (2) and let $w \in K$. We have

$$v = (1 - t)u + tw \in K \quad \forall t \in [0, 1]$$

and thus

$$|f - u| \leq |f - [(1 - t)u + tw]| = |(f - u) - t(w - u)|.$$

Therefore

$$|f - u|^2 \leq |f - u|^2 - 2t(f - u, w - u) + t^2|w - u|^2,$$

which implies that $2(f - u, w - u) \leq t|w - u|^2 \quad \forall t \in (0, 1]$. As $t \to 0$ we obtain (3).
Conversely, assume that u satisfies (3). Then we have

$$|u - f|^2 - |v - f|^2 = 2(f - u, v - u) - |u - v|^2 \leq 0 \quad \forall v \in K;$$

which implies (2).
(c) *Uniqueness.*
Assume that u_1 and u_2 satisfy (3). We have

(4) $(f - u_1, v - u_1) \leq 0 \quad \forall v \in K,$

(5) $(f - u_2, v - u_2) \leq 0 \quad \forall v \in K.$

Choosing $v = u_2$ in (4) and $v = u_1$ in (5) and adding the corresponding inequalities, we obtain $|u_1 - u_2|^2 \leq 0$.

Remark 1. It is not surprising to find that a *minimization problem* is connected with a *system of inequalities*. Let us recall a well-known example. Suppose $F : \mathbb{R} \to \mathbb{R}$ is a differentiable function and suppose $u \in [0, 1]$ is a point where F achieves its minimum on $[0, 1]$. Then either $u \in (0, 1)$ and $F'(u) = 0$, or $u = 0$ and $F'(u) \leq 0$, or $u = 1$ and $F'(u) = 1$. These three cases are summarized by saying that $u \in [0, 1]$ and $F'(u)(v - u) \leq 0 \quad \forall v \in [0, 1]$; see also Exercise 5.10.

Remark 2. Let $K \subset E$ be a nonempty closed convex set in a uniformly convex Banach space E. Then for every $f \in E$ there exists a unique element $u \in E$ such that
$$\|f - u\| = \min_{v \in K} \|f - v\| = \text{dist}(f, K);$$
see Exercise 3.32.

Proposition 5.3. *Let $K \subset H$ be a nonempty closed convex set. Then P_K does not increase distance, i.e.,*
$$|P_K f_1 - P_K f_2| \leq |f_1 - f_2| \quad \forall f_1, f_2 \in H.$$

Proof. Set $u_1 = P_K f_1$ and $u_2 = P_K f_2$. We have

(6) $\qquad\qquad (f_1 - u_1, v - u_1) \leq 0 \quad \forall v \in K$
(7) $\qquad\qquad (f_2 - u_2, v - u_2) \leq 0 \quad \forall v \in K.$

Choosing $v = u_2$ in (6) and $v = u_1$ in (5) and adding the corresponding inequalities, we obtain
$$|u_1 - u_2|^2 \leq (f_1 - f_2, u_1 - u_2).$$
It follows that $|u_1 - u_2| \leq |f_1 - f_2|$.

Corollary 5.4. *Assume that $M \subset H$ is a closed linear **subspace**. Let $f \in H$. Then $u = P_M f$ is characterized by*

(8) $\qquad\boxed{u \in M \quad and \quad (f - u, v) = 0 \quad \forall v \in M.}$

*Moreover, P_M is a linear operator, called the **orthogonal projection**.*

Proof. By (3) we have
$$(f - u, v - u) \leq 0 \quad \forall v \in M$$
and thus
$$(f - u, tv - u) \leq 0 \quad \forall v \in M, \quad \forall t \in \mathbb{R}.$$
It follows that (8) holds.

Conversely, if u satisfies (8) we have

$$(f - u, v - u) = 0 \quad \forall v \in M.$$

It is obvious that P_M is linear.

5.2 The Dual Space of a Hilbert Space

It is very easy, in a Hilbert space, to write down continuous linear functionals. Pick any $f \in H$; then the map $u \mapsto (f, u)$ is a continuous linear functional on H. It is a remarkable fact that *all* continuous linear functionals on H are obtained in this fashion:

• **Theorem 5.5 (Riesz–Fréchet representation theorem).** *Given any $\varphi \in H^\star$ there exists a unique $f \in H$ such that*

$$\langle \varphi, u \rangle = (f, u) \quad \forall u \in H.$$

Moreover,

$$|f| = \|\varphi\|_{H^\star}.$$

Proof. Once more we shall present two proofs:

1. The first one is almost identical to the proof of Theorem 4.11. Consider the map $T : H \to H^\star$ defined as follows: given any $f \in H$, the map $u \mapsto (f, u)$ is a continuous linear functional on H. It defines an element of $H^\star$, which we denote by Tf, so that

$$\langle Tf, u \rangle = (f, u) \quad \forall u \in H.$$

It is clear that $\|Tf\|_{H^\star} = |f|$. Thus T is a linear isometry from H onto $T(H)$, a closed subspace of $H^\star$. In order to conclude, it suffices to show that $T(H)$ is dense in $H^\star$. Assume that h is a continuous linear functional on $H^\star$ that vanishes on $T(H)$. Since H is *reflexive*, h belongs to H and satisfies $\langle Tf, h \rangle = 0 \, \forall f \in H$. It follows that $(f, h) = 0 \, \forall f \in H$ and thus $h = 0$.

2. The second proof is a more direct argument that avoids any use of reflexivity. Let $M = \varphi^{-1}(\{0\})$, so that M is a closed subspace of H. We may always assume that $M \neq H$ (otherwise $\varphi \equiv 0$ and the conclusion of Theorem 5.5 is obvious—just take $f = 0$). We claim that there exists some element $g \in H$ such that

$$|g| = 1 \text{ and } (g, v) = 0 \quad \forall v \in M \text{ (and thus } g \notin M).$$

Indeed, let $g_0 \in H$ with $g_0 \notin M$. Let $g_1 = P_M g_0$. Then

$$g = (g_0 - g_1)/|g_0 - g_1|$$

satisfies the required properties.

Given any $u \in H$, set

$$v = u - \lambda g \qquad \text{with } \lambda = \frac{\langle \varphi, u \rangle}{\langle \varphi, g \rangle}.$$

Note that v is well defined, since $\langle \varphi, g \rangle \neq 0$, and, moreover, $v \in M$, since $\langle \varphi, v \rangle = 0$. It follows that $(g, v) = 0$, i.e.,

$$\langle \varphi, u \rangle = \langle \varphi, g \rangle (g, u) \quad \forall u \in H,$$

which concludes the proof with $f = \langle \varphi, g \rangle g$.

• *Remark* 3. **H and $H^\star$: to identify or not to identify? The triplet $V \subset H \subset V^\star$.**
    Theorem 5.5 asserts that there is a canonical isometry from H onto $H^\star$. It is therefore "legitimate" to identify H and $H^\star$. We shall *often* do so but *not always*. Here is a typical situation—which arises in many applications—where one should be cautious with identifications. Assume that H is a Hilbert space with a scalar product $(\,,\,)$ and a corresponding norm $|\;|$. Assume that $V \subset H$ is a linear subspace that is dense in H. Assume that V has its own norm $\|\;\|$ and that V is a Banach space with $\|\;\|$. Assume that the *injection* $V \subset H$ is *continuous*, i.e.,

$$|v| \leq C\|v\| \quad \forall v \in V.$$

[For example, $H = L^2(0, 1)$ and $V = L^p(0, 1)$ with $p > 2$ or $V = C([0, 1])$.]
    There is a canonical map $T : H^\star \to V^\star$ that is simply the restriction to V of continuous linear functionals φ on H, i.e.,

$$\langle T\varphi, v \rangle_{V^\star, V} = \langle \varphi, v \rangle_{H^\star, H}.$$

It is easy to see that T has the following properties:

(i) $\|T\varphi\|_{V^\star} \leq C|\varphi|_{H^\star} \quad \forall \varphi \in H^\star$,
(ii) T is injective,
(iii) $R(T)$ is dense in $V^\star$ if V is reflexive.[1]

    Identifying $H^\star$ with H and using T as a canonical embedding from $H^\star$ into $V^\star$, one usually writes

(9) $$\boxed{V \subset H \simeq H^\star \subset V^\star}\,,$$

where all the injections are continuous and dense (provided V is reflexive). One says that H is the *pivot* space. Note that the scalar products $\langle\,,\,\rangle_{V^\star, V}$ and $(\,,\,)$ coincide whenever both make sense, i.e.,

$$\langle f, v \rangle_{V^\star, V} = (f, v) \quad \forall f \in H, \quad \forall v \in V.$$

---

[1] However, T is *not surjective* in general.

The situation becomes more delicate if V turns out to be a *Hilbert space with its own scalar product* $((\,,\,))$ associated to the norm $\|\,\|$. We could, of course, identify V and $V^\star$ with the help of $((\,,\,))$. However, (9) becomes absurd. This shows that one *cannot* identify *simultaneously* V and H with their dual spaces: one has to make a choice. The common habit is to identify $H^\star$ with H, to write (9), and *not* to identify $V^\star$ with V [naturally, there is still an isometry from V onto $V^\star$, but it is not viewed as the identity map]. Here is a *very instructive* example.

Let

$$H = \ell^2 = \left\{ u = (u_n)_{n \geq 1} ; \ \sum_{n=1}^{\infty} u_n^2 < \infty \right\}$$

equipped with the scalar product $(u, v) = \sum_{n=1}^{\infty} u_n v_n$.

Let

$$V = \left\{ u = (u_n)_{n \geq 1} ; \ \sum_{n=1}^{\infty} n^2 u_n^2 < \infty \right\}$$

equipped with the scalar product $((u, v)) = \sum_{n=1}^{\infty} n^2 u_n v_n$.

Clearly $V \subset H$ with continuous injection and V is dense in H. Here we identify $H^\star$ with H, while $V^\star$ is identified with the space

$$V^\star = \left\{ f = (f_n)_{n \geq 1} ; \ \sum_{n=1}^{\infty} \frac{1}{n^2} f_n^2 < \infty \right\},$$

which is bigger than H. The scalar product $\langle\,,\,\rangle_{V^\star, V}$ is given by

$$\langle f, v \rangle_{V^\star, V} = \sum_{n=1}^{\infty} f_n v_n,$$

and the Riesz–Fréchet isomorphism $T : V \to V^\star$ is given by

$$u = (u_n)_{n \geq 1} \mapsto Tu = (n^2 u_n)_{n \geq 1}.$$

Remark 4. It is easy to prove that Hilbert spaces are reflexive without invoking the theory of uniformly convex spaces. It suffices to use twice the Riesz–Fréchet isomorphism (from H onto $H^\star$ and then from $H^\star$ onto $H^{\star\star}$).

Remark 5. Assume that H is a Hilbert space identified with its dual space $H^\star$. Let M be a subspace of H. We have already defined $M^\perp$ (in Section 1.3) as a subspace of $H^\star$. We may now consider it as a subspace of H, namely

$$M^\perp = \{ u \in H; (u, v) = 0 \quad \forall v \in M \}.$$

Clearly we have $M \cap M^\perp = \{0\}$. Moreover, if M is closed we also have $M + M^\perp = H$. Indeed, every $f \in H$ may be written as

$$f = (P_M f) + (f - P_M f)$$

and $f - P_M f \in M^\perp$; more precisely, $f - P_M f = P_{M^\perp} f$.

It follows that in a Hilbert space every closed subspace has a complement (in the sense of Section 2.4).

5.3 The Theorems of Stampacchia and Lax–Milgram

Definition. A bilinear form $a : H \times H \to \mathbb{R}$ is said to be

(i) *continuous* if there is a constant C such that

$$|a(u, v)| \le C|u|\,|v| \quad \forall u, v \in H;$$

(ii) *coercive* if there is a constant $\alpha > 0$ such that

$$a(v, v) \ge \alpha|v|^2 \quad \forall v \in H.$$

Theorem 5.6 (Stampacchia). *Assume that $a(u, v)$ is a continuous coercive bilinear form on H. Let $K \subset H$ be a nonempty closed and convex subset. Then, given any $\varphi \in H^*$, there exists a unique element $u \in K$ such that*

$$(10) \qquad a(u, v - u) \ge \langle \varphi, v - u \rangle \quad \forall v \in K.$$

Moreover, if a is symmetric, then u is characterized by the property

$$(11) \qquad \boxed{u \in K \quad and \quad \frac{1}{2}a(u, u) - \langle \varphi, u \rangle = \min_{v \in K}\left\{\frac{1}{2}a(v, v) - \langle \varphi, v \rangle\right\}.}$$

The proof of Theorem 5.6 relies on the following very classical result.

• Theorem 5.7 (Banach fixed-point theorem—the contraction mapping principle). *Let X be a nonempty complete metric space and let $S : X \to X$ be a strict contraction, i.e.,*

$$d(Sv_1, Sv_2) \le k\,d(v_1, v_2) \quad \forall v_1, v_2 \in X \text{ with } k < 1.$$

Then S has a unique fixed point, $u = Su$.

For a proof see, e.g., T. M. Apostol [1], G. Choquet [1], A. Friedman [3].

Proof of Theorem 5.6. From the Riesz–Fréchet representation theorem (Theorem 5.5) we know that there exists a unique $f \in H$ such that

$$\langle \varphi, v \rangle = (f, v) \quad \forall v \in H.$$

On the other hand, if we *fix* $u \in H$, the map $v \mapsto a(u, v)$ is a continuous linear functional on H. Using once more the Riesz–Fréchet representation theorem we find some unique element in H, denoted by Au, such that $a(u, v) = (Au, v) \; \forall v \in H$. Clearly A is a linear operator from H into H satisfying

$$(12) \qquad\qquad |Au| \le C|u| \quad \forall u \in H,$$

$$(13) \qquad\qquad (Au, u) \ge \alpha |u|^2 \quad \forall u \in H.$$

Problem (10) amounts to finding some $u \in K$ such that

$$(14) \qquad\qquad (Au, v - u) \ge (f, v - u) \quad \forall v \in K.$$

Let $\rho > 0$ be a constant (to be determined later). Note that (14) is equivalent to

$$(15) \qquad\qquad (\rho f - \rho Au + u - u, v - u) \le 0 \quad \forall v \in K,$$

i.e.,

$$u = P_K(\rho f - \rho Au + u).$$

For every $v \in K$, set $Sv = P_K(\rho f - \rho Av + v)$. We claim that if $\rho > 0$ is properly chosen then S is a strict contraction. Indeed, since P_K does not increase distance (see Proposition 5.3) we have

$$|Sv_1 - Sv_2| \le |(v_1 - v_2) - \rho(Av_1 - Av_2)|$$

and thus

$$|Sv_1 - Sv_2|^2 = |v_1 - v_2|^2 - 2\rho(Av_1 - Av_2, v_1 - v_2) + \rho^2|Av_1 - Av_2|^2$$
$$\le |v_1 - v_2|^2(1 - 2\rho\alpha + \rho^2 C^2).$$

Choosing $\rho > 0$ in such a way that $k^2 = 1 - 2\rho\alpha + \rho^2 C^2 < 1$ (i.e., $0 < \rho < 2\alpha/C^2$) we find that S has a unique fixed point.[2]

Assume now that the form $a(u, v)$ is also *symmetric*. Then $a(u, v)$ defines a *new scalar product* on H; the corresponding norm $a(u, u)^{1/2}$ is equivalent to the original norm $|u|$. It follows that H is also a Hilbert space for this new scalar product. Using the Riesz–Fréchet theorem we may now represent the functional φ through the new scalar product, i.e., there exists some unique element $g \in H$ such that

$$\langle \varphi, v \rangle = a(g, v) \quad \forall v \in H.$$

Problem (10) amounts to finding some $u \in K$ such that

$$(16) \qquad\qquad a(g - u, v - u) \le 0 \quad \forall v \in K.$$

The solution of (16) is an old friend: u is simply the *projection onto K of g for the new scalar product a*. We also know (by Theorem 5.2) that u is the unique element K that achieves

---

[2] If one has to compute the fixed point numerically, it pays to choose $\rho = \alpha/C^2$ in order to minimize k and to accelerate the convergence of the iterates of S.

$$\min_{v \in K} a(g - v, g - v)^{1/2}.$$

This amounts to minimizing on K the function

$$v \mapsto a(g - v, g - v) = a(v, v) - 2a(g, v) + a(g, g) = a(v, v) - 2\langle \varphi, v \rangle + a(g, g),$$

or equivalently the function

$$v \mapsto \frac{1}{2} a(v, v) - \langle \varphi, v \rangle.$$

Remark 6. It is easy to check that if $a(u, v)$ is a bilinear form with the property

$$a(v, v) \geq 0 \quad \forall v \in H$$

then the function $v \mapsto a(v, v)$ is convex.

• **Corollary 5.8 (Lax–Milgram).** *Assume that $a(u, v)$ is a continuous coercive bilinear form on H. Then, given any $\varphi \in H^*$, there exists a unique element $u \in H$ such that*

(17) $a(u, v) = \langle \varphi, v \rangle \quad \forall v \in H.$

Moreover, if a is symmetric, then u is characterized by the property

(18) $\boxed{ u \in H \quad and \quad \dfrac{1}{2} a(u, u) - \langle \varphi, u \rangle = \min_{v \in H} \left\{ \dfrac{1}{2} a(v, v) - \langle \varphi, v \rangle \right\}. }$

Proof. Use Theorem 5.6 with $K = H$ and argue as in the proof of Corollary 5.4. $\blacksquare$

Remark 7. The Lax–Milgram theorem is a very *simple* and *efficient tool* for solving linear elliptic partial differential equations (see Chapters 8 and 9). It is interesting to note the connection between equation (17) and the minimization problem (18). When such questions arise in mechanics or in physics they often have a natural interpretation: least action principle, minimization of the energy, etc. In the language of the *calculus of variations* one says that (17) is the *Euler equation* associated with the minimization problem (18). Roughly speaking, (17) says that "$F'(u) = 0$," where F is the function $F(v) = \frac{1}{2} a(v, v) - \langle \varphi, v \rangle$.

Remark 8. There is a direct and elementary argument proving that (17) has a unique solution. Indeed, this amounts to showing that

$$\forall f \in H \quad \exists u \in H \quad \text{unique such that } Au = f,$$

i.e., A is bijective from H onto H. This is a trivial consequence of the following facts:

(a) A is *injective* (since A is coercive),
(b) $R(A)$ is *closed*, since $\alpha |v| \leq |Av| \; \forall v \in H$ (a consequence of the coerciveness),

(c) $R(A)$ is *dense*; indeed, suppose $v \in H$ satisfies

$$(Au, v) = 0 \quad \forall u \in H,$$

then $v = 0$.

5.4 Hilbert Sums. Orthonormal Bases

Definition. Let $(E_n)_{n \geq 1}$ be a sequence of *closed* subspaces of H. One says that H is the *Hilbert sum* of the E_n's and one writes $H = \oplus_n E_n$ if

(a) the spaces E_n are mutually orthogonal, i.e.,

$$(u, v) = 0 \quad \forall u \in E_n, \quad \forall v \in E_m, \ m \neq n,$$

(b) the linear space spanned by $\bigcup_{n=1}^{\infty} E_n$ is dense in H.[3]

• **Theorem 5.9.** *Assume that H is the Hilbert sum of the E_n's. Given $u \in H$, set*

$$u_n = P_{E_n} u$$

and

$$S_n = \sum_{k=1}^{n} u_k.$$

Then we have

(19)
$$\lim_{n \to \infty} S_n = u$$

and

(20)
$$\sum_{k=1}^{\infty} |u_k|^2 = |u|^2 \quad \text{(Bessel–Parseval's identity)}.$$

It is convenient to use the following lemma.

Lemma 5.1. *Assume that (v_n) is any sequence in H such that*

(21)
$$(v_m, v_n) = 0 \quad \forall m \neq n,$$

(22)
$$\sum_{k=1}^{\infty} |v_k|^2 < \infty.$$

Set

---

[3] The linear space spanned by the E_n's is understood in the algebraic sense, i.e., *finite* linear combinations of elements belonging to the spaces (E_n).

$$S_n = \sum_{k=1}^{n} v_k.$$

Then

$$S = \lim_{n \to \infty} S_n \quad exists$$

and, moreover,

(23) $$|S|^2 = \sum_{k=1}^{\infty} |v_k|^2.$$

Proof of Lemma 5.1. Note that for $m > n$ we have

$$|S_m - S_n|^2 = \sum_{k=n+1}^{m} |v_k|^2.$$

It follows that S_n is a Cauchy sequence and thus $S = \lim_{n \to \infty} S_n$ exists. On the other hand, we have

$$|S_n|^2 = \sum_{k=1}^{n} |v_k|^2.$$

As $n \to \infty$ we obtain (23).

Proof of Theorem 5.9. Since $u_n = P_{E_n} u$, we have (by (8))

(24) $$(u - u_n, v) = 0 \quad \forall v \in E_n,$$

and in particular,

$$(u, u_n) = |u_n|^2.$$

Adding these equalities, we find that

$$(u, S_n) = \sum_{k=1}^{n} |u_k|^2.$$

But we also have

(25) $$\sum_{k=1}^{n} |u_k|^2 = |S_n|^2,$$

and thus we obtain

$$(u, S_n) = |S_n|^2.$$

It follows that $|S_n| \le |u|$ and therefore $\sum_{k=1}^{n} |u_k|^2 \le |u|^2$.

Hence, we may apply Lemma 5.1 and conclude that $S = \lim_{n \to \infty} S_n$ exists. Let us identify S even *without assumption* (b). Let F be the linear space spanned by the E_n's. We claim that

(26) $$S = P_{\overline{F}}u.$$

Indeed, we have
$$(u - S_n, v) = 0 \qquad \forall v \in E_m, \ m \leq n$$

(just write $u - S_n = (u - u_m) - \sum_{k \neq m} u_k$). As $n \to \infty$ we obtain

$$(u - S, v) = 0 \quad \forall v \in E_m, \quad \forall m$$

and thus
$$(u - S, v) = 0 \quad \forall v \in F,$$

which implies that
$$(u - S, v) = 0 \quad \forall v \in \overline{F}.$$

On the other hand, $S_n \in F \ \forall n$, and at the limit $S \in \overline{F}$. This proves (26). Of course, if (b) holds, then $\overline{F} = H$ and thus $S = u$. Passing to the limit as $n \to \infty$ in (25) we obtain (20).

Definition. A sequence $(e_n)_{n \geq 1}$ in H is said to be an *orthonormal basis* of H (or a *Hilbert basis*[4] or simply a *basis* when there is no confusion)[5] if it satisfies the following properties:

(i) $|e_n| = 1 \ \forall n$ and $(e_m, e_n) = 0 \ \forall m \neq n$,
(ii) the linear space spanned by the e_n's is dense in H.

• **Corollary 5.10.** *Let (e_n) be an orthonormal basis. Then for every $u \in H$, we have*

$$u = \sum_{k=1}^{\infty} (u, e_k)e_k, \quad i.e., \ u = \lim_{n \to \infty} \sum_{k=1}^{n} (u, e_k)e_k$$

and

$$|u|^2 = \sum_{k=1}^{\infty} |(u, e_k)|^2.$$

Conversely, given any sequence $(\alpha_n) \in \ell^2$, the series $\sum_{k=1}^{\infty} \alpha_k e_k$ converges to some element $u \in H$ such that $(u, e_k) = \alpha_k \ \forall k$ and $|u|^2 = \sum_{k=1}^{\infty} \alpha_k^2$.

Proof. Note that H is the Hilbert sum of the spaces $E_n = \mathbb{R}e_n$ and that $P_{E_n}u = (u, e_n)e_n$. Use Theorem 5.9 and Lemma 5.1.

Remark 9. In general, the series $\sum u_k$ in Theorem 5.9 and the series $\sum (u, e_k)e_k$ in Corollary 5.10 are *not absolutely convergent*, i.e., it may happen that $\sum_{k=1}^{\infty} |u_k| = \infty$ or that $\sum_{k=1}^{\infty} |(u, e_k)| = \infty$.

• **Theorem 5.11.** *Every separable Hilbert space has an orthonormal basis.*

---

[4] Not to be confused with an *algebraic* (= *Hamel*) *basis*, which is a family $(e_i)_{i \in I}$ in H such that every $u \in H$ can be uniquely written as a *finite* linear combination of the e_i's (see Exercise 1.5).
[5] Some authors say that (e_n) is a *complete orthonormal system*.

Proof. Let (v_n) be a countable dense subset of H. Let F_k denote the linear space spanned by $\{v_1, v_2, \ldots, v_k\}$. The sequence (F_k) is a nondecreasing sequence of finite-dimensional spaces such that $\bigcup_{k=1}^{\infty} F_k$ is dense in H. Pick any unit vector e_1 in F_1. If $F_2 \neq F_1$ there is some vector e_2 in F_2 such that $\{e_1, e_2\}$ is an orthonormal basis of F_2. Repeating the same construction, one obtains an orthonormal basis of H.

Remark 10. Theorem 5.11 combined with Corollary 5.10 shows that all separable Hilbert spaces are isomorphic and isometric with the space ℓ^2. Despite this seemingly spectacular result it is still very important to consider other Hilbert spaces such as $L^2(\Omega)$ (or the Sobolev space $H^1(\Omega)$, etc.). The reason is that many nice linear (or nonlinear) operators may look dreadful when they are written in a basis.

Remark 11. If H is a *nonseparable* Hilbert space—a rather unusual situation—one may still prove (with the help of Zorn's lemma) the existence of an *uncountable* orthonormal basis $(e_i)_{i \in I}$; see, e.g., W. Rudin [2], A. E. Taylor–D. C. Lay [1], G. B. Folland [2], G. Choquet [1].

Comments on Chapter 5

1. Characterization of Hilbert spaces.
It is sometimes useful to know whether a given norm $\| \ \|$ on a vector space E is a Hilbert norm, i.e., whether there exists a scalar product $(,)$ on E such that $\|u\| = (u, u)^{1/2} \ \forall u \in E$. Various criteria are known:

(a) **Theorem 5.12 (Fréchet–von Neumann–Jordan).** *Assume that the norm $\| \ \|$ satisfies the parallelogram law* (1). *Then $\| \ \|$ is a Hilbert norm.*
For a proof see K. Yosida [1] or Exercise 5.1.

(b) **Theorem 5.13 (Kakutani [1]).** *Assume that E is a normed space with* dim $E \geq$ 3. *Assume that every subspace F of dimension 2 has a projection operator of norm 1 (i.e., there exists a bounded linear projection operator $P : E \to F$ such that $Pu = u \ \forall u \in F$ and $\|P\| \leq 1$).*[6] *Then $\| \ \|$ is a Hilbert norm.*

(c) **Theorem 5.14 (de Figueiredo–Karlovitz [1]).** *Let E be a normed space with* dim $E \geq 3$. *Consider the radial projection on the unit ball, i.e.,*

$$
Tu = \begin{cases} u & \text{if } \|u\| \leq 1, \\ u/\|u\| & \text{if } \|u\| > 1. \end{cases}
$$

Assume[7] *that*

---

[6] Let us point out that every subspace of dimension 1 has always a projection operator of norm 1. (Use Hahn–Banach.)

[7] One can show that in an *arbitrary* normed space, T satisfies

$$
\|Tu - Tv\| \leq 2 \, \|u - v\| \quad \forall u, v \in E
$$

and the constant 2 cannot be improved; see Exercise 5.6.

$$\|Tu - Tv\| \le \|u - v\| \quad \forall u, v \in E.$$

Then $\| \; \|$ *is a Hilbert norm.*

Finally, let us recall a result that has already been mentioned (Remark 2.8).

(d) **Theorem 5.15 (Lindenstrauss–Tzafriri [1]).** *Assume that E is a Banach space such that every closed subspace has a complement.*[8] *Then E is Hilbertizable, i.e., there exists an equivalent Hilbert norm.*

2. Variational inequalities.

Stampacchia's theorem is the starting point of the theory of *variational inequalities* (see, e.g., D. Kinderlehrer–G. Stampacchia [1]), which has numerous applications in mechanics and in physics (see, e.g., G. Duvaut–J. L. Lions [1]), in free boundary value problems (see, e.g., C. Baiocchi–A. Capelo [1] and A. Friedman [4]), in optimal control (see, e.g., J.-L. Lions [2] and V. Barbu [2]), in stochastic control (see A. Bensoussan–J.-L. Lions [1]).

3. Nonlinear equations associated with monotone operators.

The theorems of Stampacchia and Lax–Milgram extend to some classes of *nonlinear* operators. Let us mention the following, for example.

Theorem 5.16 (Minty–Browder). *Let E be a reflexive Banach space. Let* $A : E \to E^\star$ *be a continuous nonlinear map such that*

$$\langle Av_1 - Av_2, v_1 - v_2 \rangle > 0 \quad \forall v_1, v_2 \in E, \quad v_1 \ne v_2,$$

and

$$\lim_{\|v\| \to \infty} \frac{\langle Av, v \rangle}{\|v\|} = \infty.$$

Then for every $f \in E^\star$ *there exists a unique solution* $u \in E$ *of the equation* $Au = f$.

The interested reader will find in F. Browder [1] and J.-L. Lions [3] a proof of Theorem 5.16 as well as many extensions and applications; see also Problem 31.

4. Special orthonormal bases. Fourier series. Wavelets.

In Chapter 6 we shall present a very powerful technique for constructing orthonormal bases, namely by taking the eigenvectors of a compact self-adjoint operator. In practice one very often uses special bases of $L^2(\Omega)$ that consist of *eigenfunctions of differential operators* (see Sections 8.6 and 9.8). The orthonormal basis on $L^2(0, \pi)$ defined by

$$e_n(x) = \sqrt{2/\pi} \sin nx, n \ge 1, \quad \text{or} \quad e_n(x) = \sqrt{2/\pi} \cos nx, n \ge 0,$$

is quite beloved, since it leads to *Fourier series* and *harmonic analysis*, a major field in its own right; see, e.g., J. M. Ash [1], H. Dym–H. P. McKean [1], Y. Katznelson [1], C. S. Rees–S. M. Shah–C. V. Stanojevic [1].

---

[8] It is equivalent to say that every closed subspace has a bounded projection operator P. Note that here—in contrast to Theorem 5.13—we do not assume that $\|P\| \le 1$.

Here is a question that puzzled analysts for decades. Given $u \in L^2(0, \pi)$, consider its Fourier series $S_n = \sum_{k=1}^{n}(u, e_k)e_k$. One knows (see Corollary 5.10) that $S_n \to u$ in $L^2(0, \pi)$. It follows that a subsequence $S_{n_k} \to u$ a.e. on $(0, \pi)$ (see Theorem 4.9). But can one say that the *full sequence* $S_n \to u$ a.e. on $(0, \pi)$? The answer is given by the following very deep result:

Theorem 5.17 (Carleson [1]). *If $u \in L^2(0, \pi)$ then $S_n \to u$ a.e.*

Other classical bases of $L^2(0, 1)$ or $L^2(\mathbb{R})$ are associated with the names of *Bessel, Legendre, Hermite, Laguerre, Chebyshev, Jacobi*, etc. We refer the interested reader to R. Courant–D. Hilbert [1], Volume 1, and R. Dautray–J.-L. Lions [1], Chapter VIII; see also the comments at the end of Chapter 8 (spectral properties of the Sturm–Liouville operator). Recently, there has also been much interest in the Haar and the Walsh bases of $L^2(0, 1)$, which consist of step functions; see, e.g., Exercises 5.31, 5.32, G. Alexits [1], H. F. Harmuth [1].

The theory of *wavelets* provides a very important and beautiful new type of bases. It is a powerful tool in decomposing functions, signals, speech, images, etc. The interested reader may consult the recent books of Y. Meyer [1], [2], [3], R. Coifman and Y. Meyer [1], I. Daubechies [1], G. David [1], C. K. Chui [1], M. B. Ruskai et al. [1], J. J. Benedetto–M. W. Frazier [1], G. Kaiser [1], J. P. Kahane–P. G. Lemarié-Rieusset [1], S. Mallat [1], G. Bachman–L. Narici–E. Beckenstein [1], T. F. Chan–J. Shen [1], P. Wojtaszczyk [1], E. Hernandez–G. Weiss [1], and their references.

5. Schauder bases in Banach spaces.
Let E be a Banach space. A sequence $(e_n)_{n \geq 1}$ is said to be a *Schauder basis* if for every $u \in E$ there exists a unique sequence $(\alpha_n)_{n \geq 1}$ in $\mathbb{R}$ such that $u = \sum_{k=1}^{\infty} \alpha_k e_k$ (i.e., $u = \lim_{n \to \infty} \sum_{k=1}^{n} \alpha_k e_k$). Such bases play an important role in the geometry of Banach spaces (see, e.g., B. Beauzamy [1], J. Lindenstrauss–L. Tzafriri [2], J. Diestel [2], R. C. James [2]). All classical (separable) Banach spaces used in analysis have a Schauder basis (see, e.g., I. Singer [1]). This fact led Banach to conjecture that every separable Banach space has a basis. After a few decades of unavailing efforts a counterexample was discovered by P. Enflo [1]. One can even construct closed subspaces of ℓ^p (with $1 < p < \infty, p \neq 2$) without a Schauder basis (see J. Lindenstrauss–L. Tzafriri [2]). A. Szankowski [1] has found another surprising example: $\mathscr{L}(H)$ (with its usual norm) has no Schauder basis when H is an infinite-dimensional separable Hilbert space. In Chapter 6 we shall see that a related problem for compact operators also has a negative answer.

Exercises for Chapter 5

In what follows, H will always denote a Hilbert space equipped with the scalar product $(\ ,\ )$ and the corresponding norm $|\ |$.

5.1 *The parallelogram law.*

Suppose E is a vector space equipped with a norm $\| \ \|$ satisfying the parallelogram law, i.e.,

$$\|a+b\|^2 + \|a-b\|^2 = 2(\|a\|^2 + \|b\|^2) \quad \forall a, b \in E.$$

Our purpose is to show that the quantity defined by

$$(u, v) = \frac{1}{2}(\|u+v\|^2 - \|u\|^2 - \|v\|^2) \quad u, v \in E,$$

is a scalar product such that $(u, u) = \|u\|^2$.

1. Check that

$$(u, v) = (v, u), (-u, v) = -(u, v) \text{ and } (u, 2v) = 2(u, v) \quad \forall u, v \in E.$$

2. Prove that
$$(u+v, w) = (u, w) + (v, w) \quad \forall u, v, w \in E.$$

   [**Hint**: Use the parallelogram law successively with (i) $a = u, b = v$; (ii) $a = u+w, b = v+w$, and (iii) $a = u+v+w, b = w$.]
3. Prove that $(\lambda u, v) = \lambda(u, v) \ \forall \lambda \in \mathbb{R}, \forall u, v \in E$.
   [**Hint**: Consider first the case $\lambda \in \mathbb{N}$, then $\lambda \in \mathbb{Q}$, and finally $\lambda \in \mathbb{R}$.]
4. Conclude.

5.2 L^p *is not a Hilbert space for $p \neq 2$.*
Let Ω be a measure space and assume that there exists a measurable set $A \subset \Omega$ such that $0 < |A| < |\Omega|$.

Prove that the $\| \ \|_p$ norm does not satisfy the parallelogram law for any $1 \leq p \leq \infty$, $p \neq 2$.

[**Hint**: Use functions with disjoint supports.]

5.3 Let (u_n) be a sequence in H and let (t_n) be a sequence in $(0, \infty)$ such that

$$(t_n u_n - t_m u_m, u_n - u_m) \leq 0 \quad \forall m, n.$$

1. Assume that the sequence (t_n) is *nondecreasing* (possibly unbounded). Prove that the sequence (u_n) converges.
   [**Hint**: Show that the sequence $(|u_n|)$ is nonincreasing.]
2. Assume that the sequence (t_n) is *nonincreasing*. Prove that the following alternative holds:

   (i) either $|u_n| \to \infty$,
   (ii) or (u_n) converges.

   If $t_n \to t > 0$, prove that (u_n) converges, and if $t_n \to 0$, prove that both cases (i) and (ii) may occur.

5.4 Let $K \subset H$ be a nonempty closed convex set. Let $f \in H$ and let $u = P_K f$.
Prove that

$$|v - u|^2 \le |v - f|^2 - |u - f|^2 \quad \forall v \in K.$$

Deduce that

$$|v - u| \le |v - f| \quad \forall v \in K.$$

Give a geometric interpretation.

5.5

1. Let (K_n) be a *nonincreasing* sequence of closed convex sets in H such that $\cap_n K_n \ne \emptyset$.
   Prove that for every $f \in H$ the sequence $u_n = P_{K_n} f$ converges (strongly) to a limit and identify the limit.
2. Let (K_n) be a *nondecreasing* sequence of nonempty closed convex sets in H.

Prove that for every $f \in H$ the sequence $u_n = P_{K_n} f$ converges (strongly) to a limit and identify the limit.

Let $\varphi : H \to \mathbb{R}$ be a continuous function that is bounded from below. Prove that the sequence $\alpha_n = \inf_{K_n} \varphi$ converges and identify the limit.

5.6 *The radial projection onto the unit ball.*
Let E be a vector space equipped with the norm $\| \ \|$.
Set

$$Tu = \begin{cases} u & \text{if } \|u\| \le 1, \\ u/\|u\| & \text{if } \|u\| > 1. \end{cases}$$

1. Prove that $\|Tu - Tv\| \le 2\|u - v\| \quad \forall u, v \in E$.
2. Show that in general, the constant 2 cannot be improved.
   [**Hint**: Take $E = \mathbb{R}^2$ with the norm $\|u\| = |u_1| + |u_2|$.]
3. What happens if $\| \ \|$ is a Hilbert norm?

5.7 *Projection onto a convex cone.*
Let $K \subset H$ be a convex cone with vertex at 0, i.e.,

$$0 \in K \quad \text{and} \quad \lambda u + \mu v \in K \quad \forall \lambda, \mu > 0, \quad \forall u, v \in K;$$

assume in addition that K is closed.
Given $f \in H$, prove that $u = P_K f$ is *characterized* by the following properties:

$$u \in K, \ (f - u, v) \le 0 \quad \forall v \in K \quad \text{and} \quad (f - u, u) = 0.$$

5.8 Let Ω be a measure space and let $h : \Omega \to [0, +\infty)$ be a measurable function. Let

$$K = \{u \in L^2(\Omega); \quad |u(x)| \le h(x) \text{ a.e. on } \Omega\}.$$

Check that K is a nonempty closed convex set in $H = L^2(\Omega)$. Determine P_K.

5.9 Let $A \subset H$ and $B \subset H$ be two nonempty closed convex set such that $A \cap B = \emptyset$ and B is bounded.

Set
$$C = A - B.$$

1. Show that C is closed and convex.
2. Set $u = P_C 0$ and write $u = a_0 - b_0$ for some $a_0 \in A$ and $b_0 \in B$ (this is possible since $u \in C$).
   Prove that $|a_0 - b_0| = \text{dist}(A, B) = \inf_{a \in A, b \in B} |a - b|$.
   Determine $P_A b_0$ and $P_B a_0$.
3. Suppose $a_1 \in A$ and $b_1 \in B$ is another pair such that $|a_1 - b_1| = \text{dist}(A, B)$.
   Prove that $u = a_1 - b_1$.
   Draw some pictures where the pair $[a_0, b_0]$ is unique (resp. nonunique).
4. Find a simple proof of the Hahn–Banach theorem, second geometric form, in the case of a Hilbert space.

5.10 Let $F : H \to \mathbb{R}$ be a convex function of class C^1. Let $K \subset H$ be convex and let $u \in H$. Show that the following properties are equivalent:

(i) $F(u) \leq F(v) \quad \forall v \in K$,
(ii) $(F'(u), v - u) \geq 0 \quad \forall v \in K$.

Example: $F(v) = |v - f|^2$ with $f \in H$ given.

5.11 Let $M \subset H$ be a closed linear subspace that is not reduced to $\{0\}$. Let $f \in H$, $f \notin M^\perp$.

1. Prove that
$$m = \inf_{\substack{u \in M \\ |u|=1}} (f, u)$$
   is uniquely achieved.
2. Let $\varphi_1, \varphi_2, \varphi_3 \in H$ be given and let E denote the linear space spanned by $\{\varphi_1, \varphi_2, \varphi_3\}$. Determine m in the following cases:

   (i) $M = E$,
   (ii) $M = E^\perp$.
3. Examine the case in which $H = L^2(0, 1)$, $\varphi_1(t) = t$, $\varphi_2(t) = t^2$, and $\varphi_3(t) = t^3$.

5.12 *Completion of a pre-Hilbert space.*
   Let E be a vector space equipped with the scalar product $(\ ,\ )$. One does *not* assume that E is complete for the norm $|u| = (u, u)^{1/2}$ (E is said to be a pre-Hilbert space).
   Recall that the dual space $E^\star$, equipped with the dual norm $\|f\|_{E^\star}$, is complete. Let $T : E \to E^\star$ be the map defined by

$$\langle Tu, v \rangle_{E^\star, E} = (u, v) \quad \forall u, v \in E.$$

Check that T is a linear isometry. Is T surjective?
   Our purpose is to show that $R(T)$ is dense in $E^\star$ and that $\|\ \|_{E^\star}$ is a Hilbert norm.

1. Transfer to $R(T)$ the scalar product of E and extend it to $\overline{R(T)}$. The resulting scalar product is denoted by $((f, g))$ with $f, g \in \overline{R(T)}$.
   Check that the corresponding norm $((f, f))^{1/2}$ coincides on $\overline{R(T)}$ with $\|f\|_{E^*}$.
   Prove that
   $$\langle f, v \rangle = ((f, Tv)) \quad \forall v \in E, \quad \forall f \in \overline{R(T)}.$$

2. Prove that $\overline{R(T)} = E^*$.
   [**Hint**: Given $f \in E^*$, transfer f to a linear functional on $R(T)$ and use the Riesz–Fréchet representation theorem in $\overline{R(T)}$.]
   Deduce that E^* is a Hilbert space for the norm $\|\ \|_{E^*}$.

3. Conclude that the completion of E can be identified with E^*. (For the definition of the completion see, e.g., A. Friedman [3].)

$\boxed{5.13}$ Let E be a vector space equipped with the norm $\|\ \|_E$. The dual norm is denoted by $\|\ \|_{E^*}$. Recall that the (multivalued) duality map is defined by

$$F(u) = \{f \in E^*;\ \|f\|_{E^*} = \|u\|_E \text{ and } \langle f, u \rangle = \|u\|_E^2\}.$$

1. Assume that F satisfies the following property:
   $$F(u) + F(v) \subset F(u + v) \quad \forall u, v \in E.$$
   Prove that the norm $\|\ \|_E$ arises from a scalar product.
   [**Hint**: Use Exercise 5.1.]

2. Conversely, if the norm $\|\ \|_E$ arises from a scalar product, what can one say about F?
   [**Hint**: Use Exercise 5.12 and 1.1.]

$\boxed{5.14}$ Let $a : H \times H \to \mathbb{R}$ be a bilinear continuous form such that
$$a(v, v) \geq 0 \quad \forall v \in H.$$

Prove that the function $v \mapsto F(v) = a(v, v)$ is convex, of class C^1, and determine its differential.

$\boxed{5.15}$ Let $G \subset H$ be a linear subspace of a Hilbert space H; G is equipped with the norm of H. Let F be a Banach space. Let $S : G \to F$ be a bounded linear operator.
   Prove that there exists a bounded linear operator $T : H \to F$ that extends S and such that
$$\|T\|_{\mathscr{L}(H,F)} = \|S\|_{\mathscr{L}(G,F)}.$$

$\boxed{5.16}$ *The triplet $V \subset H \subset V^*$.*
   Let H be a Hilbert space equipped with the scalar product $(\ ,\ )$ and the corresponding norm $|\ |$. Let $V \subset H$ be a linear subspace that is dense in V. Assume that V has its own norm $\|\ \|$ and that V is a Banach space for $\|\ \|$. Assume also that the injection $V \subset H$ is continuous, i.e., $|v| \leq C\|v\| \ \forall v \in V$. Consider the operator

$T : H \to V^*$ defined by

$$\langle Tu, v \rangle_{V^*, V} = (u, v) \quad \forall u \in H, \quad \forall v \in V.$$

1. Prove that $\|Tu\|_{V^*} \le C|u| \; \forall u \in H$.
2. Prove that T is injective.
3. Prove that $R(T)$ is dense in V^* if V is reflexive.
4. Given $f \in V^*$, prove that $f \in R(T)$ iff there is a constant $a \ge 0$ such that $|\langle f, v \rangle_{V^*, V}| \le a|v| \; \forall v \in V$.

$\boxed{5.17}$ Let $M, N \subset H$ be two closed linear subspaces.
Assume that $(u, v) = 0 \; \forall u \in M, \forall v \in N$. Prove that $M + N$ is closed.

$\boxed{5.18}$ Let E be a Banach space and let H be a Hilbert space. Let $T \in \mathcal{L}(E, H)$. Show that the following properties are equivalent:

(i) T admits a left inverse,
(ii) there exists a constant C such that $\|u\| \le C|Tu| \; \forall u \in E$.

$\boxed{5.19}$ Let (u_n) be a sequence in H such that $u_n \rightharpoonup u$ weakly. Assume that $\limsup |u_n| \le |u|$. Prove that $u_n \to u$ strongly without relying on Proposition 3.32.

$\boxed{5.20}$ Assume that $S \in \mathcal{L}(H)$ satisfies $(Su, u) \ge 0 \; \forall u \in H$.

1. Prove that $N(S) = R(S)^\perp$.
2. Prove that $I + tS$ is bijective for every $t > 0$.
3. Prove that
$$\lim_{t \to +\infty} (I + tS)^{-1} f = P_{N(S)} f \quad \forall f \in H.$$

[**Hint**: Two methods are possible:

(a) Consider the cases $f \in N(S)$ and $f \in R(S)$.
(b) Use weak convergence.]

$\boxed{5.21}$ *Iterates of linear contractions. The ergodic theorem of Kakutani–Yosida.*
Let $T \in \mathcal{L}(H)$ be such that $\|T\| \le 1$. Given $f \in H$ and given an integer $n \ge 1$, set
$$\sigma_n(f) = \frac{1}{n}(f + Tf + T^2 f + \cdots + T^{n-1} f)$$
and
$$\mu_n(f) = \left(\frac{I + T}{2} \right)^n f.$$

Our purpose is to show that
$$\lim_{n \to \infty} \sigma_n(f) = \lim_{n \to \infty} \mu_n(f) = P_{N(I-T)} f.$$

1. Check that $N(I - T) = R(I - T)^\perp$.

2. Assume that $f \in R(I-T)$. Prove that there exists a constant C such that $|\sigma_n(f)| \leq C/n \; \forall n \geq 1$.
3. Deduce that for every $f \in H$, one has

$$\lim_{n \to \infty} \sigma_n(f) = P_{N(I-T)}f.$$

4. Set $S = \dfrac{1}{2}(I + T)$. Prove that

(1) $|u - Su|^2 + |Su|^2 \leq |u|^2 \quad \forall u \in H.$

Deduce that

$$\sum_{i=0}^{\infty} |S^i u - S^{i+1} u|^2 \leq |u|^2 \quad \forall u \in H$$

and that

$$|S^n(u - Su)| \leq \frac{|u|}{\sqrt{n+1}} \quad \forall u \in H \quad \forall n \geq 1.$$

5. Assume that $f \in R(I - T)$. Prove that there exists a constant C such that $|\mu_n(f)| \leq C/\sqrt{n} \; \forall n \geq 1$.
6. Deduce that for every $f \in H$, one has

$$\lim_{n \to \infty} \mu_n(f) = P_{N(I-T)}f.$$

5.22 Let $C \subset H$ be a nonempty closed convex set and let $T : C \to C$ be a nonlinear contraction, i.e.,

$$|Tu - Tv| \leq |u - v| \quad \forall u, v \in C.$$

1. Let (u_n) be a sequence in C such that

$$u_n \rightharpoonup u \text{ weakly and } (u_n - Tu_n) \to f \text{ strongly}.$$

Prove that $u - Tu = f$.
[**Hint**: Start with the case $C = H$ and use the inequality $((u-Tu)-(v-Tv), u-v) \geq 0 \; \forall u, v$.]
2. Deduce that if C is bounded and $T(C) \subset C$, then T has a fixed point.
[**Hint**: Consider $T_\varepsilon u = (1-\varepsilon)Tu + \varepsilon a$ with $a \in C$ being fixed and $\varepsilon > 0, \varepsilon \to 0$.]

5.23 *Zarantonello's inequality.*
Let $T : H \to H$ be a (nonlinear) contraction. Assume that $\alpha_1, \alpha_2, \ldots, \alpha_n \in \mathbb{R}$ are such that $\alpha_i \geq 0 \; \forall i$ and $\sum_{i=1}^{n} \alpha_i = 1$. Assume that $u_1, u_2, \ldots, u_n \in H$ and set

$$\sigma = \sum_{i=1}^{n} \alpha_i u_i.$$

Prove that

$$\left| T\sigma - \sum_{i=1}^{n} \alpha_i T u_i \right|^2 \leq \frac{1}{2} \sum_{i,j=1}^{n} \alpha_i \alpha_j \left[|u_i - u_j|^2 - |T u_i - T u_j|^2 \right].$$

[**Hint**: Write

$$\left| T\sigma - \sum_{i=1}^{n} \alpha_i T u_i \right|^2 = \sum_{i,j=1}^{n} \alpha_i \alpha_j (T\sigma - T u_i, T\sigma - T u_j)$$

and use the identity $(a, b) = \frac{1}{2}(|a|^2 + |b|^2 - |a - b|^2)$.]

What can one deduce when T is an isometry (i.e., $|Tu - Tv| = |u - v| \, \forall u, v \in H$)?

$\boxed{5.24}$ *The Banach–Saks property.*

1. Assume that (u_n) is a sequence in H such that $u_n \rightharpoonup 0$ weakly. Construct by induction a subsequence (u_{n_j}) such that $u_{n_1} = u_1$ and

$$|(u_{n_j}, u_{n_k})| \leq \frac{1}{k} \quad \forall k \geq 2 \text{ and } \forall j = 1, 2, \ldots, k-1.$$

Deduce that the sequence (σ_p) defined by $\sigma_p = \frac{1}{p} \sum_{j=1}^{p} u_{n_j}$ converges strongly to 0 as $p \to \infty$.
[**Hint**: Estimate $|\sigma_p|^2$.]

2. Assume that (u_n) is a bounded sequence in H. Prove that there exists a subsequence (u_{n_j}) such that the sequence $\sigma_p = \frac{1}{p} \sum_{j=1}^{p} u_{n_j}$ converges strongly to a limit as $p \to \infty$.
Compare with Corollary 3.8 and Exercise 3.4.

$\boxed{5.25}$ *Variations on Opial's lemma.*

Let $K \subset H$ be a nonempty closed convex set. Let (u_n) be a sequence in H such that for *each* $v \in K$ the sequence $(|u_n - v|)$ is nonincreasing.

1. Check that the sequence $(\text{dist}(u_n, K))$ is nonincreasing.
2. Prove that the sequence $(P_K u_n)$ converges strongly to a limit, denoted by ℓ.
   [**Hint**: Use Exercise 5.4.]
3. Assume here that the sequence (u_n) satisfies the property

(P) $\quad \begin{cases} \text{Whenever a subsequence } (u_{n_k}) \text{ converges weakly} \\ \text{to some limit } \overline{u} \in H, \text{ then } \overline{u} \in K. \end{cases}$

   Prove that $u_n \rightharpoonup \ell$ weakly.
4. Assume here that $\bigcup_{\lambda > 0} \lambda(K - K) = H$. Prove that there exists some $u \in H$ such that $u_n \rightharpoonup u$ weakly and $P_K u = \ell$.
5. Assume here that Int $K \neq \emptyset$. Prove that there exists some $u \in H$ such that $u_n \to u$ strongly.
   [**Hint**: Consider first the case that K is the unit ball and then the general case.]

6. Set $\sigma_n = \frac{1}{n}(u_1 + u_2 + \cdots + u_n)$ and assume that the sequence (σ_n) satisfies property (P). Prove that $\sigma_n \rightharpoonup \ell$ weakly.

5.26 Assume that (e_n) is an orthonormal basis of H.

1. Check that $e_n \rightharpoonup 0$ weakly.
   Let (a_n) be a bounded sequence in $\mathbb{R}$ and set $u_n = \frac{1}{n}\sum_{i=1}^n a_i e_i$.
2. Prove that $|u_n| \to 0$.
3. Prove that $\sqrt{n}\, u_n \rightharpoonup 0$ weakly.

5.27 Let $D \subset H$ be a subset such that the linear space spanned by D is dense in H. Let $(E_n)_{n\geq 1}$ be a sequence of closed subspaces in H that are mutually orthogonal. Assume that

$$\sum_{n=1}^{\infty} |P_{E_n} u|^2 = |u|^2 \quad \forall u \in D.$$

Prove that H is the Hilbert sum of the E_n's.

5.28 Assume that H is separable.

1. Let $V \subset H$ be a linear subspace that is dense in H. Prove that V contains an orthonormal basis of H.
2. Let $(e_n)_{n\geq 1}$ be an orthonormal sequence in H, i.e., $(e_i, e_j) = \delta_{ij}$. Prove that there exists an orthonormal basis of H that contains $\bigcup_{n=1}^{\infty}\{e_n\}$.

5.29 A lemma of Grothendieck.
   Let Ω be a measure space with $|\Omega| < \infty$. Let E be a closed subspace of $L^p(\Omega)$ with $1 \leq p < \infty$. Assume that $E \subset L^\infty(\Omega)$. Our purpose is to prove that dim $E < \infty$.

1. Prove that there exists a constant C such that

$$\|u\|_\infty \leq C\|u\|_p \quad \forall u \in E.$$

   [Hint: Use Corollary 2.8.]
2. Prove that there exists a constant M such that

$$\|u\|_\infty \leq M\|u\|_2 \quad \forall u \in E.$$

   [Hint: Distinguish the cases $1 \leq p \leq 2$ and $2 < p < \infty$.]
3. Deduce that E is a closed subspace of $L^2(\Omega)$.

   In what follows we assume that dim $E = \infty$. Let $(e_n)_{n\geq 1}$ be an orthonormal sequence of E (equipped with the L^2 scalar product).

4. Fix any integer $k \geq 1$. Prove that there exists a null set $\omega \subset \Omega$ such that

$$\sum_{i=1}^{k} \alpha_i e_i(x) \leq M\left(\sum_{i=1}^{k} \alpha_i^2\right)^{1/2} \quad \forall x \in \Omega\backslash\omega, \quad \forall \alpha = (\alpha_1, \alpha_2, \ldots, \alpha_k) \in \mathbb{R}^k.$$

[**Hint**: Start with the case $\alpha \in \mathbb{Q}^k$.]
5. Deduce that $\sum_{i=1}^{k} |e_i(x)|^2 \leq M^2 \ \forall x \in \Omega \backslash \omega$.
6. Conclude.

5.30 Let $(e_n)_{n \geq 1}$ be an orthonormal *sequence* in $H = L^2(0, 1)$. Let $p(t)$ be a given function in H.

1. Prove that for every $t \in [0, 1]$, one has

(1)
$$\sum_{n=1}^{\infty} \left| \int_0^t p(s)e_n(s)ds \right|^2 \leq \int_0^t |p(s)|^2 ds.$$

2. Deduce that

(2)
$$\sum_{n=1}^{\infty} \int_0^1 \left| \int_0^t p(s)e_n(s)ds \right|^2 dt \leq \int_0^1 |p(t)|^2(1 - t)dt.$$

3. Assume now that $(e_n)_{n \geq 1}$ is an orthonormal *basis* of H.
   Prove that (1) and (2) become equalities.
4. Conversely, assume that equality holds in (2) and that $p(t) \neq 0$ a.e. Prove that $(e_n)_{n \geq 1}$ is an orthonormal basis.

Example: $p \equiv 1$.

5.31 *The Haar basis.*
   Given an integer $n \geq 1$, write $n = k + 2^p$, where $p \geq 0$ and $k \geq 0$ are integers uniquely determined by the condition $k \leq 2^p - 1$. Consider the function defined on $(0, 1)$ by

$$\varphi_n(t) = \begin{cases} 2^{p/2} & \text{if} \quad k2^{-p} < t < (k + \dfrac{1}{2})2^{-p}, \\ -2^{p/2} & \text{if} \quad (k + \dfrac{1}{2})2^{-p} < t < (k+1)2^{-p}, \\ 0 & \text{elsewhere.} \end{cases}$$

Set $\varphi_0 \equiv 1$ and prove that $(\varphi_n)_{n \geq 0}$ is an orthonormal basis of $L^2(0, 1)$.

5.32 *The Rademacher system and the Walsh basis.*
   For every integer $i \geq 0$ consider the function $r_i(t)$ defined on $(0, 1)$ by $r_i(t) = (-1)^{[2^i t]}$ (as usual $[x]$ denotes the largest integer $\leq x$).

1. Check that $(r_i)_{i \geq 0}$ is an orthonormal sequence in $L^2(0, 1)$ (called the Rademacher system).
2. Is $(r_i)_{i \geq 0}$ an orthonormal basis?
   [**Hint**: Consider the function $u = r_1 r_2$.]
3. Given an integer $n \geq 0$, consider its binary representation $n = \sum_{i=0}^{\ell} \alpha_i 2^i$ with $\alpha_i \in \{0, 1\}$.

Set

$$w_n(t) = \prod_{i=0}^{\ell} r_{i+1}(t)^{\alpha_i}.$$

Prove that $(w_n)_{n \geq 0}$ is an orthonormal basis of $L^2(0, 1)$ (called the Walsh basis).
Note that $(r_i)_{i \geq 0}$ is a subset of $(w_n)_{n \geq 0}$.

Chapter 6
Compact Operators. Spectral Decomposition of Self-Adjoint Compact Operators

6.1 Definitions. Elementary Properties. Adjoint

Throughout this chapter, and unless otherwise specified, E and F denote two Banach spaces.

Definition. A bounded operator $T \in \mathcal{L}(E, F)$ is said to be *compact* if $T(B_E)$ has compact closure in F (in the strong topology).

The set of all compact operators from E into F is denoted by $\mathcal{K}(E, F)$. For simplicity one writes $\mathcal{K}(E) = \mathcal{K}(E, E)$.

Theorem 6.1. *The set $\mathcal{K}(E, F)$ is a closed linear subspace of $\mathcal{L}(E, F)$ (in the topology associated to the norm $\| \ \|_{\mathcal{L}(E,F)}$).*

Proof. Clearly the sum of two compact operators is a compact operator. Suppose that (T_n) is a sequence of compact operators and T is a bounded operator such that $\|T_n - T\|_{\mathcal{L}(E,F)} \to 0$. We claim that T is a compact operator. Since F is complete it suffices to check that for every $\varepsilon > 0$ there is a finite covering of $T(B_E)$ with balls of radius ε (see, e.g., J. R. Munkres [1], Section 7.3). Fix an integer n such that $\|T_n - T\|_{\mathcal{L}(E,F)} < \varepsilon/2$. Since $T_n(B_E)$ has compact closure, there is a finite covering of $T_n(B_E)$ by balls of radius $\varepsilon/2$, say $T_n(B_E) \subset \bigcup_{i \in I} B(f_i, \varepsilon/2)$. It follows that $T(B_E) \subset \bigcup_{i \in I} B(f_i, \varepsilon)$.

Definition. An operator $T \in \mathcal{L}(E, F)$ is said to be of *finite rank* if the range of T, $R(T)$, is finite-dimensional.

Clearly, any finite-rank operator is compact and thus we have the following.

Corollary 6.2. *Let (T_n) be a sequence of finite-rank operators and let $T \in \mathcal{L}(E, F)$ be such that $\|T_n - T\|_{\mathcal{L}(E,F)} \to 0$. Then $T \in \mathcal{K}(E, F)$.*

★ *Remark* 1. The celebrated "approximation problem" (Banach, Grothendieck) deals with the converse of Corollary 6.2: given a compact operator T does there always

H. Brezis, *Functional Analysis, Sobolev Spaces and Partial Differential Equations*, 157
DOI 10.1007/978-0-387-70914-7_6, © Springer Science+Business Media, LLC 2011

exist a sequence (T_n) of finite-rank operators such that $\|T_n - T\|_{\mathcal{L}(E,F)} \to 0$? The question was open for a long time until P. Enflo [1] discovered a counterexample in 1972. The original construction was quite complicated, and subsequently simpler examples were found, for example, with F being some closed subspace of ℓ^p (for any $1 < p < \infty$, $p \neq 2$). The interested reader will find a detailed discussion of the approximation problem in J. Lindenstrauss–L. Tzafriri [2]. Note that the answer to the approximation problem is *positive* in some *special cases*—for example if F is a *Hilbert space*. Indeed, set $K = \overline{T(B_E)}$. Given $\varepsilon > 0$ there is a finite covering of K with balls of radius ε, say $K \subset \bigcup_{i \in I} B(f_i, \varepsilon)$. Let G denote the vector space spanned by the f_i's and set $T_\varepsilon = P_G T$, so that T_ε is of finite rank. We claim that $\|T_\varepsilon - T\|_{\mathcal{L}(E,F)} < 2\varepsilon$. For every $x \in B_E$ there is some $i_0 \in I$ such that

$$(1) \qquad\qquad \|Tx - f_{i_0}\| < \varepsilon.$$

Thus

$$\|P_G Tx - P_G f_{i_0}\| < \varepsilon,$$

that is,

$$(2) \qquad\qquad \|P_G Tx - f_{i_0}\| < \varepsilon.$$

Combining (1) and (2), one obtains

$$\|P_G Tx - Tx\| < 2\varepsilon \quad \forall x \in B_E,$$

that is,

$$\|T_\varepsilon - T\|_{\mathcal{L}(E,F)} < 2\varepsilon.$$

[More generally, one sees that if F has a Schauder basis, then the answer to the approximation problem is positive for every space E and every compact operator from E into F.]

   In connection with the approximation problem, let us mention a technique that is very useful in nonlinear analysis to approximate a continuous map (linear or nonlinear) by *nonlinear maps* of finite rank. Let X be a topological space, let F be a Banach space, and let $T : X \to F$ be a continuous map such that $T(X)$ has compact closure in F. We claim that for every $\varepsilon > 0$ there exists a continuous map $T_\varepsilon : X \to F$ of finite rank such that

$$(3) \qquad\qquad \|T_\varepsilon(x) - T(x)\| < \varepsilon \quad \forall x \in X.$$

Indeed, since $K = \overline{T(X)}$ is compact there is a finite covering of K, say $K \subset \bigcup_{i \in I} B(f_i, \varepsilon/2)$. Set

$$T_\varepsilon(x) = \frac{\sum\limits_{i \in I} q_i(x) f_i}{\sum\limits_{i \in I} q_i(x)} \quad \text{with } q_i(x) = \max\{\varepsilon - \|Tx - f_i\|, 0\};$$

clearly T_ε satisfies (3).

This kind of approximation is very useful, for example, to deduce Schauder's fixed-point theorem from Brouwer's fixed-point theorem (see, e.g., K. Deimling [1], A. Granas–J. Dugundji [1], J. Franklin [1], and Exercise 6.26). A similar construction, combined with the Schauder fixed-point theorem, has also been used in a surprising way by Lomonosov to prove the existence of nontrivial invariant subspaces for a large class of linear operators (see, e.g., C. Pearcy [1], N. Akhiezer–I. Glazman [1], A. Granas–J. Dugundji [1], and Problem 42). Another linear result that has a simple proof based on the Schauder fixed-point theorem is the Krein–Rutman theorem (see Theorem 6.13 and Problem 41).

Proposition 6.3. *Let E, F, and G be three Banach spaces. Let $T \in \mathcal{L}(E,\ F)$ and $S \in \mathcal{K}(F,\ G)$ [resp. $T \in \mathcal{K}(E,\ F)$ and $S \in \mathcal{L}(F,\ G)$]. Then $S \circ T \in \mathcal{K}(E, G)$.*

The proof is obvious.

Theorem 6.4 (Schauder). *If $T \in \mathcal{K}(E,\ F)$, then $T^\star \in \mathcal{K}(F^\star, E^\star)$. And conversely.*

Proof. We have to show that $T^\star(B_{F^\star})$ has compact closure in $E^\star$. Let (v_n) be a sequence in $B_{F^\star}$. We claim that $(T^\star(v_n))$ has a convergent subsequence. Set $K = \overline{T(B_E)}$; this is a compact metric space. Consider the set $\mathcal{H} \subset C(K)$ defined by

$$\mathcal{H} = \{\varphi_n : x \in K \longmapsto \langle v_n, x \rangle;\ n = 1, 2, \ldots\}.$$

The assumptions of Ascoli–Arzelà's theorem (Theorem 4.25) are satisfied. Thus, there is a subsequence, denoted by φ_{n_k}, that converges uniformly on K to some continuous function $\varphi \in C(K)$. In particular, we have

$$\sup_{u \in B_E} |\langle v_{n_k}, Tu \rangle - \varphi(Tu)| \underset{k \to \infty}{\longrightarrow} 0\ .$$

Thus

$$\sup_{u \in B_E} |\langle v_{n_k}, Tu \rangle - \langle v_{n_\ell}, Tu \rangle| \underset{k, \ell \to \infty}{\longrightarrow} 0,$$

i.e., $\|T^\star v_{n_k} - T^\star v_{n_\ell}\|_{E^\star} \underset{k, \ell \to \infty}{\longrightarrow} 0$. Consequently $T^\star v_{n_k}$ converges in $E^\star$.

Conversely, assume $T^\star \in \mathcal{K}(F^\star, E^\star)$. We already know, from the first part, that $T^{\star\star} \in \mathcal{K}(E^{\star\star}, F^{\star\star})$. In particular, $T^{\star\star}(B_E)$ has compact closure in $F^{\star\star}$. But $T(B_E) = T^{\star\star}(B_E)$ and F is closed in $F^{\star\star}$. Therefore $T(B_E)$ has compact closure in F.

Remark 2. Let E and F be two Banach spaces and let $T \in \mathcal{K}(E, F)$. If (u_n) converges *weakly* to u in E, then (Tu_n) converges *strongly* to Tu. The converse is also true if E is reflexive (see Exercise 6.7).

6.2 The Riesz–Fredholm Theory

We start with some useful preliminary results.

Lemma 6.1 (Riesz's lemma). *Let E be an n.v.s. and let $M \subset E$ be a closed linear space such that $M \neq E$. Then*

$$\forall \varepsilon > 0 \ \exists u \in E \ such \ that \ \|u\| = 1 \ and \ \mathrm{dist}(u, M) \geq 1 - \varepsilon.$$

Proof. Let $v \in E$ with $v \notin M$. Since M is closed, then

$$d = \mathrm{dist}(v, M) > 0.$$

Choose any $m_0 \in M$ such that

$$d \leq \|v - m_0\| \leq d/(1 - \varepsilon).$$

Then

$$u = \frac{v - m_0}{\|v - m_0\|}$$

satisfies the required properties. Indeed, for every $m \in M$, we have

$$\|u - m\| = \left\| \frac{v - m_0}{\|v - m_0\|} - m \right\| \geq \frac{d}{\|v - m_0\|} \geq 1 - \varepsilon,$$

since $m_0 + \|v - m_0\| m \in M$.

Remark 3. If M is finite-dimensional (or more generally if M is reflexive) one can choose $\varepsilon = 0$ in Lemma 6.1. But this is not true in general (see Exercise 1.17).

• **Theorem 6.5 (Riesz).** *Let E be an n.v.s. with B_E compact. Then E is finite-dimensional.*

Proof. Assume, by contradiction, that E is infinite-dimensional. Then there is a sequence (E_n) of finite-dimensional subspaces of E such that $E_{n-1} \subset E_n$ and $E_{n-1} \neq E_n$. By Lemma 6.1 there is a sequence (u_n) with $u_n \in E_n$ such that $\|u_n\| = 1$ and $\mathrm{dist}(u_n, E_{n-1}) \geq 1/2$. In particular, $\|u_n - u_m\| \geq 1/2$ for $m < n$. Thus (u_n) has no convergent subsequence, which contradicts the assumption that B_E is compact.

• **Theorem 6.6 (Fredholm alternative).** *Let $T \in \mathcal{K}(E)$. Then*

(a) $N(I - T)$ *is finite-dimensional,*
(b) $R(I - T)$ *is closed, and more precisely* $R(I - T) = N(I - T^\star)^\perp$,
(c) $N(I - T) = \{0\} \Leftrightarrow R(I - T) = E$,
(d) $\dim N(I - T) = \dim N(I - T^\star)$.

Remark 4. The Fredholm alternative deals with the solvability of the equation $u - Tu = f$. It says that

• *either* for every $f \in E$ the equation $u - Tu = f$ has a unique solution,
• *or* the homogeneous equation $u - Tu = 0$ admits n linearly independent solutions, and in this case, the inhomogeneous equation $u - Tu = f$ is solvable if and only if f satisfies n *orthogonality conditions*, i.e.,

$$f \in N(I - T^\star)^\perp.$$

Remark 5. Property (c) is familiar in finite-dimensional spaces. If dim $E < \infty$, a linear operator from E into itself is *injective* (= one-to-one) if and only if it is *surjective* (= onto). However, *in infinite-dimensional spaces a bounded operator may be injective without being surjective and conversely*, for example the right shift (resp. the left shift) in ℓ^2 (see Remark 6). Therefore, assertion (c) is a remarkable property of the operators of the form $I - T$ with $T \in \mathcal{K}(E)$.

Proof.

(a) Let $E_1 = N(I - T)$. Then $B_{E_1} \subset T(B_E)$ and thus B_{E_1} is compact. By Theorem 6.5, E_1 must be finite-dimensional.

(b) Let $f_n = u_n - Tu_n \to f$. We have to show that $f \in R(I - T)$. Set $d_n = \operatorname{dist}(u_n, N(I - T))$. Since $N(I - T)$ is finite-dimensional, there exists $v_n \in N(I - T)$ such that $d_n = \|u_n - v_n\|$. We have

$$(4) \qquad\qquad f_n = (u_n - v_n) - T(u_n - v_n).$$

We claim that $\|u_n - v_n\|$ remains bounded. Suppose not; then there is a subsequence such that $\|u_{n_k} - v_{n_k}\| \to \infty$. Set $w_n = (u_n - v_n)/\|u_n - v_n\|$. From (4) we see that $w_{n_k} - Tw_{n_k} \to 0$. Choosing a further subsequence (still denoted by w_{n_k} for simplicity), we may assume that $Tw_{n_k} \to z$. Thus $w_{n_k} \to z$ and $z \in N(I - T)$, so that $\operatorname{dist}(w_{n_k}, N(I - T)) \to 0$. On the other hand,

$$\operatorname{dist}(w_n, N(I - T)) = \frac{\operatorname{dist}(u_n, N(I - T))}{\|u_n - v_n\|} = 1$$

(since $v_n \in N(I - T)$); a contradiction.

Thus $\|u_n - v_n\|$ remains bounded, and since T is a compact operator, we may extract a subsequence such that $T(u_{n_k} - v_{n_k})$ converges to some limit ℓ. From (4) it follows that $u_{n_k} - v_{n_k} \to f + \ell$. Letting $g = f + \ell$, we have $g - Tg = f$, i.e., $f \in R(I - T)$. This completes the proof of the fact that the operator $(I - T)$ has closed range. We may therefore apply Theorem 2.19 and deduce that

$$R(I - T) = N(I - T^\star)^\perp, \quad R(I - T^\star) = N(I - T)^\perp.$$

(c) We first prove the implication $\Rightarrow$. Assume, by contradiction, that

$$E_1 = R(I - T) \neq E.$$

Then E_1 is a Banach space and $T(E_1) \subset E_1$. Thus $T_{|E_1} \in \mathcal{K}(E_1)$ and $E_2 = (I - T)(E_1)$ is a closed subspace of E_1. Moreover, $E_2 \neq E_1$ (since $(I - T)$ is injective). Letting $E_n = (I - T)^n(E)$, we obtain a (strictly) decreasing sequence of closed subspaces. Using Riesz's lemma we may construct a sequence (u_n) such that $u_n \in E_n$, $\|u_n\| = 1$ and $\operatorname{dist}(u_n, E_{n+1}) \geq 1/2$. We have

$$Tu_n - Tu_m = -(u_n - Tu_n) + (u_m - Tu_m) + (u_n - u_m).$$

Note that if $n > m$, then $E_{n+1} \subset E_n \subset E_{m+1} \subset E_m$ and therefore

$$-(u_n - Tu_n) + (u_m - Tu_m) + u_n \in E_{m+1}.$$

It follows that $\|Tu_n - Tu_m\| \geq \text{dist}(u_m, E_{m+1}) \geq 1/2$. This is impossible, since T is a compact operator. Hence we have proved that $R(I - T) = E$.

Conversely, assume that $R(I - T) = E$. By Corollary 2.18 we know that $N(I - T^\star) = R(I - T)^\perp = \{0\}$. Since $T^\star \in \mathcal{K}(E^\star)$, we may apply the preceding step to infer that $R(I - T^\star) = E^\star$. Using Corollary 2.18 once more, we conclude that $N(I - T) = R(I - T^\star)^\perp = \{0\}$.

(d) Set $d = \dim N(I - T)$ and $d^\star = \dim N(I - T^\star)$. We will first prove that $d^\star \leq d$. Suppose not, that $d < d^\star$. Since $N(I - T)$ is finite-dimensional, it admits a complement in E (see Section 2.4, Example 1). Thus there exists a continuous projection P from E onto $N(I - T)$. On the other hand, $R(I - T) = N(I - T^\star)^\perp$ has finite codimension $d^\star$ (see Section 2.4, Example 2) and thus it has a complement (in E), denoted by F, of dimension $d^\star$. Since $d < d^\star$, there is a linear map $\Lambda : N(I - T) \to F$ that is *injective* and *not surjective*. Set $S = T + \Lambda \circ P$. Then $S \in \mathcal{K}(E)$, since $\Lambda \circ P$ has finite rank. We claim that $N(I - S) = \{0\}$. Indeed, if

$$0 = u - Su = (u - Tu) - (\Lambda \circ Pu),$$

then

$$u - Tu = 0 \quad \text{and} \quad \Lambda \circ Pu = 0,$$

i.e., $u \in N(I - T)$ and $\Lambda u = 0$. Therefore, $u = 0$.

Applying (c) to the operator S, we obtain that $R(I - S) = E$. This is absurd, since there exists some $f \in F$ with $f \notin R(\Lambda)$, and so the equation $u - Su = f$ has no solution.

Hence we have proved that $d^\star \leq d$. Applying this fact to $T^\star$, we obtain

$$\dim N(I - T^{\star\star}) \leq \dim N(I - T^\star) \leq \dim N(I - T).$$

But $N(I - T^{\star\star}) \supset N(I - T)$ and therefore $d = d^\star$.

6.3 The Spectrum of a Compact Operator

Here are some important definitions.

Definition. Let $T \in \mathcal{L}(E)$.
The *resolvent set*, denoted by $\rho(T)$, is defined by

$$\rho(T) = \{\lambda \in \mathbb{R};\ (T - \lambda I) \text{ is bijective from } E \text{ onto } E\}.$$

The *spectrum*, denoted by $\sigma(T)$, is the complement of the resolvent set, i.e., $\sigma(T) = \mathbb{R} \backslash \rho(T)$. A real number λ is said to be an *eigenvalue* of T if

$$N(T - \lambda I) \neq \{0\};$$

$N(T - \lambda I)$ is the corresponding *eigenspace*. The set of all eigenvalues is denoted by $EV(T)$.[1]

It is useful to keep in mind that if $\lambda \in \rho(T)$ then $(T - \lambda I)^{-1} \in \mathcal{L}(E)$ (see Corollary 2.7).

Remark 6. It is clear that $EV(T) \subset \sigma(T)$. In general, this inclusion can be strict:[2] there may exist some λ such that

$$N(T - \lambda I) = \{0\} \quad \text{and} \quad R(T - \lambda I) \neq E$$

(such a λ belongs to the spectrum but is not an eigenvalue). Consider, for example, in $E = \ell^2$ the *right shift*, i.e., $Tu = (0, u_1, u_2, \dots)$ with $u = (u_1, u_2, u_3, \dots)$. Then $0 \in \sigma(T)$, while $0 \notin EV(T)$. In fact, in this case $EV(T) = \emptyset$, while $\sigma(T) = [-1, +1]$ (see Exercise 6.18). It may of course happen, in finite- or infinite-dimensional spaces, that $EV(T) = \sigma(T) = \emptyset$; consider, for example, a rotation by $\pi/2$ in $\mathbb{R}^2$, or in ℓ^2 the operator $Tu = (-u_2, u_1, -u_4, u_3, \dots)$. If we work in vector spaces over $\mathbb{C}$ (see Section 11.4) the situation is *totally different*; the study of eigenvalues and spectra is much more interesting in spaces over $\mathbb{C}$. As is well known, in finite-dimensional spaces over $\mathbb{C}$, $EV(T) = \sigma(T) \neq \emptyset$ (these are the roots of the characteristic polynomial). In infinite-dimensional spaces over $\mathbb{C}$ a nontrivial result asserts that $\sigma(T)$ is *always nonempty* (see Section 11.4). However, it may happen that $EV(T) = \emptyset$ (take for example the right shift in $E = \ell^2$).

Proposition 6.7. *The spectrum $\sigma(T)$ of a bounded operator T is compact and*

$$\sigma(T) \subset [-\|T\|, +\|T\|].$$

Proof. Let $\lambda \in \mathbb{R}$ be such that $|\lambda| > \|T\|$. We will show that $T - \lambda I$ is bijective, which implies that $\sigma(T) \subset [-\|T\|, +\|T\|]$. Given $f \in E$, the equation $Tu - \lambda u = f$ has a unique solution, since it may be written as $u = \lambda^{-1}(Tu - f)$ and the contraction mapping principle (Theorem 5.7) applies.

We now prove that $\rho(T)$ is open. Let $\lambda_0 \in \rho(T)$. Given $\lambda \in \mathbb{R}$ (close to λ_0) and $f \in E$, we try to solve

(5)
$$Tu - \lambda u = f.$$

Equation (5) may be written as

$$Tu - \lambda_0 u = f + (\lambda - \lambda_0)u,$$

i.e.,

(6)
$$u = (T - \lambda_0 I)^{-1}[f + (\lambda - \lambda_0)u].$$

---

[1] Some authors write $\sigma_p(T)$ (= point spectrum) instead of $EV(T)$.

[2] Of course, if E is finite-dimensional, then $EV(T) = \sigma(T)$.

Applying the contraction mapping principle once more, we see that (6) has a solution if

$$|\lambda - \lambda_0| \|(T - \lambda_0 I)^{-1}\| < 1.$$

• **Theorem 6.8.** *Let $T \in \mathcal{K}(E)$ with dim $E = \infty$, then we have:*

(a) $0 \in \sigma(T)$,
(b) $\sigma(T) \backslash \{0\} = EV(T) \backslash \{0\}$,
(c) *one of the following cases holds:*

  - $\sigma(T) = \{0\}$,
  - $\sigma(T) \backslash \{0\}$ *is a finite set,*
  - $\sigma(T) \backslash \{0\}$ *is a sequence converging to 0.*

Proof.

(a) Suppose not, that $0 \notin \sigma(T)$. Then T is bijective and $I = T \circ T^{-1}$ is compact. Thus B_E is compact and dim $E < \infty$ (by Theorem 6.5); a contradiction.
(b) Let $\lambda \in \sigma(T)$, $\lambda \neq 0$. We shall prove that λ is an eigenvalue. Suppose not, that $N(T - \lambda I) = \{0\}$. Then by Theorem 6.6(c), we know that $R(T - \lambda I) = E$ and therefore $\lambda \in \rho(T)$; a contradiction.

For the proof of assertion (c) we shall use the following lemma.

Lemma 6.2. *Let $T \in \mathcal{K}(E)$ and let $(\lambda_n)_{n \geq 1}$ be a sequence of distinct real numbers such that*

$$\lambda_n \to \lambda$$

and

$$\lambda_n \in \sigma(T) \backslash \{0\} \quad \forall n.$$

Then $\lambda = 0$.

In other words, all the points of $\sigma(T) \backslash \{0\}$ are *isolated points*.

Proof. We know that $\lambda_n \in EV(T)$; let $e_n \neq 0$ be such that $(T - \lambda_n I)e_n = 0$. Let E_n be the space spanned by $\{e_1, e_2, \ldots, e_n\}$. We claim that $E_n \subset E_{n+1}, E_n \neq E_{n+1}$ for all n. It suffices to check that for all n, the vectors $e_1, e_2, \ldots, e_n$ are linearly independent. The proof is by induction on n. Assume that this holds up to n and suppose that $e_{n+1} = \sum_{i=1}^{n} \alpha_i e_i$. Then

$$Te_{n+1} = \sum_{i=1}^{n} \alpha_i \lambda_i e_i = \sum_{i=1}^{n} \alpha_i \lambda_{n+1} e_i.$$

It follows that $\alpha_i(\lambda_i - \lambda_{n+1}) = 0$ for $i = 1, 2, \ldots, n$ and thus $\alpha_i = 0$ for $i = 1, 2, \ldots, n$; a contradiction. Hence we have proved that $E_n \subset E_{n+1}, E_n \neq E_{n+1}$ for all n.

Applying Riesz's lemma (Lemma 6.1), we may construct a sequence $(u_n)_{n \geq 1}$ such that $u_n \in E_n$, $\|u_n\| = 1$ and dist$(u_n, E_{n-1}) \geq 1/2$ for all $n \geq 2$. For $2 \leq m < n$ we have

$$E_{m-1} \subset E_m \subset E_{n-1} \subset E_n.$$

On the other hand, it is clear that $(T - \lambda_n I)E_n \subset E_{n-1}$. Thus we have

$$\left\| \frac{Tu_n}{\lambda_n} - \frac{Tu_m}{\lambda_m} \right\| = \left\| \frac{(Tu_n - \lambda_n u_n)}{\lambda_n} - \frac{(Tu_m - \lambda_m u_m)}{\lambda_m} + u_n - u_m \right\|$$

$$\geq \mathrm{dist}(u_n, E_{n-1}) \geq 1/2.$$

If $\lambda_n \to \lambda$ and $\lambda \neq 0$ we have a contradiction, since (Tu_n) has a convergent subsequence.

Proof of Theorem 6.8, concluded. For every integer $n \geq 1$ the set

$$\sigma(T) \cap \{\lambda \in \mathbb{R}; |\lambda| \geq 1/n\}$$

is either *empty* or *finite* (if it had infinitely many distinct points we would have a subsequence that converged to some λ with $|\lambda| \geq 1/n$—since $\sigma(T)$ is compact— and this would contradict Lemma 6.2). Hence if $\sigma(T) \setminus \{0\}$ has infinitely many distinct points we may order them as a sequence tending to 0.

Remark 7. Given *any* sequence (α_n) converging to 0 there is a compact operator T such that $\sigma(T) = (\alpha_n) \cup \{0\}$. In ℓ^2 it suffices to consider the multiplication operator T defined by $Tu = (\alpha_1 u_1, \alpha_2 u_2, \ldots, \alpha_n u_n, \ldots)$, where $u = (u_1, u_2, \ldots, u_n, \ldots)$. Note that T is compact, since T is a limit of finite-rank operators. More precisely, let $T_n u = (\alpha_1 u_1, \alpha_2 u_2, \ldots, \alpha_n u_n, 0, 0, \ldots)$; then $\|T_n - T\| \to 0$. In this example, we also see that 0 may or may not belong to $EV(T)$. On the other hand, if $0 \in EV(T)$, the corresponding eigenspace, i.e., $N(T)$, may be finite- or infinite-dimensional.

6.4 Spectral Decomposition of Self-Adjoint Compact Operators

In what follows we assume that $E = H$ is a Hilbert space and that $T \in \mathcal{L}(H)$. Identifying $H^\star$ and H, we may view $T^\star$ as a bounded operator from H into itself.

Definition. A bounded operator $T \in \mathcal{L}(H)$ is said to be *self-adjoint* if $T^\star = T$, i.e.,

$$\boxed{(Tu, v) = (u, Tv) \quad \forall u, v \in H.}$$

Proposition 6.9. *Let $T \in \mathcal{L}(H)$ be a self-adjoint operator. Set*

$$m = \inf_{\substack{u \in H \\ |u|=1}} (Tu, u) \quad and \quad M = \sup_{\substack{u \in H \\ |u|=1}} (Tu, u).$$

Then $\sigma(T) \subset [m, M]$, $m \in \sigma(T)$, and $M \in \sigma(T)$. Moreover, $\|T\| = \max\{|m|, |M|\}$.

Proof. Let $\lambda > M$; we will prove that $\lambda \in \rho(T)$. We have

$$(Tu, u) \leq M|u|^2 \quad \forall u \in H,$$

and therefore

$$(\lambda u - Tu, u) \geq (\lambda - M)|u|^2 = \alpha |u|^2 \;\; \forall u \in H, \text{ with } \alpha > 0.$$

Applying Lax–Milgram's theorem (Corollary 5.8), we deduce that $\lambda I - T$ is bijective and thus $\lambda \in \rho(T)$. Similarly, any $\lambda < m$ belongs to $\rho(T)$ and therefore $\sigma(T) \subset [m, M]$.

We now prove that $M \in \sigma(T)$ (the proof that $m \in \sigma(T)$ is similar). The bilinear form $a(u, v) = (Mu - Tu, v)$ is symmetric and satisfies

$$a(v, v) \geq 0 \quad \forall v \in H.$$

Hence, it satisfies the Cauchy–Schwarz inequality

$$|a(u, v)| \leq a(u, u)^{1/2} a(v, v)^{1/2} \quad \forall u, v \in H,$$

i.e.,

$$|(Mu - Tu, v)| \leq (Mu - Tu, u)^{1/2}(Mv - Tv, v)^{1/2} \quad \forall u, v \in H.$$

It follows that

$$(7) \qquad\qquad |Mu - Tu| \leq C(Mu - Tu, u)^{1/2} \quad \forall u \in H.$$

By the definition of M there is a sequence (u_n) such that $|u_n| = 1$ and $(Tu_n, u_n) \to M$. From (7) we deduce that $|Mu_n - Tu_n| \to 0$ and thus $M \in \sigma(T)$ (since if $M \in \rho(T)$, then $u_n = (MI - T)^{-1}(Mu_n - Tu_n) \to 0$, which is impossible).

Finally, we prove that $\|T\| = \mu$, where $\mu = \max\{|m|, |M|\}$. Write $\forall u, v \in H$,

$$(T(u + v), u + v) = (Tu, u) + (Tv, v) + 2(Tu, v),$$
$$(T(u - v), u - v) = (Tu, u) + (Tv, v) - 2(Tu, v).$$

Thus

$$4(Tu, v) = (T(u + v), u + v) - (T(u - v), u - v)$$
$$\leq M|u + v|^2 - m|u - v|^2,$$

and therefore

$$4|(Tu, v)| \leq \mu(|u + v|^2 + |u - v|^2) = 2\mu(|u|^2 + |v|^2).$$

Replacing v by αv with $\alpha > 0$ yields

$$4|(Tu, v)| \leq 2\mu \left(\frac{|u|^2}{\alpha} + \alpha |v|^2 \right).$$

Next we minimize the right-hand side over α, i.e., choose $\alpha = |u|/|v|$, and then we obtain

$$|(Tu, v)| \leq \mu|u|\,|v| \quad \forall u, v, \text{ so that } \|T\| \leq \mu.$$

On the other hand, it is clear that $|(Tu, u)| \leq \|T\|\,|u|^2$, so that $|m| \leq \|T\|$ and $|M| \leq \|T\|$, and thus $\mu \leq \|T\|$.

Corollary 6.10. *Let $T \in \mathcal{L}(H)$ be a self-adjoint operator such that $\sigma(T) = \{0\}$. Then $T = 0$.*

Our last statement is a *fundamental result*. It asserts that every compact self-adjoint operator may be *diagonalized* in some suitable basis.

● **Theorem 6.11.** *Let H be a separable Hilbert space and let T be a compact self-adjoint operator. Then there exists a Hilbert basis composed of eigenvectors of T.*

Proof. Let $(\lambda_n)_{n \geq 1}$ be the sequence of all (distinct) nonzero eigenvalues of T. Set

$$\lambda_0 = 0, \quad E_0 = N(T), \quad \text{and} \quad E_n = N(T - \lambda_n I).$$

Recall that

$$0 \leq \dim E_0 \leq \infty \quad \text{and} \quad 0 < \dim E_n < \infty.$$

We claim that H is the Hilbert sum of the E_n's, $n = 0, 1, 2, \ldots$ (in the sense of Section 5.4):

(i) The spaces $(E_n)_{n \geq 0}$ are mutually orthogonal.
Indeed, if $u \in E_m$ and $v \in E_n$ with $m \neq n$, then

$$Tu = \lambda_m u \quad \text{and} \quad Tv = \lambda_n v,$$

so that

$$(Tu, v) = \lambda_m(u, v) = (u, Tv) = \lambda_n(u, v).$$

Therefore

$$(u, v) = 0.$$

(ii) Let F be the vector space spanned by the spaces $(E_n)_{n \geq 0}$. We shall prove that F is dense in H.
Clearly, $T(F) \subset F$. It follows that $T(F^\perp) \subset F^\perp$; indeed, given $u \in F^\perp$ we have

$$(Tu, v) = (u, Tv) = 0 \quad \forall v \in F,$$

so that $Tu \in F^\perp$. The operator T restricted to $F^\perp$ is denoted by T_0. This is a self-adjoint compact operator on $F^\perp$. We claim that $\sigma(T_0) = \{0\}$. Suppose not; suppose that some $\lambda \neq 0$ belongs to $\sigma(T_0)$. Since $\lambda \in EV(T_0)$, there is some $u \in F^\perp, u \neq 0$, such that $T_0 u = \lambda u$. Therefore, λ is one of the eigenvalues of T, say $\lambda = \lambda_n$ with $n \geq 1$. Thus $u \in E_n \subset F$. Since $u \in F^\perp \cap F$, we deduce that $u = 0$; a contradiction.
Applying Corollary 6.10, we deduce that $T_0 = 0$, i.e., T vanishes on $F^\perp$. It follows that $F^\perp \subset N(T)$. On the other hand, $N(T) \subset F$ and consequently $F^\perp \subset F$. This implies that $F^\perp = \{0\}$, and so F is dense in H.

Finally, we choose in each subspace $(E_n)_{n\geq 0}$ a Hilbert basis (the existence of such a basis for E_0 follows from Theorem 5.11; for the other E_n's, $n \geq 1$, this is obvious, since they are finite-dimensional). The union of these bases is clearly a Hilbert basis for H, composed of eigenvectors of T.

Remark 8. Let T be a compact self-adjoint operator. From the preceding analysis we may write any element $u \in H$ as

$$u = \sum_{n=0}^{\infty} u_n \text{ with } u_n \in E_n.$$

Then $Tu = \sum_{n=1}^{\infty} \lambda_n u_n$. Given an integer $k \geq 1$, set

$$T_k u = \sum_{n=1}^{k} \lambda_n u_n.$$

Clearly, T_k is a finite-rank operator and

$$\|T_k - T\| \leq \sup_{n \geq k+1} |\lambda_n| \to 0 \quad \text{as } k \to \infty.$$

Recall that in fact, in a Hilbert space, *every compact operator—not necessarily self-adjoint*—is the limit of a sequence of finite-rank operators (see Remark 1).

Comments on Chapter 6

⋆ 1. Fredholm operators.
Theorem 6.6 is the first step toward the theory of Fredholm operators. Given two Banach spaces E and F, one says that $A \in \mathcal{L}(E, F)$ is a *Fredholm operator* (or a *Noether operator*)—one writes $A \in \Phi(E, F)$—if it satisfies:

(i) $N(A)$ is finite-dimensional,
(ii) $R(A)$ is closed and has finite codimension.[3]

The *index* of A is defined by

$$\text{ind } A = \dim N(A) - \text{codim } R(A).$$

For example, $A = I - T$ with $T \in \mathcal{K}(E)$ is a Fredholm operator of index zero; this follows from Theorem 6.6.
The main properties of Fredholm operators are the following:

---

[3] Let $A \in \mathcal{L}(E, F)$ be such that $N(A)$ is finite-dimensional and $R(A)$ has finite codimension (i.e., there is a finite-dimensional space $G \subset F$ such that $R(A) + G = F$). Then it follows that $R(A)$ is closed (see Exercise 2.27).

(a) The class of Fredholm operators $\Phi(E, F)$ is an open subset of $\mathcal{L}(E, F)$ and the map $A \mapsto$ ind A is continuous; thus it is constant on each connected component of $\Phi(E, F)$.

(b) Every operator $A \in \Phi(E, F)$ is invertible modulo finite-rank operators, i.e., there exists an operator $B \in \mathcal{L}(F, E)$ such that

$$(A \circ B - I_F) \text{ and } (B \circ A - I_E) \text{ are finite-rank operators.}$$

Conversely, let $A \in \mathcal{L}(E, F)$ and assume that there exists $B \in \mathcal{L}(F, E)$ such that

$$A \circ B - I_F \in \mathcal{K}(F) \quad \text{and} \quad B \circ A - I_E \in \mathcal{K}(E).$$

Then $A \in \Phi(E, F)$.

(c) If $A \in \Phi(E, F)$ and $T \in \mathcal{K}(E, F)$ then $A + T \in \Phi(E, F)$ and $\text{ind}(A + T) = \text{ind} A$.

(d) If $A \in \Phi(E, F)$ and $B \in \Phi(F, G)$ then $B \circ A \in \Phi(E, G)$ and $\text{ind}(B \circ A) = \text{ind}(A) + \text{ind}(B)$.

On this question, see, e.g., T. Kato [1], M. Schechter [1], S. Lang [1], A. E. Taylor–D. C. Lay [1], P. Lax [1], L. Hörmander [2] (volume 3), and Problem 38.

⋆ **2. Hilbert–Schmidt operators.**
Let H be a separable Hilbert space. A bounded operator $T \in \mathcal{L}(H)$ is called a *Hilbert–Schmidt operator* if there is a Hilbert basis (e_n) in H such that $\|T\|_{\mathcal{HS}}^2 = \sum |Te_n|^2 < \infty$. One can prove that this definition is independent of the basis and that $\| \ \|_{\mathcal{HS}}$ is a norm. Every Hilbert–Schmidt operator is compact. Hilbert–Schmidt operators play an important role, in particular because of the following:

Theorem 6.12. *Let $H = L^2(\Omega)$ and $K(x, y) \in L^2(\Omega \times \Omega)$. Then the operator*

$$u \mapsto (Ku)(x) = \int_\Omega K(x, y)u(y)dy$$

is a Hilbert–Schmidt operator.

Conversely, every Hilbert–Schmidt operator on $L^2(\Omega)$ is of the preceding form for some unique function $K(x, y) \in L^2(\Omega \times \Omega)$.

On this question, see, e.g., A. Balakrishnan [1], N. Dunford–J. T. Schwartz [1], Volume 2, and Problem 40.

3. Multiplicity of eigenvalues.
Let $T \in \mathcal{K}(E)$ and let $\lambda \in \sigma(T) \backslash \{0\}$. One can show that the sequence $N((T - \lambda I)^k)$, $k = 1, 2, \ldots$, is strictly increasing up to some finite p and then it stays constant (see, e.g., A. E. Taylor–D. C. Lay [1], E. Kreyszig [1], and Problem 36). This integer p is called the *ascent* of $(T - \lambda I)$. The dimension of $N(T - \lambda I)$ is called by some authors the *geometric multiplicity* of λ, and the dimension of $N((T - \lambda I)^p)$ is called the *algebraic multiplicity* of λ; they coincide if E is a Hilbert space and T is self-adjoint (see Problem 36).

4. Spectral analysis.
Let H be a Hilbert space. Let $T \in \mathcal{L}(H)$ be a self-adjoint operator, possibly not compact. There is a construction called the spectral family of T that extends the spectral decomposition of Section 6.4. It allows one in particular to define a *functional calculus*, i.e., to give a sense to the quantity $f(T)$ for any continuous function f. It also extends to unbounded and non-self-adjoint operators, provided one assumes only that T is *normal*, i.e., $TT^\star = T^\star T$. *Spectral analysis* is a vast subject, especially in Banach spaces over $\mathbb{C}$ (see Section 11.4), with many applications and ramifications. For an elementary presentation see, e.g., W. Rudin [1], E. Kreyszig [1], A. Friedman [3], and K. Yosida [1]. For a more complete exposition, see, e.g., M. Reed–B. Simon [1], T. Kato [1], R. Dautray–J.-L. Lions [1], Chapters VIII and IX, N. Dunford–J. T. Schwartz [1], Volume 2, N. Akhiezer–I. Glazman [1], A. E. Taylor–D. C. Lay [1], J. Weidmann [1], J. B. Conway [1], P. Lax [1], and M. Schechter [2].

5. The min-max principle. The min-max formulas, due to Courant–Fischer, provide a very useful way of computing the eigenvalues; see, e.g., R. Courant–D. Hilbert [1], P. Lax [1], and Problem 37 . The monograph of H. Weinberger [2] contains numerous developments on this subject.

6. The Krein–Rutman theorem.
The following result has useful applications in the study of spectral properties of second-order elliptic operators (see Chapter 9).

⋆ **Theorem 6.13 (Krein–Rutman).** *Let E be a Banach space and let P be a convex cone with vertex at 0, i.e., $\lambda x + \mu y \in P \ \forall \lambda \geq 0, \ \forall \mu \geq 0, \ \forall x \in P, \ \forall y \in P$. Assume that P is closed, Int $P \neq \emptyset$, and $P \neq E$. Let $T \in \mathcal{K}(E)$ be such that $T(P \backslash \{0\}) \subset$ Int P. Then there exist some $x_0 \in$ Int P and some $\lambda_0 > 0$ such that $Tx_0 = \lambda_0 x_0$; moreover, λ_0 is the unique eigenvalue corresponding to an eigenvector of T in P, i.e., $Tx = \lambda x$ with $x \in P$ and $x \neq 0$, imply $\lambda = \lambda_0$ and $x = mx_0$ for some $m > 0$. Finally,*

$$\lambda_0 = \max\{|\lambda|; \ \lambda \in \sigma(T)\},$$

and the multiplicity (both geometric and algebraic) of λ equals one.

The proof presented in Problem 41 is due to P. Rabinowitz [2]. Variants of the above Krein–Rutman theorem may be found, e.g., in H. Schaefer [1], R. Nussbaum [1], F. F. Bonsall [1], and J. F. Toland [4].

Exercises for Chapter 6

6.1 Let $E = \ell^p$ with $1 \leq p \leq \infty$ (see Section 11.3). Let (λ_n) be a bounded sequence in $\mathbb{R}$ and consider the operator $T \in \mathcal{L}(E)$ defined by

$$Tx = (\lambda_1 x_1, \lambda_2 x_2, \ldots, \lambda_n x_n, \ldots),$$

where

$$x = (x_1, x_2, \ldots, x_n, \ldots).$$

Prove that T is a compact operator from E into E iff $\lambda_n \to 0$.

6.2 Let E and F be two Banach spaces, and let $T \in \mathcal{L}(E, F)$.

1. Assume that E is reflexive. Prove that $T(B_E)$ is closed (strongly).
2. Assume that E is reflexive and that $T \in \mathcal{K}(E, F)$. Prove that $T(B_E)$ is compact.
3. Let $E = F = C([0, 1])$ and $Tu(t) = \int_0^t u(s)ds$. Check that $T \in \mathcal{K}(E)$. Prove that $T(B_E)$ is not closed.

6.3 Let E and F be two Banach spaces, and let $T \in \mathcal{K}(E, F)$. Assume dim $E = \infty$. Prove that there exists a sequence (u_n) in E such that $\|u_n\|_E = 1$ and $\|Tu_n\|_F \to 0$.

[**Hint**: Argue by contradiction.]

6.4 Let $1 \le p < \infty$. Check that $\ell^p \subset c_0$ with continuous injection (for the definition of ℓ^p and c_0, see Section 11.3).
    Is this injection compact?

[**Hint**: Use the canonical basis (e_n) of ℓ^p.]

6.5 Let (λ_n) be a sequence of positive numbers such that $\lim_{n\to\infty} \lambda_n = +\infty$. Let V be the space of sequences $(u_n)_{n\ge1}$ such that

$$\sum_{n=1}^{\infty} \lambda_n |u_n|^2 < \infty.$$

The space V is equipped with the scalar product

$$((u, v)) = \sum_{n=1}^{\infty} \lambda_n u_n v_n.$$

Prove that V is a Hilbert space and that $V \subset \ell^2$ with compact injection.

6.6 Let $1 \le q \le p \le \infty$. Prove that the canonical injection from $L^p(0, 1)$ into $L^q(0, 1)$ is continuous but *not* compact.

[**Hint**: Use Rademacher's functions; see Exercise 4.18.]

6.7 Let E and F be two Banach spaces, and let $T \in \mathcal{L}(E, F)$. Consider the following properties:

(P) $\begin{cases} \text{For every weakly convergent sequence } (u_n) \text{ in } E, \\ u_n \rightharpoonup u, \text{ then } Tu_n \to Tu \text{ strongly in } F. \end{cases}$

(Q) $\begin{cases} T \text{ is continuous from } E \text{ equipped with the weak topology} \\ \sigma(E, E^\star) \text{ into } F \text{ equipped with the strong topology.} \end{cases}$

1. Prove that

$$(Q) \Leftrightarrow T \text{ is a finite-rank operator.}$$

2. Prove that $T \in \mathcal{K}(E, F) \Longrightarrow (P)$.
3. Assume that either $E = \ell^1$ or $F = \ell^1$. Prove that *every* operator $T \in \mathcal{L}(E, F)$ satisfies (P).

   [**Hint**: Use a result of Problem 8.]

   In what follows we assume that E is *reflexive*.

4. Prove that $T \in \mathcal{K}(E, F) \Longleftrightarrow (P)$.
5. Deduce that *every* operator $T \in \mathcal{L}(E, \ell^1)$ is compact.
6. Prove that *every* operator $T \in \mathcal{L}(c_0, E)$ is compact.

   [**Hint**: Consider the adjoint operator $T^\star$.]

6.8 Let E and F be two Banach spaces, and let $T \in \mathcal{K}(E, F)$. Assume that $R(T)$ is closed.

1. Prove that T is a finite-rank operator.

   [**Hint**: Use the open mapping theorem, i.e., Theorem 2.6.]

2. Assume, in addition, that dim $N(T) < \infty$. Prove that dim $E < \infty$.

6.9 Let E and F be two Banach spaces, and let $T \in \mathcal{L}(E, F)$.

1. Prove that the following three properties are equivalent:[4]

   (A)   dim $N(T) < \infty$ and $R(T)$ is closed.

   (B) $\begin{cases} \text{There are a finite-rank projection operator } P \in \mathcal{L}(E) \\ \text{and a constant } C \text{ such that} \\ \|u\|_E \leq C(\|Tu\|_F + \|Pu\|_E) \quad \forall u \in E. \end{cases}$

   (C) $\begin{cases} \text{There exist a Banach space } G, \text{ an operator} \\ Q \in \mathcal{K}(E, G), \text{ and a constant } C \text{ such that} \\ \|u\|_E \leq C(\|Tu\|_F + \|Qu\|_G) \quad \forall u \in E. \end{cases}$

   [**Hint**: When dim $N(T) < \infty$ consider a complement of $N(T)$; see Section 2.4.] Compare with Exercise 2.12.

2. Assume that T satisfies (A). Prove that $(T + S)$ also satisfies (A) for every $S \in \mathcal{K}(E, F)$.
3. Prove that the set of all operators $T \in \mathcal{L}(E, F)$ satisfying (A) is open in $\mathcal{L}(E, F)$.
4. Let F_0 be a closed linear subspace of F, and let $S \in \mathcal{K}(F_0, F)$.

Prove that $(I + S)(F_0)$ is a closed subspace of F.

---

[4] A *projection operator* is an operator P such that $P^2 = P$.

$\boxed{6.10}$ Let $Q(t) = \sum_{k=1}^{p} a_k t^k$ be a polynomial such that $Q(1) \neq 0$. Let E be a Banach space, and let $T \in \mathcal{L}(E)$. Assume that $Q(T) \in \mathcal{K}(E)$.

1. Prove that $\dim N(I - T) < \infty$, and that $R(I - T)$ is closed. More generally, prove that $(I - T)(E_0)$ is closed for every closed subspace $E_0 \subset E$.

   [**Hint**: Write $Q(1) - Q(t) = \widetilde{Q}(t)(1 - t)$ for some polynomial $\widetilde{Q}$ and apply Exercise 6.9.]

2. Prove that $N(I - T) = \{0\} \Leftrightarrow R(I - T) = E$.
3. Prove that $\dim N(I - T) = \dim N(I - T^\star)$.

   [**Hint for questions 2 and 3**: Use the same method as in the proof of Theorem 6.6.]

$\boxed{6.11}$ Let K be a compact metric space, and let $E = C(K; \mathbb{R})$ equipped with the usual norm $\|u\| = \max_{x \in K} |u(x)|$.

Let $F \subset E$ be a *closed* subspace. Assume that every function $u \in F$ is Hölder continuous, i.e.,

$$\begin{cases} \forall u \in F \ \exists \alpha \in (0, 1] \quad \text{and} \ \exists L \quad \text{such that} \\ |u(x) - u(y)| \leq L\, d(x, y)^\alpha \quad \forall x, y \in K. \end{cases}$$

The purpose of this exercise is to show that F is finite-dimensional.

1. Prove that there exist constants $\gamma \in (0, 1]$ and $C \geq 0$ (both independent of u) such that

$$|u(x) - u(y)| \leq C \|u\| d(x, y)^\gamma \quad \forall u \in F, \quad \forall x, y \in K.$$

   [**Hint**: Apply the Baire category theorem (Theorem 2.1) with

$$F_n = \{u \in F;\ |u(x) - u(y)| \leq n\, d(x, y)^{1/n} \quad \forall x, y \in K\}.]$$

2. Prove that B_F is compact and conclude.

$\boxed{6.12}$ *A lemma of J.-L. Lions.*

Let X, Y, and Z be three Banach spaces with norms $\|\ \|_X$, $\|\ \|_Y$, and $\|\ \|_Z$. Assume that $X \subset Y$ with *compact* injection and that $Y \subset Z$ with *continuous* injection. Prove that

$$\forall \varepsilon > 0 \ \exists C_\varepsilon \geq 0 \text{ satisfying } \|u\|_Y \leq \varepsilon \|u\|_X + C_\varepsilon \|u\|_Z \quad \forall u \in X.$$

   [**Hint**: Argue by contradiction.]

Application. Prove that $\forall \varepsilon > 0 \ \exists C_\varepsilon \geq 0$ satisfying

$$\max_{[0,1]} |u| \leq \varepsilon \max_{[0,1]} |u'| + C_\varepsilon \|u\|_{L^1} \quad \forall u \in C^1([0, 1]).$$

6.13 Let E and F be two Banach spaces with norms $\| \ \|_E$ and $\| \ \|_F$. Assume that E is reflexive. Let $T \in \mathcal{K}(E, F)$. Consider another norm $| \ |$ on E, which is weaker than the norm $\| \ \|_E$, i.e., $|u| \leq C \|u\|_E \quad \forall u \in E$. Prove that

$$\forall \varepsilon > 0 \ \exists C_\varepsilon \geq 0 \ \text{ satisfying } \|Tu\|_F \leq \varepsilon \|u\|_E + C_\varepsilon |u| \quad \forall u \in E.$$

Show that the conclusion may fail when E is *not* reflexive.

[**Hint**: Take $E = C([0, 1])$, $F = \mathbb{R}$, $\|u\| = \|u\|_{L^\infty}$ and $|u| = \|u\|_{L^1}$.]

6.14 Let E be a Banach space, and let $T \in \mathcal{L}(E)$ with $\|T\| < 1$.

1. Prove that $(I - T)$ is bijective and that

$$\|(I - T)^{-1}\| \leq 1/(1 - \|T\|).$$

2. Set $S_n = I + T + \cdots + T^{n-1}$. Prove that

$$\|S_n - (I - T)^{-1}\| \leq \|T\|^n/(1 - \|T\|).$$

6.15 Let E be a Banach space and let $T \in \mathcal{L}(E)$.

1. Let $\lambda \in \mathbb{R}$ be such that $|\lambda| > \|T\|$. Prove that

$$\|I + \lambda(T - \lambda I)^{-1}\| \leq \|T\|/(|\lambda| - \|T\|).$$

2. Let $\lambda \in \rho(T)$. Check that

$$(T - \lambda I)^{-1}T = T(T - \lambda I)^{-1},$$

and prove that
$$\text{dist}(\lambda, \sigma(T)) \geq 1/\|(T - \lambda I)^{-1}\|.$$

3. Assume that $0 \in \rho(T)$. Prove that

$$\sigma(T^{-1}) = 1/\sigma(T).$$

In what follows assume that $1 \in \rho(T)$; set

$$U = (T + I)(T - I)^{-1} = (T - I)^{-1}(T + I).$$

4. Check that $1 \in \rho(U)$ and give a simple expression for $(U - I)^{-1}$ in terms of T.
5. Prove that $T = (U + I)(U - I)^{-1}$.
6. Consider the function $f(t) = (t + 1)/(t - 1), t \in \mathbb{R}$. Prove that

$$\sigma(U) = f(\sigma(T)).$$

6.16 Let E be a Banach space and let $T \in \mathcal{L}(E)$.

1. Assume that $T^2 = I$. Prove that $\sigma(T) \subset \{-1, +1\}$ and determine $(T - \lambda I)^{-1}$ for $\lambda \neq \pm 1$.
2. More generally, assume that there is an integer $n \geq 2$ such that $T^n = I$. Prove that $\sigma(T) \subset \{-1, +1\}$ and determine $(T - \lambda I)^{-1}$ for $\lambda \neq \pm 1$.
3. Assume that there is an integer $n \geq 2$ such that $T^n = 0$. Prove that $\sigma(T) = \{0\}$ and determine $(T - \lambda I)^{-1}$ for $\lambda \neq 0$.
4. Assume that there is an integer $n \geq 2$ such that $\|T^n\| < 1$. Prove that $I - T$ is bijective and give an expression for $(I - T)^{-1}$ in terms of $(I - T^n)^{-1}$ and the iterates of T.

6.17 Let $E = \ell^p$ with $1 \leq p \leq \infty$ and let (λ_n) be a bounded sequence in $\mathbb{R}$. Consider the multiplication operator $M \in \mathcal{L}(E)$ defined by

$$Mx = (\lambda_1 x_1, \lambda_2 x_2, \ldots, \lambda_n x_n, \ldots), \quad \text{where } x = (x_1, x_2, \ldots, x_n, \ldots).$$

Determine $EV(M)$ and $\sigma(M)$.

6.18 *Spectral properties of the shifts.*
 An element $x \in E = \ell^2$ is denoted by $x = (x_1, x_2, \ldots, x_n, \ldots)$.
 Consider the operators

$$S_r x = (0, x_1, x_2, \ldots, x_{n-1}, \ldots),$$

and

$$S_\ell x = (x_2, x_3, x_4, \ldots, x_{n+1}, \ldots),$$

respectively called the *right shift* and *left shift*.

1. Determine $\|S_r\|$ and $\|S_\ell\|$. Does S_r or S_ℓ belong to $\mathcal{K}(E)$?
2. Prove that $EV(S_r) = \emptyset$.
3. Prove that $\sigma(S_r) = [-1, +1]$.
4. Prove that $EV(S_\ell) = (-1, +1)$. Determine the corresponding eigenspaces.
5. Prove that $\sigma(S_\ell) = [-1, +1]$.
6. Determine $S_r^\star$ and $S_\ell^\star$.
7. Prove that for every $\lambda \in (-1, +1)$, the spaces $R(S_r - \lambda I)$ and $R(S_\ell - \lambda I)$ are closed. Give an explicit representation of these spaces.

   [**Hint**: Apply Theorems 2.19 and 2.20.]

8. Prove that the spaces $R(S_r \pm I)$ and $R(S_\ell \pm I)$ are dense and that they are not closed.

   Consider the multiplication operator M defined by

$$Mx = (\alpha_1 x_1, \alpha_2 x_2, \ldots, \alpha_n x_n, \ldots),$$

   where (α_n) is a bounded sequence in $\mathbb{R}$.
9. Determine $EV(S_r \circ M)$.
10. Assume that $\alpha_n \to \alpha$ as $n \mapsto \infty$. Prove that

$$\sigma(S_r \circ M) = [-|\alpha|, +|\alpha|].$$

[**Hint**: Apply Theorem 6.6.]

11. Assume that for every integer n, $\alpha_{2n} = a$ and $\alpha_{2n+1} = b$ with $a \neq b$. Determine $\sigma(S_r \circ M)$.

[**Hint**: Compute $(S_r \circ M)^2$ and apply question 4 of Exercise 6.16.]

6.19 Let E be a Banach space and let $T \in \mathcal{L}(E)$.

1. Prove that $\sigma(T^\star) = \sigma(T)$.
2. Give examples showing that there is no general inclusion relation between $EV(T)$ and $EV(T^\star)$.

[**Hint**: Consider the right shift and the left shift.]

6.20 Let $E = L^p(0, 1)$ with $1 \leq p < \infty$. Given $u \in E$, set

$$Tu(x) = \int_0^x u(t)dt.$$

1. Prove that $T \in \mathcal{K}(E)$.

2. Determine $EV(T)$ and $\sigma(T)$.

3. Give an explicit formula for $(T - \lambda I)^{-1}$ when $\lambda \in \rho(T)$.

4. Determine $T^\star$.

6.21 Let V and H be two Banach spaces with norms $\| \|$ and $| |$ respectively, satisfying

$$V \subset H \text{ with compact injection.}$$

Let $p(u)$ be a seminorm on V such that $p(u) + |u|$ is a norm on V that is equivalent to $\| \|$.
Set

$$N = \{u \in V; p(u) = 0\},$$

and

$$\text{dist}(u, N) = \inf_{v \in N} \|u - v\| \text{ for } u \in V.$$

1. Prove that N is a finite-dimensional space.

[**Hint**: Consider the unit ball in N equipped with the norm $| |$.]

2. Prove that there exists a constant $K_1 > 0$ such that

$$p(u) \leq K_1 \text{dist}(u, N) \quad \forall u \in V.$$

3. Prove that there exists a constant $K_2 > 0$ such that

$$K_2 \, \text{dist}(u, N) \leq p(u) \quad \forall u \in V.$$

[**Hint**: Argue by contradiction. Assume that there is a sequence (u_n) in V such that $\text{dist}(u_n, N) = 1$ for all n and $p(u_n) \to 0$.]

$\boxed{6.22}$ Let E be a Banach space, and let $T \in \mathcal{L}(E)$. Given a polynomial $Q(t) = \sum_{k=0}^{p} a_k t^k$ with $a_k \in \mathbb{R}$, let $Q(T) = \sum_{k=0}^{p} a_k T^k$.

1. Prove that $Q(EV(T)) \subset EV(Q(T))$.

2. Prove that $Q(\sigma(T)) \subset \sigma(Q(T))$.

3. Construct an example in $E = \mathbb{R}^2$ for which the above inclusions are strict.

In what follows we assume that E is a Hilbert space (identified with its dual space $H^\star$) and that $T^\star = T$.

4. Assume here that the polynomial Q has no real root, i.e., $Q(t) \neq 0 \quad \forall t \in \mathbb{R}$. Prove that $Q(T)$ is bijective.

   [**Hint**: Start with the case that Q is a polynomial of degree 2 and more specifically, $Q(t) = t^2 + 1$.]

5. Deduce that for *every* polynomial Q, we have

   (i) $Q(EV(T)) = EV(Q(T))$,
   (ii) $Q(\sigma(T)) = \sigma(Q(T))$.

   [**Hint**: Write $Q(t) - \lambda = (t - t_1)(t - t_2) \cdots (t - t_q)\overline{Q}(t)$, where $t_1, t_2, \ldots, t_q$ are the real roots of $Q(t) - \lambda$ and $\overline{Q}$ has no real root.]

$\boxed{6.23}$ *Spectral radius.*
Let E be a Banach space and let $T \in \mathcal{L}(E)$. Set

$$a_n = \log \|T^n\|, \quad n \geq 1.$$

1. Check that

$$a_{i+j} \leq a_i + a_j \quad \forall i, j \geq 1.$$

2. Deduce that

$$\lim_{n \to +\infty} (a_n/n) \text{ exists and coincides with } \inf_{m \geq 1} (a_m/m).$$

   [**Hint**: Fix an integer $m \geq 1$. Given any integer $n \geq 1$ write $n = mq + r$, where $q = [\frac{n}{m}]$ is the largest integer $\leq n/m$ and $0 \leq r < m$. Note that $a_n \leq \frac{n}{m} a_m + a_r$.]

3. Conclude that $r(T) = \lim_{n \to \infty} \|T^n\|^{1/n}$ exists and that $r(T) \leq \|T\|$. Construct an example in $E = \mathbb{R}^2$ such that $r(T) = 0$ and $\|T\| = 1$.
   The number $r(T)$ is called the *spectral radius* of T.

4. Prove that $\sigma(T) \subset [-r(T), +r(T)]$. Deduce that if $\sigma(T) \neq \emptyset$, then

$$\max\{|\lambda|;\ \lambda \in \sigma(T)\} \leq r(T).$$

[**Hint**: Note that if $\lambda \in \sigma(T)$, then $\lambda^n \in \sigma(T^n)$; see Exercise 6.22.]

5. Construct an example in $E = \mathbb{R}^3$ such that $\sigma(T) = \{0\}$, while $r(T) = 1$.

In what follows we take $E = L^p(0, 1)$ with $1 \leq p \leq \infty$. Consider the operator $T \in \mathcal{L}(E)$ defined by

$$Tu(t) = \int_0^t u(s)ds.$$

6. Prove by induction that for $n \geq 2$,

$$(T^n u)(t) = \frac{1}{(n-1)!} \int_0^t (t-\tau)^{n-1} u(\tau)d\tau.$$

7. Deduce that $\|T^n\| \leq \frac{1}{n!}$.

[**Hint**: Use an inequality for the convolution product.]

8. Prove that the spectral radius of T is 0.

[**Hint**: Use Stirling's formula.]

9. Show that $\sigma(T) = \{0\}$. Compare with Exercise 6.20.

6.24 Assume that $T \in \mathcal{L}(H)$ is self-adjoint.

1. Prove that the following properties are equivalent:

(i) $(Tu, u) \geq 0\ \forall u \in H$,
(ii) $\sigma(T) \subset [0, \infty)$.

[**Hint**: Apply Proposition 6.9.]

2. Prove that the following properties are equivalent:

(iii) $\|T\| \leq 1$ and $(Tu, u) \geq 0\ \forall u \in H$,
(iv) $0 \leq (Tu, u) \leq |u|^2\ \forall u \in H$,
(v) $\sigma(T) \subset [0, 1]$,
(vi) $(Tu, u) \geq |Tu|^2\ \forall u \in H$.

[**Hint**: To prove that (v) $\Rightarrow$ (vi) apply Proposition 6.9 to $(T + \varepsilon I)^{-1}$ with $\varepsilon > 0$.]

3. Prove that the following properties are equivalent:

(vii) $(Tu, u) \leq |Tu|^2\ \forall u \in H$,
(viii) $(0, 1) \subset \rho(T)$.

[**Hint**: Introduce $U = 2T - I$.]

$\boxed{6.25}$ Let E be a Banach space, and let $K \in \mathcal{K}(E)$. Prove that there exist $M \in \mathcal{L}(E)$, $\widetilde{M} \in \mathcal{L}(E)$, and finite-rank projections P, $\widetilde{P}$ such that

(i) $M \circ (I + K) = I - P$,
(ii) $(I + K) \circ \widetilde{M} = I - \widetilde{P}$.

[**Hint**: Let X be a complement of $N(I + K)$ in E. Then $(I + K)_{|X}$ is bijective from X onto $R(I + K)$. Denote by M its inverse. Let Q be a projection from E onto $R(I + K)$ and set $\widetilde{M} = M \circ Q$. Show that (i) and (ii) hold.]

$\boxed{6.26}$ *From Brouwer to Schauder fixed-point theorems.*
In this exercise we assume that the following result is known (for a proof, see, e.g., K. Deimling [1], A. Granas–J. Dugundji [1], or L. Nirenberg [2]).

Theorem (Brouwer). *Let F be a finite-dimensional space, and let $Q \subset F$ be a nonempty compact convex set. Let $f : Q \to Q$ be a continuous map. Then f has a fixed point, i.e., there exists $p \in Q$ such that $f(p) = p$.*

Our goal is to prove the following.

Theorem (Schauder). *Let E be a Banach space, and let C be a nonempty closed convex set in E. Let $F : C \to C$ be a continuous map such that $F(C) \subset K$, where K is a compact subset of C. Then F has a fixed point in K.*

1. Given $\varepsilon > 0$, consider a finite covering of K, i.e., $K \subset \cup_{i \in I} B(y_i, \varepsilon/2)$, where I is finite, and $y_i \in K$ $\forall i \in I$. Define the function $q(x) = \sum_{i \in I} q_i(x)$, where

$$q_i(x) = \sum_{i \in I} \max\{\varepsilon - \|Fx - y_i\|, 0\}.$$

Check that q is continuous on C and that $q(x) \geq \varepsilon/2$ $\forall x \in C$.

2. Set

$$F_\varepsilon(x) = \frac{\sum_{i \in I} q_i(x) y_i}{q(x)}, \quad x \in C.$$

Prove that $F_\varepsilon : C \to C$ is continuous and that

$$\|F_\varepsilon(x) - F(x)\| \leq \varepsilon, \quad \forall x \in C.$$

3. Show that F_ε admits a fixed point $x_\varepsilon \in C$.

   [**Hint**: Let $Q = \text{conv}\,(\cup_{i \in I}\{y_i\})$. Check that $F_{\varepsilon|Q}$ admits a fixed point $x_\varepsilon \in Q$.]

4. Prove that (x_{ε_n}) converges to a limit $x \in C$ for some sequence $\varepsilon_n \to 0$. Show that $F(x) = x$.

Chapter 7
The Hille–Yosida Theorem

7.1 Definition and Elementary Properties of Maximal Monotone Operators

Throughout this chapter H denotes a Hilbert space.

Definition. An unbounded linear operator $A\colon D(A) \subset H \to H$ is said to be *monotone*[1] if it satisfies

$$(Av, v) \geq 0 \quad \forall v \in D(A).$$

It is called *maximal monotone* if, in addition, $R(I + A) = H$, i.e.,

$$\forall f \in H \ \exists u \in D(A) \text{ such that } u + Au = f.$$

Proposition 7.1. *Let A be a maximal monotone operator. Then*

(a) *$D(A)$ is dense in H,*
(b) *A is a closed operator,*
(c) *For every $\lambda > 0$, $(I + \lambda A)$ is bijective from $D(A)$ onto H, $(I + \lambda A)^{-1}$ is a bounded operator, and $\|(I + \lambda A)^{-1}\|_{\mathcal{L}(H)} \leq 1$.*

Proof.

(a) Let $f \in H$ be such that $(f, v) = 0 \ \forall v \in D(A)$. We claim that $f = 0$. Indeed, there exists some $v_0 \in D(A)$ such that $v_0 + Av_0 = f$. We have

$$0 = (f, v_0) = |v_0|^2 + (Av_0, v_0) \geq |v_0|^2.$$

Thus $v_0 = 0$ and hence $f = 0$.

(b) First, observe that given any $f \in H$, there exists a *unique* $u \in D(A)$ such that $u + Au = f$, since if $\bar{u}$ is another solution, we have

$$u - \bar{u} + A(u - \bar{u}) = 0.$$

---

[1] Some authors say that A is accretive or that $-A$ is dissipative.

H. Brezis, *Functional Analysis, Sobolev Spaces and Partial Differential Equations*, DOI 10.1007/978-0-387-70914-7_7, © Springer Science+Business Media, LLC 2011

Taking the scalar product with $(u - \bar{u})$ and using monotonicity, we see that $u - \bar{u} = 0$. Next, note that $|u| \leq |f|$, since $|u|^2 + (Au, u) = (f, u) \geq |u|^2$. Therefore the map $f \mapsto u$, denoted by $(I + A)^{-1}$, is a bounded linear operator from H into itself and $\|(I + A)^{-1}\|_{\mathcal{L}(H)} \leq 1$. We now prove that A is a closed operator. Let (u_n) be a sequence in $D(A)$ such that $u_n \to u$ and $Au_n \to f$. We have to check that $u \in D(A)$ and that $Au = f$. But $u_n + Au_n \to u + f$ and thus

$$u_n = (I + A)^{-1}(u_n + Au_n) \to (I + A)^{-1}(u + f).$$

Hence $u = (I + A)^{-1}(u + f)$, i.e., $u \in D(A)$ and $u + Au = u + f$.

(c) We will prove that if $R(I + \lambda_0 A) = H$ for some $\lambda_0 > 0$ then $R(I + \lambda A) = H$ for every $\lambda > \lambda_0/2$. Note first—as in part (b)—that for every $f \in H$ there is a unique $u \in D(A)$ such that $u + \lambda_0 Au = f$. Moreover, the map $f \mapsto u$, denoted by $(I + \lambda_0 A)^{-1}$, is a bounded linear operator with $\|(I + \lambda_0 A)^{-1}\|_{\mathcal{L}(H)} \leq 1$. We try to solve the equation

$$(1) \qquad\qquad u + \lambda Au = f \text{ with } \lambda > 0.$$

Equation (1) may be written as

$$u + \lambda_0 Au = \frac{\lambda_0}{\lambda} f + \left(1 - \frac{\lambda_0}{\lambda}\right) u$$

or alternatively

$$(2) \qquad\qquad u = (I + \lambda_0 A)^{-1} \left[\frac{\lambda_0}{\lambda} f + \left(1 - \frac{\lambda_0}{\lambda}\right) u\right].$$

If $|1 - \frac{\lambda_0}{\lambda}| < 1$, i.e., $\lambda > \lambda_0/2$, we may apply the contraction mapping principle (Theorem 5.7) and deduce that (2) has a solution.

Conclusion (c) follows easily by induction: since $I + A$ is surjective, $I + \lambda A$ is surjective for every $\lambda > 1/2$, and thus for every $\lambda > 1/4$, etc.

Remark 1. If A is maximal monotone then λA is also maximal monotone for every $\lambda > 0$. However, if A and B are maximal monotone operators, then $A + B$, defined on $D(A) \cap D(B)$, need not be maximal monotone.

Definition. Let A be a maximal monotone operator. For every $\lambda > 0$, set

$$\boxed{J_\lambda = (I + \lambda A)^{-1} \quad \text{and} \quad A_\lambda = \frac{1}{\lambda}(I - J_\lambda);}$$

J_λ is called the *resolvent* of A, and A_λ is the *Yosida approximation* (or *regularization*) of A. Keep in mind that $\|J_\lambda\|_{\mathcal{L}(H)} \leq 1$.

Proposition 7.2. *Let A be a maximal monotone operator. Then*

$$(a_1) \qquad\qquad A_\lambda v = A(J_\lambda v) \qquad \forall v \in H \text{ and } \forall \lambda > 0,$$

(a$_2$) $\qquad A_\lambda v = J_\lambda(Av) \qquad \forall v \in D(A)$ *and* $\forall \lambda > 0,$

(b) $\qquad |A_\lambda v| \leq |Av| \qquad \forall v \in D(A)$ *and* $\forall \lambda > 0,$

(c) $\qquad \lim_{\lambda \to 0} J_\lambda v = v \qquad \forall v \in H,$

(d) $\qquad \lim_{\lambda \to 0} A_\lambda v = Av \qquad \forall v \in D(A),$

(e) $\qquad (A_\lambda v, v) \geq 0 \qquad \forall v \in H$ *and* $\forall \lambda > 0,$

(f) $\qquad |A_\lambda v| \leq (1/\lambda)|v| \qquad \forall v \in H$ *and* $\forall \lambda > 0.$

Proof.

(a$_1$) can be written as $v = (J_\lambda v) + \lambda A(J_\lambda v)$, which is just the definition of $J_\lambda v$.

(a$_2$) By (a$_1$) we have

$$A_\lambda v + A(v - J_\lambda v) = Av,$$

i.e.,

$$A_\lambda v + \lambda A(A_\lambda v) = Av,$$

which means that $A_\lambda v = (I + \lambda A)^{-1} Av$.

(b) Follows easily from (a$_2$).

(c) Assume first that $v \in D(A)$. Then

$$|v - J_\lambda v| = \lambda |A_\lambda v| \leq \lambda |Av| \quad \text{by (b)}$$

and thus $\lim_{\lambda \to 0} J_\lambda v = v$.

Suppose now that v is a general element in H. Given any $\varepsilon > 0$ there exists some $v_1 \in D(A)$ such that $|v - v_1| \leq \varepsilon$ (since $D(A)$ is dense in H by Proposition 7.1). We have

$$|J_\lambda v - v| \leq |J_\lambda v - J_\lambda v_1| + |J_\lambda v_1 - v_1| + |v_1 - v|$$
$$\leq 2|v - v_1| + |J_\lambda v_1 - v_1| \leq 2\varepsilon + |J_\lambda v_1 - v_1|.$$

Thus

$$\limsup_{\lambda \to 0} |J_\lambda v - v| \leq 2\varepsilon \quad \forall \varepsilon > 0,$$

and so

$$\lim_{\lambda \to 0} |J_\lambda v - v| = 0.$$

(d) This is a consequence of (a$_2$) and (c).

(e) We have

$$(A_\lambda v, v) = (A_\lambda v, v - J_\lambda v) + (A_\lambda v, J_\lambda v) = \lambda |A_\lambda v|^2 + (A(J_\lambda v), J_\lambda v),$$

and thus

(3) $\qquad\qquad (A_\lambda v, v) \geq \lambda |A_\lambda v|^2.$

(f) This is a consequence of (3) and the Cauchy–Schwarz inequality.

Remark 2. Proposition 7.2 implies that $(A_\lambda)_{\lambda>0}$ is a family of *bounded* operators that "approximate" the *unbounded* operator A as $\lambda \to 0$. This approximation will be used very often. Of course, in general, $\|A_\lambda\|_{\mathcal{L}(H)}$ "blows up" as $\lambda \to 0$.

7.2 Solution of the Evolution Problem $\frac{du}{dt} + Au = 0$ on $[0, +\infty)$, $u(0) = u_0$. Existence and uniqueness

We start with a very classical result:

• **Theorem 7.3 (Cauchy, Lipschitz, Picard).** *Let E be a Banach space and let $F :$ $E \to E$ be a Lipschitz map, i.e., there is a constant L such that*

$$\|Fu - Fv\| \le L\|u - v\| \quad \forall u, v \in E.$$

Then given any $u_0 \in E$, there exists a unique solution $u \in C^1([0, +\infty); E)$ of the problem

$$(4) \qquad \begin{cases} \dfrac{du}{dt}(t) = Fu(t) & on \ [0, +\infty), \\ u(0) = u_0. \end{cases}$$

u_0 is called the *initial data*.

Proof.
Existence. Solving (4) amounts to finding some $u \in C([0, +\infty); E)$ satisfying the integral equation

$$(5) \qquad u(t) = u_0 + \int_0^t F(u(s))ds.$$

Given $k > 0$, to be fixed later, set

$$X = \left\{ u \in C([0, +\infty); E); \ \sup_{t \ge 0} e^{-kt}\|u(t)\| < \infty \right\}.$$

It is easy to check that X is a Banach space for the norm

$$\|u\|_X = \sup_{t \ge 0} e^{-kt}\|u(t)\|.$$

For every $u \in X$, the function Φu defined by

$$(\Phi u)(t) = u_0 + \int_0^t F(u(s))ds$$

also belongs to X. Moreover, we have

$$\|\Phi u - \Phi v\|_X \le \frac{L}{k} \|u - v\|_X \quad \forall u, v \in X.$$

Fixing any $k > L$, we find that Φ has a (unique) fixed point u in X, which is a solution of (5).

Uniqueness. Let u and $\bar{u}$ be two solutions of (4) and set

$$\varphi(t) = \|u(t) - \bar{u}(t)\|.$$

From (5) we deduce that

$$\varphi(t) \le L \int_0^t \varphi(s)ds \quad \forall t \ge 0$$

and consequently $\varphi \equiv 0$.

The preceding theorem is extremely useful in the study of *ordinary differential equations*. However, it is of little use in the study of partial differential equations. Our next result is a *very powerful tool* in solving *evolution partial differential equations*; see Chapter 10.

• **Theorem 7.4 (Hille–Yosida).** *Let A be a maximal monotone operator. Then, given any $u_0 \in D(A)$ there exists a unique function*[2]

$$u \in C^1([0, +\infty); H) \cap C([0, +\infty); D(A))$$

satisfying

(6)
$$\begin{cases} \dfrac{du}{dt} + Au = 0 & \text{on } [0, +\infty), \\ u(0) = u_0. \end{cases}$$

Moreover,

$$|u(t)| \le |u_0| \quad and \quad \left| \frac{du}{dt}(t) \right| = |Au(t)| \le |Au_0| \quad \forall t \ge 0.$$

Remark 3. The main interest of Theorem 7.4 lies in the fact that we reduce the study of an "*evolution problem*" to the study of the "*stationary equation*" $u + Au = f$ (assuming we already know that A is monotone, which is easy to check in practice).

Proof. It is divided into six steps.

Step 1: *Uniqueness.* Let u and $\bar{u}$ be two solutions of (6). We have

$$\left(\frac{d}{dt}(u - \bar{u}), (u - \bar{u}) \right) = -(A(u - \bar{u}), u - \bar{u}) \le 0.$$

---

[2] The space $D(A)$ is equipped with the *graph norm* $|v| + |Av|$ or with the equivalent Hilbert norm $(|v|^2 + |Av|^2)^{1/2}$.

But[3]

$$\frac{1}{2}\frac{d}{dt}|u(t) - \overline{u}(t)|^2 = \left(\frac{d}{dt}(u(t) - \overline{u}(t)), \, u(t) - \overline{u}(t)\right).$$

Thus, the function $t \mapsto |u(t) - \overline{u}(t)|$ is nonincreasing on $[0, +\infty)$. Since $|u(0) - \overline{u}(0)| = 0$, it follows that

$$|u(t) - \overline{u}(t)| = 0 \quad \forall t \geq 0.$$

The main idea in order to prove existence is to replace A by A_λ in (6), to apply Theorem 7.3 on the approximate problem, and then to pass to the limit as $\lambda \to 0$ using various *estimates that are independent of λ*. So, let u_λ be the solution of the problem

(7)
$$\begin{cases} \dfrac{du_\lambda}{dt} + A_\lambda u_\lambda = 0 \quad \text{on } [0, +\infty), \\ u_\lambda(0) = u_0 \in D(A). \end{cases}$$

Step 2: We have the estimates

(8)
$$|u_\lambda(t)| \leq |u_0| \cdot \forall t \geq 0, \quad \forall \lambda > 0,$$

(9)
$$\left|\frac{du_\lambda}{dt}(t)\right| = |A_\lambda u_\lambda(t)| \leq |Au_0| \quad \forall t \geq 0, \quad \forall \lambda > 0.$$

They follow directly from the next lemma and the fact that $|A_\lambda u_0| \leq |Au_0|$.

Lemma 7.1. *Let $w \in C^1([0, +\infty); H)$ be a function satisfying*

(10)
$$\frac{dw}{dt} + A_\lambda w = 0 \text{ on } [0, +\infty).$$

Then the functions $t \mapsto |w(t)|$ and $t \mapsto \left|\frac{dw}{dt}(t)\right| = |A_\lambda w(t)|$ are nonincreasing on $[0, +\infty)$.

Proof. We have

$$\left(\frac{dw}{dt}, \, w\right) + (A_\lambda w, \, w) = 0.$$

By Proposition 7.2(e) we know that $(A_\lambda w, w) \geq 0$ and thus $\frac{1}{2}\frac{d}{dt}|w|^2 \leq 0$, so that $|w(t)|$ is nonincreasing. On the other hand, since A_λ is a linear bounded operator, we deduce (by induction) from (10) that $w \in C^\infty([0, +\infty); H)$ and also that

$$\frac{d}{dt}\left(\frac{dw}{dt}\right) + A_\lambda\left(\frac{dw}{dt}\right) = 0.$$

Applying the preceding fact to $\frac{dw}{dt}$, we see that $\left|\frac{dw}{dt}(t)\right|$ is nonincreasing. In fact, at any order k, the function $\left|\frac{d^k w}{dt^k}(t)\right|$ is nonincreasing.

[3] Keep in mind that if $\varphi \in C^1([0, +\infty); H)$, then $|\varphi|^2 \in C^1([0, +\infty); \mathbb{R})$ and $\frac{d}{dt}|\varphi|^2 = 2(\frac{d\varphi}{dt}, \varphi)$.

Step 3: We will prove here that for every $t \geq 0$, $u_\lambda(t)$ converges, as $\lambda \to 0$, to some limit, denoted by $u(t)$. Moreover, the convergence is uniform on every bounded interval $[0, T]$.

For every λ, $\mu > 0$ we have

$$\frac{du_\lambda}{dt} - \frac{du_\mu}{dt} + A_\lambda u_\lambda - A_\mu u_\mu = 0$$

and thus

(11) $\quad \frac{1}{2}\frac{d}{dt}|u_\lambda(t) - u_\mu(t)|^2 + (A_\lambda u_\lambda(t) - A_\mu u_\mu(t),\ u_\lambda(t) - u_\mu(t)) = 0.$

Dropping t for simplicity, we write

$$(A_\lambda u_\lambda - A_\mu u_\mu, u_\lambda - u_\mu)$$

(12)
$$= (A_\lambda u_\lambda - A_\mu u_\mu, u_\lambda - J_\lambda u_\lambda + J_\lambda u_\lambda - J_\mu u_\mu + J_\mu u_\mu - u_\mu)$$
$$= (A_\lambda u_\lambda - A_\mu u_\mu, \lambda A_\lambda u_\lambda - \mu A_\mu u_\mu)$$
$$+ (A(J_\lambda u_\lambda - J_\mu u_\mu), J_\lambda u_\lambda - J_\mu u_\mu)$$
$$\geq (A_\lambda u_\lambda - A_\mu u_\mu, \lambda A_\lambda u_\lambda - \mu A_\mu u_\mu).$$

It follows from (9), (11), and (12) that

$$\frac{1}{2}\frac{d}{dt}|u_\lambda - u_\mu|^2 \leq 2(\lambda + \mu)|Au_0|^2.$$

Integrating this inequality, we obtain

$$|u_\lambda(t) - u_\mu(t)|^2 \leq 4(\lambda + \mu)t|Au_0|^2,$$

i.e.,

(13) $\quad |u_\lambda(t) - u_\mu(t)| \leq 2\sqrt{(\lambda + \mu)t}|Au_0|.$

It follows that for every fixed $t \geq 0$, $u_\lambda(t)$ is a Cauchy sequence as $\lambda \to 0$ and thus it converges to a limit, denoted by $u(t)$. Passing to the limit in (13) as $\mu \to 0$, we have

$$|u_\lambda(t) - u(t)| \leq 2\sqrt{\lambda t}|Au_0|.$$

Therefore, the convergence is uniform in t on every bounded interval $[0, T]$ and so $u \in C([0, +\infty); H)$.

Step 4: Assuming, in addition, that $u_0 \in D(A^2)$, i.e., $u_0 \in D(A)$ and $Au_0 \in D(A)$, we prove here that $\frac{du_\lambda}{dt}(t)$ converges, as $\lambda \to 0$, to some limit and that the convergence is uniform on every bounded interval $[0, T]$.

Set $v_\lambda = \frac{du_\lambda}{dt}$, so that $\frac{dv_\lambda}{dt} + A_\lambda v_\lambda = 0$. Following the same argument as in Step 3, we see that

(14) $\dfrac{1}{2}\dfrac{d}{dt}|v_\lambda - v_\mu|^2 \leq (|A_\lambda v_\lambda| + |A_\mu v_\mu|)(\lambda|A_\lambda v_\lambda| + \mu|A_\mu v_\mu|).$

By Lemma 7.1 we have

(15) $|A_\lambda v_\lambda(t)| \leq |A_\lambda v_\lambda(0)| = |A_\lambda A_\lambda u_0|$

and similarly

(16) $|A_\mu v_\mu(t)| \leq |A_\mu v_\mu(0)| = |A_\mu A_\mu u_0|.$

Finally, since $Au_0 \in D(A)$, we obtain

$$A_\lambda A_\lambda u_0 = J_\lambda A J_\lambda A u_0 = J_\lambda J_\lambda A A u_0 = J_\lambda^2 A^2 u_0$$

and thus

(17) $|A_\lambda A_\lambda u_0| \leq |A^2 u_0|, \quad |A_\mu A_\mu u_0| \leq |A^2 u_0|.$

Combining (14), (15), (16), and (17), we are led to

$$\dfrac{1}{2}\dfrac{d}{dt}|v_\lambda - v_\mu|^2 \leq 2(\lambda + \mu)|A^2 u_0|^2.$$

We conclude, as in Step 3, that $v_\lambda(t) = \frac{du_\lambda}{dt}(t)$ converges, as $\lambda \to 0$, to some limit and that the convergence is uniform on every bounded interval $[0, T]$.

Step 5: Assuming that $u_0 \in D(A^2)$ we prove here that u is a solution of (6).
   By Steps 3 and 4 we know that for all $T < \infty$,

$$\begin{cases} u_\lambda(t) \to u(t), & \text{as } \lambda \to 0, \quad \text{uniformly on } [0, T], \\ \dfrac{du_\lambda}{dt}(t) & \text{converges, as } \lambda \to 0, \text{ uniformly on } [0, T]. \end{cases}$$

It follows easily that $u \in C^1([0, +\infty); H)$ and that $\frac{du_\lambda}{dt}(t) \to \frac{du}{dt}(t)$, as $\lambda \to 0$, uniformly on $[0, T]$. Rewrite (7) as

(18) $\dfrac{du_\lambda}{dt}(t) + A(J_\lambda u_\lambda(t)) = 0.$

Note that $J_\lambda u_\lambda(t) \to u(t)$ as $\lambda \to 0$, since

$$|J_\lambda u_\lambda(t) - u(t)| \leq |J_\lambda u_\lambda(t) - J_\lambda u(t)| + |J_\lambda u(t) - u(t)|$$
$$\leq |u_\lambda(t) - u(t)| + |J_\lambda u(t) - u(t)| \to 0.$$

Applying the fact that A has a closed graph, we deduce from (18) that $u(t) \in D(A)$ $\forall t \geq 0$, and that

$$\dfrac{du}{dt}(t) + Au(t) = 0.$$

Finally, since $u \in C^1([0, +\infty); \ H)$, the function $t \mapsto Au(t)$ is continuous from $[0, +\infty)$ into H and thus $u \in C([0, +\infty); \ D(A))$. Hence we have obtained a solution of (6) satisfying, in addition,

$$|u(t)| \le |u_0| \ \forall t \ge 0 \quad \text{and} \quad \left|\frac{du}{dt}(t)\right| = |Au(t)| \le |Au_0| \ \forall t \ge 0.$$

Step 6: We conclude here the proof of the theorem.

We shall use the following lemma.

Lemma 7.2. *Let $u_0 \in D(A)$. Then $\forall \varepsilon > 0 \ \exists \overline{u}_0 \in D(A^2)$ such that $|u_0 - \overline{u}_0| < \varepsilon$ and $|Au_0 - A\overline{u}_0| < \varepsilon$. In other words, $D(A^2)$ is dense in $D(A)$ (for the graph norm).*

Proof of Lemma 7.2. Set $\overline{u}_0 = J_\lambda u_0$ for some appropriate $\lambda > 0$ to be fixed later. We have

$$\overline{u}_0 \in D(A) \quad \text{and} \quad \overline{u}_0 + \lambda A\overline{u}_0 = u_0.$$

Thus $A\overline{u}_0 \in D(A)$, i.e., $\overline{u}_0 \in D(A^2)$. On the other hand, by Proposition 7.2, we know that

$$\lim_{\lambda \to 0} |J_\lambda u_0 - u_0| = 0, \quad \lim_{\lambda \to 0} |J_\lambda Au_0 - Au_0| = 0, \quad \text{and} \quad J_\lambda Au_0 = AJ_\lambda u_0.$$

The desired conclusion follows by choosing $\lambda > 0$ small enough.

We now turn to the proof of Theorem 7.4. Given $u_0 \in D(A)$ we construct (using Lemma 7.2) a sequence (u_{0n}) in $D(A^2)$ such that $u_{0n} \to u_0$ and $Au_{0n} \to Au_0$. By Step 5 we know that there is a solution u_n of the problem

$$(19) \qquad \begin{cases} \dfrac{du_n}{dt} + Au_n = 0 \quad \text{on } [0, +\infty), \\ u_n(0) = u_{0n}. \end{cases}$$

We have, for all $t \ge 0$,

$$|u_n(t) - u_m(t)| \le |u_{0n} - u_{0m}| \underset{m,n \to \infty}{\longrightarrow} 0,$$

$$\left|\frac{du_n}{dt}(t) - \frac{du_m}{dt}(t)\right| \le |Au_{0n} - Au_{0m}| \underset{m,n \to \infty}{\longrightarrow} 0.$$

Therefore

$$u_n(t) \to u(t) \quad \text{uniformly on } [0, +\infty),$$

$$\frac{du_n}{dt}(t) \to \frac{du}{dt}(t) \quad \text{uniformly on } [0, +\infty),$$

with $u \in C^1([0, +\infty); \ H)$. Passing to the limit in (19)—using the fact that A is a closed operator—we see that $u(t) \in D(A)$ and u satisfies (6). From (6) we deduce that $u \in C([0, +\infty); \ D(A))$.

Remark 4. Let u_λ be the solution of (7):

(a) *Assume $u_0 \in D(A)$.* We know (by Step 3) that as $\lambda \to 0$, $u_\lambda(t)$ converges, for every $t \geq 0$, to some limit $u(t)$. One can prove directly that $u \in C^1([0, +\infty); H) \cap C([0, +\infty); D(A))$ and that it satisfies (6).

(b) *Assume only that $u_0 \in H$.* One can still prove that as $\lambda \to 0$, $u_\lambda(t)$ converges for every $t \geq 0$, to some limit, denoted by $u(t)$. But it may happen that this limit $u(t)$ does not belong to $D(A)$ $\forall t > 0$ and that $u(t)$ is nowhere differentiable on $[0, +\infty)$. Hence $u(t)$ is not a "classical" solution of (6). In fact, for such a u_0, problem (6) has no classical solution. Nevertheless, we may view $u(t)$ as a *"generalized" solution* of (6). We shall see in Section 7.4 that this does not happen when A is *self-adjoint*: in this case $u(t)$ is a *"classical"* solution of (6) for *every* $u_0 \in H$, even when $u_0 \notin D(A)$.

★ *Remark* 5 (*Contraction semigroups*). For each $t \geq 0$ consider the linear map $u_0 \in D(A) \mapsto u(t) \in D(A)$, where $u(t)$ is the solution of (6) given by Theorem 7.4. Since $|u(t)| \leq |u_0|$ and since $D(A)$ is dense in H, we may extend this map by continuity as a bounded operator from H into itself, denoted by $S_A(t)$.[4] It is easy to check that $S_A(t)$ satisfies the following properties:

(a) $\qquad$ for each $t \geq 0$, $S_A(t) \in \mathcal{L}(H)$ and $\|S_A(t)\|_{\mathcal{L}(H)} \leq 1$,

(b) $\qquad \begin{cases} S_A(t_1 + t_2) = S_A(t_1) \circ S_A(t_2) \quad \forall t_1, t_2 \geq 0, \\ S_A(0) = I, \end{cases}$

(c) $\qquad \lim_{\substack{t \to 0 \\ t > 0}} |S_A(t)u_0 - u_0| = 0 \quad \forall u_0 \in H$.

Such a family $\{S(t)\}_{t \geq 0}$ of operators (from H into itself) depending on a parameter $t \geq 0$ and satisfying (a), (b), (c) is called a *continuous semigroup of contractions*.

A remarkable result due to Hille and Yosida asserts that conversely, given a continuous semigroup of contractions $S(t)$ on H there exists a unique maximal monotone operator A such that $S(t) = S_A(t)$ $\forall t \geq 0$. This establishes a *bijective correspondence between maximal monotone operators and continuous semigroups of contractions*. (For a proof see the references quoted in the comments on Chapter 7.)

● *Remark* 6. Let A be a maximal monotone operator and let $\lambda \in \mathbb{R}$. The problem

$$\begin{cases} \dfrac{du}{dt} + Au + \lambda u = 0 \quad \text{on } [0, +\infty), \\ u(0) = u_0, \end{cases}$$

reduces to problem (6) using the following simple device. Set

$$v(t) = e^{\lambda t} u(t).$$

Then v satisfies

---

[4] Alternatively one may use Remark 4(b) to define $S_A(t)$ on H directly as being the map $u_0 \in H \mapsto u(t) \in H$.

$$\begin{cases} \dfrac{dv}{dt} + Av = 0 \quad \text{on } [0, +\infty), \\ v(0) = u_0. \end{cases}$$

7.3 Regularity

We shall prove here that the solution u of (6) obtained in Theorem 7.4 is more regular than just $C^1([0, +\infty);\ H) \cap C([0, +\infty);\ D(A))$ provided one makes additional assumptions on the initial data u_0. For this purpose we define by induction the space

$$D(A^k) = \{v \in D(A^{k-1});\ Av \in D(A^{k-1})\},$$

where k is any integer, $k \geq 2$. It is easily seen that $D(A^k)$ is a Hilbert space for the scalar product

$$(u,\ v)_{D(A^k)} = \sum_{j=0}^{k} (A^j u,\ A^j v);$$

the corresponding norm is

$$|u|_{D(A^k)} = \left(\sum_{j=0}^{k} |A^j u|^2 \right)^{1/2}.$$

Theorem 7.5. *Assume $u_0 \in D(A^k)$ for some integer $k \geq 2$. Then the solution u of problem* (6) *obtained in Theorem 7.4 satisfies*

$$u \in C^{k-j}([0, +\infty);\ D(A^j)) \quad \forall j = 0, 1, \ldots, k.$$

Proof. Assume first that $k = 2$. Consider the Hilbert space $H_1 = D(A)$ equipped with the scalar product $(u, v)_{D(A)}$. It is easy to check that the operator $A_1 : D(A_1) \subset H_1 \to H_1$ defined by

$$\begin{cases} D(A_1) = D(A^2), \\ A_1 u = Au \quad \text{for } u \in D(A_1), \end{cases}$$

is maximal monotone in H_1. Applying Theorem 7.4 *to the operator A_1 in the space H_1*, we see that there exists a function

$$u \in C^1([0, +\infty);\ H_1) \cap C([0, +\infty);\ D(A_1))$$

such that

$$\begin{cases} \dfrac{du}{dt} + A_1 u = 0 \quad \text{on } [0, +\infty), \\ u(0) = u_0. \end{cases}$$

In particular, u satisfies (6); by uniqueness, this u is *the* solution of (6). It remains only to check that $u \in C^2([0, +\infty);\ H)$. Since

$$A \in \mathcal{L}(H_1, H) \quad \text{and} \quad u \in C([0, +\infty); H_1),$$

it follows that $Au \in C^1([0, +\infty); H)$ and

$$(20) \qquad\qquad\qquad \frac{d}{dt}(Au) = A\left(\frac{du}{dt}\right).$$

Applying (6), we see that $\frac{du}{dt} \in C^1([0, +\infty); H)$, i.e., $u \in C^2([0, +\infty); H)$ and that

$$(21) \qquad\qquad \frac{d}{dt}\left(\frac{du}{dt}\right) + A\left(\frac{du}{dt}\right) = 0 \quad \text{on } [0, +\infty).$$

We now turn to the general case $k \geq 3$. We argue by induction on k: assume that the result holds up to order $(k-1)$ and let $u_0 \in D(A^k)$. By the preceding analysis we know that the solution u of (6) belongs to $C^2([0, +\infty); H) \cap C^1([0, +\infty); D(A))$ and that u satisfies (21). Letting

$$v = \frac{du}{dt},$$

we have

$$v \in C^1([0, +\infty); H) \cap C([0, +\infty); D(A)),$$

$$\begin{cases} \dfrac{dv}{dt} + Av = 0 \quad \text{on } [0, +\infty), \\ v(0) = -Au_0. \end{cases}$$

In other words, v is *the* solution of (6) corresponding to the initial data $v_0 = -Au_0$. Since $v_0 \in D(A^{k-1})$, we know, by the induction assumption, that

$$(22) \qquad\qquad v \in C^{k-1-j}([0, +\infty); D(A^j)) \quad \forall j = 0, 1, \ldots, k-1,$$

i.e.,

$$u \in C^{k-j}([0, +\infty); D(A^j)) \qquad \forall j = 0, 1, \ldots, k-1.$$

It remains only to check that

$$(23) \qquad\qquad\qquad u \in C([0, +\infty); D(A^k)).$$

Applying (22) with $j = k-1$, we see that

$$(24) \qquad\qquad\qquad \frac{du}{dt} \in C([0, +\infty); D(A^{k-1})).$$

It follows from (24) and equation (6) that

$$Au \in C([0, +\infty); D(A^{k-1})),$$

i.e., (23).

7.4 The Self-Adjoint Case

Let $A : D(A) \subset H \to H$ be an unbounded linear operator with $\overline{D(A)} = H$. Identifying $H^\star$ with H, we may view $A^\star$ as an unbounded linear operator in H.

Definition. One says that

- A is *symmetric* if $(Au, v) = (u, Av)$ $\forall u, v \in D(A)$,
- A is *self-adjoint* if $D(A^\star) = D(A)$ and $A^\star = A$.

Remark 7. For *bounded* operators the notions of symmetric and self-adjoint operators coincide. However, if A is *unbounded* there is a *subtle difference* between *symmetric* and *self-adjoint* operators. Clearly, any self-adjoint operator is symmetric. The converse is not true: an operator A is symmetric if and only if $A \subset A^\star$, i.e., $D(A) \subset D(A^\star)$ and $A^\star = A$ on $D(A)$. It may happen that A is symmetric and that $D(A) \neq D(A^\star)$. Our next result shows that if A is *maximal monotone*, then

$$(A \text{ is symmetric}) \Leftrightarrow (A \text{ is self-adjoint}).$$

Proposition 7.6. *Let A be a maximal monotone symmetric operator. Then A is self-adjoint.*

Proof. Let $J_1 = (I + A)^{-1}$. We will first prove that J_1 is self-adjoint. Since $J_1 \in \mathcal{L}(H)$ it suffices to check that

$$(25) \qquad\qquad (J_1 u, v) = (u, J_1 v) \quad \forall u, v \in H.$$

Set $u_1 = J_1 u$ and $v_1 = J_1 v$, so that

$$u_1 + Au_1 = u,$$
$$v_1 + Av_1 = v.$$

Since by assumption, $(u_1, Av_1) = (Au_1, v_1)$, it follows that $(u_1, v) = (u, v_1)$, i.e., (25).

Let $u \in D(A^\star)$ and set $f = u + A^\star u$. We have

$$(f, v) = (u, v + Av) \quad \forall v \in D(A),$$

i.e.,

$$(f, J_1 w) = (u, w) \quad \forall w \in H.$$

Therefore $u = J_1 f$ and thus $u \in D(A)$. This proves that $D(A^\star) = D(A)$ and hence A is self-adjoint.

Remark 8. One has to be careful that if A is a monotone operator (even a symmetric monotone operator) then $A^\star$ need not be monotone. However, one can prove that the following properties are equivalent:

A is maximal monotone $\Longleftrightarrow$ $A^\star$ is maximal monotone

$\Longleftrightarrow$ A is closed, $D(A)$ is dense, A and $A^\star$ are monotone.

A more general version of this result appears in Problem 16.

• **Theorem 7.7.** *Let A be a self-adjoint maximal monotone operator. Then for every $u_0 \in H$ there exists a unique function*[5]

$$u \in C([0, +\infty); \ H) \cap C^1((0, +\infty); \ H) \cap C((0, +\infty); \ D(A))$$

such that

$$\begin{cases} \dfrac{du}{dt} + Au = 0 \quad on\ (0, +\infty), \\ u(0) = u_0. \end{cases}$$

Moreover, we have

$$|u(t)| \le |u_0| \quad and \quad \left|\frac{du}{dt}(t)\right| = |Au(t)| \le \frac{1}{t}|u_0| \quad \forall t > 0,$$

(26) $$u \in C^k((0, +\infty); \ D(A^\ell)) \quad \forall k, \ell \ integers.$$

Proof.

Uniqueness. Let u and $\overline{u}$ be two solutions. By the monotonicity of A we see that $\varphi(t) = |u(t) - \overline{u}(t)|^2$ is nonincreasing on $(0, +\infty)$. On the other hand, φ is continuous on $[0, +\infty)$ and $\varphi(0) = 0$. Thus $\varphi \equiv 0$.

Existence. The proof is divided into two steps:

Step 1. Assume first that $u_0 \in D(A^2)$ and let u be the solution of (6) given by Theorem 7.4. We claim that

(27) $$\left|\frac{du}{dt}(t)\right| \le \frac{1}{t}|u_0| \quad \forall t > 0.$$

As in the proof of Proposition 7.6 we have

$$J_\lambda^\star = J_\lambda \quad and \quad A_\lambda^\star = A_\lambda \quad \forall \lambda > 0.$$

We go back to the approximate problem introduced in the proof of Theorem 7.4:

(28) $$\frac{du_\lambda}{dt} + A_\lambda u_\lambda = 0 \text{ on } [0, +\infty), \ u_\lambda(0) = u_0.$$

Taking the scalar product of (28) with u_λ and integrating on $[0, T]$, we obtain

---

[5] Let us emphasize the difference between Theorems 7.4 and 7.7. Here $u_0 \in H$ (instead of $u_0 \in D(A)$); the conclusion is that there is a solution of (6), which is smooth away from $t = 0$. However, $|\frac{du}{dt}(t)|$ may possibly "blow up" as $t \to 0$.

$$(29) \qquad \frac{1}{2}|u_\lambda(T)|^2 + \int_0^T (A_\lambda u_\lambda, u_\lambda)dt = \frac{1}{2}|u_0|^2.$$

Taking the scalar product of (28) with $t\frac{du_\lambda}{dt}$ and integrating over $[0, T]$, we obtain

$$(30) \qquad \int_0^T \left|\frac{du_\lambda}{dt}(t)\right|^2 t\, dt + \int_0^T \left(A_\lambda u_\lambda(t), \frac{du_\lambda}{dt}(t)\right) t\, dt = 0.$$

But

$$\frac{d}{dt}(A_\lambda u_\lambda, u_\lambda) = \left(A_\lambda \frac{du_\lambda}{dt}, u_\lambda\right) + \left(A_\lambda u_\lambda, \frac{du_\lambda}{dt}\right) = 2\left(A_\lambda u_\lambda, \frac{du_\lambda}{dt}\right),$$

since $A_\lambda^\star = A_\lambda$. Integrating the second integral in (30) by parts, we are led to

$$(31) \qquad \begin{aligned} \int_0^T \left(A_\lambda u_\lambda(t), \frac{du_\lambda}{dt}(t)\right) t\, dt &= \frac{1}{2}\int_0^T \frac{d}{dt}[(A_\lambda u_\lambda, u_\lambda)]t\, dt \\ &= \frac{1}{2}(A_\lambda u_\lambda(T), u_\lambda(T))\, T - \frac{1}{2}\int_0^T (A_\lambda u_\lambda, u_\lambda)\, dt. \end{aligned}$$

On the other hand, since the function $t \mapsto |\frac{du_\lambda}{dt}(t)|$ is nonincreasing (by Lemma 7.1), we have

$$(32) \qquad \int_0^T \left|\frac{du_\lambda}{dt}(t)\right|^2 t\, dt \geq \left|\frac{du_\lambda}{dt}(T)\right|^2 \frac{T^2}{2}.$$

Combining (29), (30), (31), and (32), we obtain

$$\frac{1}{2}|u_\lambda(T)|^2 + T(A_\lambda u_\lambda(T), u_\lambda(T)) + T^2 \left|\frac{du_\lambda}{dt}(T)\right|^2 \leq \frac{1}{2}|u_0|^2;$$

it follows, in particular, that

$$(33) \qquad \left|\frac{du_\lambda}{dt}(T)\right| \leq \frac{1}{T}|u_0| \quad \forall T > 0.$$

Finally, we pass to the limit in (33) as $\lambda \to 0$. This completes the proof of (27), since $\frac{du_\lambda}{dt} \to \frac{du}{dt}$ (see Step 5 in the proof of Theorem 7.4).

Step 2. Assume now that $u_0 \in H$. Let (u_{0n}) be a sequence in $D(A^2)$ such that $u_{0n} \to u_0$ (recall that $D(A^2)$ is dense in $D(A)$ and that $D(A)$ is dense in H; thus $D(A^2)$ is dense in H). Let u_n be the solution of

$$\begin{cases} \dfrac{du_n}{dt} + Au_n = 0 & \text{on } [0, +\infty), \\ u_n(0) = u_{0n}. \end{cases}$$

We know (by Theorem 7.4) that

$$|u_n(t) - u_m(t)| \le |u_{0n} - u_{0m}| \quad \forall m, n, \quad \forall t \ge 0,$$

and (by Step 1) that

$$\left| \frac{du_n}{dt}(t) - \frac{du_m}{dt}(t) \right| \le \frac{1}{t}|u_{0n} - u_{0m}| \quad \forall m, n, \quad \forall t > 0.$$

It follows that u_n converges uniformly on $[0, +\infty)$ to some limit $u(t)$ and that $\frac{du_n}{dt}(t)$ converges to $\frac{du}{dt}(t)$ uniformly on every interval $[\delta, +\infty), \delta > 0$. The limiting function u satisfies

$$u \in C([0, +\infty);\ H) \cap C^1((0, +\infty);\ H),$$

$$u(t) \in D(A) \quad \forall t > 0 \quad \text{and} \quad \frac{du}{dt}(t) + Au(t) = 0 \quad \forall t > 0$$

(this uses the fact that A is closed).

We now turn to the proof of (26). We will show by induction on $k \ge 2$ that

(34) $u \in C^{k-j}((0, +\infty);\ D(A^j)) \quad \forall j = 0, 1, \ldots, k.$

Assume that (34) holds up to order $k - 1$. In particular, we have

(35) $u \in C((0, +\infty);\ D(A^{k-1})).$

In order to prove (34) it suffices (in view of Theorem 7.5) to check that

(36) $u \in C((0, +\infty), D(A^k)).$

Consider the Hilbert space $\tilde{H} = D(A^{k-1})$ and the operator $\tilde{A} : D(\tilde{A}) \subset \tilde{H} \to \tilde{H}$ defined by

$$\begin{cases} D(\tilde{A}) = D(A^k), \\ \tilde{A} = A. \end{cases}$$

It is easily seen that $\tilde{A}$ is maximal monotone and symmetric in $\tilde{H}$; thus it is self-adjoint. Applying the first assertion of Theorem 7.7 in the space $\tilde{H}$ to the operator $\tilde{A}$, we obtain a unique solution v of the problem

(37) $$\begin{cases} \dfrac{dv}{dt} + Av = 0 \quad \text{on } (0, +\infty), \\ v(0) = v_0, \end{cases}$$

given any $v_0 \in \tilde{H}$. Moreover,

$$v \in C([0, +\infty);\ \tilde{H}) \cap C^1((0, +\infty);\ \tilde{H}) \cap C((0, +\infty);\ D(\tilde{A})).$$

Choosing $v_0 = u(\varepsilon)(\varepsilon > 0)$—we already know by (35) that $v_0 \in \tilde{H}$—we conclude that $u \in C((\varepsilon, +\infty);\ D(A^k))$, and this completes the proof of (36).

Comments on Chapter 7

1. The Hille–Yosida theorem in Banach spaces.
The Hille–Yosida theorem extends to Banach spaces. The precise statement is the following. Let E be a Banach space and let $A : D(A) \subset E \to E$ be an unbounded linear operator. One says that A is *m-accretive* if $\overline{D(A)} = E$ and for every $\lambda > 0$, $I + \lambda A$ is bijective from $D(A)$ onto E with $\|(I + \lambda A)^{-1}\|_{\mathcal{L}(E)} \leq 1$.

Theorem 7.8 (Hille–Yosida). *Let A be m-accretive. Then given any $u_0 \in D(A)$ there exists a unique function*

$$u \in C^1([0, +\infty); \ E) \cap C([0, +\infty); \ D(A))$$

such that

(38)
$$\begin{cases} \dfrac{du}{dt} + Au = 0 \quad on \ [0, +\infty), \\ u(0) = u_0. \end{cases}$$

Moreover,

$$\|u(t)\| \leq \|u_0\| \quad and \quad \left\| \frac{du}{dt}(t) \right\| = \|Au(t)\| \leq \|Au_0\| \quad \forall t \geq 0.$$

The map $u_0 \mapsto u(t)$ extended by continuity to all of E is denoted by $S_A(t)$. It is a continuous semigroup of contractions on E. Conversely, given any continuous semigroup of contractions $S(t)$, there exists a unique m-accretive operator A such that $S(t) = S_A(t) \ \forall t \geq 0$.

For the proof, see, e.g., P. Lax [1], A. Pazy [1], J. Goldstein [1], E. Davies [1], [2], K. Yosida [1], M. Reed–B. Simon [1], Volume 2, H. Tanabe [1], N. Dunford–J. T. Schwartz [1] Volume 1, M. Schechter [1], A. Friedman [2], R. Dautray–J.-L. Lions [1], Chapter XVII, A. Balakrishnan [1], T. Kato [1], W. Rudin [1]. These references present extensive developments on the theory of semigroups.

2. The exponential formula.
There are numerous iteration techniques for solving (38). Let us mention a basic method.

Theorem 7.9. *Assume that A is m-accretive. Then for every $u_0 \in D(A)$ the solution u of (38) is given by the "exponential formula"*

(39)
$$u(t) = \lim_{n \to +\infty} \left[\left(I + \frac{t}{n} A \right)^{-1} \right]^n u_0.$$

For a proof see, e.g., K. Yosida [1] and A. Pazy [1]. Formula (39) corresponds, in the language of *numerical analysis*, to the convergence of an *implicit time discretization* scheme for (38) (see, e.g., K. W. Morton–D. F. Mayers [1]). More precisely, one divides the interval $[0, t]$ into n intervals of equal length $\Delta t = t/n$ and one solves inductively the equations

$$\frac{u_{j+1} - u_j}{\Delta t} + A u_{j+1} = 0, \quad j = 0, 1, \dots, n - 1,$$

starting with u_0. In other words, u_n is given by

$$u_n = (I + \Delta t A)^{-n} u_0 = \left(I + \frac{t}{n} A\right)^{-n} u_0.$$

As $n \to \infty$ (i.e., $\Delta t \to 0$) it is "intuitive" that u_n converges to $u(t)$.

3. Theorem 7.7 is a first step toward the theory of *analytic semigroups*. On this subject see, e.g., K. Yosida [1], T. Kato [1], M. Reed–B. Simon [1], Volume 2, A. Friedman [2], A. Pazy [1], and H. Tanabe [1].

4. Inhomogeneous equations. Nonlinear equations.
Consider, in a Banach space E, the problem

(40)
$$\begin{cases} \dfrac{du}{dt}(t) + Au(t) = f(t) & \text{on } [0, T], \\ u(0) = u_0. \end{cases}$$

The following holds.

Theorem 7.10. *Assume that A is m-accretive. Then for every $u_0 \in D(A)$ and every $f \in C^1([0, T]; E)$ there exists a unique solution u of (40) with*

$$u \in C^1([0, T]; \ E) \cap C([0, T]; \ D(A)).$$

Moreover, u is given by the formula

(41)
$$u(t) = S_A(t)u_0 + \int_0^t S_A(t - s) f(s) ds,$$

where $S_A(t)$ is the semigroup introduced in Comment 1.

Note that if one assumes just $f \in L^1((0, T); E)$, formula (41) still makes sense and provides a generalized solution of (40). On these questions see, e.g., T. Kato [1], A. Pazy [1], R. H. Martin [1], H. Tanabe [1].

In physical applications one encounters many "semilinear" equations of the form

$$\frac{du}{dt} + Au = F(u),$$

where F is a *nonlinear* map from E into E. On these questions see, e.g., R. H. Martin [1], Th. Cazenave–A. Haraux [1], and the comments on Chapter 10.

Let us also mention that some results of Chapter 7 have nonlinear extensions. It is useful to consider nonlinear m-accretive operators $A : D(A) \subset E \rightarrow E$. On this subject, see, e.g., H. Brezis [1] and V. Barbu [1].

Chapter 8
Sobolev Spaces and the Variational Formulation of Boundary Value Problems in One Dimension

8.1 Motivation

Consider the following problem. Given $f \in C([a, b])$, find a function u satisfying

(1)
$$\begin{cases} -u'' + u = f & \text{on } [a, b], \\ u(a) = u(b) = 0. \end{cases}$$

A *classical*—or *strong*—solution of (1) is a C^2 function on $[a, b]$ satisfying (1) in the usual sense. It is well known that (1) can be solved explicitly by a very simple calculation, but we ignore this feature so as to illustrate the method on this elementary example.

Multiply (1) by $\varphi \in C^1([a, b])$ and integrate by parts; we obtain

(2)
$$\int_a^b u'\varphi' + \int_a^b u\varphi = \int_a^b f\varphi \quad \forall \varphi \in C^1([a, b]), \varphi(a) = \varphi(b) = 0.$$

Note that (2) makes sense as soon as $u \in C^1([a, b])$ (whereas (1) requires two derivatives on u); in fact, it suffices to know that $u, u' \in L^1(a, b)$, where u' has a meaning yet to be made precise. Let us say (provisionally) that a C^1 function u that satisfies (2) is a *weak* solution of (1).

The following program outlines the main steps of the *variational approach* in the theory of partial differential equations:

Step A. The notion of *weak solution* is made precise. This involves *Sobolev spaces*, which are our *basic tools*.

Step B. *Existence and uniqueness of a weak solution* is established by a variational method via the Lax–Milgram theorem.

Step C. The weak solution is proved to be of class C^2 (for example): this is a *regularity* result.

H. Brezis, *Functional Analysis, Sobolev Spaces and Partial Differential Equations*,
DOI 10.1007/978-0-387-70914-7_8, © Springer Science+Business Media, LLC 2011

Step D. A *classical* solution is recovered by showing that any weak solution that is C^2 is a classical solution.

To carry out Step D is very simple. In fact, suppose that $u \in C^2([a, b])$, $u(a) = u(b) = 0$, and that u satisfies (2). Integrating (2) by parts we obtain

$$\int_a^b (-u'' + u - f)\varphi = 0 \quad \forall \varphi \in C^1([a, b]), \ \varphi(a) = \varphi(b) = 0$$

and therefore

$$\int_a^b (-u'' + u - f)\varphi = 0 \quad \forall \varphi \in C_c^1((a, b)).$$

It follows (see Corollary 4.15) that $-u'' + u = f$ a.e. on (a, b) and thus everywhere on $[a, b]$, since $u \in C^2([a, b])$.

8.2 The Sobolev Space $W^{1,p}(I)$

Let $I = (a, b)$ be an open interval, possibly unbounded, and let $p \in \mathbb{R}$ with $1 \le p \le \infty$.

Definition. The *Sobolev space* $W^{1,p}(I)$[1] is defined to be

$$W^{1,p}(I) = \left\{ u \in L^p(I); \ \exists g \in L^p(I) \text{ such that } \int_I u\varphi' = - \int_I g\varphi \quad \forall \varphi \in C_c^1(I) \right\}.$$

We set

$$H^1(I) = W^{1,2}(I).$$

For $u \in W^{1,p}(I)$ we denote [2] $u' = g$.

Remark 1. In the definition of $W^{1,p}$ we call φ a *test function*. We could equally well have used $C_c^\infty(I)$ as the class of test functions because if $\varphi \in C_c^1(I)$, then $\rho_n \star \varphi \in C_c^\infty(I)$ for n large enough and $\rho_n \star \varphi \to \varphi$ in C^1 (see Section 4.4; of course, φ is extended to be 0 outside I).

Remark 2. It is clear that if $u \in C^1(I) \cap L^p(I)$ and if $u' \in L^p(I)$ (here u' is the usual derivative of u) then $u \in W^{1,p}(I)$. Moreover, the usual derivative of u coincides with its derivative in the $W^{1,p}$ sense—so that notation is consistent! In particular, if I is bounded, $C^1(\bar{I}) \subset W^{1,p}(I)$ for all $1 \le p \le \infty$.

Examples. Let $I = (-1, +1)$. As an exercise show the following:

(i) The function $u(x) = |x|$ belongs to $W^{1,p}(I)$ for every $1 \le p \le \infty$ and $u' = g$, where

---

[1] If there is no confusion we shall write $W^{1,p}$ instead of $W^{1,p}(I)$ and H^1 instead of $H^1(I)$.
[2] Note that this makes sense: g is well defined a.e. by Corollary 4.24.

$$g(x) = \begin{cases} +1 & \text{if } 0 < x < 1, \\ -1 & \text{if } -1 < x < 0. \end{cases}$$

More generally, a continuous function on $\bar{I}$ that is piecewise C^1 on $\bar{I}$ belongs to $W^{1,p}(I)$ for all $1 \le p \le \infty$.

(ii) The function g above does *not* belong to $W^{1,p}(I)$ for any $1 \le p \le \infty$.

★ *Remark* 3. To define $W^{1,p}$ one can also use the language of distributions (see L. Schwartz [1] or A. Knapp [2]). *All* functions $u \in L^p(I)$ admit a derivative in the sense of distributions; this derivative is an element of the huge space of distributions $\mathcal{D}'(I)$. We say that $u \in W^{1,p}$ if this distributional derivative happens to lie in L^p, which is a subspace of $\mathcal{D}'(I)$. When $I = \mathbb{R}$ and $p = 2$, Sobolev spaces can also be defined using the Fourier transform; see, e.g., J. L. Lions–E. Magenes [1], P. Mal-liavin [1], H. Triebel [1], L. Grafakos [1]. We shall not take this viewpoint here.

Notation. The space $W^{1,p}$ is equipped with the norm

$$\|u\|_{W^{1,p}} = \|u\|_{L^p} + \|u'\|_{L^p}$$

or sometimes, if $1 < p < \infty$, with the equivalent norm $(\|u\|_{L^p}^p + \|u'\|_{L^p}^p)^{1/p}$. The space H^1 is equipped with the scalar product

$$(u, v)_{H^1} = (u, v)_{L^2} + (u', v')_{L^2} = \int_a^b (uv + u'v')$$

and with the associated norm

$$\|u\|_{H^1} = (\|u\|_{L^2}^2 + \|u'\|_{L^2}^2)^{1/2}.$$

Proposition 8.1. *The space $W^{1,p}$ is a Banach space for $1 \le p \le \infty$. It is reflexive[3] for $1 < p < \infty$ and separable for $1 \le p < \infty$. The space H^1 is a separable Hilbert space.*

Proof.

(a) Let (u_n) be a Cauchy sequence in $W^{1,p}$; then (u_n) and (u'_n) are Cauchy sequences in L^p. It follows that u_n converges to some limit u in L^p and u'_n converges to some limit g in L^p. We have

$$\int_I u_n \varphi' = - \int_I u'_n \varphi \quad \forall \varphi \in C_c^1(I),$$

and in the limit

$$\int_I u \varphi' = - \int g \varphi \quad \forall \varphi \in C_c^1(I).$$

---

[3] This property is a *considerable* advantage of $W^{1,p}$. In the problems of the *calculus of variations*, $W^{1,p}$ is preferred over C^1, which is not reflexive. Existence of minimizers is easily established in reflexive spaces (see, e.g., Corollary 3.23).

Thus $u \in W^{1,p}$, $u' = g$, and $\|u_n - u\|_{W^{1,p}} \to 0$.

(b) $W^{1,p}$ is *reflexive* for $1 < p < \infty$. Clearly, the product space $E = L^p(I) \times L^p(I)$ is reflexive. The operator $T : W^{1,p} \to E$ defined by $Tu = [u, u']$ is an isometry from $W^{1,p}$ into E. Since $W^{1,p}$ is a Banach space, $T(W^{1,p})$ is a closed subspace of E. It follows that $T(W^{1,p})$ is reflexive (see Proposition 3.20). Consequently $W^{1,p}$ is also reflexive.

(c) $W^{1,p}$ is *separable* for $1 \leq p < \infty$. Clearly, the product space $E = L^p(I) \times L^p(I)$ is separable. Thus $T(W^{1,p})$ is also separable (by Proposition 3.25). Consequently $W^{1,p}$ is separable.

Remark 4. It is convenient to keep in mind the following fact, which we have used in the proof of Proposition 8.1: let (u_n) be a sequence in $W^{1,p}$ such that $u_n \to u$ in L^p and (u_n') converges to some limit in L^p; then $u \in W^{1,p}$ and $\|u_n - u\|_{W^{1,p}} \to 0$. In fact, when $1 < p \leq \infty$ it suffices to know that $u_n \to u$ in L^p and $\|u_n'\|_{L^p}$ stays *bounded* to conclude that $u \in W^{1,p}$ (see Exercise 8.2).

The functions in $W^{1,p}$ are roughly speaking the primitives of the L^p functions. More precisely, we have the following:

Theorem 8.2. *Let $u \in W^{1,p}(I)$ with $1 \leq p \leq \infty$, and I bounded or unbounded; then there exists a function $\tilde{u} \in C(\bar{I})$ such that*

$$u = \tilde{u} \quad a.e. \text{ on } I$$

and

$$\tilde{u}(x) - \tilde{u}(y) = \int_y^x u'(t)dt \quad \forall x, y \in \bar{I}.$$

Remark 5. Let us emphasize the content of Theorem 8.2. First, note that if one function u belongs to $W^{1,p}$ then all functions v such that $v = u$ a.e. on I also belong to $W^{1,p}$ (this follows directly from the definition of $W^{1,p}$). Theorem 8.2 asserts that every function $u \in W^{1,p}$ admits one (and only one) *continuous representative* on $\bar{I}$, i.e., there exists a continuous function on $\bar{I}$ that belongs to the equivalence class of u ($v \sim u$ if $v = u$ a.e.). When it is useful[4] we replace u by its continuous representative. In order to simplify the notation we also write u for its continuous representative. We finally point out that the property "u has a continuous representative" is not the same as "u is continuous a.e."

Remark 6. It follows from Theorem 8.2 that if $u \in W^{1,p}$ and if $u' \in C(\bar{I})$ (i.e., u' admits a continuous representative on $\bar{I}$), then $u \in C^1(\bar{I})$; more precisely, $\tilde{u} \in C^1(\bar{I})$, but as mentioned above, we do not distinguish u and $\tilde{u}$.

In the proof of Theorem 8.2 we shall use the following lemmas:

Lemma 8.1. *Let $f \in L^1_{loc}(I)$ be such that*

---

[4] For example, in order to give a meaning to $u(x)$ for *every* $x \in \bar{I}$.

(3)
$$\int_I f\varphi' = 0 \quad \forall \varphi \in C_c^1(I).$$

Then there exists a constant C such that $f = C$ a.e. on I.

Proof. Fix a function $\psi \in C_c(I)$ such that $\int_I \psi = 1$. For any function $w \in C_c(I)$ there exists $\varphi \in C_c^1(I)$ such that

$$\varphi' = w - \left(\int_I w \right) \psi.$$

Indeed, the function $h = w - (\int_I w)\psi$ is continuous, has compact support in I, and also $\int_I h = 0$. Therefore h has a (unique) primitive with compact support in I. We deduce from (3) that

$$\int_I f \left[w - \left(\int_I w \right) \psi \right] = 0 \quad \forall w \in C_c(I),$$

i.e.,

$$\int_I \left[f - \left(\int_I f\psi \right) \right] w = 0 \quad \forall w \in C_c(I),$$

and therefore (by Corollary 4.24) $f - (\int_I f\psi) = 0$ a.e. on I, i.e., $f = C$ a.e. on I with $C = \int_I f\psi$.

Lemma 8.2. *Let $g \in L_{loc}^1(I)$; for y_0 fixed in I, set*

$$v(x) = \int_{y_0}^x g(t)dt, \quad x \in I.$$

Then $v \in C(I)$ and

$$\int_I v\varphi' = -\int_I g\varphi \quad \forall \varphi \in C_c^1(I).$$

Proof. We have

$$\int_I v\varphi' = \int_I \left[\int_{y_0}^x g(t)dt \right] \varphi'(x)dx$$

$$= -\int_a^{y_0} dx \int_x^{y_0} g(t)\varphi'(x)dt + \int_{y_0}^b dx \int_{y_0}^x g(t)\varphi'(x)dt.$$

By Fubini's theorem,

$$\int_I v\varphi' = -\int_a^{y_0} g(t)dt \int_a^t \varphi'(x)dx + \int_{y_0}^b g(t)dt \int_t^b \varphi'(x)dx$$

$$= -\int_I g(t)\varphi(t)dt.$$

Proof of Theorem 8.2. Fix $y_0 \in I$ and set $\bar{u}(x) = \int_{y_0}^{x} u'(t)dt$. By Lemma 8.2 we have

$$\int_I \bar{u}\varphi' = -\int_I u'\varphi \quad \forall \varphi \in C_c^1(I).$$

Thus $\int_I (u - \bar{u})\varphi' = 0 \quad \forall \varphi \in C_c^1(I)$. It follows from Lemma 8.1 that $u - \bar{u} = C$ a.e. on I. The function $\tilde{u}(x) = \bar{u}(x) + C$ has the desired properties.

Remark 7. Lemma 8.2 shows that the primitive v of a function $g \in L^p$ belongs to $W^{1,p}$ provided we also know that $v \in L^p$, which is always the case when I is bounded.

Proposition 8.3. *Let $u \in L^p$ with $1 < p \leq \infty$. The following properties are equivalent:*

(i) $u \in W^{1,p}$,
(ii) *there is a constant C such that*

$$\left| \int_I u\varphi' \right| \leq C\|\varphi\|_{L^{p'}(I)} \quad \forall \varphi \in C_c^1(I).$$

Furthermore, we can take $C = \|u'\|_{L^p(I)}$ in (ii).

Proof.
   (i) $\Rightarrow$ (ii). This is obvious.
   (ii) $\Rightarrow$ (i). The linear functional

$$\varphi \in C_c^1(I) \mapsto \int_I u\varphi'$$

is defined on a dense subspace of $L^{p'}$ (since $p' < \infty$) and it is continuous for the $L^{p'}$ norm. Therefore it extends to a bounded linear functional F defined on all of $L^{p'}$ (applying the Hahn–Banach theorem, or simply extension by continuity). By the Riesz representation theorems (Theorems 4.11 and 4.14) there exists $g \in L^p$ such that

$$\langle F, \varphi \rangle = \int_I g\varphi \quad \forall \varphi \in L^{p'}.$$

In particular,

$$\int_I u\varphi' = \int_I g\varphi \quad \forall \varphi \in C_c^1$$

and thus $u \in W^{1,p}$.

$\star$ *Remark* 8 (*absolutely continuous functions and functions of bounded variation*). When $p = 1$, the implication (i) $\Rightarrow$ (ii) remains true but not the converse. To illustrate this fact, suppose that I is bounded. The functions u satisfying (i) with $p = 1$, i.e., the functions of $W^{1,1}(I)$, are called the *absolutely continuous* functions. They are also characterized by the property

$$(AC) \quad \begin{cases} \forall \varepsilon > 0, \exists \delta > 0 \text{ such that for every finite sequence} \\ \text{of disjoint intervals } (a_k, b_k) \subset I \text{ such that } \sum |b_k - a_k| < \delta, \\ \text{we have } \sum |u(b_k) - u(a_k)| < \varepsilon. \end{cases}$$

On the other hand, the functions u satisfying (ii) with $p = 1$ are called functions of *bounded variation*; these functions can be characterized in many different ways:

(a) they are the difference of two bounded nondecreasing functions (possibly discontinuous) on I,

(b) they are the functions u satisfying the property

$$(BV) \quad \begin{cases} \text{there exists a constant } C \text{ such that} \\ \sum_{i=0}^{k-1} |u(t_{i+1}) - u(t_i)| \leq C \text{ for all } t_0 < t_1 < \cdots < t_k \text{ in } I, \end{cases}$$

(c) they are the functions $u \in L^1(I)$ that have as distributional derivative a bounded measure.

Note that functions of bounded variation need not have a continuous representative. On this subject see, e.g., E. Hewitt–K. Stromberg [1], A. Kolmogorov–S. Fomin [1], S. Chae [1], H. Royden [1], G. Folland [2], G. Buttazzo–M. Giaquinta–S. Hildebrandt [1], W. Rudin [2], R. Wheeden–A. Zygmund [1], and A. Knapp [1].

Proposition 8.4. *A function u in $L^\infty(I)$ belongs to $W^{1,\infty}(I)$ if and only if there exists a constant C such that*

$$|u(x) - u(y)| \leq C|x - y| \text{ for a.e. } x, y \in I.$$

Proof. If $u \in W^{1,\infty}(I)$ we may apply Theorem 8.2 to deduce that

$$|u(x) - u(y)| \leq \|u'\|_{L^\infty}|x - y| \text{ for a.e. } x, y \in I.$$

Conversely, let $\varphi \in C_c^1(I)$. For $h \in \mathbb{R}$, with $|h|$ small enough, we have

$$\int_I [u(x + h) - u(x)]\varphi(x)dx = \int_I u(x)[\varphi(x - h) - \varphi(x)]dx$$

(these integrals make sense for h small, since φ is supported in a compact subset of I). Using the assumption on u we obtain

$$\left| \int_I u(x)[\varphi(x - h) - \varphi(x)]dx \right| \leq C|h|\|\varphi\|_{L^1}.$$

Dividing by $|h|$ and letting $h \to 0$, we are led to

$$\left| \int_I u\varphi' \right| \leq C\|\varphi\|_{L^1} \quad \forall \varphi \in C_c^1(I).$$

We may now apply Proposition 8.3 and conclude that $u \in W^{1,\infty}$.

The L^p-version of Proposition 8.4 reads as follows:

Proposition 8.5. *Let $u \in L^p(\mathbb{R})$ with $1 < p < \infty$. The following properties are equivalent:*

(i) $u \in W^{1,p}(\mathbb{R})$,
(ii) *there exists a constant C such that for all $h \in \mathbb{R}$,*

$$\|\tau_h u - u\|_{L^p(\mathbb{R})} \leq C|h|.$$

Moreover, one can choose $C = \|u'\|_{L^p(\mathbb{R})}$ in (ii).

Recall that $(\tau_h u)(x) = u(x + h)$.

Proof.
(i) $\Rightarrow$ (ii). (This implication is also valid when $p = 1$.) By Theorem 8.2 we have, for all x and h in $\mathbb{R}$,

$$u(x + h) - u(x) = \int_x^{x+h} u'(t)dt = h \int_0^1 u'(x + sh)ds.$$

Thus

$$|u(x + h) - u(x)| \leq |h| \int_0^1 |u'(x + sh)|ds.$$

Applying Hölder's inequality, we have

$$|u(x + h) - u(x)|^p \leq |h|^p \int_0^1 |u'(x + sh)|^p ds.$$

It then follows that

$$\int_{\mathbb{R}} |u(x + h) - u(x)|^p dx \leq |h|^p \int_{\mathbb{R}} dx \int_0^1 |u'(x + sh)|^p ds$$

$$\leq |h|^p \int_0^1 ds \int_{\mathbb{R}} |u'(x + sh)|^p dx.$$

But for $0 < s < 1$,

$$\int_{\mathbb{R}} |u'(x + sh)|^p dx = \int_{\mathbb{R}} |u'(y)|^p dy,$$

from which (ii) can be deduced.

(ii) $\Rightarrow$ (i). Let $\varphi \in C_c^1(\mathbb{R})$. For all $h \in \mathbb{R}$ we have

$$\int_{\mathbb{R}} [u(x + h) - u(x)]\varphi(x)dx = \int_{\mathbb{R}} u(x)[\varphi(x - h) - \varphi(x)]dx.$$

Using Hölder's inequality and (ii) one obtains

$$\left| \int_{\mathbb{R}} [u(x+h) - u(x)]\varphi(x)dx \right| \leq C|h| \|\varphi\|_{L^{p'}(\mathbb{R})}$$

and thus

$$\left| \int_{\mathbb{R}} u(x)[\varphi(x-h) - \varphi(x)]dx \right| \leq C|h| \|\varphi\|_{L^{p'}(\mathbb{R})}.$$

Dividing by $|h|$ and letting $h \to 0$, we obtain

$$\left| \int_{\mathbb{R}} u\varphi' \right| \leq C\|\varphi\|_{L^{p'}(\mathbb{R})}.$$

We may apply Proposition 8.3 once more and conclude that $u \in W^{1,p}(\mathbb{R})$.

Certain basic analytic operations have a meaning only for functions defined on all of $\mathbb{R}$ (for example convolution and Fourier transform). It is therefore useful to be able to extend a function $u \in W^{1,p}(I)$ to a function $\bar{u} \in W^{1,p}(\mathbb{R})$.[5] The following result addresses this point.

Theorem 8.6 (extension operator). *Let* $1 \leq p \leq \infty$. *There exists a bounded linear operator* $P : W^{1,p}(I) \to W^{1,p}(\mathbb{R})$, *called an* extension operator, *satisfying the following properties:*

(i) $Pu_{|I} = u \; \forall u \in W^{1,p}(I)$,
(ii) $\|Pu\|_{L^p(\mathbb{R})} \leq C\|u\|_{L^p(I)} \; \forall u \in W^{1,p}(I)$,
(iii) $\|Pu\|_{W^{1,p}(\mathbb{R})} \leq C\|u\|_{W^{1,p}(I)} \; \forall u \in W^{1,p}(I)$,

where C *depends only on* $|I| \leq \infty$.[6]

Proof. Beginning with the case $I = (0, \infty)$ we show that extension by reflexion

$$(Pu)(x) = u^{\star}(x) = \begin{cases} u(x) & \text{if } x \geq 0, \\ u(-x) & \text{if } x < 0, \end{cases}$$

works. Clearly we have

$$\|u^{\star}\|_{L^p(\mathbb{R})} \leq 2\|u\|_{L^p(I)}.$$

Setting

$$v(x) = \begin{cases} u'(x) & \text{if } x > 0, \\ -u'(-x) & \text{if } x < 0, \end{cases}$$

we easily check that $v \in L^p(\mathbb{R})$ and

$$u^{\star}(x) - u^{\star}(0) = \int_0^x v(t)dt \quad \forall x \in \mathbb{R}.$$

---

[5] If u is extended as 0 outside I then the resulting function will not, in general, be in $W^{1,p}(\mathbb{R})$ (see Remark 5 and Section 8.3).

[6] One can take $C = 4$ in (ii) and $C = 4(1 + \frac{1}{|I|})$ in (iii).

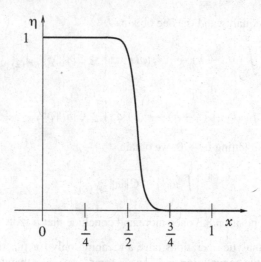

Fig. 5

It follows that $u^\star \in W^{1,p}(\mathbb{R})$ (see Remark 7) and $\|u^\star\|_{W^{1,p}(\mathbb{R})} \leq 2\|u\|_{W^{1,p}(I)}$.

Now consider the case of a *bounded interval* I; without loss of generality we can take $I = (0, 1)$. *Fix* a function $\eta \in C^1(\mathbb{R})$, $0 \leq \eta \leq 1$, such that

$$\eta(x) = \begin{cases} 1 & \text{if } x < 1/4, \\ 0 & \text{if } x > 3/4. \end{cases}$$

See Figure 5.

Given a function f on $(0, 1)$ set

$$\tilde{f}(x) = \begin{cases} f(x) & \text{if } 0 < x < 1, \\ 0 & \text{if } x > 1. \end{cases}$$

We shall need the following lemma.

Lemma 8.3. *Let* $u \in W^{1,p}(I)$. *Then*

$$\eta\tilde{u} \in W^{1,p}(0, \infty) \quad \text{and} \quad (\eta\tilde{u})' = \eta'\tilde{u} + \eta\tilde{u'}.$$

Proof. Let $\varphi \in C_c^1((0, \infty))$; then

$$\int_0^\infty \eta\tilde{u}\varphi' = \int_0^1 \eta u\varphi' = \int_0^1 u[(\eta\varphi)' - \eta'\varphi]$$

$$= -\int_0^1 u'\eta\varphi - \int_0^1 u\eta'\varphi \quad \text{since } \eta\varphi \in C_c^1((0, 1))$$

$$= -\int_0^\infty (\tilde{u'}\eta + \tilde{u}\eta')\varphi.$$

Proof of Theorem 8.6, concluded. Given $u \in W^{1,p}(I)$, write

$$u = \eta u + (1 - \eta)u.$$

The function ηu is *first* extended to $(0, \infty)$ by $\eta \tilde{u}$ (in view of Lemma 8.3) and *then* to $\mathbb{R}$ by reflection. In this way we obtain a function $v_1 \in W^{1,p}(\mathbb{R})$ that extends ηu and such that

$$\|v_1\|_{L^p(\mathbb{R})} \le 2\|u\|_{L^p(I)}, \quad \|v_1\|_{W^{1,p}(\mathbb{R})} \le C\|u\|_{W^{1,p}(I)}$$

(where C depends on $\|\eta'\|_{L^\infty}$).

Proceed in the same way with $(1 - \eta)u$, that is, *first* extend $(1 - \eta)u$ to $(-\infty, 1)$ by 0 on $(-\infty, 0)$ and *then* extend to $\mathbb{R}$ by reflection (this time about the point 1, not 0). In this way we obtain a function $v_2 \in W^{1,p}(\mathbb{R})$ that extends $(1-\eta)u$ and satisfies

$$\|v_2\|_{L^p(\mathbb{R})} \le 2\|u\|_{L^p(I)}, \quad \|v_2\|_{W^{1,p}(\mathbb{R})} \le C\|u\|_{W^{1,p}(I)}.$$

Then $Pu = v_1 + v_2$ satisfies the condition of the theorem.

Certain properties of C^1 functions remain true for $W^{1,p}$ functions (see for example Corollaries 8.10 and 8.11). It is convenient to establish these properties by a *density* argument based on the following result.

• **Theorem 8.7 (density).** *Let $u \in W^{1,p}(I)$ with $1 \le p < \infty$. Then there exists a sequence (u_n) in $C_c^\infty(\mathbb{R})$ such that $u_{n|I} \to u$ in $W^{1,p}(I)$.*

Remark 9. In general, there is no sequence (u_n) in $C_c^\infty(I)$ such that $u_n \to u$ in $W^{1,p}(I)$ (see Section 8.3). This is in contrast to L^p spaces: recall that for every function $u \in L^p(I)$ there is a sequence (u_n) in $C_c^\infty(I)$ such that $u_n \to u$ in $L^p(I)$ (see Corollary 4.23).

Proof. We can always suppose $I = \mathbb{R}$; otherwise, extend u to a function in $W^{1,p}(\mathbb{R})$ by Theorem 8.6. We use the *basic techniques of convolution* (which makes functions C^∞) and *cut-off* (which makes their support compact).

(a) Convolution.

We shall need the following lemma.

Lemma 8.4. *Let $\rho \in L^1(\mathbb{R})$ and $v \in W^{1,p}(\mathbb{R})$ with $1 \le p \le \infty$. Then $\rho \star v \in W^{1,p}(\mathbb{R})$ and $(\rho \star v)' = \rho \star v'$.*

Proof. First, suppose that ρ has compact support. We already know (Theorem 4.15) that $\rho \star v \in L^p(\mathbb{R})$. Let $\varphi \in C_c^1(\mathbb{R})$; from Propositions 4.16 and 4.20 we have

$$\int (\rho \star v)\varphi' = \int v(\check{\rho} \star \varphi') = \int v(\check{\rho} \star \varphi)' = -\int v'(\check{\rho} \star \varphi) = -\int (\rho \star v')\varphi,$$

from which it follows that

$$\rho \star v \in W^{1,p} \quad \text{and} \quad (\rho \star v)' = \rho \star v'.$$

If ρ does not have compact support introduce a sequence (ρ_n) from $C_c(\mathbb{R})$ such that $\rho_n \to \rho$ in $L^1(\mathbb{R})$ (see Corollary 4.23). From the above, we get

$$\rho_n \star v \in W^{1,p}(\mathbb{R}) \quad \text{and} \quad (\rho_n \star v)' = \rho_n \star v'.$$

But $\rho_n \star v \to \rho \star v$ in $L^p(\mathbb{R})$ and $\rho_n \star v' \to \rho \star v'$ in $L^p(\mathbb{R})$ (by Theorem 4.15). We conclude with the help of Remark 4 that

$$\rho \star v \in W^{1,p}(\mathbb{R}) \quad \text{and} \quad (\rho \star v)' = \rho \star v'.$$

(b) Cut-off.

   Fix a function $\zeta \in C_c^\infty(\mathbb{R})$ such that $0 \leq \zeta \leq 1$ and

$$\zeta(x) = \begin{cases} 1 & \text{if } |x| < 1, \\ 0 & \text{if } |x| \geq 2. \end{cases}$$

Define the sequence

(4) $\zeta_n(x) = \zeta(x/n) \quad \text{for } n = 1, 2, \dots.$

It follows easily from the dominated convergence theorem that if a function f belongs to $L^p(\mathbb{R})$ with $1 \leq p < \infty$, then $\zeta_n f \to f$ in $L^p(\mathbb{R})$.

(c) Conclusion.

   Choose a sequence of mollifiers (ρ_n). We claim that the sequence $u_n = \zeta_n(\rho_n \star u)$ converges to u in $W^{1,p}(\mathbb{R})$. First, we have $\|u_n - u\|_p \to 0$. In fact, write

$$u_n - u = \zeta_n((\rho_n \star u) - u) + (\zeta_n u - u)$$

and thus

$$\|u_n - u\|_p \leq \|\rho_n \star u - u\|_p + \|\zeta_n u - u\|_p \to 0.$$

Next, by Lemma 8.4, we have

$$u_n' = \zeta_n'(\rho_n \star u) + \zeta_n(\rho_n \star u').$$

Therefore

$$\|u_n' - u'\|_p \leq \|\zeta_n'(\rho_n \star u)\|_p + \|\zeta_n(\rho_n \star u') - u'\|_p$$
$$\leq \frac{C}{n}\|u\|_p + \|\rho_n \star u' - u'\|_p + \|\zeta_n u' - u'\|_p \to 0,$$

where $C = \|\zeta'\|_\infty$.

   The next result is an important prototype of a *Sobolev inequality* (also called a *Sobolev embedding*).

• **Theorem 8.8.** *There exists a constant C (depending only on $|I| \leq \infty$) such that*

(5) $\|u\|_{L^\infty(I)} \leq C\|u\|_{W^{1,p}(I)} \quad \forall\, u \in W^{1,p}(I), \quad \forall\, 1 \leq p \leq \infty.$

In other words, $W^{1,p}(I) \subset L^{\infty}(I)$ with continuous injection for all $1 \leq p \leq \infty$.
*Further, if I is **bounded** then*

(6)   *the **injection** $W^{1,p}(I) \subset C(\bar{I})$ is **compact** for all $1 < p \leq \infty$,*

(7)   *the **injection** $W^{1,1}(I) \subset L^q(I)$ is **compact** for all $1 \leq q < \infty$.*

Proof. We start by proving (5) for $I = \mathbb{R}$; the general case then follows from this by the extension theorem (Theorem 8.6). Let $v \in C_c^1(\mathbb{R})$; if $1 \leq p < \infty$ set $G(s) = |s|^{p-1}s$. The function $w = G(v)$ belongs to $C_c^1(\mathbb{R})$ and

$$w' = G'(v)v' = p|v|^{p-1}v'.$$

Thus, for $x \in \mathbb{R}$, we have

$$G(v(x)) = \int_{-\infty}^{x} p|v(t)|^{p-1}v'(t)dt,$$

and by Hölder's inequality

$$|v(x)|^p \leq p\|v\|_p^{p-1}\|v'\|_p,$$

from which we conclude that

(8) $$\|v\|_{\infty} \leq C\|v\|_{W^{1,p}} \quad \forall v \in C_c^1(\mathbb{R}),$$

where C is a universal constant (independent of p).[7]

Argue now by density. Let $u \in W^{1,p}(\mathbb{R})$; there exists a sequence $(u_n) \subset C_c^1(\mathbb{R})$ such that $u_n \to u$ in $W^{1,p}(\mathbb{R})$ (by Theorem 8.7). Applying (8), we see that (u_n) is a Cauchy sequence in $L^{\infty}(\mathbb{R})$. Thus $u_n \to u$ in $L^{\infty}(\mathbb{R})$ and we obtain (5).

Proof of (6). Let $\mathcal{H}$ be the unit ball in $W^{1,p}(I)$ with $1 < p \leq \infty$. For $u \in \mathcal{H}$ we have

$$|u(x) - u(y)| = \left| \int_y^x u'(t)dt \right| \leq \|u'\|_p |x-y|^{1/p'} \leq |x-y|^{1/p'} \quad \forall x, y \in I.$$

It follows then from the Ascoli–Arzelà theorem (Theorem 4.25) that $\mathcal{H}$ has a compact closure in $C(\bar{I})$.

Proof of (7). Let $\mathcal{H}$ be the unit ball in $W^{1,1}(I)$. Let P be the extension operator of Theorem 8.6 and set $\mathcal{F} = P(\mathcal{H})$, so that $\mathcal{H} = \mathcal{F}_{|I}$. We prove that $\mathcal{H}$ has a compact closure in $L^q(I)$ (for all $1 \leq q < \infty$) by applying Theorem 4.26. Clearly, $\mathcal{F}$ is bounded in $W^{1,1}(\mathbb{R})$; therefore $\mathcal{F}$ is also bounded in $L^q(\mathbb{R})$, since it is bounded both in $L^1(\mathbb{R})$ and in $L^{\infty}(\mathbb{R})$. We now check condition (22) of Chapter 4, i.e.,

$$\lim_{h \to 0} \|\tau_h f - f\|_q = 0 \quad \text{uniformly in } f \in \mathcal{F}.$$

---

[7] Noting that $p^{1/p} \leq e^{1/e} \ \forall p \geq 1$.

By Proposition 8.5 we have, for every $f \in \mathcal{F}$,

$$\|\tau_h f - f\|_{L^1(\mathbb{R})} \le |h| \|f'\|_{L^1(\mathbb{R})} \le C|h|,$$

since $\mathcal{F}$ is a bounded subset of $W^{1,1}(\mathbb{R})$. Thus

$$\|\tau_h f - f\|^q_{L^q(\mathbb{R})} \le (2\|f\|_{L^\infty(\mathbb{R})})^{q-1} \|\tau_h f - f\|_{L^1(\mathbb{R})} \le C|h|$$

and consequently

$$\|\tau_h f - f\|_{L^q(\mathbb{R})} \le C|h|^{1/q},$$

where C is independent of f. The desired conclusion follows since $q \neq \infty$.

Remark 10. The injection $W^{1,1}(I) \subset C(\bar{I})$ is continuous but it is *never compact*, even if I is a bounded interval; the reader should find an argument or see Exercise 8.2. Nevertheless, if (u_n) is a bounded sequence in $W^{1,1}(I)$ (with I bounded or unbounded) there exists a subsequence (u_{n_k}) such that $u_{n_k}(x)$ converges for *all* $x \in I$ (this is *Helly's selection theorem*; see for example A. Kolmogorov–S. Fomin [1] and Exercise 8.3). When I is *unbounded* and $1 < p \le \infty$, we know that the injection $W^{1,p}(I) \subset L^\infty(I)$ is continuous; this injection is *never compact*—again give an argument or see Exercise 8.4. However, if (u_n) is bounded in $W^{1,p}(I)$ with $1 < p \le \infty$ there exist a subsequence (u_{n_k}) and some $u \in W^{1,p}(I)$ such that $u_{n_k} \to u$ in $L^\infty(J)$ for every *bounded* subset J of I.

Remark 11. Let I be a bounded interval, let $1 \le p \le \infty$, and let $1 \le q \le \infty$. From Theorem 8.2 and (5) it can be shown easily that the norm

$$\|\|u\|\| = \|u'\|_p + \|u\|_q$$

is equivalent to the norm of $W^{1,p}(I)$.

Remark 12. Let I be an *unbounded* interval. If $u \in W^{1,p}(I)$, then $u \in L^q(I)$ for all $q \in [p, \infty]$, since

$$\int_I |u|^q \le \|u\|^{q-p}_\infty \|u\|^p_p.$$

But in general $u \notin L^q(I)$ for $q \in [1, p)$ (see Exercise 8.1).

Corollary 8.9. *Suppose that I is an unbounded interval and $u \in W^{1,p}(I)$ with $1 \le p < \infty$. Then*

$$(9) \qquad \lim_{\substack{x \in I \\ |x| \to \infty}} u(x) = 0.$$

Proof. From Theorem 8.7 there exists a sequence (u_n) in $C^1_c(\mathbb{R})$ such that $u_{n|I} \to u$ in $W^{1,p}(I)$. It follows from (5) that $\|u_n - u\|_{L^\infty(I)} \to 0$. We deduce (9) from this. Indeed, given $\varepsilon > 0$ we choose n large enough that $\|u_n - u\|_{L^\infty(I)} < \varepsilon$. For $|x|$ large enough, $u_n(x) = 0$ (since $u_n \in C^1_c(\mathbb{R})$) and thus $|u(x)| < \varepsilon$.

Corollary 8.10 (differentiation of a product).[8] *Let* $u, v \in W^{1,p}(I)$ *with* $1 \leq p \leq \infty$. *Then*

$$uv \in W^{1,p}(I)$$

and

(10)
$$(uv)' = u'v + uv'.$$

Furthermore, the formula for integration by parts holds:

(11)
$$\int_y^x u'v = u(x)v(x) - u(y)v(y) - \int_y^x uv' \quad \forall x, y \in \bar{I}.$$

Proof. First recall that $u \in L^\infty$ (by Theorem 8.8) and thus $uv \in L^p$. To show that $(uv)' \in L^p$ let us begin with the case $1 \leq p < \infty$. Let (u_n) and (v_n) be sequences in $C_c^1(\mathbb{R})$ such that $u_{n|I} \to u$ and $v_{n|I} \to v$ in $W^{1,p}(I)$. Thus $u_{n|I} \to u$ and $v_{n|I} \to v$ in $L^\infty(I)$ (again by Theorem 8.8). It follows that $u_n v_{n|I} \to uv$ in $L^\infty(I)$ and also in $L^p(I)$. We have

$$(u_n v_n)' = u_n' v_n + u_n v_n' \to u'v + uv' \text{ in } L^p(I).$$

Applying once more Remark 4 to the sequence $(u_n v_n)$, we conclude that $uv \in W^{1,p}(I)$ and that (10) holds. Integrating (10), we obtain (11).

We now turn to the case $p = \infty$; let $u, v \in W^{1,\infty}(I)$. Thus $uv \in L^\infty(I)$ and $u'v + uv' \in L^\infty(I)$. It remains to check that

$$\int_I uv\varphi' = - \int_I (u'v + uv')\varphi \quad \forall \varphi \in C_c^1(I).$$

For this, fix a bounded open interval $J \subset I$ such that $\text{supp}\, \varphi \subset J$. Thus $u, v \in W^{1,p}(J)$ for all $p < \infty$ and from the above we know that

$$\int_J uv\varphi' = - \int_J (u'v + uv')\varphi,$$

that is,

$$\int_I uv\varphi' = - \int_I (u'v + uv')\varphi.$$

Corollary 8.11 (differentiation of a composition). *Let* $G \in C^1(\mathbb{R})$ *be such that*[9] $G(0) = 0$, *and let* $u \in W^{1,p}(I)$ *with* $1 \leq p \leq \infty$. *Then*

$$G \circ u \in W^{1,p}(I) \quad and \quad (G \circ u)' = (G' \circ u)u'.$$

---

[8] Note the *contrast* of this result with the properties of L^p functions: in general, if $u, v \in L^p$, the product uv does *not* belong to L^p. We say that $W^{1,p}(I)$ is a *Banach algebra*.

[9] This restriction is unnecessary when I is bounded (or also if I is unbounded and $p = \infty$). It is essential if I is unbounded and $1 \leq p < \infty$.

Proof. Let $M = \|u\|_\infty$. Since $G(0) = 0$, there exists a constant C such that $|G(s)| \leq C|s|$ for all $s \in [-M, +M]$. Thus $|G \circ u| \leq C|u|$; it follows that $G \circ u \in L^p(I)$. Similarly, $(G' \circ u)u' \in L^p(I)$. It remains to verify that

$$(12) \qquad \int_I (G \circ u)\varphi' = -\int_I (G' \circ u)u'\varphi \quad \forall \varphi \in C_c^1(I).$$

Suppose first that $1 \leq p < \infty$. Then there exists a sequence (u_n) from $C_c^1(\mathbb{R})$ such that $u_{n|I} \to u$ in $W^{1,p}(I)$ and also in $L^\infty(I)$. Thus $(G \circ u_n)_{|I} \to G \circ u$ in $L^\infty(I)$ and $(G' \circ u_n)u'_{n|I} \to (G' \circ u)u'$ in $L^p(I)$. Clearly (by the standard rules for C^1 functions) we have

$$\int_I (G \circ u_n)\varphi' = -\int_I (G' \circ u_n)u'_n\varphi \quad \forall \varphi \in C_c^1(I),$$

from which we deduce (12). For the case $p = \infty$ proceed in the same manner as in the proof of Corollary 8.10.

The Sobolev Spaces $W^{m,p}$

Definition. Given an integer $m \geq 2$ and a real number $1 \leq p \leq \infty$ we define by induction the space

$$W^{m,p}(I) = \{u \in W^{m-1,p}(I); \; u' \in W^{m-1,p}(I)\}.$$

We also set

$$H^m(I) = W^{m,2}(I).$$

It is easily shown that $u \in W^{m,p}(I)$ if and only if there exist m functions $g_1, g_2, \dots, g_m \in L^p(I)$ such that

$$\int_I u\, D^j\varphi = (-1)^j \int_I g_j\varphi \quad \forall \varphi \in C_c^\infty(I), \quad \forall j = 1, 2, \dots, m,$$

where $D^j\varphi$ denotes the jth derivative of φ. When $u \in W^{m,p}(I)$ we may thus consider the successive derivatives of u : $u' = g_1$, $(u')' = g_2$, $\dots$, up to order m. They are denoted by $Du, D^2u, \dots, D^m u$. The space $W^{m,p}(I)$ is equipped with the norm

$$\|u\|_{W^{m,p}} = \|u\|_p + \sum_{\alpha=1}^m \|D^\alpha u\|_p,$$

and the space $H^m(I)$ is equipped with the scalar product

$$(u, v)_{H^m} = (u, v)_{L^2} + \sum_{\alpha=1}^m (D^\alpha u, D^\alpha v)_{L^2} = \int_I uv + \sum_{\alpha=1}^m \int_I D^\alpha u\, D^\alpha v.$$

One can show that the norm $\| \ \|_{W^{m,p}}$ is equivalent to the norm

$$\|\|u\|\| = \|u\|_p + \|D^m u\|_p.$$

More precisely, one proves that for every integer j, $1 \leq j \leq m-1$, and for every $\varepsilon > 0$ there exists a constant C (depending on ε and $|I| \leq \infty$) such that

$$\|D^j u\|_p \leq \varepsilon \|D^m u\|_p + C\|u\|_p \quad \forall u \in W^{m,p}(I)$$

(see, e.g., R. Adams [1], or Exercise 8.6 for the case $|I| < \infty$).

The reader can extend to the space $W^{m,p}$ all the properties shown for $W^{1,p}$; for example, if I is bounded, $W^{m,p}(I) \subset C^{m-1}(\bar{I})$ with continuous injection (resp. compact injection for $1 < p \leq \infty$).

8.3 The Space $W_0^{1,p}$

Definition. Given $1 \leq p < \infty$, denote by $W_0^{1,p}(I)$ the closure of $C_c^1(I)$ in $W^{1,p}(I)$.[10] Set

$$H_0^1(I) = W_0^{1,2}(I).$$

The space $W_0^{1,p}(I)$ is equipped with the norm of $W^{1,p}(I)$, and the space H_0^1 is equipped with the scalar product of H^1.[11]

The space $W_0^{1,p}$ is a separable Banach space. Moreover, it is reflexive for $p > 1$. The space H_0^1 is a separable Hilbert space.

Remark 13. When $I = \mathbb{R}$ we know that $C_c^1(\mathbb{R})$ is dense in $W^{1,p}(\mathbb{R})$ (see Theorem 8.7) and therefore $W_0^{1,p}(\mathbb{R}) = W^{1,p}(\mathbb{R})$.

Remark 14. Using a sequence of mollifiers (ρ_n) it is easy to check the following:

(i) $C_c^\infty(I)$ is dense in $W_0^{1,p}(I)$.
(ii) If $u \in W^{1,p}(I) \cap C_c(I)$ then $u \in W_0^{1,p}(I)$.

Our next result provides a basic characterization of functions in $W_0^{1,p}(I)$.

• **Theorem 8.12.** *Let* $u \in W^{1,p}(I)$. *Then* $u \in W_0^{1,p}(I)$ *if and only if* $u = 0$ *on* ∂I.

Remark 15. Theorem 8.12 explains the central role played by the space $W_0^{1,p}(I)$. Differential equations (or partial differential equations) are often coupled with *boundary conditions*, i.e., the value of u is prescribed on ∂I.

---

[10] We do not define $W_0^{1,p}$ for $p = \infty$.
[11] When there is no confusion we often write $W_0^{1,p}$ and H_0^1 instead of $W_0^{1,p}(I)$ and $H_0^1(I)$.

Proof. If $u \in W_0^{1,p}$, there exists a sequence (u_n) in $C_c^1(I)$ such that $u_n \to u$ in $W^{1,p}(I)$. Therefore $u_n \to u$ uniformly on $\bar{I}$ and as a consequence $u = 0$ on ∂I.

Conversely, let $u \in W^{1,p}(I)$ be such that $u = 0$ on ∂I. Fix any function $G \in C^1(\mathbb{R})$ such that

$$G(t) = \begin{cases} 0 & \text{if } |t| \le 1, \\ t & \text{if } |t| \ge 2, \end{cases}$$

and

$$|G(t)| \le |t| \quad \forall t \in \mathbb{R}.$$

Set $u_n = (1/n)G(nu)$, so that $u_n \in W^{1,p}(I)$ (by Corollary 8.11). On the other hand,

$$\text{supp}\, u_n \subset \{x \in I;\ |u(x)| \ge 1/n\},$$

and thus $\text{supp}\, u_n$ is in a compact subset of I (using the fact that $u = 0$ on ∂I and $u(x) \to 0$ as $|x| \to \infty$, $x \in I$). Therefore $u_n \in W_0^{1,p}(I)$ (see Remark 14). Finally, one easily checks that $u_n \to u$ in $W^{1,p}(I)$ by the dominated convergence theorem. Thus $u \in W_0^{1,p}(I)$.

Remark 16. Let us mention two other characterizations of $W_0^{1,p}$ functions:

(i) Let $1 \le p < \infty$ and let $u \in L^p(I)$. Define $\bar{u}$ by

$$\bar{u}(x) = \begin{cases} u(x) & \text{if } x \in I, \\ 0 & \text{if } x \in \mathbb{R}\backslash I. \end{cases}$$

Then $u \in W_0^{1,p}(I)$ if and only if $\bar{u} \in W^{1,p}(\mathbb{R})$.

(ii) Let $1 < p < \infty$ and let $u \in L^p(I)$. Then u belongs to $W_0^{1,p}(I)$ if and only if there exists a constant C such that

$$\left| \int_I u\varphi' \right| \le C\|\varphi\|_{L^{p'}(I)} \quad \forall \varphi \in C_c^1(\mathbb{R}).$$

• **Proposition 8.13 (Poincaré's inequality).** *Suppose I is a **bounded** interval. Then there exists a constant C (depending on $|I| < \infty$) such that*

(13) $$\|u\|_{W^{1,p}(I)} \le C\|u'\|_{L^p(I)} \quad \forall u \in W_0^{1,p}(I).$$

In other words, on $W_0^{1,p}$, the quantity $\|u'\|_{L^p(I)}$ is a norm equivalent to the $W^{1,p}$ norm.

Proof. Let $u \in W_0^{1,p}(I)$ (with $I = (a,b)$). Since $u(a) = 0$, we have

$$|u(x)| = |u(x) - u(a)| = \left| \int_a^x u'(t)dt \right| \le \|u'\|_{L^1}.$$

Thus $\|u\|_{L^\infty(I)} \le \|u'\|_{L^1(I)}$ and (13) then follows by Hölder's inequality.

Remark 17. If I is bounded, the expression $(u', v')_{L^2} = \int u'v'$ defines a scalar product on H_0^1 and the associated norm, i.e., $\|u'\|_{L^2}$, is equivalent to the H^1 norm.

Remark 18. Given an integer $m \geq 2$ and a real number $1 \leq p < \infty$, the space $W_0^{m,p}(I)$ is defined as the closure of $C_c^m(I)$ in $W^{m,p}(I)$. One shows (see Exercise 8.9) that

$$W_0^{m,p}(I) = \{u \in W^{m,p}(I);\ u = Du = \cdots = D^{m-1}u = 0 \quad \text{on } \partial I\}.$$

It is essential to notice the *distinction* between

$$W_0^{2,p}(I) = \{u \in W^{2,p}(I);\ u = Du = 0 \quad \text{on } \partial I\}$$

and

$$W^{2,p}(I) \cap W_0^{1,p}(I) = \{u \in W^{2,p}(I);\ u = 0 \quad \text{on } \partial I\}.$$

⋆ The Dual Space of $W_0^{1,p}(I)$

Notation. The dual space of $W_0^{1,p}(I)$ $(1 \leq p < \infty)$ is denoted by $W^{-1,p'}(I)$ and the dual space of $H_0^1(I)$ is denoted by $H^{-1}(I)$.

Following Remark 3 of Chapter 5, *we identify L^2 and its dual, but we do not identify H_0^1 and its dual*. We have the inclusions

$$H_0^1 \subset L^2 \subset H^{-1},$$

where these injections are continuous and dense (i.e., they have dense ranges).

If I is a bounded interval we have

$$W_0^{1,p} \subset L^2 \subset W^{-1,p'} \text{ for all } 1 \leq p < \infty$$

with continuous injections (and dense injections when $1 < p < \infty$).

If I is unbounded we have only

$$W_0^{1,p} \subset L^2 \subset W^{-1,p'} \text{ for all } 1 \leq p \leq 2$$

with continuous injections (see Remark 12).

The elements of $W^{-1,p'}$ can be represented with the help of functions in $L^{p'}$; to be precise, we have the following

Proposition 8.14. *Let $F \in W^{-1,p'}(I)$. Then there exist two functions $f_0, f_1 \in L^{p'}(I)$ such that*

$$\langle F, u \rangle = \int_I f_0 u + \int_I f_1 u' \quad \forall u \in W_0^{1,p}(I)$$

and

$$\|F\|_{W^{-1,p'}} = \max\{\|f_0\|_{p'}, \|f_1\|_{p'}\}.$$

When I is **bounded** we can take $f_0 = 0$.

Proof. Consider the product space $E = L^p(I) \times L^p(I)$ equipped with the norm

$$\|h\| = \|h_0\|_p + \|h_1\|_p \text{ where } h = [h_0, h_1].$$

The map $T : u \in W_0^{1,p}(I) \mapsto [u, u'] \in E$ is an isometry from $W_0^{1,p}(I)$ into E. Set $G = T(W_0^{1,p}(I))$ equipped with the norm of E and $S = T^{-1} : G \to W_0^{1,p}(I)$. The map $h \in G \mapsto \langle F, Sh \rangle$ is a continuous linear functional on G. By the Hahn–Banach theorem, it can be extended to a continuous linear functional Φ on all of E with $\|\Phi\|_{E^*} = \|F\|$. By the Riesz representation theorem we know that there exist two functions $f_0, f_1 \in L^{p'}(I)$ such that

$$\langle \Phi, h \rangle = \int_I f_0 h_0 + \int_I f_1 h_1 \quad \forall h = [h_0, h_1] \in E.$$

It is easy to check that $\|\Phi\|_{E^*} = \max\{\|f_0\|_{p'}, \|f_1\|_{p'}\}$. Also, we have

$$\langle \Phi, Tu \rangle = \langle F, u \rangle = \int_I f_0 u + \int_I f_1 u' \quad \forall u \in W_0^{1,p}.$$

When I is *bounded* the space $W_0^{1,p}(I)$ may be equipped with the norm $\|u'\|_p$ (see Proposition 8.13). We repeat the same argument with $E = L^p(I)$ and $T : u \in W^{1,p}(I) \mapsto u' \in L^p(I)$.

Remark 19. The functions f_0 and f_1 are not uniquely determined by F.

Remark 20. The element $F \in W^{-1,p'}(I)$ is usually identified with the distribution $f_0 - f_1'$ (by definition, the distribution $f_0 - f_1'$ is the linear functional $u \mapsto \int_I f_0 u + \int_I f_1 u'$, on $C_c^\infty(I)$).

Remark 21. The first assertion of Proposition 8.14 also holds for continuous linear functionals on $W^{1,p}(1 \le p < \infty)$, i.e., every continuous linear functional F on $W^{1,p}$ may be represented as

$$\langle F, u \rangle = \int_I f_0 u + \int_I f_1 u' \quad \forall u \in W^{1,p}$$

for some functions $f_0, f_1 \in L^{p'}$.

8.4 Some Examples of Boundary Value Problems

Consider the problem

(14)
$$\begin{cases} -u'' + u = f & \text{on } I = (0, 1), \\ u(0) = u(1) = 0, \end{cases}$$

where f is a given function (for example in $C(\bar{I})$ or more generally in $L^2(I)$). *The boundary condition* $u(0) = u(1) = 0$ *is called the* (homogeneous) *Dirichlet boundary condition.*

Definition. A *classical solution* of (14) is a function $u \in C^2(\bar{I})$ satisfying (14) in the usual sense. A *weak solution* of (14) is a function $u \in H_0^1(I)$ satisfying

$$(15) \qquad \int_I u'v' + \int_I uv = \int_I fv \quad \forall v \in H_0^1(I).$$

Let us "put into action" the program outlined in Section 8.1:

Step A. Every classical solution is a weak solution. This is obvious by integration by parts (as justified in Corollary 8.10).

Step B. Existence and uniqueness of a weak solution. This is the content of the following result.

• **Proposition 8.15.** *Given any* $f \in L^2(I)$ *there exists a unique solution* $u \in H_0^1$ *to* (15). *Furthermore, u is obtained by*

$$\boxed{\; \min_{v \in H_0^1} \left\{ \frac{1}{2} \int_I (v'^2 + v^2) - \int_I fv \right\};\;}$$

this is Dirichlet's principle.

Proof. We apply Lax–Milgram's theorem (Corollary 5.8) in the Hilbert space $H = H_0^1(I)$ with the bilinear form

$$a(u, v) = \int_I u'v' + \int_I uv = (u, v)_{H^1}$$

and with the linear functional $\varphi : v \mapsto \int_I fv$.

Remark 22. Given $F \in H^{-1}(I)$ we know from the Riesz–Fréchet representation theorem (Theorem 5.5) that there exists a unique $u \in H_0^1(I)$ such that

$$(u, v)_{H^1} = \langle F, v \rangle_{H^{-1}, H_0^1} \quad \forall v \in H_0^1.$$

The map $F \mapsto u$ is the Riesz–Fréchet isomorphism from H^{-1} onto H_0^1. The function u coincides with the weak solution of (14) in the sense of (15).

Steps C and D. Regularity of weak solutions. Recovery of classical solutions

First, note that if $f \in L^2$ and $u \in H_0^1$ is the weak solution of (14), then $u \in H^2$. Indeed, we have

$$\int_I u'v' = \int_I (f - u)v \quad \forall v \in C_c^1(I),$$

and thus $u' \in H^1$ (by definition of H^1 and since $f - u \in L^2$), i.e., $u \in H^2$. Furthermore, if we assume that $f \in C(\bar{I})$, then the weak solution u belongs to $C^2(\bar{I})$. Indeed, $(u')' \in C(\bar{I})$ and thus $u' \in C^1(\bar{I})$ (see Remark 6). The passage from a weak solution $u \in C^2(\bar{I})$ to a classical solution has been carried out in Section 8.1.

Remark 23. If $f \in H^k(I)$, with k an integer ≥ 1, it is easily verified (by induction) that the solution u of (15) belongs to $H^{k+2}(I)$.

The method described above is extremely flexible and can be adapted to a multitude of problems. We indicate several examples frequently encountered. *In each problem it is essential to specify precisely the function space and to find the appropriate weak formulation.*

Example 1 (*inhomogeneous Dirichlet condition*). Consider the problem

(16)
$$\begin{cases} -u'' + u = f & \text{on } I = (0, 1), \\ u(0) = \alpha, \ u(1) = \beta, \end{cases}$$

with $\alpha, \beta \in \mathbb{R}$ given and f a given function.

• **Proposition 8.16.** *Given* $\alpha, \beta \in \mathbb{R}$ *and* $f \in L^2(I)$ *there exists a unique function* $u \in H^2(I)$ *satisfying* (16). *Furthermore,* u *is obtained by*

$$\min_{\substack{v \in H^1(I) \\ v(0) = \alpha, v(1) = \beta}} \left\{ \frac{1}{2} \int_I (v'^2 + v^2) - \int_I fv \right\}.$$

If, in addition, $f \in C(\bar{I})$ *then* $u \in C^2(\bar{I})$.

Proof. We give two possible approaches:

Method 1. Fix any smooth function[12] u_0 such that $u_0(0) = \alpha$ and $u_0(1) = \beta$. Introduce as new unknown $\tilde{u} = u - u_0$. Then $\tilde{u}$ satisfies

$$\begin{cases} -\tilde{u}'' + \tilde{u} = f + u_0'' - u_0 & \text{on } I, \\ \tilde{u}(0) = \tilde{u}(1) = 0. \end{cases}$$

We are reduced to the preceding problem for $\tilde{u}$.

Method 2. Consider in the space $H^1(I)$ the closed convex set

$$K = \{v \in H^1(I); \ v(0) = \alpha \text{ and } v(1) = \beta\}.$$

If u is a classical solution of (16) we have

---

[12] Choose, for example, u_0 to be affine.

$$\int_I u'(v-u)' + \int_I u(v-u) = \int_I f(v-u) \quad \forall v \in K.$$

Then in particular,

(17) $$\int_I u'(v-u)' + \int_I u(v-u) \geq \int_I f(v-u) \quad \forall v \in K.$$

We may now invoke Stampacchia's theorem (Theorem 5.6): there exists a unique function $u \in K$ satisfying (17) and, moreover, u is obtained by

$$\min_{v \in K} \left\{ \frac{1}{2} \int_I (v'^2 + v^2) - \int_I fv \right\}.$$

To recover a classical solution of (16), set $v = u \pm w$ in (17) with $w \in H_0^1$ and obtain

$$\int_I u'w' + \int_I uw = \int_I fw \quad \forall w \in H_0^1.$$

This implies (as above) that $u \in H^2(I)$. If $f \in C(\bar{I})$ the same argument as in the homogeneous case shows that $u \in C^2(\bar{I})$.

$\star$ *Example 2* (*Sturm–Liouville problem*). Consider the problem

(18) $$\begin{cases} -(pu')' + qu = f & \text{on } I = (0, 1), \\ u(0) = u(1) = 0, \end{cases}$$

where $p \in C^1(\bar{I})$, $q \in C(\bar{I})$, and $f \in L^2(I)$ are given with

$$p(x) \geq \alpha > 0 \quad \forall x \in I.$$

If u is a classical solution of (18) we have

$$\int_I pu'v' + \int_I quv = \int_I fv \quad \forall v \in H_0^1(I).$$

We use $H_0^1(I)$ as our function space and

$$a(u, v) = \int_I pu'v' + \int_I quv$$

as symmetric continuous bilinear form on H_0^1. If $q \geq 0$ on I this form is coercive by Poincaré's inequality (Proposition 8.13). Thus, by Lax–Milgram's theorem, there exists a unique $u \in H_0^1$ such that

(19) $$a(u, v) = \int_I fv \quad \forall v \in H_0^1(I).$$

Moreover, u is obtained by

$$\min_{v \in H_0^1(I)} \left\{ \frac{1}{2} \int_I (pv'^2 + qv^2) - \int_I fv \right\}.$$

It is clear from (19) that $pu' \in H^1$; thus (by Corollary 8.10) $u' = (1/p)(pu') \in H^1$ and hence $u \in H^2$. Finally, if $f \in C(\bar{I})$, then $pu' \in C^1(\bar{I})$, and so $u' \in C^1(\bar{I})$, i.e., $u \in C^2(\bar{I})$. Step D carries over and we conclude that u is a classical solution of (18).

Consider now the more general problem

(20)
$$\begin{cases} -(pu')' + ru' + qu = f & \text{on } I = (0, 1), \\ u(0) = u(1) = 0. \end{cases}$$

The assumptions on p, q, and f are the same as above, and $r \in C(\bar{I})$. If u is a classical solution of (20) we have

$$\int_I pu'v' + \int_I ru'v + \int_I quv = \int_I fv \quad \forall v \in H_0^1.$$

We use $H_0^1(I)$ as our function space and

$$a(u, v) = \int_I pu'v' + \int_I ru'v + \int_I quv$$

as bilinear continuous form. This form is *not* symmetric. In certain cases it is coercive; for example,

(i) if $q \geq 1$ and $r^2 < 4\alpha$;
(ii) or if $q \geq 1$ and $r \in C^1(\bar{I})$ with $r' \leq 2$; here we use the fact that

$$\int rv'v = -\frac{1}{2} \int r'v^2 \quad \forall v \in H_0^1.$$

One may then apply the Lax–Milgram theorem, but there is no straightforward associated minimization problem. Here is a device that allows us to recover a symmetric bilinear form. Introduce a primitive R of r/p and set $\zeta = e^{-R}$. Equation (20) can be written, after multiplication by ζ, as

$$-\zeta pu'' - \zeta p'u' + \zeta ru' + \zeta qu = \zeta f,$$

or (since $\zeta'p + \zeta r = 0$)

$$-(\zeta pu')' + \zeta qu = \zeta f.$$

Define on H_0^1 the symmetric continuous bilinear form

$$a(u, v) = \int_I \zeta pu'v' + \int_I \zeta quv.$$

When $q \geq 0$, this form is coercive, and so there exists a unique $u \in H_0^1$ such that

$$a(u, v) = \int_I \zeta f v \quad \forall v \in H_0^1.$$

Furthermore, u is obtained by

$$\min_{v \in H_0^1(I)} \left\{ \frac{1}{2} \int_I (\zeta p v'^2 + \zeta q v^2) - \int_I \zeta f v \right\}.$$

It is easily verified that $u \in H^2$, and if $f \in C(\bar{I})$ then $u \in C^2(\bar{I})$ is a classical solution of (20).

Example 3 (*homogeneous Neumann condition*). Consider the problem

(21)
$$\begin{cases} -u'' + u = f \quad \text{on } I = (0, 1), \\ u'(0) = u'(1) = 0. \end{cases}$$

• **Proposition 8.17.** *Given* $f \in L^2(I)$ *there exists a unique function* $u \in H^2(I)$ *satisfying* (21).[13] *Furthermore, u is obtained by*

$$\min_{v \in H^1(I)} \left\{ \frac{1}{2} \int_I (v'^2 + v^2) - \int_I f v \right\}.$$

If, in addition, $f \in C(\bar{I})$, *then* $u \in C^2(\bar{I})$.

Proof. If u is a classical solution of (21) we have

(22)
$$\int_I u'v' + \int_I uv = \int_I f v \quad \forall v \in H^1(I).$$

We use $H^1(I)$ as our function space: there is no point in working in H_0^1 as above since $u(0)$ and $u(1)$ are a priori *unknown*. We apply the Lax–Milgram theorem with the bilinear form $a(u, v) = \int_I u'v' + \int_I uv$ and the linear functional $\varphi : v \mapsto \int_I f v$. In this way we obtain a unique function $u \in H^1(I)$ satisfying (22). From (22) it follows, as above, that $u \in H^2(I)$. Using (22) once more we obtain

(23) $\int_I (-u'' + u - f)v + u'(1)v(1) - u'(0)v(0) = 0 \quad \forall v \in H^1(I).$

In (23) begin by choosing $v \in H_0^1$ and obtain $-u'' + u = f$ a.e. Returning to (23), there remains

$$u'(1)v(1) - u'(0)v(0) = 0 \quad \forall v \in H^1(I).$$

Since $v(0)$ and $v(1)$ are arbitrary, we deduce that $u'(0) = u'(1) = 0$.

---

[13] Note that $u \in H^2(I) \Rightarrow u \in C^1(\bar{I})$ and thus the condition $u'(0) = u'(1) = 0$ *makes sense. It would not make sense if we knew only that* $u \in H^1$.

Example 4 (*inhomogeneous Neumann condition*). Consider the problem

$$(24) \qquad \begin{cases} -u'' + u = f \quad \text{on } I = (0, 1), \\ u'(0) = \alpha, \, u'(1) = \beta, \end{cases}$$

with $\alpha, \beta \in \mathbb{R}$ given and f a given function.

Proposition 8.18. *Given any $f \in L^2(I)$ and $\alpha, \beta \in \mathbb{R}$ there exists a unique function $u \in H^2(I)$ satisfying (24). Furthermore, u is obtained by*

$$\min_{v \in H^1(I)} \left\{ \frac{1}{2} \int_I (v'^2 + v^2) - \int_I fv + \alpha v(0) - \beta v(1) \right\}.$$

If, in addition, $f \in C(\bar{I})$ then $u \in C^2(\bar{I})$.

Proof. If u is a classical solution of (24) we have

$$\int_I u'v' + \int_I uv = \int_I fv - \alpha v(0) + \beta v(1) \quad \forall v \in H^1(I).$$

We use $H^1(I)$ as our function space and we apply the Lax–Milgram theorem with the bilinear form $a(u, v) = \int_I u'v' + \int_I uv$ and the linear functional

$$\varphi : v \mapsto \int_I fv - \alpha v(0) + \beta v(1).$$

This linear functional is continuous (by Theorem 8.8). Then proceed as in Example 3 to prove that $u \in H^2(I)$ and that $u'(0) = \alpha, u'(1) = \beta$.

Example 5 (*mixed boundary condition*). Consider the problem

$$(25) \qquad \begin{cases} -u'' + u = f \quad \text{on } I = (0, 1), \\ u(0) = 0, \, u'(1) = 0. \end{cases}$$

If u is a classical solution of (25) we have

$$(26) \qquad \int_I u'v' + \int_I uv = \int_I fv \quad \forall v \in H^1(I) \text{ with } v(0) = 0.$$

The appropriate space to work in is

$$H = \{v \in H^1(I); v(0) = 0\}$$

equipped with the H^1 scalar product. The rest is left to the reader as an exercise.

Example 6 (*Robin, or "third type," boundary condition*). Consider the problem

$$(27) \qquad \begin{cases} -u'' + u = f \quad \text{on } I = (0, 1), \\ u'(0) = ku(0), \, u(1) = 0, \end{cases}$$

where $k \in \mathbb{R}$ is given.[14]

If u is a classical solution of (27) we have

$$\int_I u'v' + \int_I uv + ku(0)v(0) = \int_I fv \quad \forall v \in H^1(I) \text{ with } v(1) = 0.$$

The appropriate space for applying Lax–Milgram is the Hilbert space

$$H = \{v \in H^1(I); v(1) = 0\}$$

equipped with the H^1 scalar product. The bilinear form

$$a(u, v) = \int_I u'v' + \int_I uv + ku(0)v(0)$$

is symmetric and continuous. It is coercive if $k \geq 0$.[15]

Example 7 (periodic boundary conditions). Consider the problem

(28)
$$\begin{cases} -u'' + u = f & \text{on } I = (0, 1), \\ u(0) = u(1), \ u'(0) = u'(1). \end{cases}$$

If u is a classical solution of (28) we have

(29) $$\int_I u'v' + \int_I uv = \int_I fv \quad \forall v \in H^1(I) \quad \text{with } v(0) = v(1).$$

The appropriate setting for applying Lax–Milgram is the Hilbert space

$$H = \{v \in H^1(I); v(0) = v(1)\}$$

with the bilinear form $a(u, v) = \int_I u'v' + \int_I uv$. When $f \in L^2(I)$ we obtain a solution $u \in H^2(I)$ of (28). If, in addition, $f \in C(\bar{I})$ then the solution is classical.

Example 8 (a boundary value problem on $\mathbb{R}$). Consider the problem

(30)
$$\begin{cases} -u'' + u = f & \text{on } \mathbb{R}, \\ u(x) \to 0 & \text{as } |x| \to \infty, \end{cases}$$

with f given in $L^2(\mathbb{R})$. A *classical solution* of (30) is a function $u \in C^2(\mathbb{R})$ satisfying (30) in the usual sense. A *weak solution* of (30) is a function $u \in H^1(\mathbb{R})$ satisfying

---

[14] More generally, one can handle the boundary condition

$$\alpha_0 u'(0) + \beta_0 u(0) = 0, \ \alpha_1 u'(1) + \beta_1 u(1) = 0,$$

with appropriate conditions on the constants α_0, β_0, α_1, and β_1.

[15] If $k < 0$ with $|k|$ small enough the form $a(u, v)$ is still coercive. On the other hand, an explicit calculation shows that there exist a negative value of k and (smooth) functions f for which (27) has no solution (see Exercise 8.21).

$$\int_{\mathbb{R}} u'v' + \int_{\mathbb{R}} uv = \int_{\mathbb{R}} fv \quad \forall v \in H^1(\mathbb{R}).$$

We have first to prove that any classical solution u is a weak solution; let us check in the first place that $u \in H^1(\mathbb{R})$. Choose a sequence (ζ_n) of cut-off functions as in the proof of Theorem 8.7. Multiplying (30) by $\zeta_n u$ and integrating by parts, we obtain

$$\int_{\mathbb{R}} u'(\zeta_n u' + \zeta_n' u) + \int_{\mathbb{R}} \zeta_n u^2 = \int_{\mathbb{R}} \zeta_n fu,$$

from which we deduce

(31) $$\int_{\mathbb{R}} \zeta_n(u'^2 + u^2) = \int_{\mathbb{R}} \zeta_n fu + \frac{1}{2} \int_{\mathbb{R}} \zeta_n'' u^2.$$

But

$$\frac{1}{2} \int_{\mathbb{R}} \zeta_n'' u^2 \le \frac{C}{n^2} \int_{n<|x|<2n} u^2 \quad \text{with } C = \|\zeta''\|_{L^\infty(\mathbb{R})}$$

and $\frac{1}{n^2} \int_{n<|x|<2n} u^2 \to 0$ as $n \to \infty$, since $u(x) \to 0$ as $|x| \to \infty$. Inserting the inequality

$$\int_{\mathbb{R}} \zeta_n fu \le \frac{1}{2} \int_{\mathbb{R}} \zeta_n u^2 + \frac{1}{2} \int_{\mathbb{R}} \zeta_n f^2$$

in (31), we see that $\int \zeta_n(u'^2 + u^2)$ remains bounded as $n \to \infty$ and therefore $u \in H^1(\mathbb{R})$.

Assuming that u is a classical solution of (30), we have

$$\int_{\mathbb{R}} u'v' + \int_{\mathbb{R}} uv = \int_{\mathbb{R}} fv \quad \forall v \in C_c^1(\mathbb{R}).$$

By density (and since $u \in H^1(\mathbb{R})$) this holds for every $v \in H^1(\mathbb{R})$. Therefore u is a weak solution of (30).

To obtain existence and uniqueness of a weak solution it suffices to apply Lax–Milgram in the Hilbert space $H^1(\mathbb{R})$. One easily verifies that the weak solution u belongs to $H^2(\mathbb{R})$ and if furthermore $f \in C(\mathbb{R})$ then $u \in C^2(\mathbb{R})$. We conclude (using Corollary 8.9) that given $f \in L^2(\mathbb{R}) \cap C(\mathbb{R})$, problem (30) has a unique classical solution (which furthermore belongs to $H^2(\mathbb{R})$).

Remark 24. The problem

$$\begin{cases} -u'' = f & \text{on } \mathbb{R}, \\ u(x) \to 0 & \text{as } |x| \to \infty, \end{cases}$$

cannot be attacked by the preceding technique because the bilinear form $a(u, v) = \int u'v'$ is *not coercive* in $H^1(\mathbb{R})$. In fact, this problem need not have a solution even if f is smooth with compact support (why?).

Remark 25. On the other hand, the same method applies to the problem

$$\begin{cases} -u'' + u = f & \text{on } I = (0, +\infty), \\ u(0) = 0 \text{ and } u(x) \to 0 \text{ as } x \to +\infty, \end{cases}$$

with f given in $L^2(0, +\infty)$.

8.5 The Maximum Principle

Here is a very useful property called the maximum principle.

• **Theorem 8.19.** *Let* $f \in L^2(I)$ *with* $I = (0, 1)$ *and let* $u \in H^2(I)$ *be the solution of the Dirichlet problem*

$$(32) \qquad \begin{cases} -u'' + u = f & \text{on } I, \\ u(0) = \alpha, u(1) = \beta. \end{cases}$$

Then we have, for every $x \in I$,[16]

$$(33) \qquad \min\{\alpha, \beta, \inf_I f\} \le u(x) \le \max\{\alpha, \beta, \sup_I f\}.$$

Proof (using Stampacchia's truncation method). We have

$$(34) \qquad \int_I u'v' + \int_I uv = \int_I fv \quad \forall v \in H_0^1(I).$$

Fix any function $G \in C^1(\mathbb{R})$ such that

 (i) G is strictly increasing on $(0, +\infty)$,
 (ii) $G(t) = 0$ for $t \in (-\infty, 0]$.

Set $K = \max\{\alpha, \beta, \sup_I f\}$ and suppose that $K < \infty$. We shall show that $u \le K$ on I. The function $v = G(u - K)$ belongs to $H^1(I)$ and even to $H_0^1(I)$, since

$$u(0) - K = \alpha - K \le 0 \quad \text{and} \quad u(1) - K = \beta - K \le 0.$$

Plugging v into (34), we obtain

$$\int_I u'^2 G'(u - K) + \int_I uG(u - K) = \int_I fG(u - K),$$

that is,

$$\int_I u'^2 G'(u - K) + \int_I (u - K)G(u - K) = \int_I (f - K)G(u - K).$$

---

[16] sup f and inf f refer respectively to the essential sup (possibly $+\infty$) and the essential inf of f (possibly $-\infty$). Recall that ess sup $f = \inf\{C; f(x) \le C \text{ a.e.}\}$ and ess inf $f = -\text{ess sup}(-f)$.

But $(f-K) \le 0$ and $G(u-K) \ge 0$, from which it follows that $(f-K)G(u-K) \le 0$, and therefore

$$\int_I (u-K)G(u-K) \le 0.$$

Since $tG(t) \ge 0 \ \forall t \in \mathbb{R}$, the preceding inequality implies $(u-K)G(u-K) = 0$ a.e. It follows that $u \le K$ a.e., and consequently everywhere on I, since u is continuous. The lower bound for u is obtained by applying this upper bound to $-u$.

Remark 26. When $f \in C(\bar{I})$, then $u \in C^2(\bar{I})$ and one can establish (33) by a different method: *the classical approach to the maximum principle.* Let $x_0 \in \bar{I}$ be the point where u attains its maximum on $\bar{I}$. If $x_0 = 0$ or if $x_0 = 1$ the conclusion is obvious. Otherwise, $0 < x_0 < 1$ and then $u'(x_0) = 0$, $u''(x_0) \le 0$. From equation (33) it follows that

$$u(x_0) = f(x_0) + u''(x_0) \le f(x_0) \le K$$

and therefore $u \le K$ on I.

Here are some immediate consequences of Theorem 8.19.

• **Corollary 8.20.** *Let u be a solution of (34).*

  (i) *If $u \ge 0$ on ∂I and if $f \ge 0$ on I, then $u \ge 0$ on I.*
  (ii) *If $u = 0$ on ∂I and if $f \in L^\infty(I)$, then $\|u\|_{L^\infty(I)} \le \|f\|_{L^\infty(I)}$.*
  (iii) *If $f = 0$ on I, then $\|u\|_{L^\infty(I)} \le \|u\|_{L^\infty(\partial I)}$.*

We have a similar result for the case of Neumann condition.

Proposition 8.21. *Let $f \in L^2(I)$ with $I = (0, 1)$ and let $u \in H^2(I)$ be the solution of the problem*

$$\begin{cases} -u'' + u = f & \text{on } I, \\ u'(0) = u'(1) = 0. \end{cases}$$

Then we have, for every $x \in \bar{I}$,

(35)
$$\inf_I f \le u(x) \le \sup_I f.$$

Proof. We have

(36)
$$\int_I u'v' + \int_I uv = \int_I fv \quad \forall v \in H^1(I).$$

Plug $v = G(u - K)$ into (36) with $K = \sup_I f$ and the same function G as above. Then proceed just as in the proof of Theorem 8.19.

Remark 27. If $f \in C(\bar{I})$, then $u \in C^2(\bar{I})$ and we can establish (35) along the same lines as in Remark 26. Note that if u achieves its maximum on ∂I, say at 0, then $u''(0) \le 0$ (extending u by reflection to the left of 0 and using the fact that $u'(0) = 0$).

Remark 28. Let $f \in L^2(\mathbb{R})$ and let $u \in H^2(\mathbb{R})$ be the solution of

$$\begin{cases} -u'' + u = f & \text{on } \mathbb{R}, \\ u(x) \to 0 & \text{as } |x| \to \infty, \end{cases}$$

discussed in Example 8. Then we have, for all $x \in \mathbb{R}$,

$$\inf_{\mathbb{R}} f \le u(x) \le \sup_{\mathbb{R}} f.$$

8.6 Eigenfunctions and Spectral Decomposition

The following is a basic result.

• **Theorem 8.22.** *Let $p \in C^1(\bar{I})$ with $I = (0, 1)$ and $p \ge \alpha > 0$ on I; let $q \in C(\bar{I})$. Then there exist a sequence (λ_n) of real numbers and a Hilbert basis (e_n) of $L^2(I)$ such that $e_n \in C^2(\bar{I})$ $\forall n$ and*

(37) $$\begin{cases} -(pe_n')' + qe_n = \lambda_n e_n & \text{on } I, \\ e_n(0) = e_n(1) = 0. \end{cases}$$

Furthermore, $\lambda_n \to +\infty$ as $n \to +\infty$.

One says that the (λ_n) are the *eigenvalues* of the differential operator $Au = -(pu')' + qu$ with Dirichlet boundary condition and that the (e_n) are the associated *eigenfunctions.*

Proof. We can always assume $q \ge 0$, for if not, pick any constant C such that $q + C \ge 0$, which amounts to replacing λ_n by $\lambda_n + C$ in (37). For every $f \in L^2(I)$ there exists a unique $u \in H^2(I) \cap H_0^1(I)$ satisfying

(38) $$\begin{cases} -(pu')' + qu = f & \text{on } I, \\ u(0) = u(1) = 0. \end{cases}$$

Denote by T the operator $f \mapsto u$ *considered as an operator from $L^2(I)$ into $L^2(I)$.*[17]

We claim that T is self-adjoint and compact. First, the compactness. Because of (38) we have

$$\int_I pu'^2 + \int_I qu^2 = \int_I fu$$

and thus $\alpha \|u'\|_{L^2}^2 \le \|f\|_{L^2} \|u\|_{L^2}$. It follows that $\|u\|_{H^1} \le C\|f\|_{L^2}$, where C is a constant depending only on α. This can be written as

$$\|Tf\|_{H^1} \le C\|f\|_{L^2} \quad \forall f \in L^2(I).$$

---

[17] We could also envisage T as an operator from H_0^1 into H_0^1 (see Section 9.8, Remark 28).

Since the injection of $H^1(I)$ into $L^2(I)$ is compact (because I is *bounded*), we deduce that T is a compact operator from $L^2(I)$ into $L^2(I)$. Next, we show that T is self-adjoint, i.e.,

$$\int_I (Tf)g = \int_I f(Tg) \quad \forall f, g \in L^2(I).$$

Indeed, setting $u = Tf$ and $v = Tg$, we have

(39) $$-(pu')' + qu = f$$

and

(40) $$-(pv')' + qv = g.$$

Multiplying (39) by v and (40) by u and then integrating, we obtain

$$\int_I pu'v' + \int_I quv = \int_I fv = \int_I gu,$$

which is the desired conclusion.

Finally, we note that

(41) $$\int_I (Tf)f = \int_I uf = \int_I pu'^2 + qu^2 \geq 0 \quad \forall f \in L^2(I)$$

and also that $N(T) = \{0\}$, since $Tf = 0$ implies $u = 0$ and so $f = 0$.

Applying Theorem 6.11, we know that $L^2(I)$ admits a Hilbert basis $(e_n)_{n \geq 1}$ consisting of eigenvectors of T with corresponding eigenvalues $(\mu_n)_{n \geq 1}$. We have $\mu_n > 0 \quad \forall n \quad (\mu_n \geq 0$ by (41) and $\mu_n \neq 0$, since $N(T) = \{0\})$. We also know that $\mu_n \to 0$. Writing that $Te_n = \mu_n e_n$, we obtain

$$\begin{cases} -(pe_n')' + qe_n = \lambda_n e_n & \text{with } \lambda_n = 1/\mu_n, \\ e_n(0) = e_n(1) = 0. \end{cases}$$

In addition, we have $e_n \in C^2(\bar{I})$, since $f = \lambda_n e_n \in C(\bar{I})$ (in fact $e_n \in C^\infty(\bar{I})$ if $p, q \in C^\infty(\bar{I})$).

Example. If $p \equiv 1$ and $q \equiv 0$ we have

$$e_n(x) = \sqrt{2} \sin(n\pi x) \quad \text{and} \quad \lambda_n = n^2\pi^2, n = 1, 2, \ldots.$$

Remark 29. For the same differential operator the eigenvalues and the eigenfunctions vary with the boundary conditions. As an exercise determine the eigenvalues of the operator $Au = -u''$ with the boundary conditions of Examples 3, 5, 6, and 7.

Remark 30. The assumption that I is *bounded* enters in an essential way in show-ing the *compactness* of the operator T. When I is not bounded the conclusion of Theorem 8.22 is in general false;[18] one encounters instead the very interesting phe-nomenon of *continuous spectrum*—on this subject, see, e.g., M. Reed–B. Simon [1]. In Exercise 8.38 we determine the eigenvalues and the spectrum of the operator $T : f \mapsto u$, where $u \in H^2(\mathbb{R})$ is the solution of problem (30): T is a self-adjoint bounded operator from $L^2(\mathbb{R})$ into itself, but it is *not compact*.

Comments on Chapter 8

1. Some further inequalities.
Let us mention some very useful inequalities involving the Sobolev norms:

(i) **Poincaré–Wirtinger's inequality.**
Let I be a bounded interval. Given $u \in L^2(I)$, set $\bar{u} = \frac{1}{|I|} \int_I u$ (this is the mean of u on I). We have

$$\|u - \bar{u}\|_\infty \le \|u'\|_1 \quad \forall u \in W^{1,1}(I)$$

(see Problem 47).

(ii) **Hardy's inequality.**
Let $I = (0, 1)$ and let $u \in W_0^{1,p}(I)$ with $1 < p < \infty$. Then the function

$$v(x) = \frac{u(x)}{x(1 - x)}$$

belongs to $L^p(I)$ and furthermore,

$$\|v\|_p \le C_p \|u'\|_p \quad \forall u \in W_0^{1,p}(I)$$

(see Exercise 8.8).

(iii) **Interpolation inequalities of Gagliardo–Nirenberg.**
Let I be a bounded interval and let $1 \le r \le \infty$, $1 \le q \le p \le \infty$. Then there exists a constant C such that

$$(42) \qquad \|u\|_p \le C\|u\|_q^{1-a}\|u\|_{W^{1,r}}^a \quad \forall u \in W^{1,r}(I),$$

where $0 \le a \le 1$ is defined by $a(\frac{1}{q} - \frac{1}{r} + 1) = \frac{1}{q} - \frac{1}{p}$ (see Exercise 8.15). In particular, it follows from inequality (42) that if $p < \infty$ (or even if $p = \infty$ but $r > 1$), then

---

[18] In certain circumstances, with some appropriate assumptions on p and q, the conclusion of Theorem 8.22 still holds on unbounded intervals (see Problem 51).

(43)
$$\begin{cases} \forall \varepsilon > 0 \ \exists C_\varepsilon > 0 \quad \text{such that} \\ \|u\|_p \le \varepsilon \|u\|_{W^{1,r}} + C_\varepsilon \|u\|_q \quad \forall u \in W^{1,r}(I). \end{cases}$$

One can also establish (43) by a direct "*compactness method*"; see Exercise 8.5. Other more general inequalities can be found in L. Nirenberg [1] (see also A. Friedman [2]). In particular, we call attention to the inequality

$$\|u'\|_p \le C\|u\|_{W^{2,r}}^{1/2}\|u\|_q^{1/2} \quad \forall u \in W^{2,r}(I),$$

where p is the *harmonic mean* of q and r, i.e., $\frac{1}{p} = \frac{1}{2}(\frac{1}{q} + \frac{1}{r})$.

2. Hilbert–Schmidt operators.
It can be shown that the operator $T : f \mapsto u$ that associates to each f in $L^2(I)$ the unique solution u of the problem

$$\begin{cases} -(pu')' + qu = f \quad \text{on } I = (0,1), \\ u(0) = u(1) = 0 \end{cases}$$

(assuming $p \ge \alpha > 0$ and $q \ge 0$) is a Hilbert–Schmidt operator from $L^2(I)$ into $L^2(I)$; see Exercise 8.37.

3. Spectral properties of Sturm–Liouville operators.
Many spectral properties of the Sturm–Liouville operator $Au = -(pu')' + qu$ with Dirichlet condition on a bounded interval I are known. Among these let us mention that:

(i) Each eigenvalue has *multiplicity one*: it is then said that each eigenvalue is *simple*.

(ii) If the eigenvalues (λ_n) are arranged in increasing order, then the eigenfunction $e_n(x)$ corresponding to λ_n possesses exactly $(n-1)$ zeros on I; in particular the *first eigenfunction* $e_1(x)$ has a *constant sign* on I, and usually one takes $e_1 > 0$ on I.

(iii) The quotient λ_n/n^2 converges as $n \to \infty$ to a positive limit.

Some of these properties are discussed in Exercises 8.33, 8.42 and Problem 49. The interested reader can also consult Weinberger [1], M. Protter–H. Weinberger [1], E. Coddington–N. Levinson [1], Ph. Hartman [1], S. Agmon [1], R. Courant–D. Hilbert [1], Vol. 1, E. Ince [1], Y. Pinchover–J. Rubinstein [1], A. Zettl [1], and G. Buttazzo–M. Giaquinta–S. Hildebrandt [1].

The celebrated Gelfand–Levitan theory deals with an important "inverse" problem: what informations on the function $q(x)$ can one retrieve purely from the knowledge of the spectrum of the Sturm–Liouville operator $Au = -u'' + q(x)u$? This question has attracted much attention because of its numerous applications; see, e.g., B. Levitan [1] and also Comment 13 in Chapter 9.

Exercises for Chapter 8

8.1 Consider the function

$$u(x) = (1 + x^2)^{-\alpha/2}(\log(2 + x^2))^{-1}, \quad x \in \mathbb{R},$$

with $0 < \alpha < 1$. Check that $u \in W^{1,p}(\mathbb{R})$ $\forall p \in [1/\alpha, \infty]$ and that $u \notin L^q(\mathbb{R})$ $\forall q \in [1, 1/\alpha)$.

8.2 Let $I = (0, 1)$.

1. Assume that (u_n) is a bounded sequence in $W^{1,p}(I)$ with $1 < p \le \infty$. Show that there exist a subsequence (u_{n_k}) and some u in $W^{1,p}(I)$ such that $\|u_{n_k} - u\|_{L^\infty} \to 0$. Moreover, $u'_{n_k} \rightharpoonup u'$ weakly in $L^p(I)$ if $1 < p < \infty$, and $u'_{n_k} \overset{\star}{\rightharpoonup} u'$ in $\sigma(L^\infty, L^1)$ if $p = \infty$.
2. Construct a bounded sequence (u_n) in $W^{1,1}(I)$ that admits no subsequence converging in $L^\infty(I)$.

[**Hint:** Consider the sequence (u_n) defined by

$$u_n(x) = \begin{cases} 0 & \text{if } x \in [0, \frac{1}{2}], \\ n(x - \frac{1}{2}) & \text{if } x \in (\frac{1}{2}, \frac{1}{2} + \frac{1}{n}), \\ 1 & \text{if } x \in [\frac{1}{2} + \frac{1}{n}, 1], \end{cases}$$

with $n \ge 2$.]

8.3 *Helly's selection theorem.*
Let (u_n) be a bounded sequence in $W^{1,1}(0, 1)$. The goal is to prove that there exists a subsequence (u_{n_k}) such that $u_{n_k}(x)$ converges to a limit for *every* $x \in [0, 1]$.

1. Show that we may always assume in addition that

(1) $\forall n, u_n$ is nondecreasing on $[0, 1]$.

[**Hint:** Consider the sequences $v_n(x) = \int_0^x |u'_n(t)|dt$ and $w_n = v_n - u_n$.]

In what follows we assume that (1) holds.

2. Prove that there exist a subsequence (u_{n_k}) and a measurable set $E \subset [0, 1]$ with $|E| = 0$ such that $u_{n_k}(x)$ converges to a limit, denoted by $u(x)$, for every $x \in [0, 1] \setminus E$.

[**Hint:** Use the fact that $W^{1,1} \subset L^1$ with compact injection.]

3. Show that u is nondecreasing on $[0, 1] \setminus E$ and deduce that there are a countable set $D \subset (0, 1)$ and a nondecreasing function $\bar{u} : (0, 1) \setminus D \to \mathbb{R}$ such that $\bar{u}(x + 0) = \bar{u}(x - 0)$ $\forall x \in (0, 1) \setminus D$ and $\bar{u}(x) = u(x)$ $\forall x \in (0, 1) \setminus (D \cup E)$.

4. Prove that $u_{n_k}(x) \to \bar{u}(x)$ $\forall x \in (0, 1) \setminus D$.

5. Construct a subsequence from the sequence (u_{n_k}) that converges for every $x \in [0, 1]$.

   [**Hint:** Use a diagonal process.]

8.4 Fix a function $\varphi \in C_c^\infty(\mathbb{R})$, $\varphi \not\equiv 0$, and set $u_n(x) = \varphi(x+n)$. Let $1 \le p \le \infty$.

1. Check that (u_n) is bounded in $W^{1,p}(\mathbb{R})$.

2. Prove that there exists no subsequence (u_{n_k}) converging strongly in $L^q(\mathbb{R})$, for any $1 \le q \le \infty$.

3. Show that $u_n \rightharpoonup 0$ weakly in $W^{1,p}(\mathbb{R})$ $\quad \forall p \in (1, \infty)$.

8.5 Let $p > 1$.

1. Prove that $\forall \varepsilon > 0 \; \exists C = C(\varepsilon, p)$ such that

   (1) $\qquad \|u\|_{L^\infty(0,1)} \le \varepsilon \|u'\|_{L^p(0,1)} + C\|u\|_{L^1(0,1)} \quad \forall u \in W^{1,p}(0, 1).$

   [**Hint:** Use Exercise 6.12 with $X = W^{1,p}(0, 1)$, $Y = L^\infty(0, 1)$, and $Z = L^1(0, 1)$.]

2. Show that (1) fails when $p = 1$.

   [**Hint:** Take $u(x) = x^n$ and let $n \to \infty$.]

3. Let $1 \le q < \infty$. Prove that $\forall \varepsilon > 0 \; \exists C = C(\varepsilon, q)$ such that

   (2) $\qquad \|u\|_{L^q(0,1)} \le \varepsilon \|u'\|_{L^1(0,1)} + C\|u\|_{L^1(0,1)} \quad \forall u \in W^{1,1}(0, 1).$

8.6 Let $I = (0, 1)$ and $p > 1$.

1. Check that $W^{2,p}(I) \subset C^1(\overline{I})$ with compact injection.

2. Deduce that $\forall \varepsilon > 0, \; \exists C = C(\varepsilon, p)$ such that

   $$\|u'\|_{L^\infty(I)} + \|u\|_{L^\infty(I)} \le \varepsilon \|u''\|_{L^p(I)} + C\|u\|_{L^1(I)} \quad \forall u \in W^{2,p}(I).$$

3. Let $1 \le q < \infty$. Prove that $\forall \varepsilon > 0 \; \exists C = C(\varepsilon, q)$ such that

   $$\|u'\|_{L^q(I)} + \|u\|_{L^\infty(I)} \le \varepsilon \|u''\|_{L^1(I)} + C\|u\|_{L^1(I)} \quad \forall u \in W^{2,1}(I).$$

   More generally, let $m \ge 2$ be an integer.

4. Show that $\forall \varepsilon > 0 \; \exists C = C(\varepsilon, m, p)$ such that

$$\sum_{j=0}^{m-1} \|D^j u\|_{L^\infty(I)} \le \varepsilon \|D^m u\|_{L^p(I)} + C\|u\|_{L^1(I)} \quad \forall u \in W^{m,p}(I).$$

5. Let $1 \le q < \infty$. Prove that $\forall \varepsilon > 0 \ \exists C = C(\varepsilon, q)$ such that

$$\|D^{(m-1)}u\|_{L^q(I)} + \sum_{j=0}^{m-2} \|D^j u\|_{L^\infty(I)} \le \varepsilon \|D^m u\|_{L^1(I)} + C\|u\|_{L^1(I)} \quad \forall u \in W^{m,1}(I).$$

8.7 Let $I = (0, 1)$. Given a function u defined on I, set

$$\overline{u}(x) = \begin{cases} u(x) & \text{if } x \in I, \\ 0 & \text{if } x \in \mathbb{R}, x \notin I. \end{cases}$$

1. Assume that $u \in W_0^{1,p}(I)$ with $1 \le p < \infty$. Prove that $\overline{u} \in W^{1,p}(\mathbb{R})$.

2. Conversely, let $u \in L^p(I)$ with $1 \le p < \infty$ be such that $\overline{u} \in W^{1,p}(\mathbb{R})$. Show that $u \in W_0^{1,p}(I)$.

3. Let $u \in L^p(I)$ with $1 < p < \infty$. Show that $u \in W_0^{1,p}(I)$ iff there exists a constant C such that

$$\left| \int_{\mathbb{R}} \overline{u}\varphi' \right| \le C\|\varphi\|_{L^{p'}(\mathbb{R})} \quad \forall \varphi \in C_c^1(\mathbb{R}).$$

8.8

1. Let $u \in W^{1,p}(0, 1)$ with $1 < p < \infty$. Show that if $u(0) = 0$, then $\frac{u(x)}{x} \in L^p(0, 1)$ and

$$\left\| \frac{u(x)}{x} \right\|_{L^p(0,1)} \le \frac{p}{p-1}\|u'\|_{L^p(0,1)}.$$

[**Hint:** Use Problem 34, part C.]

2. Conversely, assume that $u \in W^{1,p}(0, 1)$ with $1 \le p < \infty$ and that $\frac{u(x)}{x} \in L^p(0, 1)$. Show that $u(0) = 0$.

[**Hint:** Argue by contradiction.]

3. Let $u(x) = (1 + |\log x|)^{-1}$. Check that $u \in W^{1,1}(0, 1)$, $u(0) = 0$, but $\frac{u(x)}{x} \notin L^1(0, 1)$.

4. Assume that $u \in W^{1,p}(0, 1)$ with $1 \le p < \infty$ and $u(0) = 0$. Fix any function $\zeta \in C^\infty(\mathbb{R})$ such that $\zeta(x) = 0 \ \forall x \in (-\infty, 1]$ and $\zeta(x) = 1 \ \forall x \in [2, +\infty)$. Set $\zeta_n(x) = \zeta(nx)$ and $u_n(x) = \zeta_n(x)u(x)$, $n = 1, 2 \ldots$. Check that $u_n \in W^{1,p}(0, 1)$ and prove that $u_n \to u$ in $W^{1,p}(0, 1)$ as $n \to \infty$.

[**Hint:** Consider separately the cases $p = 1$ and $p > 1$.]

$\boxed{8.9}$ Set $I = (0, 1)$.

1. Let $u \in W^{2,p}(I)$ with $1 < p < \infty$. Assume that $u(0) = u'(0) = 0$. Show that $\frac{u(x)}{x^2} \in L^p(I)$ and $\frac{u'(x)}{x} \in L^p(I)$ with

   (1) $\left\| \dfrac{u(x)}{x^2} \right\|_{L^p(I)} + \left\| \dfrac{u'(x)}{x} \right\|_{L^p(I)} \leq C_p \|u''\|_{L^p(I)}.$

   [**Hint:** Look at Exercise 8.8.]

2. Deduce that $v(x) = \frac{u(x)}{x} \in W^{1,p}(I)$ with $v(0) = 0$.

3. Let u be as in question 1. Set $u_n = \zeta_n u$, where ζ_n is defined in question 4 of Exercise 8.8. Check that $u_n \in W^{2,p}(I)$ and $u_n \to u$ in $W^{2,p}(I)$ as $n \to \infty$.

4. More generally, let $m \geq 1$ be an integer, and let $1 < p < \infty$. Assume that $u \in X_m$, where

   $$X_m = \{u \in W^{m,p}(I);\ u(0) = Du(0) = \cdots = D^{m-1}u(0) = 0\}.$$

   Show that $\frac{u(x)}{x^m} \in L^p(I)$ and that $\frac{u(x)}{x^{m-1}} \in X_1$.

   [**Hint:** Use induction on m.]

5. Assume that $u \in X_m$ and prove that

   $$v = \frac{D^j u(x)}{x^{m-j-k}} \in X_k \quad \forall j, k \text{ integers}, j \geq 0, k \geq 1, j + k \leq m - 1.$$

6. Let u be as in question 4 and ζ_n as in question 3. Prove that $\zeta_n u \in W^{m,p}(I)$ and $\zeta_n u \to u$ in $W^{m,p}(I)$, as $n \to \infty$.

7. Give a proof of Remark 18 in Chapter 8 when $p > 1$.

8. Assume now that $u \in W^{2,1}(I)$ with $u(0) = u'(0) = 0$. Set

   $$v(x) = \begin{cases} \dfrac{u(x)}{x} & \text{if } x \in (0, 1], \\ 0 & \text{if } x = 0. \end{cases}$$

   Check that $v \in C([0, 1])$. Prove that $v \in W^{1,1}(I)$.

   [**Hint:** Note that $v'(x) = \frac{1}{x^2} \int_0^x u''(t)t\,dt$.]

9. Construct an example of a function $u \in W^{2,1}(I)$ satisfying $u(0) = u'(0) = 0$, but $\frac{u(x)}{x^2} \notin L^1(I)$ and $\frac{u'(x)}{x} \notin L^1(I)$.

   [**Hint:** Use question 3 in Exercise 8.8.]

10. Let u be as in question 8, and ζ_n as in question 3. Check that $u_n = \zeta_n u \in W^{2,1}(I)$, and that $u_n \to u$ in $W^{2,1}(I)$, as $n \to \infty$.

11. Give a proof of Remark 18 in Chapter 8 when $m = 2$ and $p = 1$.

12. Generalize questions 8–11 to $W^{m,1}(I)$ with $m \geq 2$.

    [The result of question 8, and its generalization to $m \geq 2$, are due to Hernan Castro and Hui Wang.]

8.10 Let $I = (0, 1)$. Let $u \in W^{1,p}(I)$ with $1 \leq p < \infty$. Our goal is to prove that $u' = 0$ a.e. on the set $E = \{x \in I; u(x) = 0\}$.

Fix a function $G \in C^1(\mathbb{R}, \mathbb{R})$ such that $|G(t)| \leq 1 \ \forall t \in \mathbb{R}$, $|G'(t)| \leq C \ \forall t \in \mathbb{R}$, for some constant C, and

$$G(t) = \begin{cases} 1 & \text{if } t \geq 1, \\ t & \text{if } |t| \leq 1/2, \\ -1 & \text{if } t \leq -1. \end{cases}$$

Set

$$v_n(x) = \frac{1}{n} G(nu(x)).$$

1. Check that $\|v_n\|_{L^\infty(I)} \to 0$ as $n \to \infty$.

2. Show that $v_n \in W^{1,p}(I)$ and compute v_n'.

3. Deduce that $|v_n'|$ is bounded by a fixed function in $L^p(I)$.

4. Prove that $v_n'(x) \to f(x)$ a.e. on I, as $n \to \infty$, and identify f.

   [**Hint:** Consider separately the cases $x \in E$ and $x \notin E$.]

5. Deduce that $v_n' \to f$ in $L^p(I)$.

6. Prove that $f = 0$ a.e. on I and conclude that $u' = 0$ a.e. on E.

8.11 Let $F \in C(\mathbb{R}, \mathbb{R})$ and assume that $F \in C^1(\mathbb{R} \setminus \{0\})$ with $|F'(t)| \leq C \ \ \forall t \in \mathbb{R} \setminus \{0\}$, for some constant C. Let $1 \leq p < \infty$.
   The goal is to prove that for every $u \in W^{1,p}(0, 1)$, $v = F(u)$ belongs to $W^{1,p}(0, 1)$ and

$$v'(x) = \begin{cases} F'(u(x))u'(x) & \text{a.e. on } [u(x) \neq 0], \\ 0 & \text{a.e. on } [u(x) = 0]. \end{cases}$$

1. Construct a sequence (F_n) in $C^\infty(\mathbb{R})$ such that $\|F_n'\|_{L^\infty(\mathbb{R})} \leq C \ \ \forall n$ and

$$\begin{cases} F_n \to F & \text{uniformly on compact subsets of } \mathbb{R}, \\ F_n' \to F' & \text{uniformly on compact subsets of } \mathbb{R} \setminus \{0\}. \end{cases}$$

2. Check that $v_n = F_n(u) \in W^{1,p}(0,1)$ and that $v_n' = F_n'(u)u'$.

3. Prove that $v_n \to v = F(u)$ in $C([0,1])$ and that v_n' converges in $L^p(0,1)$ to f, where

$$f(x) = \begin{cases} F'(u(x))u'(x) & \text{a.e. on } [u(x) \neq 0], \\ 0 & \text{a.e. on } [u(x) = 0]. \end{cases}$$

   [**Hint:** Apply dominated convergence and Exercise 8.10.]

4. Deduce that $v \in W^{1,p}(0,1)$ and $v' = f$. Show that $v_n \to v$ in $W^{1,p}(0,1)$.

5. Let (u_k) be a sequence in $W^{1,p}(0,1)$ such that $u_k \to u$ in $W^{1,p}(0,1)$. Prove that $F(u_k) \to F(u)$ in $W^{1,p}$.

   [**Hint:** Applying Theorem 4.9 and passing to a subsequence (still denoted by u_k), one may assume that $u_k' \to u'$ a.e. on $(0,1)$ and $|u_k'| \leq g$ $\forall k$, for some function $g \in L^p(0,1)$. Set $w_k = F(u_k)$ and check that $w_k' \to f$ a.e. on $(0,1)$, where f is defined in question 3. Deduce that $w_k \to F(u)$ in $W^{1,p}(0,1)$. Conclude that the full original sequence $F(u_k)$ converges to $F(u)$ in $W^{1,p}(0,1)$.]

6. Application: take $F(t) = t^+ = \max\{t,0\}$. Check that $u^+ \in W^{1,p}(0,1)$ $\forall u \in W^{1,p}(0,1)$. Compute $(u^+)'$.

8.12 Let $I = (0,1)$ and $1 \leq p \leq \infty$. Set

$$B_p = \{u \in W^{1,p}(I); \|u\|_{L^p(I)} + \|u'\|_{L^p(I)} \leq 1\}.$$

1. Prove that B_p is a closed subset of $L^p(I)$ when $1 < p \leq \infty$; more precisely, B_p is compact in $L^p(I)$.

2. Show that B_1 is not a closed subset of $L^1(I)$.

8.13 Let $1 \leq p < \infty$ and $u \in W^{1,p}(\mathbb{R})$. Set

$$D_h u(x) = \frac{1}{h}(u(x+h) - u(x)), x \in \mathbb{R}, h > 0.$$

Show that $D_h u \to u'$ in $L^p(\mathbb{R})$ as $h \to 0$.

[**Hint:** Use the fact that $C_c^1(\mathbb{R})$ is dense in $W^{1,p}(\mathbb{R})$.]

8.14 Let $u \in C^1((0,1))$. Prove that the following conditions are equivalent:

(a) $u \in W^{1,1}(0,1)$,
(b) $u' \in L^1(0,1)$ (where u' denotes the derivative of u in the usual sense),
(c) $u \in BV(0,1)$ (for the definition of BV see Remark 8).

Check that the function $u(x) = x \sin(1/x), 0 < x \leq 1$, with $u(0) = 0$, is continuous on $[0, 1]$ and that $u \notin W^{1,1}(0, 1)$.

$\boxed{8.15}$ *Gagliardo–Nirenberg's inequality (first form).*

Let $I = (0, 1)$.

1. Let $1 \leq q < \infty$ and $1 < r \leq \infty$. Prove that

(1) $$\|u\|_{L^\infty(I)} \leq C \|u\|_{W^{1,r}(I)}^a \|u\|_{L^q(I)}^{1-a} \quad \forall u \in W^{1,r}(I)$$

for some constant $C = C(q, r)$, where $0 < a < 1$ is defined by

(2) $$a\left(\frac{1}{q} + 1 - \frac{1}{r}\right) = \frac{1}{q}.$$

[**Hint:** Start with the case $u(0) = 0$ and write $G(u(x)) = \int_0^x G'(u(t))u'(t)dt$, where $G(t) = |t|^{\alpha-1}t$ and $\alpha = \frac{1}{a}$. When $u(0) \neq 0$, apply the previous inequality to ζu, where $\zeta \in C^1([0, 1]), \zeta(0) = 0$, and $\zeta(t) = 1$ for $t \in [\frac{1}{2}, 1]$.]

2. Let $1 \leq q < p < \infty$ and $1 \leq r \leq \infty$.

Prove that

(3) $$\|u\|_{L^p(I)} \leq C \|u\|_{W^{1,r}(I)}^b \|u\|_{L^q(I)}^{1-b} \quad \forall u \in W^{1,r}(I)$$

for some constant $C = C(p, q, r)$, where $0 < b < 1$ is defined by

(4) $$b\left(\frac{1}{q} + 1 - \frac{1}{r}\right) = \frac{1}{q} - \frac{1}{p}.$$

[**Hint:** Write $\|u\|_{L^p(I)}^p = \int_I |u|^q |u|^{p-q} \leq \|u\|_{L^q(I)}^q \|u\|_{L^\infty}^{p-q}$ and use (1) if $r > 1$.]

3. With the same assumptions as in question 2 show that

(5) $$\|u\|_{L^p(I)} \leq C \|u'\|_{L^r(I)}^b \|u\|_{L^q(I)}^{1-b} \quad \forall u \in W^{1,r}(I) \text{ with } \int_I u = 0.$$

$\boxed{8.16}$ Let $E = L^p(0, 1)$ with $1 \leq p < \infty$. Consider the unbounded operator $A : D(A) \subset E \to E$ defined by

$$D(A) = \{u \in W^{1,p}(0, 1), u(0) = 0\} \quad \text{and} \quad Au = u'.$$

1. Check that $D(A)$ is dense in E and that A is closed (i.e., $G(A)$ is closed in $E \times E$).

2. Determine $R(A)$ and $N(A)$.

3. Compute $A^\star$. Check that $D(A^\star)$ is dense in $E^\star = L^{p'}(0, 1)$ when $1 < p < \infty$, but $D(A^\star)$ is not dense in $E^\star = L^\infty(0, 1)$ when $p = 1$.

4. Same questions for the operator $\widetilde{A}$ defined by

$$D(\widetilde{A}) = W_0^{1,p}(0, 1) \quad \text{and} \quad \widetilde{A}u = u'.$$

$\boxed{8.17}$ Let $H = L^2(0, 1)$ and let $A : D(A) \subset H \to H$ be the unbounded operator defined by $Au = u''$, whose domain $D(A)$ will be made precise below. Determine A^*, $D(A^*)$, $N(A)$, and $N(A^*)$ in the following cases:

1. $D(A) = \{u \in H^2(0, 1); u(0) = u(1) = 0\}$.
2. $D(A) = H^2(0, 1)$.
3. $D(A) = \{u \in H^2(0, 1); u(0) = u(1) = u'(0) = u'(1) = 0\}$.
4. $D(A) = \{u \in H^2(0, 1); u(0) = u(1)\}$.

Same questions for the operator $Au = u'' - xu'$.

$\boxed{8.18}$ Check that the mapping $u \mapsto u(0)$ from $H^1(0, 1)$ into $\mathbb{R}$ is a continuous linear functional on $H^1(0, 1)$. Deduce that there exists a unique $v_0 \in H^1(0, 1)$ such that

$$u(0) = \int_0^1 (u'v_0' + uv_0) \quad \forall u \in H^1(0, 1).$$

Show that v_0 is the solution of some differential equation with appropriate boundary conditions. Compute v_0 explicitly.

[**Hint:** Consider Example 4 in Section 8.4.]

$\boxed{8.19}$ Let $H = L^2(0, 1)$ and consider the function $\varphi : H \to (-\infty, +\infty]$ defined by

$$\varphi(u) = \begin{cases} \frac{1}{2} \int_0^1 u'^2 & \text{if } u \in H^1(0, 1), \\ +\infty & \text{if } u \in L^2(0, 1) \text{ and } u \notin H^1(0, 1). \end{cases}$$

1. Check that φ is convex and l.s.c.
2. Compute $\varphi^*(f)$ for every $f \in H$.

[**Hint:** Show first that $\varphi^*(f) = +\infty$ if $\int_0^1 f \neq 0$. Assume next that $\int_0^1 f = 0$ and set $F(x) = \int_0^x f(t)dt$. Note that $\int_0^1 fv = -\int_0^1 Fv' \ \forall v \in H^1(0, 1)$.]

$\boxed{8.20}$ Set

$$V = \{v \in H^1(0, 1); v(0) = 0\}.$$

1. Given $f \in L^2(0, 1)$ such that $\frac{1}{x}f(x) \in L^2(0, 1)$, prove that there exists a unique $u \in V$ satisfying

$$(1) \quad \int_0^1 u'(x)v'(x)dx + \int_0^1 \frac{u(x)v(x)}{x^2}dx = \int_0^1 \frac{f(x)v(x)}{x^2}dx \quad \forall v \in V.$$

[**Hint:** Use question 1 in Exercise 8.8.]

2. What is the minimization problem associated with (1)?

   In what follows we assume that $\frac{1}{x^2} f(x) \in L^2(0, 1)$.

3. In (1) choose $v(x) = \frac{u(x)}{(x+\varepsilon)^2}$, $\varepsilon > 0$, and deduce that

   (2)
   $$\int_0^1 \left| \frac{d}{dx} \left(\frac{u(x)}{x+\varepsilon} \right) \right|^2 dx \leq \int_0^1 \frac{f(x)}{x^2} \frac{u(x)}{(x+\varepsilon)^2} dx.$$

4. Prove that $\frac{u(x)}{x^2} \in L^2(0, 1)$, $\frac{u(x)}{x} \in H^1(0, 1)$, and $\frac{u'(x)}{x} \in L^2(0, 1)$.

   [**Hint:** Use once more question 1 in Exercise 8.8 and pass to the limit as $\varepsilon \to 0$.]

5. Deduce that $u \in H^2(0, 1)$ and that

   (3)
   $$\begin{cases} -u''(x) + \frac{u(x)}{x^2} = \frac{f(x)}{x^2} & \text{a.e. on } (0, 1), \\ u(0) = u'(0) = 0 & \text{and} \quad u'(1) = 0. \end{cases}$$

6. Conversely, assume that a function $u \in H^2(0, 1)$ satisfies $\frac{u(x)}{x^2} \in L^2(0, 1)$ and (3). Show that (1) holds.

   8.21 Assume that $p \in C^1([0, 1])$ with $p(x) \geq \alpha > 0 \quad \forall x \in [0, 1]$ and $q \in C([0, 1])$ with $q(x) \geq 0 \quad \forall x \in [0, 1]$. Let $v_0 \in C^2([0, 1])$ be the unique solution of

   (1)
   $$\begin{cases} -(pv_0')' + qv_0 = 0 & \text{on } [0, 1], \\ v_0(0) = 1, v_0(1) = 0. \end{cases}$$

   Set $k_0 = v_0'(0)$.

1. Check that $k_0 \leq -\alpha/p(0)$.

   [**Hint:** Multiply equation (1) by v_0 and integrate by parts. Use the fact that $1 \leq \|v_0\|_1 \leq \|v_0'\|_2$.]

   We now investigate the problem

   (2)
   $$\begin{cases} -(pu')' + qu = f \text{ on } (0, 1), \\ u'(0) = ku(0), u(1) = 0, \end{cases}$$

   where $k \in \mathbb{R}$ is fixed and $f \in L^2(0, 1)$ is given.

2. Assume $k = k_0$. Show that

$$[(2) \text{ has a solution } u \in H^2(0, 1)] \iff \left[\int_0^1 f v_0 = 0 \right].$$

Is there uniqueness of u?

3. Assume now that $k \neq k_0$. Prove that for every $f \in L^2(0, 1)$, problem (2) admits a unique solution $u \in H^2(0, 1)$.

[**Hint:** Using Exercise 8.5, find a constant K such that the bilinear form $a(u, v) = \int_0^1 (pu'v' + quv + Kuv) + p(0)ku(0)v(0)$ is coercive on H^1. Write (2) in the form $u = T(f + Ku)$ for some appropriate compact operator T. Then apply assertion (c) in the Fredholm alternative.]

$\boxed{8.22}$ Set

$$K = \{\rho \in H^1(0, 1); \rho \geq 0 \text{ on } (0, 1) \text{ and } \sqrt{\rho} \in H^1(0, 1)\}.$$

1. Construct an example of a function $\rho \in H^1(0, 1)$ with $\rho \geq 0$ on $(0, 1)$ such that $\rho \notin K$.

2. Given $\rho \in H^1(0, 1)$ with $\rho \geq 0$ on $(0, 1)$, set

$$\mu = \begin{cases} \frac{1}{2} \frac{\rho'}{\sqrt{\rho}} & \text{on } [\rho > 0], \\ 0 & \text{on } [\rho = 0]. \end{cases}$$

Prove that $\rho \in K \iff \mu \in L^2$, and then $(\sqrt{\rho})' = \mu$.

[**Hint:** Consider $\rho_\varepsilon = \rho + \varepsilon$.]

3. Show that K is a convex cone with vertex at 0.

4. Prove that the function $\rho \in K \mapsto \|(\sqrt{\rho})'\|_{L^2}^2$ is convex.

$\boxed{8.23}$ Let $I = (0, 1)$ and fix a constant $k > 0$.

1. Given $f \in L^1(I)$ prove that there exists a unique $u \in H_0^1(I)$ satisfying

(1) $$\int_I u'v' + k \int_I uv = \int_I fv \quad \forall v \in H_0^1(I).$$

2. Show that $u \in W^{2,1}(I)$.

3. Prove that

$$\|u\|_{L^1(I)} \leq \frac{1}{k} \|f\|_{L^1(I)}.$$

[**Hint:** Fix a function $\gamma \in C^1(\mathbb{R}, \mathbb{R})$ such that $\gamma'(t) \geq 0 \ \forall t \in \mathbb{R}$, $\gamma(0) = 0$, $\gamma(t) = +1 \ \forall t \geq 1$, and $\gamma(t) = -1 \ \forall t \leq -1$. Take $v = \gamma(nu)$ in (1) and let $n \to \infty$.]

4. Assume now that $f \in L^p(I)$ with $1 < p < \infty$. Show that there exists a constant $\delta > 0$ independent of k and p, such that

$$\|u\|_{L^p(I)} \le \frac{1}{k + \delta/pp'} \|f\|_{L^p(I)}.$$

[**Hint:** When $2 \le p < \infty$, take $v = \gamma(u)$ in (1), where $\gamma(t) = |t|^{p-1}$ sign t. When $1 < p < 2$, use duality.]

5. Prove that if $f \in L^\infty(I)$, then

$$\|u\|_{L^\infty(I)} \le C_k \|f\|_{L^\infty(I)},$$

and find the best constant C_k.

[**Hint:** Compute explicitly the solution u of (1) corresponding to $f \equiv 1$.]

8.24 Let $I = (0, 1)$.

1. Prove that for every $\varepsilon > 0$ there exists a constant C_ε such that

$$|u(1)|^2 \le \varepsilon \|u'\|^2_{L^2(I)} + C_\varepsilon \|u\|^2_{L^2(I)} \quad \forall u \in H^1(I).$$

[**Hint:** Use Exercise 8.5 or simply write

$$u^2(1) = u^2(x) + 2 \int_x^1 u(t)u'(t)dt.]$$

2. Prove that if the constant $k > 0$ is sufficiently large, then for every $f \in L^2(I)$ there exists a unique $u \in H^2(I)$ satisfying

(1) $\qquad \begin{cases} -u'' + ku = f & \text{on } (0, 1), \\ u'(0) = 0 \quad \text{and} \quad u'(1) = u(1). \end{cases}$

What is the weak formulation of problem (1)? What is the associated minimization problem?

3. Assume that k is sufficiently large. Let T be the operator $T : f \mapsto u$, where u is the solution of (1). Prove that T is a self-adjoint compact operator in $L^2(I)$.

4. Deduce that there exist a sequence (λ_n) in $\mathbb{R}$ with $|\lambda_n| \to \infty$ and a sequence (u_n) of functions in $C^2(\overline{I})$ such that $\|u_n\|_{L^2(I)} = 1$ and

$$\begin{cases} -u_n'' = \lambda_n u_n & \text{on } (0, 1), \\ u_n'(0) = 0 \quad \text{and} \quad u_n'(1) = u_n(1). \end{cases}$$

Prove that $\lambda_n \to +\infty$.

5. Let Λ be the set of all values of $\lambda \in \mathbb{R}$ for which there exists $u \not\equiv 0$ satisfying

$$\begin{cases} -u'' = \lambda u & \text{on } (0, 1), \\ u'(0) = 0 & \text{and} \quad u'(1) = u(1). \end{cases}$$

Determine the positive elements in Λ. Show that there is exactly one negative value of λ (denoted by λ_0) in Λ.

[**Hint:** Do not try to compute Λ explicitly; use instead the intersection of two graphs.]

6. What happens in question 2 when $k = |\lambda_0|$?

8.25 Let $I = (0, 2)$ and $V = H^1(I)$. Consider the bilinear form

$$a(u, v) = \int_0^2 u'(t)v'(t)dt + \left(\int_0^1 u(t)dt \right) \left(\int_0^1 v(t)dt \right).$$

1. Check that $a(u, v)$ is a continuous symmetric bilinear form, and that $a(u, u) = 0$ implies $u = 0$.

2. Prove that a is coercive.

   [**Hint:** Argue by contradiction and assume that there exists a sequence (u_n) in $H^1(I)$ such that $a(u_n, u_n) \to 0$ and $\|u_n\|_{H^1} = 1$. Let (u_{n_k}) be a subsequence such that u_{n_k} converges weakly in $H^1(I)$ and strongly in $L^2(I)$ to a limit u. Show that $u = 0$.]

3. Deduce that for every $f \in L^2(I)$ there exists a unique $u \in H^1(I)$ satisfying

   (1) $$a(u, v) = \int_0^2 fv \quad \forall v \in H^1(I).$$

   What is the corresponding minimization problem?

4. Show that the solution of (1) belongs to $H^2(I)$ (and in particular $u \in C^1(\overline{I})$). Determine the equation and the boundary conditions satisfied by u.

   [**Hint:** It is convenient to set $g = (\int_0^1 u)\chi$, where χ is the characteristic function of $(0, 1)$.]

5. Assume that $f \in C(\overline{I})$, and let u be the solution of (1). Prove that u belongs to $W^{2,p}(I)$ for every $p < \infty$. Show that $u \in C^2(\overline{I})$ iff $\int_I f = 0$.

6. Determine explicitly the solution u of (1) when f is a constant.

7. Set $u = Tf$, where u is the solution of (1) and $f \in L^2(I)$. Check that T is a self-adjoint compact operator from $L^2(I)$ into itself.

8. Study the eigenvalues of T.

8.26 A bounded linear operator S from a Hilbert space H into itself is said to be nonnegative, written $S \geq 0$, if

$$(Sf, f) \geq 0 \quad \forall f \in H.$$

Set $I = (0, 1)$. Assume that $p \in C^1(\overline{I})$ and $q \in C(\overline{I})$ satisfy

$$p(x) \geq \alpha > 0 \quad \text{and} \quad q(x) \geq \alpha > 0 \quad \forall x \in \overline{I}.$$

Recall that given $f \in H = L^2(I)$, there exists a unique solution $u \in H^2(I)$ of the equation

(1) $$-(pu')' + qu = f \text{ on } I$$

with the Dirichlet boundary condition

(2) $$u(0) = u(1) = 0.$$

The solution is denoted by $u_D = S_D f$, where S_D is viewed as a bounded linear operator from H into itself. Similarly, there exists a unique solution $u \in H^2(I)$ of (1) with the Neumann boundary condition

(3) $$u'(0) = u'(1) = 0$$

This solution is denoted by $u_N = S_N f$, where S_N is also viewed as a bounded linear operator from H into itself.

1. Show that $S_D \geq 0$ and $S_N \geq 0$.

2. Recall the minimization principles associated with the Dirichlet and Neumann conditions. Deduce that

(4) $$\frac{1}{2} \int_I p(u'_N)^2 + qu_N^2 - \int_I fu_N \leq \frac{1}{2} \int_I p(u'_D)^2 + qu_D^2 - \int_I fu_D.$$

[**Hint:** Use the fact that $H_0^1 \subset H^1$.]

3. Prove that $S_N - S_D \geq 0$.

[**Hint:** Use (4) together with (1) multiplied, respectively, by u_D and u_N.]

Given a real number $k \geq 0$, consider the equation (1) associated to the boundary condition

(5) $$p(0)u'(0) = ku(0) \quad \text{and} \quad u(1) = 0.$$

4. Check that problem (1) with (5) admits a unique solution, denoted by $u_k = S_k f$. What is the corresponding minimization principle?

5. Show that $S_k \geq 0$.

6. Let $k_1 \geq k_2 \geq 0$. Prove that $S_{k_2} - S_{k_1} \geq 0$.

8.27 Let $I = (-1, +1)$. Consider the bilinear form defined on $H_0^1(I)$ by

$$a(u, v) = \int_I (u'v' + uv - \lambda u(0)v),$$

where $\lambda \in \mathbb{R}$ is fixed.

1. Check that $a(u, v)$ is a continuous bilinear form on $H_0^1(I)$.

2. Prove that if $|\lambda| < \sqrt{2}$, the bilinear form a is coercive.

   [**Hint:** Check that $|u(0)| \leq \|u'\|_{L^2}$ $\forall u \in H_0^1(I)$.]

3. Deduce that if $|\lambda| < \sqrt{2}$, then for every $f \in L^2(I)$ there exists a unique solution $u \in H^2(I) \cap H_0^1(I)$ of the problem

   (1) $\begin{cases} -u'' + u - \lambda u(0) = f & \text{on } I, \\ u(-1) = u(1) = 0. \end{cases}$

4. Prove that there exists a unique value $\lambda = \lambda_0 \in \mathbb{R}$, to be determined explicitly, such that the problem

   (2) $\begin{cases} -u'' + u = \lambda u(0) & \text{on } I, \\ u(-1) = u(1) = 0, \end{cases}$

   admits a solution $u \not\equiv 0$.

   [**Hint:** It is convenient to introduce the unique solution φ of the problem

   (3) $\begin{cases} -\varphi'' + \varphi = 1 & \text{on } I, \\ \varphi(-1) = \varphi(1) = 0. \end{cases}$

   Compute φ explicitly.]

5. Prove that if $\lambda \neq \lambda_0$, then for every $f \in L^2(I)$ there exists a unique solution $u \in H^2(I) \cap H_0^1(I)$ of (1).

   [**Hint:** Consider the linear operator $S : g \mapsto v$, where $g \in L^2(I)$ and $v \in H^2(I) \cap H_0^1(I)$ is the unique solution of

   (4) $\begin{cases} -v'' + v = g & \text{on } I, \\ v(-1) = v(1) = 0. \end{cases}$

   Write (1) in the form $u - \lambda u(0)\varphi = Sf$.]

6. Analyze completely problem (1) when $\lambda = \lambda_0$.

   [**Hint:** Find a simple necessary and sufficient condition on Sf such that problem (1) admits a solution.]

8.28 Let $H = L^2(0, 1)$ equipped with its usual scalar product. Consider the operator $T : H \to H$ defined by

$$(Tf)(x) = x \int_x^1 f(t)dt + \int_0^x tf(t)dt, \quad \text{for } 0 \leq x \leq 1.$$

1. Check that T is a bounded operator.

2. Prove that T is a compact operator.

3. Prove that T is self-adjoint.

4. Show that $(Tf, f) \geq 0 \ \forall f \in H$, and that $(Tf, f) = 0$ implies $f = 0$.

5. Set $u = Tf$. Prove that $u \in H^2(0, 1)$ and compute u''. Check that $u(0) = u'(1) = 0$.

6. Determine the spectrum and the eigenvalues of T. Examine carefully the case $\lambda = 0$.

In what follows, set

$$e_k(x) = \sqrt{2} \sin\left[\left(k + \frac{1}{2}\right)\pi x\right], \quad k = 0, 1, 2, \ldots.$$

7. Check that (e_k) is an orthonormal basis of H.

[**Hint:** Use question 6.]

8. Deduce that the sequence $(\tilde{e}_k)$ defined by

$$\tilde{e}_k(x) = \sqrt{2} \cos\left[\left(k + \frac{1}{2}\right)\pi x\right], k = 0, 1, 2, \ldots,$$

is also an orthonormal basis of H.

[**Hint:** Consider $e_k(1 - x)$.]

Given $f \in H$ we denote by $(\alpha_k(f))$ the components of f in the basis (e_k).

9. Compute $\alpha_k(f)$ for the following functions:

(a) $f_1(x) = \chi_{[a,b]}(x) = \begin{cases} 1 & \text{if } x \in [a, b], \\ 0 & \text{if } x \notin [a, b], \end{cases}$
where $0 \leq a < b \leq 1$.

(b) $f_2(x) = x$.

(c) $f_3(x) = x^2$.

Finally, we propose to *characterize* the functions $f \in L^2(0, 1)$ that belong to $H^1(0, 1)$, using their components $\alpha_k(f)$.

10. Assume $f \in H^1(0, 1)$. Prove that there exists a constant $a \in \mathbb{R}$ (depending on f) such that $(k\alpha_k(f) + a) \in \ell^2$, i.e.,

$$(1) \qquad \sum_{k=0}^{\infty} |k\alpha_k(f) + a|^2 < \infty.$$

[**Hint:** Use an integration by parts in the computation of $\alpha_k(f)$.]

11. Conversely, assume that $f \in L^2(0, 1)$ and that (1) holds for some $a \in \mathbb{R}$. Prove that $f \in H^1(0, 1)$.

    [**Hint:** Set $\tilde{f} = f + \frac{a\pi}{\sqrt{2}}$ and $\tilde{f}_n(x) = \sum_{k=0}^{n-1} \alpha_k(\tilde{f})e_k(x)$. Check that $\|\tilde{f}_n'\|_{L^2}$ remains bounded as $n \to \infty$.]

8.29 Set

$$a(u, v) = \int_0^1 (u'v' + uv) + (u(1) - u(0))(v(1) - v(0)) \quad \forall u, v \in H^1(0, 1).$$

1. Check that a is a continuous coercive bilinear form on $H^1(0, 1)$.

2. Deduce that for every $f \in L^2(0, 1)$, there exists a unique $u \in H^2(0, 1)$ satisfying

$$(1) \qquad a(u, v) = \int_0^1 fv \quad \forall v \in H^1(0, 1).$$

3. Check that u satisfies

$$(2) \qquad \begin{cases} -u'' + u = f & \text{on } (0, 1), \\ u'(0) = u(0) - u(1), \\ u'(1) = u(0) - u(1). \end{cases}$$

   Show that any solution $u \in H^2(0, 1)$ of (2) satisfies (1).

   Let $T: L^2(0, 1) \to L^2(0, 1)$ be the operator defined by $Tf = u$.

4. Check that T is self-adjoint and compact.

5. Show that if $f \geq 0$ a.e. on $(0, 1)$, then $u = Tf \geq 0$ on $(0, 1)$.

6. Check that $(Tf, f)_{L^2} \geq 0 \quad \forall f \in L^2(0, 1)$.

7. Determine $EV(T)$.

8.30 Let $k \in \mathbb{R}$, $k \neq 1$, and consider the space

$$V = \{v \in H^1(0, 1); \ v(0) = kv(1)\},$$

and the bilinear form

$$a(u, v) = \int_0^1 (u'v' + uv) - \left(\int_0^1 u\right)\left(\int_0^1 v\right) \quad \forall u, v \in V.$$

1. Check that V is a closed subspace of $H^1(0, 1)$. In what follows, V is equipped with the Hilbert structure induced by the H^1 scalar product.

2. Prove that a is a continuous and coercive bilinear form on V.

   [**Hint:** Show that there exists a constant C such that $\|v\|_{L^\infty(0,1)} \le C\|v'\|_{L^2(0,1)}$ $\forall v \in V$.]

3. Deduce that for every $f \in L^2(0, 1)$ there exists a unique solution of the problem

   (1) $\qquad\qquad u \in V \quad \text{and} \quad a(u, v) = \int_0^1 fv \quad \forall v \in V.$

4. Show that the solution u of (1) belongs to $H^2(0, 1)$ and satisfies

   (2) $\qquad \begin{cases} -u'' + u - \int_0^1 u = f & \text{on } (0, 1), \\ u(0) = ku(1) \text{ and } u'(1) = ku'(0). \end{cases}$

5. Conversely, prove that any function $u \in H^2(0, 1)$ satisfying (2) is a solution of (1).

6. Let $k_n \in \mathbb{R}$, $k_n \ne 1$ $\forall n$, be a sequence converging to $k \ne 1$. Set

   $$V_n = \{v \in H^1(0, 1);\ v(0) = k_n v(1)\}.$$

   Given $f \in L^2(0, 1)$, let u_n be the solution of

   $(1_n) \qquad\qquad u_n \in V_n \quad \text{and} \quad a(u_n, v) = \int_0^1 fv \quad \forall v \in V_n.$

   Prove that $u_n \to u$ in $H^1(0, 1)$ as $n \to \infty$, where u is the solution of (1). Deduce that $u_n \to u$ in $H^2(0, 1)$

   [**Hint:** Check that the function $u^{(n)}$ defined by

   $$u^{(n)}(x) = u(x) + \frac{k - k_n}{k_n - 1} u(1)$$

   belongs to V_n and converges to u in $H^1(0, 1)$. Show that $a(u_n - u^{(n)}, u_n - u^{(n)}) = (k_n - k)u'(0)(u_n(1) - u^{(n)}(1))$.]

7. What happens to the sequence (u_n) if k_n converges to 1?

8. Consider the operator $T : L^2(0, 1) \to L^2(0, 1)$ defined by $Tf = u$, where u is the solution of (1). Show that T is self-adjoint and compact. Study $EV(T)$.

8.31 Consider the Sturm–Liouville operator $Au = -u'' + u$ on $(0, 1)$ with Neumann boundary condition $u'(0) = u'(1) = 0$.

1. Compute the eigenvalues of A and the corresponding eigenfunctions.

2. Given $f \in L^2$ with $\int_0^1 f = 0$, let u be the solution of

$$\begin{cases} -u'' + u = f & \text{on } (0, 1), \\ u'(0) = u'(1) = 0. \end{cases}$$

Prove that

$$\|u\|_{L^2(0,1)} \leq \frac{1}{(1+\pi^2)} \|f\|_{L^2(0,1)}.$$

[**Hint:** Apply question 6 in Problem 49.]

3. Let (u_n) be the sequence defined inductively by

$$\begin{cases} -u_n'' + u_n = u_{n-1} & \text{on } (0, 1), \\ u_n'(0) = u_n'(1) = 0, \end{cases}$$

starting with some $u_0 \in L^2(0, 1)$. Prove that

$$\|u_n - \overline{u}_0\|_{L^2(0,1)} \leq \frac{1}{(1+\pi^2)^n} \|u_0 - \overline{u}_0\|_{L^2(0,1)} \quad \forall n,$$

where $\overline{u}_0 = \int_0^1 u_0$.

$\boxed{8.32}$ Set

$$V = \{v \in H^1(0, 1); v(1) = 0\}.$$

Let

$$H = \{f \text{ is measurable on } (0, 1) \text{ and } xf(x) \in L^2(0, 1)\}.$$

1. Show that H equipped with the scalar product

$$(f, g)_H = \int_0^1 f(x)g(x)x^2 dx$$

is a Hilbert space.

2. Given $f \in H$ and $\varepsilon > 0$, check that there exists a unique $u \in V$ satisfying

$$\int_0^1 u'(x)v'(x)(x^2 + \varepsilon)dx + \int_0^1 u(x)v(x)x^2 dx = \int_0^1 f(x)v(x)x^2 dx \quad \forall v \in V.$$

This u is denoted by u_ε.

3. Prove that $u_\varepsilon \in H^2(0, 1)$ and satisfies

(1)
$$\begin{cases} -((x^2 + \varepsilon)u_\varepsilon')' + x^2 u_\varepsilon = x^2 f & \text{on } (0, 1), \\ u_\varepsilon'(0) = 0 \quad \text{and} \quad u_\varepsilon(1) = 0. \end{cases}$$

4. Deduce that

(2) $$|u'_\varepsilon(x)| \le \frac{1}{x} \int_0^x t|f(t) - u_\varepsilon(t)|dt \quad \forall x \in [0, 1].$$

5. Prove that $xu_\varepsilon(x)$ and $u'_\varepsilon(x)$ remain bounded in $L^2(0, 1)$ as $\varepsilon \to 0$.

[**Hint:** Use question 1 in Exercise 8.8.]

6. Pass to the limit as $\varepsilon \to 0$ and conclude that there exists a unique $u \in V$ satisfying

(3) $$\int_0^1 (u'(x)v'(x) + u(x)v(x))x^2dx = \int_0^1 f(x)v(x)x^2dx \quad \forall v \in V.$$

Consider the operator $T : H \to H$ defined by $Tf = u$, where u is the solution of (3).

7. Check that T is a self-adjoint compact operator from H into itself.

8. Determine all the eigenvalues of T.

[**Hint:** Look for eigenfunctions of the form $\frac{1}{x} \sin kx$ with appropriate k.]

8.33 *Simplicity of eigenvalues.*

Consider the Sturm–Liouville operator

$$Au = -(pu')' + qu \quad \text{on } I = (0, 1),$$

where $p \in C^1([0, 1])$, $p \ge \alpha > 0$ on I, and $q \in C([0, 1])$. (No further assumptions are made; in particular, the associated bilinear form $a(u, v) = \int_0^1 (pu'v' + quv)$ need not be coercive.) Set

$$N = \{u \in H^2(0, 1); \ a(u, v) = 0 \quad \forall v \in H_0^1(0, 1)\}.$$

1. Prove that there exists a unique $U \in N$ satisfying $U(0) = 1$ and $U'(0) = 0$.

[**Hint:** Apply Theorem 7.3 (Cauchy–Lipschitz–Picard) to the equation $Au = 0$ written as a first-order differential system.]

2. Prove that dim $N = 2$.

[**Hint:** Consider the unique $V \in N$ satisfying $V(0) = 0$ and $V'(0) = 1$. Then write any $u \in N$ as $u = u(0)U + u'(0)V$.]

3. Let $N_0 = \{u \in N; u(0) = 0\}$. Check that dim $N_0 = 1$.

4. Set $N_{00} = \{u \in N; u(0) = u(1) = 0\}$. Prove that

dim $N_{00} = 1 \iff 0$ is an eigenvalue of A with zero Dirichlet condition.

Otherwise, dim $N_{00} = 0$.

5. Deduce that all the eigenvalues of A with zero Dirichlet condition are simple. (By contrast, eigenvalues associated with periodic boundary conditions can have multiplicity 2; see Exercise 8.34.)

6. Extend the above results to the case in which the condition $p \in C^1([0, 1])$ is replaced by $p \in C([0, 1])$ and N is replaced by

$$\tilde{N} = \{u \in H^1(0, 1); a(u, v) = 0 \quad \forall v \in H_0^1(0, 1)\}.$$

8.34 Consider the problem

(1)
$$\begin{cases} -u'' + u = f \quad \text{on } (0, 1), \\ u(0) = u(1) \quad \text{and} \quad u'(1) - u'(0) = k, \end{cases}$$

where $k \in \mathbb{R}$ and $f(x)$ are given.

1. Find the weak formulation of problem (1).

2. Show that for every $f \in L^2(0, 1)$ and every $k \in \mathbb{R}$ there exists a unique weak solution $u \in H^1(0, 1)$ of (1). What is the corresponding minimization problem?

3. Show that the weak solution u belongs to $H^2(0, 1)$ and satisfies (1). Check that $u \in C^2([0, 1])$ if $f \in C([0, 1])$.

4. Prove that $u \leq 0$ on $(0, 1)$ if $f \leq 0$ on $(0, 1)$ and $k \leq 0$.

5. Take $k = 0$ and consider the operator $T : L^2(0, 1) \to L^2(0, 1)$ defined by $Tf = u$. Check that T is a self-adjoint compact operator. Compute the eigenvalues of T and prove that the multiplicity of each eigenvalue λ is 2 (i.e., dim $N(T - \lambda I) = 2$), except for the first one.

Remark. Note that by contrast, each eigenvalue of a Sturm–Liouville operator with *Dirichlet* boundary condition $(u(0) = u(1) = 0)$ is *simple* (i.e., the corresponding eigenspace has dimension 1).

8.35 Fix two functions $a, b \in C([0, 1])$ and consider the problem

(1)
$$\begin{cases} -u'' + au' + bu = f \quad \text{on } (0, 1), \\ u(0) = u(1) = 0, \end{cases}$$

with $f \in L^2(0, 1)$.

Given $g \in L^2(0, 1)$, let $v \in H^2(0, 1)$ be the unique solution of

(2)
$$\begin{cases} -v'' = g \quad \text{on } (0, 1), \\ v(0) = v(1) = 0. \end{cases}$$

Set $Sg = v$, so that $S : L^2(0, 1) \to H^2(0, 1)$.

1. Check that problem (1) is equivalent to

(3) $$\begin{cases} u \in H^1(0, 1), \\ u = S(f - au' - bu). \end{cases}$$

Consider the operator $T : H^1(0, 1) \to H^1(0, 1)$ defined by

$$Tu = -S(au' + bu), u \in H^1(0, 1).$$

2. Show that T is a compact operator.

3. Prove that problem (1) has a solution for every $f \in L^2(0, 1)$ iff the only solution u of (1) with $f = 0$ is $u = 0$.

4. Assume that $b \geq 0$ on $(0, 1)$. Prove that the only solution of (1) with $f = 0$ is $u = 0$.

   [**Hint:** Fix a constant $k > 0$ such that $k^2 - ka - b > 0$ on $[0, 1]$. Set $u_\varepsilon(x) = u(x) + \varepsilon e^{kx}$, $\varepsilon > 0$. Implement on u_ε the "classical approach" to the maximum principle; see Remark 26 in Chapter 8. Deduce that $u(x) \leq \varepsilon e^k \quad \forall x \in [0, 1], \quad \forall \varepsilon > 0$.]

   Conclude that for every $f \in L^2(0, 1)$ problem (1) admits a unique solution $u \in H^2(0, 1)$.

5. Check that in general (without any assumption on a or b), the space of solutions of problem (1) with $f = 0$ has dimension 0 or 1. If this dimension is 1 prove that problem (1) has a solution iff $\int_0^1 f\varphi_0 = 0$ for some function $\varphi_0 \neq 0$ to be determined.

   [**Hint:** Use Exercise 8.33 and the Fredholm alternative.]

8.36 Let $I = (0, 1)$. Given two functions f_1, f_2 on I, consider the system

(1) $$\begin{cases} -u_1'' + u_2 = f_1 & \text{on } I, \\ -u_2'' - u_1 = f_2 & \text{on } I, \end{cases}$$

where u_1, u_2 are the unknowns.

-A-

In this part we prescribe the Dirichlet condition

(2) $$u_1(0) = u_2(0) = u_1(1) = u_2(1) = 0.$$

1. Define an appropriate concept of weak solution for the problem (1)–(2). Show that for every pair $f = [f_1, f_2] \in L^2(I) \times L^2(I)$ there exists a unique weak solution

$$u = [u_1, u_2] \in H_0^1(I) \times H_0^1(I) \quad \text{of (1)–(2)}.$$

2. Check that $u_1, u_2 \in H^2(I)$.

3. Prove that if $f = [f_1, f_2] \in C(\overline{I}) \times C(\overline{I})$, then $u = [u_1, u_2] \in C^2(\overline{I}) \times C^2(\overline{I})$ and u is a classical solution of (1)–(2).

4. Consider the operator T from $L^2(I) \times L^2(I)$ into itself defined by $Tf = u$. Check that T is compact.

5. Prove that $EV(T) = \emptyset$.

6. Is T surjective? Deduce that $\sigma(T) = \{0\}$.

7. Is T self-adjoint? Compute $T^\star$.

-B-

In this part we prescribe the Neumann condition

(3) $$u_1'(0) = u_1'(1) = u_2'(0) = u_2'(1) = 0.$$

1. Define an appropriate concept of weak solution for the problem (1)–(3). Check that the Lax–Milgram theorem does not apply.
   Given $\varepsilon > 0$, consider the system

(1_ε)
$$\begin{cases} -u_1'' + u_2 + \varepsilon u_1 = f_1 & \text{on } I, \\ -u_2'' - u_1 + \varepsilon u_2 = f_2 & \text{on } I. \end{cases}$$

2. Show that for every $f = [f_1, f_2] \in L^2(I) \times L^2(I)$ there exists a unique weak solution $u^\varepsilon = [u_1^\varepsilon, u_2^\varepsilon] \in H^1(I) \times H^1(I)$ of the problem (1_ε)–(3).

3. Prove that
$$\|u_1^\varepsilon\|_{L^2(I)}^2 + \|u_2^\varepsilon\|_{L^2(I)}^2 \leq \|f_1\|_{L^2(I)}^2 + \|f_2\|_{L^2(I)}^2.$$

4. Deduce that $u^\varepsilon = [u_1^\varepsilon, u_2^\varepsilon]$ remains bounded in $H^2(I) \times H^2(I)$ as $\varepsilon \to 0$.

5. Show that for every $f = [f_1, f_2] \in L^2(I) \times L^2(I)$ there exists a unique solution $u = [u_1, u_2] \in H^2(I) \times H^2(I)$ of (1)–(3).

8.37

1. Prove that the identity operator from $H^1(0, 1)$ into $L^2(0, 1)$ is a Hilbert–Schmidt operator (see Problem 40).

   [**Hint:** Write $u(x) = u(0) + \int_0^x u'(t)dt$ and apply questions A3, A6, and B4 in Problem 40.]

2. Consider the eigenvalues (λ_n) of the Sturm–Liouville problem

$$\begin{cases} -(pu')' + qu = \lambda u & \text{on } (0, 1), \\ u(0) = u(1) = 0. \end{cases}$$

Recall that under the asumptions of Theorem 8.22, $\lambda_n \to +\infty$. Thus, for some integer N, we have $\lambda_n > 0 \quad \forall n \geq N$. Prove that

$$\sum_{n=N}^{+\infty} \frac{1}{\lambda_n^2} < \infty.$$

Remark. A much sharper estimate is described in Exercise 8.42.

8.38 *Example of an operator with continuous spectrum.*

Given $f \in L^2(\mathbb{R})$, let $u \in H^1(\mathbb{R})$ be the unique (weak) solution of the problem

$$-u'' + u = f \quad \text{on} \quad \mathbb{R},$$

in the sense that

$$\int_{\mathbb{R}} u'v' + \int_{\mathbb{R}} uv = \int_{\mathbb{R}} fv \quad \forall v \in H^1(\mathbb{R}).$$

Set $u = Tf$ and consider T as a bounded linear operator from $H = L^2(\mathbb{R})$ into itself.

1. Check that $T^\star = T$ (H is identified with its dual space) and that $\|T\| \leq 1$.

2. Prove that $EV(T) = \emptyset$. Is T a compact operator?

3. Let $\lambda \in (-\infty, 0)$; check that $\lambda \in \rho(T)$.

4. Let $\lambda \in (1, +\infty)$; check that $\lambda \in \rho(T)$.

5. Deduce that $\sigma(T) \subset [0, 1]$.

6. Is T surjective? Deduce that $0 \in \sigma(T)$.

7. Is $(T - I)$ surjective? Deduce that $1 \in \sigma(T)$ and that $\|T\| = 1$.

8. Let $\lambda \in (0, 1)$. Is $(T - \lambda I)$ surjective?

9. Conclude that $\sigma(T) = [0, 1]$.

8.39 Given $f \in L^2(0, 1)$, consider the function $\varphi : H^1(0, 1) \to \mathbb{R}$ defined by

$$\varphi(v) = \frac{1}{2} \int_0^1 v'^2 + \frac{1}{4} \int_0^1 v^4 - \int_0^1 fv, \quad v \in H^1(0, 1).$$

1. Check that φ is convex and continuous on $H^1(0, 1)$.

2. Show that $\varphi(v) \to +\infty$ as $\|v\|_{H^1} \to \infty$.

3. Deduce that there exits a unique $u \in H^1(0, 1)$ such that

(1) $$\varphi(u) = \min_{v \in H^1(0,1)} \varphi(v).$$

4. Show that

(2)
$$\int_0^1 (u'v' + u^3 v) = \int_0^1 fv \quad \forall v \in H^1(0, 1).$$

[**Hint:** Write that $\varphi(u) \le \varphi(u + \varepsilon v)$ $\forall v \in H^1(0, 1)$ and compute $\varphi(u + \varepsilon v)$ explicitly.]

5. Prove that $u \in H^2(0, 1)$ and that u satisfies

(3)
$$\begin{cases} -u'' + u^3 = f & \text{a.e. on } (0, 1), \\ u'(0) = u'(1) = 0. \end{cases}$$

6. Conversely, show that any solution of (3) satisfies (1). Deduce that (3) admits a unique solution.

7. Show that if $f \ge 0$ a.e. on $(0, 1)$, then $u \ge 0$ on $(0, 1)$.

[**Hint:** Use the same technique as in Section 8.5.]

8. Prove that if $f \in L^\infty(0, 1)$ then

$$\|u\|^3_{L^\infty(0,1)} \le \|f\|_{L^\infty(0,1)}.$$

[**Hint:** Argue as in the proof of Theorem 8.19 using as test function $G(u - K^{1/3})$, where $K = \|f\|_{L^\infty}$.]

9. What happens when $\varphi(v)$ is replaced by

$$\psi(v) = \frac{1}{2}\int_0^1 v'^2 + \frac{1}{4}v^4(0) - \int_0^1 fv, \quad v \in H^1(0, 1)?$$

$\boxed{8.40}$ Let $j \in C^1(\mathbb{R}, \mathbb{R})$ be a convex function satisfying

(1) $j(t) \ge |t| - C \quad \forall t \in \mathbb{R}, \text{ for some } C \in \mathbb{R},$

and

(2) $-1 < j'(t) < +1 \quad \forall t \in \mathbb{R}.$

[A good example to keep in mind is $j(t) = (1 + t^2)^{1/2}$.]
Given $f \in L^2(0, 1)$, consider the function $\varphi : H^1(0, 1) \to \mathbb{R}$ defined by

$$\varphi(v) = \frac{1}{2}\int_0^1 v'^2 + \int_0^1 j(v) - \int_0^1 fv, \quad v \in H^1(0, 1).$$

1. Prove that if $|\int_0^1 f| > 1$, then

$$\inf_{v \in H^1(0,1)} \varphi(v) = -\infty.$$

[**Hint:** Take $v = $ const.]

2. Show that if $| \int_0^1 f | \leq 1$, then

$$\inf_{v \in H^1(0,1)} \varphi(v) > -\infty.$$

[**Hint:** Write $\int_0^1 fv = \int_0^1 f(v - \overline{v}) + \overline{f} \, \overline{v}$, where $\overline{v} = \int_0^1 v$, $\overline{f} = \int_0^1 f$, and use the Poincaré–Wirtinger inequality; see Comment 1 on Chapter 8, and Problem 47.]

3. Show that if $| \int_0^1 f | < 1$, then

$$\lim_{\|v\|_{H^1(0,1)} \to \infty} \varphi(v) = +\infty.$$

Deduce that $\inf_{v \in H^1(0,1)} \varphi(v)$ is achieved, and that every minimizer $u \in H^1(0, 1)$ satisfies

$$\int_0^1 u'w' + \int_0^1 j'(u)w = \int_0^1 fw \quad \forall w \in H^1(0, 1).$$

Show that $u \in H^2(0, 1)$ is a solution of the problem

(3) $\qquad \begin{cases} -u'' + j'(u) = f & \text{on } (0, 1), \\ u'(0) = u'(1) = 0. \end{cases}$

4. Suppose that $| \int_0^1 f | = 1$. Show that $\inf_{v \in H^1(0,1)} \varphi(v)$ is not achieved.

[**Hint:** Argue by contradiction. If the infimum is achieved at some u, then u satisfies (3). Integrate (3) on $(0, 1)$.]

5. What happens to a minimizing sequence (u_n) of φ when $| \int_0^1 f | = 1$?

[**Hint:** Show that $u_n = \overline{u}_n + (u_n - \overline{u}_n)$ with $|\overline{u}_n| \to \infty$ and $\|u_n - \overline{u}_n\|_{H^1} \leq C$ as $n \to \infty$.]

$\boxed{8.41}$ Let $q \in C([0, 1])$ and assume that the bilinear form

$$a(u, v) = \int_0^1 (u'v' + quv), \quad u, v \in H_0^1(0, 1),$$

is coercive on $H_0^1(0, 1)$. The space $H_0^1(0, 1)$ is equipped with the scalar product $a(u, v)$, now denoted by $(u, v)_H$, and the norm $|u|_H = (u, u)_H^{1/2}$.

1. Prove that

(1) $\qquad \alpha = \sup \left\{ \int_0^1 |u|^4; \, u \in H_0^1(0, 1) \text{ and } |u|_H = 1 \right\} > 0$

is achieved by some u_0.

[**Hint:** Consider a maximizing sequence (u_n) converging weakly in $H_0^1(0, 1)$ to some u_0. Check that $\alpha = \int_0^1 |u_0|^4$ and that $|u_0|_H \le 1$. Show that $|u_0|_H = 1$ by introducing $u_0/|u_0|_H$.]

2. Show that one can assume $u_0 \ge 0$ on $[0,1]$.

   [**Hint:** Replace u_0 by $|u_0|$ and apply Exercise 8.11.]

3. Prove that u_0 belongs to $H^2(0, 1)$ and satisfies

   (2) $$\begin{cases} -u_0'' + qu_0 = \frac{1}{\alpha}u_0^3 & \text{on } (0, 1), \\ u_0(0) = u_0(1) = 0. \end{cases}$$

   [**Hint:** Write $\|w_\varepsilon\|_{L^4(0,1)} \le \|u_0\|_{L^4(0,1)}$, where $w_\varepsilon = \frac{u_0 + \varepsilon v}{|u_0 + \varepsilon v|_H}$, $v \in H_0^1(0, 1)$, and $\varepsilon > 0$ is sufficiently small. Then use a Taylor expansion for $\|w_\varepsilon\|_{L^4(0,1)}^4$ and for $|u_0 + \varepsilon v|_H^2$ as $\varepsilon \to 0$.]

4. Deduce that $u_0(x) > 0 \; \forall x \in (0, 1)$.

   [**Hint:** Use the strong maximum principle; see Problem 45.]

5. Let u_1 be any maximizer in (1). Show that either $u_1(x) > 0 \; \forall x \in (0, 1)$ or $u_1(x) < 0 \; \forall x \in (0, 1)$.

   [**Hint:** Check that $|u_1(x)| > 0 \; \forall x \in (0, 1)$.]

6. Deduce that there exists a solution $u \in C^2([0, 1])$ of the problem

   (3) $$\begin{cases} -u'' + qu = u^3 & \text{on } (0, 1), \\ u > 0 \text{ on } (0, 1), \\ u(0) = u(1) = 0. \end{cases}$$

   [**Hint:** Take $u = ku_0$ for some appropriate constant $k > 0$.]

7. Assume now that a is not coercive and more precisely that there exists some $v_1 \in H_0^1(0, 1)$ such that $v_1 \ne 0$ and $a(v_1, v_1) \le 0$. Prove that problem (3) has no solution.

   [**Hint:** Check that $\lambda_1 \le 0$, where λ_1 is the first eigenvalue of $Au = -u'' + qu$ and multiply (3) by the corresponding eigenfunctions φ_1.]

8.42 *Asymptotic behavior of Sturm–Liouville eigenvalues.*

Consider the operator $Av = -v'' + a(x)v$ on $I = (0, L)$, with zero Dirichlet condition and $a \in C([0, L])$.

1. Let (λ_n) denote the sequence of eigenvalues of A. Prove that

   $$\left| \lambda_n - \frac{\pi^2 n^2}{L^2} \right| \le \|a\|_{L^\infty(0,L)} \quad \forall n.$$

[**Hint:** Consider the eigenvalues of the operator A_0 corresponding to $a \equiv 0$ and use the Courant–Fischer min–max principle (see Problem 49).]

Consider now the general Sturm–Liouville operator

$$Bu = -(pu')' + qu \quad \text{on } (0, 1)$$

with zero Dirichlet condition. Assume that $p \in C^2([0, 1])$, $p \ge \alpha > 0$ on $(0, 1)$, and $q \in C([0, 1])$.

Set $L = \int_0^1 p(t)^{-1/2}dt$ and introduce the new variable $x = \int_0^t p(t)^{-1/2}dt$, so that $0 < x < L$ when $0 < t < 1$. Given a function $u \in C^2([0, 1])$, set

$$v(x) = p^{1/4}(t)u(t), \quad 0 < t < 1.$$

2. Prove that u satisfies $-(pu')' + qu = \mu u$ on $(0, 1)$ iff v satisfies $-v'' + av = \mu v$ on $(0, L)$, where $a \in C([0, 1])$ depends only on p and q.

[**Hint:** Prove, after some tedious computations, that $a(x) = q(t) + \frac{1}{4}p''(t) - \frac{(p'(t))^2}{16p(t)}$.]

3. Deduce that the eigenvalues (μ_n) of the operator B satisfy

$$\left| \mu_n - \frac{\pi^2 n^2}{L^2} \right| \le C.$$

Chapter 9
Sobolev Spaces and the Variational Formulation of Elliptic Boundary Value Problems in N Dimensions

9.1 Definition and Elementary Properties of the Sobolev Spaces $W^{1,p}(\Omega)$

Let $\Omega \subset \mathbb{R}^N$ be an open set and let $p \in \mathbb{R}$ with $1 \le p \le \infty$.

Definition. The Sobolev space $W^{1,p}(\Omega)$ is defined by[1]

$$W^{1,p}(\Omega) = \left\{ u \in L^p(\Omega) \;\middle|\; \begin{array}{l} \exists g_1, g_2, \ldots, g_N \in L^p(\Omega) \quad \text{such that} \\ \displaystyle\int_\Omega u \frac{\partial \varphi}{\partial x_i} = -\int_\Omega g_i \varphi \quad \forall \varphi \in C_c^\infty(\Omega), \quad \forall i = 1, 2, \ldots, N \end{array} \right\}.$$

We set

$$H^1(\Omega) = W^{1,2}(\Omega).$$

For $u \in W^{1,p}(\Omega)$ we define[2] $\frac{\partial u}{\partial x_i} = g_i$, and we write

$$\nabla u = \operatorname{grad} u = \left(\frac{\partial u}{\partial x_1}, \frac{\partial u}{\partial x_2}, \ldots, \frac{\partial u}{\partial x_N} \right).$$

The space $W^{1,p}(\Omega)$ is equipped with the norm

$$\|u\|_{W^{1,p}} = \|u\|_p + \sum_{i=1}^{N} \left\| \frac{\partial u}{\partial x_i} \right\|_p$$

or sometimes with the equivalent norm $(\|u\|_p^p + \sum_{i=1}^{N} \|\frac{\partial u}{\partial x_i}\|_p^p)^{1/p}$ (if $1 \le p < \infty$). The space $H^1(\Omega)$ is equipped with the scalar product

---

[1] When there is no confusion we shall often write $W^{1,p}$ instead of $W^{1,p}(\Omega)$.

[2] This definition makes sense: g_i is unique (a.e.) by Corollary 4.24.

H. Brezis, *Functional Analysis, Sobolev Spaces and Partial Differential Equations*,
DOI 10.1007/978-0-387-70914-7_9, © Springer Science+Business Media, LLC 2011

$$(u, v)_{H^1} = (u, v)_{L^2} + \sum_{i=1}^{N} \left(\frac{\partial u}{\partial x_i}, \frac{\partial v}{\partial x_i} \right)_{L^2} = \int_{\Omega} uv + \sum_{i=1}^{N} \frac{\partial u}{\partial x_i} \frac{\partial v}{\partial x_i}.$$

The associated norm

$$\|u\|_{H^1} = \left(\|u\|_2^2 + \sum_{i=1}^{N} \left\| \frac{\partial u}{\partial x_i} \right\|_2^2 \right)^{1/2}$$

is equivalent to the $W^{1,2}$ norm.

• **Proposition 9.1.** $W^{1,p}(\Omega)$ *is a Banach space for every* $1 \le p \le \infty$. $W^{1,p}(\Omega)$ *is reflexive for* $1 < p < \infty$, *and it is separable for* $1 \le p < \infty$. $H^1(\Omega)$ *is a separable Hilbert space.*

Proof. Adapt the proof of Proposition 8.1 using the operator $Tu = [u, \nabla u]$.

Remark 1. In the definition of $W^{1,p}$ we could equally well have used $C_c^{\infty}(\Omega)$ as set of test functions φ (instead of $C_c^1(\Omega)$); to show this, use a sequence of mollifiers (ρ_n).

Remark 2. It is clear that if $u \in C^1(\Omega) \cap L^p(\Omega)$ and if $\frac{\partial u}{\partial x_i} \in L^p(\Omega)$ for all $i = 1, 2, \dots, N$ (here $\frac{\partial u}{\partial x_i}$ means the usual partial derivative of u), then $u \in W^{1,p}(\Omega)$. Furthermore, the usual partial derivatives coincide with the partial derivatives in the $W^{1,p}$ sense, so that notation is consistent. In particular, if Ω is *bounded*, then $C^1(\overline{\Omega}) \subset W^{1,p}(\Omega)$ for all $1 \le p \le \infty$. Conversely, one can show that if $u \in W^{1,p}(\Omega)$ for some $1 \le p \le \infty$ and if $\frac{\partial u}{\partial x_i} \in C(\Omega)$ for all $i = 1, 2, \dots, N$ (here $\frac{\partial u}{\partial x_i}$ means the partial derivative in the $W^{1,p}$ sense), then $u \in C^1(\Omega)$; more precisely, there exists a function $\tilde{u} \in C^1(\Omega)$ such that $u = \tilde{u}$ a.e.

★ *Remark* 3. For every $u \in L_{\text{loc}}^1(\Omega)$, the *theory of distributions* gives a meaning to $\frac{\partial u}{\partial x_i}$ ($\frac{\partial u}{\partial x_i}$ is an element of the huge space of distributions $\mathcal{D}'(\Omega)$, a space that contains in particular $L_{\text{loc}}^1(\Omega)$). Using the language of distributions one can say that $W^{1,p}(\Omega)$ is the set of functions $u \in L^p(\Omega)$ for which all the partial derivatives $\frac{\partial u}{\partial x_i}, 1 \le i \le N$ (in the sense of distributions), belong to $L^p(\Omega)$.

When $\Omega = \mathbb{R}^N$ and $p = 2$ one can also define the Sobolev spaces using the Fourier transform; see, e.g., J.-L. Lions–E. Magenes [1], P. Malliavin [1], H. Triebel [1], L. Grafakos [1]. We do not take this point of view here.

Remark 4. It is useful to keep in mind the following facts:

(i) Let (u_n) be a sequence in $W^{1,p}$ such that $u_n \to u$ in L^p and (∇u_n) converges to some limit in $(L^p)^N$. Then $u \in W^{1,p}$ and $\|u_n - u\|_{W^{1,p}} \to 0$. When $1 < p \le \infty$ it suffices to know that $u_n \to u$ in L^p and that (∇u_n) is *bounded* in $(L^p)^N$ to conclude that $u \in W^{1,p}$ (why?).

(ii) Given a function f defined on Ω we denote by $\bar{f}$ its extension outside Ω, that is,

$$\bar{f}(x) = \begin{cases} f(x) & \text{if } x \in \Omega, \\ 0 & \text{if } x \in \mathbb{R}^N \backslash \Omega. \end{cases}$$

Let $u \in W^{1,p}(\Omega)$ and let $\alpha \in C_c^1(\Omega)$. Then [3]

$$\overline{\alpha u} \in W^{1,p}(\mathbb{R}^N) \quad \text{and} \quad \frac{\partial}{\partial x_i}(\overline{\alpha u}) = \overline{\alpha \frac{\partial u}{\partial x_i}} + \overline{\frac{\partial \alpha}{\partial x_i} u}.$$

Indeed, let $\varphi \in C_c^1(\mathbb{R}^N)$; we have

$$\int_{\mathbb{R}^N} \overline{\alpha u} \frac{\partial \varphi}{\partial x_i} = \int_{\Omega} \alpha u \frac{\partial \varphi}{\partial x_i} = \int_{\Omega} u \left[\frac{\partial}{\partial x_i}(\alpha \varphi) - \frac{\partial \alpha}{\partial x_i} \varphi \right]$$

$$= -\int_{\Omega} \left(\frac{\partial u}{\partial x_i} \alpha \varphi + u \frac{\partial \alpha}{\partial x_i} \varphi \right) = -\int_{\mathbb{R}^N} \overline{\left(\alpha \frac{\partial u}{\partial x_i} + \frac{\partial \alpha}{\partial x_i} u \right)} \varphi.$$

The same conclusion holds if instead of assuming that $\alpha \in C_c^1(\Omega)$, we take $\alpha \in C^1(\mathbb{R}^N) \cap L^\infty(\mathbb{R}^N)$ with $\nabla \alpha \in L^\infty(\mathbb{R}^N)^N$ and $\operatorname{supp} \alpha \subset \mathbb{R}^N \backslash (\partial \Omega)$.

Here is a first density result that holds for general open sets Ω; we establish later (Corollary 9.8) a more precise result under additional assumptions on Ω. We need the following.

Definition. Let $\Omega \subset \mathbb{R}^N$ be an open set. We say that an open set ω in $\mathbb{R}^N$ is *strongly included* in Ω and we write $\omega \subset\subset \Omega$ if $\bar{\omega} \subset \Omega$ and $\bar{\omega}$ is compact. [4]

• **Theorem 9.2 (Friedrichs).** *Let $u \in W^{1,p}(\Omega)$ with $1 \le p < \infty$. Then there exists a sequence (u_n) from $C_c^\infty(\mathbb{R}^N)$ such that*

(1) $$u_{n|\Omega} \to u \quad \text{in } L^p(\Omega)$$

and

(2) $$\nabla u_{n|\omega} \to \nabla u_{|\omega} \quad \text{in } L^p(\omega)^N \text{ for all } \omega \subset\subset \Omega.$$

In case $\Omega = \mathbb{R}^N$ and $u \in W^{1,p}(\mathbb{R}^N)$ with $1 \le p < \infty$, there exists a sequence (u_n) from $C_c^\infty(\mathbb{R}^N)$ such that

$$u_n \to u \quad \text{in } L^p(\mathbb{R}^N)$$

and

$$\nabla u_n \to \nabla u \quad \text{in } L^p(\mathbb{R}^N)^N.$$

In the proof we shall use the following lemma.

---

[3] Be careful: in general, $\bar{u} \notin W^{1,p}(\mathbb{R}^N)$ (why?).
[4] $\bar{\omega}$ denotes the closure of ω in $\mathbb{R}^N$.

Lemma 9.1. *Let $\rho \in L^1(\mathbb{R}^N)$ and let $v \in W^{1,p}(\mathbb{R}^N)$ with $1 \le p \le \infty$. Then*

$$\rho \star v \in W^{1,p}(\mathbb{R}^N) \quad and \quad \frac{\partial}{\partial x_i}(\rho \star v) = \rho \star \frac{\partial v}{\partial x_i} \quad \forall i = 1, 2, \ldots, N.$$

Proof of Lemma 9.1. Adapt the proof of Lemma 8.4.

Proof of Theorem 9.2. Set

$$\bar{u}(x) = \begin{cases} u(x) & \text{if } x \in \Omega, \\ 0 & \text{if } x \in \mathbb{R}^N \backslash \Omega, \end{cases}$$

and set $v_n = \rho_n \star \bar{u}$ (where ρ_n is a sequence of mollifiers). We know (see Section 4.4) that $v_n \in C^\infty(\mathbb{R}^N)$ and $v_n \to \bar{u}$ in $L^p(\mathbb{R}^N)$. We claim that $\nabla v_{n|\omega} \to \nabla u_{|\omega}$ in $L^p(\omega)^N$ for all $\omega \subset\subset \Omega$. Indeed, given $\omega \subset\subset \Omega$, fix a function $\alpha \in C_c^1(\Omega)$, $0 \le \alpha \le 1$, such that $\alpha = 1$ on a neighborhood of ω.

If n is large enough we have

$$(3) \qquad\qquad \rho_n \star (\overline{\alpha u}) = \rho_n \star \bar{u} \quad \text{on } \omega,$$

since

$$\operatorname{supp}(\rho_n \star \overline{\alpha u} - \rho_n \star \bar{u}) = \operatorname{supp}(\rho_n \star (1 - \bar{\alpha})\bar{u})$$

$$\subset \overline{\operatorname{supp} \rho_n + \operatorname{supp}(1 - \bar{\alpha})\bar{u}} \subset \overline{B\left(0, \frac{1}{n}\right) + \operatorname{supp}(1 - \bar{\alpha})}$$

$$\subset (\omega)^c$$

for n large enough. From Lemma 9.1 and Remark 4(ii) we have

$$\frac{\partial}{\partial x_i}(\rho_n \star \overline{\alpha u}) = \rho_n \star \left(\overline{\alpha \frac{\partial u}{\partial x_i} + \frac{\partial \alpha}{\partial x_i} u} \right).$$

It follows that

$$\frac{\partial}{\partial x_i}(\rho_n \star \overline{\alpha u}) \to \overline{\alpha \frac{\partial u}{\partial x_i} + \frac{\partial \alpha}{\partial x_i} u} \quad \text{in } L^p(\mathbb{R}^N)$$

and in particular,

$$\frac{\partial}{\partial x_i}(\rho_n \star \overline{\alpha u}) \to \frac{\partial u}{\partial x_i} \quad \text{in } L^p(\omega).$$

Because of (3) we have

$$\frac{\partial}{\partial x_i}(\rho_n \star \bar{u}) \to \frac{\partial u}{\partial x_i} \quad \text{in } L^p(\omega).$$

Finally, we multiply the sequence (v_n) by a sequence of cut-off functions (ζ_n) as in the proof of Theorem 8.7.[5] It is easily verified that the sequence $u_n = \zeta_n v_n$ has the desired properties, i.e., $u_n \in C_c^\infty(\mathbb{R}^N)$, $u_n \to u$ in $L^p(\Omega)$, and $\nabla u_n \to \nabla u$ in $(L^p(\omega))^N$ for every $\omega \subset\subset \Omega$.

In case $\Omega = \mathbb{R}^N$ the sequence $u_n = \zeta_n(\rho_n \star u)$ has the desired properties.

⋆ *Remark* 5. It can be shown (Meyers–Serrin's theorem) that if $u \in W^{1,p}(\Omega)$ with $1 \leq p < \infty$ then there exists a sequence (u_n) from $C^\infty(\Omega) \cap W^{1,p}(\Omega)$ such that $u_n \to u$ in $W^{1,p}(\Omega)$; the proof of this result is fairly delicate (see, e.g., R. Adams [1] or A. Friedman [2]). In general, if Ω is an *arbitrary open* set and if $u \in W^{1,p}(\Omega)$ there need not exist a sequence (u_n) in $C_c^1(\mathbb{R}^N)$ such that $u_{n|\Omega} \to u$ in $W^{1,p}(\Omega)$. Compare the Meyers–Serrin theorem (which holds for any open set) to Corollary 9.8 (which assumes that Ω is regular).

Here is a simple characterization of $W^{1,p}$ functions:

Proposition 9.3. *Let* $u \in L^p(\Omega)$ *with* $1 < p \leq \infty$. *The following properties are equivalent:*

(i) $u \in W^{1,p}(\Omega)$,

(ii) *there exists a constant C such that*

$$\left| \int_\Omega u \frac{\partial \varphi}{\partial x_i} \right| \leq C \|\varphi\|_{L^{p'}(\Omega)} \quad \forall \varphi \in C_c^\infty(\Omega), \quad \forall i = 1, 2, \ldots, N,$$

(iii) *there exists a constant C such that for all* $\omega \subset\subset \Omega$, *and all* $h \in \mathbb{R}^N$ *with* $|h| < \mathrm{dist}(\omega, \partial\Omega)$ *we have*

$$\|\tau_h u - u\|_{L^p(\omega)} \leq C|h|.$$

(Note that $\tau_h u(x) = u(x + h)$ makes sense for $x \in \omega$ and $|h| < \mathrm{dist}(\omega, \partial\Omega)$.)
Furthermore, we can take $C = \|\nabla u\|_{L^p(\Omega)}$ in (ii) *and* (iii).

If $\Omega = \mathbb{R}^N$ we have

$$\|\tau_h u - u\|_{L^p(\mathbb{R}^N)} \leq |h| \|\nabla u\|_{L^p(\mathbb{R}^N)}.$$

Proof.

(i) $\Rightarrow$ (ii). Obvious.

(ii) $\Rightarrow$ (i). Proceed as in the proof of Proposition 8.3.

(i) $\Rightarrow$ (iii). Assume first that $u \in C_c^\infty(\mathbb{R}^N)$. Let $h \in \mathbb{R}^N$ and set

$$v(t) = u(x + th), \quad t \in \mathbb{R}.$$

---

[5] Throughout this chapter we denote *systematically* by (ζ_n) a sequence of *cut-off functions*, that is, we fix a function $\zeta \in C_c^\infty(\mathbb{R}^N)$ with $0 \leq \zeta \leq 1$ and

$$\zeta(x) = \begin{cases} 1 & \text{if } |x| \leq 1, \\ 0 & \text{if } |x| \geq 2, \end{cases}$$

and we set $\zeta_n(x) = \zeta(x/n), n = 1, 2, \ldots$.

Then $v'(t) = h \cdot \nabla u(x + th)$ and thus

$$u(x+h) - u(x) = v(1) - v(0) = \int_0^1 v'(t)dt = \int_0^1 h \cdot \nabla u(x + th)dt.$$

It then follows that for $1 \le p < \infty$,

$$|\tau_h u(x) - u(x)|^p \le |h|^p \int_0^1 |\nabla u(x + th)|^p dt$$

and

$$\begin{aligned}
\int_\omega |\tau_h u(x) - u(x)|^p dx &\le |h|^p \int_\omega dx \int_0^1 |\nabla u(x + th)|^p dt \\
&= |h|^p \int_0^1 dt \int_\omega |\nabla u(x + th)|^p dx \\
&= |h|^p \int_0^1 dt \int_{\omega + th} |\nabla u(y)|^p dy.
\end{aligned}$$

If $|h| < \text{dist}(\omega\, \partial\Omega)$, there exists an open set $\omega' \subset\subset \Omega$ such that $\omega + th \subset \omega'$ for all $t \in [0, 1]$ and thus

(4) $$\|\tau_h u - u\|_{L^p(\omega)}^p \le |h|^p \int_{\omega'} |\nabla u|^p.$$

This concludes the proof of (ii) for $u \in C_c^\infty(\mathbb{R}^N)$ and $1 \le p < \infty$. Assume now that $u \in W^{1,p}(\Omega)$ with $1 \le p < \infty$. By Theorem 9.2 there exists a sequence (u_n) in $C_c^\infty(\mathbb{R}^N)$ such that $u_n \to u$ in $L^p(\Omega)$ and $\nabla u_n \to \nabla u$ in $L^p(\omega)^N \; \forall \omega \subset\subset \Omega$. Applying (4) to (u_n) and passing to the limit, we obtain (iii) for every $u \in W^{1,p}(\Omega)$, $1 \le p < \infty$. When $p = \infty$, apply the above (for $p < \infty$) and let $p \to \infty$.

(iii) $\Rightarrow$ (ii). Let $\varphi \in C_c^\infty(\Omega)$ and consider an open set ω such that $\text{supp}\,\varphi \subset \omega \subset\subset \Omega$. Let $h \in \mathbb{R}^N$ with $|h| < \text{dist}(\omega, \partial\Omega)$. Because of (iii) we have

$$\left| \int_\Omega (\tau_h u - u)\varphi \right| \le C|h| \, \|\varphi\|_{L^{p'}(\Omega)}.$$

On the other hand, since

$$\int_\Omega (u(x+h) - u(x))\varphi(x)dx = \int_\Omega u(y)(\varphi(y - h) - \varphi(y))dy,$$

it follows that

$$\int_\Omega u(y) \frac{(\varphi(y - h) - \varphi(y))}{|h|} dy \le C\|\varphi\|_{L^{p'}(\Omega)}.$$

Choosing $h = te_i$, $t \in \mathbb{R}$, and passing to the limit as $t \to 0$, we obtain (ii).

$\star$ *Remark* 6. When $p = 1$ the following implications remain true:

$$\text{(i)} \Rightarrow \text{(ii)} \Leftrightarrow \text{(iii)}.$$

The functions that satisfy (ii) (or (iii)) with $p = 1$ are called *functions of bounded variation* (in the language of distributions a function of bounded variation is an L^1 function such that all its first derivatives, in the sense of distributions, are bounded measures). *This space plays an important role in many applications.* One encounters functions of bounded variation (or with similar properties) in the theory of *minimal surfaces* (see, e.g., E. Giusti [1] and the works of E. DeGiorgi, M. Miranda, and others cited there), in questions of *elasticity and plasticity* (functions of bounded deformation, see, e.g., R. Temam–G. Strang [2] and the cited work of P. Suquet), in *quasilinear equations of first order*, the so-called *conservation laws*, which admit *discontinuous solutions* (see, e.g., A. I. Volpert [1] and A. Bressan [1]). On this vast subject, see also the book by L. Ambrosio–N. Fusco–D. Pallara [1] and Comment 16 at the end of this chapter.

Remark 7. Proposition 9.3 ((i) $\Rightarrow$ (iii)) implies that any function $u \in W^{1,\infty}(\Omega)$ has a continuous representative on Ω. More precisely, if Ω is connected then

$$(5) \qquad |u(x) - u(y)| \le \|\nabla u\|_{L^\infty(\Omega)} \operatorname*{dist}_{\Omega}(x, y) \quad \forall x, y \in \Omega$$

(for this continuous representative u), where $\operatorname{dist}_\Omega(x, y)$ denotes the *geodesic distance* from x to y in Ω; in particular, if Ω is *convex* then $\operatorname{dist}_\Omega(x, y) = |x - y|$. From here one can also deduce that if $u \in W^{1,p}(\Omega)$ for some $1 \le p \le \infty$ (and some open set Ω), and if $\nabla u = 0$ a.e. on Ω, then u is constant on each connected component of Ω.

Proposition 9.4 (differentiation of a product). *Let* $u, v \in W^{1,p}(\Omega) \cap L^\infty(\Omega)$ *with* $1 \le p \le \infty$. *Then* $uv \in W^{1,p}(\Omega) \cap L^\infty(\Omega)$ *and*

$$\frac{\partial}{\partial x_i}(uv) = \frac{\partial u}{\partial x_i} v + u \frac{\partial v}{\partial x_i}, \quad i = 1, 2, \dots, N.$$

Proof. As in the proof of Corollary 8.10, it suffices to consider the case $1 \le p < \infty$. By Theorem 9.2 there exist sequences (u_n), (v_n) in $C_c^\infty(\mathbb{R}^N)$ such that

$$u_n \to u, \qquad v_n \to v \quad \text{in } L^p(\Omega) \text{ and a.e. on } \Omega,$$
$$\nabla u_n \to \nabla u, \quad \nabla v_n \to \nabla v \quad \text{in } L^p(\omega)^N \text{ for all } \omega \subset\subset \Omega.$$

Checking the proof of Theorem 9.2, we see easily that we have further

$$\|u_n\|_{L^\infty(\mathbb{R}^N)} \le \|u\|_{L^\infty(\Omega)} \quad \text{and} \quad \|v_n\|_{L^\infty(\mathbb{R}^N)} \le \|v\|_{L^\infty(\Omega)}.$$

On the other hand,

$$\int_\Omega u_n v_n \frac{\partial \varphi}{\partial x_i} = -\int_\Omega \left(\frac{\partial u_n}{\partial x_i} v_n + u_n \frac{\partial v_n}{\partial x_i} \right) \varphi \quad \forall \varphi \in C_c^1(\Omega).$$

Passing to the limit, by the dominated convergence theorem, this becomes

$$\int_{\Omega} uv \frac{\partial \varphi}{\partial x_i} = -\int_{\Omega} \left(\frac{\partial u}{\partial x_i} v + u \frac{\partial v}{\partial x_i} \right) \varphi \quad \forall \varphi \in C_c^1(\Omega).$$

Proposition 9.5 (differentiation of a composition). *Let $G \in C^1(\mathbb{R})$ be such that $G(0) = 0$ and $|G'(s)| \leq M$ $\forall s \in \mathbb{R}$ for some constant M. Let $u \in W^{1,p}(\Omega)$ with $1 \leq p \leq \infty$. Then*

$$G \circ u \in W^{1,p}(\Omega) \quad and \quad \frac{\partial}{\partial x_i}(G \circ u) = (G' \circ u)\frac{\partial u}{\partial x_i}, \quad i = 1, 2, \ldots, N.$$

Proof. We have $|G(s)| \leq M|s|$ $\forall s \in \mathbb{R}$ and thus $|G \circ u| \leq M|u|$; as a consequence, $G \circ u \in L^p(\Omega)$ and also $(G' \circ u)\frac{\partial u}{\partial x_i} \in L^p(\Omega)$. It remains to verify that

(6) $$\int_{\Omega} (G \circ u) \frac{\partial \varphi}{\partial x_i} = -\int_{\Omega} (G' \circ u)\frac{\partial u}{\partial x_i} \varphi \quad \forall \varphi \in C_c^1(\Omega).$$

When $1 \leq p < \infty$, one chooses a sequence (u_n) in $C_c^\infty(\mathbb{R}^N)$ such that $u_n \to u$ in $L^p(\Omega)$ and a.e. on Ω, $\nabla u_n \to \nabla u$ in $L^p(\omega)^N$ $\forall \omega \subset\subset \Omega$ (Theorem 9.2). We have

$$\int_{\Omega} (G \circ u_n) \frac{\partial \varphi}{\partial x_i} = -\int_{\Omega} (G' \circ u_n)\frac{\partial u_n}{\partial x_i} \varphi \quad \forall \varphi \in C_c^1(\Omega).$$

But $G \circ u_n \to G \circ u$ in $L^p(\Omega)$ and $(G' \circ u_n)\frac{\partial u_n}{\partial x_i} \to (G' \circ u)\frac{\partial u}{\partial x_i}$ in $L^p(\omega)$ $\forall \omega \subset\subset \Omega$ (by dominated convergence), so that (6) follows. When $p = \infty$, fix an open set Ω' such that $\operatorname{supp}\varphi \subset \Omega' \subset\subset \Omega$. Then $u \in W^{1,p}(\Omega')$ $\forall p < \infty$ and (6) follows from the above.

Proposition 9.6 (change of variables formula). *Let Ω and Ω' be two open sets in $\mathbb{R}^N$ and let $H : \Omega' \to \Omega$ be a bijective map, $x = H(y)$, such that $H \in C^1(\Omega')$, $H^{-1} \in C^1(\Omega)$, $\operatorname{Jac} H \in L^\infty(\Omega')$, $\operatorname{Jac} H^{-1} \in L^\infty(\Omega)$.[6] Let $u \in W^{1,p}(\Omega)$ with $1 \leq p \leq \infty$. Then $u \circ H \in W^{1,p}(\Omega')$ and*

$$\frac{\partial}{\partial y_j}u(H(y)) = \sum_i \frac{\partial u}{\partial x_i}(H(y))\frac{\partial H_i}{\partial y_j}(y) \quad \forall j = 1, 2, \ldots, N.$$

Proof. When $1 \leq p < \infty$, choose a sequence (u_n) in $C_c^\infty(\mathbb{R}^N)$ such that $u_n \to u$ in $L^p(\Omega)$ and $\nabla u_n \to \nabla u$ in $L^p(\omega)^N$ $\forall \omega \subset\subset \Omega$. Thus $u_n \circ H \to u \circ H$ in $L^p(\Omega')$ and

$$\left(\frac{\partial u_n}{\partial x_i} \circ H \right) \frac{\partial H_i}{\partial y_j} \to \left(\frac{\partial u}{\partial x_i} \circ H \right) \frac{\partial H_i}{\partial y_j} \quad \text{in } L^p(\omega') \quad \forall \omega' \subset\subset \Omega'.$$

---

[6] Jac H denotes the Jacobian matrix $\frac{\partial H_i}{\partial y_j}$; thus it is a function in $L^\infty(\Omega')^{N \times N}$.

Given $\psi \in C_c^1(\Omega')$, we have

$$\int_{\Omega'} (u_n \circ H) \frac{\partial \psi}{\partial y_j} dy = -\int_{\Omega'} \sum_i \left(\frac{\partial u_n}{\partial x_i} \circ H \right) \frac{\partial H_i}{\partial y_j} \psi dy.$$

In the limit we obtain the desired result. When $p = \infty$, proceed in the same way as at the end of the proof of Proposition 9.5.

The spaces $W^{m,p}(\Omega)$

Let $m \geq 2$ be an integer and let p be a real number with $1 \leq p \leq \infty$. We define by induction

$$W^{m,p}(\Omega) = \left\{ u \in W^{m-1,p}(\Omega); \ \frac{\partial u}{\partial x_i} \in W^{m-1,p}(\Omega) \quad \forall i = 1, 2, \ldots, N \right\}.$$

Alternatively, these sets could also be introduced as

$$W^{m,p}(\Omega) = \left\{ u \in L^p(\Omega) \ \middle| \ \begin{matrix} \forall \alpha \text{ with } |\alpha| \leq m, \ \exists g_\alpha \in L^p(\Omega) \text{ such that} \\ \int_\Omega u D^\alpha \varphi = (-1)^{|\alpha|} \int_\Omega g_\alpha \varphi \quad \forall \varphi \in C_c^\infty(\Omega) \end{matrix} \right\},$$

where we use the standard *multi-index notation* $\alpha = (\alpha_1, \alpha_2, \ldots, \alpha_N)$ with $\alpha_i \geq 0$ an integer,

$$|\alpha| = \sum_{i=1}^N \alpha_i \quad \text{and} \quad D^\alpha \varphi = \frac{\partial^{|\alpha|} \varphi}{\partial x_1^{\alpha_1} \partial x_2^{\alpha_2} \cdots \partial x_N^{\alpha_N}}.$$

We set $D^\alpha u = g_\alpha$. The space $W^{m,p}(\Omega)$ equipped with the norm

$$\|u\|_{W^{m,p}} = \sum_{0 \leq |\alpha| \leq m} \|D^\alpha u\|_p$$

is a Banach space.

The space $H^m(\Omega) = W^{m,2}(\Omega)$ equipped with the scalar product

$$(u, v)_{H^m} = \sum_{0 \leq |\alpha| \leq m} (D^\alpha u, D^\alpha v)_{L^2}$$

is a Hilbert space.

Remark 8. One can show that if Ω is "smooth enough" with $\Gamma = \partial\Omega$ bounded, then the norm on $W^{m,p}(\Omega)$ is equivalent to the norm

$$\|u\|_p + \sum_{|\alpha|=m} \|D^\alpha u\|_p.$$

More precisely, it is proved that for every multi-index α with $0 < |\alpha| < m$ and for every $\varepsilon > 0$ there exists a constant C (depending on $\Omega, \varepsilon, \alpha$) such that

$$\|D^\alpha u\|_p \le \varepsilon \sum_{|\beta|=m} \|D^\beta u\|_p + C\|u\|_p \quad \forall u \in W^{m,p}(\Omega)$$

(see, e.g., R. Adams [1]).

9.2 Extension Operators

It is often convenient to establish properties of functions in $W^{1,p}(\Omega)$ by beginning with the case $\Omega = \mathbb{R}^N$ (see for example the results of Section 9.3). It is therefore useful to be able to extend a function $u \in W^{1,p}(\Omega)$ to a function $\tilde{u} \in W^{1,p}(\mathbb{R}^N)$. This is not always possible (in a general domain Ω). However, if Ω is "*smooth*," such an extension can be constructed. Let us begin by making precise the notion of a smooth open set.

Notation. Given $x \in \mathbb{R}^N$, write

$$x = (x', x_N) \text{ with } x' \in \mathbb{R}^{N-1}, \quad x' = (x_1, x_2, \ldots, x_{N-1}),$$

and set

$$|x'| = \left(\sum_{i=1}^{N-1} x_i^2\right)^{1/2}.$$

We define

$$\mathbb{R}^N_+ = \{x = (x', x_N); \ x_N > 0\},$$
$$Q = \{x = (x', x_N); \ |x'| < 1 \text{ and } |x_N| < 1\},$$
$$Q_+ = Q \cap \mathbb{R}^N_+,$$
$$Q_0 = \{x = (x', 0); \ |x'| < 1\}.$$

Definition. We say that an open set Ω is of *class* C^1 if for every $x \in \partial\Omega = \Gamma$ there exist a neighborhood U of x in $\mathbb{R}^N$ and a bijective map $H : Q \to U$ such that

$$H \in C^1(\overline{Q}), \quad H^{-1} \in C^1(\overline{U}), \quad H(Q_+) = U \cap Q, \quad \text{and} \quad H(Q_0) = U \cap \Gamma.$$

The map H is called a *local chart*.

Theorem 9.7. *Suppose that Ω is of class C^1 with Γ bounded (or else $\Omega = \mathbb{R}^N_+$). Then there exists a linear extension operator*

$$P : W^{1,p}(\Omega) \to W^{1,p}(\mathbb{R}^N) \quad (1 \le p \le \infty)$$

such that for all $u \in W^{1,p}(\Omega)$,

(i) $Pu_{|\Omega} = u,$

(ii) $\|Pu\|_{L^p(\mathbb{R}^N)} \le C\|u\|_{L^p(\Omega)},$

(iii) $\|Pu\|_{W^{1,p}(\mathbb{R}^N)} \le C\|u\|_{W^{1,p}(\Omega)},$

where C depends only on Ω.

We shall begin by proving a simple but fundamental lemma concerning the *extension by reflection.*

Lemma 9.2. *Given $u \in W^{1,p}(Q_+)$ with $1 \le p \le \infty$, one defines the function $u^\star$ on Q to be the extension by reflection, that is,*

$$u^\star(x', x_N) = \begin{cases} u(x', x_N) & \text{if } x_N > 0, \\ u(x', -x_N) & \text{if } x_N < 0. \end{cases}$$

Then $u^\star \in W^{1,p}(Q)$ and

$$\|u^\star\|_{L^p(Q)} \le 2\|u\|_{L^p(Q_+)}, \quad \|u^\star\|_{W^{1,p}(Q)} \le 2\|u\|_{W^{1,p}(Q_+)}.$$

Proof. In fact, we shall prove that

(7) $$\frac{\partial u^\star}{\partial x_i} = \left(\frac{\partial u}{\partial x_i}\right)^\star \quad \text{for } 1 \le i \le N-1$$

and

(8) $$\frac{\partial u^\star}{\partial x_N} = \left(\frac{\partial u}{\partial x_N}\right)^\square,$$

where $\left(\frac{\partial u}{\partial x_i}\right)^\star$ denotes the extension by reflection of $\frac{\partial u}{\partial x_i}$ and where we set, whenever f is defined on Q_+,

$$f^\square(x', x_N) = \begin{cases} f(x', x_N) & \text{if } x_N > 0, \\ -f(x', -x_N) & \text{if } x_N < 0. \end{cases}$$

We shall use a sequence (η_k) of functions in $C^\infty(\mathbb{R})$ defined by

$$\eta_k(t) = \eta(kt), \quad t \in \mathbb{R}, \quad k = 1, 2, \ldots,$$

where η is any *fixed* function, $\eta \in C^\infty(\mathbb{R})$, such that

$$\eta(t) = \begin{cases} 0 & \text{if } t < 1/2, \\ 1 & \text{if } t > 1. \end{cases}$$

Proof of (7). Let $\varphi \in C_c^1(Q)$. For $1 \le i \le N-1$, we have

(9)
$$\int_Q u^\star \frac{\partial \varphi}{\partial x_i} = \int_{Q_+} u \frac{\partial \psi}{\partial x_i},$$

where $\psi(x', x_N) = \varphi(x', x_N) + \varphi(x', -x_N)$. The function ψ does not in general belong to $C_c^1(Q_+)$, and thus it cannot be used as a test function (in the definition of $W^{1,p}$). On the other hand, $\eta_k(x_N)\psi(x', x_N) \in C_c^1(Q_+)$ and thus

$$\int_{Q_+} u \frac{\partial}{\partial x_i}(\eta_k \psi) = -\int_{Q_+} \frac{\partial u}{\partial x_i} \eta_k \psi.$$

Since $\frac{\partial}{\partial x_i}(\eta_k \psi) = \eta_k \frac{\partial \psi}{\partial x_i}$, we have

(10)
$$\int_{Q_+} u \eta_k \frac{\partial \psi}{\partial x_i} = -\int_{Q_+} \frac{\partial u}{\partial x_i} \eta_k \psi.$$

Passing to the limit in (10) as $k \to \infty$ (by dominated convergence), we obtain

(11)
$$\int_{Q_+} u \frac{\partial \psi}{\partial x_i} = -\int_{Q_+} \frac{\partial u}{\partial x_i} \psi.$$

Combining (9) and (11), we are led to

$$\int_Q u^\star \frac{\partial \varphi}{\partial x_i} = -\int_{Q_+} \frac{\partial u}{\partial x_i} \psi = -\int_Q \left(\frac{\partial u}{\partial x_i}\right)^\star \varphi,$$

from which (7) follows.

Proof of (8). For every $\varphi \in C_c^1(Q)$ we have

(12)
$$\int_Q u^\star \frac{\partial \varphi}{\partial x_N} = \int_{Q_+} u \frac{\partial \chi}{\partial x_N},$$

where $\chi(x', x_N) = \varphi(x', x_N) - \varphi(x', -x_N)$. Note that $\chi(x', 0) = 0$ and thus there exists a constant M such that $|\chi(x', x_N)| \le M|x_N|$ on Q. Since $\eta_k \chi \in C_c^1(Q_+)$, we have

(13)
$$\int_{Q_+} u \frac{\partial}{\partial x_N}(\eta_k \chi) = -\int_{Q_+} \frac{\partial u}{\partial x_N} \eta_k \chi.$$

But

(14)
$$\frac{\partial}{\partial x_N}(\eta_k \chi) = \eta_k \frac{\partial \chi}{\partial x_N} + k\eta'(kx_N)\chi.$$

We claim that

(15)
$$\int_{Q_+} uk\eta'(kx_N)\chi \to 0 \text{ as } k \to \infty.$$

Indeed, we have

$$\left| \int_{Q_+} u k \eta'(kx_N)\chi \right| \le kMC \int_{0<x_N<1/k} |u| x_N dx \le MC \int_{0<x_N<1/k} |u| dx$$

with $C = \sup_{t\in[0,1]} |\eta'(t)|$, from which (15) follows.

We deduce from (13), (14), and (15) that

$$\int_{Q_+} u \frac{\partial\chi}{\partial x_N} = -\int_{Q_+} \frac{\partial u}{\partial x_N}\chi.$$

Finally, we have

(16)
$$\int_{Q_+} \frac{\partial u}{\partial x_N}\chi = \int_Q \left(\frac{\partial u}{\partial x_N}\right)^\square \varphi.$$

Combining (12) and (16), we obtain (8). This concludes the proof of Lemma 9.2.

The conclusion of Lemma 9.2 remains valid if Q_+ is replaced by $\mathbb{R}^N_+$ (the proof is unchanged). This establishes Theorem 9.7 for $\Omega = \mathbb{R}^N_+$.

★ *Remark* 9. Lemma 9.2 gives a very simple construction of extension operators for certain open sets Ω that are not of class C^1. Consider, for example, the square

$$\Omega = \{x \in \mathbb{R}^2;\ 0 < x_1 < 1,\ 0 < x_2 < 1\}.$$

Let $u \in W^{1,p}(\Omega)$. By four successive reflections (see Figure 6) we obtain an extension $\tilde{u} \in W^{1,p}(\tilde{\Omega})$ of u in

$$\tilde{\Omega} = \{x \in \mathbb{R}^2;\ -1 < x_1 < 3,\ -1 < x_2 < 3\}.$$

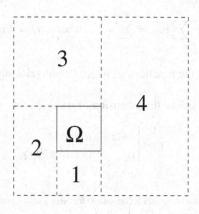

Fig. 6

Then fix any function $\psi \in C_c^1(\tilde{\Omega})$ such that $\psi = 1$ on Ω. Denote by Pu the function $\psi \tilde{u}$ extended to $\mathbb{R}^2$ by 0 outside $\tilde{\Omega}$. It is easily shown that the operator $P : W^{1,p}(\Omega) \to W^{1,p}(\mathbb{R}^2)$ satisfies (i), (ii), and (iii).

The next lemma is very useful.

Lemma 9.3 (partition of unity). *Let Γ be a compact subset of $\mathbb{R}^N$ and let U_1, U_2, ..., U_k be an open covering of Γ, i.e., $\Gamma \subset \bigcup_{i=1}^{k} U_i$. Then there exist functions θ_0, θ_1, θ_2, ..., $\theta_k \in C^\infty(\mathbb{R}^N)$ such that*

(i) $\qquad 0 \leq \theta_i \leq 1 \quad \forall i = 0, 1, 2, \ldots, k \ \text{ and } \ \sum_{i=0}^{k} \theta_i = 1 \text{ on } \mathbb{R}^N,$

(ii) $\qquad \begin{cases} \operatorname{supp} \theta_i \ \text{is compact and } \operatorname{supp} \theta_i \subset U_i \quad \forall i = 1, 2, \ldots, \\ \operatorname{supp} \theta_0 \subset \mathbb{R}^N \backslash \Gamma. \end{cases}$

If Ω is an open bounded set and $\Gamma = \partial\Omega$, then $\theta_{0|\Omega} \in C_c^\infty(\Omega)$.

Proof. This lemma is classical; similar statements can be found, for example, in S. Agmon [1], R. Adams [1], G. Folland [1], P. Malliavin [1].

Proof of Theorem 9.7. We "rectify" $\Gamma = \partial\Omega$ by local charts and use a partition of unity.[7] More precisely, since Γ is compact and of class C^1, there exist open sets $(U_i)_{1 \leq i \leq k}$ in $\mathbb{R}^N$ such that $\Gamma \subset \bigcup_{i=1}^{k} U_i$ and bijective maps $H_i = Q \to U_i$ such that

$$H_i \in C^1(\overline{Q}), \quad H_i^{-1} \in C^1(\overline{U}_i), \quad H_i(Q_+) = U_i \cap \Omega, \quad \text{and} \quad H_i(Q_0) = U_i \cap \Gamma.$$

Consider the functions θ_0, θ_1, θ_2, ..., θ_k introduced in Lemma 9.3. Given $u \in W^{1,p}(\Omega)$, write

$$u = \sum_{i=0}^{k} \theta_i u = \sum_{i=0}^{k} u_i, \quad \text{where} \ \ u_i = \theta_i u.$$

Now we extend each of the functions u_i to $\mathbb{R}^N$, distinguishing u_0 and $(u_i)_{1 \leq i \leq k}$.

(a) *Extension of u_0.* We define the extension of u_0 to $\mathbb{R}^N$ by

$$\bar{u}_0(x) = \begin{cases} u_0(x) & \text{if } x \in \Omega, \\ 0 & \text{if } x \in \mathbb{R}^N \backslash \Omega. \end{cases}$$

---

[7] In the following we shall often use this technique to transfer a result proved on $\mathbb{R}_+^N$ (or Q_+) to the same conclusion on a smooth open set Ω.

Recall that $\theta_0 \in C^1(\mathbb{R}^N) \cap L^\infty(\mathbb{R}^N)$, $\nabla\theta_0 \in L^\infty(\mathbb{R}^N)$, since $\nabla\theta_0 = -\sum_{i=1}^{k} \nabla\theta_i$ has compact support, and that $\operatorname{supp}\theta_0 \subset \mathbb{R}^N\backslash\Gamma$. It follows (by Remark 4(ii)) that

$$\bar{u}_0 \in W^{1,p}(\mathbb{R}^N) \quad \text{and} \quad \frac{\partial}{\partial x_i}\bar{u}_0 = \theta_0\frac{\overline{\partial u}}{\partial x_i} + \frac{\partial\theta_0}{\partial x_i}\bar{u}.$$

Thus

$$\|\bar{u}_0\|_{W^{1,p}(\mathbb{R}^N)} \leq C\|u\|_{W^{1,p}(\Omega)}.$$

(b) *Extension of u_i, $1 \leq i \leq k$.*

Consider the restriction of u to $U_i \cap \Omega$ and "transfer" this function to Q_+ with the help of H_i. More precisely, set $v_i(y) = u(H_i(y))$ for $y \in Q_+$. We know (Proposition 9.6) that $v_i \in W^{1,p}(Q_+)$. Then define the extension on Q by reflection of v_i (Lemma 9.2); call it $v_i^\star$. We know that $v_i^\star \in W^{1,p}(Q)$. "Retransfer" $v_i^\star$ to U_i using H_i^{-1} and call it w_i:

$$w_i(x) = v_i^\star[H_i^{-1}(x)] \quad \text{for } x \in U_i.$$

Then $w_i \in W^{1,p}(U_i)$, $w_i = u$ on $U_i \cap \Omega$, and

$$\|w_i\|_{W^{1,p}(U_i)} \leq C\|u\|_{W^{1,p}(U_i\cap\Omega)}.$$

Finally, set for $x \in \mathbb{R}^N$,

$$\hat{u}_i(x) = \begin{cases} \theta_i(x)w_i(x) & \text{if } x \in U_i, \\ 0 & \text{if } x \in \mathbb{R}^N\backslash U_i, \end{cases}$$

so that $\hat{u}_i \in W^{1,p}(\mathbb{R}^N)$ (see Remark 4(ii)), $\hat{u}_i = u_i$ on Ω, and

$$\|\hat{u}_i\|_{W^{1,p}(\mathbb{R}^N)} \leq C\|u\|_{W^{1,p}(U_i\cap\Omega)}.$$

(c) *Conclusion.* The operator $Pu = \bar{u}_0 + \sum_{i=1}^{N} \hat{u}_i$ possesses all the desired properties.

• **Corollary 9.8 (density).** *Assume that Ω is of class C^1, and let $u \in W^{1,p}(\Omega)$ with $1 \leq p < \infty$. Then there exists a sequence (u_n) from $C_c^\infty(\mathbb{R}^N)$ such that $u_{n|\Omega} \to u$ in $W^{1,p}(\Omega)$. In other words, the restrictions to Ω of $C_c^\infty(\mathbb{R}^N)$ functions form a dense subspace of $W^{1,p}(\Omega)$.*

Proof. Suppose first that Γ is *bounded*. Then there exists an extension operator P (by Theorem 9.7). The sequence[8] $\zeta_n(\rho_n \star Pu)$ converges to Pu in $W^{1,p}(\mathbb{R}^N)$ and thus it answers the problem. When Γ is *not bounded* we start by considering the sequence $\zeta_n u$. Given $\varepsilon > 0$, fix n_0 such that $\|\zeta_{n_0}u - u\|_{W^{1,p}} < \varepsilon$. One may then construct an extension $v \in W^{1,p}(\mathbb{R}^N)$ of $\zeta_{n_0}u$ (since this only involves the intersection of Γ with a large ball). We finally pick any $w \in C_c^\infty(\mathbb{R}^N)$ such that $\|w - v\|_{W^{1,p}(\mathbb{R}^N)} < \varepsilon$.

---

[8] As usual, (ρ_n) is a sequence of mollifiers and (ζ_n) is a sequence of cut-off functions as in the proof of Theorem 9.2.

9.3 Sobolev Inequalities

In Chapter 8 we saw that if Ω has dimension 1, then $W^{1,p}(\Omega) \subset L^{\infty}(\Omega)$ with continuous injection, for all $1 \leq p \leq \infty$. In dimension $N \geq 2$ this inclusion is true only for $p > N$; when $p \leq N$ one may construct functions in $W^{1,p}$ that do not belong to L^{∞} (see Remark 16). Nevertheless, an important result, essentially due to Sobolev, asserts that if $1 \leq p < N$ then $W^{1,p}(\Omega) \subset L^{p^\star}(\Omega)$ with continuous injection, for some $p^\star \in (p, +\infty)$. This result is often called the *Sobolev embedding* theorem. We begin by considering the following case:

A. The case $\Omega = \mathbb{R}^N$.

- **Theorem 9.9 (Sobolev, Gagliardo, Nirenberg).** *Let* $1 \leq p < N$. *Then*

$$W^{1,p}(\mathbb{R}^N) \subset L^{p^\star}(\mathbb{R}^N), \quad \text{where } p^\star \text{ is given by } \frac{1}{p^\star} = \frac{1}{p} - \frac{1}{N},$$

and there exists a constant[9] $C = C(p, N)$ *such that*

$$(17) \qquad \|u\|_{p^\star} \leq C\|\nabla u\|_p \quad \forall u \in W^{1,p}(\mathbb{R}^N).$$

Remark 10. The value $p^\star$ can be obtained by a very simple *scaling argument* (scaling arguments, dear to the physicists, sometimes give useful information with a minimum of effort). Indeed, *assume that there exist* constants C and q $(1 \leq q \leq \infty)$ such that

$$(18) \qquad \|u\|_q \leq C\|\nabla u\|_p \quad \forall u \in C_c^{\infty}(\mathbb{R}^N).$$

Then necessarily $q = p^\star$. To see this, fix any function $u \in C_c^{\infty}(\mathbb{R}^N)$, and plug into (18) $u_\lambda(x) = u(\lambda x)$. We obtain

$$\|u\|_q \leq C\lambda^{(1+\frac{N}{q}-\frac{N}{p})}\|\nabla u\|_p \quad \forall \lambda > 0,$$

which implies $1 + \frac{N}{q} - \frac{N}{p} = 0$, i.e., $q = p^\star$ (provided u does not vanish identically).

The proof of Theorem 9.9 relies on the following lemma:

Lemma 9.4. *Let* $N \geq 2$ *and let* $f_1, f_2, \ldots, f_N \in L^{N-1}(\mathbb{R}^{N-1})$. *For* $x \in \mathbb{R}^N$ *and* $1 \leq i \leq N$ *set*

$$\tilde{x}_i = (x_1, x_2, \ldots, x_{i-1}, x_{i+1}, \ldots, x_N) \in \mathbb{R}^{N-1},$$

i.e., x_i *is omitted from the list. Then the function*

$$f(x) = f_1(\tilde{x}_1) f_2(\tilde{x}_2) \cdots f_N(\tilde{x}_N), \quad x \in \mathbb{R}^N,$$

belongs to $L^1(\mathbb{R}^N)$ *and*

---

[9] We can take $C(p, N) = (N-1)p/(N-p)$, but this constant is not optimal. The best constant is known (but it is not simple!), see Th. Aubin [1], G. Talenti [1], and E. Lieb [1].

$$\|f\|_{L^1(\mathbb{R}^N)} \le \prod_{i=1}^{N} \|f_i\|_{L^{N-1}(\mathbb{R}^{N-1})}.$$

Proof. The case $N = 2$ is trivial (why?). Let us consider the case $N = 3$. We have

$$\int_{\mathbb{R}} |f(x)|dx_3 = |f_3(x_1, x_2)| \int_{\mathbb{R}} |f_1(x_2, x_3)||f_2(x_1, x_3)|dx_3$$

$$\le |f_3(x_1, x_2)| \left(\int_{\mathbb{R}} |f_1(x_2, x_3)|^2 dx_3 \right)^{1/2} \left(\int_{\mathbb{R}} |f_2(x_1, x_3)|^2 dx_3 \right)^{1/2}$$

(by Cauchy–Schwarz). Applying Cauchy–Schwarz once more gives

$$\int_{\mathbb{R}^3} |f(x)|dx \le \|f_3\|_{L^2(\mathbb{R}^2)} \|f_1\|_{L^2(\mathbb{R}^2)} \|f_2\|_{L^2(\mathbb{R}^2)}.$$

The general case is obtained by induction—assuming the result for N and then deducing it for $N + 1$. *Fix* $x_{N+1} \in \mathbb{R}$; because of Hölder's inequality,

$$\int_{\mathbb{R}^N} |f(x)|dx_1 dx_2 \cdots dx_N$$

$$\le \|f_{N+1}\|_{L^N(\mathbb{R}^N)} \left[\int |f_1 f_2 \cdots f_N|^{N'} dx_1 dx_2 \cdots dx_N \right]^{1/N'}$$

(with $N' = N/(N - 1)$). Applying the induction assumption to the functions $|f_1|^{N'}, |f_2|^{N'}, \ldots, |f_N|^{N'}$, we obtain

$$\int_{\mathbb{R}^N} |f_1|^{N'} \cdots |f_N|^{N'} dx_1 \cdots dx_N \le \prod_{i=1}^{N} \|f_i\|_{L^N(\mathbb{R}^{N-1})}^{N'},$$

from which it follows that

$$\int_{\mathbb{R}^N} |f(x)|dx_1 \cdots dx_N \le \|f_{N+1}\|_{L^N(\mathbb{R}^N)} \prod_{i=1}^{N} \|f_i\|_{L^N(\mathbb{R}^{N-1})}.$$

Now *vary* x_{N+1}. Each of the functions $x_{N+1} \mapsto \|f_i\|_{L^N(\mathbb{R}^{N-1})}$ belongs to $L^N(\mathbb{R})$, $1 \le i \le N$. As a consequence, their product $\prod_{i=1}^{N} \|f_i\|_{L^N(\mathbb{R}^{N-1})}$ belongs to $L^1(\mathbb{R})$ (see Remark 2 following Hölder's inequality in Chapter 4) and

$$\int_{R^{N+1}} |f(x)|dx_1 dx_2 \cdots dx_N dx_{N+1} \le \prod_{i=1}^{N+1} \|f_i\|_{L^N(\mathbb{R}^N)}.$$

Proof of Theorem 9.9. We begin with the case $p = 1$ and $u \in C_c^1(\mathbb{R}^N)$. We have

$$|u(x_1, x_2, \ldots, x_N)| = \left| \int_{-\infty}^{x_1} \frac{\partial u}{\partial x_1}(t, x_2, \ldots, x_N) dt \right|$$

$$\leq \int_{-\infty}^{+\infty} \left| \frac{\partial u}{\partial x_1}(t, x_2, \ldots, x_N) \right| dt,$$

and similarly, for each $1 \leq i \leq N$,

$$|u(x_1, x_2, \ldots, x_N)| \leq \int_{-\infty}^{+\infty} \left| \frac{\partial u}{\partial x_i}(x_1, x_2, \ldots, x_{i-1}, t, x_{i+1}, \ldots x_N) \right| dt \overset{\text{def}}{=} f_i(\tilde{x}_i).$$

Thus

$$|u(x)|^N \leq \prod_{i=1}^{N} f_i(\tilde{x}_i).$$

We deduce from Lemma 9.4 that

$$\int_{\mathbb{R}^N} |u(x)|^{N/(N-1)} dx \leq \prod_{i=1}^{N} \|f_i\|_{L^1(\mathbb{R}^{N-1})}^{1/(N-1)} = \prod_{i=1}^{N} \left\| \frac{\partial u}{\partial x_i} \right\|_{L^1(\mathbb{R}^N)}^{1/(N-1)}.$$

As a consequence, we have

(19) $$\|u\|_{L^{N/(N-1)}(\mathbb{R}^N)} \leq \prod_{i=1}^{N} \left\| \frac{\partial u}{\partial x_i} \right\|_{L^1(\mathbb{R}^N)}^{1/N}.$$

This completes the proof of (17) when $p = 1$ and $u \in C_c^1(\mathbb{R}^N)$. We turn now to the case $1 < p < N$, still with $u \in C_c^1(\mathbb{R}^N)$. Let $m \geq 1$; apply (19) to $|u|^{m-1} u$ instead of u. We obtain

(20) $$\|u\|_{mN/(N-1)}^m \leq m \prod_{i=1}^{N} \left\| |u|^{m-1} \frac{\partial u}{\partial x_i} \right\|_1^{1/N} \leq m \|u\|_{p'(m-1)}^{m-1} \prod_{i=1}^{N} \left\| \frac{\partial u}{\partial x_i} \right\|_p^{1/N}.$$

Then *choose* m such that $mN/(N - 1) = p'(m - 1)$, which gives $m = (N - 1)p^\star/N$ ($m \geq 1$ since $1 < p < N$). We obtain

$$\|u\|_{p^\star} \leq m \prod_{i=1}^{N} \left\| \frac{\partial u}{\partial x_i} \right\|_p^{1/N},$$

and thus

$$\|u\|_{p^\star} \leq C \|\nabla u\|_p \quad \forall u \in C^1(\mathbb{R}^N).$$

To complete the proof let $u \in W^{1,p}(\mathbb{R}^N)$, and let (u_n) be a sequence from $C_c^1(\mathbb{R}^N)$ such that $u_n \to u$ in $W^{1,p}(\mathbb{R}^N)$. One can also suppose, by extracting a subsequence if necesary, that $u_n \to u$ a.e. We have

$$\|u_n\|_{p^\star} \leq C \|\nabla u_n\|_p.$$

It follows from Fatou's lemma[10] that

$$u \in L^{p^\star} \quad \text{and} \quad \|u\|_{p^\star} \leq C\|\nabla u\|_p.`$$

● **Corollary 9.10.** *Let* $1 \leq p < N$. *Then*

$$W^{1,p}(\mathbb{R}^N) \subset L^q(\mathbb{R}^N) \quad \forall q \in [p, p^\star]$$

with continuous injection.

Proof. Given $q \in [p, p^\star]$, we write

$$\frac{1}{q} = \frac{\alpha}{p} + \frac{1-\alpha}{p^\star} \quad \text{for some} \quad \alpha \in [0, 1].$$

We know (see Remark 2 in Chapter 4) that

$$\|u\|_q \leq \|u\|_p^\alpha \|u\|_{p^\star}^{1-\alpha} \leq \|u\|_p + \|u\|_{p^\star}$$

(by Young's inequality). Using Theorem 9.9, we conclude that

$$\|u\|_q \leq C\|u\|_{W^{1,p}} \quad \forall u \in W^{1,p}(\mathbb{R}^N).$$

● **Corollary 9.11 (the limiting case** $p = N$**).** *We have*

$$W^{1,p}(\mathbb{R}^N) \subset L^q(\mathbb{R}^N) \quad \forall q \in [N, +\infty).$$

Proof. Assume $u \in C_c^1(\mathbb{R}^N)$; applying (20) with $p = N$, we obtain

$$\|u\|_{mN/(N-1)}^m \leq m\|u\|_{(m-1)N/(N-1)}^{m-1}\|\nabla u\|_N \quad \forall m \geq 1,$$

and thanks to Young's inequality we have

$$(21) \qquad \|u\|_{mN/(N-1)} \leq C(\|u\|_{(m-1)N/(N-1)} + \|\nabla u\|_N) \quad \forall m \geq 1.$$

In (21) we choose first $m = N$; it becomes

$$\|u\|_{N^2/(N-1)} \leq C\|u\|_{W^{1,N}},$$

and by the interpolation inequality (see Remark 2 in Chapter 4) we have

$$(22) \qquad \|u\|_q \leq C\|u\|_{W^{1,N}}$$

for all q with $N \leq q \leq N^2/(N-1)$. Reiterating this argument with $m = N + 1$, $m = N + 2$, etc., we arrive at

$$(23) \qquad \|u\|_q \leq C\|u\|_{W^{1,N}} \quad \forall u \in C^1(\mathbb{R}^N)$$

---

[10] One can also conclude by noticing that (u_n) is a Cauchy sequence in $L^{p^\star}$.

for all $q \in [N, +\infty)$, with a constant C depending on q and N.[11] Inequality (23) extends by density to $W^{1,N}$.

• **Theorem 9.12 (Morrey).** *Let $p > N$. Then*

$$(24) \qquad\qquad W^{1,p}(\mathbb{R}^N) \subset L^\infty(\mathbb{R}^N)$$

with continuous injection. Furthermore, for all $u \in W^{1,p}(\mathbb{R}^N)$, we have

$$(25) \qquad\quad |u(x) - u(y)| \leq C|x-y|^\alpha \|\nabla u\|_p \quad a.e.\ x, y \in \mathbb{R}^N,$$

where $\alpha = 1 - (N/p)$ and C is a constant (depending only on p and N).

Remark 11. Inequality (25) implies the existence of a function $\tilde{u} \in C(\mathbb{R}^N)$ such that $u = \tilde{u}$ a.e. on $\mathbb{R}^N$. (Indeed, let $A \subset \mathbb{R}^N$ be a set of measure zero such that (25) holds for $x, y \in \mathbb{R}^N \backslash A$; since $\mathbb{R}^N \backslash A$ is dense in $\mathbb{R}^N$, the function $u_{|\mathbb{R}^N \backslash A}$ admits a (unique) continuous extension to $\mathbb{R}^N$.) In other words, every function $u \in W^{1,p}(\mathbb{R}^N)$ with $p > N$ admits a *continuous representative*. When it is useful, we replace u by its continuous representative, and we also denote by u this continuous representative.

Proof. We begin by establishing (25) for $u \in C_c^1(\mathbb{R}^N)$. Let Q be an open cube, containing 0, whose sides—of length r—are parallel to the coordinate axes. For $x \in Q$ we have

$$u(x) - u(0) = \int_0^1 \frac{d}{dt} u(tx)\, dt$$

and thus

$$(26) \qquad |u(x) - u(0)| \leq \int_0^1 \sum_{i=1}^N |x_i| \left| \frac{\partial u}{\partial x_i}(tx) \right| dt \leq r \sum_{i=1}^N \int_0^1 \left| \frac{\partial u}{\partial x_i}(tx) \right| dt.$$

Set

$$\bar{u} = \frac{1}{|Q|} \int_Q u(x)\, dx = (\text{mean of } u \text{ on } Q).$$

Integrating (26) on Q, we obtain

$$|\bar{u} - u(0)| \leq \frac{r}{|Q|} \int_Q dx \sum_{i=1}^N \int_0^1 \left| \frac{\partial u}{\partial x_i}(tx) \right| dt$$

$$= \frac{1}{r^{N-1}} \int_0^1 dt \int_Q \sum_{i=1}^N \left| \frac{\partial u}{\partial x_i}(tx) \right| dx$$

$$= \frac{1}{r^{N-1}} \int_0^1 dt \int_{tQ} \sum_{i=1}^N \left| \frac{\partial u}{\partial x_i}(y) \right| \frac{dy}{t^N}.$$

---

[11] This constant "blows up" as $q \to +\infty$.

Then, from Hölder's inequality, we have

$$\int_{tQ}\left|\frac{\partial u}{\partial x_i}(y)\right|dy \le \left(\int_{Q}\left|\frac{\partial u}{\partial x_i}\right|^{p}\right)^{1/p}|tQ|^{1/p'}$$

(since $tQ \subset Q$ for $t \in (0,1)$). We deduce from this that

$$|\bar{u} - u(0)| \le \frac{1}{r^{N-1}}\|\nabla u\|_{L^p(Q)}\, r^{N/p'}\int_{0}^{1}\frac{t^{N/p'}}{t^N}dt = \frac{r^{1-(N/p)}}{1-(N/p)}\|\nabla u\|_{L^p(Q)}.$$

By translation, this inequality remains true for all cubes Q whose sides—of length r—are parallel to the coordinate axes. Thus we have

(27) $$|\bar{u} - u(x)| \le \frac{r^{1-(N/p)}}{1-(N/p)}\|\nabla u\|_{L^p(Q)}\quad \forall x \in Q.$$

By adding these (and using the triangle inequality) we obtain

(28) $$|u(x) - u(y)| \le \frac{2r^{1-(N/p)}}{1-(N/p)}\|\nabla u\|_{L^p(Q)}\quad \forall x, y \in Q.$$

Given any two points $x, y \in \mathbb{R}^N$, there exists such a cube Q with side $r = 2|x - y|$ containing x and y. This implies (25) when $u \in C_c^1(\mathbb{R}^N)$. For a general function $u \in W^{1,p}(\mathbb{R}^N)$ we use a sequence (u_n) of $C_c^1(\mathbb{R}^N)$ such that $u_n \to u$ in $W^{1,p}(\mathbb{R}^N)$ and $u_n \to u$ a.e.

We now prove (24). Let $u \in C_c^1(\mathbb{R}^N)$, $x \in \mathbb{R}^N$, and let Q be a cube of side $r = 1$ containing x. From (27) and Hölder's inequality we have

$$|u(x)| \le |\bar{u}| + C\|\nabla u\|_{L^p(Q)} \le C\|u\|_{W^{1,p}(Q)} \le C\|u\|_{W^{1,p}(\mathbb{R}^N)},$$

where C depends only on p and N. Thus

$$\|u\|_{L^\infty(\mathbb{R}^N)} \le C\|u\|_{W^{1,p}(\mathbb{R}^N)}\quad \forall u \in C_c^1(\mathbb{R}^N).$$

For a general function $u \in W^{1,p}(\mathbb{R}^N)$ we use a standard density argument.

Remark 12. We deduce from (24) that if $u \in W^{1,p}(\mathbb{R}^N)$ with $N < p < \infty$, then

$$\lim_{|x|\to\infty} u(x) = 0.$$

Indeed, there exists a sequence (u_n) in $C_c^1(\mathbb{R}^N)$ such that $u_n \to u$ in $W^{1,p}(\mathbb{R}^N)$. By (24), u is also the uniform limit on $\mathbb{R}^N$ of the u_n's.

• **Corollary 9.13.** *Let $m \ge 1$ be an integer and let $p \in [1, +\infty)$. We have*

$$W^{m,p}(\mathbb{R}^N) \subset L^q(\mathbb{R}^N), \quad where \quad \frac{1}{q} = \frac{1}{p} - \frac{m}{N} \quad if \quad \frac{1}{p} - \frac{m}{N} > 0,$$

$$W^{m,p}(\mathbb{R}^N) \subset L^q(\mathbb{R}^N) \quad \forall q \in [p, +\infty) \qquad \text{if } \frac{1}{p} - \frac{m}{N} = 0,$$

$$W^{m,p}(\mathbb{R}^N) \subset L^\infty(\mathbb{R}^N) \qquad\qquad\qquad \text{if } \frac{1}{p} - \frac{m}{N} < 0,$$

and all these injections are continuous. Moreover, if $m - (N/p) > 0$ is not an integer, set[12]

$$k = [m - (N/p)] \quad \text{and} \quad \theta = m - (N/p) - k \;\; (0 < \theta < 1).$$

We have, for all $u \in W^{m,p}(\mathbb{R}^N)$,

$$\|D^\alpha u\|_{L^\infty(\mathbb{R}^N)} \leq C\|u\|_{W^{m,p}(\mathbb{R}^N)} \quad \forall \alpha \text{ with } |\alpha| \leq k$$

and[13]

$$|D^\alpha u(x) - D^\alpha u(y)| \leq C\|u\|_{W^{m,p}(\mathbb{R}^N)}|x - y|^\theta \text{ a.e. } x, y \in \mathbb{R}^N, \quad \forall \alpha \text{ with } |\alpha| = k.$$

In particular, $W^{m,p}(\mathbb{R}^N) \subset C^k(\mathbb{R}^N)$.[14]

Proof. All of these results are obtained by repeated applications of Theorem 9.9, Corollary 9.11, and Theorem 9.12.

Remark 13. The case $p = 1$ and $m = N$ is special. We have $W^{N,1} \subset L^\infty$. (But it is not true, in general, that $W^{m,p} \subset L^\infty$ for $p > 1$ and $m = N/p$.) Indeed, for $u \in C_c^\infty(\mathbb{R}^N)$, we have

$$u(x_1, x_2, \ldots, x_N) = \int_{-\infty}^{x_1}\int_{-\infty}^{x_2}\cdots\int_{-\infty}^{x_N} \frac{\partial^N u}{\partial x_1 \partial x_2 \cdots \partial x_N}(t_1, t_2, \ldots, t_N)dt_1 dt_2 \cdots dt_N$$

and thus

(29) $$\|u\|_{L^\infty} \leq C\|u\|_{W^{N,1}} \quad \forall u \in C_c^\infty(\mathbb{R}^N).$$

The case of a general function $u \in W^{N,1}$ follows by density.

Now let us turn to the following.

B. The case $\Omega \subset \mathbb{R}^N$.

We suppose here that Ω is an open set of class C^1 with Γ bounded or else that $\Omega = \mathbb{R}^N_+$.

● **Corollary 9.14.** *Let $1 \leq p \leq \infty$. We have*

---

[12] [ ] denotes the integer part.

[13] This implies that $D^\alpha u$ is Lipschitz continuous for all α with $|\alpha| < k$, i.e.,

$$|D^\alpha u(x) - D^\alpha u(y)| \leq C\|u\|_{W^{m,p}}|x - y| \quad \text{a.e. } x, y \in \mathbb{R}^N.$$

[14] This is to be understood modulo the choice of a continuous representative.

$$W^{1,p}(\Omega) \subset L^{p^\star}(\Omega), \quad where \quad \frac{1}{p^\star} = \frac{1}{p} - \frac{1}{N}, \quad if \ p < N,$$

$$W^{1,p}(\Omega) \subset L^q(\Omega) \quad \forall q \in [p, +\infty), \quad if \ p = N,$$

$$W^{1,p}(\Omega) \subset L^\infty(\Omega), \quad if \ p > N,$$

and all these injections are continuous. Moreover, if $p > N$ we have, for all $u \in W^{1,p}(\Omega)$,

$$|u(x) - u(y)| \le C\|u\|_{W^{1,p}}|x - y|^\alpha \quad a.e. \ x, y \in \Omega,$$

with $\alpha = 1 - (N/p)$ and C depends only on Ω, p, and N. In particular, $W^{1,p}(\Omega) \subset C(\overline{\Omega})$.[15]

Proof. Consider the extension operator

$$P : W^{1,p}(\Omega) \to W^{1,p}(\mathbb{R}^N)$$

(see Theorem 9.7) and then apply Theorem 9.9, Corollary 9.11, and Theorem 9.12.

● **Corollary 9.15.** *The conclusion of Corollary 9.13 remains true if $\mathbb{R}^N$ is replaced by Ω.*[16]

Proof. By repeated application of Corollary 9.14.[17]

● **Theorem 9.16 (Rellich–Kondrachov).** *Suppose that Ω is **bounded** and of class C^1. Then we have the following **compact** injections:*

$$W^{1,p}(\Omega) \subset L^q(\Omega) \quad \forall q \in [1, p^\star), \quad where \quad \frac{1}{p^\star} = \frac{1}{p} - \frac{1}{N}, \quad if \ p < N,$$

$$W^{1,p}(\Omega) \subset L^q(\Omega) \quad \forall q \in [p, +\infty), \quad if \ p = N,$$

$$W^{1,p}(\Omega) \subset C(\overline{\Omega}), \quad if \ p > N.$$

*In particular, $W^{1,p}(\Omega) \subset L^p(\Omega)$ with **compact** injection for all p (and all N).*

Proof. The case $p > N$ follows from Corollary 9.14 and Ascoli–Arzelà's theorem. The case $p = N$ reduces to the case $p < N$. Therefore, we are left with the case $p < N$.

Let $\mathcal{H}$ be the unit ball in $W^{1,p}(\Omega)$. Let P be the extension operator of Theorem 9.7. Set $\mathcal{F} = P(\mathcal{H})$, so that $\mathcal{H} = \mathcal{F}_{|\Omega}$. In order to show that $\mathcal{H}$ has compact closure

---

[15] Once more, this is modulo the choice of a continuous representative.

[16] To be precise, if $m - (N/p) > 0$ is not an integer, then

$$W^{m,p}(\Omega) \subset C^k(\overline{\Omega}), \ where \ k = [m - (n/p)]$$

and $C^k(\overline{\Omega}) = \{u \in C^k(\Omega); \ D^\alpha u \text{ has a continuous extension on } \overline{\Omega} \text{ for all } \alpha \text{ with } |\alpha| \le k\}$.

[17] Alternatively, one could apply Corollary 9.13 together with an extension operator $P : W^{m,p}(\Omega) \to W^{m,p}(\mathbb{R}^N)$, but this would require an extra hypothesis: Ω would have to be of class C^m to construct this extension operator.

in $L^q(\Omega)$ for $q \in [1, p^\star)$ we invoke Theorem 4.26. Since Ω is bounded, we may always assume that $q \geq p$. Clearly, $\mathcal{F}$ is bounded in $W^{1,p}(\mathbb{R}^N)$ and thus it is also bounded in $L^q(\mathbb{R}^N)$ by Corollary 9.10. We have to check that

$$\lim_{|h| \to 0} \|\tau_h f - f\|_{L^q(\mathbb{R}^N)} = 0 \text{ uniformly in } f \in \mathcal{F}.$$

By Proposition 9.3 we have

$$\|\tau_h f - f\|_{L^p(\mathbb{R}^N)} \leq |h| \|\nabla f\|_{L^p(\mathbb{R}^N)} \quad \forall f \in \mathcal{F}.$$

Since $p \leq q < p^\star$, we may write

$$\frac{1}{q} = \frac{\alpha}{p} + \frac{1-\alpha}{p^\star} \quad \text{for some } \alpha \in (0, 1].$$

Thanks to the interpolation inequality (see Remark 2 in Chapter 4) we have

$$\|\tau_h f - f\|_{L^q(\mathbb{R}^N)} \leq \|\tau_h f - f\|_{L^p(\mathbb{R}^N)}^\alpha \|\tau_h f - f\|_{L^{p^\star}(\mathbb{R}^n)}^{1-\alpha}$$
$$\leq |h|^\alpha \|\nabla f\|_{L^p(\mathbb{R}^N)}^\alpha (2\|f\|_{L^{p^\star}(\mathbb{R}^N)})^{1-\alpha} \leq C|h|^\alpha,$$

where C is independent of $\mathcal{F}$ (since $\mathcal{F}$ is bounded in $W^{1,p}(\mathbb{R}^N)$). The desired conclusion follows.

Remark 14. Theorem 9.16 is "almost optimal" in the following sense:

(i) If Ω is not bounded, the injection $W^{1,p}(\Omega) \subset L^p(\Omega)$ is, in general, not compact.[18]

(ii) The injection $W^{1,p}(\Omega) \subset L^{p^\star}(\Omega)$ is never compact even if Ω is bounded and smooth.

$\star$ *Remark* 15. Let Ω be a bounded open set of class C^1. Then the norm

$$\|u\| = \|\nabla u\|_p + \|u\|_q$$

is *equivalent* to the $W^{1,p}$ norm so long as

$$1 \leq q \leq p^\star \text{ if } 1 \leq p < N,$$
$$1 \leq q < \infty \text{ if } p = N,$$
$$1 \leq q \leq \infty \text{ if } p > N.$$

$\star$ *Remark* 16 (*the limiting case* $p = N$). Let Ω be a bounded open set of class C^1 and let $u \in W^{1,N}(\Omega)$. Then in general, $u \notin L^\infty(\Omega)$. For example, if

$$\Omega = \{x \in \mathbb{R}^N; \ |x| < 1/2\},$$

the function

---

[18] See the detailed discussion in R. Adams [1], p. 167 concerning the compactness of this injection for unbounded domains.

$$u(x) = (\log 1/|x|)^{\alpha} \text{ with } 0 < \alpha < 1 - (1/N)$$

belongs to $W^{1,N}(\Omega)$, but it is not bounded because of the singularity at $x = 0$. Nevertheless, we have *Trudinger's inequality*

$$\int_{\Omega} e^{|u|^{N/(N-1)}} < \infty \quad \forall u \in W^{1,N}(\Omega)$$

(see, e.g., R. Adams [1] or D. Gilbarg–N. Trudinger [1]).

9.4 The Space $W_0^{1,p}(\Omega)$

Definition. Let $1 \le p < \infty$; $W_0^{1,p}(\Omega)$ denotes the closure of $C_c^1(\Omega)$ in $W^{1,p}(\Omega)$. Set[19]

$$H_0^1(\Omega) = W_0^{1,2}(\Omega).$$

The space $W_0^{1,p}$, equipped with the $W^{1,p}$ norm, is a separable Banach space; it is reflexive if $1 < p < \infty$. H_0^1, equipped with the H^1 scalar product, is a Hilbert space.

$\star$ *Remark 17.* Since $C_c^1(\mathbb{R}^N)$ is dense in $W^{1,p}(\mathbb{R}^N)$, we have

$$W_0^{1,p}(\mathbb{R}^N) = W^{1,p}(\mathbb{R}^N).$$

By contrast, if $\Omega \subset \mathbb{R}^N$ and $\Omega \ne \mathbb{R}^N$, then in general, $W_0^{1,p}(\Omega) \ne W^{1,p}(\Omega)$. However, if $\mathbb{R}^N \backslash \Omega$ is "sufficiently thin" and $p < N$, then $W_0^{1,p}(\Omega) = W^{1,p}(\Omega)$. For example, if $\Omega = \mathbb{R}^N \backslash \{0\}$ and $N \ge 2$ one can show that $H_0^1(\Omega) = H^1(\Omega)$.

Remark 18. It is easy to check—using a sequence of mollifiers—that $C_c^{\infty}(\Omega)$ is dense in $W_0^{1,p}(\Omega)$. In other words, $C_c^{\infty}(\Omega)$ could equally well have been used instead of $C_c^1(\Omega)$ in the definition of $W_0^{1,p}(\Omega)$.

The functions in $W_0^{1,p}(\Omega)$ are "roughly" those of $W^{1,p}(\Omega)$ that "vanish on $\Gamma = \partial\Omega$." It is delicate to make this precise, since a function $u \in W^{1,p}(\Omega)$ is defined only a.e. (and the measure of Γ is zero!) and u need not have a continuous representative.[20] The following characterizations suggest that we "really" have functions that are "zero on Γ." We begin with a simple fact:

Lemma 9.5. *Let $u \in W^{1,p}(\Omega)$ with $1 \le p < \infty$ and assume that supp u is a compact subset of Ω. Then $u \in W_0^{1,p}(\Omega)$.*

Proof. Fix an open set ω such that supp $u \subset \omega \subset\subset \Omega$ and choose $\alpha \in C_c^1(\omega)$ such that $\alpha = 1$ on supp u; thus $\alpha u = u$. On the other hand (Theorem 9.2), there exists a

---

[19] When there is ambiguity we shall write $W_0^{1,p}$, H_0^1 instead of $W_0^{1,p}(\Omega)$, $H_0^1(\Omega)$.

[20] Nevertheless, if $u \in W^{1,p}(\Omega)$ one can give a meaning to $u_{|\Gamma}$ (when Ω is regular) and one can show, for example, that $u_{|\Gamma} \in L^p(\Gamma)$. This relies on the *theory of traces* (see the comments at the end of this chapter).

sequence (u_n) in $C_c^\infty(\mathbb{R}^N)$ such that $u_n \to u$ in $L^p(\Omega)$ and $\nabla u_n \to \nabla u$ in $L^p(\omega)^N$. It follows that $\alpha u_n \to \alpha u$ in $W^{1,p}(\Omega)$. Thus αu belongs to $W_0^{1,p}(\Omega)$, and so does u.

Theorem 9.17. *Suppose that Ω is of class C^1. Let*[21]

$$u \in W^{1,p}(\Omega) \cap C(\overline{\Omega}) \quad with \ 1 \le p < \infty.$$

Then the following properties are equivalent:

(i) $u = 0$ on Γ.
(ii) $u \in W_0^{1,p}(\Omega)$.

Proof.
   (i) $\Rightarrow$ (ii). Suppose first that $\mathrm{supp}\, u$ is bounded.
Fix a function $G \in C^1(\mathbb{R})$ such that

$$|G(t)| \le |t| \quad \forall t \in \mathbb{R} \quad and \quad G(t) = \begin{cases} 0 & if \ |t| \le 1, \\ t & if \ |t| \ge 2. \end{cases}$$

Then $u_n = (1/n)G(nu)$ belongs to $W^{1,p}$ (by Proposition 9.5). It is easy to verify (using dominated convergence) that $u_n \to u$ in $W^{1,p}$. On the other hand,

$$\mathrm{supp}\, u_n \subset \{x \in \Omega; |u(x)| \ge 1/n\},$$

and thus $\mathrm{supp}\, u_n$ is a compact set contained in Ω. From Lemma 9.5, $u_n \in W_0^{1,p}$, and it follows that $u \in W_0^{1,p}$. In the general case in which $\mathrm{supp}\, u$ is not bounded, consider the sequence $(\zeta_n u)$ (where (ζ_n) is a sequence of cut-off functions as in the proof of Theorem 9.2). From the above, $\zeta_n u \in W_0^{1,p}$, and since $\zeta_n u \to u$ in $W^{1,p}$, we conclude that $u \in W_0^{1,p}$.

   (ii) $\Rightarrow$ (i). Using local charts this is reduced to the following problem. Let $u \in W_0^{1,p}(Q_+) \cap C(\overline{Q}_+)$; prove that $u = 0$ on Q_0.
   Let (u_n) be a sequence in $C_c^1(Q_+)$ such that $u_n \to u$ in $W^{1,p}(Q_+)$. We have, for $(x', x_N) \in Q_+$,

$$|u_n(x', x_N)| \le \int_0^{x_N} \left| \frac{\partial u_n}{\partial x_N}(x', t) \right| dt,$$

and thus for $0 < \varepsilon < 1$,

$$\frac{1}{\varepsilon} \int_{|x'|<1} \int_0^\varepsilon |u_n(x', x_N)| dx' dx_N \le \int_{|x'|<1} \int_0^\varepsilon \left| \frac{\partial u_n}{\partial x_N}(x', t) \right| dx' dt.$$

In the limit, when $n \to \infty$ ($\varepsilon > 0$ fixed) we obtain

$$\frac{1}{\varepsilon} \int_{|x'|<1} \int_0^\varepsilon |u(x', x_N)| dx' dx_N \le \int_{|x'|<1} \int_0^\varepsilon \left| \frac{\partial u}{\partial x_N}(x', t) \right| dx' dt.$$

[21] Recall that if $p > N$, then $u \in W^{1,p}(\Omega) \Rightarrow u \in C(\overline{\Omega})$ (see Corollary 9.14).

Finally, as $\varepsilon \to 0$, we are led to

$$\int_{|x'|<1} |u(x',0)|dx' = 0$$

(since $u \in C(\overline{Q}_+)$ and $\frac{\partial u}{\partial x_N} \in L^1(Q_+)$). Thus $u = 0$ on Q_0.

Remark 19. In the proof of (i) $\Rightarrow$ (ii) we have not used the smoothness of Ω. However, the converse (ii) $\Rightarrow$ (i) requires a smoothness hypothesis on Ω (consider for example $\Omega = \mathbb{R}^N \setminus \{0\}$ with $N \geq 2$ and $p \leq N$).

Here is another characterization of $W_0^{1,p}$.

Proposition 9.18. *Suppose Ω is of class C^1. Let $u \in L^p(\Omega)$ with $1 < p < \infty$. The following properties are equivalent:*

(i) $u \in W_0^{1,p}(\Omega)$,
(ii) *there exists a constant C such that*

$$\left| \int_\Omega u \frac{\partial \varphi}{\partial x_i} \right| \leq C \|\varphi\|_{L^{p'}(\Omega)} \quad \forall \varphi \in C_c^1(\mathbb{R}^N), \quad \forall i = 1, 2, \ldots, N,$$

(iii) *the function*

$$\bar{u}(x) = \begin{cases} u(x) & \text{if } x \in \Omega, \\ 0 & \text{if } x \in \mathbb{R}^N \setminus \Omega, \end{cases}$$

belongs to $W^{1,p}(\mathbb{R}^N)$, and in this case $\frac{\partial \bar{u}}{\partial x_i} = \overline{\frac{\partial u}{\partial x_i}}$.

Proof.
(i) $\Rightarrow$ (ii). Let (u_n) be a sequence from $C_c^1(\Omega)$ such that $u_n \to u$ in $W^{1,p}$. For $\varphi \in C_c^1(\mathbb{R}^N)$ we have

$$\left| \int_\Omega u_n \frac{\partial \varphi}{\partial x_i} \right| = \left| \int_\Omega \frac{\partial u_n}{\partial x_i} \varphi \right| \leq \left\| \frac{\partial u_n}{\partial x_i} \right\|_p \|\varphi\|_{p'}.$$

Passing to the limit, we obtain (ii).

(ii) $\Rightarrow$ (iii). Let $\varphi \in C_c^1(\mathbb{R}^N)$; we have

$$\left| \int_{\mathbb{R}^N} \bar{u} \frac{\partial \varphi}{\partial x_i} \right| = \left| \int_\Omega u \frac{\partial \varphi}{\partial x_i} \right| \leq C \|\varphi\|_{L^{p'}(\Omega)} \leq C \|\varphi\|_{L^{p'}(\mathbb{R}^N)}.$$

Thus $\bar{u} \in W^{1,p}(\mathbb{R}^N)$ (by Proposition 9.3).

(iii) $\Rightarrow$ (i). One can always assume that Ω is bounded (if not, consider the sequence $(\zeta_n u)$). By local charts and partition of unity this is reduced to the following problem. Let $u \in L^p(Q_+)$ be such that the function

$$\bar{u}(x) = \begin{cases} u(x) & \text{if } x \in Q, x_N > 0, \\ 0 & \text{if } x \in Q, x_N < 0, \end{cases}$$

belongs to $W^{1,p}(Q)$; prove that

$$\alpha u \in W_0^{1,p}(Q_+) \quad \forall \alpha \in C_c^1(Q).$$

Let (ρ_n) be a sequence of mollifiers such that

$$\operatorname{supp} \rho_n \subset \left\{ x \in \mathbb{R}^N; \ \frac{1}{2n} < x_N < \frac{1}{n} \right\};$$

one may choose, for example,

$$\rho_n(x) = n^N \rho(nx) \quad \text{and} \quad \operatorname{supp} \rho \subset \{x \in \mathbb{R}^N; \ (1/2) < x_N < 1\}.$$

Thus $\rho_n \star (\alpha \bar{u}) \to \alpha \bar{u}$ in $W^{1,p}(\mathbb{R}^N)$ (note that $\alpha \bar{u}$ extended by 0 outside Q belongs to $W^{1,p}(\mathbb{R}^N)$). On the other hand,

$$\operatorname{supp}(\rho_n \star \alpha \bar{u}) \subset \operatorname{supp} \rho_n + \operatorname{supp}(\alpha \bar{u}) \subset Q_+$$

for n large enough. It follows that

$$\rho_n \star (\alpha \bar{u}) \in C_c^1(Q_+)$$

and thus $\alpha u \in W_0^{1,p}(Q_+)$.

Remark 20. The proof of Corollary 9.14 uses the extension operator, and because of this fact one must assume that Ω is smooth. If $W^{1,p}(\Omega)$ is replaced by $W_0^{1,p}(\Omega)$ one can use the canonical extension by 0 outside Ω, which is valid for *arbitrary* domains Ω (in the proof of Proposition 9.18, the implication (i) $\Rightarrow$ (iii) does not use any smoothness hypothesis on Ω). It follows, in particular, that the conclusion of Corollary 9.14 is true for $W_0^{1,p}(\Omega)$ with an *arbitrary* open set Ω. Similarly, the conclusion of Theorem 9.16 is true for $W_0^{1,p}(\Omega)$ with an *arbitrary bounded* open set Ω. It can also be deduced from Theorem 9.9 that if Ω is an *arbitrary* open set and $1 \le p < N$, then

$$(30) \qquad \|u\|_{L^{p^\star}(\Omega)} \le C(p, N)\|\nabla u\|_{L^p(\Omega)} \quad \forall u \in W_0^{1,p}(\Omega).$$

• **Corollary 9.19 (Poincaré's inequality).** *Suppose that* $1 \le p < \infty$ *and* Ω *is a* **bounded** *open set. Then there exists a constant* C *(depending on* Ω *and* p*) such that*

$$\|u\|_{L^p(\Omega)} \le C\|\nabla u\|_{L^p(\Omega)} \quad \forall u \in W_0^{1,p}(\Omega).$$

In particular, the expression $\|\nabla u\|_{L^p(\Omega)}$ *is a norm on* $W_0^{1,p}(\Omega)$*, and it is equivalent to the norm* $\|u\|_{W^{1,p}}$*; on* $H_0^1(\Omega)$ *the expression* $\sum_{i=1}^N \int_\Omega \frac{\partial u}{\partial x_i} \frac{\partial v}{\partial x_i}$ *is a scalar product that induces the norm* $\|\nabla u\|_{L^2}$ *and it is equivalent to the norm* $\|u\|_{H^1}$*.*

Remark 21. Poincaré's inequality remains true if Ω has finite measure and also if Ω has a bounded projection on some axis.

Remark 22. For every integer $m \geq 1$ and $1 \leq p < \infty$ one defines $W_0^{m,p}(\Omega)$ as being the closure of $C_c^m(\Omega)$ in $W_0^{m,p}(\Omega)$. "Roughly," a function u belongs to $W_0^{m,p}(\Omega)$ if $u \in W^{m,p}(\Omega)$ and if $D^\alpha u = 0$ on Γ for all multi-indices α such that $|\alpha| \leq m - 1$. It is important to notice the *distinction* between $W_0^{m,p}(\Omega)$ and $W^{m,p}(\Omega) \cap W_0^{1,p}(\Omega)$ for $m \geq 2$.

The Dual Space of $W_0^{1,p}(\Omega)$

Notation. We denote by $W^{-1,p'}(\Omega)$ the dual space of $W_0^{1,p}(\Omega)$, $1 \leq p < \infty$, and by $H^{-1}(\Omega)$ the dual of $H_0^1(\Omega)$. The dual of $L^2(\Omega)$ is identified with $L^2(\Omega)$, but we do *not* identify $H_0^1(\Omega)$ with its dual (see Remark 3 in Chapter 5). We have the inclusions

$$H_0^1(\Omega) \subset L^2(\Omega) \subset H^{-1}(\Omega),$$

where these injections are continuous and dense.

If Ω is bounded then

$$W_0^{1,p}(\Omega) \subset L^2(\Omega) \subset W^{-1,p'}(\Omega) \quad \text{if} \quad 2N/(N+2) \leq p < \infty,$$

with continuous and dense injections. If Ω is not bounded, the same holds, but only for the range $2N/(N+2) \leq p \leq 2$.

The elements of $W^{-1,p'}$ are completely described by the following result:

Proposition 9.20. *Let $F \in W^{-1,p'}(\Omega)$. Then there exist functions $f_0, f_1, f_2, \ldots, f_N \in L^{p'}(\Omega)$ such that*

$$\langle F, v \rangle = \int_\Omega f_0 v + \sum_{i=1}^N \int_\Omega f_i \frac{\partial v}{\partial x_i} \quad \forall v \in W_0^{1,p}(\Omega),$$

and

$$\|F\| = \max_{0 \leq i \leq N} \|f_i\|_{p'}.$$

If Ω is bounded we can take $f_0 = 0$.

Proof. Adapt the proof of Proposition 8.14.

9.5 Variational Formulation of Some Boundary Value Problems

We are now going to use the previous setting in the study of some elliptic partial differential equations (= PDEs) of second order.

Example 1 (*homogeneous Dirichlet problem for the Laplacian*). Let $\Omega \subset \mathbb{R}^N$ be an open bounded set. We are looking for a function $u : \overline{\Omega} \to \mathbb{R}$ satisfying

(31)
$$\begin{cases} -\Delta u + u = f & \text{in } \Omega, \\ \qquad\quad u = 0 & \text{on } \Gamma = \partial\Omega, \end{cases}$$

where

$$\Delta u = \sum_{i=1}^{N} \frac{\partial^2 u}{\partial x_i^2} = \text{Laplacian of } u,$$

and f is a given function on Ω. The *boundary condition* $u = 0$ on Γ is called the (homogeneous) *Dirichlet condition*.

Definition. A *classical* solution of (31) is a function $u \in C^2(\overline{\Omega})$ satisfying (31) (in the usual sense). A *weak* solution of (31) is a function $u \in H_0^1(\Omega)$ satisfying

(32)
$$\int_\Omega \nabla u \cdot \nabla v + \int_\Omega uv = \int_\Omega fv \quad \forall v \in H_0^1(\Omega),$$

where $\nabla u \cdot \nabla v = \sum_{i=1}^{N} \frac{\partial u}{\partial x_i} \frac{\partial v}{\partial x_i}$.

We carry out the program described in Chapter 8.

Step A: Every classical solution is a weak solution.
Indeed, $u \in H^1(\Omega) \cap C(\overline{\Omega})$ and $u = 0$ on Γ, so that $u \in H_0^1(\Omega)$ by Theorem 9.17 (see also Remark 19). On the other hand, if $v \in C_c^1(\Omega)$ we have

$$\int_\Omega \nabla u \cdot \nabla v + \int_\Omega uv = \int_\Omega fv,$$

and by density this remains true for all $v \in H_0^1(\Omega)$.

Step B: Existence and uniqueness of a weak solution.
This is the content of the following basic result.

• Theorem 9.21 (Dirichlet, Riemann, Poincaré, Hilbert). *Given any $f \in L^2(\Omega)$, there exists a unique weak solution $u \in H_0^1(\Omega)$ of (31). Furthermore, u is obtained by*

$$\min_{v \in H_0^1(\Omega)} \left\{ \frac{1}{2} \int_\Omega (|\nabla v|^2 + |v|^2) - \int_\Omega fv \right\}.$$

This is *Dirichlet's principle*.

Proof. Apply Lax–Milgram in the Hilbert space $H = H_0^1(\Omega)$ with the bilinear form

$$a(u, v) = \int_\Omega (\nabla u \cdot \nabla v + uv)$$

and the linear functional $\varphi : v \mapsto \int_\Omega fv$.

Step C: Regularity of the weak solution.
This question is delicate. We shall address it in Section 9.6.

Step D: Recovery of a classical solution.
Assume that the weak solution $u \in H_0^1(\Omega)$ of (31) belongs to $C^2(\overline{\Omega})$, and assume that Ω is of class C^1. Then $u = 0$ on Γ (by Theorem 9.17). On the other hand, we have

$$\int_\Omega (-\Delta u + u)v = \int_\Omega fv \quad \forall v \in C_c^1(\Omega)$$

and thus $-\Delta u + u = f$ a.e. on Ω (by Corollary 4.24). In fact, $-\Delta u + u = f$ everywhere on Ω, since $u \in C^2(\Omega)$; thus u is a classical solution.

We describe now some other examples. *In each case it is essential to specify precisely the function space and the appropriate weak formulation.*

Example 2 (inhomogeneous Dirichlet condition). Let $\Omega \subset \mathbb{R}^N$ be a bounded open set. We look for a function $u : \overline{\Omega} \to \mathbb{R}$ satisfying

(33)
$$\begin{cases} -\Delta u + u = f & \text{in } \Omega, \\ \qquad\quad u = g & \text{on } \Gamma, \end{cases}$$

where f is given on Ω and g is given on Γ. Suppose that there exists a function $\tilde{g} \in H^1(\Omega) \cap C(\overline{\Omega})$ such that[22] $\tilde{g} = g$ on Γ and consider the set

$$K = \{v \in H^1(\Omega); \quad v - \tilde{g} \in H_0^1(\Omega)\}.$$

It follows from Theorem 9.17 that K is independent of the choice of $\tilde{g}$ and depends only on g. K is a nonempty closed convex set in $H^1(\Omega)$.

Definition. A *classical* solution of (33) is a function $u \in C^2(\overline{\Omega})$ satisfying (33). A *weak* solution of (33) is a function $u \in K$ satisfying

(34)
$$\int_\Omega (\nabla u \cdot \nabla v + uv) = \int_\Omega fv \quad \forall v \in H_0^1(\Omega).$$

As above, any classical solution is a weak solution.

● **Proposition 9.22.** *Given any* $f \in L^2(\Omega)$, *there exists a unique weak solution* $u \in K$ *of (33). Furthermore, u is obtained by*

$$\min_{v \in K} \left\{ \frac{1}{2} \int_\Omega (|\nabla v|^2 + v^2) - \int_\Omega fv \right\}.$$

Proof. We claim that $u \in K$ is a weak solution of (33) if and only if we have

(35)
$$\int_\Omega \nabla u \cdot (\nabla v - \nabla u) + \int_\Omega u(v - u) \geq \int_\Omega f(v - u) \quad \forall v \in K.$$

---

[22] This assumption is satisfied, *for example*, if Ω is of class C^1 and $g \in C^1(\Gamma)$. If Ω is regular enough it is not necessary to suppose that $\tilde{g} \in C(\overline{\Omega})$. Applying the theory of traces (see the comments at the end of this chapter), it suffices to know that $\tilde{g} \in H^1(\Omega)$, i.e., $g \in H^{1/2}(\Gamma)$.

Indeed, if u is a weak solution of (33) it is clear that (35) holds even with equality. Conversely, if $u \in K$ satisfies (35) we choose $v = u \pm w$ in (35) with $w \in H_0^1(\Omega)$, and (34) follows. We may then apply Stampacchia's theorem (Theorem 5.6) to conclude the proof.

The study of regularity and recovery of a classical solution follows the same pattern as in Example 1.

Example 3 (*general elliptic equations of second order*). Let $\Omega \subset \mathbb{R}^N$ be an open bounded set. We are given functions $a_{ij}(x) \in C^1(\overline{\Omega})$, $1 \le i, j \le N$, satisfying the *ellipticity* condition

$$(36) \qquad \sum_{i,j=1}^{N} a_{ij}(x)\xi_i\xi_j \ge \alpha|\xi|^2, \quad \forall x \in \Omega, \quad \forall \xi \in \mathbb{R}^N \text{ with } \alpha > 0.$$

A function $a_0 \in C(\overline{\Omega})$ is also given. We look for a function $u : \overline{\Omega} \to \mathbb{R}$ satisfying

$$(37) \qquad \begin{cases} -\sum_{i,j=1}^{N} \dfrac{\partial}{\partial x_j}\left(a_{ij}\dfrac{\partial u}{\partial x_i}\right) + a_0 u = f & \text{in } \Omega, \\[2mm] \qquad\qquad\qquad\qquad u = 0 & \text{on } \Gamma. \end{cases}$$

A *classical* solution of (37) is a function $u \in C^2(\overline{\Omega})$ satisfying (37) in the usual sense. A *weak* solution of (37) is a function $u \in H_0^1(\Omega)$ satisfying

$$(38) \qquad \int_\Omega \sum_{i,j=1}^{N} a_{ij}\frac{\partial u}{\partial x_i}\frac{\partial v}{\partial x_j} + \int_\Omega a_0 uv = \int_\Omega fv \quad \forall v \in H_0^1(\Omega).$$

As above, any classical solution is a weak solution. If $a_0(x) \ge 0$ on Ω then for all $f \in L^2(\Omega)$ there exists a unique weak solution $u \in H_0^1$: just apply Lax–Milgram in the space $H = H_0^1$ with the continuous bilinear form

$$a(u,v) = \int_\Omega \sum_{i,j=1}^{N} a_{ij}\frac{\partial u}{\partial x_i}\frac{\partial v}{\partial x_j} + \int_\Omega a_0 uv.$$

The coerciveness of $a(\ ,\ )$ comes from the ellipticity assumption, the assumption $a_0 \ge 0$, and Poincaré's inequality. If the matrix (a_{ij}) is also symmetric, then the form $a(\ ,\ )$ is symmetric and u is obtained by

$$\min_{v \in H_0^1}\left\{\frac{1}{2}\int_\Omega\left(\sum_{i,j=1}^{N} a_{ij}\frac{\partial v}{\partial x_i}\frac{\partial v}{\partial x_j} + a_0 v^2\right) - \int_\Omega fv\right\}.$$

We now consider a more general problem: find a function $u : \overline{\Omega} \to \mathbb{R}$ satisfying

$$(39) \quad \begin{cases} -\sum_{i,j} \dfrac{\partial}{\partial x_j}\left(a_{ij}\dfrac{\partial u}{\partial x_i}\right) + \sum_i a_i \dfrac{\partial u}{\partial x_i} + a_0 u = f & \text{in } \Omega, \\[2mm] u = 0 & \text{on } \Gamma, \end{cases}$$

where the functions $(a_{ij}) \in L^\infty(\Omega)$ satisfy the ellipticity condition and the functions $(a_i)_{0 \le i \le N}$ are given in $L^\infty(\Omega)$. A *weak* solution of (39) is a function $u \in H_0^1$ such that

$$(40) \quad \int_\Omega \sum_{i,j} a_{ij} \frac{\partial u}{\partial x_i}\frac{\partial v}{\partial x_j} + \int_\Omega \sum_i a_i \frac{\partial u}{\partial x_i} v + \int_\Omega a_0 uv = \int_\Omega fv \quad \forall v \in H_0^1.$$

The associated continuous bilinear form is

$$(41) \quad a(u,v) = \int_\Omega \sum_{i,j} a_{ij} \frac{\partial u}{\partial x_i}\frac{\partial v}{\partial x_j} + \int_\Omega \sum_i a_i \frac{\partial u}{\partial x_i} v + \int_\Omega a_0 uv.$$

In general this form is not symmetric;[23] in *certain cases* it is coercive: one may then use Lax–Milgram to obtain the existence and uniqueness of a *weak* solution. In the *general case*—even without coerciveness—one still has the following.

Theorem 9.23. *If $f = 0$, then the set of solutions $u \in H_0^1$ of (40) is a finite-dimensional vector space, say of dimension d. Moreover, there exists a subspace $F \subset L^2(\Omega)$ of dimension d such that[24]*

$$[(40) \text{ has a solution}] \iff \left[\int_\Omega fv = 0 \quad \forall v \in F\right].$$

Remark 23. Suppose that the homogeneous equation associated to (40), i.e., with $f = 0$, has $u = 0$ as its *unique* solution. Then for *every* $f \in L^2$ there *exists* a unique solution $u \in H_0^1$ of (40).[25] In particular, if $a_0 \ge 0$ on Ω one can show, by a maximum-principle-type method, that $f = 0 \Rightarrow u = 0$. We thus deduce, under only the hypothesis $a_0 \ge 0$ on Ω (and no assumption on a_i, $1 \le i \le N$), that for every $f \in L^2$ there exists a unique solution $u \in H_0^1$ of (40); see, e.g., D. Gilbarg–N. Trudinger [1].

Proof. Fix $\lambda > 0$, large enough that the bilinear form

$$a(u,v) + \lambda \int_\Omega uv$$

is coercive on H_0^1. For every $f \in L^2$ there exists a unique $u \in H_0^1$ satisfying

---

[23] In dimension N there is no known device, as there is in one dimension, to reduce it to the symmetric case.

[24] In other words, (40) has a solution iff f satisfies d orthogonality conditions.

[25] Note the close relationship between *existence* and *uniqueness* of solutions in elliptic problems. This remarkable relationship is a consequence of Fredholm's alternative (Theorem 6.6).

$$a(u, \varphi) + \lambda \int_\Omega u\varphi = \int_\Omega f\varphi \quad \forall \varphi \in H_0^1.$$

Set $u = Tf$, so that $T : L^2 \to L^2$ is a *compact* linear operator (since Ω is bounded, the injection $H_0^1 \subset L^2$ is compact; see Theorem 9.16 and Remark 20). Equation (40) is equivalent to

$$(42) \qquad\qquad u = T(f + \lambda u).$$

Set $v = f + \lambda u$ as a new unknown, and (42) becomes

$$(43) \qquad\qquad v - \lambda Tv = f.$$

The conclusion follows from Fredholm's alternative.

Example 4 (*homogeneous Neumann problem*). Let $\Omega \subset \mathbb{R}^N$ be a bounded domain of class C^1. We look for a function $u : \overline{\Omega} \to \mathbb{R}$ satisfying

$$(44) \qquad \begin{cases} -\Delta u + u = f & \text{in } \Omega, \\ \dfrac{\partial u}{\partial n} = 0 & \text{on } \Gamma, \end{cases}$$

where f is given on Ω; $\frac{\partial u}{\partial n}$ denotes the outward normal derivative of u, i.e., $\frac{\partial u}{\partial n} = \nabla u \cdot \mathbf{n}$, where $\mathbf{n}$ is the unit normal vector to Γ, pointing outward. The boundary condition $\frac{\partial u}{\partial n} = 0$ on Γ is called the (homogeneous) *Neumann condition*.

Definition. A *classical* solution of (44) is a function $u \in C^2(\overline{\Omega})$ satisfying (44). A *weak* solution of (44) is a function $u \in H^1(\Omega)$ satisfying

$$(45) \qquad \int_\Omega \nabla u \cdot \nabla v + \int_\Omega uv = \int_\Omega fv \quad \forall v \in H^1(\Omega).$$

Step A: Every classical solution is a weak solution.
Recall that by Green's formula we have

$$(46) \qquad \int_\Omega (\Delta u)v = \int_\Gamma \frac{\partial u}{\partial n} v \, d\sigma - \int_\Omega \nabla u \cdot \nabla v \quad \forall u \in C^2(\overline{\Omega}), \quad \forall v \in C^1(\overline{\Omega}),$$

where $d\sigma$ is the surface measure on Γ. If u is a classical solution of (44), then $u \in H^1(\Omega)$, and we have

$$\int_\Omega \nabla u \cdot \nabla v + \int_\Omega uv = \int_\Omega fv \quad \forall v \in C^1(\overline{\Omega}).$$

We conclude by density (Corollary 9.8) that

$$\int_\Omega \nabla u \cdot \nabla v + \int_\Omega uv = \int_\Omega fv \quad \forall v \in H^1(\Omega).$$

Step B: Existence and uniqueness of the weak solution.

Proposition 9.24. *For every* $f \in L^2(\Omega)$, *there exists a unique weak solution* $u \in H^1(\Omega)$ *of* (44). *Furthermore, u is obtained by*

$$\min_{v \in H^1(\Omega)} \left\{ \frac{1}{2} \int_\Omega (|\nabla v|^2 + v^2) - \int_\Omega fv \right\}.$$

Proof. Apply Lax–Milgram in $H = H^1(\Omega)$.

Step C: Regularity of the weak solution.
This will be discussed in Section 9.6.

Step D: Recovery of a classical solution.
If $u \in C^2(\Omega)$ is a weak solution of (44), we have from (46)

$$(47) \qquad \int_\Omega (-\Delta u + u)v + \int_\Gamma \frac{\partial u}{\partial n} v d\sigma = \int_\Omega fv \quad \forall v \in C^1(\overline{\Omega}).$$

In (47) first choose $v \in C_c^1(\Omega)$ to deduce

$$-\Delta u + u = f \text{ in } \Omega.$$

Then return to (47) with $v \in C^1(\overline{\Omega})$; one obtains

$$\int_\Gamma \frac{\partial u}{\partial n} v d\sigma = 0 \quad \forall v \in C^1(\overline{\Omega})$$

and therefore $\frac{\partial u}{\partial n} = 0$ on Γ.

Example 5 (unbounded domains). In the case that Ω is an *unbounded* open set in $\mathbb{R}^N$ one imposes—in addition to the usual boundary conditions on $\Gamma = \partial \Omega$—a *boundary condition at infinity*, for example $u(x) \to 0$ as $|x| \to \infty$. This "translates," at the level of a weak solution, by the condition $u \in H^1$. Of course, one must first *prove* that if u is a classical solution such that $u(x) \to 0$ as $|x| \to \infty$, then u *must belong* to H^1 (see the discussion in Example 8 of Chapter 8). Here are a few typical examples:

(a) $\Omega = \mathbb{R}^N$; given $f \in L^2(\mathbb{R}^N)$ the equation

$$-\Delta u + u = f \quad \text{in } \mathbb{R}^N$$

has a unique weak solution in the following sense:

$$u \in H^1(\mathbb{R}^N) \quad \text{and} \quad \int_{\mathbb{R}^N} \nabla u \cdot \nabla v + \int_{\mathbb{R}^N} uv = \int_{\mathbb{R}^N} fv \quad \forall v \in H^1(\mathbb{R}^N).$$

(b) $\Omega = \mathbb{R}^N_+$; given $f \in L^2(\mathbb{R}^N_+)$ the problem

$$\begin{cases} -\Delta u + u = f & \text{in } \mathbb{R}^N_+, \\ u(x', 0) = 0 & \text{for } x' \in \mathbb{R}^{N-1}, \end{cases}$$

has a unique weak solution in the following sense:

$$u \in H^1_0(\Omega) \quad \text{and} \quad \int_\Omega \nabla u \cdot \nabla v + \int_\Omega uv = \int_\Omega fv \quad \forall v \in H^1_0(\Omega).$$

(c) $\Omega = \mathbb{R}^N_+$; given $f \in L^2(\mathbb{R}^N_+)$ the problem

$$\begin{cases} -\Delta u + u = f & \text{in } \mathbb{R}^N_+, \\ \dfrac{\partial u}{\partial x_N}(x', 0) = 0 & \text{for } x' \in \mathbb{R}^{N-1}, \end{cases}$$

has a unique weak solution in the following sense:

$$u \in H^1(\Omega) \quad \text{and} \quad \int_\Omega \nabla u \cdot \nabla v + \int_\Omega uv = \int_\Omega fv \quad \forall v \in H^1(\Omega).$$

9.6 Regularity of Weak Solutions

Definition. We say that an open set Ω is of *class C^m*, $m \geq 1$ an integer, if for every $x \in \Gamma$ there exist a neighborhood U of x in $\mathbb{R}^N$ and a bijective mapping $H : Q \to U$ such that

$$H \in C^m(\overline{Q}), \quad H^{-1} \in C^m(\overline{U}), \quad H(Q_+) = U \cap \Omega, \quad H(Q_0) = U \cap \Gamma.$$

We say that Ω *is of class C^∞* if it is of class C^m for all m.

The main regularity results are the following.

• **Theorem 9.25 (regularity for the Dirichlet problem).** *Let Ω be an open set of class C^2 with Γ bounded (or else $\Omega = \mathbb{R}^N_+$). Let $f \in L^2(\Omega)$ and let $u \in H^1_0(\Omega)$ satisfy*

(48)
$$\int_\Omega \nabla u \nabla \varphi + \int_\Omega u\varphi = \int_\Omega f\varphi \quad \forall \varphi \in H^1_0(\Omega).$$

Then $u \in H^2(\Omega)$ and $\|u\|_{H^2} \leq C\|f\|_{L^2}$, where C is a constant depending only on Ω. Furthermore, if Ω is of class C^{m+2} and $f \in H^m(\Omega)$, then

$$u \in H^{m+2}(\Omega) \quad \text{and} \quad \|u\|_{H^{m+2}} \leq C\|f\|_{H^m}.$$

In particular, if $f \in H^m(\Omega)$ with $m > N/2$, then $u \in C^2(\overline{\Omega})$. Finally, if Ω is of class C^∞ and if $f \in C^\infty(\overline{\Omega})$, then $u \in C^\infty(\overline{\Omega})$.

Theorem 9.26 (regularity for the Neumann problem). *With the same assumptions as in Theorem 9.25 one obtains the same conclusions for the solution of the Neumann problem, i.e., for $u \in H^1(\Omega)$ such that*

$$(49) \qquad \int_\Omega \nabla u \cdot \nabla \varphi + \int_\Omega u\varphi = \int_\Omega f\varphi \quad \forall \varphi \in H^1(\Omega).$$

Remark 24. One would obtain the same conclusions for the solution of the Dirichlet (or Neumann) problem associated to a general second-order elliptic operator, i.e., if $u \in H_0^1(\Omega)$ is such that

$$\int_\Omega \sum_{i,j} a_{ij} \frac{\partial u}{\partial x_i} \frac{\partial \varphi}{\partial x_j} + \int_\Omega \sum_i a_i \frac{\partial u}{\partial x_i} \varphi + \int_\Omega a_0 u\varphi = \int_\Omega f\varphi \quad \forall \varphi \in H_0^1(\Omega);$$

then[26]

$$[f \in L^2(\Omega), \quad a_{ij} \in C^1(\overline{\Omega}) \quad \text{and} \quad a_i \in C(\overline{\Omega})] \Rightarrow u \in H^2(\Omega),$$

and for $m \geq 1$,

$$[f \in H^m(\Omega), a_{ij} \in C^{m+1}(\overline{\Omega}) \quad \text{and} \quad a_i \in C^m(\overline{\Omega})] \Rightarrow u \in H^{m+2}(\Omega).$$

We shall prove only Theorem 9.25; the proof of Theorem 9.26 is entirely analogous. The main idea of the proof is the following. We consider first the case $\Omega = \mathbb{R}^N$, then the case $\Omega = \mathbb{R}_+^N$. For a general domain Ω we proceed in two steps:

1. *Interior regularity*, i.e., u is regular on every domain $\omega \subset\subset \Omega$. Here, the proof follows the same pattern as $\Omega = \mathbb{R}^N$.
2. *Boundary regularity*, i.e., u is regular on some neighborhood of the boundary. Here, the proof resembles, in local charts, the case $\Omega = \mathbb{R}_+^N$.

We recommend that the reader study well the cases $\Omega = \mathbb{R}^N$ and $\Omega = \mathbb{R}_+^N$ before tackling the general case. The plan of this section is the following:

A. The case $\Omega = \mathbb{R}^N$.
B. The case $\Omega = \mathbb{R}_+^N$.
C. The general case:
   C_1. Interior estimates.
   C_2. Estimates near the boundary.

The essential ingredient of the proof is the *method of translations*[27] due to L. Nirenberg.

A. The case $\Omega = \mathbb{R}^N$.

Notation. Given $h \in \mathbb{R}^N$, $h \neq 0$, set

---

[26] If Ω is not bounded we must also assume that $D^\alpha a_{ij} \in L^\infty(\Omega)$ $\forall \alpha, |\alpha| \leq m+1$ and $D^\alpha a_i \in L^\infty(\Omega)$ $\forall \alpha, |\alpha| \leq m$.

[27] Also called the technique of *difference quotients*.

$$D_h u = \frac{1}{|h|}(\tau_h u - u), \quad \text{i.e.,} \quad D_h u(x) = \frac{u(x+h) - u(x)}{|h|}.$$

In (48) take $\varphi = D_{-h}(D_h u)$. This is possible, since $\varphi \in H^1(\mathbb{R}^N)$ (since $u \in H^1(\mathbb{R}^N)$); we obtain

$$\int |\nabla D_h u|^2 + \int |D_h u|^2 = \int f \, D_{-h}(D_h u)$$

and thus

(50) $$\|D_h u\|_{H^1}^2 \leq \|f\|_2 \|D_{-h}(D_h u)\|_2.$$

On the other hand, recall (Proposition 9.3) that

(51) $$\|D_{-h} v\|_2 \leq \|\nabla v\|_2 \quad \forall v \in H^1.$$

Using this with $v = D_h u$, we obtain

$$\|D_h u\|_{H^1}^2 \leq \|f\|_2 \|\nabla(D_h u)\|_2,$$

and consequently

$$\|D_h u\|_{H^1} \leq \|f\|_2.$$

In particular,

(52) $$\left\| D_h \frac{\partial u}{\partial x_i} \right\|_2 \leq \|f\|_2 \quad \forall i = 1, 2, \ldots, N.$$

Applying Proposition 9.3 once more, we see that $\frac{\partial u}{\partial x_i} \in H^1$ and thus $u \in H^2$.

We now prove that $f \in H^1 \Rightarrow u \in H^3$. We denote by Du any of the derivatives $\frac{\partial u}{\partial x_i}$, $1 \leq i \leq N$. We already know that $Du \in H^1$. We have to prove that $Du \in H^2$. For this it suffices to verify that

(53) $$\int \nabla(Du) \cdot \nabla\varphi + \int (Du)\varphi = \int (Df)\varphi \quad \forall \varphi \in H^1$$

(and then we may apply to Du the preceding analysis, which gives $Du \in H^2$).

If $\varphi \in C_c^\infty(\mathbb{R}^N)$ we may replace φ by $D\varphi$ in (48); it becomes

$$\int \nabla u \cdot \nabla(D\varphi) + \int u D\varphi = \int f D\varphi,$$

and thus

$$\int \nabla(Du) \cdot \nabla\varphi + \int (Du)\varphi = \int (Df)\varphi \quad \forall \varphi \in C_c^\infty(\mathbb{R}^N).$$

This implies (53), since $C_c^\infty(\mathbb{R}^N)$ is dense in $H^1(\mathbb{R}^N)$ (Proposition 9.2).

To show that $f \in H^m \Rightarrow u \in H^{m+2}$ it suffices to argue by induction on m and to apply (53).

B. The case $\Omega = \mathbb{R}_+^N$.
We use again *translations*, but only in the *tangential directions*, i.e., in a direction $h \in \mathbb{R}^{N-1} \times \{0\}$: we say that h is parallel to the boundary, and denote this by $h \parallel \Gamma$. It is essential to observe that

$$u \in H_0^1(\Omega) \Rightarrow \tau_h u \in H_0^1(\Omega) \quad \text{if } h \parallel \Gamma.$$

In other words, $H_0^1(\Omega)$ is *invariant under tangential translations*.
We choose $h \parallel \Gamma$ and insert $\varphi = D_{-h}(D_h u)$ in (48); we obtain

$$\int |\nabla(D_h u)|^2 + \int |D_h u|^2 = \int f \, D_{-h}(D_h u),$$

i.e.,

$$(54) \qquad \|D_h u\|_{H^1}^2 \le \|f\|_2 \|D_{-h}(D_h u)\|_2.$$

We use now the the following lemma.

Lemma 9.6. *We have*

$$\|D_h v\|_{L^2(\Omega)} \le \|\nabla v\|_{L^2(\Omega)} \quad \forall v \in H^1(\Omega), \quad \forall h \parallel \Gamma.$$

Proof. Start with $v \in C_c^1(\mathbb{R}^N)$ and follow the proof of Proposition 9.3 (note that $\Omega + th = \Omega$ for all t and all $h \parallel \Gamma$). For a general $v \in H^1(\Omega)$ argue by density.
Combining (54) and Lemma 9.6, we obtain

$$(55) \qquad \|D_h u\|_{H^1} \le \|f\|_2 \quad \forall h \parallel \Gamma.$$

Let $1 \le j \le N$, $1 \le k \le N-1$, $h = |h|e_k$, and $\varphi \in C_c^\infty(\Omega)$. We have

$$\int D_h\left(\frac{\partial u}{\partial x_j}\right)\varphi = -\int u D_{-h}\left(\frac{\partial \varphi}{\partial x_j}\right)$$

and thanks to (55),

$$\left|\int u D_{-h}\left(\frac{\partial \varphi}{\partial x_j}\right)\right| \le \|f\|_2 \|\varphi\|_2.$$

Passing to the limit as $h \to 0$, this becomes

$$(56) \qquad \left|\int u \frac{\partial^2 \varphi}{\partial x_j \partial x_k}\right| \le \|f\|_2 \|\varphi\|_2 \quad \forall 1 \le j \le N, \quad \forall 1 \le k \le N-1.$$

Finally, we claim that

(57)
$$\left| \int u \frac{\partial^2 \varphi}{\partial x_N^2} \right| \le \|f\|_2 \|\varphi\|_2 \quad \forall \varphi \in C_c^\infty(\Omega).$$

To prove (57) *we return to equation* (48) and deduce that

$$\left| \int u \frac{\partial^2 \varphi}{\partial x_N^2} \right| \le \sum_{i=1}^{N-1} \left| \int u \frac{\partial^2 \varphi}{\partial x_i^2} \right| + \left| \int (f-u)\varphi \right| \le C\|f\|_2 \|\varphi\|_2$$

from (56). Combining (56) and (57), we end up with

$$\left| \int u \frac{\partial^2 \varphi}{\partial x_j \partial x_k} \right| \le C\|f\|_2 \|\varphi\|_2 \quad \forall \varphi \in C_c^\infty(\Omega), \quad \forall 1 \le j,k \le N.$$

As a consequence, $u \in H^2(\Omega)$, since there exist functions $f_{jk} \in L^2(\Omega)$ such that

$$\int u \frac{\partial^2 \varphi}{\partial x_j \partial x_k} = \int f_{jk}\varphi \quad \forall \varphi \in C_c^\infty(\Omega)$$

(as in the proof of Proposition 8.3).

We show finally that $f \in H^m(\Omega) \Rightarrow u \in H^{m+2}(\Omega)$. By Du we mean any one of the *tangential derivatives* $Du = \frac{\partial u}{\partial x_j}, 1 \le j \le N-1$. We first establish the following result.

Lemma 9.7. *Let* $u \in H^2(\Omega) \cap H_0^1(\Omega)$ *satisfying* (48). *Then* $Du \in H_0^1(\Omega)$ *and, moreover,*

(58)
$$\int \nabla(Du) \cdot \nabla\varphi + \int (Du)\varphi = \int (Df)\varphi \quad \forall \varphi \in H_0^1(\Omega).$$

Proof. The only delicate point consists in proving that $Du \in H_0^1(\Omega)$, since (58) is derived from (48) by choosing $D\varphi$ instead of φ (with $\varphi \in C_c^\infty(\Omega)$) and then arguing by density. Let $h = |h|e_j, 1 \le j \le N-1$, so that $D_h u \in H_0^1$ (since H_0^1 is invariant under tangential translations). By Lemma 9.6 we have

$$\|D_h u\|_{H^1} \le \|u\|_{H^2}.$$

Thus there exists a sequence $h_n \to 0$ such that $D_{h_n} u$ converges weakly to some g in H_0^1 (since H_0^1 is a Hilbert space). In particular, $D_{h_n} u \rightharpoonup g$ weakly in L^2. For $\varphi \in C_c^\infty(\Omega)$ we have

$$\int (D_h u)\varphi = \int u(D_{-h}\varphi)$$

and in the limit, as $h_n \to 0$, we obtain

$$\int g\varphi = -\int u \frac{\partial \varphi}{\partial x_j} \quad \forall \varphi \in C_c^\infty(\Omega).$$

Therefore, $\frac{\partial u}{\partial x_j} = g \in H_0^1(\Omega)$.

Proof of $f \in H^m \Rightarrow u \in H^{m+2}$. This is by induction on m. Assume the claim up to order m, and let $f \in H^{m+1}$. We already know that $u \in H^{m+2}$; also Du (any tangential derivative) belongs to $H_0^1(\Omega)$ and satisfies (58). Applying the induction assumption to Du and Df, we see that $Du \in H^{m+2}$. To conclude it suffices, for example, to check that $\frac{\partial^2 u}{\partial x_N^2} \in H^{m+1}$. For this purpose we return once more to equation (48), which we write

$$\frac{\partial^2 u}{\partial x_N^2} = -\sum_{i=1}^{N-1} \frac{\partial^2 u}{\partial x_i^2} + u - f \in H^{m+1}.$$

C. The general case.
We prove only that $f \in L^2(\Omega) \Rightarrow u \in H^2(\Omega)$; the implication $f \in H^m \Rightarrow u \in H^{m+2}$ is done by induction on m as in Cases A and B. To simplify the presentation we assume that Ω is bounded. We use a partition of unity and write $u = \sum_{i=0}^k \theta_i u$ as in the proof of Theorem 9.7.

C₁. Interior estimates.
We claim that $\theta_0 u \in H^2(\Omega)$. Since $\theta_{0|\Omega} \in C_c^\infty(\Omega)$, the function $\theta_0 u$ extended by 0 outside Ω belongs to $H^1(\mathbb{R}^N)$ (see Remark 4(ii)). It is easy to verify that $\theta_0 u$ is a weak solution in $\mathbb{R}^N$ of the equation

$$-\Delta(\theta_0 u) + \theta_0 u = \theta_0 f - 2\nabla\theta_0 \cdot \nabla u - (\Delta\theta_0)u \overset{\text{def}}{=} g,$$

with $g \in L^2(\mathbb{R}^N)$. We deduce from Case A that $\theta_0 u \in H^2(\mathbb{R}^N)$ and

$$\|\theta_0 u\|_{H^2} \le C(\|f\|_2 + \|u\|_{H^1}) \le C'\|f\|_2$$

(since $\|u\|_{H^1} \le \|f\|_2$ by (48)).

C₂. Estimates near the boundary.
We claim that $\theta_i u \in H^2(\Omega)$ for $1 \le i \le k$. Recall that we have a bijective map $H : Q \to U_i$ such that

$$H \in C^2(\overline{Q}), \quad J = H^{-1} \in C^2(U_i), \quad H(Q_+) = \Omega \cap U_i, \quad \text{and} \quad H(Q_0) = \Gamma \cap U_i.$$

We write $x = H(y)$ and $y = H^{-1}(x) = J(x)$. It is easy to verify that $v = \theta_i u \in H_0^1(\Omega \cap U_i)$ and that v is a weak solution in $\Omega \cap U_i$ of the equation

$$-\Delta v = \theta_i f - \theta_i u - 2\nabla\theta_i \cdot \nabla u - (\Delta\theta_i)u \overset{\text{def}}{=} g,$$

with $g \in L^2(\Omega \cap U_i)$ and $\|g\|_2 \le C\|f\|_2$. More precisely, we have

(59)
$$\int_{\Omega \cap U_i} \nabla v \cdot \nabla\varphi \, dx = \int_{\Omega \cap U_i} g\varphi \, dx \quad \forall \varphi \in H_0^1(\Omega \cap U_i).$$

We now transfer $v_{|\Omega \cap U_i}$ to Q_+. Set

$$w(y) = v(H(y)) \text{ for } y \in Q_+,$$

i.e.,

$$w(Jx) = v(x) \text{ for } x \in \Omega \cap U_i.$$

The following lemma—which is fundamental—shows that equation (59) becomes a second-order elliptic equation for w on Q_+.[28]

Lemma 9.8. *With the above notation, w belongs to $H_0^1(Q_+)$ and satisfies*

$$(60) \qquad \sum_{k,\ell=1}^{N} \int_{Q_+} a_{k\ell} \frac{\partial w}{\partial y_k} \frac{\partial \psi}{\partial y_\ell} dy = \int_{Q_+} \tilde{g} \psi dy \quad \forall \psi \in H_0^1(Q_+),$$

where[29] $\tilde{g} = (g \circ H)|\det\text{Jac } H| \in L^2(Q_+)$ and the functions $a_{k\ell} \in C^1(\overline{Q}_+)$ satisfy the ellipticity condition (36).

Proof. Let $\psi \in H_0^1(Q_+)$ and set $\varphi(x) = \psi(Jx)$ for $x \in \Omega \cap U_i$. Then $\varphi \in H_0^1(\Omega \cap U_i)$ and

$$\frac{\partial v}{\partial x_j} = \sum_k \frac{\partial w}{\partial y_k} \frac{\partial J_k}{\partial x_j}, \qquad \frac{\partial \varphi}{\partial x_j} = \sum_\ell \frac{\partial \psi}{\partial y_\ell} \frac{\partial J_\ell}{\partial x_j}.$$

Thus

$$\int_{\Omega \cap U_i} \nabla v \cdot \nabla \varphi dx = \int_{\Omega \cap U_i} \sum_{j,k,\ell} \frac{\partial J_k}{\partial x_j} \frac{\partial J_\ell}{\partial x_j} \frac{\partial w}{\partial y_k} \frac{\partial \psi}{\partial y_\ell} dx$$

$$= \int_{Q_+} \sum_{j,k,\ell} \frac{\partial J_k}{\partial x_j} \frac{\partial J_\ell}{\partial x_j} \frac{\partial w}{\partial y_k} \frac{\partial \psi}{\partial y_\ell} |\det\text{Jac } H| dy$$

from the usual change-of-variables formulas in an integral. As a consequence,

$$(61) \qquad \int_{\Omega \cap U_i} \nabla v \cdot \nabla \varphi dx = \int_{Q_+} \sum_{k,\ell} a_{k\ell} \frac{\partial w}{\partial y_k} \frac{\partial \psi}{\partial y_\ell} dy,$$

with $a_{k\ell} = \sum_j \frac{\partial J_k}{\partial x_j} \frac{\partial J_\ell}{\partial x_j} |\det\text{Jac } H|$.

We note that $a_{k\ell} \in C^1(\overline{Q}_+)$ and that the ellipticity condition is satisfied, since for all $\xi \in \mathbb{R}^N$, we have

---

[28] More generally, if we start with an elliptic equation for v we end up with an elliptic equation for w: the *ellipticity condition is preserved under change of variables.*

[29] detJac H denotes the Jacobian determinant, i.e., the determinant of the Jacobian matrix Jac $H = (\frac{\partial H_i}{\partial y_j})$.

$$\sum_{k,\ell} a_{k\ell}\xi_k\xi_\ell = |\text{detJac } H| \sum_j \left| \sum_k \frac{\partial J_k}{\partial x_j}\xi_k \right|^2 \geq \alpha|\xi|^2$$

with $\alpha > 0$, since the Jacobian matrices Jac H and Jac J are not singular.

On the other hand, we have

(62) $$\int_{\Omega \cap U_i} g\varphi dx = \int_{Q_+} (g \circ H)\psi |\text{detJac } H| dy.$$

Combining (59), (61), and (62) we obtain (60). This completes the proof of Lemma 9.8.

We now return to the proof of the boundary estimates and show that $w \in H^2(Q_+)$ with[30] $\|w\|_{H^2} \leq C\|\tilde{g}\|_2$. This will imply, by returning to $\Omega \cap U_i$, that $\theta_i u$ belongs to $H^2(\Omega \cap U_i)$ and thus, in fact, to $H^2(\Omega)$ with $\|\theta_i u\|_{H^2} \leq C\|f\|_2$.

As in Case B ($\Omega = \mathbb{R}^N_+$), we use *tangential translations*. In (60) choose $\psi =$, $D_{-h}(D_h w)$ with $h \parallel Q_0$, and $|h|$ small enough that $\psi \in H_0^1(Q_+)$.[31] We obtain

(63) $$\sum_{k,\ell} \int_{Q_+} D_h\left(a_{k\ell}\frac{\partial w}{\partial y_k}\right)\frac{\partial}{\partial y_\ell}(D_h w) = \int_{Q_+} \tilde{g} D_{-h}(D_h w).$$

But

(64) $$\int_{Q_+} \tilde{g} D_{-h}(D_h w) \leq \|\tilde{g}\|_2 \|D_{-h}(D_h w)\|_2 \leq \|\tilde{g}\|_2 \|\nabla D_h w\|_2$$

(by Lemma 9.6).

On the other hand, write

$$D_h\left(a_{k\ell}\frac{\partial w}{\partial y_k}\right)(y) = a_{k\ell}(y+h)\frac{\partial}{\partial y_k}D_h w(y) + (D_h a_{k\ell}(y))\frac{\partial w}{\partial y_k}(y),$$

and as a consequence we have

(65) $$\sum_{k,\ell} \int_{Q_+} D_h\left(a_{k\ell}\frac{\partial w}{\partial y_k}\right)\frac{\partial}{\partial y_\ell}(D_h w) \geq \alpha\|\nabla(D_h w)\|_2^2 - C\|w\|_{H^1}\|\nabla D_h w\|_2.$$

Combining (64) and (65), we obtain

(66) $$\|\nabla D_h w\|_2 \leq C(\|w\|_{H^1} + \|\tilde{g}\|_2) \leq C\|\tilde{g}\|_2$$

(noting that because of (60) and Poincaré's inequality, $\|w\|_{H^1} \leq C\|\tilde{g}\|_2$). We deduce from (66)—as in Case B—that

---

[30] In the following we denote by C various constants depending only on $a_{k\ell}$.

[31] Recall that supp $w \subset \{(x', x_N); |x'| < 1 - \delta$ and $0 \leq x_N < 1 - \delta\}$ for some $\delta > 0$.

(67) $$\left| \int_{Q_+} \frac{\partial w}{\partial y_k} \frac{\partial \psi}{\partial y_\ell} \right| \le C\|\tilde{g}\|_2 \|\psi\|_2 \quad \forall \psi \in C_c^1(Q_+), \quad \forall (k, \ell) \ne (N, N).$$

To conclude that $w \in H^2(Q_+)$ (and $\|w\|_{H^2} \le C\|\tilde{g}\|_2$) it remains to show that

(68) $$\left| \int_{Q_+} \frac{\partial w}{\partial y_N} \frac{\partial \psi}{\partial y_N} \right| \le C\|\tilde{g}\|_2 \|\psi\|_2 \quad \forall \psi \in C_c^1(Q_+).$$

For this purpose we *return to the equation* where we replace ψ by $(1/a_{NN})\psi$ ($\psi \in C_c^1(Q_+)$); this is possible, since $a_{NN} \in C^1(\overline{Q}_+)$ and $a_{NN} \ge \alpha > 0$. It becomes

$$\int a_{NN} \frac{\partial w}{\partial y_N} \frac{\partial}{\partial y_N} \left(\frac{\psi}{a_{NN}} \right) = \int \frac{\tilde{g}}{a_{NN}} \psi - \sum_{(k,\ell) \ne (N,N)} \int a_{k\ell} \frac{\partial w}{\partial y_k} \frac{\partial}{\partial y_\ell} \left(\frac{\psi}{a_{NN}} \right),$$

that is,

(69) $$\begin{cases} \displaystyle\int \frac{\partial w}{\partial y_N} \frac{\partial \psi}{\partial y_N} = \int \frac{1}{a_{NN}} \left(\frac{\partial a_{NN}}{\partial y_N} \right) \frac{\partial w}{\partial y_N} \psi + \int \frac{\tilde{g}}{a_{NN}} \psi \\ \qquad + \displaystyle\sum_{(k,\ell) \ne (N,N)} \int \frac{\partial w}{\partial y_k} \left(\frac{\partial a_{k\ell}}{\partial y_\ell} \right) \frac{\psi}{a_{NN}} \\ \qquad - \displaystyle\sum_{(k,\ell) \ne (N,N)} \int \frac{\partial w}{\partial y_k} \frac{\partial}{\partial y_\ell} \left(\frac{a_{k\ell}\psi}{a_{NN}} \right). \end{cases}$$

Combining $(67)^{32}$ and (69), we obtain

$$\left| \int_{Q_+} \frac{\partial w}{\partial y_N} \frac{\partial \psi}{\partial y_N} \right| \le C(\|w\|_{H^1} + \|\tilde{g}\|_2)\|\psi\|_2 \quad \forall \psi \in C_c^1(Q_+).$$

This establishes (68) and completes the estimates near the boundary.

Remark 25. Let Ω be an arbitrary open set and let $u \in H^1(\Omega)$ be such that

$$\int_\Omega \nabla u \cdot \nabla \varphi = \int_\Omega f\varphi \quad \forall \varphi \in C_c^\infty(\Omega).$$

We suppose that $f \in H^m(\Omega)$. Then $\theta u \in H^{m+2}(\Omega)$ for every $\theta \in C_c^\infty(\Omega)$: we say that $u \in H_{\text{loc}}^{m+2}(\Omega)$. To prove this it suffices to proceed as in Case C_1 and to argue by induction on m. In particular, $f \in C^\infty(\Omega) \Rightarrow u \in C^\infty(\Omega)$.[33]

The same conclusion holds for a *very weak solution* in the sense that $u \in L^2(\Omega)$ is such that

---

[32] We use (67) with $(a_{k\ell}a_{NN})\psi$ instead of ψ.

[33] But in general, we *cannot* say that, for example, $u \in C(\overline{\Omega})$ (even if Ω and f are very smooth), *since no boundary condition has been prescribed.*

$$-\int_\Omega u\,\Delta\varphi = \int_\Omega f\varphi \quad \forall\varphi \in C_c^\infty(\Omega).$$

(The proof is a little more delicate; see, e.g., S. Agmon [1].) We emphasize the *local nature* of the *regularity results* in elliptic problems. More precisely, let $f \in L^2(\Omega)$ and let $u \in H_0^1(\Omega)$ be the unique weak solution of

$$\int_\Omega \nabla u \cdot \nabla\varphi + \int_\Omega u\varphi = \int_\Omega f\varphi \quad \forall\varphi \in H_0^1(\Omega).$$

Fix $\omega \subset\subset \Omega$; then $u_{|\omega}$ *depends on the values of* f *in all of* Ω—*and not only the values of* f *in* ω.[34] *By contrast, the regularity of* $u_{|\omega}$ *depends only on the regularity of* $f_{|\omega}$. For example, $f \in C^\infty(\omega) \Rightarrow u \in C^\infty(\omega)$ even if f is very irregular outside ω. This property is called *hypoellipticity*.

Remark 26. From a certain point of view, the regularity results are a little surprising. Indeed, an assumption made on Δu, i.e., on the *sum* of the derivatives $\sum_k \frac{\partial^2 u}{\partial x_k^2}$, forces a conclusion of the same nature for *all the derivatives* $\frac{\partial^2 u}{\partial x_i \partial x_j}$ *individually*.

9.7 The Maximum Principle

The maximum principle is a very useful tool, and it admits a number of formulations. We present here some simple forms.

Let Ω be a general open subset of $\mathbb{R}^N$.

- **Theorem 9.27 (maximum principle for the Dirichlet problem).** *Assume* [35] *that*

$$f \in L^2(\Omega) \quad and \quad u \in H^1(\Omega) \cap C(\overline{\Omega})$$

satisfy

(70) $$\int_\Omega \nabla u \cdot \nabla\varphi + \int_\Omega u\varphi = \int_\Omega f\varphi \quad \forall\varphi \in H_0^1(\Omega).$$

Then for all $x \in \Omega$,

$$\min\{\inf_\Gamma u, \inf_\Omega f\} \le u(x) \le \max\left\{\sup_\Gamma u, \sup_\Omega f\right\}.$$

(Here and in the following, sup = essential sup and inf = essential inf.)

Proof. We use *Stampacchia's truncation method*. Fix a function $G \in C^1(\mathbb{R})$ such that

---

[34] For example, if $f \ge 0$ in Ω, $f = 0$ in ω, and $f > 0$ in some open subset of Ω, then $u > 0$ in Ω (and thus on ω); see the strong maximum principle in the comments at the end of this chapter.

[35] If Ω is of class C^1 one can remove the assumption $u \in C(\overline{\Omega})$ by invoking the *theory of traces*, which gives a meaning to $u_{|\Gamma}$ (see comments at the end of this chapter); also if $u \in H_0^1(\Omega)$ the assumption $u \in C(\overline{\Omega})$ can be removed.

(i) $|G'(s)| \leq M \quad \forall s \in \mathbb{R}$,

(ii) G is strictly increasing on $(0, +\infty)$,

(iii) $G(s) = 0 \quad \forall s \leq 0$.

Set

$$K = \max \left\{ \sup_\Gamma u, \sup_\Omega f \right\}$$

and assume $K < \infty$ (otherwise there is nothing to prove). Let $v = G(u - K)$.

We distinguish two cases:

(a) The case $|\Omega| < \infty$.

Then $v \in H^1(\Omega)$ (from Proposition 9.5 applied to the function $t \mapsto G(t - K) - G(-K)$). On the other hand, $v \in H^1_0(\Omega)$, since $v \in C(\overline{\Omega})$ and $v = 0$ on Γ (see Theorem 9.17). Plug this v into (70) and proceed as in the proof of Theorem 8.18.

(b) The case $|\Omega| = \infty$.

We have then $K \geq 0$ (since $f(x) \leq K$ a.e. in Ω and $f \in L^2$ imply $K \geq 0$). Fix $K' > K$. By Proposition 9.5 applied to the function $t \mapsto G(t - K')$ we see that $v = G(u - K') \in H^1(\Omega)$. Moreover, $v \in C(\overline{\Omega})$ and $v = 0$ on Γ; thus $v \in H^1_0(\Omega)$. Plugging this v into (70) we have

$$\text{(71)} \qquad \int_\Omega |\nabla u|^2 G'(u - K') + \int_\Omega u G(u - K') = \int_\Omega f G(u - K').$$

On the other hand, $G(u - K') \in L^1(\Omega)$, since[36]

$$0 \leq G(u - K') \leq M|u|,$$

and on the set $[u \geq K'] = \{x \in \Omega; u(x) \geq K'\}$ we have

$$K' \int_{[u \geq K']} |u| \leq \int_\Omega u^2 < \infty.$$

We conclude from (71) that

$$\int_\Omega (u - K') G(u - K') \leq \int_\Omega (f - K') G(u - K') \leq 0.$$

It follows that $u \leq K'$ a.e. in Ω and thus $u \leq K$ a.e. in Ω (since $K' > K$ is arbitrary).

● **Corollary 9.28.** *Let $f \in L^2(\Omega)$ and $u \in H^1(\Omega) \cap C(\overline{\Omega})$[37] satisfy (70). We have*

(72) $[u \geq 0 \text{ on } \Gamma \text{ and } f \geq 0 \text{ in } \Omega] \Rightarrow [u \geq 0 \text{ in } \Omega],$

(73) $\|u\|_{L^\infty(\Omega)} \leq \max\{\|u\|_{L^\infty(\Gamma)}, \|f\|_{L^\infty(\Omega)}\}.$

---

[36] Because $G(u - K') - G(-K') \leq M|u|$ and $G(-K') = 0$ as $-K' < 0$.

[37] As above, the assumption $u \in C(\overline{\Omega})$ can be removed in certain cases.

In particular,

(74)   *if* $f = 0$ *in* Ω, *then* $\|u\|_{L^\infty(\Omega)} \leq \|u\|_{L^\infty(\Gamma)}$,

(75)   *if* $u = 0$ *on* Γ, *then* $\|u\|_{L^\infty(\Omega)} \leq \|f\|_{L^\infty(\Omega)}$.

Remark 27. If Ω is bounded and u is a *classical* solution of the equation

(76) $$-\Delta u + u = f \text{ in } \Omega$$

one can give another proof of Theorem 9.27. Indeed, let $x_0 \in \overline{\Omega}$ be a point such that $u(x_0) = \max_{\overline{\Omega}} u$.

(i) If $x_0 \in \Gamma$, then $u(x_0) \leq \sup_\Gamma u \leq K$.

(ii) If $x_0 \in \Omega$, then $\nabla u(x_0) = 0$ and $\frac{\partial^2 u}{\partial x_i^2}(x_0) \leq 0$ for all $1 \leq i \leq N$, so that $\Delta u(x_0) \leq 0$. From this, using equation (76) we have

$$u(x_0) = f(x_0) + \Delta u(x_0) \leq f(x_0) \leq K.$$

This method has the advantage that it applies to *general* second-order elliptic equations. For example, the conclusion of Theorem 9.27 holds for

(77) $$-\sum_{i,j=1}^N \frac{\partial}{\partial x_j}\left(a_{ij}\frac{\partial u}{\partial x_i}\right) + \sum_{i=1}^N a_i \frac{\partial u}{\partial x_i} + u = f \quad \text{in } \Omega.$$

Note that if $x_0 \in \Omega$, then

$$\sum_{i,j=1}^N a_{ij}(x_0)\frac{\partial^2 u}{\partial x_i \partial x_j}(x_0) \leq 0;$$

indeed, by a change of coordinates (depending on x_0) one can reduce this to the case in which the matrix $a_{ij}(x_0)$ is diagonal. The conclusion of Theorem 9.27 remains true for weak solutions of (77), but the proof is more delicate; see D. Gilbarg–N. Trudinger [1].

Proposition 9.29. *Suppose that the functions* $a_{ij} \in L^\infty(\Omega)$ *satisfy the ellipticity condition* (36), *and that* a_i, $a_0 \in L^\infty(\Omega)$ *with* $a_0 \geq 0$ *in* Ω. *Let* $f \in L^2(\Omega)$ *and* $u \in H^1(\Omega) \cap C(\overline{\Omega})$[38] *be such that*

(78) $$\int_\Omega \sum_{i,j} a_{ij}\frac{\partial u}{\partial x_i}\frac{\partial \varphi}{\partial x_j} + \int_\Omega \sum_i a_i \frac{\partial u}{\partial x_i}\varphi + \int_\Omega a_0 u\varphi = \int_\Omega f\varphi \quad \forall \varphi \in H_0^1(\Omega).$$

Then

(79) $$[u \geq 0 \text{ on } \Gamma \text{ and } f \geq 0 \text{ in } \Omega] \Rightarrow [u \geq 0 \text{ in } \Omega].$$

---

[38] As above, the assumption $u \in C(\overline{\Omega})$ can be removed in certain cases.

Suppose that $a_0 \equiv 0$ and that Ω is bounded. Then

(80)
$$[f \geq 0 \text{ in } \Omega] \Rightarrow \left[u \geq \inf_\Gamma u \ \text{ in } \Omega \right]$$

and

(81)
$$[f = 0 \text{ in } \Omega] \Rightarrow \left[\inf_\Gamma u \leq \sup_\Gamma u \ \text{ in } \Omega \right].$$

Proof. We prove this result in the case $a_i \equiv 0$, $1 \leq i \leq N$; the general case is more delicate (see D. Gilbarg–N. Trudinger [1], Theorem 8.1). To establish (79) is the same as showing that

(79′)
$$[u \leq 0 \text{ on } \Gamma \text{ and } f \leq 0 \text{ in } \Omega] \Rightarrow [u \leq 0 \text{ in } \Omega].$$

We choose $\varphi = G(u)$ in (78) with G as in the proof of Theorem 9.27; we thus obtain

$$\int_\Omega \sum_{i,j} a_{ij} \frac{\partial u}{\partial x_i} \frac{\partial u}{\partial x_j} G'(u) \leq 0,$$

and so

$$\int_\Omega |\nabla u|^2 G'(u) \leq 0.$$

Set $H(t) = \int_0^t [G'(s)]^{1/2} ds$, so that

$$H(u) \in H_0^1(\Omega) \quad \text{and} \quad |\nabla H(u)|^2 = |\nabla u|^2 G'(u) = 0.$$

It follows[39] that $H(u) = 0$ in Ω and hence $u \leq 0$ in Ω.

We now prove (80) in the following form:

(80′)
$$[f \leq 0 \text{ in } \Omega] \Rightarrow \left[u \leq \sup_\Gamma u \text{ in } \Omega \right].$$

Set $K = \sup_\Gamma u$; then $(u - K)$ satisfies (78), since $a_0 \equiv 0$ and $(u - K) \in H^1(\Omega)$, since Ω is bounded. Applying (79′), we obtain $u - K \leq 0$ in Ω, i.e., (80′). Finally, (81) follows from (80) and (80′).

Proposition 9.30 (maximum principle for the Neumann problem). *Let $f \in L^2(\Omega)$ and $u \in H^1(\Omega)$ be such that*

$$\int_\Omega \nabla u \cdot \nabla \varphi + \int_\Omega u\varphi = \int_\Omega f\varphi \quad \forall \varphi \in H^1(\Omega).$$

---

[39] Note that if $f \in W_0^{1,p}(\Omega)$ with $1 \leq p < \infty$ and $\nabla f = 0$ in Ω, then $f = 0$ in Ω. Indeed, let $\bar{f}$ be the extension of f by 0 outside Ω; then $\bar{f} \in W^{1,p}(\mathbb{R}^N)$ and $\nabla \bar{f} = \overline{\nabla f} = 0$ (see Proposition 9.18). As a consequence, $\bar{f}$ is constant (see Remark 7), and since $\bar{f} \in L^p(\mathbb{R}^N)$, $\bar{f} \equiv 0$.

Then we have, for a.e. $x \in \Omega$,

$$\inf_{\Omega} f \leq u(x) \leq \sup_{\Omega} f.$$

Proof. Analogous to the proof of Theorem 9.27.

9.8 Eigenfunctions and Spectral Decomposition

In this section we assume that Ω is a bounded open set.

• **Theorem 9.31.** *There exist a Hilbert basis $(e_n)_{n\geq 1}$ of $L^2(\Omega)$ and a sequence $(\lambda_n)_{n\geq 1}$ of reals with $\lambda_n > 0 \; \forall n$ and $\lambda_n \to +\infty$ such that*

$$(82) \qquad\qquad e_n \in H_0^1(\Omega) \cap C^\infty(\Omega),$$

$$(83) \qquad\qquad -\Delta e_n = \lambda_n e_n \quad in \; \Omega.$$

We say that the λ_n's are the eigenvalues of $-\Delta$ (with Dirichlet boundary condition) and that the e_n's are the associated eigenfunctions.

Proof. Given $f \in L^2(\Omega)$ let $u = Tf$ be the unique solution $u \in H_0^1(\Omega)$ of the problem

$$(84) \qquad\qquad \int_\Omega \nabla u \cdot \nabla \varphi = \int_\Omega f \varphi \quad \forall \varphi \in H_0^1(\Omega).$$

We consider T as an operator from $L^2(\Omega)$ into $L^2(\Omega)$. Then T is a self-adjoint compact operator (repeat the proof of Theorem 8.21 and use the fact that $H_0^1(\Omega) \subset L^2(\Omega)$ with *compact* injection). On the other hand, $N(T) = \{0\}$ and $(Tf, f)_{L^2} \geq 0$ $\forall f \in L^2$. We conclude (applying Theorem 6.11) that L^2 admits a Hilbert basis (e_n) consisting of eigenfunctions of T associated to eigenvalues (μ_n) with $\mu_n > 0 \; \forall n$ and $\mu_n \to 0$. Thus we have $e_n \in H_0^1(\Omega)$ and

$$\int_\Omega \nabla e_n \cdot \nabla \varphi = \frac{1}{\mu_n} \int_\Omega e_n \varphi \quad \forall \varphi \in H_0^1(\Omega).$$

In other words, e_n is a weak solution of (83) with $\lambda_n = 1/\mu_n$. From the regularity results of Section 9.6 (see Remark 25) we know that $e_n \in H^2(\omega)$ for every $\omega \subset\subset \Omega$. It follows that $e_n \in H^4(\omega)$ for every $\omega \subset\subset \Omega$ and then $e_n \in H^6(\omega)$ for every $\omega \subset\subset \Omega$, etc. Thus $e_n \in \cap_{m\geq 1} H^m(\omega)$ for all $\omega \subset\subset \Omega$. As a consequence, $e_n \in C^\infty(\omega)$ for all $\omega \subset\subset \Omega$, i.e., $e_n \in C^\infty(\Omega)$.

Remark 28. Under the assumptions of Theorem 9.31, the sequence $(e_n/\sqrt{\lambda_n})$ is a *Hilbert basis* of $H_0^1(\Omega)$ equipped with the scalar product $\int_\Omega \nabla u \cdot \nabla v$, and $(e_n/\sqrt{\lambda_n + 1})$ is a Hilbert basis of $H_0^1(\Omega)$ equipped with the scalar product $\int_\Omega (\nabla u \cdot \nabla v + uv)$. Indeed, it is clear that the sequence $(e_n/\sqrt{\lambda_n})$ is orthonormal in $H_0^1(\Omega)$ (use (83)). It remains to verify that the vector space spanned by the e_n's is dense in $H_0^1(\Omega)$. So, let $f \in H_0^1(\Omega)$ be such that $(f, e_n)_{H_0^1} = 0$ $\forall n$. We have to prove that $f = 0$. From (83) we have $\lambda_n \int e_n f = 0$ $\forall n$ and consequently $f = 0$ (since (e_n) is a Hilbert basis of $L^2(\Omega)$).

Remark 29. Under the hypotheses of Theorem 9.31 (for a *general* bounded domain Ω) it can be proved that $e_n \in L^\infty(\Omega)$. On the other hand, if Ω is of class C^∞ then $e_n \in C^\infty(\overline{\Omega})$; this results easily from Theorem 9.25.

Remark 30. Let $a_{ij} \in L^\infty(\Omega)$ be functions satisfying the ellipticity condition (36) and let $a_0 \in L^\infty(\Omega)$. Then there exists a Hilbert basis (e_n) of $L^2(\Omega)$ and there exists a sequence (λ_n) of reals with $\lambda_n \to +\infty$ such that $e_n \in H_0^1(\Omega)$ and

$$\int_\Omega \sum_{i,j} a_{ij} \frac{\partial e_n}{\partial x_i} \frac{\partial \varphi}{\partial x_j} + \int_\Omega a_0 e_n \varphi = \lambda_n \int_\Omega e_n \varphi \quad \forall \varphi \in H_0^1(\Omega).$$

Comments on Chapter 9

This chapter is an *introduction* to the theory of Sobolev spaces and elliptic equations. The reader who wishes to dig deeper into this *vast* subject can consult an extensive bibliography; we cite among others, S. Agmon [1], L. Bers–F. John–M. Schechter [1], J.-L. Lions [1], J.-L. Lions–E. Magenes [1], A. Friedman [2], M. Miranda [1], G. Folland [1], F. Treves [4], R. Adams [1], D. Gilbarg–N. Trudinger [1], G. Stampacchia [1], R. Courant–D. Hilbert [1] Vol. 2, H. Weinberger [1], L. Nirenberg [1], E. Giusti [2], L. C. Evans [1], M. Giaquinta [1], E. Lieb–M. Loss [1], M. Taylor [1], W. Ziemer [1], O. Ladyzhenskaya–N. Uraltseva [1], N. Krylov [1], [2], V. Maz'ja [1], C. Morrey [1], Y. Z. Chen–L. C. Wu [1], E. DiBenedetto [1], Q. Han–F. H. Lin [1], J. Jost [1], W. Strauss [1], and the references in these texts.

1. In Chapter 9 we have often supposed that Ω is of class C^1; this excludes, for example, the domains with "corners." In various situations one can weaken this hypothesis and replace it by somewhat "exotic" conditions: Ω is piecewise of class C^1, Ω is Lipschitz, Ω has the cone property, Ω has the segment property, etc.; see, for example, R. Adams [1] and S. Agmon [1].

2. Theorem 9.7 (existence of an extension operator) can be adapted to the spaces $W^{m,p}(\Omega)$ (Ω of class C^m) with the help of a *suitable generalization of the technique of extension by reflection*; see, e.g., R. Adams [1] and S. Agmon [1].

3. Some very useful inequalities involving the Sobolev norms.

• **A. Poincaré–Wirtinger's inequality.** Let Ω be a connected open set of class C^1 and let $1 \le p \le \infty$. Then there exists a constant C such that

$$\|u - \bar{u}\|_p \le C \|\nabla u\|_p \quad \forall u \in W^{1,p}(\Omega), \text{ where } \bar{u} = \frac{1}{|\Omega|} \int_\Omega u.$$

From this is deduced, because of the Sobolev inequality, that if $p < N$,

$$\|u - \bar{u}\|_{p^*} \leq C\|\nabla u\|_p \quad \forall u \in W^{1,p}(\Omega).$$

- **B. Hardy's inequality.** Let Ω be a bounded open set of class C^1 and let $1 < p < \infty$. Set $d(x) = \mathrm{dist}(x, \Gamma)$. There exists a constant C such that

$$\left\|\frac{u}{d}\right\|_p \leq C\|\nabla u\|_p \quad \forall u \in W_0^{1,p}(\Omega).$$

Conversely,

$$[u \in W^{1,p}(\Omega) \text{ and } (u/d) \in L^p(\Omega)] \Rightarrow [u \in W_0^{1,p}(\Omega)];$$

see J. L. Lions–E. Magenes [1].

- **C. Interpolation inequalities of Gagliardo–Nirenberg.** We mention only some examples that are encountered frequently in the applications. For the general case see L. Nirenberg [1] or A. Friedman [2].

To fix ideas, let $\Omega \subset \mathbb{R}^N$ be a regular bounded open set.

Example 1. Let $u \in L^p(\Omega) \cap W^{2,r}(\Omega)$ with $1 \leq p \leq \infty$ and $1 \leq r \leq \infty$. Then $u \in W^{1,q}(\Omega)$, where q is the *harmonic mean* of p and r, i.e., $\frac{1}{q} = \frac{1}{2}(\frac{1}{p} + \frac{1}{r})$, and

$$\|Du\|_{L^q} \leq C\|u\|_{W^{2,r}}^{1/2}\|u\|_{L^p}^{1/2}.$$

Particular cases:

(a) $p = \infty$, and thus $q = 2r$. We have

$$\|Du\|_{L^q} \leq C\|u\|_{W^{2,r}}^{1/2}\|u\|_{L^\infty}^{1/2}.$$

This inequality can be used, among other things, to show that $W^{2,r} \cap L^\infty$ is an *algebra*, that is to say,

$$u, v \in W^{2,r} \cap L^\infty \Rightarrow uv \in W^{2,r} \cap L^\infty$$

(this property remains true for $W^{m,r} \cap L^\infty$ with m an integer, $m \geq 2$).

(b) $p = q = r$. We have

$$\|Du\|_{L^p} \leq C\|u\|_{W^{2,p}}^{1/2}\|u\|_{L^p}^{1/2},$$

from which one deduces in particular that

$$\|Du\|_{L^p} \leq \varepsilon\|D^2 u\|_{L^p} + C_\varepsilon\|u\|_{L^p} \quad \forall \varepsilon > 0.$$

Example 2. Let $1 \leq q \leq p < \infty$. Then

(85) $\quad \|u\|_{L^p} \leq C\|u\|_{L^q}^{1-a}\|u\|_{W^{1,N}}^a \quad \forall u \in W^{1,N}(\Omega)$, where $a = 1 - (q/p)$.

We note the particular case that is *used frequently*

$$N = 2, \quad p = 4, \quad q = 2, \quad \text{and} \quad a = 1/2,$$

that is to say,

$$\|u\|_{L^4} \leq C \|u\|_{L^2}^{1/2} \|u\|_{H^1}^{1/2} \quad \forall u \in H^1(\Omega).$$

We remark, in this connection, that we have also the usual interpolation inequality (Remark 2 of Chapter 4)

$$\|u\|_{L^p} \leq \|u\|_{L^q}^{1-a} \|u\|_{L^\infty}^a \quad \text{with} \quad a = 1 - (q/p),$$

but it does not imply (85), since $W^{1,N}$ is not contained in L^∞.

Example 3. Let $1 \leq q \leq p \leq \infty$ and $r > N$. Then

(86) $$\|u\|_{L^p} \leq C \|u\|_{L^q}^{1-a} \|u\|_{W^{1,r}}^a \quad \forall u \in W^{1,r}(\Omega),$$

where $a = (\frac{1}{q} - \frac{1}{p})/(\frac{1}{q} + \frac{1}{N} - \frac{1}{r})$.

• **4.** The following property is sometimes useful. Let $u \in W^{1,p}(\Omega)$ with $1 \leq p \leq \infty$ and Ω any open set. Then $\nabla u = 0$ a.e. on the set $\{x \in \Omega; u(x) = k\}$, where k is any constant.

⋆ **5.** The functions in $W^{1,p}(\Omega)$ are *differentiable* in the usual sense a.e. in Ω when $p > N$. More precisely, let $u \in W^{1,p}(\Omega)$ with $p > N$. Then there exists a set $A \subset \Omega$ of measure zero such that

$$\lim_{h \to 0} \frac{u(x+h) - u(x) - h \cdot \nabla u(x)}{|h|} = 0 \quad \forall x \in \Omega \backslash A.$$

This property is not valid when $u \in W^{1,p}(\Omega)$ and $p \leq N (N > 1)$. On this question consult E. Stein [1] (Chapter 8).

6. Fractional Sobolev spaces.

One can define a family of spaces intermediate between $L^p(\Omega)$ and $W^{1,p}(\Omega)$. More precisely, if $0 < s < 1$ ($s \in \mathbb{R}$) and $1 \leq p < \infty$, set

$$W^{s,p}(\Omega) = \left\{ u \in L^p(\Omega); \frac{|u(x) - u(y)|}{|x - y|^{s + (N/p)}} \in L^p(\Omega \times \Omega) \right\},$$

equipped with the natural norm. Set $H^s(\Omega) = W^{s,2}(\Omega)$. For studies of these spaces, see, e.g., R. Adams [1], J.-L. Lions–E. Magenes [1], P. Malliavin [1], H. Triebel [1], and L. Grafakos [1]. The spaces $W^{s,p}(\Omega)$ can also be defined as *interpolation* spaces between $W^{1,p}$ and L^p, and also using the *Fourier transform* if $p = 2$ and $\Omega = \mathbb{R}^N$.

We define finally $W^{s,p}(\Omega)$ for s real, s not an integer, $s > 1$ as follows. Write $s = m + \sigma$ with $m = $ the integer part of s, and set

$$W^{s,p}(\Omega) = \{u \in W^{m,p}(\Omega); D^\alpha u \in W^{\sigma,p}(\Omega) \quad \forall \alpha \text{ with } |\alpha| = m\}.$$

By local charts one also defines $W^{s,p}(\Gamma)$, where Γ is a *smooth manifold* (for example the boundary of a regular open set). These spaces play an important role in the theory of traces (see Comment 7).

• 7. Theory of traces.

Let $1 \le p < \infty$. We begin with a fundamental lemma.

Lemma 9.9. *Let $\Omega = \mathbb{R}^N_+$. There exists a constant C such that*

$$\left(\int_{\mathbb{R}^{N-1}} |u(x', 0)|^p dx' \right)^{1/p} \le C \|u\|_{W^{1,p}(\Omega)} \quad \forall u \in C^1_c(\mathbb{R}^N).$$

Proof. Let $G(t) = |t|^{p-1}t$ and let $u \in C^1_c(\mathbb{R}^N)$. We have

$$G(u(x', 0)) = - \int_0^{+\infty} \frac{\partial}{\partial x_N} G(u(x', x_N)) dx_N$$
$$= - \int_0^{+\infty} G'(u(x', x_N)) \frac{\partial u}{\partial x_N}(x', x_N) dx_N.$$

Thus

$$|u(x', 0)|^p \le p \int_0^{\infty} |u(x', x_N)|^{p-1} \left| \frac{\partial u}{\partial x_N}(x', x_N) \right| dx_N$$
$$\le C \left(\int_0^{\infty} |u(x', x_N)|^p dx_N + \int_0^{\infty} \left| \frac{\partial u}{\partial x_N}(x', x_N) \right|^p dx_N \right),$$

and the conclusion follows by integration in $x' \in \mathbb{R}^{N-1}$.

It can be deduced from Lemma 9.9 that the map $u \mapsto u_{|\Gamma}$ with $\Gamma = \partial\Omega = \mathbb{R}^{N-1} \times \{0\}$ defined from $C^1_c(\mathbb{R}^N)$ into $L^p(\Gamma)$ extends, by density, to a bounded linear operator of $W^{1,p}(\Omega)$ into $L^p(\Gamma)$. This operator is, by definition, the *trace* of u on Γ; it is also denoted by $u_{|\Gamma}$.

We remark that there is a *fundamental difference* between $L^p(\mathbb{R}^N_+)$ and $W^{1,p}(\mathbb{R}^N_+)$: *the functions in $L^p(\mathbb{R}^N_+)$ do not have a trace on Γ.* One can easily imagine—using local charts—how to define the trace on $\Gamma = \partial\Omega$ for a function $u \in W^{1,p}(\Omega)$ when Ω is a *regular open* set in $\mathbb{R}^N$ (for example, Ω of class C^1 with Γ bounded). In this case $u_{|\Gamma} \in L^p(\Gamma)$ (for the surface measure $d\sigma$). The most important properties of the trace are the following:

(i) If $u \in W^{1,p}(\Omega)$, then in fact $u_{|\Gamma} \in W^{1-(1/p),p}(\Gamma)$ and

$$\|u_{|\Gamma}\|_{W^{1-(1/p),p}(\Gamma)} \le C \|u\|_{W^{1,p}(\Omega)} \quad \forall u \in W^{1,p}(\Omega).$$

Furthermore, the trace operator $u \mapsto u_{|\Gamma}$ is *surjective* from $W^{1,p}(\Omega)$ onto $W^{1-(1/p),p}(\Gamma)$.

(ii) The *kernel* of the trace operator is $W^{1,p}_0(\Omega)$, i.e.,

$$W^{1,p}_0(\Omega) = \{u \in W^{1,p}(\Omega); u_{|\Gamma} = 0\}.$$

(iii) We have *Green's formula*

$$\int_{\Omega} \frac{\partial u}{\partial x_i} v = - \int_{\Omega} u \frac{\partial v}{\partial x_i} + \int_{\Gamma} uv(\vec{n} \cdot \vec{e_i})d\sigma \quad \forall u, v \in H^1(\Omega),$$

where $\vec{n}$ is the outward unit normal vector to Γ. Note that the surface integral has a meaning, since $u, v \in L^2(\Gamma)$.

In the same way we can speak of $\frac{\partial u}{\partial n}$ for a function $u \in W^{2,p}(\Omega)$: set $\frac{\partial u}{\partial n} = (\nabla u)_{|\Gamma} \cdot \vec{n}$, which has a meaning since $(\nabla u)_{|\Gamma} \in L^p(\Gamma)^N$, and $\frac{\partial u}{\partial n} \in L^p(\Gamma)$ (in fact $\frac{\partial u}{\partial n} \in W^{1-(1/p),p}(\Gamma)$). Also Green's formula holds:

$$- \int_{\Omega} (\Delta u)v = \int_{\Omega} \nabla u \cdot \nabla v - \int_{\Gamma} \frac{\partial u}{\partial n} v d\sigma \quad \forall u, v \in H^2(\Omega).$$

(iv) The operator $u \mapsto \{u_{|\Gamma}, \frac{\partial u}{\partial n}\}$ is bounded, linear, and surjective from $W^{2,p}(\Omega)$ onto $W^{2-(1/p),p}(\Gamma) \times W^{1-(1/p),p}(\Gamma)$. On these questions, see J.-L. Lions–E. Magenes [1] for the case $p = 2$ (and the references cited therein for the case $p \neq 2$).

8. Operators of order $2m$ and elliptic systems.

The existence and regularity results proved in Chapter 9 extend to elliptic operators of order $2m$ and to elliptic systems.[40] One of the essential ingredients is Gårding's inequality. On these questions, see S. Agmon [1], J.-L. Lions–E. Magenes [1], S. Agmon–A. Douglis–L. Nirenberg [1]. The operators of order $2m$ and certain systems play an important role in mechanics and physics. We point out, in particular, the biharmonic operator Δ^2 (theory of plates), the system of elasticity, and the Stokes system (fluid mechanics); see for example Ph. Ciarlet [1], G. Duvaut–J.-L. Lions [1], R. Temam [1], J. Nečas–L. Hlaváček [1], M. Gurtin [1].

9. Regularity in L^p and $C^{0,\alpha}$ spaces.

The regularity theorems proved in Chapter 9 for $p = 2$ extend to the case $p \neq 2$.

• **Theorem 9.32 (Agmon–Douglis–Nirenberg).** *Suppose that Ω is of class C^2 with Γ bounded. Let $1 < p < \infty$. Then for all $f \in L^p(\Omega)$, there exists a unique solution $u \in W^{2,p}(\Omega) \cap W_0^{1,p}(\Omega)$ of the equation*

$$(87) \qquad\qquad -\Delta u + u = f \quad in \ \Omega.$$

Moreover, if Ω is of class C^{m+2} and if $f \in W^{m,p}(\Omega)$ ($m \geq 1$ an integer), then

$$u \in W^{m+2,p}(\Omega) \quad and \quad \|u\|_{W^{m+2,p}} \leq C\|f\|_{W^{m,p}}.$$

There is an analogous result if (87) is replaced by a second-order elliptic equation with smooth coefficients. The proof of Theorem 9.32 is *considerably more complicated* than the case $p = 2$ (Theorem 9.25). The "classical" approach rests essentially on two ingredients:

---

[40] But the *maximum principle does not*, except in very special cases.

(a) A formula for an explicit representation of u using the fundamental solution. For example, if $\Omega = \mathbb{R}^3$, then the solution of (87) is given by $u = G \star f$, where $G(x) = \frac{c}{|x|} e^{-|x|}$. So that formally, $\frac{\partial^2 u}{\partial x_i \partial x_j} = \frac{\partial^2 G}{\partial x_i \partial x_j} \star f$; "unfortunately" $\frac{\partial^2 G}{\partial x_i \partial x_j}$ does not belong to $L^1(\mathbb{R}^3)$,[41] because of the singularity at $x = 0$, and one cannot apply elementary estimates on convolution products (such as Theorem 4.15).

(b) To overcome this difficulty one uses the *theory of singular integrals in L^p* due to Calderón–Zygmund (see, for example, E. Stein [1] and L. Bers–F. John–M. Schechter [1]).

Warning: the conclusion of Theorem 9.32 is *false for $p = 1$ and $p = \infty$*.

Another basic regularity result, in the framework Hölder spaces,[42] is the following.

• **Theorem 9.33 (Schauder).** *Suppose that Ω is bounded and of class $C^{2,\alpha}$ with $0 < \alpha < 1$. Then for every $f \in C^{0,\alpha}(\overline{\Omega})$ there exists a unique solution $u \in C^{2,\alpha}(\overline{\Omega})$ of the problem*

(88)
$$\begin{cases} -\Delta u + u = f & in\ \Omega, \\ \qquad\quad u = 0 & on\ \Gamma. \end{cases}$$

Furthermore, if Ω is of class $C^{m+2,\alpha}$ ($m \geq 1$ an integer) and if $f \in C^{m,\alpha}(\overline{\Omega})$, then

$$u \in C^{m+2,\alpha}(\overline{\Omega}) \quad with\ \|u\|_{C^{m+2,\alpha}} \leq C \|f\|_{C^{m,\alpha}}.$$

An analogous result holds if (88) is replaced by a second-order elliptic operator with smooth coefficients. The proof of Theorem 9.33 rests—as does that of Theorem 9.32—on an explicit representation of u and on the *theory of singular integrals in $C^{0,\alpha}$* spaces due to Hölder, Korn, Lichtenstein, Giraud. On this subject, see S. Agmon–A. Douglis–L. Nirenberg [1], L. Bers–F. John–M. Schechter [1], C. Morrey [1], D. Gilbarg–N. Trudinger [1]. A different approach, which avoids the theory of singular integrals, has been devised by Campanato and Stampacchia (see, e.g., Y. Z. Chen–L. C. Wu [1] and E. Giusti [2]). Other elementary techniques have been developed by A. Brandt [1] (based solely on the maximum principle) and by L. Simon [2].

Let Ω be a bounded regular open set and let $f \in C(\overline{\Omega})$. From Theorem 9.32 there exists $u \in W^{2,p}(\Omega) \cap W_0^{1,p}(\Omega)$ (for all $1 < p < \infty$) that is the unique solution of (87). In particular, $u \in C^{1,\alpha}(\overline{\Omega})$ for all $0 < \alpha < 1$ (from Morrey's

---

[41] But almost!

[42] Recall that with $0 < \alpha < 1$ and m an integer,

$$C^{0,\alpha}(\overline{\Omega}) = \left\{ u \in C(\overline{\Omega});\ \sup_{\substack{x,y \in \Omega \\ x \neq y}} \frac{|u(x) - u(y)|}{|x - y|^\alpha} < \infty \right\}$$

and $C^{m,\alpha}(\overline{\Omega}) = \{u \in C(\overline{\Omega});\ D^\beta u \in C^{0,\alpha}(\overline{\Omega})\ \forall \beta\ with\ |\beta| = m\}$.

theorem (Theorem 9.12)). In general, u *does not belong to* C^2, or even to $W^{2,\infty}$. This explains why *one often avoids working in the spaces* $L^1(\Omega)$, $L^\infty(\Omega)$, and $C(\overline{\Omega})$, spaces for which we do not have optimal regularity results.

Theorems 9.32 and 9.33 extend to elliptic operators of order $2m$ and to elliptic systems; see S. Agmon–A. Douglis–L. Nirenberg [1]. We finally point out, in a different direction, that second-order elliptic equations with *discontinuous coefficients* are the subject of much work. We cite, for example, the following celebrated result.

• **Theorem 9.34 (De Giorgi, Nash, Stampacchia).** *Let* $\Omega \subset \mathbb{R}^N$, *with* $N \geq 2$, *be a bounded regular open set. Suppose that the functions* $a_{ij} \in L^\infty(\Omega)$ *satisfy the ellipticity condition* (36). *Let* $f \in L^p(\Omega)$ *with* $p > N/2$ *and let* $u \in H_0^1(\Omega)$ *be such that*

$$\int_\Omega \sum_{i,j} a_{ij} \frac{\partial u}{\partial x_i} \frac{\partial \varphi}{\partial x_j} = \int_\Omega f\varphi \quad \forall \varphi \in H_0^1(\Omega).$$

Then $u \in C^{0,\alpha}(\overline{\Omega})$ *for a certain* $0 < \alpha < 1$ *(which depends on* Ω, a_{ij} *and* p*)*.

On these questions, see G. Stampacchia [1], D. Gilbarg–N. Trudinger [1], O. Ladyzhenskaya–N. Uraltseva [1], and E. Giusti [2].

10. Some drawbacks of the variational method and how to get around them!
The variational method gives the existence of a weak solution very easily. It is not always applicable, but it can be completed. We indicate two examples. Let $\Omega \subset \mathbb{R}^N$ be a bounded regular open set.

(a) Duality method. Let $f \in L^1(\Omega)$—or even f a (Radon) measure on Ω—and look for a solution of the problem

$$(89) \qquad \begin{cases} -\Delta u + u = f & \text{in } \Omega, \\ u = 0 & \text{on } \Gamma. \end{cases}$$

As soon as $N > 1$, the linear functional $\varphi \mapsto \int_\Omega f\varphi$ is not defined for every $\varphi \in H_0^1(\Omega)$, and as a consequence the *variational method is ineffective*. On the other hand, one can use the following technique. We denote by $T : L^2(\Omega) \to L^2(\Omega)$ the operator $f \mapsto u$ (where u is the solution of (89), which exists for $f \in L^2(\Omega)$). We know that T is self-adjoint. On the other hand (Theorem 9.32), $T : L^p(\Omega) \to W^{2,p}(\Omega)$ for $2 \leq p < \infty$, and because of the theorems of Sobolev and Morrey, $T : L^p(\Omega) \to C_0(\overline{\Omega})$ if $p > N/2$. By duality we deduce that

$$T^\star : \mathcal{M}(\Omega) = C_0(\overline{\Omega})^\star \to L^{p'}(\Omega) \text{ if } p > N/2.$$

Since T is self-adjoint in L^2, $T^\star$ is an extension of T: thus one can consider $u = T^\star f$ as a generalized solution of (89). In fact, if $f \in L^1(\Omega)$, then $u = T^\star f \in L^q(\Omega)$ for all $q < N/(N-2)$; u is the unique (very) weak solution of (89) in the following sense:

$$-\int_\Omega u\Delta\varphi + \int_\Omega u\varphi = \int_\Omega f\varphi \quad \forall \varphi \in C^2(\overline{\Omega}), \quad \varphi = 0 \text{ on } \Gamma.$$

In the same spirit, one can study (89) for f given in $H^{-m}(\Omega)$; see J.-L. Lions–E. Magenes [1].

(b) Density method. Let $g \in C(\Gamma)$ and look for a solution of the problem

$$(90) \qquad \begin{cases} -\Delta u + u = 0 & \text{in } \Omega \\ \qquad\quad u = g & \text{on } \Gamma. \end{cases}$$

In general, if $g \in C(\Gamma)$, there does not exist a function $\tilde{g} \in H^1(\Omega)$ such that $\tilde{g}_{|\Gamma} = g$ (see Comment 7 and note that $C(\Gamma)$ is not contained in $H^{1/2}(\Gamma)$). It is thus not possible to look for a solution of (90) in $H^1(\Omega)$: *the variational method is ineffective*. Nevertheless, we have the following result.

• **Theorem 9.35.** *There exists a unique solution* $u \in C(\overline{\Omega}) \cap C^\infty(\Omega)$ *of* (90).

Proof. Fix $\tilde{g} \in C_c(\mathbb{R}^N)$ such that $\tilde{g}_{|\Gamma} = g$; $\tilde{g}$ exists by the Tietze–Urysohn theorem (see, e.g., J. Dieudonné [1], J. Dugundji [1], J. Munkres [1]). Let $(\tilde{g}_n)$ be a sequence in $C_c^\infty(\mathbb{R}^N)$ such that $\tilde{g}_n \to g$ uniformly on $\mathbb{R}^N$. We set $g_n = \tilde{g}_{n|\Gamma}$. Applying the *variational method and regularity results*, we see that there exists a classical solution $u_n \in C^2(\overline{\Omega})$ of the problem

$$\begin{cases} -\Delta u_n + u_n = 0 & \text{in } \Omega, \\ \qquad\quad u_n = g_n & \text{on } \Gamma. \end{cases}$$

From the maximum principle (Corollary 9.28) we have

$$\|u_m - u_n\|_{L^\infty(\Omega)} \le \|g_m - g_n\|_{L^\infty(\Gamma)}.$$

As a consequence, (u_n) is a Cauchy sequence in $C(\overline{\Omega})$ and $u_n \to u$ in $C(\overline{\Omega})$. It is clear that we have

$$\int_\Omega u(-\Delta\varphi + \varphi) = 0 \quad \forall \varphi \in C_c^\infty(\Omega)$$

and therefore $u \in C^\infty(\Omega)$ (see Remark 25). Thus $u \in C(\overline{\Omega}) \cap C^\infty(\Omega)$ satisfies (90). The uniqueness of the solution of (90) follows from the maximum principle (see Remark 27).

⋆ *Remark* 31. It is essential in Theorem 9.35 to suppose that Ω is smooth enough. When Ω has a "pathological" boundary we run into questions of potential theory (regular points, Wiener criterion, etc.).

Another approach to solving (90) is the *Perron method*, which is classical in *potential theory*. Define

$$u(x) = \sup\{v(x); v \in C(\overline{\Omega}) \cap C^2(\Omega), -\Delta v + v \le 0 \text{ in } \Omega \text{ and } v \le g \text{ on } \Gamma\},$$

and prove (directly) that u satisfies (90). A function v such that $-\Delta v + v \le 0$ in Ω and $v \le g$ on Γ is called a *subsolution* of (90).

11. The strong maximum principle.

We can strengthen the conclusion of Proposition 9.29 when u is a *classical solution*. More precisely, let Ω be a connected, bounded, regular open set. Let $a_{ij} \in C^1(\overline{\Omega})$ satisfy the ellipticity condition (36), $a_i, a_0 \in C(\overline{\Omega})$ with $a_0 \geq 0$ on Ω.

Theorem 9.36 (Hopf). *Let* $u \in C(\overline{\Omega}) \cap C^2(\Omega)$ *satisfy*

$$(91) \qquad -\sum_{i,j} \frac{\partial}{\partial x_j}\left(a_{ij}\frac{\partial u}{\partial x_i}\right) + \sum_i a_i \frac{\partial u}{\partial x_i} + a_0 u = f \quad in\ \Omega.$$

Suppose that $f \geq 0$ *in* Ω*. If there exists* $x_0 \in \Omega$ *such that* $u(x_0) = \min_{\overline{\Omega}} u$ *and if* $u(x_0) \leq 0$,[43] *then* u *is constant in* Ω *(and furthermore* $f = 0$ *in* Ω).

For the proof, see, e.g., L. Bers–F. John–M. Schechter [1], D. Gilbarg–N. Trudinger [1], M. Protter–H. Weinberger [1], and P. Pucci–J. Serrin [1].

Corollary 9.37. *Let* $u \in C(\overline{\Omega}) \cap C^2(\Omega)$ *satisfy* (91) *with* $f \geq 0$ *in* Ω*. Suppose that* $u \geq 0$ *on* Γ*. Then*

- *either* $u > 0$ *in* Ω,
- *or* $u \equiv 0$ *in* Ω.

For other results connected to the maximum principle (Harnack's inequality etc.), see, e.g., G. Stampacchia [1], D. Gilbarg–N. Trudinger [1], M. Protter–H. Weinberger [1], R. Sperb [1], and P. Pucci–J. Serrin [1].

12. Laplace–Beltrami operators.

Elliptic operators defined on *Riemannian manifolds* (with or without boundary) and in particular the Laplace–Beltrami operator play an important role in differential geometry and physics; see, for example, Y. Choquet–C. Dewitt–M. Dillard [1].

13. Spectral properties. Inverse problems.

Eigenvalues and eigenfunctions of second-order elliptic operators enjoy a number of remarkable properties. Here we cite some of them. Let $\Omega \subset \mathbb{R}^N$ be a connected, bounded, open regular set. Let $a_{ij} \in C^1(\overline{\Omega})$ satisfy the ellipticity condition (36) and $a_0 \in C(\overline{\Omega})$. Let A be the operator

$$Au = -\sum_{i,j} \frac{\partial}{\partial x_j}\left(a_{ij}\frac{\partial u}{\partial x_i}\right) + a_0 u$$

with homogeneous Dirichlet conditions ($u = 0$ on Γ). We denote by (λ_n) the sequence of eigenvalues of A arranged in increasing order, with $\lambda_n \to +\infty$ when $n \to \infty$. Then the first eigenvalue λ_1 has *multiplicity* 1 (one says that λ_1 is a *simple eigenvalue*),[44] and we can choose the associated eigenfunction e_1 to have $e_1 > 0$ in Ω; this follows from the Krein–Rutman theorem (see the comments on Chapter 6

---

[43] The hypothesis $u(x_0) \leq 0$ is unnecessary if $a_0 = 0$.

[44] In dimension $N \geq 2$ the other eigenvalues can have multiplicity > 1.

and Problem 41). Additionally, one can show that $\lambda_n \sim cn^{2/N}$ when $n \to \infty$ with $c > 0$; see S. Agmon [1].

The relations that exist between the *geometric properties*[45] of Ω and the spectrum of A are the subject of intensive research; see, e.g., M. Kac [1], Marcel Berger [1], R. Osserman [1], I. M. Singer [1], P. Bérard [1], I. Chavel [1]. The objective of *spectral geometry* is to "*recover*" *the maximum amount of information about* Ω, *purely from the knowledge of the spectrum* (λ_n).

A strikingly simple question is the following. Let Ω_1 and Ω_2 be two bounded domains in $\mathbb{R}^2$; suppose that the eigenvalues of the operator $-\Delta$ (with Dirichlet boundary conditions) are the same for Ω_1 and Ω_2. Are Ω_1 and Ω_2 isometric? This problem has been nicknamed by M. Kac: "*Can one hear the shape of a drum?*"[46] One knows that the answer is positive if Ω_1 is a disk. In 1991, C. Gordon–L. Webb–S. Wolpert [1] gave a negative answer for domains with corners. The problem of Kac is still open for smooth domains.

Another important class of "inverse problems" involves the determination of the coefficients and parameters in a PDE, or the shape and characteristics of an internal object, solely from measurements at the boundary (e.g., Dirichlet-to-Neumann map) or at "infinity" (inverse scattering). These problems arise in many areas (medical imaging, seismology, etc.); see, e.g., G. Uhlmann [1], C. B. Croke et al. [1].

14. Degenerate elliptic problems.
Consider problems of the form

$$\begin{cases} -\sum_{i,j} \frac{\partial}{\partial x_j}\left(a_{ij}\frac{\partial u}{\partial x_i}\right) + \sum_i a_i \frac{\partial u}{\partial x_i} + a_0 u = f & \text{in } \Omega \\ + \text{ boundary conditions on } \Gamma, \text{ or on part of } \Gamma, \end{cases}$$

where the functions a_{ij} do *not* satisfy the ellipticity condition (36) but only

$$(36') \qquad \sum_{i,j} a_{ij}(x)\xi_i \xi_j \geq 0 \quad \forall x \in \Omega, \quad \forall \xi \in \mathbb{R}^N.$$

Consult for example the works of J. Kohn–L. Nirenberg [1], M. S. Baouendi–C. Goulaouic [1], O. Oleinik–E. Radkevitch [1].

15. Nonlinear elliptic problems.
This is an immense field of research motivated by innumerable questions in geometry, mechanics, physics, optimal control, probability theory, etc. It has had some spectacular development since the early work of Leray and Schauder at the beginning of the 1930s. We distinguish some categories:

(a) **Semilinear problems.** This consists, for example, of problems of the form

---

[45] Particularly when Ω is a Riemannian manifold without boundary and A is the Laplace–Beltrami operator.

[46] Because the harmonics of the vibration of a membrane attached to the boundary Γ are the functions $e_n(x) \sin \sqrt{\lambda_n} t$, where (λ_n, e_n) are the eigenvalues and eigenfunctions of $-\Delta$ with Dirichlet boundary conditions.

$$(92) \qquad \begin{cases} -\Delta u = f(x, u) & \text{in } \Omega, \\ u = 0 & \text{on } \Gamma, \end{cases}$$

where $f(x, u)$ is a given function.

This category includes, among others, *bifurcation problems*, in which one studies the structure of the set of solutions (λ, u) of the problem

$$(91') \qquad \begin{cases} -\Delta u = f_\lambda(x, u) & \text{in } \Omega, \\ u = 0 & \text{on } \Gamma, \end{cases}$$

with λ a variable parameter.

(b) Quasilinear problems. Consider problems of the form

$$(93) \qquad \begin{cases} -\displaystyle\sum_{i,j} \frac{\partial}{\partial x_j}\left(a_{ij}(x, u, \nabla u)\frac{\partial u}{\partial x_i}\right) = f(x, u, \nabla u) & \text{in } \Omega, \\ u = 0 & \text{on } \Gamma, \end{cases}$$

where the functions $a_{ij}(x, u, p)$ are elliptic, but possibly degenerate; we have for example

$$\sum_{i,j} a_{ij}(x, u, p)\xi_i\xi_j \geq \alpha(u, p)|\xi|^2 \quad \forall x \in \Omega, \; \forall \xi \in \mathbb{R}^N, \; \forall u \in \mathbb{R}, \; \forall p \in \mathbb{R}^N,$$

with $\alpha(u, p) > 0 \; \forall u \in \mathbb{R}, \forall p \in \mathbb{R}^N$, but $\alpha(u, p)$ is not uniformly bounded below by a constant $\alpha > 0$. In particular, the celebrated *equation of minimal surfaces* falls in this category with $a_{ij}(x, u, p) = \delta_{ij}(1 + |p|^2)^{-1/2}$. More generally, one considers *fully nonlinear* elliptic problems of the form

$$(94) \qquad F(x, u, Du, D^2u) = 0,$$

where the matrix $\frac{\partial F}{\partial q_{ij}}(x, u, p, q)$ is elliptic (possibly degenerate). For example, the *Monge–Ampère equation* fits into this category.

(c) Free boundary problems. It is a question of solving a linear elliptic equation in an *open set Ω that is not given a priori*. The fact that Ω is unknown is often "compensated for" by having *two boundary conditions* on Γ; for example Dirichlet *and* Neumann. The problem consists in finding simultaneously an open set Ω and a function u such that. . . .

Techniques:

(a) There are several techniques used for the problems (92) or (92'):

- *Monotonicity methods*, see F. Browder [1] and J. L. Lions [3].
- *Topological methods* (Schauder's fixed-point theorem, Leray–Schauder degree theory, etc.); see J. T. Schwartz [1], M. Krasnoselskii [1], and L. Nirenberg [2], [3].

- *Variational methods* (critical point theory, min-max techniques, Morse theory, etc.); see P. Rabinowitz [1], [2], Melvyn Berger [1], M. Krasnoselskii [1], L. Nirenberg [3], J. Mawhin–M. Willem [1], M. Willem [1], M. Struwe [1].

For a general survey, see, e.g., the books of A. Ambrosetti–G. Prodi [1] and E. Zeidler [1].

(b) Solving problems of type (93) may involve elaborate techniques of estimates;[47] see the works of E. De Giorgi, O. Ladyzhenskaya–N. Uraltseva [1], J. Serrin [1], E. Bombieri [1] and D. Gilbarg–N. Trudinger [1]. Important progress on the fully nonlinear equations and in particular on the Monge–Ampère equation has also been made recently; see, e.g., S. T. Yau [1], L. Caffarelli–L. Nirenberg–J. Spruck [1] and X. Cabré–L. Caffarelli [1].

(c) On the *free boundary problems* many new results have appeared in recent years, often in connection with the *theory of variational inequalities*; see, e.g., D. Kinderlehrer–G. Stampacchia [1], C. Baiocchi–A. Capelo [1], A. Friedman [4], J. Crank [1] and L. Caffarelli–S. Salsa [1].

16. Geometric measure theory.
At the interface between geometry and PDE, this area has been extensively developed since the 1960s, starting with basic contributions by H. Federer, E. De Giorgi, A. I. Volpert, and F. Almgren, in connection with questions arising in the calculus of variations, isoperimetric inequalities, etc. It has numerous applications to physical problems, such as phase transitions, fractures in mechanics, edge detection in image processing, line vortices in liquid crystals, superconductors and superfluids. The space BV (functions of bounded variation) plays a distinguished role in these questions. We refer, e.g., to L. Ambrosio–N. Fusco–D. Pallara [1], L. Simon [1], L. C. Evans–R. Gariepy [1], and F. H. Lin–X. P. Yang [1].

---

[47] This is the case, for example, for the minimal surface equation.

Chapter 10
Evolution Problems: The Heat Equation and the Wave Equation

10.1 The Heat Equation: Existence, Uniqueness, and Regularity

Notation. Let $\Omega \subset \mathbb{R}^N$ be an open set with boundary Γ. Set

$$Q = \Omega \times (0, +\infty)$$
$$\Sigma = \Gamma \times (0, +\infty);$$

Σ is called the *lateral* boundary of the cylinder Q. See Figure 7.

Consider the following problem: find a function $u(x, t) : \overline{\Omega} \times [0, +\infty) \to \mathbb{R}$ such that

(1)
$$\frac{\partial u}{\partial t} - \Delta u = 0 \quad \text{in } Q,$$

(2)
$$u = 0 \quad \text{on } \Sigma,$$

(3)
$$u(x, 0) = u_0(x) \quad \text{on } \Omega,$$

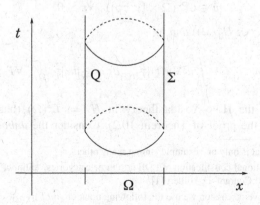

Fig. 7

H. Brezis, *Functional Analysis, Sobolev Spaces and Partial Differential Equations*,
DOI 10.1007/978-0-387-70914-7_10, © Springer Science+Business Media, LLC 2011

where $\Delta = \sum_{i=1}^{N} \frac{\partial^2}{\partial x_i^2}$ denotes the *Laplacian in the space variables* x, t is the *time variable*, and $u_0(x)$ is a given function called the *initial* (or Cauchy) *data*.

Equation (1) is called the *heat equation* because it models the temperature distribution u in the domain Ω at time t. The heat equation and its variants occur in many *diffusion phenomena*[1] (see the comments at the end of this chapter). The heat equation is the simplest example of a *parabolic equation*.[2]

Equation (2) is the (homogeneous) *Dirichlet boundary condition*; it could be replaced by the Neumann condition

(2′)
$$\frac{\partial u}{\partial n} = 0 \quad \text{on} \quad \Sigma$$

(n is the outward unit normal vector to Γ) or any of the boundary conditions encountered in Chapters 8 and 9. Condition (2) corresponds to the assumption that the boundary Γ is kept at zero temperature; condition (2′) corresponds to the assumption that the heat flux across Γ is zero. We solve problem (1), (2), (3) by viewing $u(x, t)$ as a *function defined on* $[0, +\infty)$ *with values in a space H*, where H is a space of functions depending only on x: for example $H = L^2(\Omega)$, or $H = H_0^1(\Omega)$. When we write just $u(t)$, we mean that $u(t)$ is an element of H, namely the function $x \mapsto u(x, t)$. This viewpoint allows us to solve *very easily* problem (1), (2), (3) by combining the theorem of Hille–Yosida with the results of Chapters 8 and 9.

To simplify matters, we assume throughout Chapter 10 that Ω *is of class C^∞ with* Γ *bounded* (but this assumption may be considerably weakened if we are interested only in weak solutions).

• **Theorem 10.1.** *Assume $u_0 \in L^2(\Omega)$. Then there exists a unique function $u(x, t)$ satisfying* (1), (2), (3) *and*

(4) $u \in C([0, \infty); \ L^2(\Omega)) \cap C((0, \infty); \ H^2(\Omega) \cap H_0^1(\Omega))$,

(5) $u \in C^1((0, \infty); \ L^2(\Omega))$.

Moreover,
$$u \in C^\infty(\overline{\Omega} \times [\varepsilon, \infty)) \quad \forall \varepsilon > 0.$$

Finally, $u \in L^2(0, \infty; H_0^1(\Omega))$ and [3]

(6) $$\frac{1}{2}|u(T)|^2_{L^2(\Omega)} + \int_0^T |\nabla u(t)|^2_{L^2(\Omega)} dt = \frac{1}{2}|u_0|^2_{L^2(\Omega)} \quad \forall T > 0.$$

Proof. We apply the Hille–Yosida theory in $H = L^2(\Omega)$ (but other choices are possible; see the proof of Theorem 10.2). Consider the *unbounded* operator

---

[1] The diffusion of heat is only one example among many others.

[2] Regarding the traditional classification of PDE into three categories, "elliptic," "parabolic," "hyperbolic," see, e.g., R. Courant–D. Hilbert [1].

[3] In line with the above discussion we use the following notation: $|u(T)|_{L^2(\Omega)} = \int_\Omega |u(x, T)|^2 dx$ and $|\nabla u(t)|^2_{L^2(\Omega)} = \sum_{i=1}^{N} \int_\Omega |\frac{\partial u}{\partial x_i}(x, t)|^2 dx.$

$A: D(A) \subset H \to H$ defined by

$$\begin{cases} D(A) = H^2(\Omega) \cap H^1_0(\Omega), \\ Au = -\Delta u. \end{cases}$$

It is important to note that the *boundary condition* (2) *has been incorporated in the definition of the domain of A*. We claim that A is a *self-adjoint maximal monotone operator*. We may then apply Theorem 7.7 and deduce the existence of a unique solution of (1), (2), (3) satisfying (4) and (5).

(i) *A is monotone.* For every $u \in D(A)$ we have

$$(Au, u)_{L^2} = \int_\Omega (-\Delta u)u = \int_\Omega |\nabla u|^2 \geq 0.$$

(ii) *A is maximal monotone.* We have to check that $R(I + A) = H = L^2$. But we already know (see Theorem 9.25) that for every $f \in L^2$ there exists a unique solution $u \in H^2 \cap H^1_0$ of the equation $u - \Delta u = f$.

(iii) *A is self-adjoint.* In view of Proposition 7.6 it suffices to verify that A is symmetric. For every $u, v \in D(A)$ we have

$$(Au, v)_{L^2} = \int_\Omega (-\Delta u)v = \int_\Omega \nabla u \cdot \nabla v$$

and

$$(u, Av)_{L^2} = \int_\Omega u(-\Delta v) = \int_\Omega \nabla u \cdot \nabla v,$$

so that $(Au, v) = (u, Av)$.

Next, it follows from Theorem 9.25 that $D(A^\ell) \subset H^{2\ell}(\Omega)$, for every integer ℓ, with continuous injection. More precisely,

$$D(A^\ell) = \{u \in H^{2\ell}(\Omega); u = \Delta u = \cdots = \Delta^{\ell-1}u = 0 \quad \text{on } \Gamma\}.$$

We know by Theorem 7.7 that the solution u of (1), (2), (3) satisfies

$$u \in C^k((0, \infty); D(A^\ell)) \quad \forall k, \quad \forall \ell$$

and therefore

$$u \in C^k((0, \infty); H^{2\ell}(\Omega)) \quad \forall k, \quad \forall \ell.$$

It follows (thanks to Corollary 9.15) that

$$u \in C^k((0, \infty); C^k(\overline{\Omega})) \quad \forall k.$$

We now turn to the proof of (6). Formally, we multiply (1) by u and integrate on $\Omega \times (0, T)$. However, one has to be careful, since $u(t)$ is differentiable on $(0, \infty)$ but *not* on $[0, \infty)$. Consider the function $\varphi(t) = \frac{1}{2}|u(t)|^2_{L^2(\Omega)}$. It is of class C^1 on

$(0, \infty)$ (by (5)) and, for $t > 0$,

$$\varphi'(t) = \left(u(t), \frac{du}{dt}(t) \right)_{L^2} = (u(t), \Delta u(t))_{L^2} = - \int_{\Omega} |\nabla u(t)|^2.$$

Therefore, for $0 < \varepsilon < T < \infty$, we obtain

$$\varphi(T) - \varphi(\varepsilon) = \int_{\varepsilon}^{T} \varphi'(t)dt = - \int_{\varepsilon}^{T} |\nabla u(t)|^2_{L^2}dt.$$

Finally we let $\varepsilon \to 0$. Since $\varphi(\varepsilon) \to \frac{1}{2}|u_0|^2$ (because $u \in C([0, \infty]; L^2(\Omega)))$, we find that $u \in L^2(0, \infty; H_0^1(\Omega))$ and that (6) holds.

If we make additional assumptions on u_0 the solution u becomes more regular up to $t = 0$ (recall that away from $t = 0$, Theorem 10.1 *always* guarantees that u is smooth, i.e., $u \in C^{\infty}(\overline{\Omega} \times [\varepsilon, \infty)) \, \forall \varepsilon > 0$).

Theorem 10.2.

(a) *If $u_0 \in H_0^1(\Omega)$ then the solution u of (1), (2), (3) satisfies*

$$u \in C([0, \infty); H_0^1(\Omega)) \cap L^2(0, \infty; H^2(\Omega))$$

and

$$\frac{\partial u}{\partial t} \in L^2(0, \infty; L^2(\Omega)).$$

Moreover, we have

(7) $$\int_0^T \left| \frac{\partial u}{\partial t}(t) \right|^2_{L^2(\Omega)} dt + \frac{1}{2}|\nabla u(T)|^2_{L^2(\Omega)} = \frac{1}{2}|\nabla u_0|^2_{L^2(\Omega)}.$$

(b) *If $u_0 \in H^2(\Omega) \cap H_0^1(\Omega)$, then*

$$u \in C([0, \infty); H^2(\Omega)) \cap L^2(0, \infty; H^3(\Omega))$$

and

$$\frac{\partial u}{\partial t} \in L^2(0, \infty; H_0^1(\Omega)).$$

(c) *If $u_0 \in H^k(\Omega) \, \forall k$ and satisfies the so-called* **compatibility conditions**

(8) $$u_0 = \Delta u_0 = \cdots = \Delta^j u_0 = \cdots = 0 \quad on \, \Gamma$$

for every integer j, then $u \in C^{\infty}(\overline{\Omega} \times [0, \infty))$.

Proof of (a). We work here in the space $H_1 = H_0^1(\Omega)$ equipped with the scalar product

$$(u, v)_{H_1} = \int_{\Omega} \nabla u \cdot \nabla v + \int_{\Omega} uv.$$

In H_1 consider the unbounded operator $A_1 : D(A_1) \subset H_1 \to H_1$ defined by

$$\begin{cases} D(A_1) = \{u \in H^3(\Omega) \cap H_0^1(\Omega);\ \Delta u \in H_0^1(\Omega)\}, \\ A_1 u = -\Delta u. \end{cases}$$

We claim that A_1 is maximal monotone and self-adjoint.

(i) A_1 *is monotone.* For every $u \in D(A_1)$ we have

$$(A_1 u, v)_{H_1} = \int_\Omega \nabla(-\Delta u) \cdot \nabla u + \int_\Omega (-\Delta u)u = \int_\Omega |\Delta u|^2 + \int_\Omega |\nabla u|^2 \geq 0.$$

(ii) A_1 *is maximal monotone.* We know (by Theorem 9.25) that for every $f \in H^1(\Omega)$ the solution $u \in H_0^1(\Omega)$ of the problem

$$\begin{cases} u - \Delta u = f & \text{in } \Omega, \\ u = 0 & \text{on } \Gamma, \end{cases}$$

belongs to $H^3(\Omega)$. If, in addition, $f \in H_0^1(\Omega)$ then $\Delta u \in H_0^1(\Omega)$, and so $u \in D(A_1)$.

(iii) A_1 *is symmetric.* For every $u, v \in D(A_1)$ we have

$$\begin{aligned} (A_1 u, v)_{H_1} &= \int_\Omega \nabla(-\Delta u) \cdot \nabla v + \int_\Omega (-\Delta u)v \\ &= \int_\Omega \Delta u\, \Delta v + \int_\Omega \nabla u \cdot \nabla v = (u, A_1 v)_{H_1}. \end{aligned}$$

Applying Theorem 7.7, we see that if $u_0 \in H_0^1(\Omega)$ there exists a solution u of (1), (2), (3) (which coincides with the one obtained in Theorem 10.1 because of uniqueness) such that

$$u \in C([0, \infty);\ H_0^1(\Omega)).$$

Finally, set $\varphi(t) = \frac{1}{2}|\nabla u(t)|^2_{L^2(\Omega)}$. This function is C^∞ on $(0, \infty)$ and

$$\varphi'(t) = \left(\nabla u(t), \nabla \frac{du}{dt}(t)\right)_{L^2} = \left(-\Delta u(t), \frac{du}{dt}(t)\right)_{L^2} = -\left|\frac{du}{dt}(t)\right|^2_{L^2}.$$

It follows that for $0 < \varepsilon < T < \infty$, we have

$$\varphi(T) - \varphi(\varepsilon) + \int_\varepsilon^T \left|\frac{du}{dt}(t)\right|^2_{L^2} dt = 0.$$

As $\varepsilon \to 0$, $\varphi(\varepsilon) \to \frac{1}{2}|\nabla u_0|^2_{L^2}$, and we conclude easily.

Proof of (b). We work here in the space $H_2 = H^2(\Omega) \cap H_0^1(\Omega)$ equipped with the scalar product

$$(u, v)_{H_2} = (\Delta u, \Delta v)_{L^2} + (u, v)_{L^2}$$

(the corresponding norm is equivalent to the usual H^2 norm; why?). In H_2 consider the unbounded operator $A_2 : D(A_2) \subset H_2 \to H_2$ defined by

$$
\begin{cases}
D(A_2) = \{u \in H^4(\Omega); \ u \in H_0^1(\Omega) \text{ and } \Delta u \in H_0^1(\Omega)\}, \\
A_2 u = -\Delta u.
\end{cases}
$$

It is easy to show that A_2 is a self-adjoint maximal monotone operator in H_2.[4] We may therefore apply Theorem 7.7 to A_2 in H_2. Finally, we set $\varphi(t) = \frac{1}{2}|\Delta u(t)|_{L^2}^2$. This function is C^∞ on $(0, \infty)$ and

$$
\varphi'(t) = \left(\Delta u(t), \ \Delta \frac{du}{dt}(t) \right)_{L^2} = (\Delta u(t), \ \Delta^2 u(t))_{L^2} = -|\nabla \Delta u(t)|_{L^2}^2.
$$

Thus, for $0 < \varepsilon < T < \infty$, we have

$$
\frac{1}{2}|\Delta u(T)|_{L^2}^2 - \frac{1}{2}|\Delta u(\varepsilon)|_{L^2}^2 + \int_\varepsilon^T |\nabla \Delta u(t)|_{L^2}^2 dt = 0.
$$

In the limit, as $\varepsilon \to 0$, we see that $u \in L^2(0, \infty; H^3(\Omega))$ (why?) and (because of equation (1)), $\frac{du}{dt} \in L^2(0, \infty; H^1(\Omega))$.

Proof of (c). In the space $H = L^2(\Omega)$, consider the operator $A : D(A) \subset H \to H$ defined by

$$
\begin{cases}
D(A) = H^2(\Omega) \cap H_0^1(\Omega), \\
Au = -\Delta u.
\end{cases}
$$

Applying Theorem 7.5, we know that if $u_0 \in D(A^k)$, $k \geq 1$, then

$$
u \in C^{k-j}([0, \infty); \ D(A^j)) \quad \forall j = 0, 1, \ldots, k.
$$

Assumption (8) says precisely that $u_0 \in D(A^k)$ for every integer $k \geq 1$. Therefore we have

$$
u \in C^{k-j}([0, \infty); \ D(A^j)) \quad \forall k \geq 1, \quad \forall j = 0, 1, \ldots, k.
$$

It follows (as in the proof of Theorem 10.1) that $u \in C^\infty(\overline{\Omega} \times [0, \infty))$.

• *Remark* 1. Theorem 10.1 shows that the heat equation has a *strong smoothing effect on the initial data* u_0. Note that the solution $u(x, t)$ is C^∞ in x for every $t > 0$ even if the initial data is discontinuous. This effect implies, in particular, that the heat equation is *time irreversible*. In general one cannot solve the problem

$$
(9) \qquad\qquad \frac{\partial u}{\partial t} - \Delta u = 0 \qquad \text{in } \Omega \times (0, T),
$$

---

[4] More generally, if $A : D(A) \subset H \to H$ is a self-adjoint maximal monotone operator one may consider the Hilbert space $\tilde{H} = D(A)$ equipped with the scalar product $(u, v)_{\tilde{H}} = (Au, Av) + (u, v)$. Then the operator $\tilde{A} : D(\tilde{A}) \subset \tilde{H} \to \tilde{H}$ defined by $D(\tilde{A}) = D(A^2)$ and $\tilde{A}u = Au$ is a self-adjoint maximal monotone operator in $\tilde{H}$.

(10) $$u = 0 \qquad \text{on } \Gamma \times (0, T),$$

with "final" data

(11) $$u(x, T) = u_T(x) \quad \text{on } \Omega.$$

We would necessarily have to assume that

$$u_T \in C^\infty(\overline{\Omega}) \quad \text{and} \quad \Delta^j u_T = 0 \text{ on } \Gamma \quad \forall j \geq 0.$$

But even with this assumption there need not be a solution of the *backward* problem (9), (10), (11). This problem should not be confused with the problem (9'), (10), (11), where

(9') $$-\frac{\partial u}{\partial t} - \Delta u = 0 \quad \text{in } \Omega \times (0, T),$$

which *always* has a unique solution for any data $u_T \in L^2(\Omega)$ (change t into $T - t$ and apply Theorem 10.1).

Remark 2. The preceding results are also true—with some slight modifications—if we replace the Dirichlet condition by the Neumann condition.

Remark 3. When Ω is *bounded*, problem (1), (2), (3) can also be solved by a *decomposition in a Hilbert basis* of $L^2(\Omega)$. For this purpose it is very convenient to choose a basis $(e_i(x))_{i \geq 1}$ of $L^2(\Omega)$ composed of *eigenfunctions* of $-\Delta$ (with zero Dirichlet condition), i.e.,

$$\begin{cases} -\Delta e_i = \lambda_i e_i & \text{in } \Omega, \\ \quad e_i = 0 & \text{on } \Gamma \end{cases}$$

(see Section 9.8). We seek a solution u of (1), (2), (3) in the form of a series [5]

(12) $$u(x, t) = \sum_{i=1}^{\infty} a_i(t) e_i(x).$$

We see immediately that the functions $a_i(t)$ must satisfy

$$a_i'(t) + \lambda_i a_i(t) = 0,$$

so that $a_i(t) = a_i(0) e^{-\lambda_i t}$. The constants $a_i(0)$ are determined by the relation

(13) $$u_0(x) = \sum_{i=1}^{\infty} a_i(0) e_i(x).$$

In other words, the solution u of (1), (2), (3) is given by

---

[5] For obvious reasons this method is also called the method of "*separation of variables*," or Fourier method. In fact, Fourier discovered the Fourier series while studying the heat equation in one space variable.

$$(14) \qquad u(x,t) = \sum_{i=1}^{\infty} a_i(0)e^{-\lambda_i t}e_i(x),$$

where the constants $a_i(0)$ are the components of $u_0(x)$ in the basis (e_i), i.e., $a_i(0) = \int_\Omega u_0 e_i$.

For the study of the convergence of this series (and also the regularity of u obtained in this way) we refer to H. Weinberger [1]. Note the analogy between this method and the standard technique used in solving the *linear system of differential equations*

$$\frac{d\mathbf{u}}{dt} + M\mathbf{u} = 0,$$

where $\mathbf{u}(t)$ takes its values in a finite-dimensional vector space, and M is a symmetric matrix. Of course, the main difference comes from the fact that problem (1), (2), (3) is associated with an *infinite-dimensional system*.

Remark 4. The *compatibility conditions* (8) look perhaps mysterious, but in fact they are natural. These are *necessary conditions* in order to have a solution u of (1), (2), (3) that is smooth up to $t = 0$, i.e., $u \in C^\infty(\overline{\Omega} \times [0, \infty))$ (the assumption $u_0 \in C^\infty(\overline{\Omega})$ with $u_0 = 0$ on $\partial\Omega$ does not guarantee smoothness up to $t = 0$). Indeed, suppose $u \in C^\infty(\overline{\Omega} \times [0, \infty))$ satisfies (1), (2), (3). Then clearly,

$$(15) \qquad \frac{\partial^j u}{\partial t^j} = 0 \quad \text{on } \Gamma \times (0, \infty) \quad \forall j,$$

and by continuity, we also have

$$\frac{\partial^j u}{\partial t^j} = 0 \quad \text{on } \Gamma \times [0, \infty) \quad \forall j.$$

On the other hand,

$$\frac{\partial^2 u}{\partial t^2} = \Delta\left(\frac{\partial u}{\partial t}\right) = \Delta^2 u \quad \text{in } Q,$$

and by induction,

$$\frac{\partial^j u}{\partial t^j} = \Delta^j u \quad \text{in } Q \quad \forall j.$$

By continuity once more we have

$$(16) \qquad \frac{\partial^j u}{\partial t^j} = \Delta^j u \quad \text{in } \overline{\Omega} \times [0, \infty).$$

Comparing (15) and (16) on $\Gamma \times \{0\}$, we obtain (8).

Remark 5. Of course, there are many variants of the regularity results for u near $t = 0$ if we make assumptions that are *intermediate* between the cases (b) and (c) of Theorem 10.2.

10.2 The Maximum Principle

The main result is the following.

• **Theorem 10.3.** *Assume $u_0 \in L^2(\Omega)$ and let u be the solution of (1), (2), (3). Then we have, for all $(x, t) \in Q$,*

$$\min\left\{0, \inf_\Omega u_0\right\} \leq u(x, t) \leq \max\left\{0, \sup_\Omega u_0\right\}.$$

Proof. As in the elliptic case we use Stampacchia's truncation method. Set

$$K = \max\left\{0, \sup_\Omega u_0\right\}$$

and assume that $K < +\infty$. Fix a function G as in the proof of Theorem 9.27 and let

$$H(s) = \int_0^s G(\sigma)d\sigma, \quad s \in \mathbb{R}.$$

It easily checked that the function φ defined by

$$\varphi(t) = \int_\Omega H(u(x, t) - K)dx$$

has the following properties:

(17) $\varphi \in C([0, \infty); \mathbb{R}), \quad \varphi(0) = 0, \quad \varphi \geq 0 \quad \text{on } [0, \infty),$

(18) $\varphi \in C^1((0, \infty); \mathbb{R}),$

and

$$\varphi'(t) = \int_\Omega G(u(x, t) - K)\frac{\partial u}{\partial t}(x, t)dx = \int_\Omega G(u(x, t) - K)\Delta u(x, t)dx$$

$$= -\int_\Omega G'(u - K)|\nabla u|^2 dx \leq 0,$$

since $G(u(x, t) - K) \in H_0^1(\Omega)$ for every $t > 0$. It follows that $\varphi \equiv 0$ and thus, for every $t > 0$, $u(x, t) \leq K$ a.e. on Ω.

Corollary 10.4. *Let $u_0 \in L^2(\Omega)$. The solution u of (1), (2), (3) has the following properties:*

(i) *If $u_0 \geq 0$ a.e. on Ω, then $u \geq 0$ in Q.*
(ii) *If $u_0 \in L^\infty(\Omega)$, then $u \in L^\infty(Q)$ and*

(19) $\|u\|_{L^\infty(Q)} \leq \|u_0\|_{L^\infty(\Omega)}.$

Corollary 10.5. *Let $u_0 \in C(\overline{\Omega}) \cap L^2(\Omega)$ with $u_0 = 0$ on Γ.[6] Then the solution u of (1), (2), (3) belongs to $C(\overline{Q})$.*

Proof of Corollary 10.5. Let (u_{0n}) be a sequence of functions in $C_c^\infty(\Omega)$ such that $u_{0n} \to u_0$ in $L^\infty(\Omega)$ and in $L^2(\Omega)$ (the existence of such a sequence is easily established). By Theorem 10.2 the solution u_n of (1), (2), (3) corresponding to the initial data u_{0n} belongs to $C^\infty(\overline{Q})$. On the other hand (Theorem 7.7), we know that

$$|u_n(t) - u(t)|_{L^2(\Omega)} \le |u_{0n} - u_0|_{L^2(\Omega)} \quad \forall t \ge 0.$$

Because of (19) we have

$$\|u_n - u_m\|_{L^\infty(Q)} \le \|u_{0n} - u_{0m}\|_{L^\infty(\Omega)}.$$

Therefore, the sequence (u_n) converges to u uniformly on $\overline{Q}$, and so $u \in C(\overline{Q})$.

As in the elliptic case, there is another approach to the maximum principle. For simplicity we assume here that Ω is bounded. Let $u(x, t)$ be a function satisfying[7]

(20) $$u \in C(\overline{\Omega} \times [0, T]),$$

(21) $\quad u$ is of class C^1 in t and of class C^2 in x in $\Omega \times (0, T)$,

(22) $$\frac{\partial u}{\partial t} - \Delta u \le 0 \quad \text{in } \Omega \times (0, T).$$

Theorem 10.6. *Assume (20), (21), and (22). Then*

(23) $$\max_{\overline{\Omega} \times [0,T]} u = \max_P u,$$

*where $P = (\overline{\Omega} \times \{0\}) \cup (\Gamma \times [0, T])$ is called the "**parabolic boundary**" of the cylinder $\Omega \times (0, T)$.*

Proof. Set $v(x, t) = u(x, t) + \varepsilon|x|^2$ with $\varepsilon > 0$, so that

(24) $$\frac{\partial v}{\partial t} - \Delta v \le -2\varepsilon N < 0 \quad \text{in } \Omega \times (0, T).$$

We claim that

$$\max_{\overline{\Omega} \times [0,T]} v = \max_P v.$$

Suppose not. Then there is some point $(x_0, t_0) \in \overline{\Omega} \times [0, T]$ such that $(x_0, t_0) \notin P$ and

$$\max_{\overline{\Omega} \times [0,T]} v = v(x_0, t_0).$$

Since $x_0 \in \Omega$ and $0 < t_0 \le T$, we have

[6] If Ω is not bounded we also assume that $u_0(x) \to 0$ as $|x| \to \infty$.
[7] Note that we do not prescribe any boundary condition or any initial data.

(25)
$$\Delta v(x_0, t_0) \leq 0$$

and

(26)
$$\frac{\partial v}{\partial t}(x_0, t_0) \geq 0 .$$

(if $t_0 < T$ we have $\frac{\partial v}{\partial t}(x_0, t_0) = 0$, and if $t_0 = T$ we have $\frac{\partial v}{\partial t}(x_0, t_0) \geq 0$).[8]
Combining (25) and (26), we obtain $\left(\frac{\partial v}{\partial t} - \Delta v\right)(x_0, t_0) \geq 0$, a contradiction with
(24). Therefore we have

$$\max_{\overline{\Omega} \times [0,T]} v = \max_{P} v \leq \max_{P} u + \varepsilon C,$$

where $C = \sup_{x \in \Omega} |x|^2$. Since $u \leq v$, we conclude that

$$\max_{\overline{\Omega} \times [0,T]} u \leq \max_{P} u + \varepsilon C \quad \forall \varepsilon > 0.$$

This completes the proof of (23).

10.3 The Wave Equation

Let $\Omega \subset \mathbb{R}^N$ be an open set. As above, we set

$$Q = \Omega \times (0, \infty) \quad \text{and} \quad \Sigma = \Gamma \times (0, \infty).$$

Consider the following problem: find a function $u(x, t) : \overline{\Omega} \times [0, \infty) \to \mathbb{R}$ satisfying

(27)
$$\boxed{\frac{\partial^2 u}{\partial t^2} - \Delta u = 0 \quad \text{in } Q,}$$

(28)
$$\boxed{u = 0 \quad \text{on } \Sigma,}$$

(29)
$$\boxed{u(x, 0) = u_0(x) \quad \text{on } \Omega,}$$

(30)
$$\boxed{\frac{\partial u}{\partial t}(x, 0) = v_0(x) \quad \text{on } \Omega,}$$

where $\Delta = \sum_{i=1}^{N} \frac{\partial^2}{\partial x_i^2}$ denotes the *Laplacian in the space variables* x, t is the *time variable*, and u_0, v_0 are given functions.

Equation (27) is called the *wave equation*. The operator $(\frac{\partial^2}{\partial t^2} - \Delta)$ is often denoted by $\square$ and is called the *d'Alembertian*. The wave equation is a typical example of a *hyperbolic equation*.

---

[8] To be safe one should work in $\Omega \times (0, T')$ with $T' < T$ and then let $T' \to T$, since v is of class C^1 in t only in $\Omega \times (0, T)$.

When $N = 1$ and $\Omega = (0, 1)$, equation (27) models the small[9] *vibrations of a string* in the absence of any exterior force. For each t, the graph of the function $x \in \Omega \mapsto u(x, t)$ represents the configuration of the string at time t. When $N = 2$ equation (27) models the small *vibrations of an elastic membrane*. For each t, the graph of the function $x \in \Omega \mapsto u(x, t)$ represents the configuration of the membrane at time t. More generally, equation (27) models the *propagation of a wave* (acoustic, electromagnetic, etc.) in some homogeneous elastic medium $\Omega \subset \mathbb{R}^N$.

Equation (28) is the (homogeneous) *Dirichlet boundary condition*; it could be replaced by the Neumann condition or any of the boundary conditions encountered in Chapter 8 or 9. The condition $u = 0$ on Σ means that the string (or the membrane) is *fixed* on Γ, while the Neumann condition says that the string is free at its endpoints.

Equations (29) and (30) represent the initial state of the system: the *initial configuration* (one also says initial displacement) is described by u_0, and the *initial velocity* is described by v_0. The data (u_0, v_0) are usually called the *Cauchy data*.

To simplify matters we assume throughout this section that Ω is of class C^∞, with Γ bounded.

• **Theorem 10.7 (existence and uniqueness).** *Assume* $u_0 \in H^2(\Omega) \cap H_0^1(\Omega)$ *and* $v_0 \in H_0^1(\Omega)$. *Then there exists a unique solution u of* (27), (28), (29), (30) *satisfying*

(31) $\quad u \in C([0, \infty); H^2(\Omega) \cap H_0^1(\Omega)) \cap C^1([0, \infty); H_0^1(\Omega)) \cap C^2([0, \infty); L^2(\Omega))$.

Moreover,[10]

$$(32) \quad \left| \frac{\partial u}{\partial t}(t) \right|^2_{L^2(\Omega)} + |\nabla u(t)|^2_{L^2(\Omega)} = |v_0|^2_{L^2(\Omega)} + |\nabla u_0|^2_{L^2(\Omega)} \quad \forall t \geq 0.$$

Remark 6. Equation (32) is a *conservation law* that asserts that the energy of the system is *invariant* in time.

Before proving Theorem 10.7 let us mention a regularity result.

Theorem 10.8 (regularity). *Assume that the initial data satisfy*

$$u_0 \in H^k(\Omega), \quad v_0 \in H^k(\Omega) \quad \forall k,$$

*and the **compatibility conditions***

$$\Delta^j u_0 = 0 \quad on \ \Gamma \quad \forall j \geq 0, \ j \ integer,$$

$$\Delta^j v_0 = 0 \quad on \ \Gamma \quad \forall j \geq 0, \ j \ integer.$$

Then the solution u of (27), (28), (29), (30) *belongs to* $C^\infty(\overline{\Omega} \times [0, \infty))$.

---

[9] The full equation is a very difficult *nonlinear* equation; equation (27) is a *linearized* version of this near an equilibrium.

[10] We use the same notation as in the preceding sections, that is, $\left| \frac{\partial u}{\partial t}(t) \right|^2_{L^2(\Omega)} = \int_\Omega \left| \frac{\partial u}{\partial t}(x, t) \right|^2 dx$, $|\nabla u(t)|^2_{L^2(\Omega)} = \sum_{i=1}^N \int_\Omega \left| \frac{\partial u}{\partial x_i}(x, t) \right|^2 dx$.

Proof of Theorem 10.7. As in Section 10.1 we consider $u(x, t)$ as a vector-valued function defined on $[0, \infty)$; more precisely, for each $t \geq 0$, $u(t)$ denotes the map $x \mapsto u(x, t)$. We write (27) in the form of a system of first-order equations:[11]

$$
(33) \qquad
\begin{cases}
\dfrac{\partial u}{\partial t} - v = 0 & \text{in } Q, \\[2mm]
\dfrac{\partial v}{\partial t} - \Delta u = 0 & \text{in } Q,
\end{cases}
$$

and we set $U = \binom{u}{v}$, so that (33) becomes

$$
(34) \qquad \frac{dU}{dt} + AU = 0,
$$

where

$$
(35) \qquad AU = \begin{pmatrix} 0 & -I \\ -\Delta & 0 \end{pmatrix} U = \begin{pmatrix} 0 & -I \\ -\Delta & 0 \end{pmatrix} \begin{pmatrix} u \\ v \end{pmatrix} = \begin{pmatrix} -v \\ -\Delta u \end{pmatrix}.
$$

We now apply the Hille–Yosida theory in the space $H = H_0^1(\Omega) \times L^2(\Omega)$ equipped with the scalar product

$$
(U_1, \ U_2) = \int_\Omega \nabla u_1 \cdot \nabla u_2 dx + \int_\Omega u_1 u_2 dx + \int_\Omega v_1 v_2 dx,
$$

where $U_1 = \binom{u_1}{v_1}$ and $U_2 = \binom{u_2}{v_2}$.

Consider the unbounded operator $A : D(A) \subset H \to H$ defined by (35) with

$$
D(A) = (H^2(\Omega) \cap H_0^1(\Omega)) \times H_0^1(\Omega).
$$

Note that the boundary condition (28) has been incorporated in the space H. The condition $v = \frac{\partial u}{\partial t} = 0$ on Σ is a direct consequence of (28).

We claim that $A + I$ is maximal monotone in H:

(i) $A + I$ is *monotone*; indeed, if $U = \binom{u}{v} \in D(A)$ we have

$$
\begin{aligned}
(AU, U)_H + |U|_H^2 \\
= -\int_\Omega \nabla v \cdot \nabla u - \int_\Omega uv + \int_\Omega (-\Delta u) v + \int_\Omega u^2 + \int_\Omega |\nabla u|^2 + \int_\Omega v^2 \\
= -\int_\Omega uv + \int_\Omega u^2 + \int_\Omega v^2 + \int_\Omega |\nabla u|^2 \geq 0.
\end{aligned}
$$

(ii) $A + I$ is *maximal monotone*. This amounts to proving that $A + 2I$ is surjective. Given $F = \binom{f}{g} \in H$, we must solve the equation $AU + 2U = F$, i.e., the system

---

[11] This is the standard device, which consists in writing a differential equation of order k as a system of k first-order equations.

$$(36) \quad \begin{cases} -v + 2u = f & \text{in } \Omega, \\ -\Delta u + 2v = g & \text{in } \Omega, \end{cases}$$

with

$$u \in H^2(\Omega) \cap H_0^1(\Omega) \quad \text{and} \quad v \in H_0^1(\Omega).$$

It follows from (36) that

$$(37) \qquad -\Delta u + 4u = 2f + g.$$

Equation (37) has a unique solution $u \in H^2(\Omega) \cap H_0^1(\Omega)$ (by Theorem 9.25). Then we obtain $v \in H_0^1(\Omega)$ simply by taking $v = 2u - f$. This solves (36).

Applying Hille–Yosida's theorem (Theorem 7.4) and Remark 7.7, we see that there exists a unique solution of the problem

$$(38) \quad \begin{cases} \dfrac{dU}{dt} + AU = 0 & \text{on } [0, \infty), \\ U(0) = U_0, \end{cases}$$

with

$$(39) \qquad U \in C^1([0, \infty); H) \cap C([0, \infty); D(A)),$$

since $U_0 = \binom{u_0}{v_0} \in D(A)$. From (39) we deduce (31).

In order to prove (32) it suffices to multiply (27) by $\frac{\partial u}{\partial t}$ and to integrate on Ω. Note that

$$\int_\Omega \frac{\partial^2 u}{\partial t^2} \frac{\partial u}{\partial t} dx = \frac{1}{2} \frac{\partial}{\partial t} \int_\Omega \left| \frac{\partial u}{\partial t}(x, t) \right|^2 dx$$

and

$$\int_\Omega (-\Delta u) \frac{\partial u}{\partial t} dx = \int_\Omega \nabla u \cdot \frac{\partial}{\partial t}(\nabla u) dx = \frac{1}{2} \frac{\partial}{\partial t} \int_\Omega |\nabla u|^2 dx.$$

Remark 7. When Ω is *bounded* we may use on $H_0^1(\Omega)$ the scalar product $\int \nabla u_1 \cdot \nabla u_2$ (see Corollary 9.19), and on $H = H_0^1(\Omega) \times L^2(\Omega)$ the scalar product

$$(U_1, U_2) = \int_\Omega \nabla u_1 \cdot \nabla u_2 + \int_\Omega v_1 v_2, \quad \text{where } U_1 = \binom{u_1}{v_1} \text{ and } U_2 = \binom{u_2}{v_2}.$$

With this scalar product we have

$$(AU, U) = -\int_\Omega \nabla v \cdot \nabla u + \int_\Omega (-\Delta u) v = 0 \quad \forall U = \binom{u}{v} \in D(A).$$

It is easy to check that:

(i) A and $-A$ are maximal monotone,
(ii) $A^* = -A$.

As a consequence we may also solve the problem

$$\frac{dU}{dt} - AU = 0 \text{ on } [0, +\infty), \quad U(0) = U_0,$$

or equivalently

$$\frac{dU}{dt} + AU = 0 \text{ on } (-\infty, 0], \quad U(0) = U_0$$

(just change t into $-t$).[12] Relation (32) may be written as

$$|U(t)|_H = |U_0|_H \quad \forall t \in \mathbb{R}.$$

One says that the one-parameter family $\{U(t)\}_{t \in \mathbb{R}}$ is a *group of isometries on H*.

• **Remark 8.** The wave equation has *no smoothing effect on the initial data*, in contrast with the heat equation. To convince oneself of this it suffices consider the case $\Omega = \mathbb{R}$. Then there is a very simple *explicit solution* of (27), (28), (29), (30), namely

$$(40) \qquad u(x, t) = \frac{1}{2}(u_0(x + t) + u_0(x - t)) + \frac{1}{2}\int_{x-t}^{x+t} v_0(s)ds.$$

In particular, if $v_0 = 0$, we have

$$u(x, t) = \frac{1}{2}(u_0(x + t) + u_0(x - t)).$$

Clearly *u is not more regular than* u_0. We can be even more precise. Assume $u_0 \in C^\infty(\mathbb{R}\backslash\{x_0\})$. Then $u(x, t)$ is C^∞ on $\mathbb{R} \times \mathbb{R}$, except on the lines $x + t = x_0$ and $x - t = x_0$. These are called the *characteristics* passing through the point $(x_0, 0)$. One says that *singularities propagate along the characteristics*.

Remark 9. When Ω is bounded, problem (27), (28), (29), (30) can be solved by *decomposition in a Hilbert basis*, as was done for the heat equation. It is very convenient to work in the basis (e_i) of $L^2(\Omega)$ composed of *eigenfunctions* of $-\Delta$ (with Dirichlet condition), i.e., $-\Delta e_i = \lambda_i e_i$ in Ω, $e_i = 0$ on Γ; recall that $\lambda_i > 0$. We seek a solution of (27), (28), (29), (30) in the form of a series

$$(41) \qquad u(x, t) = \sum_i a_i(t)e_i(x).$$

We see immediately that the functions $a_i(t)$ must satisfy

$$a_i''(t) + \lambda_i a_i(t) = 0,$$

so that

$$a_i(t) = a_i(0) \cos(\sqrt{\lambda_i}t) + \frac{a_i'(0)}{\sqrt{\lambda_i}} \sin(\sqrt{\lambda_i}t).$$

---

[12] In other words, time is *reversible*; from this viewpoint there is a basic difference between the wave equation and the heat equation (for which time is not reversible).

The constants $a_i(0)$ and $a'_i(0)$ are determined by the relations

$$u_0(x) = \sum_i a_i(0)e_i(x) \quad \text{and} \quad v_0(x) = \sum_i a'_i(0)e_i(x).$$

In other words, $a_i(0)$ and $a'_i(0)$ are the components of u_0 and v_0 in the basis (e_i). For the study of the convergence of this series see, e.g., H. Weinberger [1].

Proof of Theorem 10.8. We use the same notation as in the proof of Theorem 10.7. It is easy to see, by induction on k, that

$$D(A^k) = \left\{ \begin{pmatrix} u \\ v \end{pmatrix} \middle| \begin{array}{l} u \in H^{k+1}(\Omega) \text{ and } \Delta^j u = 0 \text{ on } \Gamma \ \forall j, \ 0 \le j \le [k/2] \\ v \in H^k(\Omega) \text{ and } \Delta^j v = 0 \text{ on } \Gamma \ \forall j, \ 0 \le j \le [(k+1)/2] - 1 \end{array} \right\}.$$

In particular, $D(A^k) \subset H^{k+1}(\Omega) \times H^k(\Omega)$ with continuous injection. Applying Theorem 7.5, we see that if $U_0 = \begin{pmatrix} u_0 \\ v_0 \end{pmatrix} \in D(A^k)$, then the solution U of (38) satisfies

$$U \in C^{k-j}([0, \infty); D(A^j)) \quad \forall j = 0, 1, \ldots, k.$$

Thus $u \in C^{k-j}([0, \infty); H^{j+1}(\Omega)) \ \forall j = 0, 1, \ldots, k$. We conclude with the help of Corollary 9.15 that under the assumptions of Theorem 10.8 (i.e., $U_0 \in D(A^k) \ \forall k$), $u \in C^k(\overline{\Omega} \times [0, \infty)) \ \forall k$.

Remark 10. The compatibility conditions introduced in Theorem 10.8 are *necessary and sufficient* in order to have a solution $u \in C^\infty(\overline{\Omega} \times [0, \infty))$ of the problem (27), (28), (29), (30). The proof is the same as in Remark 4.

Remark 11. The techniques presented in Section 10.3 may also be used for solving the *Klein–Gordon equation*

$$(27') \qquad\qquad \frac{\partial^2 u}{\partial t^2} - \Delta u + m^2 u = 0 \text{ in } Q, \quad m > 0.$$

Note that $(27')$ *cannot be reduced* to (27) by a change of unknown such as $v(x, t) = e^{\lambda t} u(x, t)$.

Comments on Chapter 10

Comments on the heat equation

1. The approach of J.-L. Lions.

The following result allows us to prove, in a very general framework, the existence and uniqueness of a *weak solution* for parabolic problems. *This theorem can be viewed as a parabolic counterpart of the Lax–Milgram theorem.* Let H be a Hilbert space with scalar product $(\,,\,)$ and norm $|\,|$. The dual space H^* is identified with H.

Let V be another Hilbert space with norm $\| \ \|$. We assume that $V \subset H$ with dense and continuous injection, so that

$$V \subset H \subset V^\star$$

(see Remark 5.1).

Let $T > 0$ be fixed; for a.e. $t \in [0, T]$ we are given a bilinear form $a(t; u, v)$: $V \times V \to \mathbb{R}$ satisfying the following properties:

(i) For every $u, v \in V$ the function $t \mapsto a(t; u, v)$ is measurable,
(ii) $|a(t; u, v)| \le M\|u\|\|v\|$ for a.e. $t \in [0, T]$, $\forall u, v \in V$,
(iii) $a(t; v, v) \ge \alpha\|v\|^2 - C|v|^2$ for a.e. $t \in [0, T]$, $\forall v \in V$,

where $\alpha > 0$, M and C are constants.

Theorem 10.9 (J.-L. Lions). *Given $f \in L^2(0, T; V^\star)$ and $u_0 \in H$, there exists a unique function u satisfying*

$$u \in L^2(0, T; V) \cap C([0, T]; H), \quad \frac{du}{dt} \in L^2(0, T; V^\star)$$

$$\left\langle \frac{du}{dt}(t), v \right\rangle + a(t; u(t), v) = \langle f(t), v \rangle \quad \text{for a.e. } t \in (0, T), \quad \forall v \in V,$$

and

$$u(0) = u_0.$$

For a proof see, e.g., J.-L. Lions–E. Magenes [1].
Application. $H = L^2(\Omega)$, $V = H_0^1(\Omega)$ and

$$a(t; u, v) = \sum_{i,j} \int_\Omega a_{ij}(x, t) \frac{\partial u}{\partial x_i} \frac{\partial v}{\partial x_j} dx + \sum_i \int_\Omega a_i(x, t) \frac{\partial u}{\partial x_i} v + \int_\Omega a_0(x, t) uv \, dx$$

with $a_{ij}, a_i, a_0 \in L^\infty(\Omega \times (0, T))$ and

$$(42) \quad \sum_{i,j} a_{ij}(x, t)\xi_i\xi_j \ge \alpha|\xi|^2 \quad \text{for a.e. } (x, t) \in \Omega \times (0, T), \quad \forall \xi \in \mathbb{R}^N, \ \alpha > 0.$$

In this way we obtain a weak solution of the problem

$$(43) \quad \begin{cases} \dfrac{\partial u}{\partial t} - \displaystyle\sum_{i,j} \frac{\partial}{\partial x_j}\left(a_{ij}\frac{\partial u}{\partial x_i}\right) + \sum_i a_i \frac{\partial u}{\partial x_i} + a_0 u = f & \text{in } \Omega \times (0, T), \\[2mm] \hspace{4.5cm} u = 0 & \text{on } \Gamma \times (0, T), \\[2mm] \hspace{3.5cm} u(x, 0) = u_0(x) & \text{on } \Omega. \end{cases}$$

Under additional assumptions on the data, the solution of (43) has greater regularity; see the following comments.

2. C^∞- regularity.

We assume here that Ω is bounded and of class C^∞. Let $a_{ij}, a_i, a_0 \in C^\infty(\overline{\Omega} \times [0, T])$ satisfy (42).

Theorem 10.10. *Assume $u_0 \in L^2(\Omega)$ and $f \in C^\infty(\overline{\Omega} \times [0, T])$. Then the solution u of (43) belongs to $C^\infty(\overline{\Omega} \times [\varepsilon, T])$ for every $\varepsilon > 0$. If in addition $u_0 \in C^\infty(\overline{\Omega})$ and $\{f, u_0\}$ satisfy the appropriate* **compatibility conditions**[13] *on $\Gamma \times \{0\}$, then $u \in C^\infty(\overline{\Omega} \times [0, T])$.*

For a proof, see, e.g., J.-L. Lions–E. Magenes [1], A. Friedman [1], [2], and O. Ladyzhenskaya–V. Solonnikov–N. Uraltseva [1]; it is based on estimates very similar to those presented in Chapter 7 and in Section 10.1.

Let us mention that there is also an *abstract theory that extends the Hille–Yosida theory* to problems of the form $\frac{du}{dt}(t) + A(t)u(t) = f(t)$, where for each t, $A(t)$ is a maximal monotone operator. This theory has been developed by T. Kato, H. Tanabe, P. E. Sobolevski, and others. It is technically more complicated to handle than the Hille–Yosida theory; see A. Friedman [2], H. Tanabe [1], and K. Yosida [1].

3. L^p and $C^{0,\alpha}$-regularity.

Consider the problem[14]

$$
(44) \qquad
\begin{cases}
\dfrac{\partial u}{\partial t} - \Delta u = f & \text{in } \Omega \times (0, T), \\[2mm]
u = 0 & \text{on } \Gamma \times (0, T), \\[2mm]
u(x, 0) = u_0(x) & \text{on } \Omega.
\end{cases}
$$

Assume, for convenience, that Ω is bounded and of class C^∞. Let us start with a simple result.

Theorem 10.11 (L^2-regularity). *Given $f \in L^2(\Omega \times (0, T))$ and $u_0 \in H_0^1(\Omega)$, there is a unique solution of (44) satisfying*

$$
u \in C([0, T]; \; H_0^1(\Omega)) \cap L^2(0, T; \; H^2(\Omega) \cap H_0^1(\Omega))
$$

and

$$
\frac{\partial u}{\partial t} \in L^2(0, T; \; L^2(\Omega)).
$$

The proof is easy; see, e.g., J.-L. Lions–E. Magenes [1]. More generally, in L^p spaces, we have the following.

Theorem 10.12 (L^p-regularity). *Given $f \in L^p(\Omega \times (0, T))$ with $1 < p < \infty$ and $u_0 = 0$,[15] there exists a unique solution of (44) satisfying*

---

[13] We do not write down explicitly these relations; they are the natural extensions of (8) (see also Remark 4).

[14] Of course, we could also prescribe an inhomogeneous Dirichlet condition $u(x, t) = g(x, t)$ on $\Gamma \times (0, T)$, but for simplicity we deal only with the case $g = 0$.

[15] To simplify matters.

$$u, \frac{\partial u}{\partial t}, \frac{\partial u}{\partial x_i}, \frac{\partial^2 u}{\partial x_i \partial x_j} \in L^p(\Omega \times (0, T)) \quad \forall i, j.$$

Theorem 10.13 (Hölder regularity). *Let* $0 < \alpha < 1$. *Assume that* [16] $f \in C^{\alpha, \alpha/2}(\overline{\Omega} \times [0, T])$ *and* $u_0 \in C^{2+\alpha}(\overline{\Omega})$ *satisfy the natural compatibility conditions*

$$u_0 = 0 \text{ on } \Gamma \quad \text{and} \quad -\Delta u_0 = f(x, 0) \text{ on } \Gamma.$$

Then (44) *has a unique solution u such that*

$$u, \frac{\partial u}{\partial t}, \frac{\partial u}{\partial x_i}, \frac{\partial^2 u}{\partial x_i \partial x_j} \in C^{\alpha, \alpha/2}(\overline{\Omega} \times [0, T]) \quad \forall i, j.$$

The proofs of Theorems 10.12 and 10.13 are delicate, except for the case $p = 2$ of Theorem 10.12. As in the elliptic case (see the comments at the end of Chapter 9) they rely on the following:

(i) an *explicit representation formula* for u involving the fundamental solution of $\frac{\partial}{\partial t} - \Delta$. For example, if $\Omega = \mathbb{R}^N$ and $f = 0$ then

$$(45) \qquad u(x, t) = \int_{\mathbb{R}^N} E(x - y, t) u_0(y) dy = E \star u_0,$$

where $\star$ refers to *convolution solely in the space variables* x, and E is the *heat kernel*, $E(x, t) = (4\pi t)^{-N/2} e^{-|x|^2/4t}$; see, e.g., G. Folland [1].

(ii) a technique of *singular integrals*.

On this topic see, e.g., O. Ladyzhenskaya–V. Solonnikov–N. Uraltseva [1], A. Friedman [1], N. Krylov [1], [2], P. Grisvard [1] (Section 9), D. Stroock–S. Varadhan [1]. A. Brandt [2], B. Knerr [1], and L. Simon [2] have devised more elementary arguments for the Hölder regularity.

The general "philosophy" to keep in mind is the following: if u is the solution of (44) with $u_0 = 0$ then $\frac{\partial u}{\partial t}$ and Δu both have the same regularity as f.

Finally, we mention that the conclusions of Theorems 10.11, 10.12, and 10.13 still hold if Δ is replaced by

$$\sum_{i,j} \frac{\partial}{\partial x_j} \left(a_{ij}(x, t) \frac{\partial u}{\partial x_i} \right) + \sum_i a_i(x, t) \frac{\partial u}{\partial x_i} + a_0(x, t) u$$

with smooth coefficients such that

$$(46) \qquad \sum_{i,j} a_{ij}(x, t) \xi_i \xi_j \geq \nu |\xi|^2 \quad \forall x, t, \quad \forall \xi \in \mathbb{R}^N, \quad \nu > 0.$$

---

[16] That is, $|f(x_1, t_1) - f(x_2, t_2)| \leq C(|x_1 - x_2|^2 + |t_1 - t_2|)^{\alpha/2} \ \forall x_1, x_2, t_1, t_2.$

In the case of *irregular coefficients* (i.e., $a_{ij} \in L^\infty(\Omega \times (0, T))$ satisfying (46)) a difficult result of Nash–Moser asserts that there exists some $\alpha > 0$ such that $u \in C^{\alpha, \alpha/2}(\overline{\Omega} \times [0, T])$; see, e.g., O. Ladyzhenskaya–V. Solonnikov–N. Uraltseva [1].

4. Some examples of parabolic equations.

Linear and *nonlinear* parabolic equations (and systems) occur in many fields: mechanics, physics, chemistry, biology, optimal control, probability, finance, image processing etc. Let us mention some examples:

(i) *The Navier–Stokes system*:

$$(47) \quad \frac{\partial u_i}{\partial t} - \Delta u_i + \sum_j u_j \frac{\partial u_i}{\partial x_j} = f_i + \frac{\partial p}{\partial x_i} \quad \text{in } \Omega \times (0, T), 1 \le i \le N,$$

$$(48) \quad \operatorname{div} u = \sum_{i=1}^{N} \frac{\partial u_i}{\partial x_i} = 0 \qquad \text{on } \Omega \times (0, T),$$

$$(49) \quad u = 0 \qquad \text{on } \Gamma \times (0, T),$$

$$(50) \quad u(x, 0) = u_0(x) \qquad \text{on } \Omega,$$

plays a central role in fluid mechanics; see, e.g., R. Temam [1] and its references.

(ii) *Reaction–diffusion systems*. These are nonlinear parabolic equations or systems of the form

$$\begin{cases} \dfrac{\partial \mathbf{u}}{\partial t} - M\Delta \mathbf{u} = f(\mathbf{u}) & \text{in } \Omega \times (0, T) \\ + \text{ boundary conditions and initial data,} \end{cases}$$

where $\mathbf{u}(x, t)$ takes its values in $\mathbb{R}^m$, M is an $m \times m$ (diagonal) matrix, and f is a nonlinear map from $\mathbb{R}^m$ into $\mathbb{R}^m$. These systems are used to model phenomena occurring in various fields: chemistry, biology, neurophysiology, epidemiology, combustion, population genetics, ecology, geology, etc.; see, e.g., P. Fife [1] and its numerous references. The solutions of reaction–diffusion equations display a wide range of behaviors, including the formation of traveling waves and self-organized patterns.

(iii) *Free boundary problems*. For example, the *Stefan problem* describes the evolution of a mixture of ice and water; see, e.g., the expository paper of E. Magenes [1] and the book of A. Friedman [4].

(iv) *Diffusion* equations play a central role in *probability* (Brownian motion, Markov processes, diffusion processes, stochastic differential equations, etc.); see, e.g., D. Stroock–S. Varadhan [1].

(v) Many other examples of semilinear parabolic problems are presented in D. D. Henry [1], Th. Cazenave–A. Haraux [1].

(vi) An interesting use of the heat equation has been made in connection with the Atiyah–Singer index; see, e.g., P. Gilkey [1].

(vii) More sophisticated nonlinear diffusion equations are used in image processing (variants of the Perona–Malik model). The recent solution by G. Perelman of the celebrated Poincaré conjecture relies on R. Hamilton's careful study of the Ricci flow, which is a kind of nonlinear heat equation.

5. For further results concerning the *maximum principle for parabolic equations*, see, e.g., A. Friedman [1], M. Protter–H. Weinberger-[1], R. Sperb [1]. For example, if u is the solution of (1), (2), (3) with $u_0 \geq 0$ and $u_0 \not\equiv 0$, then $u(x, t) > 0$ $\forall x \in \Omega, \forall t > 0$. When $\Omega = \mathbb{R}^N$ this follows easily from the explicit representation formula (45).

Comments on the wave equation

6. Weak solutions of the wave equation.
There is a general abstract setting for the existence and uniqueness of a weak solution of the wave equation. Let V and H be two Hilbert spaces such that $V \subset H \subset V^*$ (as in Comment 1). For each $t \in [0, T]$ we are given a symmetric continuous bilinear form $a(t; u, v) : V \times V \to \mathbb{R}$ such that

(i) the function $t \mapsto a(t; u, v)$ is of class C^1 $\forall u, v \in V$,
(ii) $a(t; v, v) \geq \alpha \|v\|^2 - C|v|^2$ $\forall t \in [0, T]$, $\forall v \in V, \alpha > 0$.

Theorem 10.14 (J.-L. Lions). *Given $f \in L^2(0, T; H)$, $u_0 \in V$, and $v_0 \in H$, there exists a unique function u satisfying*

$$u \in C([0, T]; V), \quad \frac{du}{dt} \in C([0, T]; H), \quad \frac{d^2u}{dt^2} \in L^2(0, T; V^*),$$

$$\left\langle \frac{d^2u}{dt^2}(t), v \right\rangle + a(t; u(t), v) = \langle f(t), v \rangle \quad \text{for a.e. } t \in (0, T), \quad \forall v \in V,$$

$$u(0) = u_0 \quad \text{and} \quad \frac{du}{dt}(0) = v_0.$$

For a proof, see, e.g., J.-L. Lions–E. Magenes [1].

Application. Let $H = L^2(\Omega)$, $V = H_0^1(\Omega)$,

$$a(t; u, v) = \int_\Omega \sum_{i,j} a_{ij}(x, t) \frac{\partial u}{\partial x_i} \frac{\partial v}{\partial x_j} dx + \int_\Omega a_0(x, t) uv dx$$

with (42) and

$$a_{ij}, \frac{\partial a_{ij}}{\partial t}, a_0, \frac{\partial a_0}{\partial t} \in L^\infty(\Omega \times (0, T)), \quad a_{ij} = a_{ji} \quad \forall i, j.$$

Then there is a unique weak solution of the problem

$$\begin{cases} \dfrac{\partial^2 u}{\partial t^2} - \sum_{i,j} \dfrac{\partial}{\partial x_j}\left(a_{ij}\dfrac{\partial u}{\partial x_i}\right) + a_0 u = f \quad \text{in } \Omega \times (0, T), \\ (28), (29), (30). \end{cases}$$

Note that *the assumptions on the initial data* ($u_0 \in H_0^1(\Omega)$ and $v_0 \in L^2(\Omega)$) *are weaker than those made in Theorem* 10.7. Under additional assumptions on f, u_0, and v_0 (regularity *and* compatibility conditions) as well as on a_{ij}, a_0 one gains regularity on u.

7. The L^p-theory for the wave equation is delicate and had been extensively studied over the past 30 years. The *Strichartz estimates* are an important tool; see, e.g., S. Klainerman [1].

8. Maximum principle.

Some *very special forms* of the maximum principle hold for the wave equation; see, e.g., M. Protter–H. Weinberger [1]. For example, let u be the solution of (27), (28), (29), (30).

(i) If $\Omega = \mathbb{R}$, $u_0 \geq 0$ and $v_0 \geq 0$, then $u \geq 0$.
(ii) If $\Omega = \mathbb{R}^2$, $u_0 = 0$ and $v_0 \geq 0$, then $u \geq 0$.

Assertion (i) follows from the representation formula (40). A similar but *more complicated* formula holds in $\mathbb{R}^N$; see, e.g., S. Mizohata [1], G. Folland [1], H. Weinberger [1], R. Courant–D. Hilbert [1], and S. Mikhlin [1]. It implies (ii).

However, the reader is warned of the following:

(iii) If $\Omega = (0, 1)$, $u_0 \geq 0$, and $v_0 = 0$, then in general one *cannot* infer that $u \geq 0$.
(iv) If $\Omega = \mathbb{R}^2$, $u_0 \geq 0$, and $v_0 = 0$, then in general one *cannot* infer that $u \geq 0$.

An unusual form of maximum principle for the telegraph equation (which resembles the wave equation) has recently been devised by J. Mawhin–R. Ortega–A. M. Robles–Perez [1].

9. Domain of dependence. Wave propagation. Huygens' principle.

There is a *fundamental difference* between the heat equation and the wave equation:

(i) For *the heat equation, a small perturbation*[17] *of the initial data is immediately felt everywhere*, i.e., $\forall x \in \Omega$, $\forall t > 0$. For example, we have seen that if $u_0 \geq 0$ and $u_0 \not\equiv 0$, then $u(x, t) > 0$ $\forall x \in \Omega$, $\forall t > 0$. One says that *the heat propagates at infinite speed.* [18]

(ii) For the *wave equation*, the situation is *completely different*. Assume for example $\Omega = \mathbb{R}$. The explicit formula (40) shows that $u(\bar{x}, \bar{t})$ depends *solely* on the values of u_0 and v_0 in the interval $[\bar{x} - \bar{t}, \bar{x} + \bar{t}]$; see Figure 8.

One says that the interval $[\bar{x} - \bar{t}, \bar{x} + \bar{t}]$ on the x-axis is the *domain of dependence* of the point $(\bar{x}, \bar{t})$. The same holds for $\Omega = \mathbb{R}^N (N \geq 2) : u(\bar{x}, \bar{t})$ depends only on the values of u_0 and v_0 in the *ball* $\{x \in \mathbb{R}^N; |x - \bar{x}| \leq \bar{t}\}$. This ball in the hyperplane $\mathbb{R}^N \times \{0\}$ is called the *domain of dependence* of the point $(\bar{x}, \bar{t})$. Geometrically it is the intersection of the cone

---

[17] That is, localized in a small region.

[18] Physically this is not realistic! However, the representation formula (45) shows that a perturbation on the initial data localized near $x = x_0$ has *negligible effects* at the point (x, t) if t is small and $|x - x_0|$ is large.

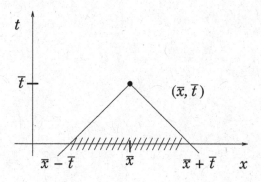

Fig. 8

$$\{(x, t) \in \mathbb{R}^N \times \mathbb{R}; \ |x - \bar{x}| \leq \bar{t} - t \text{ and } t \leq \bar{t}\}$$

with the hyperplane $\mathbb{R}^N \times \{0\}$. The physical interpretation is that waves propagate at speed less than 1.[19] A signal localized in the domain[20] D at time $t = 0$ is felt at the point $x \in \mathbb{R}^N$ only *after time* $t \geq \mathrm{dist}(x, D)$ ($u(x, t) = 0$ for $t < \mathrm{dist}(x, D)$).

When $N > 1$ is *odd*, for example $N = 3$, there is an even more striking effect: $u(\bar{x}, \bar{t})$ depends only on the values of u_0 and v_0[21] on the *sphere* $\{x \in \mathbb{R}^N; |x - \bar{x}| = t\}$. This is *Huygens' principle*. Physically, it says that a signal localized in the domain D at time $t = 0$ is observed at the point $x \in \mathbb{R}^N$ only during the time $[t_1, t_2]$ with $t_1 = \inf_{y \in D} \mathrm{dist}(x, y)$ and $t_2 = \sup_{y \in D} \mathrm{dist}(x, y)$. After the time t_2 the signal is not felt at the point x.

On the other hand, if the dimension N is *even* (for example $N = 2$) the signal persists at x for all time $t > t_1$.[22]

An application to music. A listener placed in $\mathbb{R}^3$ at distance d from a musical intrument[23] hears at time t the note played at time $(t - d)$ and nothing else![24] For more details on Huygens' principle the reader may consult R. Courant–D. Hilbert [1], G. Folland [1], P. Garabedian [1], and S. Mikhlin [1].

---

[19] The speed 1 comes in because we have normalized the wave equation. Some readers may prefer to work with the equation $\frac{\partial^2 u}{\partial t^2} - c^2 \Delta u = 0$.

[20] That is, u_0 and v_0 have their supports in D.

[21] And of some of their derivatives.

[22] The effect is damped out with time but it does not vanish completely.

[23] Of small dimension.

[24] While in $\mathbb{R}^2$ he would hear a weighted average of all notes played during the time $[0, t - d]$.

Chapter 11
Miscellaneous Complements

This chapter contains various complements that have not been incorporated in the main body of the book in order to keep the presentation more compact. They are connected to Chapters 1–7. Some of the proofs are very sketchy. Several proofs have been omitted, and the interested reader is invited to consult the references.

11.1 Finite-Dimensional and Finite-Codimensional Spaces

As is well known, every finite-dimensional space X of dimension p is isomorphic to $\mathbb{R}^p$. In particular, X is complete, all norms on X are equivalent, and the closed unit ball B_X is compact.

Proposition 11.1. *Let E be a Banach space and let $X \subset E$ be a finite-dimensional space. Then X is closed.*

Proof. Assume that (x_n) is a sequence in X such that $x_n \to x$ in E. Then (x_n) is a Cauchy sequence in X and thus (x_n) converges to a limit in X. Hence $x \in X$.

Proposition 11.2. *Assume that X is finite-dimensional and F is a Banach space. Then every linear operator $T : X \to F$ must be bounded.*

Proof. Let (e_i) be a basis in X and write $x = \sum_{i=1}^{p} x_i e_i$. Then $Tx = \sum_{i=1}^{p} x_i T e_i$, so that $\|Tx\| \leq \sum_{i=1}^{p} |x_i| \, \|T e_i\| \leq (\max_i \|T e_i\|) \sum_{i=1}^{p} |x_i| \leq C \|x\|$.

In particular, all linear functionals on X are continuous. The dual space $X^\star$ of a finite-dimensional space X is also finite-dimensional, and $\dim X^\star = \dim X$. More precisely, if (e_i) is a basis of X then write $x \in E$ as $x = \sum_{i=1}^{p} x_i e_i$ and set $f_i(x) = x_i$, $i = 1, 2, \ldots, p$. Clearly the functionals (f_i) are linearly independent in $X^\star$ and they generate $X^\star$. Thus they form a basis of $X^\star$. What is less obvious is the following:

Proposition 11.3. *Assume that X is a Banach space (with $\dim X \leq \infty$) such that $X^\star$ is finite-dimensional. Then X is finite-dimensional and $\dim X = \dim X^\star$.*

H. Brezis, *Functional Analysis, Sobolev Spaces and Partial Differential Equations,*
DOI 10.1007/978-0-387-70914-7_11, © Springer Science+Business Media, LLC 2011

Proof. We need Hahn–Banach, or more precisely Corollary 1.4. Let $J : X \to X^{**}$ be the canonical injection defined in Section 1.3. Since $\dim X^* < \infty$, we deduce from the above discussion that $\dim X^{**} < \infty$. But X is isomorphic to $J(X) \subset X^{**}$, and thus $\dim X \le \dim X^{**} = \dim X^*$. Therefore $\dim X < \infty$, and we deduce (again from the above discussion) that $\dim X^* = \dim X$.

Proposition 11.4. *Let E be a Banach space and let $M \subset E$ be a closed subspace. Assume that $X \subset E$ is a finite-dimensional subspace. Then $(M + X)$ is closed. Moreover, $(M + X)$ admits a complement in E if and only if M does.*

[**Warning:** Recall that in general, the sum of two closed subspaces need not be closed; see, e.g., Exercise 1.14.]

Proof. First, assume in addition that $M \cap X = \{0\}$. Write $u_n = x_n + y_n$ with $x_n \in X$, $y_n \in M$, and $u_n \to u$ in E. We claim that (x_n) is bounded. If not, then $\|x_{n_k}\| \to \infty$ for some subsequence $n_k \to \infty$. Passing to a further subsequence, we may assume that $\frac{x_{n_k}}{\|x_{n_k}\|} \to \xi$ in X, with $\|\xi\| = 1$ (here we use the fact that $\dim X < \infty$). Thus $\frac{y_{n_k}}{\|x_{n_k}\|} = \frac{u_{n_k}}{\|x_{n_k}\|} - \frac{x_{n_k}}{\|x_{n_k}\|} \to -\xi$; moreover, $\xi \in M$ (since M is closed). Thus $\xi \in M \cap X$ and we must have $\xi = 0$. Impossible. Hence we have shown that (x_n) is bounded. Passing to a subsequence, we may assume that $x_{n_k} \to x$ in X. Then $y_n \to u - x \in M$ (since M is closed). Therefore $u \in (M + X)$, and this completes the proof that $(M + X)$ is closed when $M \cap X = \{0\}$.

In the general case, let $\widetilde{X}$ be a complement of $(M \cap X)$ in X (this is finite-dimensional stuff). Clearly $\widetilde{X}$ is finite-dimensional, $M \cap \widetilde{X} = \{0\}$, and $M + \widetilde{X} = M + X$. We have already proved that $(M + \widetilde{X})$ is closed, and so is $(M + X)$.

Suppose now that M admits a complement, say N, in E. Let P_M and P_N be the projections onto M and N. Since $P_N(X)$ has finite dimension, it has a complement, say $\widetilde{N}$, in N (see Section 2.4). We claim that $\widetilde{N}$ is a complement of $(M + X)$ in E.

First we have
$$(M + X) \cap \widetilde{N} = \{0\}.$$

Indeed, if $\tilde{n} = m + x$ with $\tilde{n} \in \widetilde{N}$, $m \in M$, and $x \in X$, then
$$\tilde{n} = P_N \tilde{n} = P_N(m + x) = P_N x \in P_N(X),$$

and thus $\tilde{n} \in \widetilde{N} \cap P_N(X) = \{0\}$.

Next, we have
$$(M + X) + \widetilde{N} = E.$$

Indeed, any $\xi \in E$ may be written as
$$\xi = P_M \xi + P_N \xi,$$

and $P_N \xi$ may be further decomposed as
$$P_N \xi = P_N x + \tilde{n},$$

for some $x \in X$ and some $\tilde{n} \in \widetilde{N}$. But $x = P_M x + P_N x$, so that

$$P_N \xi = (x - P_M x) + \tilde{n},$$

and therefore

$$\xi = P_M \xi + x - P_M x + \tilde{n} \in (M + X) + \widetilde{N}.$$

Conversely, assume that $(M + X)$ admits a complement, say W, in E. Let $\widetilde{X}$ be, as above, a complement of $(M \cap X)$ in X. We claim that $(W + \widetilde{X})$ is a complement of M.

First we have

$$(W + \widetilde{X}) \cap M = \{0\}.$$

Indeed, if $m \in M$ can be written as $m = w + \tilde{x}$ with $w \in W$ and $\tilde{x} \in \widetilde{X}$, then $w = m - \tilde{x}$, so that $w \in (M + X) \cap W = \{0\}$. Therefore $m = \tilde{x} \in (M \cap X) \cap \widetilde{X} = \{0\}$.

Finally, we verify that

$$(W + \widetilde{X}) + M = E,$$

Indeed, it suffices to check that

$$(M + \widetilde{X}) = (M + X)$$

(since $W + (M + X) = E$). Clearly $M + \widetilde{X} \subset M + X$ (since $\widetilde{X} \subset X$). Conversely, any $x \in X$ can be written as $x = x_1 + \tilde{x}$ with $x_1 \in M \cap X$ and $\tilde{x} \in \widetilde{X}$. Therefore $M + X \subset M + \widetilde{X}$.

Let M be a subspace of a Banach space E. Recall that M has *finite codimension* if there exists a finite-dimensional space $X \subset E$ such that $M + X = E$. We may always assume that $M \cap X = \{0\}$ (otherwise choose a complement of $M \cap X$ in X). The codimension M, codim M, is by definition the dimension of such X (and is independent of the special choice of X); it coincides with $\dim(E/M)$.

[**Warning:** A subspace of finite codimension need not be closed. For example, if $\dim E = \infty$, take any linear functional f on E that is *not* continuous (see Exercise 1.5). Then $M = f^{-1}(\{0\})$ has codimension 1 but M is not closed (by Proposition 1.5); in fact, M is dense in E.]

Proposition 11.5. *Let E be a Banach space and let M be a closed subspace of E of finite codimension. Then any subspace $\widetilde{M}$ of E containing M must be closed.*

Proof. The space M admits an algebraic complement in $\widetilde{M}$, say X. Clearly $\dim X < \infty$, and $\widetilde{M} = X + M$. Applying Proposition 11.4, we see that $\widetilde{M}$ is closed.

Proposition 11.6. *Let E be a Banach space and let M be a closed subspace of E of finite codimension. Let D be a dense subspace of E. Then there exists a complement X of M with $X \subset D$.*

Proof. Let d be the codimension of M in E. If $d = 0$, we have $M = E$ and we may take $X = \{0\}$. Hence we may assume that $d \geq 1$. Fix any $x_1 \in D$ with $x_1 \notin M$; this is possible, for otherwise $D \subset M$ implies $E = \overline{D} \subset M \neq E$; a contradiction. Let

$M_1 = M + \mathbb{R}x_1$. Then M_1 is closed (by Proposition 11.4) and codim $M_1 = d - 1$. Repeating this construction $(d - 1)$ times yields a subspace $X \subset D$, of dimension d, such that $M + X = E$ and $M \cap X = \{0\}$.

Proposition 11.7. *Let E be a Banach space and let $G, L \subset E$ be closed subspaces. Assume that there exist finite-dimensional spaces $X_1, X_2 \subset E$ such that*

(1) $G + L + X_1 = E,$

(2) $G \cap L \subset X_2.$

Then G (resp. L) admits a complement.

Proof. We divide the proof into two steps.

Step 1: The conclusion of Proposition 11.7 holds when $X_2 = \{0\}$.

Let $\widetilde{X}_1$ be a complement of $(G + L) \cap X_1$ in X_1. We already know by Proposition 11.4 that $(L + \widetilde{X}_1)$ is closed. We claim that $(L + \widetilde{X}_1)$ is a complement of G.

First, we have

$$G + (L + \widetilde{X}_1) = E.$$

Indeed, any $\xi \in E$ may be written as $\xi = g + \ell + h$ with $g \in G, \ell \in L, h \in X_1$, and h may be further decomposed as $h = h_1 + h_2$ with $h_1 \in (G + L) \cap X_1$ and $h_2 \in \widetilde{X}_1$. Hence $\xi \in G + L + \widetilde{X}_1$.

Next we have

$$G \cap (L + \widetilde{X}_1) = \{0\}.$$

Indeed, suppose that $g = \ell + \tilde{x}_1$ with $g \in G, \ell \in L$, and $\tilde{x}_1 \in X_1$. Then $\tilde{x}_1 = g - \ell$, so that $\tilde{x}_1 \in (G + L) \cap \widetilde{X}_1 = \{0\}$. Hence $g = \ell \in G \cap L = \{0\}$ (this is assumed in Step 1).

Step 2: The general case.

Let $\widetilde{G}$ be a complement of $(G \cap L)$ in G and let $\widetilde{L}$ be a complement of $(G \cap L)$ in L (note that $\widetilde{G}$ and $\widetilde{L}$ exist, since $G \cap L$ is finite-dimensional; see Section 2.4).

We claim that

(3) $(\widetilde{G} + \widetilde{L}) + (X_1 + X_2) = E$

and

(4) $(\widetilde{G} \cap \widetilde{L}) = \{0\}.$

This will complete the proof of the proposition. Indeed, from Step 1 we deduce that $\widetilde{G}$ admits a complement. Therefore $G = \widetilde{G} + (G \cap L)$ also admits a complement by Proposition 11.4.

Verification of (3). Any $\xi \in E$ may be written as

$$\xi = g + \ell + x_1 \quad \text{with} \quad g \in G, \ \ell \in L, \text{ and } x_1 \in X_1.$$

But $g = \tilde{g} + h_1$ with $\tilde{g} \in \widetilde{G}$ and $h_1 \in G \cap L$; similarly $\ell = \tilde{\ell} + h_2$ with $\tilde{\ell} \in \widetilde{L}$ and $h_2 \in G \cap L$. Therefore

$$\xi = (\tilde{g} + \tilde{\ell}) + x_1 + (h_1 + h_2) \in (\tilde{G} + \tilde{L}) + (X_1 + X_2).$$

Verification of (4). Assume that $g \in \tilde{G} \cap \tilde{L}$. Then $g \in (G \cap L) \cap \tilde{L}$ (since $\tilde{G} \subset G$ and $\tilde{L} \subset L$). But $(G \cap L) \cap \tilde{L} = \{0\}$.

11.2 Quotient Spaces

Let E be a Banach space and let M be a closed subspace. We consider an equivalence relation on E defined by $x \sim y$ if $x - y \in M$. The set of all equivalence classes is a vector space, denoted by E/M, and is called the *quotient space* of E (mod M). The canonical map that associates to every $x \in E$ its equivalence class $[x]$ is denoted by $\pi : E \to E/M$. Clearly π is a surjective linear operator. The quotient space E/M is equipped with the quotient norm

$$\|[x]\|_{E/M} = \|\pi(x)\|_{E/M} = \inf_{\substack{y \in E \\ y \in [x]}} \|y\| = \inf_{m \in M} \|x - m\|.$$

It is clear that $\|[x]\|_{E/M}$ is a norm on E/M (to check that $\|[x]\|_{E/M} = 0$ implies $[x] = 0$, one uses the fact that M is closed). Moreover, $\pi : E \to E/M$ is a bounded operator and $\|\pi\| \leq 1$. When there is no confusion we simply write $\| \ \|$ instead of $\| \ \|_{E/M}$.

Proposition 11.8. *The quotient space E/M equipped with the norm $\| \ \|_{E/M}$ is a Banach space.*

Proof. Let $(\pi(x_k))$ be a Cauchy sequence in E/M. We have to show that $(\pi(x_k))$ converges, and since $(\pi(x_k))$ is Cauchy, it suffices to prove that a subsequence converges. Passing to a subsequence (still denoted by (x_k)), we may assume that $\|\pi(x_{k+1}) - \pi(x_k)\| < \frac{1}{2^k}$ $\forall k$ (see the proof of Theorem 4.8). Hence there exists a sequence (m_k) in M such that $\|x_{k+1} - x_k - m_k\| < \frac{1}{2^k}$. Write $m_k = \mu_{k+1} - \mu_k$ with $\mu_1 = 0$ and $\mu_k \in M$ $\forall k$. Since $(x_k - \mu_k)$ is a Cauchy sequence in E, it converges to a limit ℓ in E. Therefore $\pi(x_k) = \pi(x_k - \mu_k)$ also converges (to $\pi(\ell)$) in E/M.

Proposition 11.9. *Let M be a closed subspace of E and let $\pi^\star : (E/M)^\star \to E^\star$ be the adjoint of $\pi : E \to E/M$. Then $R(\pi^\star) = M^\perp$, and more precisely, $\pi^\star$ is bijective from $(E/M)^\star$ onto $M^\perp$, with*

$$\|\pi^\star(\xi)\|_{E^\star} = \|\xi\|_{(E/M)^\star} \quad \forall \xi \in (E/M)^\star.$$

In particular, $(E/M)^\star$ is isomorphic and isometric to $M^\perp$.

Proof. With $\xi \in (E/M)^\star$ and $x \in E$, write

$$\langle \pi^\star(\xi), x \rangle = \langle \xi, \pi(x) \rangle.$$

If $x \in M$ we have $\pi(x) = 0$ and thus $\langle \pi^\star(\xi), x \rangle = 0$ $\forall x \in M$, i.e., $\pi^\star(\xi) \in M^\perp$.

Conversely, let $f \in M^\perp$; we need to show that $f = \pi^\star(\xi)$ for some $\xi \in (E/M)^\star$. Given $y \in E/M$, write $y = \pi(x)$ for some $x \in E$ and then define $\xi(y) = \langle f, x \rangle$.

Note that this definition does not depend on the special choice of x, since $f \in M^{\perp}$. Clearly ξ is linear in y and we have $|\xi(y)| \leq \|f\|_{E^\star}\|x - m\|$ $\forall m \in M$. Taking the inf over all $m \in M$ gives $|\xi(y)| \leq \|f\|_{E^\star}\|\pi(x)\|_{E/M} = \|f\|_{E^\star}\|y\|_{E/M}$. Hence $\xi \in (E/M)^\star$, and clearly, $\langle \pi^\star(\xi), x \rangle = \langle \xi, \pi(x) \rangle = \langle f, x \rangle$ $\forall x \in E$, i.e., $\pi^\star(\xi) = f$. Moreover, $\|\xi\|_{(E/M)^\star} \leq \|f\|_{E^\star} = \|\pi^\star(\xi)\|_{E^\star}$.

On the other hand, we know that

$$\|\pi^\star(\xi)\|_{E^\star} \leq \|\xi\|_{(E/M)^\star},$$

since $\|\pi^\star\| = \|\pi\| \leq 1$. Consequently,

$$\|\pi^\star(\xi)\|_{E^\star} = \|\xi\|_{(E/M)^\star} \quad \forall \xi \in (E/M)^\star.$$

Let F, G be Banach spaces and let $T \in \mathcal{L}(F, G)$. Consider the closed subspace $N(T)$ of F, the quotient space $F/N(T)$, and the canonical map $\pi : F \to F/N(T)$. The operator T can be factored as $T = \tilde{T} \circ \pi$, where $\tilde{T} : F/N(T) \to G$; indeed, given $y \in F/N(T)$, write $y = \pi(x)$ for some $x \in F$ and set $\tilde{T}y = Tx$. Clearly $\tilde{T}$ is well defined independently of the choice of x, and bijective from $F/N(T)$ onto $R(T)$; moreover, $\|\tilde{T}\| = \|T\|$.

Consider now a special case of this setting. Let M be a closed subspace of a Banach space E. Let $T : E^\star \to M^\star$ be defined by

$$T(f) = f_{|M} \quad \forall f \in E^\star.$$

Then $N(T) = M^{\perp}$ and (by Hahn–Banach) $R(T) = M^\star$. Applying the above to $F = E^\star$ and $G = M^\star$, we obtain an operator $\tilde{T} : E^\star/M^{\perp} \to M^\star$ that is bijective, and such that $\tilde{T} \circ \pi = T$.

Proposition 11.10. *For any Banach space E and any closed subspace M of E, the operator $\tilde{T}$ is a bijective isometry from $E^\star/M^{\perp}$ onto $M^\star$.*

Proof. We have only to show that $\tilde{T}$ is an isometry. Given any $f \in E^\star$, consider the functional $f_{|M}$ on M. By Corollary 1.2 we know that there exists a functional $\tilde{f} \in E^\star$ such that $\tilde{f}_{|M} = f_{|M}$ and $\|\tilde{f}\|_{E^\star} = \|f_{|M}\|_{M^\star} = \|T(f)\|_{M^\star}$.

Since $f - \tilde{f} \in M^{\perp}$, we have

$$\|\pi(f)\|_{E^\star/M^{\perp}} = \text{dist}(f, M^{\perp}) \leq \|f - (f - \tilde{f})\|_{E^\star} = \|\tilde{f}\|_{E^\star} = \|Tf\|_{M^\star}.$$

Hence we have proved that

$$\|\pi(f)\|_{E^\star/M^{\perp}} \leq \|Tf\|_{M^\star} \quad \forall f \in E^\star.$$

But $T = \tilde{T} \circ \pi$, so that

$$\|\pi(f)\|_{E^\star/M^{\perp}} \leq \|\tilde{T}(\pi(f))\|_{M^\star} \quad \forall f \in E^\star,$$

i.e.,

$$\|y\|_{E^\star/M^\perp} \leq \|\widetilde{T}(y)\|_{M^\star} \quad \forall y \in E^\star/M^\perp.$$

On the other hand, it is clear that

$$\|(\widetilde{T} \circ \pi)(f)\|_{M^\star} = \|T(f)\|_{M^\star} \leq \|f\|_{E^\star} \quad \forall f \in E^\star.$$

Replacing f by $(f - g)$ with $g \in M^\perp$ and taking the infimum over $g \in M^\perp$ yields

$$\|\widetilde{T}(\pi(f))\|_{M^\star} \leq \|\pi(f)\|_{E^\star/M^\perp} \quad \forall f \in E^\star,$$

i.e.,

$$\|\widetilde{T}(y)\|_{M^\star} \leq \|y\|_{E^\star/M^\perp} \quad \forall y \in E^\star/M^\perp.$$

We conclude that $\widetilde{T}$ is an isometry.

The quotient space E/M inherits many of the properties of the space E, e.g., reflexivity and uniform convexity.

Proposition 11.11. *Assume that E is a reflexive Banach space and M is a closed subspace. Then E/M is reflexive.*

Proof. We know that $E^\star$ is reflexive (see Corollary 3.21) and thus $M^\perp$ is also reflexive (being a closed subspace of $E^\star$; see Proposition 3.20). On the other hand, $M^\perp$ is isomorphic to $(E/M)^\star$ (by Proposition 11.9). Therefore $(E/M)^\star$ is reflexive, and so is E/M, again by Corollary 3.21.

Proposition 11.12. *Assume that E is a uniformly convex Banach space and M is a closed subspace. Then E/M is uniformly convex.*

Proof. Let $\pi(x), \pi(y) \in E/M$ be such that $\|\pi(x)\| \leq 1$, $\|\pi(y)\| \leq 1$, and $\|\pi(x) - \pi(y)\| > \varepsilon$. Since E is reflexive, we know (see Corollary 3.23) that there exist $m_1 \in M$ and $m_2 \in M$ such that $\|x - m_1\| \leq 1$ and $\|x - m_2\| \leq 1$. Moreover, $\|(x - y) - m\| > \varepsilon \quad \forall m \in M$. The uniform convexity of E yields

$$\left\| \frac{(x - m_1) + (y - m_2)}{2} \right\| < 1 - \delta,$$

and thus

$$\left\| \frac{\pi(x) + \pi(y)}{2} \right\| < 1 - \delta.$$

Proposition 11.13. *Let E be a Banach space and let $M \subset E$ be a closed subspace. Then*

(a) *$\dim M < \infty$ if and only if $\operatorname{codim} M^\perp < \infty$, and in that case*

$$\dim M = \operatorname{codim} M^\perp,$$

(b) *$\operatorname{codim} M < \infty$ if and only if $\dim M^\perp < \infty$, and in that case*

$$\operatorname{codim} M = \dim M^\perp.$$

Proof.

(a) We know by Proposition 11.10 that $E^\star/M^\perp$ is always isomorphic to $M^\star$. Thus $\dim M^\star < \infty \Leftrightarrow \dim(E^\star/M^\perp) < \infty$. By Proposition 11.3 we know that $\dim M < \infty \Leftrightarrow \dim M^\star < \infty$ and then $\dim M = \dim M^\star$. On the other hand, $\dim(E^\star/M^\perp) < \infty \Leftrightarrow \operatorname{codim} M^\perp < \infty$. Hence $\dim M < \infty \Leftrightarrow \operatorname{codim} M^\perp < \infty$ and $\dim M = \dim M^\star = \dim(E^\star/M^\perp) = \operatorname{codim} M^\perp$.

(b) Proposition 11.9 yields that $\dim M^\perp < \infty \Leftrightarrow \dim(E/M)^\star < \infty$. Using once more Proposition 11.3, this is equivalent to $\dim(E/M) < \infty$, i.e., $\operatorname{codim} M < \infty$. Then $\dim M^\perp = \dim(E/M)^\star = \dim(E/M) = \operatorname{codim} M$.

A "dual" statement is partially true.

Proposition 11.14. *Let $N \subset E^\star$ be a closed subspace. Then* $\dim N < \infty$ *if and only if* $\operatorname{codim} N^\perp < \infty$, *and in that case* $\dim N = \operatorname{codim} N^\perp$. *It is also true that* $\dim N^\perp \leq \operatorname{codim} N$, *but it may happen that* $\dim N^\perp < \operatorname{codim} N < \infty$.

Proof. Recall that

$$N^\perp = \{x \in E;\ \langle f, x \rangle = 0 \quad \forall f \in N\}.$$

Clearly $\overline{N} \subset N^{\perp\perp}$; but it may happen that $\overline{N} \neq N^{\perp\perp}$ (see Remark 6 in Chapter 1). For example, take $\xi \in E^{\star\star}$ with $\xi \notin E$ and let $N = \xi^{-1}(\{0\}) = \{f \in E^\star;\ \langle \xi, f \rangle = 0\}$. Then N is a closed subspace of $E^\star$ of codimension 1 (i.e., N is a hyperplane). However, $N^\perp = \{0\}$ (because the orthogonal of N in $E^{\star\star}$ is $\mathbb{R}\xi$ by Lemma 3.2 and thus $N^\perp$, the orthogonal of N in E, is reduced to $\{0\}$). In this case $N = \overline{N} \neq N^{\perp\perp} = E^\star$, and $\dim N^\perp = 0$, while $\operatorname{codim} N = 1$.

We now return to the general case. Since $N \subset N^{\perp\perp}$, we have $\operatorname{codim} N^{\perp\perp} \leq \operatorname{codim} N \leq \infty$. Set $M = N^\perp \subset E$. By Proposition 11.10 we have

$$\operatorname{codim} M^\perp = \dim(E^\star/M^\perp) = \dim M^\star,$$

and thus $\operatorname{codim} N^{\perp\perp} = \dim M \leq \infty$. Therefore

$$\dim N^\perp \leq \operatorname{codim} N \leq \infty.$$

We now prove that $\dim N < \infty \Rightarrow \operatorname{codim} N^\perp < \infty$ and $\operatorname{codim} N^\perp = \dim N$. We first claim that $N^{\perp\perp} = N$. We already know that $N \subset N^{\perp\perp}$. Let $f_1, f_2, \ldots, f_p$ be a basis of N and let $f \in N^{\perp\perp}$. Since $f = 0$ on $N^\perp = \{x \in E;\ \langle f_i, x \rangle = 0 \ \forall i\}$, we may apply Lemma 3.2 and conclude that $f = \sum \lambda_i f_i$. Therefore $N^{\perp\perp} \subset N$. As above, set $M = N^\perp$. Since $\dim M^\perp < \infty$, we deduce from Proposition 11.13 that $\operatorname{codim} M < \infty$ and that $\operatorname{codim} M = \dim M^\perp$, i.e., $\operatorname{codim} N^\perp = \dim N$.

Conversely, assume $\operatorname{codim} N^\perp < \infty$, and set again $M = N^\perp$, so that $\operatorname{codim} M < \infty$. Applying Proposition 11.13 once more yields $\dim M^\perp < \infty$, i.e., $\dim N^{\perp\perp} < \infty$. Since $N \subset N^{\perp\perp}$, we deduce that $\dim N < \infty$ and we are back to the previous situation. Hence $\dim N = \operatorname{codim} N^\perp$.

11.3 Some Classical Spaces of Sequences

Given a sequence $x = (x_1, x_2, \ldots, x_k, \ldots)$, set

$$\|x\|_p = \left(\sum_{k=1}^{\infty} |x_k|^p \right)^{1/p}, \quad 1 \le p < \infty,$$

$$\|x\|_\infty = \sup_k |x_k|$$

and consider the corresponding spaces

$$\ell^p = \{x; \|x\|_p < \infty\}, 1 \le p < \infty,$$
$$\ell^\infty = \{x; \|x\|_\infty < \infty\},$$

which are Banach spaces for the ℓ^p (resp. ℓ^∞) norms. This can be established directly (and is quite easy); or one can rely on Theorem 4.8 applied to $\Omega = \mathbb{N}$ equipped with the counting measure, $\mu(E) =$ the number of points in a set $E \subset \mathbb{N}$. Many properties mentioned below are consquences of general results from Chapter 4. For the convenience of the reader, we also present some direct proofs.

There are two interesting subspaces of ℓ^∞:

$$c = \left\{ x; \lim_{k \to \infty} x_k \text{ exists} \right\}$$

and

$$c_0 = \left\{ x; \lim_{k \to \infty} x_k = 0 \right\}.$$

They are both equipped with the ℓ^∞ norm. Clearly $c_0 \subset c \subset \ell^\infty$ with c_0 closed in c, and c closed in ℓ^∞.

Hölder's inequality takes the form

$$(5) \quad \left| \sum_{k=1}^{\infty} x_k\, y_k \right| \le \|x\|_p \|y\|_{p'} \quad \forall x \in \ell^p, \quad \forall y \in \ell^{p'} \text{ with } \frac{1}{p} + \frac{1}{p'} = 1.$$

The space ℓ^2 is a Hilbert space equipped with the scalar product

$$(x, y) = \sum_{k=1}^{\infty} x_k\, y_k.$$

It is clear that $\ell^p \subset c_0$ with

$$\|x\|_\infty \le \|x\|_p \quad \forall p, \quad 1 \le p < \infty, \quad \forall x \in \ell^p,$$

and this yields $\ell^p \subset \ell^q$ when $1 \le p \le q \le \infty$, with

$$\|x\|_q \le \|x\|_p \quad \forall x \in \ell^p.$$

Proposition 11.15. *The space ℓ^p is reflexive, and even uniformly convex, for $1 < p < \infty$.*

Proof. Apply Theorem 4.10 and Exercise 4.12 with $\Omega = \mathbb{N}$.

Proposition 11.16. *The spaces c, c_0, and ℓ^p, with $1 \le p < \infty$, are separable.*

Proof. Let

$$D = \{x = (x_k); x_k \in \mathbb{Q} \quad \forall k, \text{ and } x_k = 0 \text{ for } k \text{ sufficiently large}\}.$$

It is clear that D is countable; moreover, D is dense in ℓ^p when $1 \le p < \infty$ and in c_0. The set $D + \lambda(1, 1, 1, \dots)$, with $\lambda \in \mathbb{Q}$, is countable and dense in c.

Proposition 11.17. *The space ℓ^∞ is not separable.*

Proof. Assume that $A \subset \ell^\infty$ is countable. We will check that A cannot be dense in ℓ^∞. Write $A = (a^k)$, where each $a^k \in \ell^\infty$, so that $a^k = (a_1^k, a_2^k, \dots)$. For each integer k set

$$b_k = \begin{cases} a_k^k + 1 & \text{if } |a_k^k| \le 1, \\ 0 & \text{if } |a_k^k| > 1. \end{cases}$$

Note that $b = (b_k) \in \ell^\infty$ and $|b_k - a_k^k| \ge 1 \; \forall k$. Therefore,

$$\|b - a^k\|_\infty \ge |b_k - a_k^k| \ge 1 \quad \forall k,$$

and thus $b \notin \overline{A}$.

Proposition 11.18. *Let $1 \le p < \infty$. Given any $\phi \in (\ell^p)^\star$, there exists a unique $u \in \ell^{p'}$ such that*

$$\langle \phi, x \rangle = \sum_{k=1}^\infty u_k x_k \quad \forall x \in \ell^p.$$

Moreover,

$$\|u\|_{p'} = \|\phi\|_{(\ell^p)^\star}.$$

Proof. Let $e_k = (0, 0, \dots, \underset{(k)}{1}, 0, 0, \dots)$. Set $u_k = \phi(e_k)$. We claim that $u = (u_k) \in \ell^{p'}$ and

(6) $$\|u\|_{p'} \le \|\phi\|_{(\ell^p)^\star}.$$

Inequality (6) is clear when $p = 1$, since

$$|u_k| \le \|\phi\|_{(\ell^1)^\star} \|e_k\|_1 \le \|\phi\|_{(\ell^1)^\star} \quad \forall k.$$

We now turn to the case $1 < p < \infty$. Fix an integer N. Then for every $x = (x_1, x_2, \dots, x_N, 0, 0, \dots)$ we have

(7)
$$\sum_{k=1}^{N} u_k x_k = \phi\left(\sum_{k=1}^{N} x_k e_k\right) \le \|\phi\|_{(\ell^p)^\star} \|x\|_p.$$

Choosing $x_k = |u_k|^{p'-2} u_k$ yields

$$\left(\sum_{k=1}^{N} |u_k|^{p'}\right)^{1/p'} \le \|\phi\|_{(\ell^p)^\star}.$$

As $N \to \infty$ we see that $u \in \ell^{p'}$ and (6) holds. Moreover,

$$\phi(x) = \sum_{k=1}^{\infty} u_k \, x_k \quad \forall x \in D,$$

where D is defined in the proof of Proposition 11.16. Since D is dense in ℓ^p we obtain

$$\phi(x) = \sum_{k=1}^{\infty} u_k \, x_k \quad \forall x \in \ell^p.$$

Hölder's inequality yields

$$|\phi(x)| \le \|u\|_{p'} \|x\|_p \quad \forall x \in \ell^p,$$

and therefore $\|\phi\|_{(\ell^p)^\star} \le \|u\|_{p'}$. Combining with (6), we obtain

$$\|\phi\|_{(\ell^p)^\star} = \|u\|_{p'}.$$

The uniqueness of u is obvious.

Proposition 11.19. *Given any $\phi \in (c_0)^\star$, there exists a unique $u \in \ell^1$ such that*

$$\langle \phi, x \rangle = \sum_{k=1}^{\infty} u_k \, x_k \quad \forall x \in c_0.$$

Moreover,
$$\|u\|_1 = \|\phi\|_{(c_0)^\star}.$$

Proof. This is an easy adaptation of the proof of Proposition 11.18 (with $p = \infty$ and $p' = 1$); the last part of the proof holds since D is dense in c_0 (but not in ℓ^∞).

Proposition 11.20. *Given $\phi \in (c)^\star$, there exists a unique pair $(u, \lambda) \in \ell^1 \times \mathbb{R}$ such that*

$$\langle \phi, x \rangle = \sum_{k=1}^{\infty} u_k \, x_k + \lambda \lim_{k\to\infty} x_k \quad \forall x \in c.$$

Moreover,
$$\|u\|_1 + |\lambda| = \|\phi\|_{(c)^\star}.$$

Proof. Applying Proposition 11.19 to $\phi_{|c_0}$, we find some $u \in \ell^1$ such that

$$\phi(y) = \sum_{k=1}^{\infty} u_k y_k \quad \forall y \in c_0.$$

If $x \in c$ write $x = y + ae$, where $e = (1, 1, 1, \ldots)$, $a = \lim_{k \to \infty} x_k$, and $y \in c_0$. Then

$$\phi(x) = \sum_{k=1}^{\infty} u_k \, y_k + a\phi(e) = \sum_{k=1}^{\infty} u_k(x_k - a) + a\phi(e) = \sum_{k=1}^{\infty} u_k \, x_k + \lambda a,$$

where $\lambda = \phi(e) - \sum_{k=1}^{\infty} u_k$.

Conversely, given any $u \in \ell^1$ and $\lambda \in \mathbb{R}$, the functional

$$(8) \qquad \phi(x) = \sum_{k=1}^{\infty} u_k \, x_k + \lambda \lim_{k \to \infty} x_k, \quad x \in c,$$

defines an element of $(c)^\star$. We claim that

$$(9) \qquad \|\phi\|_{(c)^\star} = \|u\|_1 + |\lambda|.$$

It is clear that

$$(10) \qquad \|\phi\|_{(c)^\star} \leq \|u\|_1 + |\lambda|.$$

Choosing $x = (x_k)$ in (8), where N is a fixed integer and

$$x_k = \begin{cases} \text{sign}\,(u_k), & 1 \leq k \leq N, \\ \text{sign}\,(\lambda), & k > N, \end{cases}$$

yields

$$\phi(x) = \sum_{k=1}^{N} |u_k| + \text{sign}\,(\lambda) \sum_{k=N+1}^{\infty} u_k + |\lambda| \leq \|\phi\|_{(c)^\star}.$$

As $N \to \infty$ we obtain

$$\|u\|_1 + |\lambda| \leq \|\phi\|_{(c)^\star},$$

which, together with (10), gives (9).

Proposition 11.21. *The spaces ℓ^1, ℓ^∞, c, and c_0 are not reflexive.*

Proof. From Propositions 11.19 and 11.18 we know that $(c_0)^\star$ is ℓ^1 and $(\ell^1)^\star$ is ℓ^∞. Therefore the identity map from c_0 into ℓ^∞ corresponds to the canonical injection $J : c_0 \to (c_0)^{\star\star}$ defined in Section 1.3. Since it is not surjective, we conclude that c_0 is not reflexive. Applying Corollary 3.21, we deduce that ℓ^1 and ℓ^∞ are not reflexive. Moreover, c cannot be reflexive; otherwise, c_0, which is a closed subspace of c, would be reflexive by Proposition 3.20.

The following table summarizes the main properties discussed above:

	Reflexive	Separable	Dual Space
ℓ^p with $1 < p < \infty$	YES	YES	$\ell^{p'}$
ℓ^1	NO	YES	ℓ^∞
c_0	NO	YES	ℓ^1
c	NO	YES	$\ell^1 \times \mathbb{R}$
ℓ^∞	NO	NO	Strictly bigger than ℓ^1

11.4 Banach Spaces over $\mathbb{C}$: What Is Similar and What Is Different?

Throughout this section we assume that E is a vector space over $\mathbb{C}$. Of course we may associate to E a vector space over $\mathbb{R}$ simply by considering the product λx with λ restricted to $\mathbb{R}$, and $x \in E$; the corresponding vector space over $\mathbb{R}$ will often be denoted by $E_{\mathbb{R}}$ to distinguish it from E.

A linear subspace $M \subset E$ is a subset M satisfying $\lambda x \in M$ and $x + y \in M$ $\forall \lambda \in \mathbb{C}, \forall x, y \in M$. Of course a linear subspace M of E is also a linear subspace of $E_{\mathbb{R}}$. But the converse is *not* true. For example, a line L in $\mathbb{R}^2$ containing 0 is a linear subspace of $\mathbb{R}^2$. However, if we identify $\mathbb{C}$ with $\mathbb{R}^2$, the line L is no longer a linear subspace of $\mathbb{C}$ because $iL = L$ rotated by $\pi/2$, is not contained in L.

A norm on E is by definition a function E with values in $[0, +\infty)$ such that $\|x\| = 0 \Leftrightarrow x = 0$, $\|\lambda x\| = |\lambda| \, \|x\| \; \forall \lambda \in \mathbb{C}, \forall x \in E$, and $\|x + y\| \leq \|x\| + \|y\|$. Clearly $\| \; \|$ is also a norm on $E_{\mathbb{R}}$, but the converse is not true.

A linear functional on E is a map $f : E \to \mathbb{C}$ such that $f(\lambda x) = \lambda f(x)$ and $f(x + y) = f(x) + f(y) \; \forall \lambda \in \mathbb{C}, \forall x, y \in E$. The dual space $E^\star$ is the space of all continuous linear functionals on E; $E^\star$ is a vector space over $\mathbb{C}$ and is equipped with the norm

$$\|f\|_{E^\star} = \sup_{\substack{x \in E \\ \|x\| \leq 1}} |f(x)|.$$

The complex number $f(x)$ is also denoted by $\langle f, x \rangle$, and we clearly have $\langle \lambda f, \mu x \rangle = \lambda \mu \langle f, x \rangle \; \forall \lambda, \mu \in \mathbb{C}, \forall x \in E$. The correspondence between the complex dual $E^\star$ and the real dual $E_{\mathbb{R}}^\star$ is given by the following simple but illuminating result.

Proposition 11.22. *The map*

$$I : f \in E^\star \mapsto \operatorname{Re} f \in E_{\mathbb{R}}^\star$$

is a bijective isometry from $E^\star$ onto $E_{\mathbb{R}}^\star$.

Proof. Clearly,

$$|\operatorname{Re}\langle f, x \rangle| \leq |\langle f, x \rangle| \leq \|f\|_{E^\star} \|x\|$$

and thus

(11)
$$\|I(f)\|_{E_{\mathbb{R}}^\star} \leq \|f\|_{E^\star}.$$

It is also clear that I is injective because $\mathrm{Re}\langle f, x \rangle = 0 \; \forall x \in E$ implies $\mathrm{Re}\langle f, ix \rangle = 0$ $\forall x \in E$, i.e., $\mathrm{Im}\langle f, x \rangle = 0 \; \forall x \in E$, and thus $f = 0$. Next we claim that I is surjective. Indeed, given $\varphi \in E_\mathbb{R}^\star$ set

$$(12) \qquad\qquad f(x) = \varphi(x) - i\varphi(ix) \quad \forall x \in E$$

[**warning:** $\varphi(ix)$ is not equal to $i\varphi(x)$, because both $\varphi(x)$ and $\varphi(ix)$ belong to $\mathbb{R}$]. It is easy to check that $f \in E^\star$, i.e., $f(\lambda x) = \lambda f(x) \; \forall \lambda \in \mathbb{C}, \forall x \in E$ (please verify!) and that $I(f) = \mathrm{Re}\, f = \varphi$. From (11) we have $\|\varphi\|_{E_\mathbb{R}^\star} \leq \|f\|_{E^\star}$. It is also clear from (12) that

$$|f(x)| \leq \left(|\varphi(x)|^2 + |\varphi(ix)|^2 \right)^{1/2} \leq \sqrt{2}\|\varphi\|_{E_\mathbb{R}^\star}\|x\|$$

(since $\|ix\| = \|x\|$). But we can do better. Assume $f(x) \neq 0$ and set $\lambda = \frac{f(x)}{|f(x)|} \in \mathbb{C}$. Then

$$|f(x)| = \frac{1}{\lambda} f(x) = f\left(\frac{x}{\lambda}\right) = \varphi\left(\frac{x}{\lambda}\right) - i\varphi\left(\frac{ix}{\lambda}\right).$$

Since $|f(x)| \in \mathbb{R}, \varphi\left(\frac{x}{\lambda}\right) \in \mathbb{R}$, and $\varphi\left(\frac{ix}{\lambda}\right) \in \mathbb{R}$, we see that $\varphi\left(\frac{ix}{\lambda}\right) = 0$ and thus $|f(x)| = \varphi\left(\frac{x}{\lambda}\right)$. Therefore

$$|f(x)| \leq \|\varphi\|_{E_\mathbb{R}^\star}\left\|\frac{x}{\lambda}\right\| = \frac{1}{|\lambda|}\|\varphi\|_{E_\mathbb{R}^\star}\|x\| = \|\varphi\|_{E_\mathbb{R}^\star}\|x\|.$$

Hence $\|f\|_{E_\mathbb{R}^\star} \leq \|\varphi\|_{E_\mathbb{R}^\star} = \|I(f)\|_{E_\mathbb{R}^\star}$. Combining this with (11), we conclude that I is an isometry.

Proposition 11.22 implies that there are very few changes in Chapters 1–5 when we are dealing with vector spaces over $\mathbb{C}$, except that we need to be a little careful with Hahn–Banach (see below). A *major* change occurs in Chapter 6 when we deal with eigenvalues and spectrum. This is already visible in finite dimension: any $n \times n$ matrix M with entries in $\mathbb{C}$ admits eigenvalues in $\mathbb{C}$; but it may have no eigenvalues in $\mathbb{R}$, even if the entries of M belong to $\mathbb{R}$. We now describe chapter by chapter the changes to be made.

Chapter 1. We select a few examples showing that some statements remain unchanged while some others need slight modifications.

Proposition 11.23. *Let $G \subset E$ be a linear subspace. If $g : G \to \mathbb{C}$ is a continuous linear functional, then there exists $f \in E^\star$ that extends g, and such that*

$$\|f\|_{E^\star} = \|g\|_{G^\star}.$$

Proof. Set $\psi = \mathrm{Re}\, g$, so that ψ is an element of $G_\mathbb{R}^\star$ and $\|\psi\|_{G_\mathbb{R}^\star} = \|g\|_{G^\star}$. By Corollary 1.2 there exists some $\varphi \in E_\mathbb{R}^\star$ that extends ψ, and such that

$$\|\varphi\|_{E_\mathbb{R}^\star} = \|\psi\|_{G_\mathbb{R}^\star}.$$

Applying Proposition 11.22, we see that there exists $f \in E^\star$ such that $\varphi = \text{Re } f$ and $\|f\|_{E^\star} = \|\varphi\|_{E^\star_{\mathbb{R}}} = \|\psi\|_{G^\star_{\mathbb{R}}} = \|g\|_{G^\star}$. In addition, we have $\varphi = \text{Re } f = \psi = \text{Re } g$ on G, i.e., $\text{Re } f(x) = \text{Re } g(x) \ \forall x \in G$; taking ix instead of x yields $\text{Im } f(x) = \text{Im } g(x) \ \forall x \in G$, and thus $f = g$ on G.

Next, we state one of the geometric forms of Hahn–Banach. A closed *real* hyperplane H in E is a set of the form

$$H = \{x \in E; \ \text{Re}\langle f, x \rangle = \alpha\} = [\text{Re } f = \alpha],$$

for some $f \in E^\star$, $f \neq 0$, and some $\alpha \in \mathbb{R}$. We again warn the reader that if $\alpha = 0$, then H is a linear subspace of $E_{\mathbb{R}}$, but it is *not* a linear subspace of E over $\mathbb{C}$; for example, in $E = \mathbb{C}$, H is a line (and a line is not a linear subspace of E). We say that H separates $A, B \subset E$ if

$$\text{Re}\langle f, x \rangle \leq \alpha \quad \forall x \in A \quad \text{and} \quad \text{Re}\langle f, x \rangle \geq \alpha \quad \forall x \in B.$$

Proposition 11.24. *Let $A, B \subset E$ be two nonempty convex subsets of E such that $A \cap B = \emptyset$. Assume that one of them is open. Then there exists a closed real hyperplane that separates A and B.*

Proof. Applying Theorem 1.6 to $E_{\mathbb{R}}$ yields a hyperplane $H = [\varphi = \alpha]$ for some $\varphi \in E^\star_{\mathbb{R}}$ that separates A and B in the usual sense. Then use Proposition 11.23 to assert that $\varphi = \text{Re } f$ for some $f \in E^\star$.

The definition of the orthogonal $M^\perp$ of a linear subspace M of E is unchanged,

$$M^\perp = \{f \in E^\star; \ \langle f, x \rangle = 0 \ \ \forall x \in M\},$$

and clearly we have $M^\perp = \{f \in E^\star; \ \text{Re}\langle f, x \rangle = 0 \ \forall x \in M\}$ (since we may take ix in place of x). It is easily seen that $M^{\perp\perp} = \overline{M}$.

Given a function $\varphi : E \to (-\infty, +\infty]$, we define its conjugate $\varphi^\star$ on $E^\star$ by

$$\varphi^\star(f) = \sup_{x \in E}\{\text{Re}\langle f, x \rangle - \varphi(x)\}.$$

With obvious notation we have

$$\varphi^\star(f) = \varphi^\star_{\mathbb{R}}(If) \quad \forall f \in E^\star.$$

Proposition 11.25. *Assume that $\varphi : E \to (-\infty, +\infty]$ is convex, l.s.c., and $\varphi \not\equiv +\infty$. Then $\varphi^{\star\star} = \varphi$.*

Proof. There are two methods. Either one can apply Theorem 1.11 to $\tilde{\varphi} = \varphi$ viewed on $E_{\mathbb{R}}$, in conjunction with Proposition 11.22. Or one can repeat the proof of Theorem 1.11; when Hahn–Banach is used, one can separate the convex sets A and B using a real hyperplane as above.

The definition of the indicator function I_K is unchanged. If M is a linear subspace of E and $\varphi = I_M$, then

$$\varphi^{\star}(f) = \sup_{x \in M} \mathrm{Re}\langle f, x \rangle = I_{M^{\perp}}.$$

Indeed if $f \in M^{\perp}$ we have $\langle f, x \rangle = 0 \ \forall x \in M$ and thus $\varphi^{\star}(f) = 0$. Otherwise, if $f \notin M^{\perp}$ there exists some $x_0 \in M$ such that $\langle f, x_0 \rangle \neq 0$. Replacing x_0 by ix_0 if needed we may assume that $\mathrm{Re}\langle f, x_0 \rangle \neq 0$. Replacing x_0 by $-x_0$ if needed we may assume that $\mathrm{Re}\langle f, x_0 \rangle > 0$ and then $\sup_{\lambda>0}\langle f, \lambda x_0 \rangle = +\infty$.

Chapter 2. All the statements are unchanged (in Corollaries 2.4 and 2.5 replace $\mathbb{R}$ by $\mathbb{C}$). Some proofs rely on the $\mathbb{R}$-structure (e.g., formula (21) in the proof Theorem 2.16). They can easily be adapted to $\mathbb{C}$; alternatively, the $\mathbb{C}$-statement can be established by applying the $\mathbb{R}$-version to $E_{\mathbb{R}}$.

Chapter 3. All the statements are unchanged (in Lemmas 3.2 and 3.3 replace $\mathbb{R}$ by $\mathbb{C}$). Some proofs require obvious modifications (e.g., the proof of Proposition 3.11).

Chapter 4. Totally unchanged.

Chapter 5. A Hilbert space over $\mathbb{C}$ is a vector space over $\mathbb{C}$ equipped with a scalar product $(u, v) \in \mathbb{C}$. This is a map from $H \times H$ into $\mathbb{C}$ satisfying

$$(u, v) = \overline{(v, u)} \quad \forall (u, v) \in H,$$
$$\text{for every } v \in H, u \mapsto (u, v) \text{ is linear,}$$
$$(u, u) > 0 \quad \forall u \neq 0.$$

In particular, we have

$$(\lambda u, \mu v) = \lambda \overline{\mu}(u, v) \quad \forall \lambda, \mu \in \mathbb{C}, \quad \forall u, v \in H.$$

The quantity $|u| = (u, u)^{1/2}$ is a norm; we have

$$|u + v|^2 = |v|^2 + 2\,\mathrm{Re}(u, v) + |v|^2 \quad \forall u, v \in H,$$

and the Cauchy–Schwarz inequality becomes

$$|(u, v)| \leq |u||v| \quad \forall u, v \in H.$$

A typical example is $L^2(\Omega; \mathbb{C})$ equipped with the scalar product

$$(u, v) = \int_{\Omega} u(x)\overline{v(x)}d\mu.$$

The connection between Hilbert spaces over $\mathbb{R}$ and over $\mathbb{C}$ goes as follows. Suppose H is a Hilbert space over $\mathbb{C}$. Then $H_{\mathbb{R}}$ equipped with the scalar product $\mathrm{Re}(u, v)$ becomes a Hilbert space over $\mathbb{R}$. Therefore all the statements of Chapter 5 apply to $H_{\mathbb{R}}$. Here are some examples.

Proposition 11.26. *Let $K \subset H$ be a nonempty closed convex set. Then for every $f \in H$ there exists a unique element $u \in K$ such that*

$$|f - u| = \min_{v \in K} |f - v| = \text{dist}(f, K).$$

Moreover, u is characterized by the property

$$u \in K \quad and \quad \text{Re}(f - u, v - u) \le 0 \quad \forall v \in K.$$

Proposition 11.27. *Given any $\varphi \in H^\star$ there exists a unique $f \in H$ such that*

$$\varphi(u) = (u, f) \quad \forall u \in H.$$

Moreover,

$$|f| = \|\varphi\|_{H^\star}.$$

Proof. Applying Theorem 5.5 to $\text{Re}\,\varphi$ in $H_\mathbb{R}$, we find some $f \in H$ such that

$$\text{Re}\,\varphi(u) = \text{Re}(u, f) \quad \forall u \in H.$$

Applying this to iu yields $\text{Im}\,\varphi(u) = \text{Im}(u, f)$ and thus $\varphi(u) = (u, f)\,\forall u \in H$.

Consider now a function $a(u, v) : H \times H \to \mathbb{C}$ satisfying

(13) $\quad \forall v \in H, u \mapsto a(u, v)$ is linear and $\forall u \in H, v \mapsto \overline{a(u, v)}$ is linear,

(14) $\qquad a$ is continuous, i.e., $|a(u, v)| \le C|u||v| \quad \forall uv \in H$,

(15) $\qquad a$ is coercive, i.e., $\text{Re}\,a(u, u) \ge \alpha|u|^2 \quad \forall u \in H$, for some $\alpha > 0$.

Proposition 11.28. *Assume that a satifies (13), (14), and (15). Let K be a nonempty closed convex set in H. Then given any $\varphi \in H^\star$ there exists a unique $u \in K$ such that*

(16) $\qquad \text{Re}\,a(u, v - u) \ge \text{Re}\langle\varphi, v - u\rangle \quad \forall v \in K.$

Moreover, if $a(v, w) = \overline{a(w, v)}\,\forall v, w \in H$, then u is characterized by the property

$$u \in K \quad and \quad \frac{1}{2}a(u, u) - \text{Re}\langle\varphi, u\rangle = \min_{v \in K}\left\{\frac{1}{2}a(v, v) - \text{Re}\langle\varphi, v\rangle\right\}.$$

When $K = H$, (16) becomes $a(u, v) = \overline{\langle\varphi, v\rangle}\,\forall v \in H$. In particular, we deduce that any operator $T \in \mathcal{L}(H)$ satisfying

(17) $\qquad \text{Re}(Tu, u) \ge \alpha|u|^2 \quad \forall u \in H$, for some $\alpha > 0$,

is bijective from H onto itself. There is a variant that looks slightly more general (see, however, Remark 1 below).

Proposition 11.29 (Lax–Milgram). *Assume that $T \in \mathcal{L}(H)$ satisfies*

(18) $\qquad |(Tu, u)| \ge \alpha|u|^2 \quad \forall u \in H$, *for some $\alpha > 0$.*

Then T is bijective.

Proof. See Remark 8 in Chapter 5.

Remark 1. Clearly (17) implies (18). Conversely, assume that (18) holds. Then there exists some $\xi \in \mathbb{C}$ with $|\xi| = 1$ such that

$$(19) \qquad \qquad \text{Re}(\xi Tu, u) \geq \alpha |u|^2 \quad \forall u \in H.$$

Indeed, the numerical range

$$W(T) = \{(Tu, u); u \in H, |u| = 1\}$$

is a convex set (by Proposition 11.33 below). Moreover, by (18) we know that $0 \notin \overline{W(T)}$, and in fact $\text{dist}(0, W(T)) \geq \alpha$. Let p denote the projection of 0 onto $\overline{W(T)}$ (in $\mathbb{C} \simeq \mathbb{R}^2$). After a rotation in the plane (i.e., a multiplication by $\xi \in \mathbb{C}, |\xi| = 1$) bringing p to the point $(0, |p|)$ on the x-axis, we conclude that (19) holds.

Chapter 6. Sections 6.1 and 6.2 are totally unchanged. The main difference occurs in Section 6.3.

Let E be a Banach space over $\mathbb{C}$ and let $T \in \mathcal{L}(E)$. The *resolvent set* is defined by

$$\rho(T) = \{\lambda \in \mathbb{C}; (T - \lambda I) \text{ is bijective from } E \text{ onto } E\}.$$

The *spectrum* is the complement of $\rho(T)$, i.e., $\sigma(T) = \mathbb{C} \setminus \rho(T)$. A number $\lambda \in \mathbb{C}$ is an *eigenvalue* if the corresponding eigenspace $N(T - \lambda I) \neq \{0\}$ and the set of all eigenvalues is denoted by $EV(T)$. Clearly $EV(T) \subset \sigma(T)$. It may happen that $EV(T) = \emptyset$ (e.g., the right shift $Tu = (0, u_1, u_2, \dots)$). However, $\sigma(T)$ is *never* empty.

Proposition 11.30. *The spectrum $\sigma(T)$ is a nonempty compact set and*

$$\sigma(T) \subset \{\lambda \in \mathbb{C}; |\lambda| \leq \|T\|\}.$$

Proof. The main novelty is that $\sigma(T)$ is nonempty. The proof relies on the theory of analytic functions on $\mathbb{C}$ (more precisely Liouville's theorem) and we will not present it here. The interested reader may consult A. Taylor–D. Lay [1], W. Rudin [2], or A. Knaap [2].

The estimate $|\lambda| \leq \|T\| \ \forall \lambda \in \sigma(T)$ is usually not sharp. For example, in $\mathbb{C}^2$ the operator $T(u_1, u_2) = (u_2, 0)$ satisfies $\sigma(T) = \{0\}$ and $\|T\| = 1$. The optimal bound is given in terms of the spectral radius. We already know (see Exercise 6.23) that for every operator $T \in \mathcal{L}(E)$,

$$r(T) = \lim_{n \to \infty} \|T^n\|^{1/n} \text{ exists}$$

and clearly $r(T) \leq \|T\|$; $r(T)$ is called the *spectral radius*.

Proposition 11.31. *For every $T \in \mathcal{L}(E)$ we have*

$$r(T) = \max\{|\lambda|; \lambda \in \sigma(T)\}.$$

For the proof we refer again to A. Taylor–D. Lay [1], W. Rudin [2], or A. Knaap [2]. The argument relies heavily on the fact that E is a Banach space over $\mathbb{C}$ through the theory of power series on $\mathbb{C}$. When E is a Banach space over $\mathbb{R}$ we can say only that $\max\{|\lambda|; \lambda \in \sigma(T)\} \leq r(T)$, and the inequality can be strict even if $\sigma(T)$ is nonempty (see Exercise 6.23).

Another interesting difference between real and complex spaces concerns the so-called *spectral mapping theorem*. Consider first the real case: let $Q(t) = \sum_{k=0}^{p} a_k t^k$ be a polynomial with coefficients $a_k \in \mathbb{R}$ and let $T \in \mathcal{L}(E)$, where E is a Banach space over $\mathbb{R}$. We know (see Exercise 6.22) that

(20) $$Q(EV(T)) \subset EV(Q(T)) \quad \text{and} \quad Q(\sigma(T)) \subset \sigma(Q(T)),$$

and these inclusions might be strict (except, e.g., in the case of a Hilbert space when $T^\star = T$). In the complex case these inclusions become *equalities*: Suppose $Q(t)$ is a polynomial with coefficients $a_k \in \mathbb{C}$ and let $T \in \mathcal{L}(E)$, where E is a Banach space over $\mathbb{C}$.

Proposition 11.32. *We have*

(21) $$Q(EV(T)) = EV(Q(T))$$

and

(22) $$Q(\sigma(T)) = \sigma(Q(T)).$$

Proof. We already know that (20) holds (the argument is the same as in Exercise 6.22). Assume by contradiction that the inclusions are strict. Then there exists $\mu \in EV(Q(T))$ such that $\mu \notin Q(EV(T))$. Write

$$Q(t) - \mu = \alpha(t - t_1)(t - t_2) \cdots (t - t_p),$$

with $\alpha \neq 0$ and $t_i \notin EV(T)$ $\forall i$. In addition, we have some $x \neq 0$ such that $Q(T)x = \mu x$. Since $(T - t_1 I)$ is injective, we deduce that $(T - t_2 I) \cdots (T - t_p I)x = 0$, and repeating the same argument yields $x = 0$. Impossible.

Similarly, suppose $\mu \in \sigma(Q(T))$ is such that $\mu \notin Q(\sigma(T))$. Write $Q(t) - \mu$ as above with $t_i \notin \sigma(T)$ $\forall i$. Then $Q(T) - \mu I$ can be written as a product of bijective operators. Therefore $Q(T) - \mu I$ is bijective, i.e., $\mu \in \rho(Q(T))$. Impossible.

In Hilbert spaces, a useful tool in the study of the spectrum is the *numerical range*. Let H be a Hilbert space over $\mathbb{C}$; the numerical range of an operator $T \in \mathcal{L}(H)$ is defined by

$$W(T) = \{(Tu, u); u \in H \text{ and } |u| = 1\}.$$

Proposition 11.33. *We have*

$$\sigma(T) \subset \overline{W(T)},$$

and more precisely, if $\lambda \notin \overline{W(T)}$, then $\lambda \in \rho(T)$ with

(23) $\|(T - \lambda I)^{-1}\| \leq 1/\operatorname{dist}(\lambda, W(T))$.

In addition, $W(T)$ is convex.

Proof. Assume that $\lambda \notin \overline{W(T)}$ and set $\alpha = \operatorname{dist}(\lambda, W(T))$. We have

$$|(Tu, u) - \lambda| \geq \alpha \quad \forall u \in H \text{ with } |u| = 1.$$

Thus
$$|(Tu - \lambda u, u)| \geq \alpha|u|^2 \quad \forall u \in H.$$

Applying Lax–Milgram (Proposition 11.29), we conclude that $(T - \lambda I)$ is bijective and that $|Tu - \lambda u| \geq \alpha|u| \; \forall u \in H$, i.e., $\|(T - \lambda I)^{-1}\| \leq 1/\alpha$.

The convexity of $W(T)$ is a counterintuitive fact due to Toeplitz and Hausdorff. For the proof we refer to P. R. Halmos [2].

In general, the numerical range $W(T)$ can be much larger than the spectrum. For example, with $H = \mathbb{C}^2$ and $T(u_1, u_2) = (u_2, 0)$ we have $EV(T) = \sigma(T) = \{0\}$, while $W(T) = \{\lambda \in \mathbb{C}; |\lambda| \leq 1/2\}$. However, if T is self-adjoint, or more generally normal (see below), then $\overline{W(T)} = \operatorname{conv} \sigma(T)$, the convex hull of $\sigma(T)$ (see P. R. Halmos [2] and Remark 2 below).

When H is a Hilbert space over $\mathbb{C}$ and $T \in \mathcal{L}(H)$, a word of caution about the concept of adjoint $T^\star$ is necessary. Following a general procedure, the adjoint of an operator $T \in \mathcal{L}(H)$ is defined via the relation

$$\langle T^\star f, u \rangle_{H^\star, H} = \langle f, Tu \rangle_{H^\star, H} \quad \forall f \in H^\star, \quad \forall u \in H,$$

and then $T^\star \in \mathcal{L}(H^\star)$ (we emphasize that $T^\star(\lambda f) = \lambda T^\star f \; \forall \lambda \in \mathbb{C}$ and $\forall f \in H^\star$). Moreover, $(\lambda T)^\star = \lambda T^\star \; \forall \lambda \in \mathbb{C}$ (because $f : H \to \mathbb{C}$ is linear).

On the other hand, we may also identify $H^\star$ with H (via the isomorphism in Proposition 11.27), and view $T^\star$ as an operator from H into itself defined through the relation
$$(Tu, v) = (u, T^\star v) \quad \forall u, v \in H,$$

and we have $T^\star \in \mathcal{L}(H)$ (we emphasize that $T^\star(\lambda v) = \lambda T^\star v \; \forall \lambda \in \mathbb{C}$ and $\forall v \in H$). *However*, we now have

(24) $(\lambda T)^\star = \bar{\lambda} T^\star \quad \forall \lambda \in \mathbb{C}$

(as can be easily checked). This convention is commonly used, so that $T^\star$ and T live in the same world: one can compare $T^\star$ and T, compose $T^\star$ and T, etc.

We say that an operator $T \in \mathcal{L}(H)$ is *self-adjoint* (or Hermitian) if $T^\star = T$, i.e.,

$$(Tu, v) = (u, Tv) \quad \forall u, v \in H.$$

If T is self-adjoint, then $(Tu, u) = (u, Tu) = \overline{(Tu, u)} \; \forall u \in H$, so that $(Tu, u) \in \mathbb{R}$ $\forall u \in H$. In particular, the numerical range $W(T)$ is a subset of $\mathbb{R}$ and thus $\sigma(T) \subset \mathbb{R}$.

The spectral decomposition of compact, self-adjoint operators is exactly the same as in Chapter 6.

Proposition 11.34. *Let H be a separable Hilbert space over $\mathbb{C}$ and let T be a compact self-adjoint operator. Then there exists a Hilbert basis composed of eigenvectors of T (and the corresponding eigenvalues are real).*

We say that an operator $T \in \mathcal{L}(H)$ is *normal* if it satisfies $T^\star \circ T = T \circ T^\star$. Various properties of normal operators are discussed in Problem 43 when the underlying space H is a Hilbert space over $\mathbb{R}$; they still remain valid when H is a Hilbert space over $\mathbb{C}$. But we have now much more:

Proposition 11.35. *Let H be a Hilbert space over $\mathbb{C}$ and let T be a normal operator. Then*

(25) $$\max\{|\lambda|; \ \lambda \in \sigma(T)\} = \|T\|.$$

Proof. Since T is normal, we have

$$\|T^p\| = \|T\|^p \text{ for every integer } p \geq 1.$$

This is proved in Problem 43 when H is a Hilbert space over $\mathbb{R}$, and the same argument remains valid when H is a Hilbert space over $\mathbb{C}$ (alternatively apply the real result to T on $H_{\mathbb{R}}$). Therefore $r(T) = \lim_{n\to\infty}\|T^n\|^{1/n} = \|T\|$. Combining this with Proposition 11.31 yields (25).

Proposition 11.36. *Let H be a separable Hilbert space over $\mathbb{C}$ and let T be a compact normal operator, then there exists a Hilbert basis composed of eigenvectors of T (but the corresponding eigenvalues need not be real).*

Proof. If T is normal, so is $(T - \lambda I)$ for any $\lambda \in \mathbb{C}$. Therefore (as in Problem 43) we have $N(T - \lambda I) = N((T - \lambda I)^\star) = N(T^\star - \bar{\lambda} I)$. It follows that $N(T - \lambda I)$ and $N(T - \mu I)$ are orthogonal when $\lambda \neq \mu$. We may then proceed exactly as in the proof of Theorem 6.11. We obtain a compact normal operator T_0 on $F^\perp$ with $\sigma(T_0) = \{0\}$. Instead of invoking Corollary 6.10 to conclude that $T_0 = 0$, we apply instead Proposition 11.35 and derive that $T_0 = 0$. It is here that we make use of the fact that H is a space over $\mathbb{C}$ (the same conclusion fails in real spaces).

Remark 2. It is easy to deduce from Proposition 11.36 that $\overline{W(T)} = \text{conv}\,\sigma(T)$ when T is a compact normal operator. Indeed, choose a basis (e_i) as in Proposition 11.36. Given $u \in H$ with $|u| = 1$ write $u = \sum u_i e_i$ and $\sum |u_i|^2 = 1$. Then $Tu = \sum \lambda_i u_i e_i$ and $(Tu, u) = \sum \lambda_i |u_i|^2$. It is still true that $\overline{W(T)} = \text{conv}\,\sigma(T)$ for any normal operator T (not necessarily compact); see P. R. Halmos [2].

Let H be a Hilbert space over $\mathbb{C}$. We say that an operator $T \in \mathcal{L}(H)$ is an *isometry* if $|Tu| = |u| \ \forall u \in H$, and T is a *unitary* operator if T is an isometry that is also surjective. Various properties of isometries and unitary operators are discussed in Problem 44 when the underlying space H is a Hilbert space over $\mathbb{R}$; most of them remain valid when H is a Hilbert space over $\mathbb{C}$, except a statement about the spectrum, which needs to be modified as follows:

Proposition 11.37. *Let T be an isometry. Then*

$$EV(T) \subset S^1 = \{\lambda \in \mathbb{C}; |\lambda| = 1\}.$$

If T is a unitary operator, then
$$\sigma(T) \subset S^1,$$

*and if T is **not** a unitary operator, then*

$$\sigma(T) = \{\lambda \in \mathbb{C}; |\lambda| \leq 1\}.$$

The proof is an easy adaptation of the one given in the solution of Problem 44, question 6.

An operator $T \in \mathcal{L}(H)$ is said to be *skew-adjoint* (or antisymmetric) if $T^* = -T$. Clearly, T is skew-adjoint if and only if iT is self-adjoint (this follows from (24)). Thus, for any skew-adjoint operator we have $EV(T) \subset \sigma(T) \subset \overline{W(T)} \subset i\mathbb{R}$.

Chapter 7. Very little needs to be changed. In the definition of a monotone operator replace the assumption $(Av, v) \geq 0 \ \forall v \in D(A)$ by $\mathrm{Re}(Av, v) \geq 0 \ \forall v \in D(A)$. Many computations in Sections 7.2, 7.3, and 7.4 rely on the following identity: if $\varphi \in C^1([0, +\infty); H)$, then $|\varphi|^2 \in C^1([0, +\infty); \mathbb{R})$ and $\frac{d}{dt}|\varphi|^2 = 2\,\mathrm{Re}(\frac{d\varphi}{dt}, \varphi)$, since

$$\frac{d}{dt}|\varphi|^2 = \frac{d}{dt}(\varphi, \varphi) = \left(\frac{d\varphi}{dt}, \varphi\right) + \left(\varphi, \frac{d\varphi}{dt}\right)$$

$$= \left(\frac{d\varphi}{dt}, \varphi\right) + \overline{\left(\frac{d\varphi}{dt}, \varphi\right)} = 2\,\mathrm{Re}\left(\frac{d\varphi}{dt}, \varphi\right).$$

Chapters 8 and 9. Interesting properties of the spectrum of second-order elliptic operators that are *not* self-adjoint may be found in S. Agmon [1] (Section 16).

Solutions of Some Exercises

In this section the formulas are numbered (S1), (S2), etc, in order to avoid any confusion with formulas from the previous sections.

1.1

1. The equality $\langle f, x \rangle = \|x\|^2$ implies that $\|x\| \leq \|f\|$. Corollary 1.3 implies that $F(x)$ is nonempty. It is clear from the second form of $F(x)$ that $F(x)$ is closed and convex.
2. In a strictly convex normed space any nonempty convex set that is contained in a sphere is reduced to a single point.
3. Note that

$$\langle f, y \rangle \leq \|f\| \, \|y\| \leq \frac{1}{2}\|f\|^2 + \frac{1}{2}\|y\|^2.$$

Conversely, assume that f satisfies

(S1) $$\frac{1}{2}\|y\|^2 - \frac{1}{2}\|x\|^2 \geq \langle f, y - x \rangle \quad \forall y \in E.$$

First choose $y = \lambda x$ with $\lambda \in \mathbb{R}$ in (S1); by varying λ one sees that $\langle f, x \rangle = \|x\|^2$. Next choose y in (S1) such that $\|y\| = \delta > 0$; it follows that

$$\langle f, y \rangle \leq \frac{1}{2}\delta^2 + \frac{1}{2}\|x\|^2.$$

Therefore we obtain

$$\delta\|f\| = \sup_{\substack{y \in E \\ \|y\|=\delta}} \langle f, y \rangle \leq \frac{1}{2}\delta^2 + \frac{1}{2}\|x\|^2.$$

The conclusion follows by choosing $\delta = \|x\|$.
4. If $f \in F(x)$ one has

H. Brezis, *Functional Analysis, Sobolev Spaces and Partial Differential Equations*, DOI 10.1007/978-0-387-70914-7, © Springer Science+Business Media, LLC 2011

$$\frac{1}{2}\|y\|^2 - \frac{1}{2}\|x\|^2 \geq \langle f, y - x \rangle$$

and if $g \in F(y)$ one has

$$\frac{1}{2}\|x\|^2 - \frac{1}{2}\|y\|^2 \geq \langle g, x - y \rangle.$$

Adding these inequalities leads to $\langle f - g, x - y \rangle \geq 0$. On the other hand, note that

$$\langle f - g, x - y \rangle = \|x\|^2 + \|y\|^2 - \langle f, y \rangle - \langle g, x \rangle$$
$$\geq \|x\|^2 + \|y\|^2 - 2\|x\| \, \|y\|.$$

5. By question 4 we already know that $\|x\| = \|y\|$. On the other hand, we have

$$\langle F(x) - F(y), x - y \rangle = [\|x\|^2 - \langle F(x), y \rangle] + [\|y\|^2 - \langle F(y), x \rangle],$$

and both terms in brackets are ≥ 0. It follows that $\|x\|^2 = \|y\|^2 = \langle F(x), y \rangle = \langle F(y), x \rangle$, which implies that $F(x) \in F(y)$ and thus $F(x) = F(y)$ by question 2.

$\boxed{1.2}$

1(a).
$$\|f\|_{E^\star} = \max_{1 \leq i \leq n} |f_i|.$$

1(b). $f \in F(x)$ iff for every $1 \leq i \leq n$ one has

$$f_i = \begin{cases} (\text{sign } x_i)\|x\|_1 & \text{if } x_i \neq 0, \\ \text{anything in the interval } [-\|x\|_1, +\|x\|_1] & \text{if } x_i = 0. \end{cases}$$

2(a).
$$\|f\|_{E^\star} = \sum_{i=1}^{n} |f_i|.$$

2(b). Given $x \in E$ consider the set

$$I = \{1 \leq i \leq n; |x_i| = \|x\|_\infty\}.$$

Then $f \in F(x)$ iff one has

(i) $f_i = 0 \quad \forall i \notin I$,
(ii) $f_i x_i \geq 0 \, \forall i \in I$ and $\sum_{i \in I} |f_i| = \|x\|_\infty$.

3.
$$\|f\|_{E^\star} = \left(\sum_{i=1}^{n} |f_i|^2 \right)^{1/2}$$

and $f \in F(x)$ iff one has $f_i = x_i \, \forall i = 1, 2, \ldots, n$. More generally,

$$\|f\|_{E^*} = \left(\sum_{i=1}^{n} |f_i|^{p'} \right)^{1/p'},$$

where $1/p + 1/p' = 1$, and $f \in F(x)$ iff one has $f_i = |x_i|^{p-2} x_i / \|x\|_p^{p-2}$ $\forall i = 1, 2, \ldots, n$.

1.3

1. $\|f\|_{E^*} = 1$ (note that $f(t^\alpha) = 1/(1 + \alpha)$ $\forall \alpha > 0$).
2. If there exists such a u we would have $\int_0^1 (1 - u) dt = 0$ and thus $u \equiv 1$; absurd.

1.5

1. Let P denote the family of all linearly independent subsets of E. It is easy to see that P (ordered by the usual inclusion) is inductive. Zorn's lemma implies that P has a maximal element, denoted by $(e_i)_{i \in I}$, which is clearly an algebraic basis. Since $e_i \neq 0$ $\forall i \in I$, one may assume, by normalization, that $\|e_i\| = 1$ $\forall i \in I$.
2. Since E is infinite-dimensional one may assume that $\mathbb{N} \subset I$. There exists a (unique) linear functional on E such that $f(e_i) = i$ if $i \in \mathbb{N}$ and $f(e_i) = 0$ if $i \in I \backslash \mathbb{N}$.
3. Assume that I is countable, i.e., $I = \mathbb{N}$. Consider the vector space F_n spanned by $(e_i)_{1 \le i \le n}$. F_n is closed (see Section 11.1) and, moreover, $\bigcup_{n=1}^{\infty} F_n = E$. It follows from the Baire category theorem that there exists some n_0 such that $\text{Int}(F_{n_0}) \neq \emptyset$. Thus $E = F_{n_0}$; absurd.

1.7

1. Let $x, y \in \overline{C}$, so that $x = \lim x_n$ and $y = \lim y_n$ with $x_n, y_n \in C$. Thus $tx + (1 - t)y = \lim[tx_n + (1 - t)y_n]$ and therefore $tx + (1 - t)y \in \overline{C}$ $\forall t \in [0, 1]$. Assume $x, y \in \text{Int } C$, so that there exists some $r > 0$ such that $B(x, r) \subset C$ and $B(y, r) \subset C$. It follows that

$$t B(x, r) + (1 - t) B(y, r) \subset C \ \forall t \in [0, 1].$$

But $t B(x, r) + (1 - t) B(y, r) = B(tx + (1 - t)y, r)$ (why?).
2. Let $r > 0$ be such that $B(y, r) \subset C$. One has

$$tx + (1 - t) B(y, r) \subset C \quad \forall t \in [0, 1],$$

and therefore $B(tx + (1 - t)y, (1 - t)r) \subset C$. It follows that $tx + (1 - t)y \in \text{Int } C$ $\forall t \in [0, 1)$.
3. Fix any $y_0 \in \text{Int } C$. Given $x \in C$ one has $x = \lim_{n \to \infty}[(1 - \frac{1}{n})x + \frac{1}{n}y_0]$. But $(1 - \frac{1}{n})x + \frac{1}{n}y_0 \in \text{Int } C$ and therefore $x \in \overline{\text{Int } C}$. This proves that $C \subset \overline{\text{Int } C}$ and hence $\overline{C} \subset \overline{\text{Int } C}$.

1.8

1. We already know that

$$p(\lambda x) = \lambda p(x) \; \forall \lambda > 0, \; \forall x \in E \quad \text{and} \quad p(x+y) \leq p(x) + p(y) \; \forall x, y \in E.$$

It remains to check that

(i) $p(-x) = p(x) \; \forall x \in E$, which follows from the symmetry of C.
(ii) $p(x) = 0 \Rightarrow x = 0$, which follows from the fact that C is bounded. More precisely, let $L > 0$ be such that $\|x\| \leq L \; \forall x \in C$. It is easy to see that

$$p(x) \geq \frac{1}{L} \|x\| \; \forall x \in E.$$

2. C is *not* bounded. Consider for example the sequence $u_n(t) = \sqrt{n}/(1 + nt)$ and check that $u_n \in C$, while $\|u_n\| = \sqrt{n}$. Here $p(u) = \left(\int_0^1 |u(t)|^2 dt \right)^{1/2}$ is a norm that is not equivalent to $\|u\|$.

1.9

1. Let

$$P = \left\{ \lambda = (\lambda_1, \lambda_2, \ldots, \lambda_n) \in \mathbb{R}^n; \; \lambda_i \geq 0 \; \forall i \text{ and } \sum_{i=1}^{n} \lambda_i = 1 \right\},$$

so that P is a compact subset of $\mathbb{R}^n$ and C_n is the image of P under the continuous map $\lambda \mapsto \sum_{i=1}^{n} \lambda_i x_i$.
2. Apply Hahn–Banach, second geometric form, to C_n and $\{0\}$. Normalize the linear functional associated to the hyperplane that separates C_n and $\{0\}$.
4. Apply the above construction to $C = A - B$.

1.10

(A) $\Rightarrow$ (B) is obvious.
(B) $\Rightarrow$ (A). Let G be the vector space spanned by the x_i's $(i \in I)$. Given $x \in G$ write $x = \sum_{i \in J} \beta_i x_i$ and set $g(x) = \sum_{i \in J} \beta_i \alpha_i$. Assumption (B) implies that this definition makes sense and that $|g(x)| \leq M \|x\| \; \forall x \in G$. Next, extend g to all of E using Corollary 1.2.

1.11

(A) $\Rightarrow$ (B) is again obvious.
(B) $\Rightarrow$ (A). Assume first that the f_i's are linearly independent $(1 \leq i \leq n)$. Set $\alpha = (\alpha_1, \alpha_2, \ldots, \alpha_n) \in \mathbb{R}^n$. Consider the map $\varphi : E \to \mathbb{R}^n$ defined by

$$\varphi(x) = (\langle f_1, x \rangle, \ldots, \langle f_n, x \rangle).$$

Let $C = \{x \in E; \|x\| \leq M + \varepsilon\}$. One has to show that $\alpha \in \varphi(C)$. Suppose, by contradiction, that $\alpha \notin \varphi(C)$ and separate $\varphi(C)$ and $\{\alpha\}$ (see Exercise 1.9). Hence, there exists some $\beta = (\beta_1, \beta_2, \ldots, \beta_n) \in \mathbb{R}^n, \; \beta \neq 0$, such that

$$\beta \cdot \varphi(x) \leq \beta \cdot \alpha \quad \forall x \in C, \text{ i.e., } \left\langle \sum \beta_i f_i, x \right\rangle \leq \sum \beta_i \alpha_i \quad \forall x \in C.$$

It follows that $(M+\varepsilon)\|\sum \beta_i f_i\| \leq \sum \beta_i \alpha_i$. Using asumption (B) one finds that $\sum \beta_i f_i = 0$. Since the f_i's are linearly independent one concludes that $\beta = 0$; absurd.

In the general case, apply the above result to a maximal linearly independent subset of $(f_i)_{1 \leq i \leq n}$.

$\boxed{1.15}$

1. It is clear that $C \subset C^{**}$ and that C^{**} is closed. Conversely, assume that $x_0 \in C^{**}$ and $x_0 \notin \overline{C}$. One may strictly separate $\{x_0\}$ and $\overline{C}$, so that there exist some $f_0 \in E^*$ and some $\alpha_0 \in \mathbb{R}$ such that

$$\langle f_0, x \rangle < \alpha_0 < \langle f_0, x_0 \rangle \ \forall x \in \overline{C}.$$

Since $0 \in C$ it follows that $\alpha_0 > 0$; letting $f = (1/\alpha_0) f_0$, one has

$$\langle f, x \rangle < 1 < \langle f, x_0 \rangle \ \forall x \in C.$$

Thus $f \in C^*$ and we are led to a contradiction, since $x_0 \in C^{**}$.

2. If C is a linear subspace then

$$C^* = \{f \in E^*; \langle f, x \rangle = 0 \ \forall x \in C\} = C^{\perp}.$$

$\boxed{1.18}$

(a) $$\varphi^*(f) = \begin{cases} -b & \text{if } f = a, \\ +\infty & \text{if } f \neq a. \end{cases}$$

(b) $$\varphi^*(f) = \begin{cases} f \log f - f & \text{if } f > 0, \\ 0 & \text{if } f = 0, \\ +\infty & \text{if } f < 0. \end{cases}$$

(c) $$\varphi^*(f) = |f|.$$

(d) $$\varphi^*(f) = 0.$$

(e) $$\varphi^*(f) = \begin{cases} +\infty & \text{if } f \geq 0, \\ -1 - \log|f| & \text{if } f < 0. \end{cases}$$

(f) $$\varphi^*(f) = (1 + f^2)^{1/2}.$$

(g) $$\varphi^*(f) = \begin{cases} \frac{1}{2}f^2 & \text{if } |f| \leq 1, \\ +\infty & \text{if } |f| > 1. \end{cases}$$

(h) $$\varphi^*(f) = \frac{1}{p'}|f|^{p'} \quad \text{with } \frac{1}{p} + \frac{1}{p'} = 1.$$

(i) $$\varphi^*(f) = \begin{cases} 0 & \text{if } 0 \leq f \leq 1, \\ +\infty & \text{otherwise.} \end{cases}$$

(j)
$$\varphi^\star(f) = \begin{cases} \frac{1}{p'} f^{p'} & \text{if } f \geq 0, \\ 0 & \text{if } f < 0. \end{cases}$$

(k)
$$\varphi^\star(f) = \begin{cases} +\infty & \text{if } f \geq 0, \\ -\frac{1}{p'}|f|^{p'} & \text{if } f < 0. \end{cases}$$

(l)
$$\varphi^\star(f) = |f| + \frac{1}{p'}|f|^{p'}.$$

1.20 The conjugate functions are defined on $\ell^{p'}$ with $\frac{1}{p} + \frac{1}{p'} = 1$ by

(a)
$$\varphi^\star(f) = \begin{cases} \frac{1}{4}\sum_{k=1}^{\infty}\frac{1}{k}|f_k|^2 & \text{if } \sum_{k=1}^{\infty}\frac{1}{k}|f_k|^2 < +\infty, \\ +\infty & \text{otherwise.} \end{cases}$$

(b)
$$\varphi^\star(f) = \begin{cases} \sum_{k=2}^{+\infty} a_k|f_k|^{k/(k-1)} & \text{if } \sum_{k=2}^{+\infty} a_k|f_k|^{k/(k-1)} < +\infty, \\ +\infty & \text{otherwise,} \end{cases}$$

with $a_k = \dfrac{(k-1)}{k^{k/(k-1)}}$.

(c)
$$\varphi^\star(f) = \begin{cases} 0 & \text{if } \|f\|_{\ell^\infty} \leq 1, \\ +\infty & \text{otherwise.} \end{cases}$$

1.21

2. $\varphi^\star = I_A$, where $A = \{[f_1, f_2]; f_1 \leq 0, f_2 \leq 0, \text{ and } 4f_1 f_2 \geq 1\}$.
3. One has
$$\inf_{x \in E} \{\varphi(x) + \psi(x)\} = 0$$
and
$$\varphi^\star = I_{D^\perp}, \quad \text{where} \quad D^\perp = \{[f_1, f_2]; f_2 = 0\}.$$
It follows that
$$\varphi^\star(-f) + \psi^\star(f) = +\infty \quad \forall f \in E^\star,$$
and thus
$$\sup_{f \in E^\star} \{-\varphi^\star(-f) - \psi^\star(f)\} = -\infty.$$

4. The assumptions of Theorem 1.12 are not satisfied: there is *no* element $x_0 \in E$ such that $\varphi(x_0) < +\infty$, $\psi(x_0) < +\infty$, and φ is continuous at x_0.

1.22

1. Write that
$$\|x - a\| \leq \|x - y\| + \|y - a\|.$$
Taking $\inf_{a \in A}$ leads to $\varphi(x) \leq \|x - y\| + \varphi(y)$. Then exchange x and y.

2. Let $x, y \in E$ and $t \in [0, 1]$ be fixed. Given $\varepsilon > 0$ there exist some $a \in A$ and some $b \in A$ such that

$$\|x - a\| \le \varphi(x) + \varepsilon \quad \text{and} \quad \|y - b\| \le \varphi(y) + \varepsilon.$$

Therefore

$$\|tx + (1 - t)y - [ta + (1 - t)b]\| \le t\varphi(x) + (1 - t)\varphi(y) + \varepsilon.$$

But $ta + (1 - t)b \in A$, so that

$$\varphi(tx + (1 - t)y) \le t\varphi(x) + (1 - t)\varphi(y) + \varepsilon \quad \forall \varepsilon > 0.$$

3. Since A is closed, one has $A = \{x \in E; \varphi(x) \le 0\}$, and therefore A is convex if φ is convex.
4. One has
$$\varphi^\star(f) = \sup_{x \in E}\{\langle f, x \rangle - \inf_{a \in A} \|x - a\|\}$$
$$= \sup_{x \in E}\sup_{a \in A}\{\langle f, x \rangle - \|x - a\|\}$$
$$= \sup_{a \in A}\sup_{x \in E}\{\langle f, x \rangle - \|x - a\|\}$$
$$= (I_A)^\star(f) + I_{B_{E^\star}}(f).$$

1.23

1. Let $f \in D(\varphi^\star) \cap D(\psi^\star)$. For every $x, y \in E$ one has

$$\langle f, x - y \rangle - \varphi(x - y) \le \varphi^\star(f),$$
$$\langle f, y \rangle - \psi(y) \le \psi^\star(f).$$

Adding these inequalities leads to

$$(\varphi \nabla \psi)(x) \ge \langle f, x \rangle - \varphi^\star(f) - \psi(f).$$

In particular, $(\varphi \nabla \psi)(x) > -\infty$. Also, we have

$$(\varphi \nabla \psi)^\star(f) = \sup_{x \in E}\{\langle f, x \rangle - \inf_{y \in E}[\varphi(x - y) + \psi(y)]\}$$
$$= \sup_{x \in E}\sup_{y \in E}\{\langle f, x \rangle - \varphi(x - y) - \psi(y)\}$$
$$= \sup_{y \in E}\sup_{x \in E}\{\langle f, x \rangle - \varphi(x - y) - \psi(y)\}$$
$$= \varphi^\star(f) + \psi^\star(f).$$

2. One has to check that $\forall f, g \in E^\star$ and $\forall x \in E$,

$$\langle f, x \rangle - \varphi(x) - \psi(x) \le \varphi^\star(f - g) + \psi^\star(g).$$

This becomes obvious by writing

$$\langle f, x \rangle = \langle f - g, x \rangle + \langle g, x \rangle.$$

3. Given $f \in E^\star$, one has to prove that

(S1) $$\sup_{x \in E} \{\langle f, x \rangle - \varphi(x) - \psi(x)\} = \inf_{g \in E^\star} \{\varphi^\star(f - g) + \psi^\star(g)\}.$$

Note that

$$\sup_{x \in E} \{\langle f, x \rangle - \varphi(x) - \psi(x)\} = - \inf_{x \in E} \{\tilde{\varphi}(x) + \psi(x)\}$$

with $\tilde{\varphi}(x) = \varphi(x) - \langle f, x \rangle$. Applying Theorem 1.12 to the functions $\tilde{\varphi}$ and ψ leads to

$$\inf_{x \in E} \{\tilde{\varphi}(x) + \psi(x)\} = \sup_{g \in E^\star} \{-\tilde{\varphi}^\star(-g) - \psi^\star(g)\},$$

which corresponds precisely to (S1).

4. Clearly one has

$$\begin{aligned}
(\varphi^\star \nabla \psi^\star)^\star(x) &= \sup_{f \in E^\star} \{\langle f, x \rangle - \inf_{g \in E^\star} [\varphi^\star(f - g) + \psi^\star(g)]\} \\
&= \sup_{f \in E^\star} \sup_{g \in E^\star} \{\langle f, x \rangle - \varphi^\star(f - g) - \psi^\star(g)\} \\
&= \sup_{g \in E^\star} \sup_{f \in E^\star} \{\langle f, x \rangle - \varphi^\star(f - g) - \psi^\star(g)\} \\
&= \varphi^{\star\star}(x) + \psi^{\star\star}(x).
\end{aligned}$$

1.24

1. One knows (Proposition 1.10) that there exist some $f \in E^\star$ and a constant C such that $\varphi(y) \geq \langle f, y \rangle - C \quad \forall y \in E$. Choosing $n \geq \|f\|$, one has $\varphi_n(x) \geq -n\|x\| - C > -\infty$.

2. The function φ_n is the inf-convolution of two convex functions; thus φ_n is convex (see question 7 in Exercise 1.23). In order to prove that $|\varphi_n(x_1) - \varphi_n(x_2)| \leq n\|x_1 - x_2\|$, use the same argument as in question 1 of Exercise 1.22.

3. $(\varphi_n)^\star = I_{nB_{E^\star}} + \varphi^\star$ (by question 1 of Exercise 1.23).

5. By question 1 we have $\varphi(y) \geq -\|f\| \, \|y\| - C \quad \forall y \in E$, which leads to

$$n\|x - y_n\| \leq \|f\| \, \|y_n\| + C + \varphi(x) + 1/n.$$

It follows that $\|y_n\|$ remains bounded as $n \to \infty$, and therefore $\lim_{n \to \infty} \|x - y_n\| = 0$. On the other hand, we have $\varphi_n(x) \geq \varphi(y_n) - 1/n$, and since φ is l.s.c. we conclude that $\liminf_{n \to \infty} \varphi_n(x) \geq \varphi(x)$.

6. Suppose, by contradiction, that there exists a constant C such that $\varphi_n(x) \leq C$ along a subsequence still denoted by $\varphi_n(x)$. Choosing y_n as in question 5 we see

that $y_n \to x$. Moreover, $\varphi(y_n) \leq C + 1/n$ and thus $\varphi(x) \leq \liminf_{n\to\infty} \varphi(y_n) \leq C$; absurd.

1.25

4. For each fixed $t > 0$ the function

$$y \mapsto \frac{1}{2t}\left[\|x + ty\|^2 - \|x\|^2\right]$$

is convex. Thus the function $y \mapsto [x, y]$ is convex as a *limit* of convex functions. On the other hand, $G(x, y) = \sup_{t>0}\{-\frac{1}{2t}[\|x + ty\|^2 - \|x\|^2]\}$ is l.s.c. as a *supremum* of continuous functions.

5. One already knows (see question 3 of Exercise 1.1) that

$$\frac{1}{2}\|x + ty\|^2 - \frac{1}{2}\|x\|^2 \geq \langle f, ty \rangle$$

and therefore

$$[x, y] \geq \langle f, y \rangle \quad \forall x, y \in E, \ \forall f \in F(x).$$

On the other hand, one has

$$\varphi^*(f) = \frac{1}{2}\|f\|^2 - \langle f, x \rangle + \frac{1}{2}\|x\|^2$$

and

$$\psi^*(f) = \begin{cases} 0 & \text{if } \langle f, y \rangle + \alpha \leq 0, \\ +\infty & \text{if } \langle f, y \rangle + \alpha > 0. \end{cases}$$

It is easy to check that $\inf_{z\in E}\{\varphi(z) + \psi(z)\} = 0$. It follows from Theorem 1.12 that there exists some $f_0 \in E^*$ such that $\varphi^*(f_0) + \psi^*(-f_0) = 0$, i.e., $\langle f_0, y \rangle \geq \alpha$ and $\frac{1}{2}\|f_0\|^2 - \langle f_0, x \rangle + \frac{1}{2}\|x\|^2 = 0$. Consequently, we have $\|f_0\| = \|x\|$ and $\langle f_0, x \rangle = \|x\|^2$, i.e., $f_0 \in F(x)$.

6. (a) $1 < p < \infty$, $[x, y] = \dfrac{\sum |x_i|^{p-2} x_i y_i}{\|x\|_p^{p-2}}$.

   (b) $p = 1$, $[x, y] = \|x\|_1\left[\sum_{x_i \neq 0}(\text{sign } x_i)y_i + \sum_{x_i = 0}|y_i|\right]$.

   (c) $p = \infty$, $[x, y] = \max_{i\in I}\{x_i y_i\}$, where $I = \{1 \leq i \leq n; |x_i| = \|x\|_\infty\}$.

1.27 Let $\widetilde{T} : E \to F$ be a continuous linear extension of T. It is easy to check that $E = N(\widetilde{T}) + G$ and $N(\widetilde{T}) \cap G = \{0\}$, so that $N(\widetilde{T})$ is a complement of G; absurd.

2.1 Without loss of generality we may assume that $x_0 = 0$.

1. Let $X = \{x \in E; \|x\| \leq \rho\}$ with $\rho > 0$ small enough that $X \subset D(\varphi)$. The sets F_n are closed and $\bigcup_{n=1}^{\infty} F_n = X$. By the Baire category theorem there is some n_0 such that $\text{Int}(F_{n_0}) \neq \emptyset$. Let $x_1 \in E$ and $\rho_1 > 0$ be such that $B(x_1, \rho_1) \subset F_{n_0}$. Given any $x \in E$ with $\|x\| < \rho_1/2$ write $x = \frac{1}{2}(x_1 + 2x) + \frac{1}{2}(-x_1)$ to conclude that $\varphi(x) \leq \frac{1}{2}n_0 + \frac{1}{2}\varphi(-x_1)$.

2. There exist some $\xi \in E$ and some constant $t \in [0, 1]$ such that $\|\xi\| = R$ and $x_2 = tx_1 + (1 - t)\xi$. It follows that

$$\varphi(x_2) \le t\varphi(x_1) + (1 - t)M$$

and consequently $\varphi(x_2) - \varphi(x_1) \le (1 - t)[M - \varphi(x_1)]$. But $x_2 - x_1 = (1 - t)(\xi - x_1)$ and thus $\|x_2 - x_1\| \ge (1 - t)(R - r)$. Hence we have

$$\varphi(x_2) - \varphi(x_1) \le \frac{\|x_2 - x_1\|}{R - r}[M - \varphi(x_1)].$$

On the other hand, if $x_2 = 0$ one obtains $t\|x_1\| = (1 - t)R$ and therefore

$$(1 - t) = \frac{\|x_1\|}{\|x_1\| + R} \le \frac{1}{2}.$$

It follows that $\varphi(0) - \varphi(x_1) \le \frac{1}{2}[M - \varphi(x_1)]$, so that $M - \varphi(x_1) \le 2[M - \varphi(0)]$.

$\boxed{2.2}$ We have $p(0) \le p(x_n) + p(-x_n) \to 0$, so that $p(0) \le 0$. On the other hand $p(0) \le 2p(0)$ by (i). Thus $p(0) = 0$.

Next we prove that $p(\alpha_n x_n) \to 0$. Argue by contradiction and assume that $|p(\alpha_n x_n)| > 2\varepsilon$ along a subsequence, for some $\varepsilon > 0$. Passing to a further subsequence we may assume that $\alpha_n \to \alpha$ for some $\alpha \in \mathbb{R}$. For simplicity we still denote (x_n) and (α_n) the corresponding sequences.

The sets F_n are closed and $\bigcup_{n \ge 1} F_n = \mathbb{R}$. Applying the Baire category theorem, we find some n_0 such that $\text{Int } F_{n_0} \ne \emptyset$. Hence, there exist some $\lambda_0 \in \mathbb{R}$ and some $\delta > 0$ such that $|p((\lambda_0 + t)x_k)| \le \varepsilon \ \forall k \ge n_0, \ \forall t$ with $|t| < \delta$. On the other hand, note that

$$p(\alpha_k x_k) \le p((\lambda_0 + \alpha_k - \alpha)x_k) + p((\alpha - \lambda_0)x_k),$$
$$-p(\alpha_k x_k) \le -p((\lambda_0 + \alpha_k - \alpha)x_k) + p((\lambda_0 - \alpha)x_k).$$

Hence we obtain $|p(\alpha_k x_k)| \le 2\varepsilon$ for k large enough. A contradiction.

Finally, write

$$p(\alpha_n x_n) - p(\alpha x) \le p(\alpha_n(x_n - x)) + p(\alpha_n x) - p(\alpha x) \to 0$$

and

$$p(\alpha_n x) \le p(\alpha_n(x - x_n)) + p(\alpha_n x_n),$$

so that

$$p(\alpha_n x_n) - p(\alpha x) \ge -p(\alpha_n(x_n - x)) + p(\alpha_n x) - p(\alpha x) \to 0.$$

$\boxed{2.4}$ By (i) there exists a linear operator $T : E \to F^\star$ such that $a(x, y) = \langle Tx, y \rangle_{F^\star, F} \ \forall x, y$. The aim is to show that T is a bounded operator, i.e., $T(B_E)$ is bounded in $F^\star$. In view of Corollary 2.5 it suffices to fix $y \in F$ and to check that $\langle T(B_E), y \rangle$ is bounded. This follows from (ii).

2.6

1. One has $\langle Ax_n - A(x_0+x), x_n - x_0 - x \rangle \geq 0$ and thus $\langle Ax_n, x \rangle \leq \varepsilon_n \|Ax_n\| + C(x)$ with $\varepsilon_n = \|x_n - x_0\|$ and $C(x) = \|A(x_0 + x)\|(1 + \|x\|)$ (assuming $\varepsilon_n \leq 1\ \forall n$). It follows from Exercise 2.5 that (Ax_n) is bounded; absurd.
2. Assume that there is a sequence (x_n) in $D(A)$ such that $x_n \to x_0$ and $\|Ax_n\| \to \infty$. Choose $r > 0$ such that $B(x_0, r) \subset \operatorname{conv} D(A)$. For every $x \in E$ with $\|x\| < r$ write

$$x_0 + x = \sum_{i=1}^{m} t_i y_i \text{ with } t_i \geq 0\ \forall i, \quad \sum_{i=1}^{m} t_i = 1, \quad \text{and} \quad y_i \in D(A)\ \forall i$$

(of course t_i, y_i, and m depend on x). We have

$$\langle Ax_n - Ay_i, x_n - y_i \rangle \geq 0$$

and thus $t_i \langle Ax_n, x_n - y_i \rangle \geq t_i \langle Ay_i, x_n - y_i \rangle$. It follows that

$$\langle Ax_n, x_n - x_0 - x \rangle \geq \sum_{i=1}^{m} t_i \langle Ay_i, x_n - y_i \rangle,$$

which leads to

$$\langle Ax_n, x \rangle \leq \varepsilon_n \|Ax_n\| + C(x)$$

with $\varepsilon_n = \|x_n - x_0\|$ and $C(x) = \sum_{i=1}^{m} t_i \|Ay_i\|(1 + \|x_0 - y_i\|)$.
3. Let $x_0 \in \operatorname{Int} D(A)$. Following the same argument as in question 1, one shows that there exist two constants $R > 0$ and C such that

$$\|f\| \leq C \quad \forall x \in D(A) \text{ with } \|x - x_0\| < R \text{ and } \forall f \in Ax.$$

2.7
For every $x \in \ell^p$ set $T_n x = \sum_{i=1}^{n} \alpha_i x_i$, so that $T_n x$ converges to a limit for every $x \in \ell^p$. It follows from Corollary 2.3 that there exists a constant C such that

$$|T_n x| \leq C \|x\|_{\ell^p} \quad \forall x \in \ell^p, \quad \forall n.$$

Choosing x appropriately, one sees that $\alpha \in \ell^{p'}$ and $\|\alpha\|_{p'} \leq C$.

2.8
Method (ii). Let us check that the graph of T is closed. Let (x_n) be a sequence in E such that $x_n \to x$ and $Tx_n \to f$. Passing to the limit in the inequality $\langle Tx_n - Ty, x_n - y \rangle \geq 0$ leads to

$$\langle f - Ty, x - y \rangle \geq 0 \quad \forall y \in E.$$

Choosing $y = x + tz$ with $t \in \mathbb{R}$ and $z \in E$, one sees that $f = Tx$.

2.10

1. If $T(M)$ is closed then $M + N(T) = T^{-1}(T(M))$ is also closed. Conversely, assume that $M+N(T)$ is closed. Since T is surjective, one has $T((M+N(T))^c) =$

$(T(M))^c$. The open mapping theorem implies that $T((M + N(T))^c)$ is open and thus $T(M)$ is closed.

2. If M is any closed subspace and N is any finite-dimensional space then $M + N$ is closed (see Section 11.1).

$\boxed{2.11}$ By the open mapping theorem there is a constant $c > 0$ such that $T(B_E) \supset c B_F$. Let (e_n) denote the canonical basis of ℓ^1, i.e.,

$$e_n = (0, 0, \ldots, 0, \underset{(n)}{1}, 0, \ldots).$$

There exists some $u_n \in E$ such that $\|u_n\| \leq 1/c$ and $T(u_n) = e_n$. Given $y = (y_1, y_2, \ldots, y_n, \ldots) \in \ell^1$, set $Sy = \sum_{i=1}^{\infty} y_i e_i$. Clearly the series converges and S has all the required properties.

$\boxed{2.12}$ Without loss of generality we may assume that T is *surjective* (otherwise, replace E by $R(T)$). Assume by contradiction that there is a sequence (x_n) in E such that

$$\|x_n\|_E = 1 \quad \text{and} \quad \|Tx_n\|_F + |x_n| < 1/n.$$

By the open mapping theorem there is a constant $c > 0$ such that $T(B_E) \supset c B_F$. Since $\|Tx_n\|_F < 1/n$, there exists some $y_n \in E$ such that

$$Tx_n = Ty_n \quad \text{and} \quad \|y_n\|_E < 1/nc.$$

Write $x_n = y_n + z_n$ with $z_n \in N(T)$, $\|y_n\|_E \to 0$ and $\|z_n\|_E \to 1$. On the other hand, $|x_n| < 1/n$; hence $|z_n| < (1/n) + |y_n| \leq (1/n) + M\|y_n\|_E$, and consequently $|z_n| \to 0$. This is impossible, since the norms $\| \ \|_E$ and $| \ |$ are equivalent on the finite-dimensional space $N(T)$.

$\boxed{2.13}$ First, let $T \in \mathcal{O}$ so that $T^{-1} \in \mathcal{L}(F, E)$ (by Corollary 2.7). Then $T + U \in \mathcal{O}$ for every $U \in \mathcal{L}(E, F)$ with $\|U\|$ small enough. Indeed, the equation $Tx + Ux = f$ may be written as $x = T^{-1}(f - Ux)$; it has a unique solution (for every $f \in F$) provided $\|T^{-1}\| \|U\| < 1$ (by Banach's fixed-point theorem; see Theorem 5.7).

Next, let $T \in \Omega$. In view of Theorem 2.13, $R(T)$ is closed and has a complement in F. Let $P : F \to R(T)$ be a continuous projection. The operator PT is bijective from E onto $R(T)$ and hence the above analysis applies. Let $U \in \mathcal{L}(E, F)$ be such that $\|U\| < \delta$; the operator $(PT + PU) : E \to R(T)$ is bijective if δ is small enough and thus $(PT + PU)^{-1}$ is well-defined as an element of $\mathcal{L}(R(T), E)$. Set $S = (PT + PU)^{-1} P$. Clearly $S \in \mathcal{L}(F, E)$ and $S(T + U) = I_E$.

$\boxed{2.14}$

1. Consider the quotient space $\widetilde{E} = E/N(T)$ and the canonical surjection $\pi : E \to \widetilde{E}$, so that $\|\pi x\|_{\widetilde{E}} = \text{dist}(x, N(T)) \ \forall x \in E$. T induces an injective operator $\widetilde{T}$ on $\widetilde{E}$. More precisely, write $T = \widetilde{T} \circ \pi$ with $\widetilde{T} \in \mathcal{L}(\widetilde{E}, F)$, so that $R(T) = R(\widetilde{T})$.

On the other hand, Corollary 2.7 shows that $R(\widetilde{T})$ is closed iff there is a constant C such that

$$\|y\|_{\widetilde{E}} \leq C\|\widetilde{T}y\| \quad \forall y \in \widetilde{E},$$

or equivalently

$$\|\pi x\|_{\widetilde{E}} \leq C\|\widetilde{T}\pi x\| \quad \forall x \in E.$$

The last inequality reads

$$\text{dist}(x, N(T)) \leq C\|Tx\| \quad \forall x \in E.$$

$\boxed{2.15}$ The operator $T : E_1 \times E_2 \to F$ is linear, bounded, and surjective. Moreover, $N(T) = N(T_1) \times N(T_2)$ (since $R(T_1) \cap R(T_2) = \{0\}$). Applying Exercise 2.10 with $M = E_1 \times \{0\}$, one sees that $T(M) = R(T_1)$ is closed provided $M + N(T)$ is closed. But $M + N(T) = E_1 \times N(T_2)$ is indeed closed.

$\boxed{2.16}$ Let π denote the canonical surjection from E onto E/L (see Section 11.2). Consider the operator $T : G \to E/L$ defined by $Tx = \pi x$ for $x \in G$. We have

$$\text{dist}(x, N(T)) = \text{dist}(x, G \cap L) \leq C \text{ dist}(x, L) = C\|Tx\| \quad \forall x \in G.$$

It follows (see Exercise 2.14) that $R(T) = \pi(G)$ is closed. Therefore $\pi^{-1}[\pi(G)] = G + L$ is closed.

$\boxed{2.19}$ Recall that $N(A^\star) = R(A)^\perp$.

1. Let $u \in N(A)$ and $v \in D(A)$; we have

$$\langle A(u + tv), u + tv \rangle \geq -C\|A(u + tv)\|^2 \quad \forall t \in \mathbb{R},$$

which implies that $\langle Av, u \rangle = 0$. Thus $N(A) \subset R(A)^\perp$.
2. $D(A)$ equipped with the graph norm is a Banach space. $R(A)$ equipped with the norm of $E^\star$ is a Banach space. The operator $A : D(A) \to R(A)$ satisfies the assumptions of the open mapping theorem. Hence there is a constant C such that

$$\forall f \in R(A), \exists v \in D(A) \text{ with } Av = f \text{ and } \|v\|_{D(A)} \leq C\|f\|.$$

In particular, $\|v\| \leq C\|f\|$. Given $u \in D(A)$, the above result applied to $f = Au$ shows that there is some $v \in D(A)$ such that $Au = Av$ and $\|v\| \leq C\|Au\|$. Since $u - v \in N(A) \subset R(A)^\perp$, we have

$$\langle Au, u \rangle = \langle Av, u \rangle = \langle Av, v \rangle \geq -\|Av\| \, \|v\| \geq -C\|Au\|^2.$$

$\boxed{2.21}$

1. Distinguish two cases:

   Case (i): $f(a) = 1$. Then $N(A) = \mathbb{R}a$ and $R(A) = N(f)$.
   Case (ii): $f(a) \neq 1$. Then $N(A) = \{0\}$ and $R(A) = E$.
2. A is not closed. Otherwise the closed graph theorem would imply that A is bounded and consequently that f is continuous.
3. $D(A^\star) = \{u \in E^\star; \langle u, a \rangle = 0\}$ and $A^\star u = u \quad \forall u \in D(A^\star)$.

4. $N(A^\star) = \{0\}$ and $R(A^\star) = \{u \in E^\star; \langle u, a \rangle = 0\}$.
5. $R(A)^\perp = \{0\}$ and $R(A^\star)^\perp = \mathbb{R}a$ (note that $N(f)$ is dense in E; see Exercise 1.6).
   It follows that $N(A^\star) = R(A)^\perp$ and $N(A) \subset R(A^\star)^\perp$.
   Observe that in Case (ii), $N(A) \neq R(A^\star)^\perp$.
6. If A is not closed it may happen that $N(A) \neq R(A^\star)^\perp$.

$\boxed{2.22}$

1. Clearly $D(A)$ is dense in E. In order to check that A is closed let (u^j) be a
   sequence in $D(A)$ such that $u^j \to u$ in E and $Au^j \to f$ in E. It follows that

$$u_n^j \xrightarrow[j\to\infty]{} u_n \quad \forall n \quad \text{and} \quad nu_n^j \xrightarrow[j\to\infty]{} f_n \quad \forall n.$$

   Thus $nu_n = f_n \ \forall n$, so that $u \in D(A)$ and $Au = f$.
2.
$$D(A^\star) = \{v = (v_n) \in \ell^\infty; (nv_n) \in \ell^\infty\},$$
$$A^\star v = (nv_n) \text{ and } \overline{D(A^\star)} = c_0.$$

$\boxed{2.24}$

1. We have $D(B^\star) = \{v \in G^\star; T^\star v \in D(A^\star)\}$ and $B^\star = A^\star T^\star$.
2. If $D(A) \neq E$ and $T = 0$, then B is not closed. Indeed, let (u_n) be a sequence in
   $D(A)$ such that $u_n \to u$ with $u \notin D(A)$. Then $Bu_n \to 0$ but $u \notin D(B)$.

$\boxed{2.25}$

2. By Corollary 2.7, $T^{-1} \in \mathcal{L}(F, E)$. Since $T^{-1}T = I_E$ and $TT^{-1} = I_F$, it follows
   that $T^\star(T^{-1})^\star = I_{E^\star}$ and $(T^{-1})^\star T^\star = I_{F^\star}$.

$\boxed{2.26}$ We have

$$\varphi^\star(T^\star f) = \sup_{x \in E}\{\langle T^\star f, x \rangle - \varphi(x)\} = \sup_{y \in R(T)}\{\langle f, y \rangle - \psi(y)\} = -\inf_{y \in F}\{\psi(y) + \zeta(y)\},$$

where $\zeta(y) = -\langle f, y \rangle + I_{R(T)}(y)$. Applying Theorem 1.12, we obtain

$$\varphi^\star(T^\star f) = \min_{g \in F^\star}\{\zeta^\star(g) + \psi^\star(-g)\}.$$

But

$$\zeta^\star(g) = \begin{cases} 0 & \text{if } f + g \in R(T)^\perp, \\ +\infty & \text{if } f + g \notin R(T)^\perp, \end{cases}$$

and thus

$$\varphi^\star(T^\star f) = \min_{f+g \in N(T^\star)} \psi^\star(-g) = \min_{h \in N(T^\star)} \psi^\star(f - h).$$

$\boxed{2.27}$ Let $G = E \times X$ and consider the operator

$$S(x, y) = Tx + y : G \to F.$$

Applying the open mapping theorem, we know that S is an open map, and thus $S(E \times (X \setminus \{0\})) = R(T) + (X \setminus \{0\})$ is open in F. Hence its complement, $R(T)$, is closed.

$\boxed{3.1}$ Apply Corollary 2.4.

$\boxed{3.2}$ Note that $\langle f, \sigma_n \rangle = \frac{1}{n} \sum_{i=1}^{n} \langle f, x_i \rangle$ $\quad \forall f \in E^\star$. Since $\langle f, x_n \rangle \to \langle f, x \rangle$, it follows that $\langle f, \sigma_n \rangle \to \langle f, x \rangle$.

$\boxed{3.4}$

1. Set $G_n = \text{conv}\left(\bigcup_{i=n}^{\infty}\{x_i\}\right)$. Since $x_n \rightharpoonup x$ for the topology $\sigma(E, E^\star)$ it follows that $x \in \overline{G_n}^{\sigma(E,E^\star)}$ $\forall n$. On the other hand, G_n being convex, its closure for the weak topology $\sigma(E, E^\star)$ and that for the strong topology are the same (see Exercise 3.3). Hence $x \in \overline{G_n}$ $\forall n$ (the strong closure of G_n) and there exists a sequence (y_n) such that $y_n \in G_n$ $\forall n$ and $y_n \to x$ strongly.
2. There exists a sequence (u_k) in E such that $u_k \to x$ and $u_k \in \text{conv}\left(\bigcup_{i=1}^{\infty}\{x_i\}\right)$ $\forall k$. Hence there exists an increasing sequence of integers (n_k) such that

$$ u_k \in \text{conv}\left(\bigcup_{i=1}^{n_k}\{x_i\}\right) \quad \forall k. $$

The sequence (z_n) defined by $z_n = u_k$ for $n_k \leq n < n_{k+1}$ (and $z_n = x_1$ for $1 \leq n < n_1$) has all the required properties.

$\boxed{3.7}$

1. Let $x \notin A + B$. We shall construct a neighborhood W of 0 for $\sigma(E, E^\star)$ such that

$$ (x + W) \cap (A + B) = \emptyset. $$

For every $y \in B$ there exists a convex neighborhood $V(y)$ of 0 such that

$$ (x + V(y)) \cap (A + y) = \emptyset $$

(since $A + y$ is closed and $x \notin A + y$).
Clearly

$$ B \subset \bigcup_{y \in B} \left(y - \frac{1}{2}V(y)\right), $$

and since B is compact, there is some finite set I such that

$$ B \subset \bigcup_{i \in I} \left(y_i - \frac{1}{2}V(y_i)\right) \text{ with } y_i \in B. $$

Set

$$ W = \frac{1}{2}\bigcap_{i \in I} V(y_i). $$

Solutions of Some Exercises

We claim that $(x + W) \cap (A + B) = \emptyset$. Indeed, suppose by contradiction that there exists some $w \in W$ such that $x + w \in (A + B)$. Hence there is some $i \in I$ such that

$$x + w \in A + y_i - \frac{1}{2}V(y_i).$$

Since $V(y_i)$ is convex it follows that there exists some $w' \in V(y_i)$ such that $x + w' \in A + y_i$. Consequently $(x + V(y_i)) \cap (A + y_i) \neq \emptyset$; absurd.

Remark. If $E^\star$ is *separable* and A is *bounded* one may use *sequences* in order to prove that $A + B$ is closed, since the weak topology is metrizable on bounded sets (see Theorem 3.29). This makes the argument somewhat easier. Indeed, let $x_n = a_n + b_n$ be a sequence such that $x_n \rightharpoonup x$ weakly $\sigma(E, E^\star)$ with $a_n \in A$ and $b_n \in B$. Since B is weakly compact (and metrizable), there is a subsequence such that $b_{n_k} \rightharpoonup b$ weakly $\sigma(E, E^\star)$ with $b \in B$. Thus $a_{n_k} \rightharpoonup x - b$ weakly $\sigma(E, E^\star)$. But A is weakly closed and therefore $x - b \in A$, i.e., $x \in A + B$.
2. By question 1, $(A - B)$ is weakly closed and therefore it is strongly closed. Hence one may strictly separate $\{0\}$ and $(A - B)$.

3.8

1. Since V_k is a neighborhood of 0 for $\sigma(E, E^\star)$, one may assume (see Proposition 3.4) that V_k has the form

$$V_k = \{x \in E; |\langle f, x \rangle| < \varepsilon_k \ \forall f \in F_k\},$$

where $\varepsilon_k > 0$ and F_k is a *finite* subset of $E^\star$. Hence the set $F = \bigcup_{k=1}^\infty F_k$ is countable. We claim that any $g \in E^\star$ can be written as a finite linear combination of elements in F. Indeed, set

$$V = \{x \in E; |\langle g, x \rangle| < 1\}.$$

Since V is neighborhood of 0 for $\sigma(E, E^\star)$, there exists some integer m such that $\{x \in E; d(x, 0) < 1/m\} \subset V$ and consequently $V_m \subset V$. Suppose $x \in E$ is such that $\langle f, x \rangle = 0 \ \forall f \in F_m$. Then $tx \in V_m \ \forall t \in \mathbb{R}$ and thus $tx \in V \ \forall t \in \mathbb{R}$, i.e., $\langle g, x \rangle = 0$. Applying Lemma 3.2, we see that g is a linear combination of elements in F_m.
2. Use the same method as in question 3 of Exercise 1.5.
3. If $\dim E^\star < \infty$, then $\dim E^{\star\star} < \infty$; consequently $\dim E < \infty$ (since there is a canonical injection from E into $E^{\star\star}$).
4. Apply the following lemma (which is an easy consequence of Lemma 3.2): Assume that $x_1, x_2, \ldots, x_k, y \in E$ satisfy

$$[f \in E^\star; \langle f, x_i \rangle = 0 \ \forall i] \Rightarrow [\langle f, y \rangle = 0].$$

Then there exist constants $\lambda_1, \lambda_2, \ldots, \lambda_k$ such that $y = \sum_{i=1}^k \lambda_i x_i$.

3.9

1. Apply Theorem 1.12 with

$$\varphi(x) = \langle f_0, x \rangle + I_{B_E}(x) \quad \text{and} \quad \psi(x) = I_M(x).$$

2. Note that $B_{E^\star}$ is compact for $\sigma(E^\star, E)$, while $M^\perp$ is closed for $\sigma(E^\star, E)$ (why?).

3.11 It suffices to argue on sequences (why?). Assume $x_n \to x$ strongly in E and $Ax_n \not\rightharpoonup Ax$ for $\sigma(E^\star, E)$, i.e., there exists some $y \in E$ such that $\langle Ax_n, y \rangle \not\to \langle Ax, y \rangle$. We already know (by Exercise 2.6) that (Ax_n) is bounded. Hence, there is a subsequence such that $\langle Ax_{n_k}, y \rangle \to \ell \neq \langle Ax, y \rangle$. Applying the monotonicity of A, we have

$$\langle Ax_{n_k} - A(x + ty), x_{n_k} - x - ty \rangle \geq 0.$$

Passing to the limit, we obtain

$$-t\ell + t\langle A(x + ty), y \rangle \geq 0,$$

which implies that $\ell = \langle Ax, y \rangle$; absurd.

3.12

1. Assumption (A) implies that $\varphi^\star(f) \geq R\|f\| + \langle f, x_0 \rangle - M \quad \forall f \in E^\star$. Conversely, assume that (B) holds and set $\psi(f) = \varphi^\star(f) - \langle f, x_0 \rangle$. We claim that there exist constants $k > 0$ and C such that

(S1) $$\psi(f) \geq k\|f\| - C \quad \forall f \in E^\star.$$

After a translation we may always assume that $\psi(0) < \infty$ (see Proposition 1.10). Fix $\alpha > \psi(0)$. Using assumption (B) we may find some $r > 0$ such that

$$\psi(g) \geq \alpha \quad \forall g \in E^\star \text{ with } \|g\| \geq r.$$

Given $f \in E^\star$ with $\|f\| \geq r$ write

$$\psi(tf) \leq t\psi(f) + (1 - t)\psi(0) \text{ with } t = r/\|f\|.$$

Since $\|tf\| = r$, this leads to $\alpha - \psi(0) \leq \frac{r}{\|f\|}(\psi(f) - \psi(0))$, which establishes claim (S1). Passing to the conjugate of (S1) we obtain (A).

2. The function ψ is convex and l.s.c. for the weak* topology (why?). Assumption (B) says that for every $\lambda \in \mathbb{R}$ the set $\{f \in E^\star; \psi(f) \leq \lambda\}$ is bounded. Hence, it is weak* compact (by Theorem 3.16), and thus $\inf_{E^\star} \psi$ is achieved. On the other hand,

$$\inf_{E^\star} \psi = - \sup_{f \in E^\star} \{\langle f, x_0 \rangle - \varphi^\star(f)\} = -\varphi^{\star\star}(x_0) = -\varphi(x_0).$$

Alternatively, one could also apply Theorem 1.12 to the functions φ and $I_{\{x_0\}}$ (note that φ is continuous at x_0; see Exercise 2.1).

3.13

1. For every fixed p we have $x_{p+n} \in K_p \ \forall n$. Passing to the limit (as $n \to \infty$) we see that $x \in K_p$ since K_p is weakly closed (see Theorem 3.7). On the other hand,

let V be a convex neighborhood of x for the topology $\sigma(E, E^\star)$. There exists an integer N such that $x_n \in V$ $\forall n \geq N$. Thus $K_n \subset \overline{V}$ $\forall n \geq N$ and consequently $\bigcap_{n=1}^{\infty} K_n \subset \overline{V}$. Since this is true for any convex neighborhood V of x, it follows that $\bigcap_{n=1}^{\infty} K_n \subset \{x\}$ (why?).

2. Let V be an open neighborhood of x for the topology $\sigma(E, E^\star)$. Set $K'_n = K_n \cap (V^c)$. Since K_n is compact for $\sigma(E, E^\star)$ (why?), it follows that K'_n is also compact for $\sigma(E, E^\star)$. On the other hand, $\bigcap_{n=1}^{\infty} K'_n = \emptyset$, and hence there is some integer N such that $K'_N = \emptyset$, i.e., $K_N \subset V$.

3. We may assume that $x = 0$. Consider the recession cone

$$C_n = \bigcap_{\lambda > 0} \lambda K_n.$$

Since $C_n \subset K_n$ we deduce that $\bigcap_{n=1}^{\infty} C_n = \{0\}$. Let $S_E = \{x \in E; \|x\| = 1\}$. The sequence $(C_n \cap S)$ is decreasing and $\bigcap_{n=1}^{\infty}(C_n \cap S) = \emptyset$. Thus, by compactness, $C_{n_0} \cap S = \emptyset$ for some n_0. Therefore $C_{n_0} = \{0\}$ and consequently K_{n_0} is bounded (why?). Hence (x_n) is bounded and we are reduced to question 2.

4. Consider the sequence $x_n = (0, 0, \ldots, \underset{(n)}{n}, 0, \ldots)$, when n is odd, and $x_n = 0$ when n is even.

3.18

2. Suppose, by contradiction, that $e^{n_k} \underset{k \to \infty}{\rightharpoonup} a$ in ℓ^1 for the topology $\sigma(\ell^1, \ell^\infty)$. Thus we have $\langle \xi, e^{n_k} \rangle \underset{k \to \infty}{\longrightarrow} \langle \xi, a \rangle$ $\forall \xi \in \ell^\infty$. Consider the element $\xi \in \ell^\infty$ defined by

$$\xi = (0, 0, \ldots, \underset{(n_1)}{-1}, 0, \ldots, \underset{(n_2)}{1}, 0, \ldots, \underset{(n_3)}{-1}, 0, \ldots).$$

Note that $\langle \xi, e^{n_k} \rangle = (-1)^k$ does not converge as $k \to \infty$; a contradiction.

3. Let $E = \ell^\infty$, so that $\ell^1 \subset E^\star$. Set $f_n = e^n$, considered as a sequence in $E^\star$. We claim that (f_n) has no subsequence that converges for $\sigma(E^\star, E)$. Suppose, by contradiction, that $f_{n_k} \overset{\star}{\rightharpoonup} f$ in $E^\star$ for $\sigma(E^\star, E)$, i.e., $\langle f_{n_k}, \eta \rangle \to \langle f, \eta \rangle$ $\forall \eta \in E$. Choosing $\eta = \xi$ as in question 2, we see that $\langle f_{n_k}, \xi \rangle = (-1)^k$ does not converge; a contradiction. Here, the set $B_{E^\star}$, equipped with the topology $\sigma(E^\star, E)$ is compact (by Theorem 3.16), but it is *not metrizable*. Applying Theorem 3.28, we may also say that $E = \ell^\infty$ is *not separable* (for another proof see Remark 8 in Chapter 4 and Proposition 11.17).

3.19

1. Note that if $x^n \to x$ strongly in ℓ^p, then

$$\forall \varepsilon > 0 \quad \exists I \text{ such that } \sum_{i=I}^{\infty} |x_i^n|^p \leq \varepsilon^p \ \forall n.$$

2. Apply Exercise 3.17.

3. The space B_E is metrizable for the topology $\sigma(E, E^\star)$ (by Theorem 3.29). Thus it suffices to check the continuity of A on *sequences*.

3.20

1. Consider the map $T : E \to C(K)$ defined by

$$(Tx)(t) = \langle t, x \rangle \text{ with } x \in E \text{ and } t \in B_{E^\star} = K.$$

Clearly $\|Tx\| = \sup_{t \in K} |(Tx)(t)| = \|x\|$.

2. Since $K = B_{E^\star}$ is metrizable and compact for $\sigma(E^\star, E)$, there is a dense countable subset (t_n) in K. Consider the map $S : E \to \ell^\infty$ defined by

$$Sx = (\langle t_1, x \rangle, \langle t_2, x \rangle, \ldots, \langle t_n, x \rangle, \ldots).$$

Check that $\|Sx\|_{\ell^\infty} = \|x\|$.

3.21

Let (a_i) be a dense countable subset of E. Choose a first subsequence such that $\langle f_{n_k}, a_1 \rangle$ converges to a limit as $k \to \infty$. Then, pick a subsequence out of (n_k) such that $\langle f_{n'_k}, a_2 \rangle$ converges, etc.

By a standard diagonal process we may extract a sequence (g_k) out of the sequence (f_n) such that $\langle g_k, a_i \rangle \xrightarrow[k \to \infty]{} \ell_i \; \forall i$. Since the set (a_i) is dense in E, we easily obtain that $\langle g_k, a \rangle \to \ell_a \; \forall a \in E$. It follows that g_k converges for $\sigma(E^\star, E)$ to some g (see Exercise 3.16).

3.22

(a) B_E is metrizable for $\sigma(E, E^\star)$ (by Theorem 3.29) and 0 belongs to the closure of $S = \{x \in E; \|x\| = 1\}$ for $\sigma(E, E^\star)$ (see Remark 2 in Chapter 3).

(b) Since $\dim E = \infty$ there is a closed subspace E_0 in E that is separable and such that $\dim E_0 = \infty$ (why?). Note that E_0 is reflexive and apply Case (a) (in conjunction with Corollary 3.27).

3.25

Suppose, by contradiction, that $C(K)$ is reflexive. Then $E = \{u \in C(K); u(a) = 0\}$ is also reflexive and $\sup_{u \in B_E} f(u)$ is achieved.

On the other hand, we claim that $\sup_{u \in B_E} f(u) = 1$. Indeed, $\forall N, \exists u \in E$ such that $0 \le u \le 1$ on K and $u(a_i) = 1 \; \forall i = 1, 2, \ldots, N$. (Apply, for example, the Tietze–Urysohn theorem; see, e.g., J. Munkres [1].) Hence there exists some $u \in B_E$ such that $f(u) = 1$. This leads to $u(a_n) = 1 \; \forall n$ and $u(a) = 0$; absurd.

3.26

1. Given $y \in B_F$, there is some integer n_1 such that $\|y - a_{n_1}\| < 1/2$. Since the set $\frac{1}{2}(a_i)_{i > n_1}$ is dense in $\frac{1}{2} B_F$, there is some $n_2 > n_1$ such that

$$\left\| y - a_{n_1} - \frac{1}{2} a_{n_2} \right\| < \frac{1}{4}.$$

Construct by induction an increasing sequence $n_k \uparrow \infty$ of integers such that

$$y = a_{n_1} + \frac{1}{2}a_{n_2} + \frac{1}{4}a_{n_3} + \cdots + \frac{1}{2^{k-1}}a_{n_k} + \cdots .$$

2. Suppose, by contradiction, that $S \in \mathcal{L}(F, \ell^1)$ is such that $TS = I_F$. Let (y_n) be any sequence in F such that $\|y_n\| = 1 \; \forall n$ and $y_n \rightharpoonup 0$ weakly $\sigma(F, F^*)$. Thus $Sy_n \rightharpoonup 0$ for $\sigma(\ell^1, \ell^\infty)$ and consequently $Sy_n \to 0$ strongly in ℓ^1 (see Problem 8). It follows that $y_n = TSy_n \to 0$; absurd.

3. Use Theorem 2.12.

4. $T^* : F^* \to \ell^\infty$ is defined by

$$T^* v = (\langle v, a_1 \rangle, \langle v, a_2 \rangle, \ldots, \langle v, a_n \rangle, \ldots).$$

3.27 B_{E^*} is compact and metrizable for $\sigma(E^*, E)$. Hence there exists a countable subset of B_{E^*} that is dense for $\sigma(E^*, E)$.

1. Clearly $\|f\| \le \|f\|_1 \le \sqrt{2}\|f\| \; \forall f \in E^*$.

2. Set $|f|^2 = \sum_{n=1}^\infty \frac{1}{2^n}|\langle f, a_n \rangle|^2$. Note that the *norm* $|\;|$ is associated to a scalar product (why?), and thus it is strictly convex, i.e., the *function* $f \mapsto |f|^2$ is strictly convex. More precisely, we have $\forall t \in [0, 1], \forall f, g \in E^*$,

(S1) $|tf + (1 - t)g|^2 + t(1 - t)|f - g|^2 = t|f|^2 + (1 - t)|g|^2.$

Consequently, the function $f \mapsto \|f\|^2 + |f|^2$ is also strictly convex.

3. Same method as in question 2. Note that if $\langle b_n, x \rangle = 0 \; \forall n$, then $x = 0$ (why?).

4. Given $x \in E$ set $[x] = \{\sum_{n=1}^\infty \frac{1}{2^n}|\langle b_n, x \rangle|^2\}^{1/2}$, and let $[f]$ denote the dual norm of $[\;]$ on E^*. Note that $[f]$ also satisfies the identity (S1). Indeed, we have

$$\frac{1}{2}[tf + (1 - t)g]^2 = \sup_{x \in E} \left\{ \langle tf + (1 - t)g, x \rangle - \frac{1}{2}[x]^2 \right\},$$

$$\frac{1}{2}[f - g]^2 = \sup_{y \in E} \left\{ \langle f - g, y \rangle - \frac{1}{2}[y]^2 \right\},$$

and thus

$$\frac{1}{2}[tf + (1 - t)g]^2 + \frac{1}{2}t(1 - t)[f - g]^2$$

$$= \sup_{x, y} \left\{ \langle tf + (1 - t)g, x \rangle + t(1 - t)\langle f - g, y \rangle - \frac{1}{2}[x]^2 - \frac{1}{2}t(1 - t)[y]^2 \right\}.$$

We conclude that (S1) holds by a change of variables $x = t\xi + (1 - t)\eta$ and $y = \xi - \eta$. Applying question 3 of Exercise 1.23, we see that

$$\|f\|_2^2 = \inf_{h \in E^*} \left\{ \|f - h\|_1^2 + [h]^2 \right\} = \min_{h \in E^*} \left\{ \|f - h\|_1^2 + [h]^2 \right\}.$$

We claim that the function $f \mapsto \|f\|_2^2$ is strictly convex. Indeed, given $f, g \in E^*$, fix $h_1, h_2 \in E^*$ such that

$$\|f\|_2^2 = \|f - h_1\|_2^2 + [h_1]^2,$$

$$\|g\|_2^2 = \|g - h_2\|_1^2 + [h_2]^2.$$

For every $t \in (0, 1)$ we have

$$\|tf + (1-t)g\|_2^2 \le \|tf + (1-t)g - (th_1 + (1-t)h_2)\|_1^2 + [th_1 + (1-t)h_2]^2$$
$$< t\|f\|_2^2 + (1-t)\|g\|_2^2,$$

unless $f - h_1 = g - h_2$ and $h_1 = h_2$, i.e., $f = g$.

3.28 Since E is reflexive, $\sup_{x \in B_E} \langle f, x \rangle$ is achieved by some unique point $x_0 \in B_E$. Then $x = x_0 \|f\|$ satisfies $f \in F(x)$.

Alternatively, we may also consider the duality map F^* from E^* into E^{**}. The set $F^*(f)$ is nonempty (by Corollary 1.3). Fix any $\xi \in F^*(f)$. Since E is reflexive there exists some $x \in E$ such that $Jx = \xi$ (J is the canonical injection from E into E^{**}). We have

$$\|\xi\| = \|f\| = \|x\| \quad \text{and} \quad \langle \xi, f \rangle = \|f\|^2 = \langle f, x \rangle.$$

Thus $f \in F(x)$.

Uniqueness. Let x_1 and x_2 be such that $f \in F(x_1)$ and $f \in F(x_2)$. Then $\|x_1\| = \|x_2\| = \|f\|$, and therefore, if $x_1 \ne x_2$ we have

$$\left\| \frac{x_1 + x_2}{2} \right\| < \|f\|.$$

On the other hand, $\langle f, x_1 \rangle = \langle f, x_2 \rangle = \|f\|^2$ and hence

$$\|f\|^2 = \left\langle f, \frac{x_1 + x_2}{2} \right\rangle < \|f\|^2 \quad \text{if } x_1 \ne x_2.$$

3.29

1. Assume, by contradiction, that there exist $M_0 > 0$, $\varepsilon_0 > 0$, and two sequences (x_n), (y_n) such that

$$\|x_n\| \le M, \|y_n\| \le M, \quad \|x_n - y_n\| > \varepsilon_0,$$

and

(S1) $$\left\| \frac{x_n + y_n}{2} \right\|^2 > \frac{1}{2}\|x_n\|^2 + \frac{1}{2}\|y_n\|^2 - \frac{1}{n}.$$

Consider subsequences, still denoted by (x_n) and (y_n), such that $\|x_n\| \to a$ and $\|y_n\| \to b$. We find that $a + b \ge \varepsilon_0$ and $\frac{1}{2}a^2 + \frac{1}{2}b^2 \le \left(\frac{a+b}{2}\right)^2$. Therefore $a = b \ne 0$.
Set

$$x'_n = \frac{x_n}{\|x_n\|} \quad \text{and} \quad y'_n = \frac{y_n}{\|y_n\|}.$$

For n large enough we have $\|x'_n - y'_n\| \geq (\varepsilon_0/a) + o(1)$ (as usual, we denote by $o(1)$ various quantities—positive or negative—that tend to zero as $n \to \infty$). By uniform convexity there exists $\delta_0 > 0$ such that

$$\left\| \frac{x'_n + y'_n}{2} \right\| \leq 1 - \delta_0.$$

Thus

$$\left\| \frac{x_n + y_n}{2} \right\| \leq a(1 - \delta_0) + o(1).$$

By (S1) we have

$$\left\| \frac{x_n + y_n}{2} \right\|^2 \geq a^2 + o(1).$$

Hence $a^2 \leq a^2(1 - \delta_0)^2 + o(1)$; absurd.

$\boxed{3.32}$

1. The infimum is achieved since E is reflexive and we may apply Corollary 3.23. The uniqueness comes from the fact that the *space* E is strictly convex and thus the *function* $y \mapsto \|y - x\|^2$ is strictly convex.
2. Let (y_n) be a minimizing sequence; set $d_n = \|x - y_n\|$ and $d = \inf_{y \in C} \|x - y\|$, so that $d_n \to d$. Let (y_{n_k}) be a sequence such that $y_{n_k} \rightharpoonup z$ weakly. Thus $z \in C$ and $\|x - z\| \leq d$ (why?). It follows that

$$x - y_{n_k} \rightharpoonup x - z \text{ weakly} \quad \text{and} \quad \|x - y_{n_k}\| \to d = \|x - z\|,$$

and therefore (see Proposition 3.32) $y_{n_k} \to z$ strongly. The uniqueness of the limit implies that the *whole sequence* (y_n) converges strongly to $P_C x$. The argument is standard and we will use it many times. We recall it for the convenience of the reader. Assume, by contradiction, that (y_n) does *not* converge to $y = P_C x$. Then there exist $\varepsilon > 0$ and a subsequence, (y_{m_j}), such that $\|y_{m_j} - y\| \geq \varepsilon \quad \forall j$. From (y_{m_j}) we extract (by the argument above) a further subsequence, denoted by (y_{n_k}), such that $y_{n_k} \to P_C x$. Since (y_{n_k}) is a subsequence of (y_{m_j}), we have $\|y_{n_k} - y\| \geq \varepsilon \quad \forall k$ and thus $\|P_C x - y\| \geq \varepsilon$. Absurd.

3 and 4. Assume, by contradiction, that there exist some $\varepsilon_0 > 0$ and sequences (x_n) and (y_n) such that

$$\|x_n\| \leq M, \quad \|y_n\| \leq M, \quad \|x_n - y_n\| \to 0, \quad \text{and} \quad \|P_C x_n - P_C y_n\| \geq \varepsilon_0 \quad \forall n.$$

We have

$$\|x_n - P_C x_n\| \leq \left\| x_n - \frac{P_C x_n + P_C y_n}{2} \right\| \leq \left\| \frac{x_n + y_n}{2} - \frac{P_C x_n + P_C y_n}{2} \right\| + o(1),$$

and similarly

$$\|y_n - P_C y_n\| \le \left\| \frac{x_n + y_n}{2} - \frac{P_C x_n + P_C y_n}{2} \right\| + o(1).$$

It follows that

(S1) $\dfrac{1}{2}\|x_n - P_C x_n\|^2 + \dfrac{1}{2}\|y_n - P_C y_n\|^2 \le \left\| \dfrac{x_n + y_n}{2} - \dfrac{P_C x_n + P_C y_n}{2} \right\|^2 + o(1).$

On the other hand, if we set $a_n = x_n - P_C x_n$ and $b_n = y_n - P_C y_n$, then $\|a_n - b_n\| \ge \varepsilon_0 + o(1)$ and $\|a_n\| \le M'$, $\|b_n\| \le M'$. Using Exercise 3.29, we know that there is some $\delta_0 > 0$ such that

$$\left\| \frac{a_n + b_n}{2} \right\|^2 \le \frac{1}{2}\|a_n\|^2 + \frac{1}{2}\|b_n\|^2 - \delta_0,$$

that is,

(S2) $\left\| \dfrac{x_n + y_n}{2} - \dfrac{P_C x_n + P_C y_n}{2} \right\|^2 \le \dfrac{1}{2}\|x_n - P_C x_n\|^2 + \dfrac{1}{2}\|y_n - P_C y_n\|^2 - \delta_0.$

Combining (S1) and (S2) leads to a contradiction.

5. Same argument as in question 1.
6. We have

(S3) $\qquad n\|y_n - x\|^2 + \varphi(y_n) \le n\|y - x\|^2 + \varphi(y) \quad \forall y \in D(\varphi).$

Since φ is bounded below by an affine continuous function (see Proposition 1.10), we see that (y_n) remains bounded as $n \to \infty$ (check the details). Let (y_{n_k}) be a subsequence such that $y_{n_k} \rightharpoonup z$ weakly. Note that $z \in \overline{D(\varphi)}$ (why?). From (S3) we obtain $\|z - x\| \le \|y - x\|$ $\forall y \in D(\varphi)$, and thus $\forall y \in \overline{D(\varphi)}$. Hence $z = P_C x$, where $C = \overline{D(\varphi)}$. Using (S3) once more leads to

$$\limsup_{n \to \infty}\|y_n - x\| \le \|y - x\| \ \forall y \in \overline{D(\varphi)}, \text{ and in particular for } y = z.$$

We conclude that $y_{n_k} \to z$ strongly, and finally the uniqueness of the limit shows (as above) that the *whole sequence* (y_n) converges strongly to $P_C x$.

4.3

2. Note that $h_n = \frac{1}{2}\left(|f_n - g_n| + f_n + g_n\right).$
3. Note that $f_n g_n - fg = (f_n - f)g_n + f(g_n - g)$ and that $f(g_n - g) \to 0$ in $L^p(\Omega)$ by dominated convergence.

4.5

1. Recall that $L^1(\Omega) \cap L^\infty(\Omega) \subset L^p(\Omega)$ and more precisely $\|f\|_p^p \le \|f\|_\infty^{p-1}\|f\|_1$. Since Ω is σ-finite, we may write $\Omega = \bigcup_n \Omega_n$ with $|\Omega_n| < \infty$ $\forall n$. Given $f \in L^p(\Omega)$, check that $f_n = \chi_{\Omega_n} T_n f \in L^1(\Omega) \cap L^\infty(\Omega)$ and that $f_n \xrightarrow[n \to \infty]{} f$ in $L^p(\Omega)$.

2. Let (f_n) be sequence in $L^p(\Omega) \cap L^q(\Omega)$ such that $f_n \to f$ in $L^p(\Omega)$ and $\|f_n\|_q \leq 1$. We assume (by passing to a subsequence) that $f_n \to f$ a.e. (see Theorem 4.9). It follows from Fatou's lemma that $f \in L^q(\Omega)$ and that $\|f\|_q \leq 1$.

3. We already know, by question 2, that $f \in L^q(\Omega)$ and thus $f \in L^r(\Omega)$ for every r between p and q. On the other hand, we may write $\frac{1}{r} = \frac{\alpha}{p} + \frac{1-\alpha}{q}$ with $0 < \alpha \leq 1$, and we obtain

$$\|f_n - f\|_r \leq \|f_n - f\|_p^\alpha \|f_n - f\|_q^{1-\alpha} \leq \|f_n - f\|_p^\alpha (2C)^{1-\alpha}.$$

| 4.6 |

1. We have $\|f\|_p \leq \|f\|_\infty |\Omega|^{1/p}$ and thus $\limsup_{p\to\infty} \|f\|_p \leq \|f\|_\infty$. On the other hand, fix $0 < k < \|f\|_\infty$, and let

$$A = \{x \in \Omega; |f(x)| > k\}.$$

Clearly $|A| \neq 0$ and $\|f\|_p \geq k|A|^{1/p}$. It follows that $\liminf_{p\to\infty} \|f\|_p \geq k$ and therefore $\liminf_{p\to\infty} \|f\|_p \geq \|f\|_\infty$.

2. Fix $k > C$ and let A be defined as above. Then $k^p|A| \leq \|f\|_p^p \leq C^p$ and thus $|A| \leq (C/k)^p \ \forall p \geq 1$. Letting $p \to \infty$, we see that $|A| = 0$.

3. $f(x) = \log|x|$.

| 4.7 | Consider the operator $T : L^p(\Omega) \to L^q(\Omega)$ defined by $Tu = au$. We claim that the graph of T is closed. Indeed, let (u_n) be a sequence in $L^p(\Omega)$ such that $u_n \to u$ in $L^p(\Omega)$ and $au_n \to f$ in $L^q(\Omega)$. Passing to a subsequence we may assume that $u_n \to u$ a.e. and $au_n \to f$ a.e. Thus $f = au$ a.e., and so $f = Tu$. It follows from the closed graph theorem (Theorem 2.9) that T is bounded and so there is a constant C such that

(S1) $\|au\|_q \leq C\|u\|_p \quad \forall u \in L^p(\Omega).$

Case 1: $p < \infty$. It follows from (S1) that

$$\int |a|^q |v| \leq C^q \|v\|_{p/q} \quad \forall v \in L^{p/q}(\Omega).$$

Therefore the map $v \mapsto \int |a|^q v$ is a continuous linear functional on $L^{p/q}(\Omega)$ and thus $|a|^q \in L^{(p/q)'}(\Omega)$.

Case 2: $p = \infty$. Choose $u \equiv 1$ in (S1).

| 4.8 |

1. X equipped with the norm $\|\ \|_1$ is a Banach space. For every n, X_n is a closed subset of X (see Exercise 4.5). On the other hand, $X = \bigcup_n X_n$. Indeed, for every $f \in X$ there is some $q > 1$ such that $f \in L^q(\Omega)$. Thus $f \in L^{1+1/n}(\Omega)$ provided $1 + (1/n) \leq q$, and, moreover,

$$\|f\|_{1+1/n} \le \|f\|_1^{\alpha_n} \|f\|_q^{1-\alpha_n} \quad \text{with } \frac{1}{1+(1/n)} = \frac{\alpha_n}{1} + \frac{1-\alpha_n}{q}.$$

It follows from the Baire category theorem that there is some integer n_0 such that Int $X_{n_0} \ne \emptyset$. Thus $X \subset L^{1+1/n_0}(\Omega)$.

2. The identity map $I : X \to L^p(\Omega)$ is a linear operator whose graph is closed. Thus it is a bounded operator.

$\boxed{4.9}$ For every $t \in \mathbb{R}$ we have

$$f(x)t \le j(f(x)) + j^\star(t) \text{ a.e. on } \Omega,$$

and by integration we obtain

$$\left(\frac{1}{|\Omega|}\int_\Omega f\right)t \le \frac{1}{|\Omega|}\int_\Omega j(f) + j^\star(t).$$

Therefore

$$j\left(\frac{1}{|\Omega|}\int_\Omega f\right) = \sup_{t\in\mathbb{R}}\left\{\left(\frac{1}{|\Omega|}\int_\Omega f\right)t - j^\star(t)\right\} \le \frac{1}{|\Omega|}\int_\Omega j(f).$$

$\boxed{4.10}$

1. Let $u_1, u_2 \in D(J)$ and let $t \in [0, 1]$. The function $x \mapsto j(tu_1(x)+(1-t)u_2(x))$ is measurable (since j is continuous). On the other hand, $j(tu_1 + (1 - t)u_2) \le tj(u_1) + (1 - t)j(u_2)$. Recall that there exist constants a and b such that $j(s) \ge as+b \; \forall s \in \mathbb{R}$ (see Proposition 1.10). It follows that $j(tu_1 +(1+t)u_2) \in L^1(\Omega)$ and that $J(tu_1 + (1 - t)u_2) \le tJ(u_1) + (1 - t)J(u_2)$.

2. Assume first that $j \ge 0$. We claim that for every $\lambda \in \mathbb{R}$ the set $\{u \in L^p(\Omega); J(u) \le \lambda\}$ is closed. Indeed, let (u_n) be a sequence in $L^p(\Omega)$ such that $u_n \to u$ in $L^p(\Omega)$ and $\int j(u_n) \le \lambda$. Passing to a subsequence we may assume that $u_n \to u$ a.e. It follows from Fatou's lemma that $j(u) \in L^1(\Omega)$ and that $\int j(u) \le \lambda$. Therefore J is l.s.c. In the general case, let $\tilde{j}(s) = j(s) - (as + b) \ge 0$. We already know that $\tilde{J}$ is l.s.c., and so is $J(u) = \tilde{J}(u) + a\int u + b|\Omega|$.

3. We first claim that

$$J^\star(f) \le \int j^\star(f) \quad \forall f \in L^{p'}(\Omega) \text{ such that } j^\star(f) \in L^1(\Omega).$$

Indeed, we have $fu - j(u) \le j^\star(f)$ a.e. on Ω, $\forall u \in L^p(\Omega)$, and thus

$$\sup_{u\in D(J)}\left\{\int fu - J(u)\right\} \le \int j^\star(f).$$

The proof of the reverse inequality is more delicate and requires some "regularization" process. Assume first that $1 < p < \infty$ and set

$$j_n(t) = j(t) + \frac{1}{n}|t|^p, \quad t \in \mathbb{R}.$$

We claim that

(S1) $$J_n^\star(f) = \int j_n^\star(f) \quad \forall f \in L^{p'}(\Omega).$$

Indeed, let $f \in L^{p'}(\Omega)$. For a.e. fixed $x \in \Omega$,

$$\sup_{u \in \mathbb{R}} \left\{ f(x)u - j(u) - \frac{1}{n}|u|^p \right\}$$

is achieved by some unique element $u = u(x)$. Clearly we have

$$j(u(x)) + \frac{1}{n}|u(x)|^p - f(x)u(x) \le j(0).$$

It follows that $u \in L^p(\Omega)$ and that $j(u) \in L^1(\Omega)$ (why?).
We conclude that

$$J_n^\star(f) = \sup_{v \in D(J)} \left\{ \int fv - J_n(v) \right\} \ge \int \left\{ fu - j(u) - \frac{1}{n}|u|^p \right\} = \int j_n^\star(f).$$

Since we have already established the reverse inequality, we see that (S1) holds.
Next we let $n \uparrow \infty$. Clearly, $J \le J_n$, so that $J_n^\star \le J^\star$, i.e., $\int j_n^\star(f) \le J^\star(f)$.
We claim that for every $s \in \mathbb{R}$, $j_n^\star(s) \uparrow j^\star(s)$ as $n \uparrow \infty$. Indeed, we know that
$j_n^\star = j^\star \nabla \left(\frac{1}{n}| |^p\right)^\star$ (see Exercise 1.23), and we may then argue as in Exercise 1.24.
We conclude by monotone convergence that if $f \in D(J^\star)$, then

$$j^\star(f) \in L^1(\Omega) \quad \text{and} \quad \int j^\star(f) \le J^\star(f).$$

Finally, if $p = 1$, the above method can be modified using, for example, $j_n(t) = j(t) + \frac{1}{n}t^2$.

4. Assuming first that $f(x) \in \partial j(u(x))$ a.e. on Ω, we have

$$j(v) - j(u(x)) \ge f(x)(v - u(x)) \quad \forall v \in \mathbb{R}, \text{ a.e. on } \Omega.$$

Choosing $v = 0$, we see that $j(u) \in L^1(\Omega)$ and thus

$$J(v) - J(u) \ge \int f(v - u) \quad \forall v \in D(J).$$

Conversely, assume that $f \in \partial J(u)$. Then we have $J(u) + J^\star(f) = \int fu$.
Thus $j(u) \in L^1(\Omega)$, $j^\star(f) \in L^1(\Omega)$, and $\int \{j(u) + j^\star(f) - fu\} = 0$. Since
$j(u) + j^\star(f) - fu \ge 0$ a.e., we find that $j(u) + j^\star(f) - fu = 0$ a.e., i.e.,
$f(x) \in \partial j(u(x))$ a.e.

$\boxed{4.11}$ Set $f = u^\alpha$, $g = v^\alpha$, and $p = 1/\alpha$. We have to show that

(S1)
$$\left(\int f \right)^p + \left(\int g \right)^p \le \left[\int (f^p + g^p)^{1/p} \right]^p.$$

Set $a = \int f$ and $b = \int g$, so that we have

$$a^p + b^p = \int a^{p-1} f + b^{p-1} g \le \int (a^p + b^p)^{1/p'} (f^p + g^p)^{1/p}$$

$$= (a^p + b^p)^{1/p'} \int (f^p + g^p)^{1/p}.$$

It follows that $(a^p + b^p)^{1/p} \le \int (f^p + g^p)^{1/p}$, i.e., (S1).

$\boxed{4.12}$

1. It suffices to show that

$$\inf_{t \in [-1, +1]} \left\{ \frac{(|t|^p + 1)^{1-s}(|t|^p + 1 - 2|\frac{t+1}{2}|^p)^s}{|t-1|^p} \right\} > 0,$$

or equivalently that

$$\inf_{t \in [-1, +1]} \left\{ \frac{|t|^p + 1 - 2|\frac{t+1}{2}|^2}{|t-1|^2} \right\} > 0.$$

But the function $\varphi(t) = |t|^p + 1 - 2|\frac{t+1}{2}|^p$ satisfies

$$\varphi(t) > 0 \quad \forall t \in [-1, +1), \quad \varphi(1) = \varphi'(1) = 0 \quad \text{and} \quad \varphi''(1) > 0.$$

$\boxed{4.14}$

1 and 2. Let $g_n = \chi_{S_n(\alpha)}$, so that $g_n \to 0$ a.e. and $|g_n| \le 1$. It follows—by dominated convergence (since $|\Omega| < \infty$)—that $\int g_n \to 0$, i.e., $|S_n(\alpha)| \to 0$.

3. Given any integer $m \ge 1$, we may apply question 2 with $\alpha = 1/m$ to find an integer N_m such that $|S_{N_m}(1/m)| < \delta/2^m$. Letting $\Sigma_m = S_{N_m}(1/m)$, we obtain

$$|f_k(x) - f(x)| \le \frac{1}{m} \quad \forall k \ge N_m, \quad \forall x \in \Omega \backslash \Sigma_m.$$

Finally, set $A = \bigcup_{m=1}^\infty \Sigma_m$, so that $|A| < \delta$. We claim that $f_n \to f$ uniformly on $\Omega \backslash A$. Indeed, given $\varepsilon > 0$, fix an integer m_0 such that $m_0 > 1/\varepsilon$. Clearly,

$$|f_k(x) - f(x)| < \varepsilon \quad \forall k \ge N_{m_0}, \quad \forall x \in \Omega \backslash \Sigma_{m_0},$$

and consequently

$$|f_k(x) - f(x)| < \varepsilon \quad \forall k \ge N_{m_0}, \quad \forall x \in \Omega \backslash A.$$

4. Given $\varepsilon > 0$, first fix some $\delta > 0$ using (i) and then fix some A using question 3. We obtain that $\int_A |f_n|^p \leq \varepsilon$ $\forall n$ and $f_n \to f$ uniformly on $\Omega \backslash A$. It follows from Fatou's lemma that $\int_A |f|^p \leq \varepsilon$ and thus

$$\int_\Omega |f_n - f|^p = \int_A |f_n - f|^p + \int_{\Omega \backslash A} |f_n - f|^p \leq 2^p \varepsilon + |\Omega| \, \|f_n - f\|_{L^\infty(\Omega \backslash A)}^p.$$

4.15

1(iv). Note that $\int f_n \varphi \to 0$ $\forall \varphi \in C_c(\Omega)$. Suppose, by contradiction, that $f_{n_k} \rightharpoonup f$ weakly $\sigma(L^1, L^\infty)$. It follows that $\int f \varphi = 0$ $\forall \varphi \in C_c(\Omega)$ and thus (by Corollary 4.24) $f = 0$ a.e. Also $\int f_{n_k} \to \int f = 0$; but $\int f_{n_k} = \int_0^{n_k} e^{-t} dt \to 1$; a contradiction.

2(iv). Note that $\int g_n \varphi \to 0$ $\forall \varphi \in C_c(\Omega)$ and use the fact that $C_c(\Omega)$ is dense in $L^{p'}(\Omega)$ (since $p' < \infty$).

4.16

1. Let us first check that if a sequence (f_n) satisfies

   (S1) $f_n \rightharpoonup \tilde{f}$ weakly $\sigma(L^p, L^{p'})$

   and

   (S2) $f_n \to f$ a.e.

   then $f = \tilde{f}$ a.e.

   Indeed, we know from Exercise 3.4 that there exists a sequence (g_n) in $L^p(\Omega)$ such that

   (S3) $g_n \in \operatorname{conv} \{f_n, f_{n+1}, \dots\}$,

   and

   (S4) $g_n \to \tilde{f}$ strongly in $L^p(\Omega)$.

   It follows from (S2) and (S3) that $g_n \to f$ a.e. On the other hand (by Theorem 4.9), there is a subsequence (g_{n_k}) such that $g_{n_k} \to \tilde{f}$ a.e. Therefore $f = \tilde{f}$ a.e.

   Let us now check, under the assumptions (i) and (ii), that $f_n \rightharpoonup f$ weakly $\sigma(L^p, L^{p'})$. There exists a subsequence (f_{n_k}) converging weakly $\sigma(L^p, L^{p'})$ to some limit, say $\tilde{f}$. From the preceding discussion we know that $f = \tilde{f}$ a.e. The "uniqueness of the limit" implies that the *whole* sequence (f_n) converges weakly to f (fill in the details using a variant of the argument in Exercise 3.32).

3. *First method.* Write

(S5) $\|f_n - f\|_q \leq \|f_n - T_k f_n\|_q + \|T_k f_n - T_k f\|_q + \|T_k f - f\|_q.$

Note that for every $k > 0$ we have

$$\int |f_n - T_k f_n|^q \leq \int_{[|f_n| \geq k]} |f_n|^q.$$

On the other hand, we have $\int |f_n|^p \leq C^p$ and thus $k^{p-q} \int_{[|f_n| \geq k]} |f_n|^q \leq C^p$.

It follows that

(S6) $\|f_n - T_k f_n\|_q \leq \left(\dfrac{C^p}{k^{p-q}} \right)^{1/q} \quad \forall n.$

Passing to the limit (as $n \to \infty$), with the help of Fatou's lemma we obtain

(S7) $\|f - T_k f\|_q \leq \left(\dfrac{C^p}{k^{p-q}} \right)^{1/q}.$

Given $\varepsilon > 0$, *fix k large enough* that $(C^p / k^{p-q})^{1/q} < \varepsilon$. It is clear (by dominated convergence) that $\|T_k f_n - T_k f\|_q \xrightarrow[n \to \infty]{} 0$, and hence there is some integer N such that

(S8) $\|T_k f_n - T_k f\|_q < \varepsilon \quad \forall n \geq N.$

Combining (S5), (S6), (S7), and (S8), we see that $\|f_n - f\|_q < 3\varepsilon \ \forall n \geq N$.

Second method. By Egorov's theorem we know that given $\delta > 0$ there exists some $A \subset \Omega$ such that $|A| < \delta$ and $f_n \to f$ uniformly on $\Omega \backslash A$. Write

$$\int_\Omega |f_n - f|^q = \int_{\Omega \backslash A} + \int_A$$
$$\leq \|f_n - f\|^q_{L^\infty(\Omega \backslash A)} |\Omega| + \|f_n - f\|^q_p |A|^{1-(q/p)}$$
$$\leq \|f_n - f\|^q_{L^\infty(\Omega \backslash A)} |\Omega| + (2C)^q \delta^{1-(q/p)},$$

which leads to

$$\limsup_{n \to \infty} \int |f_n - f|^q \leq (2C)^q \delta^{1-(q/p)} \quad \forall \delta > 0.$$

$\boxed{4.17}$

1. By homogeneity it suffices to check that

$$\sup_{t \in [-1,+1]} \left\{ \frac{\big| |t+1|^p - |t|^p - 1 \big|}{|t|^{p-1} + |t|} \right\} < \infty.$$

4.18

1. First, it is easy to check that $\int_a^b u_n(t)dt \to (b-a)\overline{f}$ (for every $a, b \in (0, 1)$). This implies that $u_n \rightharpoonup \overline{f}$ weakly $\sigma(L^p, L^{p'})$ whenever $1 < p \le \infty$ (since $p' < \infty$, step functions are dense in $L^{p'}$). When $p = 1$, i.e., $f \in L^1_{loc}(\mathbb{R})$, there is a T-periodic function $g \in L^\infty(\mathbb{R})$ such that $\frac{1}{T}\int_0^T |f - g| < \varepsilon$ (where $\varepsilon > 0$ is fixed arbitrarily).

   Set $v_n(x) = g(nx)$, $x \in (0, 1)$ and let $\varphi \in L^\infty(0, 1)$. We have

   $$\left| \int u_n\varphi - \overline{f}\int \varphi \right| \le 3\varepsilon\|\varphi\|_\infty + \left| \int v_n\varphi - \overline{g}\int\varphi \right|$$

   and thus $\limsup_{n\to\infty} \left| \int u_n\varphi - \overline{f}\int\varphi \right| \le 3\varepsilon\|\varphi\|_\infty$ $\forall\varepsilon > 0$. It follows that $u_n \rightharpoonup \overline{f}$ weakly $\sigma(L^1, L^\infty)$.

2. $\lim_{n\to\infty}\|u_n - \overline{f}\|_p = \left[\frac{1}{T}\int_0^T |f - \overline{f}|^p \right]^{1/p}$.

3. (i) $u_n \overset{\star}{\rightharpoonup} 0$ for $\sigma(L^\infty, L^1)$.
   (ii) $u_n \overset{\star}{\rightharpoonup} \frac{1}{2}(\alpha + \beta)$ for $\sigma(L^\infty, L^1)$.

4.20

1. Let (u_n) be a sequence in $L^p(\Omega)$ such that $u_n \to u$ strongly in $L^p(\Omega)$. There exists a subsequence such that $u_{n_k}(x) \to u(x)$ a.e. and $|u_{n_k}| \le v$ $\forall k$ with $v \in L^p(\Omega)$ (see Theorem 4.9). It follows by dominated convergence that $Au_{n_k} \to Au$ strongly in $L^q(\Omega)$. The "uniqueness of the limit" implies that the *whole* sequence (Au_n) converges to Au strongly in $L^q(\Omega)$ (as in the solution to Exercise 3.32).

2. Consider the sequence (u_n) defined in Exercise 4.18, question 3(ii). Note that $u_n \rightharpoonup \frac{1}{2}(\alpha + \beta)$, while $Au_n \rightharpoonup \frac{1}{2}(a(\alpha) + a(\beta))$. It follows that

   $$a\left(\frac{\alpha + \beta}{2}\right) = \frac{1}{2}(a(\alpha) + a(\beta)) \quad \forall\alpha, \beta \in \mathbb{R},$$

   and thus a must be an *affine* function.

4.21

1. Check that $\int_I u_n(t)dt \to 0$ for every bounded interval I. Then use the density of step functions (with compact support) in $L^{p'}(\mathbb{R})$.

2. We claim once more that $\int_I u_n(t)dt \to 0$ for every bounded interval I. Indeeed, given $\varepsilon > 0$, fix $\delta > 0$ such that $\delta(\|u_0\|_\infty + |I|) < \varepsilon$. Set $E = [|u_0| > \delta]$ and write

   $$\int_I u_n(t)dt = \int_{(I+n)} u_0(t)dt = \int_{(I+n)\cap E} u_0 + \int_{(I+n)\cap E^c} u_0.$$

   Choose N large enough that $|(I + n) \cap E| < \delta$ $\forall n > N$ (why is it possible?).

We obtain

$$\left| \int_I u_n(t)dt \right| \leq \delta \|u_0\|_\infty + \delta |I| < \varepsilon \quad \forall n \geq N.$$

Then use the density of step functions (with compact support) in $L^1(\mathbb{R})$.
3. Suppose, by contradiction, that $u_{n_k} \rightharpoonup u$ weakly $\sigma(L^1, L^\infty)$. Consider the function $f \in L^\infty(\mathbb{R})$ defined by

$$f = \sum_i (-1)^i \chi_{(-n_i, -n_i+1)}.$$

Note that $\int u_{n_k} f = (-1)^k$ does not converge.

4.22

1. In order to prove that (B) $\Rightarrow$ (A) use the fact that the vector space spanned by the functions χ_E with E measurable and $|E| < \infty$ is dense in $L^{p'}(\Omega)$ provided $p' < \infty$ (why?).
2. Use the fact that the vector space spanned by the functions χ_E (with $E \subset \Omega$ and E measurable) is dense in $L^\infty(\Omega)$ (why?).
4. Given $\varepsilon > 0$, fix some measurable subset $\omega \subset \Omega$ such that $|\omega| < \infty$ and

(S1) $$\int_{\omega^c} f < \varepsilon.$$

We have

$$\int_{\omega^c} f_n = \int_{\omega^c} f + \left(\int_\omega f - \int_\omega f_n \right) + \left(\int_\Omega f_n - \int_\Omega f \right)$$

and therefore

(S2) $$\int_{\omega^c} f_n = \int_{\omega^c} f + o(1) \quad \text{(by (b) and (c))}.$$

On the other hand, we have

$$\int_F f_n = \int_{F\cap\omega} f_n + \int_{F\cap(\omega^c)} f_n = \int_{F\cap\omega} f + \int_{F\cap(\omega^c)} f_n + o(1)$$

and thus

(S3) $$\int_F f_n - \int_F f = \int_{F\cap(\omega^c)} (f_n - f) + o(1).$$

Combining (S1), (S2), and (S3), we obtain

$$\left| \int_F f_n - \int_F f \right| \leq 2\varepsilon + o(1).$$

It follows that $\int_F f_n \to \int_F f$. Finally, we use the fact that the vector space spanned by the functions χ_F with $F \subset \Omega$, F measurable and $|F| \leq \infty$, is dense in $L^\infty(\Omega)$ (why?).

4.23

1. Let (u_n) be a sequence in C such that $u_n \to u$ strongly in $L^p(\Omega)$. There exists a subsequence (u_{n_k}) such that $u_{n_k} \to u$ a.e. Thus $u \geq f$ a.e.
2. Assume that $u \in L^\infty(\Omega)$ satisfies

$$\int u\varphi \geq \int f\varphi \quad \forall \varphi \in L^1(\Omega) \text{ such that } f\varphi \in L^1(\Omega) \text{ and } \varphi \geq 0.$$

We claim that $u \geq f$ a.e. Indeed, write $\Omega = \bigcup_n \Omega_n$ with $|\Omega_n| < \infty$ and set $\Omega'_n = \Omega_n \cap [|f| < n]$, so that $\bigcup_n \Omega'_n = \Omega$. Let $A = [u < f]$. Choosing $\varphi = \chi_{A \cap \Omega'_n}$, we find that $\int_{A \cap \Omega'_n} |f - u| \leq 0$ and thus $|A \cap \Omega'_n| = 0 \ \forall n$. It follows that $|A| = 0$.
3. Note that if $\varphi \in L^1(\Omega)$ is *fixed* with $f\varphi \in L^1(\Omega)$ then the set $\{u \in L^\infty(\Omega); \int u\varphi \geq \int f\varphi\}$ is closed for the topology $\sigma(L^\infty, L^1)$.

4.24

1. For every $\varphi \in L^1(\mathbb{R}^N)$ we have

$$\int v_n\varphi = \int u\zeta_n(\check{\rho}_n \star \varphi) = \int u\zeta_n(\check{\rho}_n \star \varphi - \varphi) + \int u\zeta_n\varphi$$

and thus

$$\left| \int v_n\varphi - \int v\varphi \right| \leq \|u\|_\infty \|\check{\rho}_n \star \varphi - \varphi\|_1 + \|u\|_\infty \|(\zeta_n - \zeta)\varphi\|_1.$$

The first term on the right side tends to zero by Theorem 4.22, while the second term on the right side tends to zero by dominated convergence.
2. Let $B = B(x_0, R)$ and let χ denote the characteristic function of $B(x_0, R+1)$. Set $\tilde{v}_n = \rho_n \star (\zeta_n \chi u)$. Note that $\tilde{v}_n = v_n$ on $B(x_0, R)$, since

$$\text{supp}(\tilde{v}_n - v_n) \subset \overline{B(0, 1/n) + B(x_0, R+1)^c}.$$

On the other hand, we have

$$\int_B |v_n - v| = \int_B |\tilde{v}_n - \chi v| \leq \int_{\mathbb{R}^N} |\tilde{v}_n - \chi v|$$

$$\leq \int_{\mathbb{R}^N} |\rho_n \star (\zeta_n - \zeta)\chi u| + \int_{\mathbb{R}^N} |(\rho_n \star \chi v) - \chi v|$$

$$\leq \int_{\mathbb{R}^N} |(\zeta_n - \zeta)\chi u| + \int_{\mathbb{R}^N} |(\rho_n \star \chi v) - \chi v| \to 0.$$

4.25

1. Let $\bar{u}$ denote the extension of u by 0 outside Ω. Let

$$\Omega_n = \{x \in \Omega; \operatorname{dist}(x, \partial\Omega) > 2/n \text{ and } |x| < n\}.$$

Let ζ_n (resp. ζ) denote the characteristic function of Ω_n (resp. Ω), so that $\zeta_n \to \zeta$ on $\mathbb{R}^N$. Let $v_n = \rho_n \star (\zeta_n \bar{u})$. We know that $v_n \in C_c^\infty(\Omega)$ and that $\int_B |v_n - \bar{u}| \to 0$ for every ball B (by Exercise 4.24). Thus, for every ball B there is a subsequence (depending on B) that converges to $\bar{u}$ a.e. on B. By a *diagonal process* we may construct a subsequence (v_{n_k}) that converges to $\bar{u}$ a.e. on $\mathbb{R}^N$.

4.26

1. Assume that $A < \infty$. Let us prove that $f \in L^1(\Omega)$ and that $\|f\|_1 \leq A$. We have

$$\left| \int f\varphi \right| \leq A \|\varphi\|_\infty \quad \forall \varphi \in C_c(\Omega).$$

Let $K \subset \Omega$ be any compact subset and let $\psi \in C_c(\Omega)$ be a function such that $0 \leq \psi \leq 1$ and $\psi = 1$ on K. Let u be any function in $L^\infty(\Omega)$. Using Exercise 4.25 we may construct a sequence (u_n) in $C_c(\Omega)$ such that $\|u_n\|_\infty \leq \|u\|_\infty$ and $u_n \to u$ a.e. on Ω. We have

$$\left| \int f\psi u_n \right| \leq A \|u\|_\infty.$$

Passing to the limit as $n \to \infty$ (by dominated convergence) we obtain

$$\left| \int f\psi u \right| \leq A \|u\|_\infty \quad \forall u \in L^\infty(\Omega).$$

Choosing $u = \operatorname{sign}(f)$ we find that $\int_K |f| \leq A$ for every compact subset $K \subset \Omega$. It follows that $f \in L^1(\Omega)$ and that $\|f\|_1 \leq A$.

2. Assume that $B < \infty$. We have

$$\int f\varphi \leq B \|\varphi\|_\infty \quad \forall \varphi \in C_c(\Omega), \ \varphi \geq 0.$$

Using the same method as in question 1, we obtain

$$\int f\psi u \leq B \|u\|_\infty \quad \forall u \in L^\infty(\Omega), \ u \geq 0.$$

Choosing $u = \chi_{[f>0]}$ we find that $\int_K f^+ \leq B$.

4.27 Let us first examine an abstract setting. Let E be a vector space and let f, g be two linear functionals on E such that $f \not\equiv 0$. Assume that

$$[\varphi \in E \text{ and } f(\varphi) > 0] \Rightarrow [g(\varphi) \geq 0].$$

We claim that there exists a constant $\lambda \geq 0$ such that $g = \lambda f$. Indeed, fix any $\varphi_0 \in E$ such that $f(\varphi_0) = 1$. For every $\varphi \in E$ and every $\varepsilon > 0$, we have

$$f(\varphi - f(\varphi)\varphi_0 + \varepsilon\varphi_0) = \varepsilon > 0$$

and thus $g(\varphi - f(\varphi)\varphi_0 + \varepsilon\varphi_0) \geq 0$. It follows that $g(\varphi) \geq \lambda f(\varphi) \ \forall \varphi \in E$, and thus $g = \lambda f$, with $\lambda = g(\varphi_0) \geq 0$.

Application. $E = C_c^\infty(\Omega)$, $f(\varphi) = \int u\varphi$, and $g(\varphi) = \int v\varphi$.

<hr>

4.30

1 and 2. Note that $\frac{1}{p'} + \frac{1}{q'} + \frac{1}{r} = 1$ and that $(1 - \alpha)r = p$, $(1 - \beta)r = q$. For a.e. $x \in \mathbb{R}^N$ write

$$|f(x - y)g(y)| = \varphi_1(y)\varphi_2(y)\varphi_3(y)$$

with $\varphi_1(y) = |f(x - y)|^\alpha$, $\varphi_2(y) = |g(y)|^\beta$, and $\varphi_3(y) = |f(x - y)|^{1-\alpha}|g(y)|^{1-\beta}$. Clearly, $\varphi_1 \in L^{q'}(\mathbb{R}^N)$ and $\varphi_2 \in L^{p'}(\mathbb{R}^N)$. On the other hand, $|\varphi_3(y)|^r = |f(x - y)|^p|g(y)|^q$. We deduce from Theorem 4.15 that for a.e. $x \in \mathbb{R}^N$ the function $y \mapsto |\varphi_3(y)|^r$ is integrable. It follows from Hölder's inequality (see Exercise 4.4) that for a.e. $x \in \mathbb{R}^N$, the function $y \mapsto |f(x - y)g(y)|$ is integrable and that

$$\int |f(x - y)| \, |g(y)|dy \leq \|f\|_p^\alpha \|g\|_q^\beta \left(\int |f(x - y)|^p|g(y)|^q dy \right)^{1/r}.$$

Thus

$$|(f \star g)(x)|^r \leq \|f\|_p^{\alpha r} \|g\|_q^{\beta r} \int |f(x - y)|^p|g(y)|^q dy,$$

and consequently

$$\int |(f \star g)(x)|^r dx \leq \|f\|_p^{\alpha r} \|g\|_q^{\beta r} \|f\|_p^p \|g\|_q^q = \|f\|_p^r \|g\|_q^r.$$

3. If $1 < p < \infty$ and $1 < q < \infty$, there exist sequences (f_n) and (g_n) in $C_c(\mathbb{R}^N)$ such that $f_n \to f$ in $L^p(\mathbb{R}^N)$ and $g_n \to g$ in $L^q(\mathbb{R}^N)$. Then $f_n \star g_n \in C_c(\mathbb{R}^N)$, and, moreover, $\|(f_n \star g_n) - (f \star g)\|_\infty \to 0$. It follows that $(f \star g)(x) \to 0$ as $|x| \to \infty$.

<hr>

4.34 Given any $\varepsilon > 0$ there is a finite covering of $\mathcal{F}$ by balls of radius ε in $L^p(\mathbb{R}^N)$, say $\mathcal{F} \subset \bigcup_{i=1}^k B(f_i, \varepsilon)$.

2. For each i there is some $\delta_i > 0$ such that

$$\|\tau_h f_i - f_i\|_{L^p(\mathbb{R}^N)} < \varepsilon \quad \forall h \in \mathbb{R}^N \text{ with } |h| < \delta_i$$

(see Lemma 4.3). Set $\delta = \min_{1 \leq i \leq k} \delta_i$. It is easy to check that

$$\|\tau_h f - f\|_p < 3\varepsilon \quad \forall f \in \mathcal{F}, \ \forall h \in \mathbb{R}^N \text{ with } |h| < \delta.$$

3. For each i there is some bounded open set $\Omega_i \subset \mathbb{R}^N$ such that

$$\|f_i\|_{L^p(\mathbb{R}^N \setminus \Omega_i)} < \varepsilon.$$

Set $\Omega = \bigcup_{i=1}^k \Omega_i$ and check that $\|f\|_{L^p(\mathbb{R}^N \setminus \Omega)} < 2\varepsilon \quad \forall f \in \mathcal{F}$.

$\boxed{4.37}$

1. Write

$$\int_I u_n(x)\varphi(x)dx = \int_{-n}^{+n} f(t)\left(\varphi\left(\frac{t}{n}\right) - \varphi(0)\right)dt + \varphi(0)\int_{-n}^{+n} f(t)dt$$

$$= A_n + B_n;$$

$A_n \to 0$ by Lebesgue's theorem and $B_n \to 0$ since $\int_{-\infty}^{+\infty} f(t)dt = 0$.

2. Note that, for all $\delta > 0$,

$$\int_0^\delta |u_n(x)|dx = \int_0^{n\delta} |f(t)|dt \to \int_0^\infty |f(t)|dt > 0.$$

3. Argue by contradiction. We would have

$$\int_I u\varphi = 0 \quad \forall \varphi \in C([-1, +1])$$

and thus $u \equiv 0$ (by Corollary 4.24). On the other hand, if we choose $\varphi = \chi_{(0,1)}$ we obtain

$$\int_I u_n\varphi = \int_0^n f(t)dt \to \int_0^{+\infty} f(t)dt > 0.$$

Impossible.

$\boxed{4.38}$

2. Check that, $\forall \varphi \in C^1([0, 1])$,

$$\int_I u_n\varphi = \int_I \varphi + O\left(\frac{1}{n}\right), \quad \text{as } n \to \infty.$$

Then use the facts that $\|u_n\|_1$ is bounded and $C^1([0, 1])$ is dense in $C([0, 1])$.

3. The sequence (u_n) cannot be equi-integrable since $|\operatorname{supp} u_n| \to 0$ and

$$1 = \int_I u_n = \int_{\operatorname{supp} u_n} |u_n|.$$

4. If $u_{n_k} \rightharpoonup u$ weakly $\sigma(L^1, L^\infty)$ we would have, by question 2 and Corollary 4.24, $u \equiv 1$. Choose a further subsequence $(u_{n'_k})$ such that $\sum_k |\operatorname{supp} u_{n'_k}| < 1$. Let $\varphi = \chi_A$ where

$$A = I \setminus \left(\bigcup_k \operatorname{supp} u_{n'_k} \right),$$

so that $|A| > 0$. We have

$$\int_I u_{n'_k} \varphi = 0 \quad \forall k$$

and thus $0 = \int_I \varphi = |A|$. Impossible.

5. Consider a subsequence (u_{n_k}) such that

$$\sum_k |\operatorname{supp} u_{n_k}| < \infty.$$

Let $B_k = \bigcup_{j \geq k} (\operatorname{supp} u_{n_j})$ and $B = \bigcap_k B_k$. Clearly $|B_k| \to 0$ as $k \to \infty$, and thus $|B| = 0$. If $x \notin B$ there exists some k_0 such that $u_{n_k}(x) = 0 \; \forall k \geq k_0$.

5.1

1. Using the parallelogram law with $a = u + v$ and $b = v$ leads to $(u, 2v) = 2(u, v)$.

2. Compute (i) $-$ (ii) $+$ (iii).

3. Note that by definition of $(\,,\,)$, the map $\lambda \in \mathbb{R} \mapsto (\lambda u, v)$ is continuous.

5.2 Let A be a measurable set such that $0 < |A| < |\Omega|$, and choose a measurable set B such that $A \cap B = \emptyset$ and $0 < |B| < |\Omega|$, Let $u = \chi_A$ and $v = \chi_B$. Assume first that $1 \leq p < \infty$. We have $\|u + v\|_p^p = \|u - v\|_p^p = |A| + |B|$ and thus $\|u + v\|_p^2 + \|u - v\|_p^2 = 2(|A| + |B|)^{2/p}$. On the other hand, we have $2(\|u\|_p^2 + \|v\|_p^2) = 2(|A|^{2/p} + |B|^{2/p})$. Finally, note that

$$(\alpha + \beta)^{2/p} > \alpha^{2/p} + \beta^{2/p} \; \forall \alpha, \beta > 0 \text{ if } p < 2,$$
$$(\alpha + \beta)^{2/p} < \alpha^{2/p} + \beta^{2/p} \; \forall \alpha, \beta > 0 \text{ if } p > 2.$$

Examine the case $p = \infty$ with the same functions u and v.

5.3 Check that

(S1) $2(t_n u_n - t_m u_m, u_n - u_m) = (t_n + t_m)|u_n - u_m|^2 + (t_n - t_m)(|u_n|^2 - |u_m|^2)$,

which implies that

$$(t_n - t_m)(|u_n|^2 - |u_m|^2) \leq 0 \quad \forall m, n.$$

1. Let $n > m$, so that $t_n \geq t_m$ and thus $|u_n| \leq |u_m|$. (Note that if $t_n = t_m$, then $u_n = u_m$ in view of (S1)). On the other hand, we have for $n > m$,

$$(t_n + t_m)|u_n - u_m|^2 \leq (t_n - t_m)(|u_m|^2 - |u_n|^2) \leq t_n(|u_m|^2 - |u_n|)^2)$$

and thus

$$|u_n - u_m|^2 \leq |u_m|^2 - |u_n|^2.$$

It follows that $|u_n| \downarrow \ell$ as $n \uparrow \infty$ and that (u_n) is a Cauchy sequence.

2. Let $n > m$, so that $t_m \geq t_n$ and $|u_m| \leq |u_n|$. For $n > m$ we have

$$(t_n + t_m)|u_n - u_m|^2 \leq (t_m - t_n)(|u_n|^2 - |u_m|^2) \leq t_m(|u_n|^2 - |u_m|^2)$$

and thus

$$|u_n - u_m|^2 \leq |u_n|^2 - |u_m|^2.$$

We now have the following alternative:

(i) either $|u_n| \uparrow \infty$ as $n \uparrow \infty$,

(ii) or $|u_n| \uparrow \ell < \infty$ as $n \uparrow \infty$ and then (u_n) is a Cauchy sequence.

On the other hand, letting $v_n = t_n u_n$ and $s_n = 1/t_n$, we obtain

$$(s_n v_n - s_m v_m, v_n - v_m) \leq 0,$$

and thus (v_n) converges to a limit by question 1. It follows that if $t_n \to t > 0$ then (u_n) also converges to a limit. Finally if $t_n \to 0$, both cases (i) and (ii) may occur. Take, for example, $H = \mathbb{R}$, $u_n = C/t_n$ for (i), $u_n = C$ for (ii).

$\boxed{5.4}$ Note that

$$|v - u|^2 = |v - f|^2 - |u - f|^2 + 2(f - u, v - u).$$

$\boxed{5.5}$

1. Let $K = \bigcap_n K_n$. We claim that $u_n \to u = P_K f$. First, note that the sequence $d_n = |f - u_n| = \text{dist}(f, K_n)$ is nondecreasing and bounded above. Thus $d_n \uparrow \ell < \infty$ as $n \uparrow \infty$. Next, using the parallelogram law (with $a = f - u_n$ and $b = f - u_m$), we obtain

$$\left| f - \frac{u_n + u_m}{2} \right|^2 + \left| \frac{u_n - u_m}{2} \right|^2 = \frac{1}{2} \left(|f - u_n|^2 + |f - u_m|^2 \right).$$

It follows that $|u_n - u_m|^2 \leq 2(d_m^2 - d_n^2)$ if $m \geq n$. Thus (u_n) converges to a limit, say u, and clearly $u \in K$. On the other hand, we have $|f - u_n| \leq |f - v| \; \forall v \in K_n$ and in particular $|f - u_n| \leq |f - v| \; \forall v \in K$. Passing to the limit, we obtain $|f - u| \leq |f - v| \; \forall v \in K$.

2. Clearly $K = \overline{\bigcup_n K_n}$ is convex (why?). We claim that $u_n \to u = P_K f$. First, note that the sequence $d_n = |f - u_n| = \text{dist}(f, K_n)$ is nonincreasing and

thus $d_n \to \ell$. Next, we have (with the same method as above) $|u_n - u_m|^2 \le 2\left(d_n^2 - d_m^2\right)$ if $m \ge n$. Thus (u_n) converges to a limit, say u, and clearly $u \in K$. Finally, note that $|f - u_m| \le |f - v| \ \forall v \in K_n$ provided $m \ge n$. Passing to the limit (as $m \to \infty$) leads to $|f - u| \le |f - v| \ \forall v \in \bigcup_n K_n$, and by density $\forall v \in K$.

The sequence (α_n) is nonincreasing and thus it converges to a limit, say α. We claim that $\alpha = \inf_K \varphi$. First, it is clear that $\inf_K \varphi \le \alpha_n$ and thus $\inf_K \varphi \le \alpha$. On the other hand, let u be any element in K and let $u_n = P_{K_n} u$. Passing to the limit in the inequality $\alpha_n \le \varphi(u_n)$, we obtain $\alpha \le \varphi(u)$ (since $u_n \to u$). It follows that $\alpha \le \inf_K \varphi$.

5.6

1. Consider, for example, the case that $\|u\| \ge 1$ and $\|v\| \le 1$. We have

$$\|Tu - Tv\| = \left\| \frac{u}{\|u\|} - v \right\| = \frac{\|(u-v)+(v-v\|u\|)\|}{\|u\|}$$
$$\le \|u-v\| + \|u\| - 1 \le 2\|u-v\|,$$

since $\|u\| \le \|u-v\| + \|v\| \le \|u-v\| + 1$.
2. Let $u = (1,0)$ and $v = (1,\alpha)$. Then we have $\|Tu - Tv\| = 2|\alpha|/(1+|\alpha|)$, while $\|u-v\| = |\alpha|$. We conclude by choosing $\alpha \ne 0$ and arbitrarily small.
3. T coincides with P_{B_E}. Just check that if $\|u\| \ge 1$. then

$$\left(u - \frac{u}{\|u\|}, v - \frac{u}{\|u\|} \right) \le 0 \quad \forall v \in B_E.$$

5.10

(i) $\Rightarrow$ (ii). Write that

$$F(u) \le F((1-t)u + tv) \quad \forall t \in (0,1), \ \forall v \in K,$$

which implies that

$$\frac{1}{t}[F(u+t(v-u)) - F(u)] \ge 0.$$

Passing to the limit as $t \to 0$ we obtain (ii).
(ii) $\Rightarrow$ (i). We claim that

$$F(v) - F(u) \ge \left(F'(u), v-u\right) \quad \forall u, v \in H.$$

Indeed, the function $t \in \mathbb{R} \mapsto \varphi(t) = F(u+t(v-u))$ is of class C^1 and convex. Thus $\varphi(1) - \varphi(0) \ge \varphi'(0)$.

5.12 T is surjective iff E is complete.

1. Transfer onto $R(T)$ the scalar product of E by letting

$$((T(u), T(v))) = (u, v) \quad \forall u, v \in E.$$

Note that $|((f, g))| \leq \|f\|_{E^*}\|g\|_{E^*} \ \forall f, g \in R(T)$. The scalar product $((\, ,))$ can be extended by continuity and density to $\overline{R(T)}$, which is now equipped with the structure of a *Hilbert space*.

2. Fix any $f \in E^*$. The map $g \in R(T) \mapsto \langle f, T^{-1}(g) \rangle$ is a continuous linear functional on $R(T)$. It may be extended (by continuity) to $\overline{R(T)}$. Using the Riesz–Fréchet representation theorem in $\overline{R(T)}$ we obtain some element $h \in \overline{R(T)}$ such that $((h, g)) = \langle f, T^{-1}(g) \rangle \ \forall g \in R(T)$. Thus we have $((h, T(v))) = \langle f, v \rangle$ $\forall v \in E$. On the other hand, we have $((h, Tv)) = \langle h, v \rangle \ \forall h \in \overline{R(T)}, \ \forall v \in E$ (this is obvious when $h \in R(T)$). It follows that $f = h$ and consequently $f \in \overline{R(T)}$, i.e., $\overline{R(T)} = E^*$.

3. We have constructed an isometry $T : E \to E^*$ with $R(T)$ dense in E^*. Since E^* is complete, we conclude that (up to an isomorphism) E^* is the completion of E.

$\boxed{5.13}$

1. We claim that the parallelogram law holds. Indeed, let $f \in F(u)$ and let $g \in F(v)$. Then $f \pm g \in F(u \pm v)$ and so we have

$$\langle f + g, u + v \rangle = \|u + v\|^2 \quad \text{and} \quad \langle f - g, u - v \rangle = \|u - v\|^2.$$

Adding these relations leads to

$$2(\|u\|^2 + \|v\|^2) = \|u + v\|^2 + \|u - v\|^2.$$

2. Let $T : E \to E^*$ be the map introduced in Exercise 5.12. We claim that $F(u) = \{T(u)\}$. Clearly, $T(u) \in F(u)$. On the other hand, we know that E^* is a Hilbert space for the dual norm $\| \ \|_{E^*}$. In particular, E^* is strictly convex and thus (see Exercise 1.1) $F(u)$ is reduced to a single element.

$\boxed{5.14}$ The convexity inequality $a(tu + (1-t)v, tu + (1-t)v) \leq ta(u, u) + (1-t)a(v, v)$ is equivalent to $t(1-t)a(u - v, u - v) \geq 0$.

Consider the operator $A \in \mathcal{L}(H)$ defined by $a(u, v) = (Au, v) \ \forall u, v \in H$. Then $F'(u) = Au + A^*u$, since we have

$$F(u + h) - F(u) = (Au + A^*u, h) + a(h, h).$$

$\boxed{5.15}$ First, extend S by continuity into an operator $\widetilde{S} : \overline{G} \to F$. Next, let $T = \widetilde{S} \circ P_{\overline{G}}$, where $P_{\overline{G}}$ denotes the projection from H onto $\overline{G}$.

$\boxed{5.18}$

(ii) $\Rightarrow$ (i). Assumption (ii) implies that T is injective and that $R(T)$ is closed. Thus $R(T)$ has a complement (since H is a Hilbert space). We deduce from Theorem 2.13 that T has a left inverse.

(i) $\Rightarrow$ (ii). Assumption (i) implies that T is injective and that $R(T)$ is closed. Then, use Theorem 2.21.

5.19 Note that $\lim \sup_{n\to\infty} |u_n - u|^2 = \lim \sup_{n\to\infty} (|u_n|^2 - 2(u_n, u) + |u|^2) \leq 0$.

5.20

1. If $u \in N(S)$ we have $(Sv, v - u) \geq 0 \; \forall v \in H$; replacing v by tv, we see that $(Sv, u) = 0 \; \forall v \in H$. Conversely, if $u \in R(S)^\perp$ we have $(Sv - Su, v) \geq 0 \; \forall v \in H$; replacing v by tv, we see that $(Su, v) = 0 \; \forall v \in H$. (See also Problem 16.)
2. Apply Corollary 5.8 (Lax–Milgram).
3. *Method* (a). Set $u_t = (I + tS)^{-1} f$.

   If $f \in N(S)$, then $u_t = f \; \forall t > 0$.

   If $f \in R(S)$, write $f = Sv$, so that $u_t + S(tu_t - v) = 0$. It follows that

   $(u_t, tu_t - v) \leq 0$ and thus $|u_t| \leq (1/t)|v|$. Consequently $u_t \to 0$ as $t \to \infty$. By density, one can still prove that $u_t \to 0$ as $t \to \infty$ for every $f \in \overline{R(S)}$ (fill in the details).

   In the general case $f \in H$, write $f = f_1 + f_2$ with $f_1 = P_{N(S)} f$ and $f_2 = P_{\overline{R(S)}} f$.

   Method (b). We have $u_t + tSu_t = f$ and thus $|u_t| \leq |f|$. Passing to a subsequence $t_n \to \infty$ we may assume that $u_{t_n} \rightharpoonup u$ weakly and that $Su = 0$ (why?), i.e., $u \in N(S)$. From question 1 we know that $(Su_t, v) = 0 \; \forall v \in N(S)$ and thus $(f - u_t, v) = 0 \; \forall v \in N(S)$. Passing to the limit, we find that $(f - u, v) = 0 \; \forall v \in N(S)$. Thus $u = P_{N(S)} f$ and the "uniqueness of the limit" implies that $u_t \rightharpoonup u$ weakly as $t \to \infty$. On the other hand, we have $(Su_t, u_t) \geq 0$, i.e., $(f - u_t, u_t) \geq 0$ and consequently $\lim \sup_{t\to\infty} |u_t|^2 \leq (f, u) = |u|^2$. It follows that $u_t \to u$ strongly as $t \to \infty$.

5.21

1. Set $S = I - T$ and apply question 1 of Exercise 5.20.
2. Write $f = u - Tu$ and note that $\sigma_n(f) = \frac{1}{n}(u - T^n u)$.
3. First, check that $\lim_{n\to\infty} \sigma_n(f) = 0 \; \forall f \in R(I - T)$. Next, split a general $f \in H$ as $f = f_1 + f_2$ with $f_1 \in N(I - T)$ and $f_2 \in N(I - T)^\perp = \overline{R(I - T)}$. We then have $\sigma_n(f) = \sigma_n(f_1) + \sigma_n(f_2) = f_1 + \sigma_n(f_2)$.
4. Apply successively inequality (1) to $u, Su, S^2 u, \ldots, S^i u, \ldots$, and add the resulting inequalities. Note that

$$|S^n u - S^{n+1} u| \leq |S^i u - S^{i+1} u| \quad \forall i = 0, 1, \ldots, n.$$

5. Writing $f = u - Tu = 2(u - Su)$, we obtain $|\mu_n(f)| \leq 2|u|/\sqrt{n + 1}$.
6. Use the same method as in question 3.

$\boxed{5.25}$

2. Let $m > n$. Applying Exercise 5.4 with $f = u_m$ and $v = P_K u_n$, one obtains

$$|P_K u_n - P_K u_m|^2 \le |P_K u_n - u_m|^2 - |P_K u_m - u_m|^2$$
$$\le |P_K u_n - u_n|^2 - |P_K u_m - u_m|^2.$$

Therefore, $(P_K u_n)$ is a Cauchy sequence.

3. We may assume that $u_{n_k} \rightharpoonup \overline{u}$ weakly. Recall now that $(u_n - P_K u_n, v - P_K u_n) \le 0 \; \forall v \in K$. Passing to the limit (along the sequence n_k) leads to $(\overline{u} - \ell, v - \ell) \le 0 \; \forall v \in K$. Since $\overline{u} \in K$, we may take $v = \overline{u}$ and conclude that $\overline{u} = \ell$. Once more, the "uniqueness of the limit" implies that $u_n \rightharpoonup \ell$ weakly.

4. For every $v \in K$, $\lim_{n \to \infty} |u_n - v|^2$ exists and thus $\lim_{n \to \infty}(u_n, v - w)$ also exists for every $v, w \in K$. It follows that $\varphi(z) = \lim_{n \to \infty}(u_n, z)$ exists for every $z \in H$. Using the Riesz–Fréchet representation theorem we may write $\varphi(z) = (u, z)$ for some $u \in H$. Finally, note that $(u - \ell, v - \ell) \le 0 \; \forall v \in K$ and thus $\ell = P_K u$.

5. By translation and dilation we may always assume that $K = B_H$. Thus $|u_n| \downarrow \alpha$.

If $\alpha < 1$, then $u_n = P_K u_n$ for n large enough (and we already know that $P_K u_n$ converges strongly).

If $\alpha \ge 1$, then $P_K u_n = u_n/|u_n|$ converges strongly and so does u_n.

6. Recall that $(u_n - P_K u_n, v - P_K u_n) \le 0 \; \forall v \in K$ and thus $(u_n - \ell, v - \ell) \le \varepsilon_n$ $\forall v \in K$, with $\varepsilon_n \to 0$ (ε_n depends on v). Adding these inequalities leads to $(\sigma_n - \ell, v - \ell) \le \varepsilon_n' \; \forall v \in K$, with $\varepsilon_n' \to 0$. Assuming that $\sigma_{n_k} \rightharpoonup \overline{\sigma}$ weakly, then $\overline{\sigma} \in K$ satisfies $(\overline{\sigma} - \ell, v - \ell) \le 0 \; \forall v \in K$. Therefore $\overline{\sigma} = \ell$ and the "uniqueness of the limit" implies that $\sigma_n \rightharpoonup \ell$ weakly.

$\boxed{5.26}$

3. Note that $\sqrt{n}u_n$ is bounded, and that for each fixed j, $(\sqrt{n}u_n, e_j) \to 0$ as $n \to \infty$.

$\boxed{5.27}$ Let $F^\bullet$ be the closure of the vector space spanned by the E_n's. We know (see the proof of Theorem 5.9) that $\sum_{n=1}^\infty |P_{E_n} u|^2 = |P_F u|^2 \; \forall u \in H$, and thus $|P_F u| = |u| \; \forall u \in D$. It follows that $|P_F u|^2 = |u|^2 \; \forall u \in D$ and therefore $P_{F^\perp} u = 0 \; \forall u \in D$. Consequently $P_{F^\perp} u = 0 \; \forall u \in H$, i.e., $F^\perp = \{0\}$, and so $F = H$.

$\boxed{5.28}$

1. V is separable by Proposition 3.25. Consider a dense countable subset (v_n) of V and conclude as in the proof of Theorem 5.11.

$\boxed{5.29}$

2. If $2 < p < \infty$ use the inequality $\|u\|_p \le \|u\|_\infty^{1-2/p} \|u\|_2^{2/p}$. Note that *every* infinite-dimensional Hilbert space (separable or not) admits an infinite orthonormal sequence.

6. Integrating over Ω, we find that $k \le M^2 |\Omega|$, which provides an upper bound for the dimension of E.

5.30

1. For every fixed $t \in [0, 1]$ consider the function $u_t(s) = p(s)\chi_{[0,t]}(s)$ and write that $\sum_{n=1}^{\infty} |(u_t, e_n)| \leq \|u_t\|_2^2$.
2. Equality in (2) implies equality in (1) for a.e. $t \in [0, 1]$. Thus $u_t = \sum_{n=1}^{\infty}(u_t, e_n)e_n$ for a.e. $t \in [0, 1]$, and hence $u_t \in E =$ the closure of the vector space spanned by the e_n's. It remains to check that the space spanned by the functions (u_t) is dense in L^2. Let $f \in L^2$ be such that $\int_0^1 f u_t = 0$ for a.e. t. It follows that $\int_0^t fp = 0 \; \forall t \in [0, 1]$, and so $fp = 0$ a.e.

5.31 It is easy to check that $(\varphi_i, \varphi_j) = 0$ for $i \neq j$. Let $n = 2^{p+1} - 1$. Let E denote the space spanned by $\{\varphi_0, \varphi_1, \ldots, \varphi_n\}$ and let F denote the space spanned by the characteristic functions of the intervals $(\frac{i}{2^{p+1}}, \frac{i+1}{2^{p+1}})$, where i is an integer with $0 \leq i \leq 2^{p+1} - 1$. Clearly $E \subset F$, $\dim E = n + 1 = 2^{p+1}$, and $\dim F = 2^{p+1}$. Thus $E = F$.

5.32

2. The function $u = r_1 r_2$ is orthogonal to all the functions $(r_i)_{i \geq 0}$ and $u \neq 0$. Thus $(r_i)_{i \geq 0}$ is not a basis.
3. It is easy to check that $(w_n)_{n \geq 0}$ is an orthonormal system and that $w_0 = r_0$, $w_{2\ell} = r_{\ell+1} \; \forall \ell \geq 0$. In order to prove that $(w_n)_{n \geq 0}$ is a basis one can use the same argument as in Exercise 5.31.

6.2

3. Consider the sequence of functions defined on $[0, 1]$ by

$$u_n(t) = \begin{cases} 0 & \text{if } 0 \leq t \leq \frac{1}{2}, \\ n(t - \frac{1}{2}) & \text{if } \frac{1}{2} < t \leq \frac{1}{2} + \frac{1}{n}, \\ 1 & \text{if } \frac{1}{2} + \frac{1}{n} < t \leq 1. \end{cases}$$

Note that $T(u_n) \to f$, but $f \notin T(B_E)$, since $f \notin C^1([0, 1])$.

6.3 Argue by contradiction. If the conclusion fails, there exists some $\delta > 0$ such that $\|Tu\|_F \geq \delta \|u\|_E \; \forall u \in E$. Hence $R(T)$ is closed. Consider the operator $T_0 : E \to R(T)$ defined by $T_0 = T$. Clearly T_0 is bijective. By Corollary 2.6, $T_0^{-1} \in \mathcal{L}(R(T), E)$. On the other hand, $T_0 \in \mathcal{K}(E, R(T))$. Hence B_E is compact and $\dim E < \infty$.

6.5 Let $T : V \to \ell^2$ be the operator defined by

$$Tu = (\sqrt{\lambda_1}u_1, \sqrt{\lambda_2}u_2, \ldots, \sqrt{\lambda_n}u_n, \ldots).$$

Clearly $|Tu|_{\ell^2} = \|u\|_V \; \forall u \in V$, and T is surjective from V onto ℓ^2. Since ℓ^2 is complete, it follows that V is also complete.
Consider the operator $J_n : V \to \ell^2$ defined by

$$J_n u = (u_1, u_2, \ldots, u_n, 0, 0, \ldots).$$

It is easy to check that $\|J_n - I\|_{\mathcal{L}(V, \ell^2)} \to 0$ and thus the canonical injection from V into ℓ^2 is compact.

6.7

1. Assume that T is continuous from E weak into F strong. Then for every $\varepsilon > 0$ there exists a neighborhood V of 0 in E weak such that $x \in V \Rightarrow \|Tx\| < \varepsilon$. We may assume that V has the form

$$V = \{x \in E; \, |\langle f_i, x \rangle| < \delta \quad \forall i = 1, 2, \ldots, n\},$$

where $f_1, f_2, \ldots, f_n \in E^*$ and $\delta > 0$.
Let $M = \{x \in E; \langle f_i, x \rangle = 0 \; \forall i = 1, 2, \ldots, n\}$, so that $Tx = 0 \; \forall x \in M$. On the other hand, M has finite codimension (see Example 2 in Section 2.4). Thus $E = M + N$ with $\dim N < \infty$. It follows that $R(T) = T(N)$ is finite-dimensional.
2. Note that if $u_n \rightharpoonup u$ weakly in E then $Tu_n \rightharpoonup Tu$ weakly in F. On the other hand, (Tu_n) has compact closure in F (for the strong topology). Thus $Tu_n \to Tu$ (see, e.g., Exercise 3.5).
6. Note that $T^* \in \mathcal{L}(E^*, (c_0)^*)$. But $(c_0)^* = \ell^1$ (see Section 11.3). Since E^* is reflexive, it follows from question 5 that T^* is compact. Hence (by Theorem 6.4) T is compact.

6.8

1. There is a constant c such that $B_{R(T)} \subset cT(B_E)$ and thus the unit ball of $R(T)$ is compact.
2. Let E_0 be a complement of $N(T)$. Then $T_0 = T_{|E_0}$ is bijective from E_0 onto $R(T)$. Thus $\dim E_0 = \dim R(T) < \infty$.

6.9

1. (A) $\Rightarrow$ (B):
Let E_0 be a complement of $N(T)$ and let $P : E \to N(T)$ be an associated projection operator. Then $T_0 = T_{|E_0}$ is bijective from E_0 onto $R(T)$. By the open mapping theorem there exists a constant C such that

$$\|u\|_E \leq C\|Tu\|_F \quad \forall u \in E_0.$$

It follows that $\forall u \in E$,

$$\|u\|_E \leq \|u - Pu\|_E + \|Pu\|_E \leq C\|Tu\|_F + \|Pu\|_E.$$

(C) $\Rightarrow$ (A):
(i) To check that the unit ball in $N(T)$ is compact, let (u_n) be a sequence in $N(T)$ such that $\|u_n\|_E \leq 1$. Since $(Q(u_n))$ has compact closure in G, one may extract a subsequence $(Q(u_{n_k}))$ converging in G. Applying (C), we see that (u_{n_k}) is Cauchy.

(ii) Introducing a complement of $N(T)$ we may assume in addition that T is injective. Let (u_n) be a sequence in E such that $Tu_n \to f$. Let us first check that (u_n) is bounded. If not, set $v_n = u_n/\|u_n\|$. Applying (C), we see that a subsequence (v_{n_k}) is Cauchy. Let $v_{n_k} \to v$ with $v \in N(T)$ and $\|v\| = 1$; impossible. Therefore (u_n) is bounded and we may extract a subsequence $(Q(u_{n_k}))$ converging in G. Applying (C) once more, we find that (u_{n_k}) is Cauchy.

To recover the result in Exercise 2.12 write

$$\|u\|_E \le C(\|Tu\|_F + \|Pu\|_E) \le C(\|Tu\|_F + |Pu|),$$

since all norms on $N(T)$ are equivalent. Moreover,

$$|Pu| \le |u - Pu| + |u| \le C\|u - Pu\|_E + |u| \le C\|Tu\|_F + |u|.$$

2. Note that

$$\|u\|_E \le C(\|Tu\|_F + \|Pu\|_E) \le C(\|(T+S)u\|_F + \|Pu\|_E + \|Su\|_F)$$

and consider the compact operator $Q : E \to E \times F$ defined by $Qu = [Pu, Su]$.

6.10

1. Note that $\forall u \in E$,

$$
\begin{aligned}
|Q(1)|\|u\| &\le \|Q(1)u - Q(T)u\| + \|Q(T)u\| \\
&= \|\tilde{Q}(T)(u - Tu)\| + \|Q(T)u\| \\
&\le C(\|u - Tu\| + \|Q(T)u\|).
\end{aligned}
$$

2. Proof of the implication $N(I - T) = \{0\} \Rightarrow R(I - T) = E$. Suppose by contradiction that $R(I - T) = E_1 \ne E$. Set $E_n = (I - T)^n E$. Then (E_n) is a decreasing sequence of closed subspaces. Choose $u_n \in E_n$ such that $\|u_n\| = 1$ and $\text{dist}(u_n, E_{n+1}) \ge 1/2$. Write

$$Q(T)u_n - Q(T)u_m = Q(T)u_n - Q(1)u_n + Q(1)u_n - Q(1)u_m + Q(1)u_m - Q(T)u_m.$$

Thus, for $m > n$, we have

$$\|Q(T)u_n - Q(T)u_m\| \ge |Q(1)|/2,$$

and this is impossible.

For the converse, follow the argument described in the proof of Theorem 6.6.

3. Using the same notation as in the proof of Theorem 6.6, write $S = T + \Lambda \circ P$. Here $S \notin \mathcal{K}(E)$, but $\Lambda \circ P \in \mathcal{K}(E)$. Thus $Q(S) \in \mathcal{K}(E)$ (why?). Then continue as in the proof of Theorem 6.6.

$\boxed{6.11}$

1. There exists an integer $n_0 \geq 1$ such that Int $F_{n_0} \neq \emptyset$ and thus $B(u_0, \rho) \subset F_{n_0}$. For every $u \in F$ and $|\lambda| < \rho/\|u\|$ we have $u_0 + \lambda u \in F_{n_0}$. Therefore

$$|\lambda| \, |u(x) - u(y)| \leq |u_0(x) - u_0(y)| + n_0 d(x, y)^{1/n_0} \leq 2n_0 d(x, y)^{1/n_0}.$$

It follows that

$$|u(x) - u(y)| \leq \frac{2n_0}{\rho} \|u\| d(x, y)^{1/n_0} \quad \forall x, y \in K.$$

2. The theorem of Ascoli–Arzelà implies that B_F is compact.

$\boxed{6.13}$ Suppose, by contradiction, that there exist some $\varepsilon_0 > 0$ and a sequence (u_n) such that $\|u_n\|_E = 1$ and $\|T u_n\|_F \geq \varepsilon_0 + n|u_n|$. Then $|u_n| \to 0$ and we may assume that $u_{n_k} \rightharpoonup u$ weakly. But the function $u \mapsto |u|$ is convex and continuous. Thus it is l.s.c. for the weak topology and hence $u = 0$. It follows that $T u_n \to 0$. Impossible.

$\boxed{6.15}$

1. If $u = f + \lambda(T - \lambda I)^{-1} f$, we have $\lambda u = T(u - f)$ and hence $|\lambda| \, \|u\| \leq \|T\|(\|u\| + \|f\|)$.
2. By the proof of Proposition 6.7 we know that if $\mu \in \mathbb{R}$ is such that $|\mu - \lambda| \, \|(T - \lambda I)^{-1}\| < 1$, then $\mu \in \rho(T)$. Thus $\text{dist}(\lambda, \sigma(T)) \geq 1/\|(T - \lambda I)^{-1}\|$.
4. $(U - I)^{-1} = \frac{1}{2}(T - I)$.
6. Note that the relation $Uu - \lambda u = f$ is equivalent to

$$Tu - \frac{(\lambda + 1)}{(\lambda - 1)} u = \frac{1}{(\lambda - 1)}(f - Tf).$$

$\boxed{6.16}$

2. $(T - \lambda I)^{-1} = \frac{1}{1 - \lambda^n} \sum_{i=0}^{n-1} \lambda^{n-i-1} T^i$.
3. $(T - \lambda I)^{-1} = -\sum_{i=0}^{n-1} \lambda^{-i-1} T^i$.
4. $(I - T)^{-1} = (I - T^n)^{-1} \sum_{i=0}^{n-1} T^i$.

$\boxed{6.18}$

1. $\|S_r\| = \|S_\ell\| = 1$. Note that $S_\ell \circ S_r = I$ and thus $S_r \notin \mathcal{K}(E)$, $S_\ell \notin \mathcal{K}(E)$.
3. For every $\lambda \in [-1, +1]$ the operator $(S_r - \lambda I)$ is not surjective: for example, if $f = (-1, 0, 0, \dots)$ the equation $S_r x - \lambda x = f$ has no solution $x \in \ell^2$.
4. $N(S_\ell - \lambda I) = \mathbb{R}(1, \lambda, \lambda^2, \dots)$.
6. $S_r^\star = S_\ell$ and $S_\ell^\star = S_r$.
7. Writing $S_r x - \lambda x = f$, we have

$$|x| = |S_r x| = |\lambda x + f| \leq |\lambda| |x| + |f|.$$

Thus

$$|S_r x - \lambda x| \geq (1 - |\lambda|)|x|,$$

and hence $R(S_r - \lambda I)$ is closed. Applying Theorem 2.19 yields

$$R(S_r - \lambda I) = N(S_\ell - \lambda I)^\perp = \left\{ x \in \ell^2; \ \sum_{i=1}^\infty \lambda^{i-1} x_i = 0 \right\}$$

and

$$R(S_\ell - \lambda I) = N(S_r - \lambda I)^\perp = E.$$

8. We have $\overline{R(S_r \pm I)} = N(S_\ell \pm I)^\perp = E$ and $\overline{R(S_\ell \pm I)} = N(S_r \pm I)^\perp = E$. We already know (see question 3) that $R(S_r \pm I) \neq E$. On the other hand, $R(S_\ell \pm I) \neq E$; otherwise, since $S_\ell \pm I$ is injective, we would have $\pm 1 \in \rho(S_\ell)$. Impossible.

9. $EV(S_r \circ M) = \emptyset$ if $\alpha_n \neq 0 \ \forall n$ and $EV(S_r \circ M) = \{0\}$ if $\alpha_n = 0$ for some n.

10. We may always assume that $\alpha \neq 0$; otherwise $S_r \circ M$ is compact and the conclusion is obvious.

Let us show that $(T - \lambda I)$ is bijective for every λ with $|\lambda| > |\alpha|$. Note that $M = \alpha I + K$, where K is a compact operator. Letting $T = S_r \circ M$, we obtain $T = \alpha S_r + K_1$ and $(T - \lambda I) = (\alpha S_r - \lambda I) + K_1 = J \circ (I + K_2)$, where $J = (\alpha S_r - \lambda I)$ is bijective and K_1, K_2 are compact. Applying Theorem 6.6 (c), it suffices to check that $N(T - \lambda I) = \{0\}$. This has already been established in question 9.

Let us show that $(T - \lambda I)$ is not bijective for $|\lambda| \leq |\alpha|$. Assume by contradiction that $(T - \lambda I)$ is bijective. Write $(S_r - \frac{\lambda}{\alpha} I) = \frac{1}{\alpha}(T - \lambda I) - \frac{1}{\alpha} K_1 = J' \circ (I + K_3)$, where J' is bijective and K_3 is compact. Applying once more Theorem 6.6 (c), we see that

$$\left(S_r - \frac{\lambda}{\alpha} I \right) \text{ injective } \Leftrightarrow \left(S_r - \frac{\lambda}{\alpha} I \right) \text{ surjective.}$$

But we already know (from questions 2 and 3) that $(S_r - \frac{\lambda}{\alpha} I)$ is injective and not surjective, for $|\lambda| \leq |\alpha|$. Impossible.

11. $\sigma(S_r \circ M) = [-\sqrt{|ab|}, +\sqrt{|ab|}]$. Indeed, if $|\lambda| \leq \sqrt{|ab|}$, the operator $(S_r \circ M - \lambda I)$ is not surjective, since (for example)

$$f = (-1, 0, 0, \ldots) \notin R(S_r \circ M - \lambda I).$$

On the other hand, if $|\lambda| > \sqrt{|ab|}$, the operator $(S_r \circ M - \lambda I)$ is bijective, since $(S_r \circ M)^2 = ab S_r^2$. Thus $\|(S_r \circ M)^2\| \leq |ab|$ and we may apply Exercise 6.16, question 4.

| 6.20 |

1. Note that

$$|Tu(x) - Tu(y)| \leq |x - y|^{1/p'} \|u\|_p.$$

If $1 < p < \infty$ we may apply Ascoli–Arzelà to conclude that $T(B_E)$ has compact closure in $C([0, 1])$ and a fortiori in $L^p(0, 1)$. If $p = 1$, apply Theorem 4.26.

2. $EV(T) = \emptyset$. Note first that $0 \notin EV(T)$. Indeed, the equation $Tu = 0$ implies

$$\int_0^1 u\chi_{[a,b]} = 0 \quad \forall a, b \in [0, 1].$$

If $1 < p < 1$ we may use the density of step functions in $L^{p'}$ to conclude that $u \equiv 0$. When $p = 1$, we prove that

$$\int_0^1 u\varphi = 0 \quad \forall \varphi \in C([0, 1])$$

by approximating uniformly φ by step functions. We conclude with the help of Corollary 4.24 that $u \equiv 0$.

3. For $\lambda \neq 0$ and for $f \in C([0, 1])$, set $u = (T - \lambda I)^{-1} f$. Then $v(x) = \int_0^x u(t)dt$ satisfies:

$$v \in C^1([0, 1]) \quad \text{and} \quad v - \lambda v' = f \quad \text{with } v(0) = 0.$$

Therefore

$$u(x) = -\frac{1}{\lambda} f(x) - \frac{1}{\lambda^2} \int_0^x e^{(x-t)/\lambda} f(t)dt.$$

The same formula remains valid for $f \in L^p$ (argue by density).

4. $(T^\star v)(x) = \int_x^1 v(t)dt.$

6.22

2. Suppose, by contradiction, that there exists some $\mu \in Q(\sigma(T))$ such that $\mu \notin \sigma(Q(T))$. Then $\mu = Q(\lambda)$ with $\lambda \in \sigma(T)$, and $Q(T) - Q(\lambda)I = S$ is bijective. We may write

$$Q(t) - Q(\lambda) = (t - \lambda)\overline{Q}(t) \quad \forall t \in \mathbb{R},$$

and thus

$$(T - \lambda I)\overline{Q}(T) = \overline{Q}(T)(T - \lambda I) = S.$$

Hence $T - \lambda I$ is bijective and $\lambda \in \rho(T)$; impossible.

3. Take $E = \mathbb{R}^2$, $T = \left(\begin{smallmatrix} 0 & 1 \\ -1 & 0 \end{smallmatrix}\right)$ and $Q(t) = t^2$.

Then $EV(T) = \sigma(T) = \emptyset$ and $EV(T^2) = \sigma(T^2) = \{-1\}$.

4. $T^2 + I$ is bijective by Lax–Milgram. Every polynomial of degree 2 without real roots may be written (modulo a nonzero factor) as

$$Q(t) = t^2 + at + b = \left(t + \frac{a}{2}\right)^2 + b - \frac{a^2}{4}$$

with $b - a^2/4 > 0$, and we may apply Lax–Milgram once more.

If a polynomial $Q(t)$ has no real root, then its roots are complex conjugates. We may then write $Q(t) = Q_1(t)Q_2(t)\ldots Q_\ell(t)$, where each $Q_i(t)$ is a polynomial of degree 2 without real roots. Since $Q_i(T)$ is bijective, the same holds for $Q(T)$.

5. (i) Suppose, by contradiction, that $\mu \in EV(Q(T))$ and $\mu \notin Q(EV(T))$. Then there exists $u \neq 0$ such that $Q(T)u = \mu u$. Write

$$Q(t) - \mu = (t - t_1)(t - t_2) \cdots (t - t_q)\overline{Q}(t),$$

where the t_i's are the real roots of the polynomial $Q(t) - \mu$ and $\overline{Q}$ has no real root. Then $t_i \notin EV(T)\, \forall i$, since $\mu \notin Q(EV(T))$. We have

$$(T - t_1 I)(T - t_2 I) \cdots (T - t_k I)\overline{Q}(T)u = 0.$$

Since each factor in this product is injective, we conclude that $u = 0$. Impossible.

(ii) Argue as in (i).

6.23

3. In $E = \mathbb{R}^2$ take $T(u_1, u_2) = (u_2, 0)$. Then $T^2 = 0$, so that $r(T) = 0$, while $\|T\| = 1$.

5. In $E = \mathbb{R}^3$ take $T(u_1, u_2, u_3) = (u_2, -u_1, 0)$. Then $\sigma(T) = \{0\}$. Using the fact that $T^3 = -T$ it is easy to see that $r(T) = 1$.

Comment. If we work in Banach spaces over $\mathbb{C}$ the situation is totally different; see Section 11.4. There, we always have $r(T) = \max\{|\lambda|; \lambda \in \sigma(T)\}$. Taking $E = \mathbb{C}^3$ in the current example we have $\sigma(T) = \{0, +i, -i\}$ and then $r(T) = \max\{|\lambda|; \lambda \in \sigma(T)\} = 1$.

6. Assuming that the formula holds for T^n, we have

$$(T^{n+1}u)(t) = \frac{1}{(n-1)!} \int_0^t ds \int_0^s (s - \tau)^{n-1} u(\tau) d\tau$$

$$= \frac{1}{(n-1)!} \int_0^t u(\tau) \left[\int_\tau^t (s - \tau)^{n-1} ds \right] d\tau$$

$$= \frac{1}{n!} \int_0^t (t - \tau)^n u(\tau) d\tau.$$

7. Consider the functions f and g defined on $\mathbb{R}$ by

$$f(t) = \begin{cases} \frac{1}{(n-1)!} t^{n-1} & \text{if } 0 \le t \le 1, \\ 0 & \text{otherwise,} \end{cases}$$

$$g(t) = \begin{cases} u(t) & \text{if } 0 \le t \le 1, \\ 0 & \text{otherwise,} \end{cases}$$

so that for $0 \le t \le 1$, we have

$$(f \star g)(t) = \int_0^1 (t - \tau) u(\tau) d\tau = (T^n u)(t).$$

We deduce that

$$\|f \star g\|_{L^p(0,1)} \le \|f \star g\|_{L^p(\mathbb{R})} \le \|f\|_{L^1(\mathbb{R})} \|g\|_{L^p(\mathbb{R})} = \frac{1}{n!} \|u\|_{L^p(0,1)}.$$

8. Apply Stirling's formula.

6.24

2. (v) $\Rightarrow$ (vi). For every $\varepsilon > 0$, $T_\varepsilon = T + \varepsilon I$ is bijective and $\sigma(T_\varepsilon) \subset [\varepsilon, 1 + \varepsilon]$. Thus $\sigma(T_\varepsilon^{-1}) \subset [\frac{1}{1+\varepsilon}, \frac{1}{\varepsilon}]$. Applying Proposition 6.9 to T_ε^{-1} yields

$$(T_\varepsilon^{-1}v, v) \geq \frac{1}{1+\varepsilon}|v|^2 \quad \forall v \in H,$$

i.e.,

$$(T_\varepsilon u, u) \geq \frac{1}{1+\varepsilon}|T_\varepsilon u|^2 \quad \forall u \in H.$$

3. Set $U = 2T - I$. Clearly (vii) is equivalent to

(vii') $\qquad\qquad\qquad |u| \leq |Uu| \quad \forall u \in H.$

Applying Theorem 2.20, we see that (vii) $\Rightarrow$ $(-1, +1) \subset \rho(U) = 2\rho(T) - 1$. Thus (vii) $\Rightarrow$ (viii).

Conversely, (viii) $\Rightarrow$ $(-1, +1) \subset \rho(U)$. Thus $\sigma(U) \subset (-\infty, -1] \cup [1, +\infty)$ and $\sigma(U^{-1}) \subset [-1, +1]$. By Proposition 6.9 we know that $\|U^{-1}\| \leq 1$, i.e., (vii') holds.

6.25 By construction we have

$$M \circ (I + K) = I \quad \text{on } X,$$
$$(I + K) \circ M = I \quad \text{on } R(I + K).$$

Given any $x \in E$, write $x = x_1 + x_2$ with $x_1 \in X$ and $x_2 \in N(I + K)$. Then

$$M \circ (I + K)(x) = M \circ (I + K)(x_1) = x_1 = x - Px$$

where P is a projection onto $N(I + K)$.
For any $x \in E$ we have

$$(I + K) \circ \tilde{M}(x) = (I + K) \circ M \circ Q(x) = Qx = x - \tilde{P}x,$$

where $\tilde{P}$ is a finite-rank projection onto a complement of $R(I + K)$ in E.

8.8

4. We have

$$u_n' = \zeta_n u' + \zeta_n' u.$$

Clearly $\zeta_n u' \to u'$ in L^p by dominated convergence. It remains to show that $\zeta_n' u \to 0$ in L^p. Note that

$$\|\zeta_n' u\|_p^p \le C \int_{1/n}^{2/n} n^p |u(x)|^p dx,$$

where $C = \|\zeta'\|_{L^\infty}^p$.
When $p = 1$ we have, since $u \in C([0, 1])$ and $u(0) = 0$,

$$n \int_{1/n}^{2/n} |u(x)| dx \le \max_{x \in [\frac{1}{n}, \frac{2}{n}]} |u(x)| \to 0 \text{ as } n \to \infty.$$

When $p > 1$ we have

$$n^p \int_{1/n}^{2/n} |u(x)|^p dx = n^p \int_{1/n}^{2/n} x^p \frac{|u(x)|^p}{x^p} dx \le 2^p \int_{1/n}^{2/n} \frac{|u(x)|^p}{x^p} dx \to 0$$

by question 1.

8.9

1. By question 1 in Exercise 8.8 we know that $\frac{u'(x)}{x} \in L^p$. On the other hand,

$$u(x) = \int_0^x u'(t) dt = x u'(x) - \int_0^x u''(t) t \, dt,$$

and thus

$$\frac{u(x)}{x^2} = \frac{u'(x)}{x} - \frac{1}{x^2} \int_0^x u''(t) t \, dt.$$

But

$$\frac{1}{x^2} \left| \int_0^x u''(t) t \, dt \right| \le \frac{1}{x} \int_0^x |u''(t)| dt \in L^p,$$

as above.
2. We have $v \in C^1((0, 1))$ and

$$v'(x) = -\frac{u(x)}{x^2} + \frac{u'(x)}{x} \in L^p,$$

by question 1.

Moreover,

$$v(x) = \frac{u(x)}{x} = \frac{1}{x} \int_0^x u'(t) dt \to 0 \quad \text{as } x \to 0,$$

since $u \in C^1([0, 1])$ and $u'(0) = 0$.
3. We need only to show that

$$\|\zeta_n' u'\|_p + \|\zeta_n'' u\|_p \to 0 \text{ as } n \to 0.$$

But

$$\|\zeta_n' u'\|_p^p \leq Cn^p \int_{1/n}^{2/n} |u'(x)|^p dx \leq 2^p C \int_{1/n}^{2/n} \frac{|u'(x)|^p}{x^p} dx$$

and

$$\|\zeta_n'' u\|_p^p \leq Cn^{2p} \int_{1/n}^{2/n} |u(x)|^p dx \leq 4^p C \int_{1/n}^{2/n} \frac{|u(x)|^p}{x^{2p}} dx,$$

and the conclusion follows since $\frac{u'(x)}{x} \in L^p$ and $\frac{u(x)}{x^2} \in L^p$.

4. Let $u \in X_m$. Then $u' \in X_{m-1}$ and $\frac{u'(x)}{x^{m-1}} \in L^p(I)$ by the induction assumption.
   Next, observe that

$$\frac{u(x)}{x^m} = \frac{1}{x^m} \int_0^x \frac{u'(t)}{t^{m-1}} t^{m-1} dt.$$

Applying once more Hardy's inequality (see Problem 34, part C) we obtain

$$\frac{|u(x)|}{x^m} \leq \frac{1}{x} \int_0^x \frac{|u'(t)|}{t^{m-1}} dt \in L^p(I).$$

In order to prove that $\frac{u(x)}{x^{m-1}} \in X_1$, note that

$$D\left(\frac{u(x)}{x^{m-1}}\right) = \frac{Du(x)}{x^{m-1}} - (m-1)\frac{u(x)}{x^m} \in L^p(I),$$

and that

$$\frac{|u(x)|}{x^{m-1}} \leq \frac{1}{x^{m-1}} \int_0^x \frac{|u'(t)|}{t^{m-1}} t^{m-1} dt \leq \int_0^x \frac{|u'(t)|}{t^{m-1}} dt \rightarrow 0 \quad \text{as } x \rightarrow 0,$$

since $\frac{u'(t)}{t^{m-1}} \in L^p(I)$.

5. It suffices to check that $D^\ell v \in X_1$ for every integer ℓ such that $0 \leq \ell \leq k - 1$.
   But $D^\ell v$ is a linear combination of functions of the form $\frac{D^{j+\alpha} u(x)}{x^{m-j-k+\ell-\alpha}}$, where α is
   an integer such that $0 \leq \alpha \leq \ell$. Then use question 4.

6. It suffices to show that $(D^\alpha \zeta_n)(D^\beta u) \rightarrow 0$ in $L^p(I)$ when $\alpha + \beta = m$ and
   $1 \leq \alpha \leq m$. But $|D^\alpha \zeta_n(x)| \leq Cn^\alpha$ and thus

$$\int_0^1 |D^\alpha \zeta_n(x)|^p |D^\beta u(x)|^p dx \leq Cn^{\alpha p} \int_{1/n}^{2/n} \left|\frac{D^\beta u(x)}{x^\alpha}\right|^p x^{\alpha p} dx$$

$$\leq C \int_{1/n}^{2/n} \left|\frac{D^\beta u(x)}{x^\alpha}\right|^p dx \rightarrow 0$$

since $\frac{D^\beta u(x)}{x^\alpha} \in L^p(I)$ by question 4.

8. To prove that $v \in C([0,1])$, note that $v(x) = \frac{1}{x} \int_0^x u'(t) dt$ and that $u' \in C([0,1])$
   with $u'(0) = 0$.
   Next, we prove that $v \in W^{1,1}(I)$. Integrating by parts, we see that

$$v'(x) = \frac{1}{x^2} \int_0^x u''(t) t\, dt,$$

and a straightforward computation gives

$$\|v'\|_1 \le \int_0^1 |u''(t)|(1-t)\, dt \le \|u''\|_1.$$

9. Set

$$u(x) = \int_0^x (1 + |\log t|)^{-1}\, dt.$$

It is clear that $u \in W^{2,1}(I)$ with $u(0) = u'(0) = 0$, and, moreover, $\frac{u'(x)}{x} \notin L^1(I)$. The relation

$$\frac{u(x)}{x^2} = \frac{u'(x)}{x} - v'(x),$$

combined with question 8 shows that $\frac{u(x)}{x^2} \notin L^1(I)$.

$\boxed{8.10}$

4. Clearly, as $n \to \infty$,

$$v'_n(x) = G'(nu(x))u'(x) \to f(x) \quad \text{a.e.,}$$

where

$$f(x) = \begin{cases} 0 & \text{if} \quad u(x) \ne 0, \\ u'(x) & \text{if} \quad u(x) = 0. \end{cases}$$

6. We have

$$\int_0^1 v_n \varphi' = -\int_0^1 v'_n \varphi \quad \forall \varphi \in C_c^1(I).$$

Passing to the limit as $n \to \infty$ yields

$$\int_0^1 f\varphi = 0 \quad \forall \varphi \in C_c^1(I),$$

and therefore $f = 0$ a.e. on I, i.e., $u'(x) = 0$ a.e. on $[u = 0]$.

$\boxed{8.12}$

1. Use Exercise 8.2 and the fact that

$$\liminf_{n \to \infty} \|u'_n\|_{L^p} \ge \|u'\|_{L^p}.$$

2. Consider the sequence (u_n) in Exercise 8.2. We have $\|u_n\|_{L^1} \le \frac{1}{2}$ and $\|u'_n\|_{L^1} = 1$. Thus $\frac{2}{3} u_n \in B_1$. On the other hand, $\frac{2}{3} u_n \to \frac{2}{3} u$ in L^1, where

$$u(x) = \begin{cases} 0 & \text{if } x \in (0, 1/2), \\ 1 & \text{if } x \in (1/2, 1). \end{cases}$$

But $u \notin W^{1,1}$. Thus B_1 is not closed in L^1.

8.16

2. $R(A) = L^p(0, 1)$ and $N(A) = \{0\}$.
3. $v \in D(A^\star)$ iff $v \in L^{p'}$ and there is a constant C such that

$$\left| \int_0^1 vu' \right| \le C \|u\|_p \quad \forall u \in D(A).$$

In particular, $v \in D(A^\star) \Rightarrow v \in W^{1,p'}$, and then

(S1) $$\left| u(1)v(1) - \int_0^1 uv' \right| \le C \|u\|_p \quad \forall u \in D(A).$$

We deduce from (S1) that

$$|u(1)| \, |v(1)| \le (C + \|v'\|_{p'}) \|u\|_p \quad \forall u \in D(A).$$

It follows that $v(1) = 0$, since there exists a sequence (u_n) in $D(A)$ such that $u_n(1) = 1$ and $\|u_n\|_p \to 0$. Hence we have proved that

$$v \in D(A^\star) \Rightarrow v \in W^{1,p'} \quad \text{and} \quad v(1) = 0.$$

It follows easily that

$$D(A^\star) = \{v \in W^{1,p'} \text{ and } v(1) = 0\},$$

with $A^\star v = -v'$.
4. We have

$$N(\tilde{A}) = \{0\}, \quad R(\tilde{A}) = \left\{ f \in L^p; \int_0^1 f(t)dt = 0 \right\},$$

and

$$(\tilde{A})^\star v = -v' \quad \text{with } D((\tilde{A})^\star) = W^{1,p'}.$$

8.17 In the determination of $D(A^\star)$ it is useful to keep in mind the following fact.
Let $I = (0, 1)$ and $1 < p \le \infty$. Assume that $u \in L^p(I)$ satisfies

(S1) $$\left| \int_I u\varphi' \right| \le C \|\varphi\|_{p'} \quad \forall \varphi \in C_c^1(I) \text{ such that } \int_I \varphi = 0,$$

then $u \in W^{1,p}(I)$.

Indeed, fix a function $\psi_0 \in C_c^1(I)$ such that $\int_I \psi_0 = 1$. Let ζ be any function in $C_c^1(I)$. Inserting $\varphi = \zeta - (\int_I \zeta)\psi_0$ into (S1), we obtain

$$\left| \int_I u\zeta' \right| \le C\|\zeta\|_{p'} + C' \left| \int_I \zeta \right|,$$

where C' depends only on u and ψ_0. Therefore $u \in W^{1,p}(I)$ by Proposition 8.3.

When $Au = u'' - xu'$ we have

$$A^\star v = v'' + xv' + v.$$

Note the following identity

$$A^\star(e^{-\frac{x^2}{2}}u) = e^{-\frac{x^2}{2}}Au \quad \forall u \in H^2(I),$$

which allows to compute $N(A^\star)$ under the various boundary conditions.

$\boxed{8.19}$ Given $f \in L^2(0, 1)$, set $F(x) = \int_0^x f(t)dt$. Then

$$\varphi^\star(f) = \begin{cases} \frac{1}{2} \int_0^1 F^2(x)dx & \text{if } \int_0^1 f(t)dt = 0, \\ +\infty & \text{otherwise.} \end{cases}$$

Indeed, if $\int_0^1 f(t)dt = 0$, then $\int_0^1 fv = \int_0^1 F'v = -\int_0^1 Fv' \quad \forall v \in H^1(0, 1)$, and

$$\varphi^\star(f) = \sup_{v \in H^1} \left\{ \int_0^1 fv - \frac{1}{2} \int_0^1 v'^2 \right\} = \sup_{v \in H^1} \left\{ -\int_0^1 Fv' - \frac{1}{2} \int_0^1 v'^2 \right\}$$

$$= \sup_{w \in L^2} \left\{ -\int_0^1 Fw - \frac{1}{2} \int_0^1 w^2 \right\} = \frac{1}{2} \int_0^1 F^2.$$

$\boxed{8.21}$

2. Let U be any function satisfying

$$\begin{cases} -(pU')' + qU = f & \text{on } (0, 1), \\ \qquad\qquad U(1) = 0. \end{cases}$$

Then

$$\int_0^1 fv_0 = p(0)(U'(0) - k_0 U(0)).$$

Therefore, if $\int_0^1 fv_0 = 0$, any such function U satisfies $U'(0) = k_0 U(0)$. Since $U(0)$ can be chosen arbitrarily we see that the set of solutions is one-dimensional.

$\boxed{8.22}$

1. The function $\rho(x) = x$ belongs to $H^1(0, 1)$, but $\sqrt{\rho(x)} = \sqrt{x} \notin H^1(0, 1)$.

2. For every $\rho \in H^1(0, 1)$, with $\rho \geq 0$ on $(0, 1)$, set $\gamma_\varepsilon = \sqrt{\rho + \varepsilon}$. Since the function $t \mapsto \sqrt{t + \varepsilon}$ is C^1 on $[0, +\infty)$, we deduce that $\gamma_\varepsilon \in H^1(0, 1)$ and, moreover,

$$\gamma_\varepsilon' = \frac{1}{2} \frac{\rho'}{\sqrt{\rho + \varepsilon}},$$

so that $|\gamma_\varepsilon'| \leq \mu$ on the set $[\rho > 0]$. On the other hand, we know that $\rho' = 0$ a.e. on the set $[\rho = 0]$ (see Exercise 8.10) and thus $|\gamma_\varepsilon'| \leq \mu$ a.e. on $[\rho = 0]$. Therefore $|\gamma_\varepsilon'| \leq \mu$ a.e. on $(0, 1)$.

Consequently, if $\mu \in L^2$ we deduce that $\|\gamma_\varepsilon'\|_{L^2} \leq C$ as $\varepsilon \to 0$. Since $\gamma_\varepsilon \to \sqrt{\rho}$, as $\varepsilon \to 0$, in $C([0, 1])$ and $\gamma_\varepsilon' \to \mu$ in $L^2(0, 1)$, we conclude (see Exercise 8.2) that $\sqrt{\rho} \in H^1(0, 1)$ and $(\sqrt{\rho})' = \mu$.

Conversely, if $\sqrt{\rho} \in H^1(0, 1)$, set $\gamma = \sqrt{\rho}$, so that $\rho = \gamma^2$ and $\rho' = 2\gamma\gamma'$. Hence $\mu = \gamma'$ a.e. on $[\rho > 0]$ and, moreover, $\mu = \gamma'$ a.e. on $[\rho = 0]$ since $\gamma' = 0$ a.e. on $[\gamma = 0] = [\rho = 0]$.

8.24

1. One may choose $C_\varepsilon = 1 + 1/\varepsilon$.
2. The weak formulation is

$$\begin{cases} u \in H^1(I), \\ a(u, v) = \int_0^1 (u'v' + kuv) - u(1)v(1) = \int_0^1 fv \quad \forall v \in H^1(I). \end{cases}$$

Clearly $a(u, v)$ is a continuous bilinear form on $H^1(0, 1)$. By question 1 it is coercive, e.g., if $k > 2$.

The corresponding minimization problem is

$$\min_{v \in H^1} \left\{ \frac{1}{2} \int_0^1 (v'^2 + kv^2) - \frac{1}{2} v(1)^2 - \int_0^1 fv \right\}.$$

3. Let $g \in L^2(I)$ and let $v \in H^2(I)$ be the corresponding solution of (1) (with f replaced by g). We have

$$(Tf, g)_{L^2} = \int_0^1 ug = \int_0^1 u(-v'' + kv)$$
$$= -u(1)v'(1) + u(0)v'(0) + u'(1)v(1) - u'(0)v(0)$$
$$\quad + \int_0^1 (-u'' + ku)v$$
$$= -u(1)v(1) + u(1)v(1) + \int_0^1 fv = (f, Tg)_{L^2}.$$

Therefore T is self-adjoint. It is compact since it is a bounded operator from $L^2(I)$ into $H^1(I)$, and $H^1(I) \subset L^2(I)$ with compact injection (see Theorem 8.8).

4. By the results of Section 6.4 we know that there exists a sequence (u_n) in $L^2(I)$ satisfying $Tu_n = \mu_n u_n$ with $\|u_n\|_{L^2} = 1$, $\mu_n > 0$ $\forall n$, and $\mu_n \to 0$. Thus we have $-u_n'' + ku_n = \frac{1}{\mu_n}u_n$, so that $-u_n'' = (\frac{1}{\mu_n} - k)u_n$ on I.

5. The value $\lambda = 0$ is excluded (why?). If $\lambda > 0$ we have $u(x) = A\cos\sqrt{\lambda}x + B\sin\sqrt{\lambda}x$, where the constants A and B are adjusted to satisfy the boundary condition, i.e., $B = 0$ and $A(\cos\sqrt{\lambda} + \sqrt{\lambda}\sin\sqrt{\lambda}) = 0$, so that $A \neq 0$ iff $\sqrt{\lambda}$ is a solution of the equation $\tan t = -1/t$ (which has an infinite sequence of positive solutions $t_n \to \infty$, as can be seen by inspection of the graphs). If $\lambda < 0$ we have $u(x) = Ae^{\sqrt{|\lambda|}x} + Be^{-\sqrt{|\lambda|}x}$. Putting this together with the boundary conditions gives $A = B$ and $A\sqrt{|\lambda|}e^{\sqrt{|\lambda|}} - B\sqrt{|\lambda|}e^{-\sqrt{|\lambda|}} = Ae^{\sqrt{|\lambda|}} + Be^{-\sqrt{|\lambda|}}$. In order to have some $u \not\equiv 0$, λ must satisfy $\sqrt{|\lambda|}(e^{\sqrt{|\lambda|}} - e^{-\sqrt{|\lambda|}}) = e^{\sqrt{|\lambda|}} + e^{-\sqrt{|\lambda|}}$, i.e., $t = \sqrt{|\lambda|}$ is a solution of the equation $e^{2t} = \frac{t+1}{t-1}$. An inspection of the graphs shows that there is a unique solution $t_0 > 1$ and then $\lambda = -t_0^2$.

8.25

2. Assume by contradiction that there is a sequence (u_n) in $H^1(I)$ such that $a(u_n, u_n) \to 0$ and $\|u_n\|_{H^1(I)} = 1$. Passing to a subsequence (u_{n_k}) we may assume that $u_{n_k}' \rightharpoonup u'$ weakly in L^2 and $u_n \to u$ strongly in L^2. By lower semicontinuity (see Proposition 3.5) we have $\liminf \int_I (u_{n_k}')^2 \geq \int_I (u')^2$ and therefore $a(u, u) = 0$, so that $u = 0$. But $\int_I (u_{n_k}')^2 = 1 - \int_I u_{n_k}^2$ and thus $a(u_{n_k}, u_{n_k}) = \int_I (u_{n_k}')^2 + (\int_0^1 u_{n_k})^2 = 1 - \int_I u_{n_k}^2 + (\int_0^1 u_{n_k})^2 \to 1$. Impossible.

4. We have
$$\int_I u'v' = \int_I gv \quad \forall v \in H^1(I),$$
where $g = f - (\int_0^1 u)\chi_{(0,1)}$. Therefore $u \in H^2(I)$ and satisfies
$$\begin{cases} -u'' + (\int_0^1 u)\chi_{(0,1)} = f & \text{on } I, \\ u'(0) = u'(2) = 0. \end{cases}$$

5. We have $u \in C^2(\bar{I})$ iff $\int_0^1 u = 0$. This happens iff $\int_I f = 0$.

8. The eigenvalues of T are positive and if $1/\lambda$ is an eigenvalue, we must have a function $u \not\equiv 0$ satisfying
$$\begin{cases} -u'' + \int_0^1 u = \lambda u & \text{on } (0,1), \\ -u'' = \lambda u & \text{on } (1,2), \\ u'(0) = u'(2) = 0, \\ u(1-) = u(1+) \text{ and } u'(1-) = u'(1+). \end{cases}$$

Therefore
$$u(x) = \frac{k}{\lambda} + A\cos(\sqrt{\lambda}x) \quad \text{on } (0,1),$$

$$u(x) = A' \cos(\sqrt{\lambda}(x - 2)) \quad \text{on } (1, 2),$$

where the constants k, A and A' are determined using the relations

$$\begin{cases} u(1-) = u(1+) \text{ and } u'(1-) = u(1+), \\ k = \int_0^1 u. \end{cases}$$

We conclude that either $\sin(\sqrt{\lambda}) = 0$, i.e., $\lambda = n^2 \pi^2$ with $n = 1, 2, \ldots$, or λ is a solution of the equation $\tan(\sqrt{\lambda}) = 2\sqrt{\lambda}(1 - \lambda)$.

8.26

3. Set $a(v, v) = \int_I p v'^2 + q v^2$. We have $(S_N f - S_D f, f) = \int_I f(u_N - u_D)$. We already know that $\frac{1}{2} a(u_N, u_N) - \int_I f u_N \leq \frac{1}{2} a(u_D, u_D) - \int_I f u_D$. On the other hand, $a(u_N, u_N) = \int_I f u_N$ and $a(u_D, u_D) = \int_I f u_D$. Therefore $\int_I f(u_N - u_D) \geq 0$.

6. Set $a_i(v, v) = a(v, v) + k_i v^2(0)$, $i = 1, 2$, and $u_{k_1} = u_1$, $u_{k_2} = u_2$. Since u_i is a minimizer of $\left(\frac{1}{2} a_i(v, v) - \int_I f v\right)$ on $V = \{v \in H^1(I); v(1) = 0\}$, we have

$$\frac{1}{2} a(u_2, u_2) + \frac{1}{2} k_2 u_2^2(0) - \int_I f u_2 \leq \frac{1}{2} a(u_1, u_1) + \frac{1}{2} k_2 u_1^2(0) - \int_I f u_1.$$

On the other hand, we have

$$a(u_1, u_1) + k_1 u_1^2(0) = \int_I f u_1,$$

and

$$a(u_2, u_2) + k_2 u_2^2(0) = \int_I f u_2.$$

Therefore

$$-\frac{1}{2} \int_I f u_2 \leq \frac{1}{2} \int_I f u_1 + \frac{1}{2}(k_2 - k_1) u_1^2(0) - \int_I f u_1,$$

so that

$$(S_{k_2} f - S_{k_1} f, f) = \int_I f(u_2 - u_1) \geq (k_1 - k_2) u_1^2(0) \geq 0.$$

8.27

4. The solution φ of

$$\begin{cases} -\varphi'' + \varphi = 1 \text{ on } I, \\ \varphi(-1) = \varphi(1) = 0, \end{cases}$$

is given by $\varphi(x) = 1 + A(e^x + e^{-x})$, where $A = -e/(e^2 + 1)$. By uniqueness of φ we must have $u = \lambda u(0)\varphi$. Therefore

$$\lambda_0 = \frac{1}{\varphi(0)} = \frac{e^2 + 1}{(e-1)^2}.$$

5. Equation (1) becomes $u = S(f + \lambda u(0)) = Sf + \lambda u(0)S1 = Sf + \lambda u(0)\varphi$. Thus $u(0)(1 - \lambda\varphi(0)) = (Sf)(0)$, i.e., $u(0) = \frac{\lambda_0(Sf)(0)}{\lambda_0 - \lambda}$ and $u = Sf + \frac{\lambda\lambda_0(Sf)(0)\varphi}{\lambda_0 - \lambda}$ is the desired solution.

6. When $\lambda = \lambda_0$, the existence of a solution u implies $(Sf)(0) = 0$ (just follow the computation in question 5). Conversely, assume that $(Sf)(0) = 0$. A solution of (1) must have the form $u = Sf + A\varphi$ for some constant A. A direct computation shows that any such u satisfies $-u'' + u = f + A$. But $u(0) = (Sf)(0) + \frac{A}{\lambda_0} = \frac{A}{\lambda_0}$. Thus we have $-u'' + u = f + \lambda_0 u(0)$, i.e., (1) holds for any A. Therefore the set of all solutions of (1) when $\lambda = \lambda_0$ is $Sf + \mathbb{R}\varphi$.

$\boxed{8.29}$

2. The existence and uniqueness of a solution $u \in H^1(0, 1)$ comes from Lax–Milgram. In particular, u satisfies

$$\int_0^1 u'v' = \int_0^1 (f - u)v \quad \forall v \in H_0^1(0, 1),$$

and therefore $u' \in H^1(0, 1)$, i.e., $u \in H^2(0, 1)$; moreover, $-u'' + u = f$ on $(0, 1)$. Using the information that $u \in H^2(0, 1)$, we may now write

$$a(u, v) = \int_0^1 (-u'' + u)v + u'(1)v(1) - u'(0)v(0)$$
$$+ (u(1) - u(0))(v(1) - v(0))$$
$$= \int_0^1 fv \quad \forall v \in H^1(0, 1).$$

Consequently,

$$(u'(1) + u(1) - u(0))v(1) - (u'(0) + u(1) - u(0))v(0) = 0 \quad \forall v \in H^1(0, 1).$$

Since $v(0)$ and $v(1)$ are arbitrary, we conclude that

$$u'(1) + u(1) - u(0) = 0 \quad \text{and} \quad u'(0) + u(1) - u(0) = 0.$$

5. Using the same function G as in the proof of Theorem 8.19 we have, taking $v = G(-u)$, $a(u, G(-u)) = \int_0^1 fG(-u) \geq 0$ since $f \geq 0$ and $G \geq 0$. On the other hand,

$$a(u, G(-u)) = -\int_0^1 u'^2 G'(-u) - \int_0^1 (-u) G(-u)$$
$$+ (u(1) - u(0))(G(-u(1)) - G(-u(0)))$$
$$\leq -\int_0^1 (-u) G(-u),$$

since G is nondecreasing. It follows that

$$\int_0^1 (-u) G(-u) \leq 0,$$

and consequently $-u \leq 0$.

7. Let $1/\lambda$ be an eigenvalue and let u be a corresponding eigenfunction. Then

$$\begin{cases} -u'' + u = \lambda u & \text{on } (0, 1), \\ u'(0) = u(0) - u(1), \\ u'(1) = u(0) - u(1). \end{cases}$$

Since $a(u, u) = \lambda \int_0^1 u^2 \geq \int_0^1 u^2$, we see that $\lambda \geq 1$. Moreover, $\lambda = 1$ is an eigenvalue corresponding to $u = $ const. Assume now $\lambda > 1$ and set $\alpha = \sqrt{\lambda - 1}$. We must have

$$u(x) = A \cos \alpha x + B \sin \alpha x.$$

In order to satisfy the boundary condition we need to impose

$$\begin{cases} B\alpha = A - A \cos \alpha - B \sin \alpha, \\ -A\alpha \sin \alpha + B\alpha \cos \alpha = A - A \cos \alpha - B \sin \alpha. \end{cases}$$

This system admits a nontrivial solution iff $2(1 - \cos \alpha) + \alpha \sin \alpha = 0$, i.e., $\sin(\alpha/2) = 0$ or $(\alpha/2) + \tan(\alpha/2) = 0$.

8.34

1. Let u be a classical solution. Then we have

$$-u'(1)v(1) + u'(0)v(0) + \int_0^1 (u'v' + uv) = \int_0^1 fv \quad \forall v \in H^1(0, 1).$$

Let $V = \{v \in H^1(0, 1); \ v(0) = v(1)\}$. If $v \in V$ we obtain

$$a(u, v) = \int_0^1 (u'v' + uv) = \int_0^1 fv + kv(0).$$

The weak formulation is

$$u \in V \quad \text{and} \quad a(u, v) = \int_0^1 fv + kv(0) \quad \forall v \in V.$$

2. By Lax–Milgram there exists a unique weak solution $u \in V$, and the corresponding minimization problem is

$$\min_{v \in V} \left\{ \frac{1}{2} \int_0^1 (v'^2 + v^2) - \int_0^1 fv - kv(0) \right\}.$$

3. Clearly, any weak solution u belongs to $H^2(0,1)$ and satisfies

$$-u'' + u = f \text{ a.e. on } (0,1),$$
$$u'(1)v(1) - u'(0)v(0) = kv(0), \quad \forall v \in V,$$

i.e.,
$$u'(1) - u'(0) = k.$$

5. The eigenvalues of T are given by $\lambda_k = 1/\mu_k$, where μ_k corresponds to a nontrivial solution of

$$\begin{cases} -u'' + u = \mu_k u & \text{a.e. on } (0,1), \\ u(1) = u(0), \ u'(1) = u'(0). \end{cases}$$

Therefore $\mu_k \geq 1$ and u is given by

$$u(x) = A \sin\left(\sqrt{\mu_k - 1}\,x\right) + B \cos\left(\sqrt{\mu_k - 1}\,x\right)$$

with $\sqrt{\mu_k - 1} = 2\pi k, k = 0, 1, \ldots$.

8.38

2. Suppose that $Tu = \lambda u$ with $u \in H^2(\mathbb{R})$ and $u \not\equiv 0$. Clearly $\lambda \neq 0$ and u satisfies

$$-u'' + u = \frac{1}{\lambda} u \quad \text{on } \mathbb{R}.$$

If $\lambda = 1$, we have $u(x) = Ax + B$ for some constants A, B. Since $u \in L^2(\mathbb{R})$ we deduce that $A = B = 0$. Therefore $1 \notin EV(T)$.

If $(\frac{1}{\lambda} - 1) > 0$ we have $u(x) = A \sin \alpha x + B \cos \alpha x$, with $\alpha = \sqrt{\frac{1}{\lambda} - 1}$. The condition $u \in L^2(\mathbb{R})$ yields again $A = B = 0$. Similarly, if $(\frac{1}{\lambda} - 1) < 0$ we have no solution, except $u \equiv 0$. Hence $EV(T) = \emptyset$. T cannot be a compact operator. Otherwise we would have $\sigma(T) = \{0\}$ by Theorem 6.8 and then $T \equiv 0$ by Corollary 6.10. But obviously $T \not\equiv 0$ (otherwise any f in $L^2(\mathbb{R})$ would be $\equiv 0$).

3. If $\lambda < 0$, $(T - \lambda I)$ is bijective from $H = L^2(\mathbb{R})$ onto itself, for example by Lax–Milgram and the fact that $(Tf, f) \geq 0 \ \forall f \in H$. Thus $\lambda \in \rho(T)$.

4. If $\lambda > 1 \geq \|T\|$ we have $\lambda \in \rho(T)$ by Proposition 6.7.

6. T is not surjective, since $R(T) \subset H^2(\mathbb{R})$.

7. $T - I$ is not surjective. Indeed, if we try to solve $Tf - f = \varphi$ for a given φ in $L^2(\mathbb{R})$ we are led to $-u'' + u = f$ (letting $u = Tf$) and $u = f + \varphi$. Therefore $u'' = \varphi$ admits a solution $u \in H^2$. Suppose, for example, that supp $\varphi \subset [0, 1]$. An immediate computation yields $u(x) = 0 \; \forall x \leq 0$ and $u(x) = 0 \; \forall x \geq 1$. Thus $u'(0) = u'(1) = 0$. It follows that $0 = u'(1) - u'(0) = \int_0^1 \varphi$. Therefore the equation $Tf - f = \varphi$ has no solution $f \in L^2(\mathbb{R})$ when $\int_0^1 \varphi \neq 0$. Hence $T - I$ cannot be surjective.

8. $T - \lambda I$ is not surjective. Indeed, if we try to solve $Tf - \lambda f = \varphi$ we are led to $-u'' + u = f$ (letting $u = Tf$) and $u = \lambda f + \varphi$. Therefore $-u'' + u = \frac{1}{\lambda}(u - \varphi)$. Assume again that supp $\varphi \subset [0, 1]$. We would have $u'' = -\mu^2 u$ outside $[0, 1]$, with $\mu = \sqrt{\frac{1}{\lambda} - 1}$. Therefore $u \equiv 0$ outside $[0, 1]$ and consequently $u(0) = u'(0) = u(1) = u'(1) = 0$. The equation $-u'' + (1 - \frac{1}{\lambda})u = -\frac{1}{\lambda}\varphi$ implies that $\int_0^1 \varphi v = 0$ for any solution v of $-v'' = \mu^2 v$ on $(0, 1)$; for example $\int_0^1 \varphi(x) \sin \mu x = 0$. Therefore the equation $Tf - \lambda f = \varphi$ has no solution $f \in L^2(\mathbb{R})$ when $\int_0^1 \varphi(x) \sin \mu x \neq 0$. Consequently $(T - \lambda I)$ is not surjective.

| 8.39 |

2. We have $v^2 \leq \frac{1}{2}v^4 + \frac{1}{2} \; \forall v \in \mathbb{R}$, and thus

$$\varphi(v) \geq \frac{1}{2}\|v\|_{H^1}^2 - \frac{1}{4} - \|f\|_{L^2}\|v\|_{H^1}.$$

Therefore $\varphi(v) \to \infty$ as $\|v\|_{H^1} \to \infty$.

3. The uniqueness follows from the fact that φ is strictly convex on $H^1(0, 1)$; this is a consequence of the strict convexity of the function $t \mapsto t^4$ on $\mathbb{R}$.

4. We have

$$\varphi(u + \varepsilon v) = \frac{1}{2}\int_0^1 (u'^2 + 2\varepsilon u'v' + \varepsilon^2 v'^2)$$
$$+ \frac{1}{4}\int_0^1 (u^4 + 4\varepsilon u^3 v + 6\varepsilon^2 u^2 v^2 + 4\varepsilon^3 uv^3 + \varepsilon^4 v^4)$$
$$- \int_0^1 f(u + \varepsilon v).$$

Writing that $\varphi(u) \leq \varphi(u + \varepsilon v)$ gives

$$\int_0^1 (u'v' + u^3 v - fv) + A_\varepsilon \geq 0,$$

where $A_\varepsilon \to 0$ as $\varepsilon \to 0$. Passing to the limit as $\varepsilon \to 0$ and choosing $\pm v$ yields

$$\int_0^1 (u'v' + u^3 v - fv) = 0 \quad \forall v \in H^1(0, 1).$$

6. From the convexity of the function $t \mapsto t^4$ we have

$$\frac{1}{4}v^4 - \frac{1}{4}u^4 \geq u^3(v-u) \quad \forall u, v \in \mathbb{R}.$$

On the other hand, we clearly have

$$\frac{1}{2}v'^2 - \frac{1}{2}u'^2 \geq u'(v'-u') \text{ a.e. on } (0,1) \quad \forall u, v \in H^1(0,1).$$

Thus $\forall u, v \in H^1(0,1)$

$$\varphi(v) - \varphi(u) \geq \int_0^1 u'(v'-u') + \int_0^1 u^3(v-u) - \int_0^1 f(v-u).$$

If u is a solution of (3) we have

$$\int_0^1 u'(v'-u') + \int_0^1 u^3(v-u) = \int_0^1 f(v-u) \quad \forall v \in H^1(0,1),$$

and therefore $\varphi(u) \leq \varphi(v) \ \forall v \in H^1(0,1)$.

9. We claim that $\psi(v) \to +\infty$ as $\|v\|_{H^1} \to \infty$. Indeed, this boils down to showing that for every constant C the set $\{v \in H^1(0,1); \psi(v) \leq C\}$ is bounded in $H^1(0,1)$. If $\psi(v) \leq C$ write

$$\int_0^1 fv = \int_0^1 f(v - v(0)) + v(0) \leq \|f\|_{L^2} \left(\|v'\|_{L^2} + |v(0)|\right),$$

so that $\|v'\|_{L^2}$ and $|v(0)|$ are bounded (why?). Hence $\|v\|_{L^2} \leq \|v'\|_{L^2} + |v(0)|$ is also bounded, so that $\|v\|_{H^1}$ is bounded. For the uniqueness of the minimizer check that $\psi(\frac{u_1+u_2}{2}) \leq \frac{1}{2}(\psi(u_1) + \psi(u_2))$, and equality holds iff $u_1' = u_2'$, and $u_1(0) = u_2(0)$, i.e., $u_1 = u_2$.
We have

$$\psi(u + \varepsilon v) = \frac{1}{2}\int_0^1 (u'^2 + 2\varepsilon u'v' + \varepsilon^2 v'^2)$$

$$+ \frac{1}{4}\left(u^4(0) + 4\varepsilon u^3(0)v(0) + \cdots + \varepsilon^4 v^4(0)\right) - \int_0^1 f(u + \varepsilon v).$$

If u is a minimizer of ψ we write $\psi(u) \leq \psi(u + \varepsilon v)$, and obtain

$$\int_0^1 (u'v' - fv) + u^3(0)v(0) + B_\varepsilon \geq 0,$$

where $B_\varepsilon \to 0$ as $\varepsilon \to 0$. Passing to the limit as $\varepsilon \to 0$, and choosing $\pm v$ yields

(S1) $$\int_0^1 (u'v' - fv) + u^3(0)v(0) = 0 \quad \forall v \in H^1(0,1).$$

Consequently, $u \in H^2(0, 1)$ satisfies

(S2) $\qquad\qquad\qquad -u'' = f \quad$ a.e. on $(0, 1)$.

Returning to (S1) and using (S2) yields

$$u'(1)v(1) - u'(0)v(0) + u^3(0)v(0) = 0 \quad \forall v \in H^1(0, 1),$$

so that

(S3) $\qquad\qquad\qquad u'(1) = 0, \quad u'(0) = u^3(0).$

Conversely, any function u satisfying (S2) and (S3) is a minimizer of ψ: the argument is the same as in question 6. In this case we have an explicit solution. The general solution of (S2) is given by

$$u(x) = -\int_0^x (x - t)f(t)dt + Ax + B,$$

and then (S3) is equivalent to

$$A = \int_0^1 f(t)dt, \quad \text{with } A = B^3.$$

$\boxed{8.42}$

2. Differentiating the equation

(S1) $\qquad\qquad\qquad v(x) = p^{1/4}(t)u(t)$

with respect to t gives

$$v'(x)p^{-1/2}(t) = \frac{1}{4}p^{-3/4}(t)p'(t)u(t) + p^{1/4}(t)u'(t).$$

Thus

(S2)
$$p(t)u'(t) = v'(x)p^{1/4}(t) - \frac{1}{4}p'(t)u(t)$$
$$= v'(x)p^{1/4}(t) - \frac{1}{4}p'(t)p^{-1/4}(t)v(x).$$

Differentiating (S2) with respect to t gives

(S3) $(pu')' = v''(x)p^{-1/4}(t) - \frac{1}{4}p''(t)p^{-1/4}(t)v(x) + \frac{1}{16}p'(t)^2 p^{-5/4}(t)v(x).$

Combining (S3) with the equation $-(pu')' + qu = \mu u$ on $(0, 1)$ yields

$$\text{(S4)} \quad v''(x)p^{-1/4}(t) - \frac{1}{4}p''(t)p^{-1/4}(t)v(x) + \frac{1}{16}p'(t)^2 p^{-5/4}(t)v(x)$$
$$= (q(t) - \mu)p^{-1/4}(t)v(x).$$

Hence v satisfies

$$-v'' + a(x)v = \mu v \text{ on } (0, L),$$

where

$$a(x) = q(t) + \frac{1}{4}p''(t) - \frac{1}{16}p'(t)^2 p^{-1}(t).$$

Problems

The numbers in parentheses refer to the chapters in the book whose knowledge is needed to solve the problem.

PROBLEM 1 (1, 4 only for question 9)

Extreme points; the Krein–Milman theorem

Let E be an n.v.s. and let $K \subset E$ be a convex subset. A point $a \in K$ is said to be an *extreme point* if

$$tx + (1 - t)y \neq a \quad \forall t \in (0, 1), \quad \forall x, y \in K \text{ with } x \neq y.$$

1. Check that $a \in K$ is an extreme point iff the set $K \setminus \{a\}$ is convex.

2. Let a be an extreme point of K. Let $(x_i)_{1 \leq i \leq n}$ be a finite sequence in K and let $(\alpha_i)_{1 \leq i \leq n}$ be a finite sequence of real numbers such that $\alpha_i > 0 \ \forall i$, $\sum \alpha_i = 1$, and $\sum \alpha_i x_i = a$. Prove that $x_i = a \ \forall i$.

   In what follows we assume that $K \subset E$ is a nonempty compact convex subset of E. A *subset* $M \subset K$ is said to be an *extreme set* if M is nonempty, closed, and whenever $x, y \in K$ are such that $tx + (1 - t)y \in M$ for some $t \in (0, 1)$, then $x \in M$ and $y \in M$.

3. Let $a \in K$. Check that a is an extreme point iff $\{a\}$ is an extreme set.

   Our first goal is to show that every extreme set contains at least one extreme point.

4. Let $A \subset K$ be an extreme set and let $f \in E^\star$. Set

$$B = \left\{ x \in A; \ \langle f, x \rangle = \max_{y \in A} \langle f, y \rangle \right\}.$$

   Prove that B is an extreme subset of K.

H. Brezis, *Functional Analysis, Sobolev Spaces and Partial Differential Equations,*
DOI 10.1007/978-0-387-70914-7, © Springer Science+Business Media, LLC 2011

5. Let $M \subset K$ be an extreme set of K. Consider the collection $\mathcal{F}$ of all the extreme sets of K that are contained in M; $\mathcal{F}$ is equipped with the following ordering:

$$A \leq B \quad \text{if} \quad B \subset A.$$

Prove that $\mathcal{F}$ has a maximal element M_0.

6. Prove that M_0 is reduced to a single point.

   [**Hint**: Use Hahn–Banach and question 4.]

7. Conclude.

8. Prove that K coincides with the closed convex hull of all its extreme points.

   [**Hint**: Argue by contradiction and use Hahn–Banach.]

9. Determine the set $\mathcal{E}$ of all the extreme points of B_E (= the closed unit ball of E) in the following cases:

   (a) $E = \ell^\infty$,
   (b) $E = c$,
   (c) $E = c_0$,
   (d) $E = \ell^1$,
   (e) $E = \ell^p$ with $1 < p < \infty$,
   (f) $E = L^1(\mathbb{R})$.

   [For the notation see Section 11.3].

PROBLEM 2 (1, 2 only for question B4)

Subdifferentials of convex functions

Let E be an n.v.s. and let $\varphi : E \to (-\infty, +\infty]$ be a convex function such that $\varphi \not\equiv +\infty$. For every $x \in E$ the *subdifferential* of φ is defined by

$$\begin{cases} \partial\varphi(x) = \{f \in E^*; \varphi(y) - \varphi(x) \geq \langle f, y - x \rangle \ \ \forall y \in E\} & \text{if } x \in D(\varphi), \\ \partial\varphi(x) = \emptyset & \text{if } x \notin D(\varphi), \end{cases}$$

and we set

$$D(\partial\varphi) = \{x \in E; \ \partial\varphi(x) \neq \emptyset\},$$

so that $D(\partial\varphi) \subset D(\varphi)$. Construct an example for which this inclusion is strict.

- A -

1. Show that $\partial\varphi(x)$ is a closed convex subset of E^*.

2. Let $x_1, x_2 \in D(\partial\varphi)$, $f_1 \in \partial\varphi(x_1)$, and $f_2 \in \partial\varphi(x_2)$. Prove that

$$\langle f_1 - f_2, x_1 - x_2 \rangle \geq 0.$$

3. Prove that

$$f \in \partial\varphi(x) \Longleftrightarrow \varphi(x) + \varphi^*(f) = \langle f, x \rangle.$$

4. Determine $\partial\varphi$ in the following cases:

   (a) $\varphi(x) = \frac{1}{2}\|x\|^2$,
   (b) $\varphi(x) = \|x\|$,
   (c) $\varphi(x) = I_K(x)$ (the indicator function of K), where $K \subset E$ is a nonempty convex set (resp. a linear subspace),
   (d) $\varphi(x)$ is a differentiable convex function on E.

   [**Hint**: In the cases (a), (b), $\partial\varphi$ is related to the duality map F defined in Remark 2 of Chapter 1; see also Exercise 1.1.]

5. Let $\psi : E \to (-\infty, +\infty]$ be another convex function such that $\psi \not\equiv +\infty$. Assume that $D(\varphi) \cap D(\psi) \neq \emptyset$. Prove that

$$\partial\varphi(x) + \partial\psi(x) \subset \partial(\varphi + \psi)(x) \quad \forall x \in E$$

(with the convention that $A + B = \emptyset$ if either $A = \emptyset$ or $B = \emptyset$). Construct an example for which this inclusion is strict.

- B -

Throughout part B we assume that $x_0 \in E$ satisfies the assumption

(1) $\exists M \in \mathbb{R}$ and $\exists R > 0$ such that $\varphi(x) \leq M \quad \forall x \in E$ with $\|x - x_0\| \leq R$.

1. Prove that $\partial\varphi(x_0) \neq \emptyset$.

   [**Hint**: Use Hahn–Banach in $E \times \mathbb{R}$.]

2. Prove that $\|f\| \leq \frac{1}{R}(M - \varphi(x_0)) \quad \forall f \in \partial\varphi(x_0)$.

3. Deduce that $\forall r < R, \exists L \geq 0$ such that

$$|\varphi(x_1) - \varphi(x_2)| \leq L\|x_1 - x_2\| \quad \forall x_1, x_2 \in E \text{ with } \|x_i - x_0\| \leq r, \ i = 1, 2.$$

[See also Exercise 2.1 for an alternative proof.]

4. Assume here that E is a Banach space and that φ is l.s.c. Prove that

$$\text{Int } D(\partial\varphi) = \text{Int } D(\varphi).$$

5. Prove that for every $y \in E$ one has

$$\lim_{t\downarrow 0} \frac{\varphi(x_0 + ty) - \varphi(x_0)}{t} = \max_{f \in \partial\varphi(x_0)} \langle f, y \rangle.$$

[**Hint**: Look at Exercise 1.25, question 5.]

6. Let $\psi : E \to (-\infty, +\infty]$ be a convex function such that $x_0 \in D(\psi)$. Prove that

$$\partial\varphi(x) + \partial\psi(x) = \partial(\varphi + \psi)(x) \quad \forall x \in E.$$

[**Hint**: Given $f_0 \in \partial(\varphi + \psi)(x)$, apply Theorem 1.12 to the functions $\widetilde{\varphi}(y) = \varphi(y) - \varphi(x) - \langle f_0, y - x \rangle$ and $\widetilde{\psi}(y) = \psi(y) - \psi(x)$.]

- C -

1. Let $\varphi : E \to \mathbb{R}$ be a convex function such that $\varphi(x) \le k\|x\| + C \; \forall x \in E$, for some constants $k \ge 0$ and C. Prove that

$$|\varphi(x_1) - \varphi(x_2)| \le k\|x_1 - x_2\| \quad \forall x_1, x_2 \in E.$$

What can one say about $D(\varphi^\star)$?

2. Let $A \subset \mathbb{R}^n$ be open and convex. Let $\varphi : A \to \mathbb{R}$ be a convex function. Prove that φ is continuous on A.

- D -

Let $\varphi : E \to \mathbb{R}$ be a continuous convex function and let

$$C = \{x \in E; \; \varphi(x) \le 0\}.$$

Assume that there exists some $x_0 \in E$ such that $\varphi(x_0) < 0$. Given $x \in C$ prove that $f \in \partial I_C(x)$ iff there exists some $\lambda \in \mathbb{R}$ such that $f \in \lambda \partial\varphi(x)$ with $\lambda = 0$ if $\varphi(x) < 0$, and $\lambda \ge 0$ if $\varphi(x) = 0$.

PROBLEM 3 (1)

The theorems of Ekeland, Brönsted–Rockafellar,
and Bishop–Phelps; the ε-subdifferential

- A -

Let M be a nonempty complete metric space equipped with the distance $d(x, y)$. Let $\psi : M \to (-\infty, +\infty]$ be an l.s.c. function that is bounded below and such that $\psi \not\equiv +\infty$. Our goal is to prove that there exists some $a \in M$ such that

$$\psi(x) - \psi(a) + d(x, a) \ge 0 \quad \forall x \in M.$$

Given $x \in M$ set

$$S(x) = \{y \in M; \; \psi(y) - \psi(x) + d(x, y) \le 0\}.$$

1. Check that $x \in S(x)$, and that $y \in S(x) \Rightarrow S(y) \subset S(x)$.

2. Fix any sequence of real numbers (ε_n) with $\varepsilon_n > 0$ $\forall n$ and $\varepsilon_n \to 0$. Given $x_0 \in M$, one constructs by induction a sequence (x_n) as follows: once x_n is known, pick any element x_{n+1} satisfying

$$\begin{cases} x_{n+1} \in S(x_n), \\ \psi(x_{n+1}) \leq \inf_{x \in S(x_n)} \psi(x) + \varepsilon_{n+1}. \end{cases}$$

Check that $S(x_{n+1}) \subset S(x_n)$ $\forall n$ and that

$$\psi(x_{n+p}) - \psi(x_n) + d(x_n, x_{n+p}) \leq 0 \quad \forall n, \quad \forall p.$$

Deduce that (x_n) is a Cauchy sequence, and so it converges to a limit, denoted by a.

3. Prove that a satisfies the required property.

[**Hint**: Given $x \in M$, consider two cases: either $x \in S(x_n)$ $\forall n$, or $\exists N$ such that $x \notin S(x_N)$.]

4. Give a geometric interpretation.

- B -

Let E be a Banach space and let $\varphi : E \to (-\infty, +\infty]$ be a convex l.s.c. function such that $\varphi \not\equiv +\infty$. Given $\varepsilon > 0$ and $x \in D(\varphi)$, set

$$\partial_\varepsilon \varphi(x) = \{f \in E^\star; \ \varphi(x) + \varphi^\star(f) - \langle f, x \rangle \leq \varepsilon\}.$$

Check that $\partial_\varepsilon \varphi(x) \neq \emptyset$.

Our purpose is to show that given any $x_0 \in D(\varphi)$ and any $f_0 \in \partial_\varepsilon \varphi(x_0)$ the following property holds:

$$\begin{cases} \forall \lambda > 0, \ \exists x_1 \in D(\varphi) \text{ and } \exists f_1 \in E^\star \text{ with } f_1 \in \partial \varphi(x_1) \\ \text{such that } \|x_1 - x_0\| \leq \varepsilon/\lambda \text{ and } \|f_1 - f_0\| \leq \lambda. \end{cases}$$

(The *subdifferential* $\partial \varphi$ is defined in Problem 2; it is recommended to solve Problem 2 before this one.)

1. Consider the function ψ defined by

$$\psi(x) = \varphi(x) + \varphi^\star(f_0) - \langle f_0, x \rangle.$$

Prove that there exists some $x_1 \in E$ such that $\|x_1 - x_0\| \leq \varepsilon/\lambda$ and

$$\psi(x) - \psi(x_1) + \lambda\|x - x_1\| \geq 0 \quad \forall x \in E.$$

[**Hint**: Use the result of part A on the set

$$M = \{x \in E;\ \psi(x) \le \psi(x_0) - \lambda\|x - x_0\|\}.]$$

2. Conclude.

   [**Hint**: Use the result of Problem 2, question B6.]

3. Deduce that

$$\overline{D(\partial\varphi)} = \overline{D(\varphi)} \quad \text{and} \quad \overline{R(\partial\varphi)} = \overline{D(\varphi^\star)},$$

   where $R(\partial\varphi) = \{f \in E^\star;\ \exists x \in D(\partial\varphi) \text{ such that } f \in \partial\varphi(x)\}$.

- C -

Let E be a Banach space and let $C \subset E$ be a nonempty closed convex set.

1. Assuming that C is also bounded, prove that the set

$$\left\{ f \in E^\star;\ \sup_{x \in C}\langle f, x\rangle \text{ is achieved} \right\}$$

is dense in $E^\star$.

   [**Hint**: Apply the results of part B to the function $\varphi = I_C$.]

2. One says that a closed hyperplane H of E is a *supporting hyperplane* to C at a point $x \in C$ if H separates C and $\{x\}$. Prove that the set of points in C that admit a supporting hyperplane is dense in the boundary of $C(= C \setminus \text{Int}\, C)$.

PROBLEM 4 (1)

Asplund's theorem and strictly convex norms

Let E be an n.v.s. and let $\varphi_0, \psi_0 : E \to [0, \infty)$ be two convex functions such that $\varphi_0(0) = \psi_0(0) = 0$ and $0 \le \psi_0(x) \le \varphi_0(x)\ \forall x \in E$. Starting with φ_0 and ψ_0 one defines by induction two sequences of functions (φ_n) and (ψ_n) as follows:

$$\varphi_{n+1}(x) = \frac{1}{2}(\varphi_n(x) + \psi_n(x))$$

and

$$\psi_{n+1}(x) = \frac{1}{2}\inf_{y \in E}\{\varphi_n(x + y) + \psi_n(x - y)\} = \frac{1}{2}(\varphi_n \nabla \psi_n)(2x).$$

[Before starting this problem solve Exercise 1.23, which deals with the inf-convolution ∇.]

- A -

1. Check that $0 \le \psi_n(x) \le \varphi_n(x)\ \forall x \in E,\ \forall n$ and that $\varphi_n(0) = \psi_n(0) = 0$.

2. Check that φ_n and ψ_n are convex.

3. Prove that the sequence (φ_n) is nonincreasing and that the sequence (ψ_n) is nondecreasing. Deduce that (φ_n) and (ψ_n) have a common limit, denoted by θ, with $\psi_0 \leq \theta \leq \varphi_0$, and that θ is convex.

4. Prove that $\varphi_n^\star \uparrow \theta^\star$.

5. Prove that $\psi_{n+1}^\star = \frac{1}{2}(\varphi_n^\star + \psi_n^\star)$, and deduce that $\psi_n^\star \downarrow \theta^\star$ when $D(\psi_0^\star) = E^\star$.

6. Assume that there exists some $x_0 \in D(\varphi_0)$ such that φ_0 is continuous at x_0. Prove that φ_n and ψ_n are also continuous at x_0.

   [**Hint**: Apply question 2 of Exercise 2.1.]

   Deduce that

   $$\varphi_{n+1}^\star(f) = \frac{1}{2} \inf_{g \in E^\star} \{\varphi_n^\star(f + g) + \psi_n^\star(f - g)\}.$$

- B -

Let $\varphi : E \to [0, +\infty)$ be a convex function that is *homogeneous of order two*, i.e., $\varphi(\lambda x) = \lambda^2 \varphi(x)\ \forall \lambda \in \mathbb{R},\ \forall x \in E$. Prove that

$$\varphi(x + y) \leq \frac{1}{t}\varphi(x) + \frac{1}{1-t}\varphi(y) \quad \forall x, y \in E, \quad \forall t \in (0, 1).$$

Deduce that the function $x \mapsto \sqrt{\varphi(x)}$ is a seminorm and conversely. Establish also that

(1) $\qquad 4\varphi(x) \leq \frac{1}{t}\varphi(x + y) + \frac{1}{1-t}\varphi(x - y) \quad \forall x, y \in E, \quad \forall t \in (0, 1).$

In what follows we assume, in addition, that φ_0 and ψ_0 are homogeneous of order two and that there is a constant $C > 0$ such that

$$\varphi_0(x) \leq (1 + C)\psi_0(x) \quad \forall x \in E.$$

1. Check that φ_n, ψ_n, and θ are homogeneous of order two.

2. Prove that for every n, one has

   $$\varphi_n(x) \leq \left(1 + \frac{C}{4^n}\right)\psi_n(x) \quad \forall x \in E.$$

   [**Hint**: Argue by induction and use (1).]

3. Assuming that either φ_0 or ψ_0 is strictly convex, prove that θ is strictly convex (for the definition of a strictly convex function, see Exercise 1.26).

[**Hint**: Use the inequality established in question B2. It is convenient to split φ_n as $\varphi_n = \theta_n + \frac{1}{2^n}\varphi_0$, where θ_n is some convex function that one should not try to write down explicitly. Note that

$$\theta_n + \left(\frac{1}{2^n} - \frac{C}{4^n}\right)\varphi_0 \leq \theta \leq \theta_n + \frac{1}{2^n}\varphi_0.]$$

- C -

Assume that there exist on E two equivalent norms denoted by $\| \ \|_1$ and $\| \ \|_2$. Let $\| \ \|_1^\star$ and $\| \ \|_2^\star$ denote the corresponding dual norms on $E^\star$. Assume that the norms $\| \ \|_1$ and $\| \ \|_2^\star$ are strictly convex. Using the above results, prove that there exists a third norm $\| \ \|$, equivalent to $\| \ \|_1$ (and to $\| \ \|_2$), that is strictly convex as well as its dual norm $\| \ \|^\star$.

PROBLEM 5 (1, 2)

Positive linear functionals

Let E be an n.v.s. and let P be a convex cone with vertex at 0, i.e., $\lambda x + \mu y \in P$, $\forall x, y \in P, \forall \lambda, \mu > 0$. Set $F = P - P$, so that F is a linear subspace. Consider the following two properties:

(i) Every linear functional f on E such that $f(x) \geq 0 \ \forall x \in P$, is continuous on E.
(ii) F is a closed subspace of finite codimension.

The goal of this problem is to show that (i) $\Rightarrow$ (ii) and that conversely, (ii) $\Rightarrow$ (i) when E is a Banach space and P is closed.

- A -

Throughout part A we assume (i).

1. Prove that F is closed.

   [**Hint**: Given any $x_0 \notin F$, construct a linear functional f on E such that $f(x_0) = 1$ and $f = 0$ on F.]

2. Let M be any linear subspace of E such that $M \cap F = \{0\}$. Prove that $\dim M < +\infty$.

   [**Hint**: Use Exercise 1.5.]

3. Deduce that (i) $\Rightarrow$ (ii).

- B -

Throughout part B we assume that E is a Banach space and that P is closed.

1. Assume here in addition that

   (iii) $P - P = E$.

Prove that there exists a constant $C > 0$ such that every $x \in E$ has a decomposition $x = y - z$ with $y, z \in P$, $\|y\| \leq C\|x\|$ and $\|z\| \leq C\|x\|$.

[**Hint**: Consider the set

$$K = \{x = y - z \text{ with } y, z \in P, \|y\| \leq 1 \text{ and } \|z\| \leq 1\}$$

and follow the idea of the proof of the open mapping theorem (Theorem 2.6).]

2. Deduce that (iii) $\Rightarrow$ (i).

[**Hint**: Argue by contradiction and consider a sequence (x_n) in E such that $\|x_n\| \leq 1/2^n$ and $f(x_n) \geq 1$. Then, use the result of question B1.]

3. Prove that (ii) $\Rightarrow$ (i).

- C -

In the following examples determine $F = P - P$ and examine whether (i) or (ii) holds:

(a) $E = C([0, 1])$ with its usual norm and

$$P = \{u \in E; u(t) \geq 0 \quad \forall t \in [0, 1]\},$$

(b) $E = C([0, 1])$ with its usual norm and

$$P = \{u \in E; u(t) \geq 0 \quad \forall t \in [0, 1], \text{ and } u(0) = u(1) = 0\},$$

(c) $E = \{u \in C^1([0, 1]); u(0) = u(1) = 0\}$ with its usual norm and

$$P = \{u \in E; u(t) \geq 0 \quad \forall t \in [0, 1]\}.$$

PROBLEM 6 (1, 2)

Let E be a Banach space and let $A : D(A) \subset E \to E^\star$ be a closed unbounded operator satisfying

$$\langle Ax, x \rangle \geq 0 \quad \forall x \in D(A).$$

- A -

Our purpose is to show that the following properties are equivalent:

(i) $\forall x \in D(A)$, $\exists C(x) \in \mathbb{R}$ such that $\langle Ay, y - x \rangle \geq C(x) \quad \forall y \in D(A)$,

(ii) $\exists k \geq 0$ such that

$$|\langle Ay, x \rangle| \leq k(\|x\| + \|Ax\|)\sqrt{\langle Ay, y \rangle} \quad \forall x, y \in D(A).$$

1. Prove that (ii) $\Rightarrow$ (i).

Conversely, assume (i).

2. Prove that there exist two constants $R > 0$ and $M \geq 0$ such that

$$\langle Ay, x - y \rangle \leq M \quad \forall y \in D(A) \quad \text{and} \quad \forall x \in D(A) \text{ with } \|x\| + \|Ax\| \leq R.$$

[**Hint**: Consider the function $\varphi(x) = \sup_{y \in D(A)} \langle Ay, x - y \rangle$ and apply Exercise 2.1.]

3. Deduce that

$$|\langle Ay, x \rangle|^2 \leq 4M \langle Ay, y \rangle \quad \forall y \in D(A) \quad \text{and} \quad \forall x \in D(A) \text{ with } \|x\| + \|Ax\| \leq R.$$

4. Conclude.

- B -

In what follows assume that $D(A) = E$. Let $\alpha > 0$.

1. Prove that the following properties are equivalent:

(iii) $$\|Ay\| \leq \alpha \sqrt{\langle Ay, y \rangle} \quad \forall y \in E,$$

(iv) $$\langle Ay, y - x \rangle \geq -\frac{1}{4} \alpha^2 \|x\|^2 \quad \forall x, y \in E.$$

[**Hint**: Use the same method as in part A.]

2. Let $A^\star \in \mathcal{L}(E^{\star\star}, E^\star)$ be the adjoint of A. Prove that (iv) is equivalent to

(iv*) $$\langle A^\star y, y - x \rangle \geq -\frac{1}{4} \alpha^2 \|x\|^2 \quad \forall x, y \in E.$$

3. Deduce that (iii) is equivalent to

(iii*) $$\|A^\star y\| \leq \alpha \sqrt{\langle A^\star y, y \rangle} \quad \forall y \in E.$$

PROBLEM 7 (1, 2)

The adjoint of the sum of two unbounded linear operators

Let E be a Banach space. Given two closed linear subspace M and N in E, set

$$\rho(M, N) = \sup_{\substack{x \in M \\ \|x\| \leq 1}} \text{dist}(x, N).$$

- A -

1. Check that $\rho(M, N) \leq 1$; if, in addition, $N \subset M$ with $N \neq M$, prove that $\rho(M, N) = 1$.

[**Hint**: Use Lemma 6.1.]

2. Let L, M, and N be three closed linear subspaces.
   Set $a = \rho(M, N)$ and $b = \rho(N, L)$. Prove that $\rho(M, L) \leq a + b + ab$.
   Deduce that if $L \subset M$, $a \leq 1/3$, and $b \leq 1/3$, then $L = M$.

3. Prove that $\rho(M, N) = \rho(N^\perp, M^\perp)$.

   [**Hint**: Check with the help of Theorem 1.12 that $\forall x \in E$ and $\forall f \in E^*$

   $$\text{dist}(x, N) = \sup_{\substack{g \in N^\perp \\ \|g\| \leq 1}} \langle g, x \rangle \quad \text{and} \quad \text{dist}(f, M^\perp) = \sup_{\substack{y \in M \\ \|y\| \leq 1}} \langle f, y \rangle.]$$

- B -

Let E and F be two Banach spaces; $E \times F$ is equipped with the norm $\|[u, v]\|_{E \times F} = \|u\|_E + \|v\|_F$. Given two unbounded operators $A : D(A) \subset E \to F$ and $B : D(B) \subset E \to F$ that are densely defined and closed, set

$$\rho(A, B) = \rho(G(A), G(B)).$$

1. Prove that $\rho(A, B) = \rho(B^*, A^*)$.

2. Prove that if $D(A) \cap D(B)$ is dense in E, then

$$A^* + B^* \subset (A + B)^*.$$

[Recall that $D(A + B) = D(A) \cap D(B)$ and $D(A^* + B^*) = D(A^*) \cap D(B^*)$.]

It may happen that the inclusion $A^* + B^* \subset (A + B)^*$ is strict—construct such an example. Our purpose is to prove that equality holds under some additional assumptions.

3. Assume

   (H) $\quad \begin{cases} D(A) \subset D(B) \text{ and there exist constants } k \in [0, 1) \text{ and } C \geq 0 \\ \text{such that } \|Bu\| \leq k\|Au\| + C\|u\| \quad \forall u \in D(A). \end{cases}$

   Prove that $A + B$ is closed and that $\rho(A, A + B) \leq k + C$.

4. In addition to (H) assume also

   (H*) $\quad \begin{cases} D(A^*) \subset D(B^*) \text{ and there exist constants } k^* \in [0, 1) \text{ and} \\ C^* \geq 0 \text{ such that } \|B^* v\| \leq k^*\|A^* v\| + C^*\|v\| \quad \forall v \in D(A^*). \end{cases}$

   Let $\varepsilon > 0$ be such that $\varepsilon(k + C) \leq 1/3$ and $\varepsilon(k^* + C^*) \leq 1/3$.

   Prove that $A + \varepsilon B^* = A^* + \varepsilon B^*$.

5. Assuming (H) and (H*) prove that $(A + B)^\star = A^\star + B^\star$.

   [**Hint**: Use successive steps. Check that the following inequality holds $\forall t \in [0, 1]$:

$$\|Bu\| \leq \frac{k}{1-k}\|Au + tBu\| + \frac{C}{1-k}\|u\| \quad \forall u \in D(A).]$$

PROBLEM 8 (2, 3, 4 only for question 6)

Weak convergence in ℓ^1. Schur's theorem.

Let $E = \ell^1$, so that $E^\star = \ell^\infty$ (see Section 11.3). Given $x \in E$ write

$$x = (x_1, x_2, \ldots, x_i, \ldots) \quad \text{and} \quad \|x\|_1 = \sum_{i=1}^{\infty} |x_i|,$$

and given $f \in E^\star$ write

$$f = (f_1, f_2, \ldots, f_i, \ldots) \quad \text{and} \quad \|f\|_\infty = \sup_i |f_i|.$$

Let (x^n) be a sequence in E such that $x^n \rightharpoonup 0$ weakly $\sigma(E, E^\star)$. Our goal is to show that $\|x^n\|_1 \to 0$.

1. Given $f, g \in B_{E^\star}$ (i.e., $\|f\|_\infty \leq 1$ and $\|g\|_\infty \leq 1$) set

$$d(f, g) = \sum_{i=1}^{\infty} \frac{1}{2^i}|f_i - g_i|.$$

   Check that d is a metric on $B_{E^\star}$ and that $B_{E^\star}$ is compact for the corresponding topology.

2. Given $\varepsilon > 0$ set

$$F_k = \{f \in B_{E^\star}; \ |\langle f, x^n \rangle| \leq \varepsilon \quad \forall n \geq k\}.$$

   Prove that there exist some $f^0 \in B_{E^\star}$, a constant $\rho > 0$, and an integer k_0 such that

$$[f \in B_{E^\star} \text{ and } d(f, f^0) < \rho] \Rightarrow [f \in F_{k_0}].$$

   [**Hint**: Use Baire category theorem.]

3. Fix an integer N such that $(1/2^{N-1}) < \rho$. Prove that

$$\|x^n\|_1 \leq \varepsilon + 2 \sum_{i=1}^{N} |x_i^n| \quad \forall n \geq k_0.$$

4. Conclude.

5. Using a similar method prove that if (x^n) is a sequence in ℓ^1 such that for every $f \in \ell^\infty$ the sequence $(\langle f, x^n \rangle)$ converges to some limit, then (x^n) converges to a limit strongly in ℓ^1.

6. Consider $E = L^1(0, 1)$, so that $E^* = L^\infty(0, 1)$. Construct a sequence (u^n) in E such that $u^n \rightharpoonup 0$ weakly $\sigma(E, E^*)$ and such that $\|u^n\|_1 = 1 \ \forall n$.

PROBLEM 9 (1, 2, 3)

Hahn–Banach for the weak* topology and applications

Let E be a Banach space.

- A -

1. Let $A \subset E^*$ and $B \subset E^*$ be two nonempty convex sets such that $A \cap B = \emptyset$. Assume that A is open in the topology $\sigma(E^*, E)$. Prove that there exist some $x \in E, x \neq 0$, and a constant α such that the hyperplane $\{f \in E^*; \langle f, x \rangle = \alpha\}$ separates A and B.

2. Assume that $A \subset E^*$ is closed in $\sigma(E^*, E)$ and $B \subset E^*$ is compact in $\sigma(E^*, E)$. Prove that $A + B$ is closed in $\sigma(E^*, E)$.

3. Let $A \subset E^*$ and $B \subset E^*$ be two nonempty convex sets such that $A \cap B = \emptyset$. Assume that A is closed in $\sigma(E^*, E)$ and B is compact in $\sigma(E^*, E)$. Prove that there exist some $x \in E, x \neq 0$, and a constant α such that the hyperplane $\{f \in E^*; \langle f, x \rangle = \alpha\}$ strictly separates A and B.

4. Let $A \subset E^*$ be convex. Prove that $\overline{A}^{\sigma(E^*, E)}$, the closure of A in $\sigma(E^*, E)$, is convex.

- B -

Here are various applications of the above results:

1. Let $N \subset E^*$ be a linear subspace. Recall that

$$N^\perp = \{x \in E; \ \langle f, x \rangle = 0 \quad \forall f \in N\}$$

and

$$N^{\perp\perp} = \{f \in E^*; \ \langle f, x \rangle = 0 \quad \forall x \in N^\perp\}.$$

Prove that $N^{\perp\perp} = \overline{N}^{\sigma(E^*, E)}$.

What can one say if E is reflexive?

Deduce that c_0 is dense in ℓ^∞ in the topology $\sigma(\ell^\infty, \ell^1)$.

2. Let $\varphi : E \to (-\infty, +\infty]$ be a convex l.s.c. function, $\varphi \not\equiv +\infty$. Prove that $\psi = \varphi^*$ is l.s.c. in the topology $\sigma(E^*, E)$.

Conversely, given a convex function $\psi : E^\star \to (-\infty, +\infty]$ that is l.s.c. for the topology $\sigma(E^\star, E)$ and such that $\psi \not\equiv +\infty$, prove that there exists a convex l.s.c. function $\varphi : E \to (-\infty, +\infty]$, $\varphi \not\equiv +\infty$, such that $\psi = \varphi^\star$.

3. Let F be another Banach space and let $A : D(A) \subset E \to F$ be an unbounded linear operator that is densely defined and closed. Prove that

   (i) $\overline{R(A^\star)}^{\sigma(E^\star, E)} = N(A)^\perp$,

   (ii) $\overline{D(A^\star)}^{\sigma(F^\star, F)} = F^\star$.

   What can one say if E (resp. F) is reflexive?

4. Prove—without the help of Lemma 3.3—that $J(B_E)$ is dense in $B_{E^{\star\star}}$ in the topology $\sigma(E^{\star\star}, E^\star)$ (see Lemma 3.4).

5. Let $A : B_E \to E^\star$ be a monotone map, that is,

$$\langle Ax - Ay, x - y \rangle \geq 0 \quad \forall x, y \in B_E.$$

Set $S_E = \{x \in E; \|x\| = 1\}$. Prove that $A(B_E) \subset \overline{\mathrm{conv}\, A(S_E)}^{\sigma(E^\star, E)}$.

PROBLEM 10 (3)

The Eberlein–Šmulian theorem

Let E be a Banach space and let $A \subset E$. Set $B = \overline{A}^{\sigma(E, E^\star)}$. The goal of this problem is to show that the following properties are equivalent:

(P) B is compact in the topology $\sigma(E, E^\star)$.

(Q)   Every sequence (x_n) in A has a weakly convergent subsequence.

Moreover, (P) (or (Q)) implies the following property:

(R) $\begin{cases} \text{For every } y \in B \text{ there exists a } \textit{sequence } (y_n) \subset A \\ \text{such that } y_n \rightharpoonup y \text{ weakly } \sigma(E, E^\star). \end{cases}$

- A -

Proof of the claim (P) $\Rightarrow$ (Q).

1. Prove that (P) $\Rightarrow$ (Q) under the additional assumption that E is separable.

   [**Hint**: Consider a set (b_k) in $B_{E^\star}$ that is countable and dense in $B_{E^\star}$ for the topology $\sigma(E^\star, E)$ (why does such a set exist?). Check that the quantity $d(x, y) = \sum_{k=1}^\infty \frac{1}{2^k} |\langle b_k, x - y \rangle|$ is a metric and deduce that B is metrizable for $\sigma(E, E^\star)$.]

2. Show that (P) $\Rightarrow$ (Q) in the general case.

   [**Hint**: Use question A1.]

- B -

For later purpose we shall need the following:

Lemma. *Let F be an n.v.s. and let $M \subset F^*$ be a finite-dimensional vector space. Then there exists a finite subset $(a_i)_{1 \leq i \leq k}$ in B_F such that*

$$\max_{1 \leq i \leq k} \langle g, a_i \rangle \geq \frac{1}{2} \|g\| \quad \forall g \in M.$$

[Hint: First choose points $(g_i)_{1 \leq i \leq k}$ in S_M such that $S_M \subset \bigcup_{i=1}^{k} B(g_i, 1/4)$, where $S_M = \{g \in M;\ \|g\| = 1\}$.]

- C -

Let $\xi \in E^{**}$ be such that $\xi \in \overline{A}^{\sigma(E^{**}, E^*)}$. Using assumption (Q) we shall prove that $\xi \in B$ and that there exists a sequence $(y_k) \subset A$ such that $y_k \rightharpoonup \xi$ in $\sigma(E, E^*)$.

1. Set $n_1 = 1$ and fix any $f_1 \in B_{E^*}$. Prove that there exists some $x_1 \in A$ such that

$$|\langle \xi, f_1 \rangle - \langle f_1, x_1 \rangle| < 1.$$

2. Let $M_1 = [\xi, x_1]$ be the linear space spanned by ξ and x_1. Prove that there exist $(f_i)_{1 < i \leq n_2}$ in B_{E^*} such that

$$\max_{1 < i \leq n_2} \langle \eta, f_i \rangle \geq \frac{1}{2} \|\eta\| \quad \forall \eta \in M_1.$$

Prove that there exists some $x_2 \in A$ such that

$$|\langle \xi, f_i \rangle - \langle f_i, x_2 \rangle| < \frac{1}{2} \quad \forall i, 1 \leq i \leq n_2.$$

3. Iterating the above construction, we obtain two sequences $(x_k) \subset A$ and $(f_i) \subset B_{E^*}$, and an increasing sequence of integers (n_k) such that

   (a) $$\max_{n_k < i \leq n_{k+1}} \langle \eta, f_i \rangle \geq \frac{1}{2} \|\eta\| \quad \forall \eta \in M_k = [\xi, x_1, x_2, \ldots, x_k],$$

   (b) $$|\langle \xi, f_i \rangle - \langle f_i, x_{k+1} \rangle| < \frac{1}{k+1} \quad \forall i, 1 \leq i \leq n_{k+1}.$$

4. Deduce from (a) that

$$\sup_{i \geq 1} \langle \eta, f_i \rangle \geq \frac{1}{2} \|\eta\| \quad \forall \eta \in \bigcup_{k=1}^{\infty} M_k = M$$

and then that

$$\sup_{i \geq 1} \langle \eta, f_i \rangle \geq \frac{1}{2} \|\eta\| \quad \forall \eta \in \overline{M},$$

where $\overline{M}$ denotes the closure of M in $E^{\star\star}$, in the strong topology.

5. Using (b) and assumption (Q), prove that there exists some $x \in B \cap \overline{M}$ such that

$$\langle \xi, f_i \rangle = \langle f_i, x \rangle \quad \forall i \geq 1.$$

Deduce that $\xi = x$ and conclude.

- D -

1. Prove that (Q) $\Rightarrow$ (P).
2. Prove that (Q) $\Rightarrow$ (R).

PROBLEM 11 (3)

A theorem of Banach–Dieudonné–Krein–Šmulian

Let E be a Banach space and let $C \subset E^{\star}$ be a convex set. Assume that for each integer n, the set $C \cap (nB_{E^{\star}})$ is closed in the topology $\sigma(E^{\star}, E)$. The goal of this problem is to show that C is closed in the topology $\sigma(E^{\star}, E)$.

- A -

Suppose, in addition, that $0 \notin C$. We shall prove that there exists a sequence (x_n) in E such that

(1) $\|x_n\| \to 0 \quad \text{and} \quad \sup_n \langle f, x_n \rangle > 1 \quad \forall f \in C.$

Let $d = \mathrm{dist}(0, C)$ and consider a sequence $d_n \uparrow +\infty$ such that $d_1 > d$. Set

$$C_k = \{ f \in C; \ \|f\| \leq d_k \}.$$

1. Check that the sets C_k are compact in the topology $\sigma(E^{\star}, E)$. Prove that there exists some $f_0 \in C$ such that $d = \|f_0\| > 0$.

2. Prove that there exists some $x_1 \in E$ such that

$$\langle f, x_1 \rangle > 1 \quad \forall f \in C_1.$$

[**Hint**: Use Hahn–Banach for the weak* topology; see question A3 of Problem 9.]

3. Set $A_1 = \{x_1\}$. Prove that there exists a finite subset $A_2 \subset E$ such that $A_2 \subset \frac{1}{d_1} B_E$ and $\sup_{x \in A_1 \cup A_2} \langle f, x \rangle > 1 \ \forall f \in C_2$.
   [**Hint**: For each finite subset $A \subset E$ such that $A \subset \frac{1}{d_1} B_E$ consider the set

$$Y_A = \left\{ f \in C_2; \ \sup_{x \in A_1 \cup A} \langle f, x \rangle \leq 1 \right\},$$

and prove first that $\cap_A Y_A = \emptyset$.]

4. Construct, by induction, a finite subset $A_k \subset E$ such that

$$A_k \subset \frac{1}{d_{k-1}} B_E \quad \text{and} \quad \sup_{x \in \cup_{i=1}^k A_i} \langle f, x \rangle > 1 \quad \forall f \in C_k.$$

5. Construct a sequence (x_n) satisfying (1).

- B -

1. Assume once more that $0 \notin C$. Prove that there exists some $x \in E$ such that

$$\langle f, x \rangle \geq 1 \quad \forall f \in C.$$

[**Hint**: Let (x_n) be a sequence satisfying (1). Consider the operator $T : E^\star \to c_0$ defined by $T(f) = (\langle f, x_n \rangle)_n$ and separate (in c_0) $T(C)$ and the open unit ball of c_0.]

2. Conclude.

PROBLEM 12 (1, 2, 3)

Before starting this problem it is necessary to solve Exercise 1.23.

Let E be a reflexive Banach space and let $\varphi, \psi : E \to (-\infty, +\infty]$ be convex l.s.c. functions such that $D(\varphi) \cap D(\psi) \neq \emptyset$. Set $\theta = \varphi^\star \nabla \psi^\star$.

- A -

We claim that

$$\overline{D((\varphi + \psi)^\star)} = \overline{D(\varphi^\star) + D(\psi^\star)}.$$

1. Prove that $D(\varphi^\star) + D(\psi^\star) \subset D((\varphi + \psi)^\star)$.

2. Prove that θ maps $E^\star$ into $(-\infty, +\infty]$, θ is convex, $D(\theta) = D(\varphi^\star) + D(\psi^\star)$ and $\theta^\star = \varphi + \psi$.

3. Deduce that $\overline{D((\varphi + \psi)^\star)} = \overline{D(\theta)}$ and conclude.

- B -

Assume, in addition, that φ and ψ satisfy

(H) $$\bigcup_{\lambda \geq 0} \lambda(D(\varphi) - D(\psi)) = E.$$

We claim that

(i) $$(\varphi + \psi)^\star = \varphi^\star \nabla \psi^\star,$$

(ii) $$\inf_{x\in E}\{\varphi(x)+\psi(x)\}=\max_{g\in E^*}\{-\varphi^*(-g)-\psi^*(g)\},$$

(iii) $$D((\varphi+\psi)^*)=D(\varphi^*)+D(\psi^*).$$

1. Prove that for every fixed $f\in E^*$ and $\alpha\in\mathbb{R}$ the set

$$M=\{g\in E^*;\ \varphi^*(f-g)+\psi^*(g)\le\alpha\}$$

   is bounded.
   [**Hint**: Use assumption (H) and Corollary 2.5.]

2. Let $\alpha\in\mathbb{R}$ be fixed. Let (f_n) and (g_n) be two sequences in E^* such that (f_n) is bounded and $\varphi^*(f_n-g_n)+\psi^*(g_n)\le\alpha$. Prove that (g_n) is bounded.

3. Deduce that θ is l.s.c.

4. Prove (i), (ii), and (iii).
   Compare these results with question 3 of Exercise 1.23 and with Theorem 1.12.

PROBLEM 13 (1, 3)

Properties of the duality map. Uniform
convexity. Differentiability of the norm

Let E be a Banach space. Recall the definition of the duality map (see Remark 2 in Chapter 1): For every $x\in E$,

$$F(x)=\{f\in E^*;\ \|f\|=\|x\|\ \text{and}\ \langle f,x\rangle=\|x\|^2\}.$$

Before starting this problem it is useful to solve Exercises 1.1 and 1.25.

- A -

Assume that E^* is strictly convex, so that $F(x)$ consists of a single element.

1. Check that

$$\lim_{\lambda\to0}\frac{1}{2\lambda}(\|x+\lambda y\|^2-\|x\|^2)=\langle Fx,y\rangle\quad\forall x,y\in E.$$

   [**Hint**: Apply a result of Exercise 1.25; distinguish the cases $\lambda>0$ and $\lambda<0$.]

2. Prove that for every $x,y\in E$, the map $t\in\mathbb{R}\mapsto\langle F(x+ty),y\rangle$ is continuous at $t=0$.

   [**Hint**: Use the inequality $\frac{1}{2}(\|v\|^2-\|u\|^2)\ge\langle Fu,v-u\rangle$ with $u=x+ty$ and $v=x+\lambda y$.]

3. Deduce that F is continuous from E strong into E^* weak*.

   [**Hint**: Use the result of Exercise 3.11.]

Prove the same result by a simple direct method in the case that E is reflexive or separable.

4. Check that

$$\langle Fx + Fy, \, x + y \rangle + \langle Fx - Fy, \, x - y \rangle = 2(\|x\|^2 + \|y\|^2) \quad \forall x, y \in E.$$

Deduce that

$$\|Fx + Fy\| + \|x - y\| \geq 2 \quad \forall x, y \in E \text{ with } \|x\| = \|y\| = 1.$$

5. Assume, in addition, that E is reflexive and strictly convex. Prove that F is bijective from E onto $E^\star$. Check that F^{-1} coincides with the duality map of $E^\star$.

- B -

In this part we assume that $E^\star$ is uniformly convex.

1. Prove that F is continuous from E strong into $E^\star$ strong.

2. More precisely, prove that F is uniformly continuous on bounded sets of E.

[**Hint**: Argue by contradiction and apply question A4.]

3. Deduce that the function $\varphi(x) = \frac{1}{2}\|x\|^2$ is differentiable and that its differential is F, i.e., for every $x_0 \in E$ we have

$$\lim_{\substack{x \to x_0 \\ x \neq x_0}} \frac{\varphi(x) - \varphi(x_0) - \langle Fx_0, \, x - x_0 \rangle}{\|x - x_0\|} = 0.$$

- C -

Conversely, assume that for every $x \in E$, the set Fx consists of a single element and that F is uniformly continuous on bounded sets of E. Prove that $E^\star$ is uniformly convex.

[**Hint**: Prove first the inequality

$$\|f + g\| \leq \frac{1}{2}\|f\|^2 + \frac{1}{2}\|g\|^2 - \langle f - g, y \rangle + \sup_{\substack{x \in E \\ \|x\| \leq 1}} \{\varphi(x + y) + \varphi(x - y)\}$$

$$\forall y \in E, \quad \forall f, g \in E^\star.]$$

PROBLEM 14 (1, 3)

Regularization of convex functions by inf-convolution

Let E be a Banach space such that $E^\star$ is uniformly convex. Assume that $\varphi : E \to (-\infty, +\infty]$ is convex l.s.c. and $\varphi \not\equiv +\infty$. The goal of this problem is to show that there exists a sequence (φ_n) of differentiable convex functions such that $\varphi_n \uparrow \varphi$ as $n \uparrow +\infty$.

- **A** -

For each fixed $x \in E$ consider the function $\Phi_x : E^\star \to (-\infty, +\infty]$ defined by

$$\Phi_x(f) = \frac{1}{2}\|f\|^2 + \varphi^\star(f) - \langle f, x \rangle, \quad f \in E^\star.$$

1. Check that there exists a unique element $f_x \in E^\star$ such that

$$\Phi_x(f_x) = \inf_{f \in E^\star} \Phi_x(f).$$

Set $Sx = f_x$.

2. Prove that the map $x \mapsto Sx$ is continuous from E strong into $E^\star$ strong.

   [**Hint**: Prove first that S is continuous from E strong into $E^\star$ weak*.]

- **B** -

Consider the function $\psi : E \to \mathbb{R}$ defined by

$$\psi(x) = \psi(x) = \inf_{y \in E} \left\{ \frac{1}{2}\|x - y\|^2 + \varphi(y) \right\}.$$

We claim that ψ is convex, differentiable, and that its differential coincides with S.

1. Check that ψ is convex and that

   (i) $\qquad \psi(x) = -\min_{f \in E^\star}\left\{ \frac{1}{2}\|f\|^2 + \varphi^\star(f) - \langle f, x \rangle \right\} \quad \forall x \in E,$

   (ii) $\qquad \psi^\star(f) = \frac{1}{2}\|f\|^2 + \varphi^\star(f) \quad \forall f \in E^\star.$

   [**Hint**: Apply Theorem 1.12.]

2. Deduce that

   $$\psi(x) + \psi^\star(Sx) = \langle Sx, x \rangle \quad \forall x \in E$$

   and that

   $$|\psi(y) - \psi(x) - \langle Sx, y - x \rangle| \leq \|Sy - Sx\| \, \|y - x\| \quad \forall x, y \in E.$$

3. Conclude.

- **C** -

For each integer $n \geq 1$ and every $x \in E$ set

$$\varphi_n(x) = \inf_{y \in E} \left\{ \frac{n}{2} \|x - y\|^2 + \varphi(y) \right\}.$$

Prove that φ_n is convex, differentiable, and that for every $x \in E$, $\varphi_n(x) \uparrow \varphi(x)$ as $n \uparrow +\infty$.

[**Hint**: Use the same method as in Exercise 1.24.]

PROBLEM 15 (1, 5 for question B6)

Center of a set in the sense of Chebyshev. Normal structure.
Asymptotic center of a sequence in the sense of Edelstein. Fixed points
of contractions following Kirk, Browder, Göhde, and Edelstein.

Before starting this problem it is useful to solve Exercise 3.29.

Let E be a uniformly convex Banach space and let $C \subset E$ be a nonempty closed convex set.

- A -

Let $A \subset C$ be a nonempty *bounded set*. For every $x \in E$ define

$$\varphi(x) = \sup_{y \in A} \|x - y\|.$$

1. Check that φ is a convex function and that

$$|\varphi(x_1) - \varphi(x_2)| \leq \|x_1 - x_2\| \quad \forall x_1, x_2 \in E.$$

2. Prove that there exists a unique element $c \in C$ such that

$$\varphi(c) = \inf_{x \in C} \varphi(x).$$

The point c is called the *center* of A and is denoted by $c = \sigma(A)$.

3. Prove that if A is not reduced to a single point then

$$\varphi(\sigma(A)) < \operatorname{diam} A = \sup_{x, y \in A} \|x - y\|.$$

- B -

Let (a_n) be a *bounded sequence* in C; set

$$A_n = \bigcup_{i=n}^{\infty} \{a_i\} \quad \text{and} \quad \varphi_n(x) = \sup_{y \in A_n} \|x - y\| \quad \text{for } x \in E.$$

1. For every $x \in E$, consider $\varphi(x) = \lim_{n \to +\infty} \varphi_n(x)$. Prove that this limit exists and that φ is convex and continuous on E.

2. Prove that there exists a unique element $\sigma \in C$ such that

$$\varphi(\sigma) = \inf_{x \in C} \varphi(x).$$

The point σ is called the *asymptotic center* of the sequence (a_n).

3. Let $\sigma_n = \sigma(A_n)$ be the center of the set A_n in the sense of question A2. Prove that

$$\lim_{n \to \infty} \varphi_n(\sigma_n) = \lim_{n \to \infty} \varphi(\sigma_n) = \varphi(\sigma),$$

and that $\sigma_n \rightharpoonup \sigma$ weakly $\sigma(E, E^*)$.

4. Deduce that $\sigma_n \to \sigma$ strongly.

[**Hint**: Argue by contradiction and apply the result of Exercise 3.29.]

5. Assume $a_n \to a$ strongly. Determine the asymptotic center of the sequence (a_n).

6. Assume here that E is a Hilbert space and that $a_n \rightharpoonup a$ weakly $\sigma(E, E^*)$. Compute $\varphi(x)$ and determine the asymptotic center of the sequence (a_n).
[**Hint**: Expand squares of norms.]

- C -

Assume that $T : C \to C$ is a contraction, that is,

$$\|Tx - Ty\| \le \|x - y\| \quad \forall x, y \in C.$$

Let $a \in C$ be given and let $a_n = T^n a$ be the sequence of its iterates. Assume that the sequence (a_n) is *bounded*. Let σ be the asymptotic center of the sequence (a_n).

1. Prove that σ is a fixed point of T, i.e., $T\sigma = \sigma$.

2. Check that the set of fixed points of T is closed and convex.

PROBLEM 16 (2, 3)

Characterization of linear maximal monotone operators

Let E be a Banach space and let $A : D(A) \subset E \to E^*$ be an unbounded linear operator satisfying the *monotonicity* condition

(M) $\langle Au, u \rangle \ge 0 \quad \forall u \in D(A).$

We denote by (P) the following property:

(P) $\begin{cases} \text{If } x \in E \text{ and } f \in E^* \text{ are such that} \\ \langle Au - f, u - x \rangle \ge 0 \quad \forall u \in D(A), \\ \text{then } x \in D(A) \text{ and } Ax = f. \end{cases}$

- A -

1. Prove that if (P) holds then $D(A)$ is dense in E.

   [**Hint**: Show that if $f \in E^\star$ and $\langle f, u \rangle = 0 \; \forall u \in D(A)$, then $f = 0$.]

2. Prove that if (P) holds then A is closed.

3. Prove that the function $u \in D(A) \mapsto \langle Au, u \rangle$ is convex.

4. Prove that $N(A) \subset R(A)^\perp$. Deduce that if $D(A)$ is dense in E then $N(A) \subset N(A^\star)$.

5. Prove that if $D(A) = E$, then (P) holds.

   Throughout the rest of this problem we assume, in addition, that

   (i) E is reflexive, E and $E^\star$ are strictly convex,
   (ii) $D(A)$ is dense in E and A is closed,

   so that $A^\star : D(A^\star) \subset E \to E^\star$ and $D(A^\star)$ is dense in E (why?).

   The goal of this problem is to establish the equivalence (P) $\Leftrightarrow$ (M*), where (M*) denotes the following property:

   (M*) $\langle A^\star v, v \rangle \geq 0 \quad \forall v \in D(A^\star)$.

- B -

In this section we assume that (P) holds.

1. Prove that
$$\langle A^\star v, v \rangle \geq 0 \quad \forall v \in D(A) \cap D(A^\star).$$

2. Let $v \in D(A^\star)$ with $v \notin D(A)$. Prove that $\forall f \in E^\star$, $\exists u \in D(A)$ such that

$$\langle Au - f, u - v \rangle < 0.$$

   Choosing $f = -A^\star v$, prove that $\langle A^\star v, v \rangle > 0$. Deduce that (M*) holds.

3. Prove that $N(A) = N(A^\star)$ and $\overline{R(A)} = \overline{R(A^\star)}$.

- C -

In this part we assume that (M*) holds.

1. Check that the space $D(A)$ equipped with the graph norm $\|u\|_{D(A)} = \|u\|_E + \|Au\|_{E^\star}$ is reflexive.

2. Given $x \in E$ and $f \in E^\star$, consider the function φ defined on $D(A)$ by

$$\varphi(u) = \frac{1}{2}\|Au - f\|^2 + \frac{1}{2}\|u - x\|^2 + \langle Au - f, u - x \rangle.$$

Prove that φ is convex and continuous on $D(A)$.
Prove that $\varphi(u) \to +\infty$ as $\|u\|_{D(A)} \to \infty$.

3. Deduce that there exists some $u_0 \in D(A)$ such that $\varphi(u_0) \le \varphi(u)\ \forall u \in D(A)$. What equation (involving A and $A^\star$) does one obtain by choosing $u = u_0 + tv$ with $v \in D(A), t > 0$, and letting $t \to 0$?

[**Hint**: Apply the result of Problem 13, part A.]

4. Prove that $(M^\star) \Rightarrow (P)$.

5. Deduce that $A^\star$ also satisfies property (P).

PROBLEM 17 (1, 3, 4)

- A -

Let E be a reflexive Banach space and let M be a closed linear subspace of E. Let C be a convex subset of $E^\star$. For every $u \in E$ set

$$\varphi(u) = \sup_{g \in C} \langle g, u \rangle.$$

1. Prove that for every $f \in \overline{M^\perp + C}$ we have

$$\varphi(u) \ge \langle f, u \rangle \quad \forall u \in M.$$

[**Hint**: Start with the case $f \in M^\perp + C$.]

2. Conversely, let $f \in E^\star$ be such that

$$\varphi(u) \ge \langle f, u \rangle \quad \forall u \in M.$$

Prove that $f \in \overline{M^\perp + C}$.

[**Hint**: Use Hahn–Banach.]

3. Assuming that C is closed and bounded, prove that $M^\perp + C$ is closed.

- B -

In this section we assume that $E = L^p(\Omega)$ with $1 < p < \infty$,

$$M = \left\{ u \in L^p(\Omega);\ \int ju = 0 \right\},$$

and

$$C = \{ g \in L^{p'}(\Omega);\ |g(x)| \le k(x)\ \text{a.e.}\ x \in \Omega \},$$

where j and $k \ge 0$ are given functions in $L^{p'}(\Omega)$.

1. Check that M is a closed linear subspace and that C is convex, closed, and bounded.

2. Determine $M^\perp$.
3. Determine $\varphi(u)$ for every $u \in L^p(\Omega)$.
4. Deduce that if $f \in L^{p'}(\Omega)$ satisfies

$$\int k|u| \geq \int fu \quad \forall u \in M,$$

then there exist a constant $\lambda \in \mathbb{R}$ and a function $g \in C$ such that $f = \lambda j + g$.
5. Prove that the converse also holds.

- C -

Let $M \subset L^1(\Omega)$ be a linear subspace. Let $f, g \in L^\infty(\Omega)$ be such that $f \leq g$ a.e. on Ω. Prove that the following properties are equivalent:

(i) $\qquad\qquad \exists \varphi \in M^\perp$ such that $f \leq \varphi \leq g$ a.e. on Ω,

(ii) $\qquad\qquad \int (fu^+ - gu^-) \leq 0 \quad \forall u \in M,$

where $u^+ = \max\{u, 0\}$ and $u^- = \max\{-u, 0\}$.
[**Hint**: Assuming (ii), check that $\int(g + f)u \leq \int(g - f)|u| \; \forall u \in M$ and apply Theorem I.12 to find some $\psi \in L^\infty(\Omega)$ with $|\psi| \leq g - f$, such that $\psi - (g + f) \in M^\perp$. Take $\varphi = \frac{1}{2}(g + f - \psi)$.]

PROBLEM 18 (3, 4)

Let Ω be a measure space with finite measure. Let $1 < p < \infty$. Let $g : \mathbb{R} \to \mathbb{R}$ be a continuous nondecreasing function such that

$$|g(t)| \leq C(|t|^{p-1} + 1) \quad \forall t \in \mathbb{R}, \quad \text{for some constant } C.$$

Set $G(t) = \int_0^t g(s)ds$.

1. Check that for every $u \in L^p(\Omega)$, we have $g(u) \in L^{p'}(\Omega)$ and $G(u) \in L^1(\Omega)$.
Let (u_n) be a sequence in $L^p(\Omega)$ and let $u \in L^p(\Omega)$ be such that

(i) $\qquad\qquad u_n \rightharpoonup u \quad \text{weakly } \sigma(L^p, L^{p'})$

and

(ii) $\qquad\qquad \limsup \int G(u_n) \leq \int G(u).$

The purpose of this problem is to establish the following properties:

(1) $\qquad\qquad g(u_n) \to g(u) \text{ strongly in } L^q \text{ for every } q \in [1, p'),$

(2) $\qquad\begin{cases} \text{Assuming, in addition, that } g \text{ is increasing (strictly),} \\ \text{then } u_n \to u \text{ strongly in } L^q \text{ for every } q \in [1, p). \end{cases}$

2. Check that $G(a) - G(b) - g(b)(a - b) \geq 0 \ \forall a, b \in \mathbb{R}$.

   What can one say if $G(a) - G(b) - g(b)(a - b) = 0$?

3. Let (a_n) be a sequence in $\mathbb{R}$ and let $b \in \mathbb{R}$ be such that

   $$\lim[G(a_n) - G(b) - g(b)(a_n - b)] = 0.$$

   Prove that $g(a_n) \to g(b)$.

4. Prove that $\int |G(u_n) - G(u) - g(u)(u_n - u)| \to 0$.
   Deduce that there exists a subsequence (u_{n_k}) such that

   $$G(u_{n_k}) - G(u) - g(u)(u_{n_k} - u) \to 0 \text{ a.e. on } \Omega$$

   and therefore $g(u_{n_k}) \to g(u)$ a.e. on Ω.

5. Prove that (1) holds. (Check (1) for the whole sequence and not only for a subsequence.)

6. Prove that (2) holds.

   In what follows, we assume, in addition, that there exist constants $\alpha > 0$ and C such that

   (3) $$|g(t)| \geq \alpha |t|^{p-1} - C \quad \forall t \in \mathbb{R}.$$

7. Prove that $g(u_n) \to g(u)$ strongly in $L^{p'}$.

8. Can one reach the same conclusion without assumption (3)?

9. If, in addition, g is increasing (strictly) prove that $u_n \to u$ strongly in L^p.

PROBLEM 19 (3, 4)

Let E be the space $L^1(\mathbb{R}) \cap L^2(\mathbb{R})$ equipped with the norm

$$\|u\|_E = \|u\|_1 + \|u\|_2.$$

1. Check that E is a Banach space. Let $f(x) = f_1(x) + f_2(x)$ with $f_1 \in L^\infty(\mathbb{R})$ and $f_2 \in L^2(\mathbb{R})$. Check that the mapping $u \mapsto \int_{\mathbb{R}} f(x)u(x)dx$ is a continuous linear functional on E.

2. Let $0 < \alpha < 1/2$; check that the mapping

   $$u \mapsto \int_{\mathbb{R}} \frac{1}{|x|^\alpha} u(x)dx$$

   is a continuous linear functional on E.

   [**Hint**: Split the integral into two parts: $[|x| > M]$ and $[|x| \leq M]$.]

3. Set

$$K = \left\{ u \in E; \ u \geq 0 \text{ a.e. on } \mathbb{R} \text{ and } \int_{\mathbb{R}} u(x)dx \leq 1 \right\}.$$

Check that K is a closed convex subset of E.

4. Let (u_n) be a sequence in K and let $u \in K$ be such that $u_n \rightharpoonup u$ weakly in $L^2(\mathbb{R})$. Check that $u \in K$ and prove that

$$\int_{\mathbb{R}} \frac{1}{|x|^\alpha} u_n(x)dx \to \int_{\mathbb{R}} \frac{1}{|x|^\alpha} u(x)dx.$$

Consider the function J defined, for every $u \in E$, by

$$J(u) = \int_{\mathbb{R}} u^2(x)dx - \int_{\mathbb{R}} \frac{1}{|x|^\alpha} u(x)dx.$$

5. Check that there is a constant C such that $J(u) \geq C \ \forall u \in K$.

We claim that $m = \inf_{u \in K} J(u)$ is achieved.

6. Let (u_n) be a sequence in K such that $J(u_n) \to m$. Prove that $\|u_n\|_E$ is bounded.

7. Let (u_{n_k}) be a subsequence such that $u_{n_k} \rightharpoonup u$ weakly in $L^2(\mathbb{R})$. Prove that $J(u) = m$.

8. Is E a reflexive space?

PROBLEM 20 (4)

Clarkson's inequalities. Uniform convexity of L^p

- A -

In this part we assume that $2 \leq p < \infty$ and we shall establish the following inequalities:

(1) $\qquad |x + y|^p + |x - y|^p \leq 2(|x|^{p'} + |y|^{p'})^{p/p'} \quad \forall x, y \in \mathbb{R},$

(2) $\qquad 2(|x|^{p'} + |y|^{p'})^{p/p'} \leq 2^{p-1}(|x|^p + |y|^p) \quad \forall x, y \in \mathbb{R}.$

1. Prove (2).

[**Hint**: Use the convexity of the function $g(t) = |t|^{p/p'}$.]

2. Set

$$f(x) = (1 + x^{1/p})^p + (1 - x^{1/p})^p, \quad x \in (0, 1).$$

Prove that

$$f''(x) \leq 0 \quad \forall x \in (0, 1).$$

Deduce that

(3) $\qquad f(x) \le f(y) + (x - y)f'(y) \quad \forall x, y \in (0, 1).$

3. Prove that

$$f(x) \le 2(1 + x^{p'/p})^{p/p'} \quad \forall x \in (0, 1).$$

[**Hint**: Use (3) with $y = x^{p'}$.]

4. Deduce (1).

In what follows Ω denotes a σ-finite measure space.

- B -

In this part we assume again that $2 \le p < \infty$.

1. Prove the following inequalities:

(4) $\quad \|f + g\|_p^p + \|f - g\|_p^p \le 2(\|f\|_p^{p'} + \|g\|_p^{p'})^{p/p'} \quad \forall f, g \in L^p(\Omega),$

(5) $\quad 2(\|f\|_p^{p'} + \|g\|_p^{p'})^{p/p'} \le 2^{p-1}(\|f\|_p^p + \|g\|_p^p) \quad \forall f, g \in L^p(\Omega).$

2. Deduce *Clarkson's first inequality* (see Theorem 4.10).

- C -

In this part we assume that $1 < p \le 2$.

1. Establish the following inequality:

(6) $\quad \|f + g\|_p^{p'} + \|f - g\|_p^{p'} \le 2(\|f\|_p^p + \|g\|_p^p)^{p'/p} \quad \forall f, g \in L^p(\Omega).$

Inequality (6) is called *Clarkson's second inequality*.

[**Hint**: There are two different methods:

(i) By duality from (4), observing that

$$\sup_{\varphi, \psi \in L^{p'}} \left\{ \frac{\int (u\varphi + v\psi)}{[\|\varphi\|_{p'}^p + \|\psi\|_{p'}^p]^{1/p}} \right\} = (\|u\|_p^{p'} + \|v\|_p^{p'})^{1/p'}.$$

(ii) Directly from (1) combined with the result of Exercise 4.11.]

2. Deduce that $L^p(\Omega)$ is uniformly convex for $1 < p \le 2$.

PROBLEM 21 (4)

The distribution function. Marcinkiewicz spaces

Throughout this problem Ω denotes a measure space with finite measure μ. Given a measurable function $f : \Omega \to \mathbb{R}$, we define its *distribution function* α to be

$$\alpha(t) = |[|f| > t]| = \text{meas}\{x \in \Omega; \ |f(x)| > t\} \quad \forall t \geq 0.$$

- A -

1. Check that α is nonincreasing. Prove that $\alpha(t + 0) = \alpha(t) \ \forall t \geq 0$. Construct a simple example in which $\alpha(t - 0) \neq \alpha(t)$ for some $t > 0$.

2. Let (f_n) be a sequence of measurable functions such that $f_n \to f$ a.e. on Ω. Let (α_n) and α denote the corresponding distribution functions. Prove that

$$\alpha(t) \leq \lim_{n \to \infty} \inf \alpha_n(t) \leq \lim_{n \to \infty} \sup \alpha_n(t) \leq \alpha(t - 0) \quad \forall t > 0.$$

Deduce that $\alpha_n(t) \to \alpha(t)$ a.e.

- B -

1. Let $g \in L^1_{\text{loc}}(\mathbb{R})$ be a function such that $g \geq 0$ a.e. Set

$$G(t) = \int_0^t g(s)ds.$$

Prove that for every measurable function f,

$$\int_\Omega G(|f(x)|)d\mu < \infty \iff \int_0^\infty \alpha(t)g(t)dt < \infty$$

and that

$$\int_\Omega G(|f(x)|)d\mu = \int_0^\infty \alpha(t)g(t)dt.$$

[**Hint**: Use Fubini and Tonelli.]

2. More generally, prove that

$$\int_{[|f|>\lambda]} G(|f(x)|)d\mu = \alpha(\lambda)G(\lambda) + \int_\lambda^\infty \alpha(t)g(t)dt \quad \forall \lambda \geq 0.$$

3. Deduce that for $1 \leq p < \infty$,

$$f \in L^p(\Omega) \iff \int_0^\infty \alpha(t)t^{p-1} \, dt < \infty$$

and that

$$\int_{[|f|>\lambda]} |f(x)|^p d\mu = \alpha(\lambda)\lambda^p + p \int_\lambda^\infty \alpha(t)t^{p-1}dt \quad \forall \lambda \geq 0.$$

Check that if $f \in L^p(\Omega)$, then $\lim_{t \to +\infty} \alpha(t)t^p = 0$.

- C -

Let $1 < p < \infty$. For every $f \in L^1(\Omega)$ define

$$[f]_p = \sup\left\{|A|^{-1/p'}\int_A |f|; \ A \subset \Omega \text{ measurable}, |A| > 0\right\} \le \infty,$$

and consider the set

$$M^p(\Omega) = \{f \in L^1(\Omega); \ [f]_p < \infty\},$$

called the *Marcinkiewicz space* of order p. The space M^p is also called the *weak L^p* space, but this terminology is confusing because the word "weak" is already used in connection with the weak topology.

1. Check that $M^p(\Omega)$ is a linear space and that $[\ ]_p$ is a norm. Prove that

   $$L^p(\Omega) \subset M^p(\Omega) \text{ and that } [f]_p \le \|f\|_p \text{ for every } f \in L^p(\Omega).$$

2. Prove that $M^p(\Omega)$, equipped with the norm $[\ ]_p$, is a Banach space.
   Check that $M^p(\Omega) \subset M^q(\Omega)$ with continuous injection for $1 < q \le p$.

   We claim that

   $$[f \in M^p(\Omega)] \iff \left[f \text{ is measurable and } \sup_{t>0} t^p \alpha(t) < \infty\right].$$

3. Prove that if $f \in M^p(\Omega)$ then $t^p \alpha(t) \le [f]_p^p \ \forall t > 0$.

4. Conversely, let f be a measurable function such that

   $$\sup_{t>0} t^p \alpha(t) < \infty.$$

   Prove that there exists a constant C_p (depending only on p) such that

   $$[f]_p^p \le C_p \sup_{t>0} t^p \alpha(t).$$

   [**Hint**: Use question B3 and write

   $$\int_A |f| = \int_{A \cap [|f|>\lambda]} |f| + \int_{A \cap [|f|\le\lambda]} |f|;$$

   then vary λ.]

5. Prove that $M^p(\Omega) \subset L^q(\Omega)$ with continuous injection for $1 \le q < p$.

6. Let $1 < q < r < \infty$ and $\theta \in (0, 1)$; set

   $$\frac{1}{p} = \frac{\theta}{q} + \frac{1-\theta}{r}.$$

Prove that there is a constant C—depending only on q, r, and θ—such that

$$\|f\|_p \le C[f]_q^\theta [f]_r^{1-\theta} \quad \forall f \in M^r(\Omega).$$

7. Set $\Omega = \{x \in \mathbb{R}^N; |x| < 1\}$, equipped with the Lebesgue measure, and let $f(x) = |x|^{-N/p}$ with $1 < p < \infty$. Check that $f \in M^p(\Omega)$, while $f \notin L^p(\Omega)$.

PROBLEM 22 (4)

An interpolation theorem (Schur, Riesz, Thorin, Marcinkiewicz)

Let Ω be a measure space with finite measure. Let

$$T : L^1(\Omega) \to L^1(\Omega)$$

be a bounded linear operator whose norm is denoted by $N_1 = \|T\|_{\mathcal{L}(L^1, L^1)}$. We assume that

$$T(L^\infty(\Omega)) \subset L^\infty(\Omega).$$

1. Prove that T is a bounded operator from $L^\infty(\Omega)$ into itself. Set

$$N_\infty = \|T\|_{\mathcal{L}(L^\infty, L^\infty)}.$$

The goal of this problem is to show that

$$T(L^p(\Omega)) \subset L^p(\Omega) \quad \text{for every } 1 < p < \infty$$

and that $T : L^p(\Omega) \to L^p(\Omega)$ is a bounded operator whose norm $N_p = \|T\|_{\mathcal{L}(L^p, L^p)}$ satisfies the inequality $N_p \le 2N_1^{1/p} N_\infty^{1/p'}$.

For simplicity, we assume first that $N_\infty = 1$. Given a function $u \in L^1(\Omega)$, we set, for every $\lambda > 0$,

$$u = v_\lambda + w_\lambda \quad \text{with} \quad v_\lambda = u\chi_{[|u|>\lambda]} \quad \text{and} \quad w_\lambda = u\chi_{[|u|\le\lambda]},$$

$$f = Tu, \quad g_\lambda = Tv_\lambda, \quad \text{and} \quad h_\lambda = Tw_\lambda, \quad \text{so that} \quad f = g_\lambda + h_\lambda.$$

2. Check that

$$\|g_\lambda\|_1 \le N_1 \int_{[|u|>\lambda]} |u(x)|d\mu \quad \text{and} \quad \|h_\lambda\|_\infty \le \lambda \quad \forall \lambda > 0.$$

3. Consider the distribution functions

$$\alpha(t) = |[|u| > t]|, \beta(t) = |[|f| > t]|, \gamma_\lambda(t) = [|g_\lambda| > t].$$

Prove that

$$\int_0^\infty \gamma_\lambda(t)dt \le N_1[\alpha(\lambda)\lambda + \int_\lambda^\infty \alpha(t)dt] \quad \forall \lambda > 0,$$

and that

$$\beta(t) \le \gamma_\lambda(t - \lambda) \quad \forall \lambda > 0, \quad \forall t > \lambda.$$

[**Hint**: Apply the results of Problem 21, part B.]

4. Assuming $u \in L^p(\Omega)$, prove that $f \in L^p(\Omega)$ and that

$$\|f\|_p \le 2N_1^{1/p}\|u\|_p.$$

5. Conclude in the general case, in which $N_\infty \ne 1$.

Remark. By a different argument one can prove in fact that $N_p \le N_1^{1/p} N_\infty^{1/p'}$; see, e.g., Bergh–Löfström [1] and the references in the Notes on Chapter 1 of their book.

PROBLEM 23 (3, 4)

Weakly compact subsets of L^1 and equi-integrable families.
The theorems of Hahn–Vitali–Saks, Dunford–Pettis, and de la Vallée-Poussin.

Let Ω be a σ-finite measure space. We recall (see Exercise 4.36) that a subset $\mathcal{F} \subset L^1(\Omega)$ is said to be *equi-integrable* if it satisfies the following properties:

(a) $\qquad \mathcal{F}$ is bounded in $L^1(\Omega)$,

(b) $\begin{cases} \forall \varepsilon > 0 \ \exists \delta > 0 \quad \text{such that } \int_A |f| < \varepsilon \quad \forall f \in \mathcal{F}, \\ \forall A \subset \Omega \text{ with } A \text{ measurable and } |A| < \delta, \end{cases}$

(c) $\begin{cases} \forall \varepsilon > 0 \ \exists \omega \subset \Omega \quad \text{measurable with } |\omega| < \infty \\ \text{such that } \int_{\Omega \setminus \omega} |f| < \varepsilon. \end{cases}$

The first goal of this problem is to establish the equivalence of the following properties for a given set $\mathcal{F}$ in $L^1(\Omega)$:

(i) $\mathcal{F}$ is contained in a weakly ($\sigma(L^1, L^\infty)$) compact set of $L^1(\Omega)$,

(ii) $\mathcal{F}$ is equi-integrable.

- **A** -

The implication (i) $\Rightarrow$ (ii).

1. Let (f_n) be a sequence in $L^1(\Omega)$ such that

$$\int_A f_n \to 0 \quad \forall A \subset \Omega \text{ with } A \text{ measurable and } |A| < \infty.$$

Prove that (f_n) satisfies property (b).

[**Hint**: Consider the subset $X \subset L^1(\Omega)$ defined by

$$X = \{\chi_A \text{ with } A \subset \Omega, A \text{ measurable and } |A| < \infty\}.$$

Check that X is closed in $L^1(\Omega)$ and apply the Baire category theorem to the sequence

$$X_n = \left\{\chi_A \in X; \left|\int_A f_k\right| \le \varepsilon \quad \forall k \ge n\right\},$$

where $\varepsilon > 0$ is fixed.]

2. Let (f_n) be a sequence in $L^1(\Omega)$ such that

$$\int_A f_n \to 0 \quad \forall A \subset \Omega \text{ with } A \text{ measurable and } |A| \le \infty.$$

Prove that (f_n) satisfies property (c).

[**Hint**: Let (Ω_i) be a nondecreasing sequence of measurable sets with finite measure such that $\Omega = \bigcup_i \Omega_i$. Consider on $L^\infty(\Omega)$ the metric d defined by

$$d(f, g) = \sum_i \frac{1}{2^i |\Omega_i|} \int_{\Omega_i} |f - g|.$$

Set $Y = \{\chi_A \text{ with } A \subset \Omega, A \text{ measurable}\}$. Check that Y is complete for the metric d and apply the Baire category theorem to the sequence

$$Y_n = \left\{\chi_A \in Y; \left|\int_A f_k\right| \le \varepsilon \quad \forall k \ge n\right\},$$

where $\varepsilon > 0$ is fixed.]

3. Deduce that if (f_n) is a sequence in $L^1(\Omega)$ such that $f_n \rightharpoonup f$ weakly $\sigma(L^1, L^\infty)$, then (f_n) is equi-integrable.

4. Prove that (i) $\Rightarrow$ (ii).

   [**Hint**: Argue by contradiction and apply the theorem of Eberlein–Šmulian; see Problem 10.]

5. Take up again question 1 (resp. question 2) assuming only that $\int_A f_n$ converges to a finite limit $\ell(A)$ for every $A \subset \Omega$ with A measurable and $|A| < \infty$ (resp. $|A| \le \infty$).

- **B** -

The implication (ii) $\Rightarrow$ (i).

1. Let E be a Banach space and let $\mathcal{F} \subset E$. Assume that

$$\forall \varepsilon > 0 \; \exists \mathcal{F}_\varepsilon \subset E, \; \mathcal{F}_\varepsilon \text{ weakly } (\sigma(E, E^\star)) \text{ compact such that } \mathcal{F} \subset \mathcal{F}_\varepsilon + \varepsilon B_E.$$

Prove that $\mathcal{F}$ is contained in a weakly compact subset of E.

[**Hint**: Consider $\mathcal{G} = \overline{\mathcal{F}}^{\sigma(E^{**},E^*)}$.]

2. Deduce that (ii) $\Rightarrow$ (i).

[**Hint**: Consider the family $(\chi_\omega T_n f)_{f\in\mathcal{F}}$ with $|\omega| < \infty$ and T_n is the truncation as in the proof of Theorem 4.12.]

- C -

Some applications.

1. Let (f_n) be a sequence in $L^1(\Omega)$ such that $f_n \rightharpoonup f$ weakly $\sigma(L^1, L^\infty)$ and $f_n \to f$ a.e. Prove that $\|f_n - f\|_1 \to 0$.

   [**Hint**: Apply Exercise 4.14.]

2. Let $u_1, u_2 \in L^1(\Omega)$ with $u_1 \le u_2$ a.e. Prove that the set $K = \{f \in L^1(\Omega);\ u_1 \le f \le u_2$ a.e.$\}$ is compact in the weak topology $\sigma(L^1, L^\infty)$.

3. Let (f_n) be an equi-integrable sequence in $L^1(\Omega)$. Prove that there exists a subsequence (f_{n_k}) such that $f_{n_k} \rightharpoonup f$ weakly $\sigma(L^1, L^\infty)$.

4. Let (f_n) be a bounded[1] sequence in $L^1(\Omega)$ such that $\int_A f_n$ converges to a finite limit, $\ell(A)$, for every measurable set $A \subset \Omega$. Prove that there exists some $f \in L^1(\Omega)$ such that $f_n \rightharpoonup f$ weakly $\sigma(L^1, L^\infty)$.

5. Let $g : \mathbb{R} \to \mathbb{R}$ be a continuous increasing function such that

$$|g(t)| \le C \quad \forall t \in \mathbb{R}.$$

Set $G(t) = \int_0^t g(s)ds$. Let (f_n) be a sequence in $L^1(\Omega)$ such that $f_n \rightharpoonup f$ weakly $\sigma(L^1, L^\infty)$ and $\limsup \int G(f_n) \le \int G(f)$. Prove that $\|f_n - f\|_1 \to 0$.

[**Hint**: Look at Problem 18.]

- D -

In this part, we assume that $|\Omega| < \infty$. Let $\mathcal{F} \subset L^1(\Omega)$.

1. Let $G : [0, +\infty) \to [0, +\infty)$ be a continuous function such that $\lim_{t\to+\infty} G(t)/t = +\infty$. Assume that there exists a constant C such that

$$\int G(|f|) \le C \quad \forall f \in \mathcal{F}.$$

Prove that $\mathcal{F}$ is equi-integrable.

---

[1] In fact, it is not necessary to assume that (f_n) is bounded, but then the proof is more complicated; see, e.g., R. Edwards [1] p. 276–277.

2. Conversely, assume that $\mathcal{F}$ is equi-integrable. Prove that there exists a convex increasing function $G : [0, +\infty) \to [0, +\infty)$ such that $\lim_{t \to +\infty} G(t)/t = +\infty$ and $\int G(|f|) \leq C \ \forall f \in \mathcal{F}$, for some constant C.

[**Hint**: Use the distribution function; see Problem 21.]

PROBLEM 24 (1, 3, 4)

Radon measures

Let K be a compact metric space, with distance d, and let $E = C(K)$ equipped with its usual norm

$$\|f\| = \max_{x \in K} |f(x)|.$$

The dual space $E^\star$, denoted by $\mathcal{M}(K)$, is called the space of *Radon measures* on K. The space $\mathcal{M}(K)$ is equipped with the dual norm, denoted by $\| \ \|_{\mathcal{M}}$ or simply $\| \ \|$. The purpose of this problem is to present some properties of $\mathcal{M}(K)$.

- A -

We prove here that $C(K)$ is separable. Given $\delta > 0$, let $\bigcup_{j \in J} B(a_j, \delta/2)$ be a finite covering of K. Set

$$q_j(x) = \max\{0, \delta - d(x, a_j)\}, \quad j \in J, \ x \in K,$$

and

$$q(x) = \sum_{j \in J} q_j(x).$$

1. Check that the functions $(q_j)_{j \in J}$ and q are continuous on K. Show that

$$q(x) \geq \delta/2 \quad \forall x \in K.$$

2. Set

$$\theta_j(x) = \frac{q_j(x)}{q(x)}, \quad j \in J, \ x \in K.$$

Show that the functions $(\theta_j)_{j \in J}$ are continuous on K,

$$[\theta_j(x) \neq 0] \iff [d(x, a_j) < \delta],$$

and

$$\sum_{j \in J} \theta_j(x) = 1 \quad \forall x \in K.$$

The collection of functions $(\theta_j)_{j \in J}$ is called a *partitition of unity* (subordinate to the open covering $\bigcup_{j \in J} B(a_j, 2\delta)$, because $\operatorname{supp} \theta_j \subset B(a_j, 2\delta)$).

3. Given $f \in C(K)$, set

$$\tilde{f}(x) = \sum_{j \in J} f(a_j)\theta_j(x).$$

Prove that

$$\|f - \tilde{f}\| \le \sup_{\substack{x,y \in K \\ d(x,y) < \delta}} |f(x) - f(y)|.$$

4. Choosing $\delta = 1/n$, the above construction yields a finite set J, now denoted by J_n, a finite collection of points $(a_j)_{j \in J_n}$, and a finite collection of functions $(\theta_j)_{j \in J_n}$. Show that the vector space spanned by the functions (θ_j), $j \in J_n$, $n = 1, 2, 3 \ldots$, is dense in $C(K)$.
5. Deduce that $C(K)$ is separable.

- B -

In this part we assume that $K = \overline{\Omega}$, where Ω is a bounded open set in $\mathbb{R}^N$. It is convenient to identify $L^1(\Omega)$ with a subspace of $\mathcal{M}(\overline{\Omega})$ through the embedding $T : L^1(\Omega) \to \mathcal{M}(\overline{\Omega})$ defined by

$$\langle Tu, f \rangle = \int_\Omega uf \quad \forall u \in L^1(\Omega), \quad \forall f \in C(\overline{\Omega}).$$

1. Check that $\|Tu\|_{\mathcal{M}} = \|u\|_{L^1} \ \forall u \in L^1(\Omega)$.
   [**Hint**: Use Exercise 4.26.]
2. Let (v_n) be a bounded sequence in $L^1(\Omega)$. Show that there exist a subsequence (v_{n_k}) and some $\mu \in \mathcal{M}(\overline{\Omega})$ such that $v_{n_k} \overset{\star}{\rightharpoonup} \mu$ in $\mathcal{M}(\overline{\Omega})$ and

$$\|\mu\|_{\mathcal{M}} \le \liminf_{k \to \infty} \|v_{n_k}\|_{L^1}.$$

[**Hint**: Use Corollary 3.30.]

The aim is now to prove that given any $\mu_0 \in \mathcal{M}(\overline{\Omega})$ there exists a sequence (u_n) in $C_c^\infty(\Omega)$ such that

(1) $$\int_\Omega u_n f \to \langle \mu_0, f \rangle \quad \forall f \in C(\overline{\Omega})$$

and

(2) $$\|u_n\|_{L^1} = \|\mu_0\|_{\mathcal{M}} \quad \forall n.$$

Without loss of generality we may assume that $\|\mu_0\|_{\mathcal{M}} = 1$ (why?).

Set
$$A = \{u \in C_c^\infty(\Omega); \ \|u\|_{L^1} \le 1\}.$$

3. Prove that $\mu_0 \in \overline{A}^{\sigma(E^\star, E)}$.
   [**Hint**: Apply Hahn–Banach in $E^\star$ for the weak* topology $\sigma(E^\star, E)$; see Problem 9. Then use Corollary 4.23.]
4. Deduce that there exists a sequence (v_n) in A such that $v_n \overset{\star}{\rightharpoonup} \mu_0$ in $\sigma(E^\star, E)$. Check that $\lim_{n \to \infty} \|v_n\|_{L^1} = 1$.

5. Conclude that the sequence $u_n = v_n/\|v_n\|_{L^1}$ satifies (1) and (2).

We say that $\mu \geq 0$ if

$$\langle \mu, f \rangle \geq 0 \quad \forall f \in C(\overline{\Omega}), \, f \geq 0 \text{ on } \overline{\Omega}.$$

6. Check that if $\mu \geq 0$, then $\langle \mu, 1 \rangle = \|\mu\|$, where 1 denotes the function $f \equiv 1$.
7. Assume $\mu_0 \in \mathcal{M}(\overline{\Omega})$, with $\mu_0 \geq 0$ and $\|\mu_0\| = 1$ (such measures are called *probability measures*). Construct a sequence (u_n) in $C_c^\infty(\Omega)$ satisfying (1), (2), and, moreover,

$$(3) \qquad\qquad u_n(x) \geq 0 \quad \forall n, \, \forall x \in \overline{\Omega}.$$

8. Compute $\|u + \delta_a\|_{\mathcal{M}}$, where $u \in L^1$, and δ_a, with $a \in \overline{\Omega}$, is defined below.

- C -

We now return to the general setting and denote by δ_a the *Dirac mass* at a point $a \in K$, i.e., the measure defined by

$$\langle \delta_a, f \rangle = f(a) \quad \forall f \in C(K).$$

Set

$$D = \left\{ \mu = \sum_{j \in J} \alpha_j \delta_{a_j}; \, J \text{ is finite, } \alpha_j \in \mathbb{R}, \text{ and the points } a_j\text{'s are all distinct} \right\}.$$

1. Show that if $\mu \in D$ then

$$\|\mu\| = \sum_{j \in J} |\alpha_j|$$

and

$$[\mu \geq 0] \Leftrightarrow [\alpha_j \geq 0 \quad \forall j].$$

Set

$$D_1 = \{\mu \in D; \, \|\mu\| \leq 1\}.$$

2. Show that any measure $\mu_0 \in \mathcal{M}(K)$ with $\|\mu_0\| \leq 1$ belongs to $\overline{D_1}^{\sigma(E^\star, E)}$.
   [**Hint**: Use the same technique as in part B.]
3. Deduce that given any measure $\mu_0 \in \mathcal{M}(\overline{\Omega})$ there exists a sequence (v_n) in D such that $v_n \overset{\star}{\rightharpoonup} \mu_0$ and $\|v_n\| = \|\mu_0\| \, \forall n$.
4. Let μ_0 be a probability measure. Prove that there exists a sequence (v_n) of probability measures in D such that $v_n \overset{\star}{\rightharpoonup} \mu_0$.

Remark. An alternative approach to question 4 is to show that the Dirac masses are the extremal points of the convex set of probability measures; then apply Krein–Milman (see Problem 1) in the weak* topology; for more details see, e.g., R. Edwards [1].

- D -

The goal of this part is to show that every $\mu \in \mathcal{M}(K)$ admits a unique decomposition $\mu = \mu_1 - \mu_2$ with $\mu_1, \mu_2 \in \mathcal{M}(K)$, $\mu_1, \mu_2 \geq 0$, and $\|\mu_1\| + \|\mu_2\| = \|\mu\|$. (The measures μ_1 and μ_2 are often denoted by μ^+ and μ^-.)

Given $f \in C(K)$ with $f \geq 0$, set

$$L(f) = \sup\{\langle \mu, g \rangle; g \in C(K) \text{ and } 0 \leq g \leq f \text{ on } K\}.$$

1. Check that $0 \leq L(f) \leq \|\mu\| \|f\|$, $L(\lambda f) = \lambda L(f) \ \forall \lambda \geq 0$, and

$$L(f_1 + f_2) = L(f_1) + L(f_2) \quad \forall f_1, f_2 \in C(K) \text{ with } f_1 \geq 0 \text{ and } f_2 \geq 0.$$

Given any $f \in C(K)$, set

$$\mu_1(f) = L(f^+) - L(f^-), \quad \text{where } f^+ = \max\{f, 0\} \text{ and } f^- = \max\{-f, 0\}.$$

2. Show that the mapping $f \mapsto \mu_1(f)$ is linear on $C(K)$ and that $|\mu_1(f)| \leq \|\mu\| \|f\| \ \forall f \in C(K)$, so that $\mu_1 \in \mathcal{M}(K)$. Check that $\mu_1 \geq 0$.
3. Set $\mu_2 = \mu_1 - \mu$ and check that $\mu_2 \geq 0$. Show that $\|\mu\| = \|\mu_1\| + \|\mu_2\|$.
4. Let $\nu \in \mathcal{M}(K)$ be such that $\nu \geq 0$ and $\nu \geq \mu$ (i.e., $\nu - \mu \geq 0$). Show that $\nu \geq \mu_1$. Similarly if $\nu \in \mathcal{M}(K)$ and $\nu \geq -\mu$, show that $\nu \geq \mu_2$. Deduce the uniqueness of the decomposition.

- E -

Show that all the above results (except question B6) remain valid when the space $E = C(\overline{\Omega})$ is replaced by the subspace

$$E_0 = \{f \in C(\overline{\Omega}); f = 0 \text{ on the boundary of } \overline{\Omega}\}.$$

The dual of E_0 is often denoted by $\mathcal{M}(\Omega)$ (as opposed to $\mathcal{M}(\overline{\Omega})$).

- F -

Dunford–Pettis revisited

Let (f_n) be a sequence in $L^1(\Omega)$. Recall that (f_n) is said to be equi-integrable if it satisfies the property

(4)
$$\begin{cases} \forall \varepsilon > 0 \ \exists \delta > 0 \text{ such that } \int_A |f_n| < \varepsilon \quad \forall n, \\ \text{and } \forall A \subset \Omega \text{ with } A \text{ measurable and } |A| < \delta. \end{cases}$$

The goal is to prove that every equi-integrable sequence (f_n) admits a subsequence (f_{n_j}) such that $f_{n_j} \rightharpoonup f$ weakly $\sigma(L^1, L^\infty)$, for some function $f \in L^1(\Omega)$.

1. Show that (f_n) is bounded in $L^1(\Omega)$.
2. Check that

$$\int_{\Omega} |f_n - T_k f_n| \leq \int_{\substack{\Omega \\ [|f_n| > k]}} |f_n| \quad \forall n, \forall k,$$

where T_k denotes the truncation operation.

3. Deduce that $\forall \varepsilon > 0 \; \exists k > 0$ such that

$$\int_{\Omega} |f_n - T_k f_n| \leq \varepsilon \quad \forall n.$$

[**Hint**: Use (4); see also Exercise 4.36.]

Passing to a subsequence, still denoted by (f_n), we may assume that $f_n \overset{\star}{\rightharpoonup} \mu$ weak* in $\mathcal{M}(\overline{\Omega})$, for some measure $\mu \in \mathcal{M}(\overline{\Omega})$.

4. Prove that $\forall \varepsilon > 0 \; \exists g = g_\varepsilon \in L^\infty(\Omega)$ such that

$$\|\mu - g_\varepsilon\|_{\mathcal{M}} \leq \varepsilon.$$

[**Hint**: For fixed k, a subsequence of $(T_k f_n)$ converges to some limit g weak* in $\sigma(L^\infty, L^1)$.]

5. Deduce that $\mu \in L^1(\Omega)$.

[**Hint**: Use a Cauchy sequence argument in $L^1(\Omega)$.]

6. Prove that $f_n \rightharpoonup \mu$ weakly $\sigma(L^1, L^\infty)$.

[**Hint**: Given $u \in L^\infty(\Omega)$, consider a sequence (u_m) in $C_c^\infty(\Omega)$ such that $u_m \to u$ a.e. on Ω and $\|u_m\|_\infty \leq \|u\|_\infty \; \forall m$ (see Exercise 4.25); then use Egorov's theorem (see Theorem 4.29 and Exercise 4.14).]

PROBLEM 25 (1, 5)

Let H be a Hilbert space and let $C \subset H$ be a convex cone with vertex at 0, that is, $0 \in C$ and $\lambda u + \mu v \in C \; \forall \lambda, \mu > 0, \forall u, v \in C$. We assume that C is nonempty, open, and that $C \neq H$.

Check that $0 \notin C$ and that $0 \in \overline{C}$. Consider the set

$$\Sigma = \{u \in H; \; (u, v) \leq 0 \quad \forall v \in C\}.$$

1. Check that Σ is a convex cone with vertex at 0, Σ is closed, and $0 \in \Sigma$. Prove that $C = \{v \in H; \; (u, v) < 0 \; \forall u \in \Sigma \setminus \{0\}\}$ and deduce that Σ is not reduced to $\{0\}$.

[**Hint**: Use Hahn–Banach.]

2. Let $\omega \in C$ be fixed and consider the set

$$K = \{u \in \Sigma; \; (u, \omega) = -1\}.$$

Prove that K is a nonempty, bounded, closed, convex set such that $0 \notin K$ and

$$\Sigma \setminus \{0\} = \bigcup_{\lambda > 0} \lambda K.$$

Draw a figure.

[**Hint**: Consider a ball centered at ω of radius $\rho > 0$ contained in C.]

3. Let $a = P_K 0$. Prove that $a \in (-C) \cap \Sigma$.

4. Prove directly, by a simple argument, that $(-C) \cap \Sigma \neq \emptyset$.

5. Let $D \subset H$ be a nonempty, open, convex set and let $x_0 \notin D$. Prove that there exists some $w_0 \in D$ such that

$$(w_0 - x_0, w - x_0) > 0 \quad \forall w \in D.$$

Give a geometric interpretation.

[**Hint**: Consider the set $C = \cup_{\mu > 0} \mu(D - x_0)$.]

PROBLEM 26 (1, 5)

The Prox map in the sense of Moreau

Let H be a Hilbert space and let $\varphi : H \to (-\infty, +\infty]$ be a convex l.s.c. function such that $\varphi \not\equiv +\infty$.

1. Prove that for every $f \in H$, there exists some $u \in D(\varphi)$ such that

(P) $\qquad \dfrac{1}{2}|f - u|^2 + \varphi(u) = \inf_{v \in H} \left\{ \dfrac{1}{2}|f - v|^2 + \varphi(v) \right\} \equiv I.$

[**Hint**: Check first that $I > -\infty$. Then use either a Cauchy sequence argument or the fact that H is reflexive.]

2. Check that u satisfies (P) iff

(Q) $\quad u \in D(\varphi) \quad$ and $\quad (u, v - u) + \varphi(v) - \varphi(u) \geq (f, v - u) \quad \forall v \in D(\varphi)$.

3. Prove that if u and $\bar{u}$ are solutions of (P) corresponding to f and $\bar{f}$, then $|u - \bar{u}| \leq |f - \bar{f}|$. Deduce the uniqueness of the solution of (P).

4. Investigate the special case in which $\varphi = I_K$ is the indicator function of a closed convex set K.

5. Let φ^* be the conjugate function of φ and consider the problem

(P*) $\qquad \dfrac{1}{2}|f - u^\star|^2 + \varphi^\star(u^\star) = \inf_{v \in H} \left\{ \dfrac{1}{2}|f - v|^2 + \varphi^\star(v) \right\} = I^\star.$

Prove that the solutions u of (P) and $u^\star$ of (P*) satisfy

$$u + u^* = f \quad \text{and} \quad I + I^* = \frac{1}{2}|f|^2.$$

6. Given $f \in H$ and $\lambda > 0$ let u_λ denote the solution of the problem

$$(P_\lambda) \qquad \frac{1}{2}|f - u_\lambda|^2 + \lambda\varphi(u_\lambda) = \inf_{v \in H} \left\{ \frac{1}{2}|f - v|^2 + \lambda\varphi(v) \right\}.$$

Prove that $\lim_{\lambda \to 0} u_\lambda = P_{\overline{D(\varphi)}} f = $ the projection of f on $\overline{D(\varphi)}$.

[**Hint**: Either start with weak convergence or use Exercise 5.3.]

7. Let $\dot{K} = \{v \in D(\varphi); \varphi(v) = \inf_H \varphi\}$ and assume $K \neq \emptyset$.
Check that K is a closed convex set and prove that $\lim_{\lambda \to +\infty} u_\lambda = P_K f$.
What happens to (u_λ) as $\lambda \to +\infty$ when $K = \emptyset$?

8. Prove that $\lim_{\lambda \to +\infty} \frac{1}{\lambda} u_\lambda = -P_{\overline{D(\varphi^*)}} 0$.

[**Hint**: Start with the case where $f = 0$ and apply questions 5 and 6.]

PROBLEM 27 (5)

Alternate projections

Let H be a Hilbert space and let $K \subset H$ be a nonempty closed convex set. Check that

$$|P_K u - P_K v|^2 \leq (P_K u - P_K v, u - v) \leq |u - v|^2 \quad \forall u, v \in H.$$

Let $K_1 \subset H$ and $K_2 \subset H$ be two nonempty closed convex sets. Set $P_1 = P_{K_1}$ and $P_2 = P_{K_2}$. Given $u \in H$, define by induction the sequence (u_n) as follows:

$$u_0 = u, \quad u_1 = P_1 u_0, \quad u_2 = P_2 u_1, \quad \ldots, u_{2n-1} = P_1 u_{2n-2}, \quad u_{2n} = P_2 u_{2n-1}, \quad \ldots$$

- A -

The purpose of this part is to prove that the sequence $(u_{2n} - u_{2n-1})$ converges to $P_K 0$, where $K = \overline{K_2 - K_1}$ (note that K is convex, why?).

1. Given $v \in H$ consider the sequence (v_n) defined by the same iteration as above starting with $v_0 = v$. Check that

$$|u_{2n} - v_{2n}|^2 \leq (u_{2n} - v_{2n}, u_{2n-1} - v_{2n-1}) \leq |u_{2n-1} - v_{2n-1}|^2$$

and that

$$|u_{2n+1} - v_{2n+1}|^2 \leq (u_{2n+1} - v_{2n+1}, u_{2n} - v_{2n}) \leq |u_{2n} - v_{2n}|^2.$$

2. Deduce that the sequence $(|u_n - v_n|)$ is nonincreasing and thus converges to a limit, denoted by ℓ.
Prove that $\lim_{n \to \infty} |(u_{2n} - v_{2n}) - (u_{2n-1} - v_{2n-1})|^2 = 0$.

3. Check that the sequence $(|u_{2n} - u_{2n-1}|)$ is nonincreasing.

Set $d = \text{dist}(K_1, K_2) = \inf\{|a_1 - a_2|; a_1 \in K_1 \text{ and } a_2 \in K_2\}$.
We claim that $\lim_{n\to\infty} |u_{2n} - u_{2n-1}| = d$.

4. Given $\varepsilon > 0$, choose $v \in K_2$ such that $\text{dist}(v, K_1) \leq d + \varepsilon$.

Prove that $|v_{2n} - v_{2n-1}| \leq d + \varepsilon \; \forall n$.

5. Deduce that $\lim_{n\to\infty} |u_{2n} - u_{2n-1}| = d$.

Set $z = P_K 0$.

6. Check that $|z| = d$ and that $|z|^2 \leq (z, w) \; \forall w \in K_2 - K_1$.

7. Prove that the sequence $(u_{2n} - u_{2n-1})$ converges to z.

[**Hint**: Estimate $|z - (u_{2n} - u_{2n-1})|^2$ using the above results.]

8. Give a geometric interpretation.

- B -

Throughout the rest of this problem we assume that $z = P_K 0 \in K_2 - K_1$. (This assumption holds, for example, if one of the sets K_1 or K_2 is bounded, why?)

We claim that there exist $a_1 \in K_1$ and $a_2 \in K_2$ with $a_2 - a_1 = z$ such that $u_{2n} \rightharpoonup a_2$ and $u_{2n-1} \rightharpoonup a_1$ weakly. Note that a_1 and a_2 may depend on the choice of $u_0 = u$. Draw a figure.

1. Consider the Hilbert space $\mathcal{H} = H \times H$ equipped with its natural scalar product. Set $\mathcal{K} = \{[b_1, b_2] \in \mathcal{H} \; ; b_1 \in K_1, b_2 \in K_2 \text{ and } b_2 - b_1 = z\}$.
   Check that $\mathcal{K}$ is a nonempty closed convex set.

2. Let $b = [b_1, b_2] \in \mathcal{K}$. Determine the sequence (v_n) corresponding to $v_0 = b_1$.
   Deduce that the sequences $(|u_{2n-1} - b_1|)$ and $(|u_{2n} - b_2|)$ are nonincreasing.

3. Set $x_n = [u_{2n-1}, u_{2n}]$ and prove that the sequence (x_n) satisfies the following property:

   (P) $\quad \begin{cases} \text{For every subsequence } (x_{n_k}) \text{ that converges weakly to some} \\ \text{element } \bar{x} \in \mathcal{H}, \text{ then } \bar{x} \in \mathcal{K}. \end{cases}$

4. Apply Opial's lemma (see Exercise 5.25, question 3) and conclude.

PROBLEM 28 (5)

Projections and orthogonal projections

Let H be a Hilbert space. An operator $P \in \mathcal{L}(H)$ such that $P^2 = P$ is called a *projection*. Check that a projection satisfies the following properties:

(a) $I - P$ is a projection,

(b) $N(I - P) = R(P)$ and $N(P) = R(I - P)$,
(c) $N(P) \cap N(I - P) = \{0\}$,
(d) $H = N(P) + N(I - P)$.

- A -

An operator $P \in \mathcal{L}(H)$ is called an *orthogonal projection* if there exists a closed linear subspace M such that $P = P_M$ (where P_M is defined in Corollary 5.4). Check that every orthogonal projection is a projection.

1. Given a projection P, prove that the following properties are equivalent:

   (a) P is an orthogonal projection,
   (b) $P^\star = P$,
   (c) $\|P\| \leq 1$,
   (d) $N(P) \perp N(I - P)$,

   where the notation $X \perp Y$ means that $(x, y) = 0 \; \forall x \in X, \forall y \in Y$.

2. Let $T \in \mathcal{L}(H)$ be an operator such that

$$T^\star = T \quad \text{and} \quad T^2 = I.$$

   Prove that $P = \frac{1}{2}(I - T)$ is an orthogonal projection. Prove the converse.

   Assuming, in addition, that $(Tu, u) \geq 0 \; \forall u \in H$, prove that $T = I$.

- B -

Throughout this part, M and N denote two closed linear subspaces of H. Set $P = P_M$ and $Q = P_N$.

1. Prove that the following properties are equivalent:

   (a) $PQ = QP$,
   (b) PQ is a projection,
   (c) QP is a projection.

   In this case, check that

   (i) PQ is the orthogonal projection onto $M \cap N$,
   (ii) $(P + Q - PQ)$ is the orthogonal projection onto $\overline{M + N}$.

2. Prove that the following properties are equivalent:

   (a) $M \perp N$,
   (b) $PQ = 0$,
   (c) $QP = 0$,
   (d) $|Pu|^2 + |Qu|^2 \leq |u|^2 \; \forall u \in H$,
   (e) $|Pu| \leq |u - Qu| \; \forall u \in H$,
   (f) $|Qu| \leq |u - Pu| \; \forall u \in H$,

(g) $P + Q$ is a projection.

In this case, check that $(P + Q)$ is the orthogonal projection onto $M + N$ (note that $M + N$ is closed; why?).

3. Prove that the following properties are equivalent:

   (a) $M \subset N$,
   (b) $PQ = P$,
   (c) $QP = P$,
   (d) $|Pu| \leq |Qu| \; \forall u \in H$,
   (e) $Q - P$ is a projection.

In this case, check that $Q - P$ is the orthogonal projection onto $M^{\perp} \cap N$.

PROBLEM 29 (5)

Iterates of nonlinear contractions.
The ergodic theorems of Opial and Baillon

Let H be a Hilbert space and let $T : H \to H$ be a nonlinear contraction, that is,

$$|Tu - Tv| \leq |u - v| \quad \forall u, v \in H.$$

We assume that the set

$$K = \{u \in H; \; Tu = u\}$$

of fixed points is nonempty. Check that K is closed and convex. Given $f \in H$ set

$$\sigma_n = \frac{1}{n}(f + Tf + T^2 f + \cdots + T^{n-1} f)$$

and

$$\mu_n = \left(\frac{I + T}{2}\right)^n f.$$

The goal of this problem is to prove the following:

(A) Each of the sequences (σ_n) and (μ_n) converges *weakly* to a fixed point of T.
(B) If, in addition, T is *odd*, that is, $T(-v) = -Tv \; \forall v \in H$, then (σ_n) and (μ_n) converge *strongly*.

It is advisable to solve Exercises 5.22 and 5.25 before starting this problem. In the special case that T is linear, see also Exercise 5.21.

- A -

Set

$$u_n = T^n f.$$

1. Check that for every $v \in K$, the sequence $(|u_n - v|)$ is nonincreasing. Deduce that the sequences (σ_n) and $(T\sigma_n)$ are bounded.

2. Prove that

$$|\sigma_n - T\sigma_n| \le \frac{1}{\sqrt{n}}|f - T\sigma_n| \quad \forall n \ge 1.$$

[**Hint**: Note that $|T\sigma_n - Tu_i|^2 \le |\sigma_n - u_i|^2$ and add these inequalities for $0 \le i \le n-1$.]

3. Deduce that the sequence (σ_n) satisfies property (P) of Exercise 5.25. Conclude that $\sigma_n \rightharpoonup \sigma$ weakly, with $\sigma \in K$.

Set

$$S = \frac{1}{2}(I + T).$$

4. Prove that

$$|(u - Su) - (v - Sv)|^2 + |Su - Sv|^2 \le |u - v|^2 \quad \forall u, v \in H.$$

5. Deduce that for every $v \in K$,

$$\sum_{n=0}^{\infty} |\mu_n - \mu_{n+1}|^2 \le |f - v|^2$$

and consequently

$$|\mu_n - S\mu_n| \le \frac{1}{\sqrt{n+1}}|f - v| \quad \forall n.$$

6. Conclude that $\mu_n \rightharpoonup \mu$ weakly, with $\mu \in K$.

- B -

Throughout the rest of this problem we assume that T is *odd*, that is,

$$T(-v) = -Tv \quad \forall v \in H.$$

1. Prove that for every integer p,

$$2|(u, v) - (T^p u, T^p v)| \le |u|^2 + |v|^2 - |T^p u|^2 - |T^p v|^2 \quad \forall u, v \in H.$$

[**Hint**: Start with the inequality $|T^p u - T^p v|^2 \le |u - v|^2 \ \forall u, v \in H.$]

2. Deduce that for every fixed integer $i \ge 0$,

$$\ell(i) = \lim_{n \to \infty} (u_n, u_{n+i}) \quad \text{exists.}$$

Prove that this convergence holds uniformly in i, that is,

(1) $|(u_n, u_{n+i}) - \ell(i)| \le \varepsilon_n \quad \forall i \ \text{ and } \ \forall n, \quad \text{with } \lim_{n \to \infty} \varepsilon_n = 0.$

3. Similarly, prove that for every fixed integer $i \geq 0$,

$$m(i) = \lim_{n \to \infty} (\mu_n, \mu_{n+i}) \quad \text{exists.}$$

Prove that $m(0) = m(1) = m(2) = \cdots$.

[**Hint**: Use the result of question A5.]

4. Deduce that $\mu_n \to \mu$ strongly.

   We now claim that $\sigma_n \to \sigma$ strongly.

5. Set

$$X_p = \frac{1}{p} \sum_{i=0}^{p-1} \ell(i).$$

Prove that

$$|(u_n, \sigma_{n+p}) - X_p| \leq \varepsilon_n + \frac{2n}{p}|f|^2 \quad \forall n, \quad \forall p.$$

[**Hint**: Use (1).]

6. Deduce that

   (i) $X = \lim_{p \to \infty} X_p$ exists,
   (ii) $|(u_n, \sigma) - X| \leq \varepsilon_n \ \forall n$,
   (iii) $|\sigma|^2 = X$.

7. Prove that

$$|\sigma_n|^2 \leq \frac{2}{n^2} \sum_{i=0}^{n-1} (n-i)\ell(i) + \frac{2}{n} \sum_{i=0}^{n-1} \varepsilon_i.$$

8. Deduce that $\lim \sup_{n \to \infty} |\sigma_n|^2 \leq X$ and conclude.

PROBLEM 30 (3, 5)

Variants of Stampacchia's theorem. The min–max theorem of von Neumann
Let H be a Hilbert space.

- A -

Let $a(u, v) : H \times H \to \mathbb{R}$ be a continuous bilinear form such that

$$a(v, v) \geq 0 \ \forall v \in H.$$

Let $K \subset H$ be a nonempty closed convex set. Let $f \in H$. Assume that there exists some $v_0 \in K$ such that the set

$$\{u \in K; \ a(u, v_0 - u) \geq (f, \ v_0 - u)\}$$

is bounded.

1. Prove that there exists some $u \in K$ such that

$$a(u, v - u) \geq (f, v - u) \quad \forall v \in K.$$

[**Hint**: Set $f_\varepsilon = f + \varepsilon v_0$ and consider the bilinear form $a_\varepsilon(u, v) = a(u, v) + \varepsilon(u, v)$, $\varepsilon > 0$. Then, pass to the limit as $\varepsilon \to 0$ using Exercise 5.14.]

2. Recover Stampacchia's theorem.

3. Give a geometric interpretation in the case that K is bounded and $a(u, v) = 0$ $\forall u, v \in H$.

- B -

Let $b(u, v) : H \times H \to \mathbb{R}$ be a bilinear form that is continuous and *coercive*. Let $\varphi : H \to (-\infty, +\infty]$ be a convex l.s.c. function such that $\varphi \not\equiv +\infty$.

1. Prove that there exists a unique $u \in D(\varphi)$ such that

$$b(u, v - u) + \varphi(v) - \varphi(u) \geq 0 \quad \forall v \in D(\varphi).$$

[**Hint**: Apply the result of question A1 in the space $H \times \mathbb{R}$ with $K = \text{epi } \varphi$, $f = [0, -1]$, $a(U, V) = b(u, v)$ with $U = [u, \lambda]$ and $V = [v, \mu]$. Note that a is *not coercive*.]

2. Recover Stampacchia's theorem.

- C -

Let H_1 and H_2 be two Hilbert spaces and let $A \subset H_1$, $B \subset H_2$ be two nonempty, bounded, closed convex sets.

1. Let $F(\lambda, \mu) : H_1 \times H_2 \to \mathbb{R}$ be a continuous bilinear form. Prove that there exist $\overline{\lambda} \in A$ and $\overline{\mu} \in B$ such that

(1) $\qquad F(\overline{\lambda}, \mu) \leq F(\overline{\lambda}, \overline{\mu}) \leq F(\lambda, \overline{\mu}) \quad \forall \lambda \in A, \quad \forall \mu \in B.$

[**Hint**: Apply question A1 with $H = H_1 \times H_2$, $K = A \times B$, and $a(u, v) = F(\lambda, \overline{\mu}) - F(\overline{\lambda}, \mu)$, where $u = [\overline{\lambda}, \overline{\mu}]$, $v = [\lambda, \mu]$.]

2. Deduce that

(2) $\qquad \min_{\lambda \in A} \max_{\mu \in B} F(\lambda, \mu) = \max_{\mu \in B} \min_{\lambda \in A} F(\lambda, \mu).$

Note that all min and max are achieved (why?).

[**Hint**: Check that without any further assumptions, $\max \min \leq \min \max$; use (1) to prove the reverse inequality.]

3. Prove that (2) implies the existence of some $\overline{\lambda} \in A$ and $\overline{\mu} \in B$ satisfying (1).

- D -

Let E and F be two reflexive Banach spaces; let $A \subset E$ and $B \subset F$ be two nonempty, bounded, closed convex sets. Let $K : E \times F \to \mathbb{R}$ be a function satisfying the following assumptions:

(a) For every fixed $v \in B$ the function $u \mapsto K(u, v)$ is convex and l.s.c. .
(b) For every fixed $u \in A$ the function $v \mapsto K(u, v)$ is concave and u.s.c., i.e., the function $v \mapsto -K(u, v)$ is convex and l.s.c.

Our goal is to prove that

$$\min_{u \in A} \max_{v \in B} K(u, v) = \max_{v \in B} \min_{u \in A} K(u, v).$$

We shall argue by contradiction and assume that there exists a constant γ such that

$$\max_{v \in B} \min_{u \in A} K(u, v) < \gamma < \min_{u \in A} \max_{v \in B} K(u, v).$$

1. For every $u \in A$, set

$$B_u = \{v \in B; \ K(u, v) \geq \gamma\}$$

and for every $v \in B$, set

$$A_v = \{u \in A; \ K(u, v) \leq \gamma\}.$$

Check that $\cap_{u \in A} B_u = \emptyset$ and $\cap_{v \in B} A_v = \emptyset$.

2. Choose $u_1, u_2, \ldots, u_n \in A$ and $v_1, v_2, \ldots, v_m \in B$ such that $\cap_{i=1}^n B_{u_i} = \emptyset$ and $\cap_{j=1}^m A_{v_j} = \emptyset$ (justify). Apply the result of C1 with $H_1 = \mathbb{R}^n$, $H_2 = \mathbb{R}^m$,

$$A' = \left\{\lambda = (\lambda_1, \lambda_2, \ldots, \lambda_n); \ \lambda_i \geq 0 \quad \forall i \text{ and } \sum_{i=1}^n \lambda_i = 1\right\},$$

$$B' = \left\{\mu = (\mu_1, \mu_2, \ldots, \mu_m); \ \mu_j \geq 0 \quad \forall j \text{ and } \sum_{j=1}^m \mu_j = 1\right\},$$

and $F(\lambda, \mu) = \sum_{i,j} \lambda_i \mu_j K(u_i, v_j)$. Set $\overline{u} = \sum_i \overline{\lambda}_i u_i$ and $\overline{v} = \sum_j \overline{\mu}_j v_j$. Prove that

$$K(\overline{u}, v_\ell) \leq K(u_k, \overline{v}) \quad \forall k = 1, 2, \ldots, n, \quad \forall \ell = 1, 2, \ldots, m.$$

3. Check that on the other hand,

$$\min_k K(u_k, \overline{v}) < \gamma \quad \text{and} \quad \max_\ell K(\overline{u}, v_\ell) > \gamma.$$

[**Hint**: Argue by contradiction.]

4. Conclude.

PROBLEM 31 (3, 5)

Monotone operators. The theorem of Minty–Browder

Let E be a reflexive Banach space. A (nonlinear) mapping

$$A : D(A) \subset E \to E^\star$$

is said to be *monotone* if it satisfies

$$\langle Au - Av, \ u - v \rangle \geq 0 \quad \forall u, v \in D(A)$$

(here $D(A)$ denotes any subset of E).

- A -

Let $A : D(A) \subset E \to E^\star$ be a monotone mapping and let $K \subset E$ be a nonempty, bounded, closed convex set. Our goal is to prove that there exists some $u \in K$ such that

$$\langle Av, u - v \rangle \geq 0 \quad \forall v \in D(A) \cap K.$$

For this purpose, set, for each $v \in D(A) \cap K$,

$$K_v = \{u \in K; \ \langle Av, v - u \rangle \geq 0\}.$$

We have to prove that $\cap_{v \in D(A) \cap K} K_v \neq \emptyset$; we shall argue by contradiction and assume that

$$\underset{v \in D(A) \cap K}{\cap} K_v = \emptyset.$$

1. Check that K_v is closed and convex.

2. Deduce that there exist $v_1, v_2, \ldots, v_n \in D(A) \cap K$ such that

$$\overset{n}{\underset{i=1}{\cap}} K_{v_i} = \emptyset.$$

Set $B = \{\lambda = (\lambda_1, \lambda_2, \ldots, \lambda_n); \lambda_i \geq 0 \ \forall i \text{ and } \sum_{i=1}^{n} \lambda_i = 1\}$, and consider the bilinear form

$$F : \mathbb{R}^n \times \mathbb{R}^n \to \mathbb{R}$$

defined by

$$F(\lambda, \mu) = \sum_{i,j=1}^{n} \lambda_i \mu_j \langle Av_j, \ v_i - v_j \rangle.$$

3. Check that $F(\lambda, \lambda) \leq 0 \ \forall \lambda \in \mathbb{R}^n$.

4. Prove that there exists some $\overline{\lambda} \in B$ such that $F(\overline{\lambda}, \mu) \leq 0 \ \forall \mu \in B$.

[**Hint**: Apply question C1 of Problem 30.]

5. Set $\bar{u} = \sum_{i=1} \bar{\lambda}_i v_i$ and prove that

$$\langle Av_j, \bar{u} - v_j \rangle \le 0 \quad \forall j = 1, 2, \dots, n.$$

6. Conclude.

- B -

Throughout the rest of this problem, we assume that $D(A) = E$, $A : E \to E^*$ is monotone, and A is continuous.

1. Let $K \subset E$ be a nonempty, bounded, closed convex set. Prove that there exists some $u \in K$ such that $\langle Au, w - u \rangle \ge 0 \ \forall w \in K$.

   [**Hint**: Consider $v_t = (1 - t)u + tw$ with $t \in (0, 1)$ and $w \in K$.]

2. Let K be a closed convex set containing 0 (K need not be bounded). Assume that the set $\{u \in K; \ \langle Au, u \rangle \le 0\}$ is bounded. Prove that there exists some $u \in K$ such that
   $$\langle Au, v - u \rangle \ge 0 \quad \forall v \in K.$$

   [**Hint**: Apply B1 to the set $K_R = \{v \in K; \ \|v\| \le R\}$ with R large enough.]

3. Assume here that
   $$\lim_{\|v\| \to \infty} \frac{\langle Av, v \rangle}{\|v\|} = +\infty.$$

   Prove that A is surjective.

4. Assume here that E is a Hilbert space identified with E^*. Prove that $I + A$ is bijective from E onto itself.

PROBLEM 32 (5)

Extension of contractions. The theorem of Kirszbraun–Valentine
via the method of Schoenberg

Let H be a Hilbert space and let I be a finite set of indices.

- A -

Let $(y_i)_{i \in I}$ be elements of H and let $(c_i)_{i \in I}$ be elements of $\mathbb{R}$. Set

$$\varphi(u) = \max_{i \in I}\{|u - y_i|^2 - c_i\}, \quad u \in H,$$

and

$$J(u) = \{i \in I; \ |u - y_i|^2 - c_i = \varphi(u)\}.$$

1. Check that $\inf_{u \in H} \varphi(u)$ is achieved by some unique element $u_0 \in H$.
2. Prove that $\max_{i \in J(u_0)} (v, u_0 - y_i) \ge 0 \ \forall v \in H$.
3. Deduce that

(1)
$$u_0 \in \text{conv} \left(\bigcup_{i \in J(u_0)} \{y_i\} \right).$$

4. Conversely, if $u_0 \in H$ satisfies (1), prove that $\varphi(u_0) = \inf_{u \in H} \varphi(u)$.
5. Extend this result to the case in which $\varphi(u) = \max_{i \in I} \{f_i(u)\}$ and each $f_i : H \to \mathbb{R}$ is a convex C^1 function.

- B -

Let $(x_i)_{i \in I}$ and $(y_i)_{i \in I}$ be elements of H such that

$$|y_i - y_j| \le |x_i - x_j| \quad \forall i, j \in I.$$

We claim that given any $p \in H$, there exists some $q \in \text{conv} \left(\bigcup_{i \in I} \{y_i\} \right)$ such that

$$|q - y_i| \le |p - x_i| \quad \forall i \in I.$$

1. Set $P = \{\lambda = (\lambda_i)_{i \in I}; \lambda_i \ge 0 \ \forall i \text{ and } \sum_{i \in I} \lambda_i = 1\}$.
   Prove that for every $p \in H$ and for every $\lambda \in P$,

$$\sum_{j \in I} \lambda_j \left| \left(\sum_{i \in I} \lambda_i y_i \right) - y_j \right|^2 \le \sum_{j \in I} \lambda_j |p - x_j|^2.$$

· [**Hint:** Check that $\sum_{j \in I} \lambda_j \left| \left(\sum_{i \in I} \lambda_i y_i \right) - y_j \right|^2 = \frac{1}{2} \sum_{i,j \in I} \lambda_i \lambda_j |y_i - y_j|^2$.]

2. Consider the function

$$\varphi(u) = \max_{i \in I} \{|u - y_i|^2 - |p - x_i|^2\}.$$

   Let $u_0 \in H$ be such that $\varphi(u_0) = \inf_{u \in H} \varphi(u)$. Prove that $\varphi(u_0) \le 0$.

   [**Hint:** Apply questions A3 and B1.]

3. Conclude.

- C -

1. Extend the result of part B to the case that I is an infinite set of indices.

2. Let $D \subset H$ by any subset of H and let $S : D \to H$ be a contraction, i.e.,

$$|Su - Sv| \le |u - v| \quad \forall u, v \in D.$$

   Prove that there exists a contraction T defined on all of H that extends S and such that
   $$T(H) \subset \overline{\text{conv } S(D)}.$$

   [**Hint:** Use Zorn's lemma and question C1.]

PROBLEM 33 (4, 6)

Multiplication operator in L^p

Let Ω be a measure space (having finite or infinite measure). Set $E = L^p(\Omega)$ with $1 < p < \infty$. Let $q : \Omega \to \mathbb{R}$ be a measurable function. Consider the unbounded linear operator $A : D(A) \subset E \to E$ defined by

$$D(A) = \{u \in L^p(\Omega); au \in L^p(\Omega)\} \quad \text{and} \quad Au = au.$$

1. Prove that $D(A)$ is dense in E.

   [**Hint**: Given $u \in E$, consider the sequence $u_n(x) = (1 + n^{-1}|a(x)|)^{-1}u(x)$.]

2. Show that A closed.

3. Prove that $D(A) = E$ iff $a \in L^\infty(\Omega)$.

   [**Hint**: Apply the closed graph theorem.]

4. Determine $N(A)$ and $N(A)^\perp$.

5. Determine $D(A^\star)$, $A^\star$, $N(A^\star)$, and $N(A^\star)^\perp$.

6. Prove that A is surjective iff there exists $\alpha > 0$ such that $|a(x)| \geq \alpha$ a.e. on Ω.

   [**Hint**: Use question 3.]

   In what follows we assume that $a \in L^\infty(\Omega)$.

7. Determine the eigenvalues and the spectrum of A. Check that $\sigma(A) \subset [\inf_\Omega a, \sup_\Omega a]$ and that $\inf_\Omega a \in \sigma(A)$, $\sup_\Omega a \in \sigma(A)$. Here $\inf_\Omega$ and $\sup_\Omega$ refer to the ess $\inf_\Omega$ and ess $\sup_\Omega$ (defined in Section 8.5).

8. In case Ω is an open set in $\mathbb{R}^N$ (equipped with the Lebesgue measure) and $a \in C(\Omega) \cap L^\infty(\Omega)$, prove that $\sigma(A) = \overline{a(\Omega)}$.

9. Prove that $\sigma(A) = \{0\}$ iff $a = 0$ a.e. on Ω.

10. Assume that Ω has no atoms. Prove that A is compact iff $a = 0$ a.e. on Ω.

PROBLEM 34 (4, 6)

Spectral analysis of the Hardy operator $Tu(x) = \frac{1}{x}\int_0^x u(t)dt$

- A -

Let $E = C([0, 1])$ equipped with the norm $\|u\| = \sup_{t \in [0,1]} |u(t)|$. Given $u \in E$ define the function Tu on $[0, 1]$ by

$$Tu(x) = \begin{cases} \frac{1}{x}\int_0^x u(t)dt & \text{if } x \in (0, 1], \\ u(0) & \text{if } x = 0. \end{cases}$$

Check that $Tu \in E$ and that $\|Tu\| \le \|u\| \ \forall u \in E$, so that $T \in \mathcal{L}(E)$.

1. Prove that $EV(T) = (0, 1]$ and determine the corresponding eigenfunctions.

2. Check that $\|T\|_{\mathcal{L}(E)} = 1$. Is T a compact operator from E into itself?

3. Show that $\sigma(T) = [0, 1]$. Give an explicit formula for $(T - \lambda I)^{-1}$ when $\lambda \in \rho(T)$. Prove that $(T - \lambda I)$ is surjective from E onto E for every $\lambda \in \mathbb{R}, \lambda \notin \{0, 1\}$. Check that T and $(T - I)$ are not surjective.

4. In this question we consider T as a bounded operator from $E = C([0, 1])$ into $F = L^q(0, 1)$ with $1 \le q < \infty$. Prove that $T \in \mathcal{K}(E, F)$.

   [**Hint**: Consider the operator $(T_\varepsilon u)(x) = \frac{1}{x+\varepsilon} \int_0^x u(t)dt$ with $\varepsilon > 0$ and estimate $\|T_\varepsilon - T\|_{\mathcal{L}(E,F)}$ as $\varepsilon \to 0$.]

- B -

In this part we set $E = C^1([0, 1])$ equipped with the norm

$$\|u\| = \sup_{t \in [0,1]} |u(t)| + \sup_{t \in [0,1]} |u'(t)|.$$

Given $u \in C^1([0, 1])$ we define Tu as in part A.

1. Check that if $u \in C^1([0, 1])$, then $Tu \in C^1([0, 1])$ and $\|Tu\| \le \|u\| \ \forall u \in E$. .

2. Prove that $EV(T) = (0, \frac{1}{2}] \cup \{1\}$.

3. Prove that $\sigma(T) = [0, \frac{1}{2}] \cup \{1\}$.

- C -

In this part we set $E = L^p(0, 1)$ with $1 < p < \infty$. Given $u \in L^p(0, 1)$ define Tu by

$$Tu(x) = \frac{1}{x} \int_0^x u(t)dt \quad \text{for } x \in (0, 1].$$

Check that $Tu \in C((0, 1])$ and that $Tu \in L^q(0, 1)$ for every $q < p$. Our goal is to prove that $Tu \in L^p(0, 1)$ and that

(1) $$\|Tu\|_{L^p(0,1)} \le \frac{p}{p-1} \|u\|_{L^p(0,1)} \quad \forall u \in E.$$

1. Prove that (1) holds when $u \in C_c((0, 1))$.

   [**Hint**: Set $\varphi(x) = \int_0^x u(t)dt$; check that $|\varphi|^p \in C^1([0, 1])$ and compute its derivative. Estimate $\|Tu\|_{L^p}$ using the formula

   $$\int_0^1 |Tu(x)|^p dx = \int_0^1 |\varphi(x)|^p \frac{dx}{x^p} = \frac{1}{p-1} \int_0^1 |\varphi(x)|^p d\left(-\frac{1}{x^{p-1}}\right)$$

   and integrating by parts.]

2. Prove that (1) holds for every $u \in E$.

   In what follows we consider T as a bounded operator from E into itself.

3. Show that $EV(T) = (0, \frac{p}{p-1})$.

4. Deduce that $\|T\|_{\mathcal{L}(E)} = \frac{p}{p-1}$. Is T a compact operator from E into itself?

5. Prove that $\sigma(T) = [0, \frac{p}{p-1}]$.

6. Determine $T^\star$.

7. In this question we consider T as a bounded operator from $E = L^p(0, 1)$ into $F = L^q(0, 1)$ with $1 \le q < p < \infty$. Show that $T \in \mathcal{K}(E, F)$.

PROBLEM 35 (6)

Cotlar's lemma

Let H be a Hilbert space identified with its dual space.

- A -

Assume $T \in \mathcal{L}(H)$, so that $T^\star \in \mathcal{L}(H)$.

1. Prove that $\|T^\star T\| = \|T\|^2$.

2. Assume in this question that T is self-adjoint.
   Show that
   $$\|T^N\| = \|T\|^N \text{ for every integer } N.$$

3. Deduce that (for a general $T \in \mathcal{L}(H)$),
   $$\|(T^\star T)^N\| = \|T\|^{2N} \text{ for every integer } N.$$

- B -

Let (T_j), $1 \le j \le m$, be a finite collection of operators in $\mathcal{L}(H)$. Assume that $\forall j, k \in \{1, 2, \ldots, m\}$,

(1) $$\|T_j^\star T_k\|^{1/2} \le \omega(j - k),$$

(2) $$\|T_k T_j^\star\|^{1/2} \le \omega(k - j),$$

where $\omega : \mathbb{Z} \to [0, \infty)$.
   Set
   $$\sigma = \sum_{i=-(m-1)}^{m-1} \omega(i).$$

The goal of this problem is to show that

(3)
$$\left\| \sum_{j=1}^{m} T_j \right\| \le \sigma.$$

Set

$$U = \sum_{j=1}^{m} T_j$$

and fix an integer N.

1. Show that

$$\|T_{j_1}^\star T_{k_1} T_{j_2}^\star T_{k_2} \cdots T_{j_N}^\star T_{k_N}\|$$
$$\le \sigma\omega(j_1 - k_1)\omega(k_1 - j_2)\omega(j_2 - k_2)\cdots\omega(k_{N-1} - j_N)\omega(j_N - k_N),$$

for any choice of the integers $j_1, k_1, \ldots, j_N, k_N \in \{1, 2, \ldots, m\}$.

2. Deduce that

$$\sum_{j_1}\sum_{k_1}\cdots\sum_{j_N}\sum_{k_N}\|T_{j_1}^\star T_{k_1}\cdots T_{j_N}^\star T_{k_N}\| \le m\sigma^{2N},$$

where the summation is taken over all possible choices of the integers $j_i, k_i \in \{1, 2, \ldots, m\}$.

3. Prove that

$$\|(U^\star U)^N\| \le m\sigma^{2N}$$

and deduce that (3) holds.

PROBLEM 36 (6)

More on the Riesz–Fredholm theory

Let E be a Banach space and let $T \in \mathcal{K}(E)$. For every integer $k \ge 1$ set

$$N_k = N((I - T)^k) \quad \text{and} \quad R_k = R((I - T)^k).$$

1. Check that $\forall k \ge 1$, $R_{k+1} \subset R_k$, R_k is closed, $T(R_k) \subset R_k$, and $(I - T)R_k \subset R_{k+1}$.

2. Prove that there exists an integer $p \ge 1$ such that

$$\begin{cases} R_{k+1} \ne R_k & \forall k < p \text{ (no condition if } p = 1), \\ R_{k+1} = R_k & \forall k \ge p. \end{cases}$$

3. Check that $\forall k \ge 1$, $N_k \subset N_{k+1}$, $\dim N_k < \infty$, $T(N_k) \subset N_k$, and $(I - T)N_{k+1} \subset N_k$.

4. Show that

$$\text{codim } R_k = \dim N_k \quad \forall k \geq 1,$$

and deduce that

$$\begin{cases} N_{k+1} \neq N_k & \forall k < p \text{ (no condition if } p = 1\text{)}, \\ N_{k+1} = N_k & \forall k \geq p. \end{cases}$$

5. Prove that

$$\begin{cases} R_p \cap N_p = \{0\}, \\ R_p + N_p = E. \end{cases}$$

6. Prove that $(I - T)$ restricted to R_p is bijective from R_p onto itself.

7. Assume here in addition that E is a Hilbert space and that T is self-adjoint. Prove that $p = 1$.

PROBLEM 37 (6)

Courant–Fischer min–max principle. Rayleigh–Ritz method

Let H be an infinite-dimensional separable Hilbert space. Let T be a self-adjoint compact operator from H into itself such that $(Tx, x) \geq 0 \ \forall x \in H$. Denote by (μ_k), $k \geq 1$, its eigenvalues, repeated with their multiplicities, and arranged in nonincreasing order:

$$\mu_1 \geq \mu_2 \geq \cdots \geq 0.$$

Let (e_j) be an associated orthonormal basis composed of eigenvectors. Let E_k be the space spanned by $\{e_1, e_2, \ldots, e_k\}$. For $x \neq 0$ we define the *Rayleigh quotient*

$$R(x) = \frac{(Tx, x)}{|x|^2}.$$

1. Prove that $\forall k \geq 1$,

$$\min_{\substack{x \in E_k \\ x \neq 0}} R(x) = \mu_k.$$

2. Prove that $\forall k \geq 2$,

$$\max_{\substack{x \in E_{k-1}^\perp \\ x \neq 0}} R(x) = \mu_k,$$

and

$$\max_{\substack{x \in H \\ x \neq 0}} R(x) = \mu_1.$$

3. Let Σ be any k-dimensional subspace of H with $k \geq 1$. Prove that

$$\min_{\substack{x \in \Sigma \\ x \neq 0}} R(x) \leq \mu_k.$$

[**Hint:** If $k \geq 2$, show that $\Sigma \cap E_{k-1}^{\perp} \neq \{0\}$ and apply question 2.]

4. Deduce that $\forall k \geq 1$,

$$\max_{\substack{\Sigma \subset H \\ \dim \Sigma = k}} \min_{\substack{x \in \Sigma \\ x \neq 0}} R(x) = \mu_k.$$

5. Let Σ be any $(k-1)$-dimensional subspace of H with $k \geq 2$. Prove that

$$\max_{\substack{x \in \Sigma^{\perp} \\ x \neq 0}} R(x) \geq \mu_k.$$

[**Hint:** Prove that $\Sigma^{\perp} \cap E_k \neq \{0\}$.]

6. Deduce that $\forall k \geq 1$,

$$\min_{\substack{\Sigma \subset H \\ \dim \Sigma = k-1}} \max_{\substack{x \in \Sigma^{\perp} \\ x \neq 0}} R(x) = \mu_k.$$

7. Assume here that $N(T) = \{0\}$, so that $R(x) \neq 0 \ \forall x \neq 0$, or equivalently $\mu_k > 0 \ \forall k$. Show that $\forall k \geq 1$,

$$\min_{\substack{\Sigma \subset H \\ \dim \Sigma = k}} \max_{\substack{x \in \Sigma \\ x \neq 0}} \frac{1}{R(x)} = \frac{1}{\mu_k},$$

and

$$\max_{\substack{\Lambda \subset H \\ \operatorname{codim} \Lambda = k-1}} \min_{\substack{x \in \Lambda \\ x \neq 0}} \frac{1}{R(x)} = \frac{1}{\mu_k},$$

where Σ and Λ are closed subspaces of H.
In particular, for $k = 1$,

$$\min_{\substack{x \in H \\ x \neq 0}} \frac{1}{R(x)} = \frac{1}{\mu_1};$$

and, moreover, $\forall k \geq 2$,

$$\min_{\substack{x \in E_{k-1}^{\perp} \\ x \neq 0}} \frac{1}{R(x)} = \frac{1}{\mu_k}.$$

8. Let V be a closed subspace of H (finite- or infinite-dimensional). Let P_V be the orthogonal projection from H onto V and consider the operator $S : V \to V$ defined by $S = P_V \circ T_{|V}$. Check that S is a self-adjoint compact operator from V into itself such that $(Sx, x) \geq 0 \ \forall x \in V$.

9. Denote by (ν_k), $k \geq 1$, the eigenvalues of S, repeated with their multiplicities and arranged in nonincreasing order. Prove that $\forall k$ with $1 \leq k \leq \dim V$,

$$\max_{\substack{\Sigma \subset V \\ \dim \Sigma = k}} \min_{\substack{x \in \Sigma \\ x \neq 0}} R(x) = \nu_k.$$

Deduce that $v_k \leq \mu_k$ $\forall k$ with $1 \leq k \leq \dim V$.

10. Consider now an increasing sequence $V^{(n)}$ of closed subspaces of H such that

$$\overline{\bigcup_n V^{(n)}} = H.$$

Set $S^{(n)} = P_{V^{(n)}} \circ T_{|V^{(n)}}$ and let $(v_k^{(n)})$ denote the eigenvalues of $S^{(n)}$ arranged as in question 9. Prove that for each fixed k the sequence $n \mapsto v_k^{(n)}$ is nondecreasing and converges, as $n \to \infty$, to μ_k.

PROBLEM 38 (2, 6, 11)

Fredholm–Noether operators

Let E and F be Banach spaces and let $T \in \mathcal{L}(E, F)$.

- A -

The goal of part A is to prove that the following conditions are equivalent:

(1) $\begin{cases} \text{(a)} \quad R(T) \text{ is closed and has finite codimension in } F, \\ \text{(b)} \quad N(T) \text{ admits a complement in } E. \end{cases}$

(2) $\begin{cases} \text{There exist } S \in \mathcal{L}(F, E) \text{ and } K \in \mathcal{K}(F, F) \text{ such that} \\ T \circ S = I_F + K. \end{cases}$

(3) $\begin{cases} \text{There exist } U \in \mathcal{L}(F, E) \text{ and a finite-rank} \\ \text{projection } P \text{ in } F \text{ such that } T \circ U = I_F - P. \end{cases}$

Moreover, one can choose U and P such that $\dim R(P) = \operatorname{codim} R(T)$.

1. Prove that $(1) \Rightarrow (3)$

   [**Hint**: Let X be a complement of $N(T)$ in E. Then $T_{|X}$ is bijective from X onto $R(T)$. Denote by U_0 its inverse. Let Q be a projection from F onto $R(T)$ and set $U = U_0 \circ Q$.]

2. Prove that $(2) \Rightarrow (3)$.

   [**Hint**: Use Exercise 6.25.]

3. Prove that $(3) \Rightarrow (1)$.

   [**Hint**: To establish part (a) of (1) note that $R(T) \supset R(I_F - P)$ and apply Proposition 11.5. Similarly, show that $R(U^\star)$ is closed and thus $R(U)$ is also closed. Finally, prove that there exist finite-dimensional spaces Σ_1 and Σ_2 in E such that $N(T) + R(U) + \Sigma_1 = E$ and $N(T) \cap R(U) \subset \Sigma_2$. Then apply Proposition 11.7.]

4. Conclude.

- B -

Prove that the following conditions are equivalent:

(4)
$\begin{cases} \text{(a)} & R(T) \text{ is closed and admits a complement,} \\ \text{(b)} & \dim N(T) < \infty. \end{cases}$

(5)
$\begin{cases} \text{There exists } \widetilde{S} \in \mathcal{L}(F, E) \text{ and } \widetilde{K} \in \mathcal{K}(E, E) \text{ such that} \\ \widetilde{S} \circ T = I_E + \widetilde{K}. \end{cases}$

(6)
$\begin{cases} \text{There exist } \widetilde{U} \in \mathcal{L}(F, E) \text{ and a finite-rank} \\ \text{projection } \widetilde{P} \text{ in } E \text{ such that } \widetilde{U} \circ T = I_E - \widetilde{P}. \end{cases}$

- C -

One says that an operator $T \in \mathcal{L}(E, F)$ is *Fredholm* (or *Noether*) if it satisfies

(FN)
$\begin{cases} \text{(a)} & R(T) \text{ is closed and has finite codimension,} \\ \text{(b)} & \dim N(T) < \infty. \end{cases}$

(The property that $R(T)$ is closed can be deduced from the other assumptions; see Exercise 2.27.)

The class of operators satisfying (FN) is denoted by $\Phi(E, F)$. The *index* of T is by definition

$$\text{ind } T = \dim N(T) - \text{codim } R(T).$$

1. Assume that $T \in \Phi(E, F)$. Show that there exist $U \in \mathcal{L}(F, E)$ and finite-rank projections P in F (resp. $\widetilde{P}$ in E) such that

(7)
$\begin{cases} \text{(a)} & T \circ U = I_F - P, \\ \text{(b)} & U \circ T = I_E - \widetilde{P}, \end{cases}$

with $\dim R(P) = \text{codim } R(T)$, $\dim R(\widetilde{P}) = \dim N(T)$.
[**Hint**: Use the operator U constructed in question A1.]

An operator $V \in \mathcal{L}(F, E)$ satisfying

(8)
$\begin{cases} \text{(a)} & T \circ V = I_F + K, \\ \text{(b)} & V \circ T = I_E + \widetilde{K}, \end{cases}$

with $K \in \mathcal{K}(F)$ and $\widetilde{K} \in \mathcal{K}(E)$, is called a *pseudoinverse* of T (or an *inverse modulo compact operators*).

2. Show that any pseudoinverse V belongs to $\Phi(F, E)$.

3. Prove that an operator $T \in \mathcal{L}(E, F)$ belongs to $\Phi(E, F)$ iff $R(T)$ is closed, $\dim N(T) < \infty$, and $\dim N(T^\star) < \infty$. Moreover,

$$\text{ind } T = \dim N(T) - \dim N(T^*).$$

[**Hint**: Apply Propositions 11.14 and 2.18.]

4. Let $T \in \Phi(E, F)$. Prove that $T^* \in \Phi(F^*, E^*)$ and that

$$\text{ind } T^* = - \text{ind } T.$$

[**Hint**: Apply Proposition 11.13 and Theorem 2.19.]

5. Conversely, let $T \in \mathcal{L}(E, F)$ be such that $T^* \in \Phi(F^*, E^*)$. Prove that $T \in \Phi(E, F)$.

6. Assume that $J \in \mathcal{L}(E, F)$ is bijective and $K \in \mathcal{K}(E, F)$. Show that $T = J + K$ belongs to $\Phi(E, F)$ and ind $T = 0$. Conversely, if $T \in \Phi(E, F)$ and ind $T = 0$, prove that T can be written as $T = J + K$ with J and K as above (one may even choose K to be of finite rank).

[**Hint**: Applying Theorem 6.6, prove that $I_E + J^{-1} \circ K$ belongs to $\Phi(E, E)$ and has index zero. For the converse, consider an isomorphism from $N(T)$ onto a complement Y of $R(T)$.]

7. Let $T \in \Phi(E, F)$ and $K \in \mathcal{K}(E, F)$. Prove that $T + K \in \Phi(E, F)$.

8. Under the assumptions of the previous question, show that

$$\text{ind}(T + K) = \text{ind } T.$$

[**Hint**: Set $\widetilde{E} = E \times Y$, $\widetilde{F} = F \times N(T)$, and $\widetilde{T} : \widetilde{E} \to \widetilde{F}$ defined by $\widetilde{T}(x, y) = (Tx + Kx, 0)$. Show that $\widetilde{T} = \widetilde{J} + \widetilde{K}$, where $\widetilde{J}$ is bijective from $\widetilde{E}$ onto $\widetilde{F}$ and $\widetilde{K} \in \mathcal{K}(\widetilde{E}, \widetilde{F})$. Then apply question 6.]

9. Let $T \in \Phi(E, F)$. Prove that there exists $\varepsilon > 0$ (depending on T) such that for every $M \in \mathcal{L}(E, F)$ with $\|M\| < \varepsilon$, we have $T + M \in \Phi(E, F)$. Show that

$$\text{ind}(T + M) = \text{ind } T.$$

[**Hint**: Let V be a pseudoinverse of T. Then $W = I_E + (V \circ M)$ is bijective if $\|M\| < \|V\|^{-1}$. Check that $T + M = (T \circ W) +$ compact; then apply the previous question.]

10. Let (H_t), $t \in [0, 1]$, be a family of operators in $\mathcal{L}(E, F)$. Assume that $t \mapsto H_t$ is continuous from $[0, 1]$ into $\mathcal{L}(E, F)$, and that $H_t \in \Phi(E, F)$ $\forall t \in [0, 1]$. Prove that ind H_t is constant on $[0, 1]$.

11. Let E_1, E_2, and E_3 be Banach spaces and let $T_1 \in \Phi(E_1, E_2)$, $T_2 \in \Phi(E_2, E_3)$. Prove that $T_2 \circ T_1 \in \Phi(E_1, E_3)$.

12. With the same notation as above, show that

$$\text{ind}(T_2 \circ T_1) = \text{ind } T_1 + \text{ind } T_2.$$

[**Hint:** Consider the family of operators $H_t : E_1 \times E_2 \to E_2 \times E_3$ defined in matrix notation, for $t \in [0, 1]$, by

$$H_t = \begin{pmatrix} I & 0 \\ 0 & T_2 \end{pmatrix} \begin{pmatrix} (1-t)I & tI \\ -tI & (1-t)I \end{pmatrix} \begin{pmatrix} T_1 & 0 \\ 0 & I \end{pmatrix},$$

where I is the identity operator in E_2. Check that $t \mapsto H_t$ is continuous from $[0, 1]$ into $\mathcal{L}(E_1 \times E_2, E_2 \times E_3)$. Using the previous question, show that for each t, $H_t \in \Phi(E_1 \times E_2, E_2 \times E_3)$. Compute ind H_0 and ind H_1.]

13. Let $T \in \Phi(E, F)$. Compute the index of any pseudoinverse V of T.

- D -

In this part we study two simple examples.

1. Assume dim $E < \infty$ and dim $F < \infty$. Show that any linear operator T from E into F belongs to $\Phi(E, F)$ and compute its index.

2. Let $E = F = \ell^2$. Consider the shift operators S_r and S_ℓ defined in Exercise 6.18. Prove that for every $\lambda \in \mathbb{R}, \lambda \neq +1, \lambda \neq -1$, we have $S_r - \lambda I \in \Phi(\ell^2, \ell^2)$, and $S_\ell - \lambda I \in \Phi(\ell^2, \ell^2)$. Compute their indices.
Show that $S_r \pm I$, $S_\ell \pm I$ do not belong to $\Phi(\ell^2, \ell^2)$.

[**Hint:** Use the results of Exercise 6.18.]

PROBLEM 39 (5, 6)

Square root of a self-adjoint nonnegative operator

Let H be a Hilbert space. Let $S \in \mathcal{L}(H)$; we say that S is *nonnegative*, and we write $S \geq 0$, if $(Sx, x) \geq 0 \; \forall x \in H$. When $S_1, S_2 \in \mathcal{L}(H)$, we write $S_1 \geq S_2$ (or $S_2 \leq S_1$) if $S_1 - S_2 \geq 0$.

- A -

1. Let $S \in \mathcal{L}(H)$ be such that $S^\star = S$ and $0 \leq S \leq I$. Show that $\|S^2\| = \|S\|^2 \leq 1$, and that $0 \leq S^2 \leq S \leq I$.

[**Hint:** Use Exercise 6.24.]

2. Let $S \in \mathcal{L}(H)$ be such that $S^\star = S$ and $S \geq 0$. Let $P(t) = \sum a_k t^k$ be a polynomial such that $a_k \geq 0 \; \forall k$. Prove that $[P(S)]^\star = P(S)$ and $P(S) \geq 0$.

3. Let (S_n) be a sequence in $\mathcal{L}(H)$ such that $S_n^\star = S_n \; \forall n$ and $S_{n+1} \leq S_n \; \forall n$. Assume that $\|S_n\| \leq M \; \forall n$, for some constant M. Prove that for every $x \in H$, $S_n x$ converges as $n \to \infty$ to a limit, denoted by Sx, and that $S \in \mathcal{L}(H)$ with $S^\star = S$.

[Hint: Let $n \geq m$. Use Exercise 6.24 to prove that $|S_n x - S_m x|^2 \leq 2M(S_m x - S_n x, x)$.]

- B -

Assume that $T \in \mathcal{L}(H)$ satisfies $T^\star = T$, $T \geq 0$, and $\|T\| \leq 1$. Consider the sequence (S_n) defined by

$$S_{n+1} = S_n + \frac{1}{2}(T - S_n^2), \quad n \geq 0,$$

starting with $S_0 = I$.

1. Show that $S_n^\star = S_n \ \forall n \geq 0$.

2. Show that
$$I - S_{n+1} = \frac{1}{2}(I - S_n)^2 + \frac{1}{2}(I - T),$$

   and deduce that $I - S_n \geq 0 \ \forall n$.

3. Prove that $S_n \geq 0 \ \forall n$.

   [Hint: Show by induction that $I - S_n \leq I$ using questions A.1 and B.2.]

4. Deduce that $\|S_n\| \leq 1 \ \forall n$.

5. Prove that

$$S_n - S_{n+1} = \frac{1}{2} \left[(I - S_n) + (I - S_{n-1}) \right] \circ (S_{n-1} - S_n) \quad \forall n$$

   and deduce that $S_{n-1} - S_n \geq 0 \ \forall n$.

   [Hint: Show by induction that $(I - S_n) = P_n(I - T)$ and $(S_{n-1} - S_n) = Q_n(I - T)$, where P_n and Q_n are polynomials with nonnegative coefficients.]

6. Show that $\lim_{n \to \infty} S_n x = Sx$ exists. Prove that $S \in \mathcal{L}(H)$ satisfies $S^\star = S$, $S \geq 0$, $\|S\| \leq 1$, and $S^2 = T$.

- C -

1. Let $U \in \mathcal{L}(H)$ be such that $U^\star = U$ and $U \geq 0$. Prove that there exists $V \in \mathcal{L}(H)$ such that $V^\star = V$, $V \geq 0$, and $V^2 = U$.

   [Hint: Apply the construction of part B to $T = U/\|U\|$.]

   Next, we prove the uniqueness of V. More precisely, if W is any operator $W \in \mathcal{L}(H)$ such that $W^\star = W$, $W \geq 0$, and $W^2 = U$, then $W = V$. The operator V is called the *square root* of U and is denoted by $U^{1/2}$.

2. Prove that the operator V constructed above commutes with every operator X that commutes with U (i.e., $X \circ U = U \circ X$ implies $X \circ V = V \circ X$).

3. Prove that W commutes with U and deduce that V commutes with W.

4. Check that $(V - W) \circ (V + W) = 0$ and deduce that $V = W$ on $R(V + W)$. Show that $N(V) = N(W) = N(U) = N(V + W)$. Conclude that $V = W$ on H.

   [**Hint**: Note that $V = W$ on $\overline{R(V + W)} = N(U)^{\perp}$, and that $V = W = 0$ on $N(U)$.]

5. Show that $\|U^{1/2}\| = \|U\|^{1/2}$.

6. Let $U_1, U_2 \in \mathcal{L}(H)$ be such that $U_1^{\star} = U_1$, $U_2^{\star} = U_2$, $U_1 \geq 0$, $U_2 \geq 0$, and $U_1 \circ U_2 = U_2 \circ U_1$. Prove that $U_1 \circ U_2 \geq 0$.

   [**Hint**: Introduce $U_1^{1/2}$ and $U_2^{1/2}$.]

- D -

Let $U \in \mathcal{K}(H)$ be such that $U^{\star} = U$ and $U \geq 0$. Prove that its square root V belongs to $\mathcal{K}(H)$. Assuming that H is separable, compute V on a Hilbert basis composed of eigenvectors of U. Find the eigenvalues of V.

PROBLEM 40 (4, 5, 6)

Hilbert–Schmidt operators

- A -

Let E and F be separable Hilbert spaces, both identified with their dual spaces. The norms on E and on F are denoted by the same symbol $| \; |$. Let $T \in \mathcal{L}(E, F)$, so that $T^{\star} \in \mathcal{L}(F, E)$.

1. Let (e_k) (resp. (f_k)) be any orthonormal basis of E (resp. F). Show that $\sum_{k=1}^{\infty} |T(e_k)|^2 < \infty$ iff $\sum_{k=1}^{\infty} |T^{\star}(f_k)|^2 < \infty$, and that

$$\sum_{k=1}^{\infty} |T(e_k)|^2 = \sum_{k=1}^{\infty} |T^{\star}(f_k)|^2.$$

2. Let (e_k) and $(\tilde{e}_k)$ be two orthonormal bases of E. Show that $\sum_{k=1}^{\infty} |T(e_k)|^2 < \infty$ iff $\sum_{k=1}^{\infty} |T(\tilde{e}_k)|^2 < \infty$ and that

$$\sum_{k=1}^{\infty} |T(e_k)|^2 = \sum_{k=1}^{\infty} |T(\tilde{e}_k)|^2.$$

   One says that $T \in \mathcal{L}(E, F)$ is a *Hilbert–Schmidt* operator and one writes $T \in \mathcal{HS}(E, F)$ if there exists some orthonormal basis (e_k) of E such that

$$\sum_{k=1}^{\infty} |T(e_k)|^2 < \infty.$$

3. Prove that $\mathcal{HS}(E, F)$ is a linear subspace of $\mathcal{L}(E, F)$ and that

$$\|T\|_{\mathcal{HS}} = \left(\sum_{k=1}^{\infty} |T(e_k)|^2 \right)^{1/2}$$

defines a norm on $\mathcal{HS}(E, F)$. Let $\| \ \|$ denote the standard norm or $\mathcal{L}(E, F)$. Show that

$$\|T\| \le \|T\|_{\mathcal{HS}} \quad \forall T \in \mathcal{HS}(E, F).$$

4. Prove that $\mathcal{HS}(E, F)$ equipped with the norm $\| \ \|_{\mathcal{HS}}$ is a Banach space. Show that in fact, it is a Hilbert space.

5. Show that $\mathcal{HS}(E, F) \subset \mathcal{K}(E, F)$.

   [**Hint**: Given $x \in E$, write $x = \sum_{k=1}^{\infty} x_k e_k$ and set $T_n(x) = \sum_{k=1}^{n} x_k T(e_k)$. Show that $\|T_n - T\| \to 0$ as $n \to \infty$.]

6. Show that any finite-rank operator from E into F belongs to $\mathcal{HS}(E, F)$.

7. Let $T \in \mathcal{L}(E, F)$. Prove that $T \in \mathcal{HS}(E, F)$ iff $T^\star \in \mathcal{HS}(F, E)$ and that

$$\|T^\star\|_{\mathcal{HS}(F,E)} = \|T\|_{\mathcal{HS}(E,F)}.$$

8. Assume that $T \in \mathcal{K}(E, E)$ with $T^\star = T$, and let (λ_k) denote the sequence of eigenvalues of T. Show that $T \in \mathcal{HS}(E, E)$ iff $\sum_{k=1}^{\infty} \lambda_k^2 < \infty$ and that

$$\|T\|_{\mathcal{HS}}^2 = \sum_{k=1}^{\infty} \lambda_k^2.$$

   Construct an example of an operator $T \in \mathcal{K}(E, E)$ with $E = \ell^2$ such that $T \notin \mathcal{HS}(E, E)$.

9. Let G be another separable Hilbert space. Let $T_1 \in \mathcal{L}(E, F)$ and $T_2 \in \mathcal{L}(F, G)$. Show that $T_2 \circ T_1 \in \mathcal{HS}(E, G)$ if either T_1 or T_2 belongs to $\mathcal{HS}$.

10. Let $T \in \mathcal{HS}(E, E)$ and assume $N(I + T) = \{0\}$. Show that $(I + T)$ is bijective and that $(I + T)^{-1} = I + S$ with $S \in \mathcal{HS}(E, E)$.

11. Let (e_k) (resp. (f_k)) be an orthonormal basis of E (resp. F). Consider the operator $T_{k,\ell} : E \to F$ defined by

$$T_{k,\ell}(x) = (x, e_k) f_\ell.$$

Show that $(T_{k,\ell})$ is an orthonormal basis of $\mathcal{HS}(E, F)$.

- B -

Assume that Ω is an open subset of $\mathbb{R}^N$. In what follows we take $E = F = L^2(\Omega)$. Let $K \in L^2(\Omega \times \Omega)$, and consider the operator

(1) $$(Tu)(x) = \int_\Omega K(x, y)u(y)dy.$$

1. Show that $T \in \mathcal{L}(E, E)$ and that

$$\|T\|_{\mathcal{L}(E,E)} \leq \|K\|_{L^2(\Omega \times \Omega)}.$$

2. Show that $T \in \mathcal{HS}(E, E)$ and that

$$\|T\|_{\mathcal{HS}(E,E)} \leq \|K\|_{L^2(\Omega \times \Omega)}.$$

[**Hint:** Let (e_j) be an orthonormal basis of $L^2(\Omega)$. Check that the family $e_{j,k} = e_j \otimes e_k$, where $(e_j \otimes e_k)(x, y) = e_j(x)e_k(y)$, is an orthonormal basis of $L^2(\Omega \times \Omega)$. Then write

$$\|T(e_k)\|^2_{L^2(\Omega)} = \sum_{j=1}^{\infty} |(T(e_k), e_j)|^2 = \sum_{j=1}^{\infty} |(K, e_j \otimes e_k)|.]$$

3. Conversely, let $T \in \mathcal{HS}(E, E)$. Prove that there exists a unique function $K \in L^2(\Omega \times \Omega)$ such that (1) holds. K is called the *kernel* of T.

[**Hint:** Let $t_{j,k} = (Te_k, e_j)$ and check that $\sum_{j,k=1}^{\infty} |t_{j,k}|^2 < \infty$. Define $K = \sum_{j,k=1}^{\infty} t_{j,k} e_j \otimes e_k$ and prove that (1) holds.]

4. Assume that $\Omega = (0, 1)$, $E = L^2(\Omega)$, and consider the operator

$$(Tu)(x) = \int_0^x u(t)dt.$$

Show that $T \in \mathcal{HS}(E, E)$ and compute $\|T\|_{\mathcal{HS}}$.

PROBLEM 41 (1, 6)

The Krein–Rutman theorem

Let E be a Banach space and let $P \subset E$ be a closed convex set containing 0. Assume that P is a convex cone with vertex at 0, i.e., $\lambda x + \mu y \in P$ $\forall \lambda > 0$, $\mu > 0, x \in P$, and $y \in P$.

Assume that

(1) $\text{Int } P \neq \emptyset$

and

(2) $P \neq E$.

Let $T \in \mathcal{K}(E)$ be such that

(3) $T(P \setminus \{0\}) \subset \text{Int } P$.

- A -

1. Show that $(\text{Int } P) \cap (-P) = \emptyset$.

   [**Hint**: Use Exercise 1.7.]

   In what follows we *fix* some $u \in \text{Int } P$.

2. Show that there exists $\alpha > 0$ such that

$$\|x + u\| \geq \alpha \quad \forall x \in P.$$

   [**Hint**: Argue by contradiction and deduce that $-u \in P$.]

3. Check that there exists $r > 0$ such that

$$Tu - ru \in P.$$

4. Assume that some $x \in P$ satisfies

$$T(x + u) = \lambda x \quad \text{for some } \lambda \in \mathbb{R}.$$

   Prove that $\lambda \geq r$.

   [**Hint**: It is convenient to introduce an order relation on E defined by $y \geq z$ if $y - z \in P$. Show by induction that $\left(\frac{\lambda}{r}\right)^n x \geq u$, $n = 1, 2, \ldots$.]

5. Consider the nonlinear map

$$F(x) = T\left(\frac{x + u}{\|x + u\|}\right), \quad x \in P.$$

   Show that $F : P \to P$ is continuous and $F(P) \subset K$ for some compact set $K \subset E$. Deduce that there exists some $x_1 \in P$ such that

$$T(x_1 + u) = \lambda_1 x_1$$

   with $\lambda_1 = \|x_1 + u\| \geq r$.

   [**Hint**: Apply the Schauder fixed-point theorem; see Exercise 6.26.]

6. Deduce that for every $\varepsilon > 0$ there exists $x_\varepsilon \in P$ such that

$$T(x_\varepsilon + \varepsilon u) = \lambda_\varepsilon x_\varepsilon$$

   with $\lambda_\varepsilon = \|x_\varepsilon + \varepsilon u\| \geq r$.

7. Prove that there exist $x_0 \in \text{Int } P$ and $\mu_0 > 0$ such that

$$Tx_0 = \mu_0 x_0.$$

[**Hint**: Show that (x_ε) is bounded. Deduce that there exists a sequence $\varepsilon_n \to 0$ such that $x_{\varepsilon_n} \to x_0$ and $\lambda_{\varepsilon_n} \to \mu_0$ with the required properties.]

- **B** -

1. Given two points $a \in \text{Int}\, P$ and $b \in E$, $b \notin P$, prove that there exists a unique $\sigma \in (0, 1)$ such that

$$(1 - t)a + tb \in \text{Int}\, P \quad \forall t \in [0, \sigma),$$
$$(1 - \sigma)a + \sigma b \in P,$$
$$(1 - t)a + tb \notin P \qquad \forall t \in (\sigma, 1].$$

Then we set $\tau(a, b) = \sigma/(1 - \sigma)$, with $0 < \tau(a, b) < \infty$.

2. Let $x \in P \setminus \{0\}$ be such that

$$Tx = \mu x \quad \text{for some } \mu \in \mathbb{R}.$$

Prove that $\mu = \mu_0$ and $x = mx_0$ for some $m > 0$, where μ_0 and x_0 have been constructed in question A7.

[**Hint**: Suppose by contradiction that $x \neq mx_0$, $\forall m > 0$. Show that $\mu > 0$, $x \in \text{Int}\, P$, and $-x \notin P$. Set $y = x_0 - \tau_0 x$, where $\tau_0 = \tau(x_0, -x)$. Compute Ty and deduce that $\mu < \mu_0$. Then reverse the roles of x_0 and x.]

3. Let $x \in E \setminus \{0\}$ be such that

$$Tx = \mu x \quad \text{for some } \mu \in \mathbb{R}.$$

Prove that *either* $\mu = \mu_0$ and $x = mx_0$ with $m \in \mathbb{R}$, $m \neq 0$, *or* $|\mu| < \mu_0$.

[**Hint**: In view of question 2 one may assume that $x \notin P$ and $-x \notin P$. If $\mu > 0$ consider $\tau(x_0, x)$, and if $\mu < 0$ consider both $\tau(x_0, x)$ and $\tau(x_0, -x)$.]

4. Deduce that $N(T - \mu_0 I) = \mathbb{R}x_0$. In other words, the geometric multiplicity of the eigenvalue μ_0 is one.

5. Prove that $N((T - \mu_0 I)^k) = \mathbb{R}x_0$ for all $k \geq 2$. In other words, the algebraic multiplicity of the eigenvalue μ_0 is also one.

[**Hint**: In view of Problem 36, it suffices to show that $N((T - \mu_0 I)^2) = \mathbb{R}x_0$.]

PROBLEM 42 (6)

Lomonosov's theorem on invariant subspaces

Let E be an infinite-dimensional Banach space and let $T \in \mathcal{K}(E)$, $T \neq 0$. The goal of part A is to prove that there exists a nontrivial, closed, invariant subspace Z of T, i.e., $T(Z) \subset Z$, with $Z \neq \{0\}$, and $Z \neq E$.

- A -

Set

$$\mathcal{A} = \text{span}\{I, T, T^2, \dots\}$$

$$= \left\{ \sum_{i \in I} \lambda_i T^i, \text{ with } \lambda_i \in \mathbb{R} \text{ and } I \text{ is a finite subset of } \{0, 1, 2, \dots\} \right\}.$$

For every $y \in E$, set $\mathcal{A}_y = \{Sy; \ S \in \mathcal{A}\}$. Clearly, $y \in \mathcal{A}_y$ and thus $\mathcal{A}_y \neq \{0\}$ for every $y \neq 0$. Moreover, $\mathcal{A}_y$ is a subspace of E and $T(\mathcal{A}_y) \subset \mathcal{A}_y$, so that $T(\overline{\mathcal{A}_y}) \subset \overline{\mathcal{A}_y}$. If $\overline{\mathcal{A}_y} \neq E$ for some $y \neq 0$, then $\overline{\mathcal{A}_y}$ is a nontrivial, closed, invariant subspace of T. Therefore we can assume that

(1) $\overline{\mathcal{A}_y} = E \quad \forall y \in E, \ y \neq 0.$

Since $T \neq 0$, we may *fix* some $x_0 \in E$ such that $Tx_0 \neq 0$, and some r such that

$$0 < r \leq \frac{\|Tx_0\|}{2\|T\|} \leq \frac{\|x_0\|}{2}.$$

Set

$$C = \{x \in E; \ \|x - x_0\| \leq r\}.$$

1. Check that $0 \notin C$ and that

$$\|Tx - Tx_0\| \leq \frac{1}{2} \|Tx_0\| \quad \forall x \in C,$$

   so that

$$\|Tx\| \geq \frac{1}{2} \|Tx_0\| \quad \forall x \in C.$$

   Deduce that $0 \notin \overline{T(C)}$.

2. Prove that for every $y \in E$, $y \neq 0$, there exists some $S \in \mathcal{A}$, denoted by S_y, such that
$$\|Sy - x_0\| \leq \frac{r}{2}.$$

   [**Hint**: Use assumption (1).]

3. Deduce that for every $y \in E$, $y \neq 0$, there exists some $\varepsilon > 0$ (depending on y), denoted by ε_y, such that

$$\|Sz - x_0\| \leq r \quad \forall z \in B(y, \varepsilon),$$

   where S is as in question 2.

4. Consider a finite covering of $\overline{T(C)}$ by balls $B(y_j, \frac{1}{2}\varepsilon_{y_j})$ with $j \in J$, J finite. Set, for $j \in J$ and $x \in E$,

$$q_j(x) = \max\{0, \varepsilon_{y_j} - \|Tx - y_j\|\} \quad \text{and} \quad q(x) = \sum_{j \in J} q_j(x).$$

Check that the functions q_j, $j \in J$, and q are continuous on E. Show that $\forall x \in C$,

$$q(x) \geq \min_{j \in J} \left\{ \frac{1}{2} \varepsilon_{y_j} \right\} > 0.$$

Set

$$F(x) = \frac{1}{q(x)} \sum_{j \in J} q_j(x) S_{y_j}(Tx), \quad x \in C.$$

5. Prove that F is continuous from C into E and that

$$\|F(x) - x_0\| \leq r \quad \forall x \in C.$$

[**Hint**: Use question 3.]

6. Prove that $F(C) \subset K$, where K is a compact subset of C. Deduce that there exists $\xi \in C$ such that $F(\xi) = \xi$.

[**Hint**: Apply the Schauder fixed-point theorem; see Exercise 6.26.]

7. Set

$$U = \frac{1}{q(x)} \sum_{j \in J} q_j(\xi)(S_{y_j} \circ T),$$

with ξ as in question 6. Show that $U \in \mathcal{K}(E)$. Deduce that $Z = N(I - U)$ is finite-dimensional; check that $\xi \in Z$.

8. Prove that $T(Z) \subset Z$ and conclude.

[**Hint**: Show that $U \in \mathcal{A}$ and deduce that $T \circ U = U \circ T$.]

9. Construct a linear operator $T : \mathbb{R}^2 \to \mathbb{R}^2$ that has no invariant subspaces except the trivial ones.

- B -

We now establish a stronger version of the above result. Assume that $T \in \mathcal{K}(E)$ and $T \neq 0$. Let $R \in \mathcal{L}(E)$ be such that $R \circ T = T \circ R$. Prove that R admits a nontrivial, closed, invariant subspace.

[**Hint**: Set $\mathcal{B} = \text{span} \{I, R, R^2, \dots\}$ and $\mathcal{B}_y = \{Sy; S \in \mathcal{B}\}$. Check that all the steps in part A still hold with $\mathcal{A}$ replaced by $\mathcal{B}$ and $\mathcal{A}_y$ by $\mathcal{B}_y$.]

PROBLEM 43 (2, 4, 5, 6)

Normal operators

Let H be a Hilbert space identified with its dual space. An operator $T \in \mathcal{L}(H)$ is said to be *normal* if it satisfies

$$T \circ T^\star = T^\star \circ T.$$

1. Prove that T is normal iff it satisfies

$$|Tu| = |T^\star u| \quad \forall u \in H.$$

   [**Hint**: Compute $|T(u + v)|^2$.]

   Throughout the rest of this problem we assume that T is normal.

2. Assume that $u \in N(T - \lambda I)$ and $v \in N(T - \mu I)$ with $\lambda \neq \mu$. Show that $(u, v) = 0$.

   [**Hint**: Prove, using question 1, that $N(T^\star - \mu I) = N(T - \mu I)$, and compute (Tu, v).]

3. Prove that $\overline{R(T)} = \overline{R(T^\star)} = N(T)^\perp = N(T^\star)^\perp$.

4. Let $f \in R(T)$. Check that there exists $u \in \overline{R(T^\star)}$ satisfying $f = Tu$.

   [**Hint**: Note that $H = \overline{R(T)} \oplus N(T)$.]

5. Consider a sequence $u_n \in R(T^\star)$ such that $u_n \to u$ as $n \to \infty$. Write $u_n = T^\star y_n$ for some $y_n \in H$. Show that Ty_n converges as $n \to \infty$ to a limit $z \in H$ that satisfies $T^\star z = f$.

   [**Hint**: Use question 1 and a Cauchy sequence argument.]

6. Deduce that $R(T) = R(T^\star)$.

   [**Hint**: Use the fact that $N(T) = N(T^\star)$.]

7. Show that $\|T^2\| = \|T\|^2$.

   [**Hint**: Write $|Tu|^2 \leq |T^\star T u| \, |u| = |T^2 u| \, |u|$.]

8. Deduce that $\|T^p\| = \|T\|^p$ for every integer $p \geq 1$.

   [**Hint**: Consider first the case $p = 2^k$. For a general integer p, choose any k such that $2^k \geq p$ and write $\|T\|^{2^k} = \|T^{2^k}\| = \|T^{2^k - p} T^p\|$.]

9. Prove that $N(T^2) = N(T)$ and deduce that $N(T^p) = N(T)$ for every integer $p \geq 1$.

   [**Hint**: Note that if $T^2 u = 0$, then $Tu \in N(T) \cap R(T)$.]

PROBLEM 44 (5, 6)

Isometries and unitary operators. Skew-adjoint operators.
Polar decomposition and Cayley transform.

Let H be a Hilbert space identified with its dual space and let $T \in \mathcal{L}(H)$. One says that

(i) T is an *isometry* if $|Tu| = |u| \ \forall u \in H$,
(ii) T is a *unitary* operator if T is an isometry that is also surjective,
(iii) T is *skew-adjoint* (or *antisymmetric*) if $T^* = -T$.

- A -

1. Assume that T is an isometry. Check that $\|T\| = 1$.

2. Prove that $T \in \mathcal{L}(H)$ is an isometry iff $T^* \circ T = I$.

3. Assume that $T \in \mathcal{L}(H)$ is an isometry. Prove that the following conditions are equivalent:

   (a) T is a unitary operator,
   (b) T^* is injective,
   (c) $T \circ T^* = I$,
   (d) T^* is an isometry,
   (e) T^* is a unitary operator.

4. Give an example of an isometry that is *not* a unitary operator.

   [**Hint**: Use Exercise 6.18.]

5. Assume that T is an isometry. Prove that $R(T)$ is closed and that $T \circ T^* = P_{R(T)} = $ the orthogonal projection on $R(T)$.

6. Assume that T is an isometry. Prove that
   either T is a unitary operator and then $\sigma(T) \subset \{-1, +1\}$,
   or T is *not* a unitary operator and then $\sigma(T) = [-1, +1]$.

7. Assume that $T \in \mathcal{K}(H)$ is an isometry. Show that dim $H < \infty$.

8. Prove that $T \in \mathcal{L}(H)$ is skew-adjoint iff $(Tu, u) = 0 \ \forall u \in H$.

9. Assume that $T \in \mathcal{L}(H)$ is skew-adjoint. Show that $\sigma(T) \subset \{0\}$.

   [**Hint**: Use Lax–Milgram.]

10. Assume that $T \in \mathcal{L}(H)$ is skew-adjoint. Set

$$U = (T + I) \circ (T - I)^{-1}.$$

Check that U is well defined, that $U = (T-I)^{-1} \circ (T+I)$, and that $U \circ T = T \circ U$. Prove that U is a unitary operator (U is called the *Cayley transform* of T).

11. Conversely, let $T \in \mathcal{L}(H)$ be such that $1 \notin \sigma(T)$. Assume that $U = (T + I) \circ (T - I)^{-1}$ is an isometry. Prove that T is skew-adjoint.

- B -

We will say that an operator $T \in \mathcal{L}(H)$ satisfies property (1) if

(1)   there exists an isometry J from $N(T)$ into $N(T^\star)$.

The goal of part B is to prove that every operator $T \in \mathcal{L}(H)$ satisfying property (1) can be factored as

$$T = U \circ P,$$

where $U \in \mathcal{L}(H)$ is an isometry and $P \in \mathcal{L}(H)$ is a self-adjoint nonnegative operator (recall that nonnegative means $(Pu, u) \geq 0 \; \forall u \in H$). Such a factorization is called a *polar decomposition* of T. In addition, P is uniquely determined on H, and U is uniquely determined on $N(T)^\perp$ (but not on H).

1. Check that assumption (1) is satisfied in the following cases:

   (i) T is injective,
   (ii) $\dim H < \infty$,
   (iii) T is normal (see Problem 43),
   (iv) $T = I - K$ with $K \in \mathcal{K}(H)$.

2. Give an example in which (1) is *not* satisfied.

   [**Hint**: Use Exercise 6.18.]

3. Assume that we have a polar decomposition $T = U \circ P$. Prove that $P^2 = T^\star \circ T$.

4. Deduce that P is uniquely determined on H.

   [**Hint**: Use Problem 39.]

5. Let $T = U \circ P$ be a polar decomposition of T. Show that U is uniquely determined on $N(T)^\perp$.

6. Assume that T admits a polar decomposition. Show that (1) holds.

   [**Hint**: Set $J = U_{|N(T)}$.]

7. Prove that every operator $T \in \mathcal{L}(H)$ satisfying (1) admits a polar decomposition.

8. Assume that T satisfies the stronger assumption

   (2)   there exists an isometry J from $N(T)$ *onto* $N(T^\star)$.

   Show that T admits a polar decomposition $T = U \circ P$, where U is a unitary operator.

9. Deduce that every normal $T \in \mathcal{L}(H)$ admits a polar decomposition $T = U \circ P$ where U is a unitary operator and $U \circ P = P \circ U$.

10. Show that every operator $T \in \mathcal{L}(H)$ satisfying (2) can be factored as $T = P \circ U$, where $U \in \mathcal{L}(H)$ is a unitary operator and $P \in \mathcal{L}(H)$ is a self-adjoint nonnegative operator.

    [**Hint**: Apply question 8 to $T^\star$.]

11. Show that every operator $T \in \mathcal{K}(H)$ satisfying (1) admits a polar decomposition $T = U \circ P$, where $P \in \mathcal{K}(H)$.

12. Assume that H is separable and $T \in \mathcal{K}(H)$ (but T does not necessarily satisfy (1)). Prove that there exist two orthonormal bases (e_n) and (f_n) of H such that

$$Tu = \sum_{n=1}^{\infty} \alpha_n (u, e_n) f_n \quad \forall u \in H,$$

    where (α_n) is a sequence such that $\alpha_n \geq 0 \ \forall n$ and $\alpha_n \to 0$ as $n \to \infty$. Compute $T^\star$. Conversely, show that any operator of this form must be compact.

PROBLEM 45 (8)

Strong maximum principle

Consider the bilinear form

$$a(u, v) = \int_0^1 pu'v' + quv,$$

where $p \in C^1([0, 1])$, $p \geq \alpha > 0$ on $(0, 1)$, and $q \in C([0, 1])$. We assume that a is coercive on $H_0^1(0, 1)$ (but we make no sign assumption on q).

Given $f \in L^2(0, 1)$, let $u \in H^2(0, 1)$ be the solution of

(1)
$$\begin{cases} -(pu')' + qu = f & \text{on } (0, 1), \\ u(0) = u(1) = 0. \end{cases}$$

Assume that $f \geq 0$ a.e. on $(0, 1)$ and $f \not\equiv 0$. Our goal is to prove that

(2)
$$u'(0) > 0, \ u'(1) < 0$$

and

(3)
$$u(x) > 0 \quad \forall x \in (0, 1).$$

1. Assume that $\psi \in H^1(0, 1)$ satisfies

(4)
$$\begin{cases} a(\psi, v) \leq 0 \quad \forall v \in H_0^1(0, 1), v \geq 0 \text{ on } (0, 1), \\ \psi(0) \leq 0, \ \psi(1) \leq 0. \end{cases}$$

Prove that $\psi \leq 0$ on $(0, 1)$.

[**Hint:** Take $v = \psi^+$ in (4) and use Exercise 8.11.]

Consider the problem

(5)
$$\begin{cases} -(p\zeta')' + q\zeta = 0 & \text{on } (0, 1), \\ \zeta(0) = 0, \; \zeta(1) = 1. \end{cases}$$

2. Show that (5) has a unique solution ζ and that $\zeta \geq 0$ on $(0, 1)$.
3. Check that $u \geq 0$ on $(0, 1)$ and deduce that $u'(0) \geq 0$ and $u'(1) \leq 0$.
4. Prove that

(6)
$$p(1)|u'(1)| = \int_0^1 f\zeta.$$

[**Hint:** Multiply (1) by ζ and (5) by u.]

Set $\varphi(x) = (e^{Bx} - 1)$, $B > 0$.

5. Check that if B is sufficiently large (depending only on p and q), then

(7)
$$-(p\varphi')' + q\varphi \leq 0 \quad \text{on } (0, 1).$$

In what follows we fix B such that (7) holds.
6. Let $A = (e^B - 1)^{-1}$. Prove that

$$\zeta \geq A\varphi \quad \text{on } (0, 1).$$

[**Hint:** Apply question 1 to $\psi = A\varphi - \zeta$.]
7. Deduce that $u'(1) < 0$.

[**Hint:** Apply question 4.]
8. Check that $u'(0) > 0$.

[**Hint:** Change t into $(1 - t)$.]
9. Fix $\delta \in (0, \frac{1}{2})$ so small that

$$\frac{u(x)}{x} \geq \frac{1}{2}u'(0) \; \setminus \forall x \in (0, \delta) \quad \text{and} \quad \frac{u(x)}{1-x} \geq \frac{1}{2}|u'(1)| \quad \forall x \in (1 - \delta, 1).$$

Why does such δ exist? Let v be the solution of the problem

$$\begin{cases} -(pv')' + qv = 0 & \text{on } (\delta, 1 - \delta), \\ v(\delta) = v(1 - \delta) = \gamma, \end{cases}$$

where
$$\gamma = \frac{\delta}{2} \min\{u'(0), |u'(1)|\}.$$

Show that $u \geq v \geq 0$ on $(\delta, 1 - \delta)$.
10. Prove that $v > 0$ on $(\delta, 1 - \delta)$.

[**Hint:** Assume by contradiction that $v(x_0) = 0$ for some $x_0 \in (\delta, 1 - \delta)$, and apply Theorem 7.3 (Cauchy–Lipschitz–Picard) as in Exercise 8.33.]

11. Deduce that $u(x) > 0 \; \forall x \in (0, 1)$.

Finally, we present a sharper form of the strong maximum principle.

12. Prove that there is a constant $a > 0$ (depending only on p and q) such that

$$u(x) \geq ax(1 - x) \int_0^1 f(t)t(1 - t)dt.$$

[**Hint:** Start with the case where $p \equiv 1$ and $q \equiv k^2$ is a positive constant; use an explicit solution of (1). Next, consider the case where $p \equiv 1$ and no further assumption is made on q. Finally, reduce the general case to the previous situation, using a change of variable.]

PROBLEM 46 (8)

The method of subsolutions and supersolutions

Let $h(t) : [0, +\infty) \to [0, +\infty)$ be a continuous nondecreasing function. Assume that there exist two functions $v, w \in C^2([0, 1])$ satisfying

(1)
$$\begin{cases} 0 \leq v \leq w & \text{on } I = (0, 1), \\ -v'' + v \leq h(v) & \text{on } I, \quad v(0) = v(1) = 0, \\ -w'' + w \geq h(w) & \text{on } I, \quad w(0) \geq 0, w(1) \geq 0, \end{cases}$$

(v is called a *subsolution* and w a *supersolution*). The goal is to prove that there exists a solution $u \in C^2([0, 1])$ of the problem

(2)
$$\begin{cases} -u'' + u = h(u) & \text{on } I, \\ u(0) = u(1) = 0, \\ v \leq u \leq w & \text{on } I. \end{cases}$$

Consider the sequence $(u_n)_{n \geq 1}$ defined inductively by

(3)
$$\begin{cases} -u_n'' + u_n = h(u_{n-1}) & \text{on } I, \; n \geq 1, \\ u_n(0) = u_n(1) = 0, \end{cases}$$

starting from $u_0 = w$.

1. Show that $v \leq u_1 \leq w$ on I.

[**Hint:** Apply the maximum principle to $(u_1 - w)$ and to $(u_1 - v)$.]

2. Prove by induction that for every $n \geq 1$,

$$v \leq u_n \text{ on } I \quad \text{and} \quad u_{n+1} \leq u_n \text{ on } I.$$

3. Deduce that the sequence (u_n) converges in $L^2(I)$ to a limit u and that $h(u_n) \to h(u)$ in $L^2(I)$.
4. Show that $u \in H_0^1(I)$, and that

$$\int_0^1 u'\varphi' + \int_0^1 u\varphi = \int_0^1 h(u)\varphi \quad \forall \varphi \in H_0^1(I).$$

5. Conclude that $u \in C^2([0, 1])$ is a classical solution of (2).

In what follows we choose $h(t) = t^\alpha$, where $0 < \alpha < 1$. The goal is to prove that there exists a unique function $u \in C^2([0, 1])$ satisfying

(4)
$$\begin{cases} -u'' + u = u^\alpha & \text{on } I, \\ u(0) = u(1) = 0, \\ u(x) > 0 & \forall x \in I. \end{cases}$$

6. Let $v(x) = \varepsilon \sin(\pi x)$ and $w(x) \equiv 1$. Show that if ε is sufficiently small, assumption (1) is satisfied. Deduce that there exists a solution of (4).

We now turn to the question of uniqueness. Let u be the solution of (4) obtained by the above method, starting with $u_0 \equiv 1$. Let $\tilde{u} \in C^2([0, 1])$ be another solution of (4).

7. Show that $\tilde{u} \leq 1$ on I.

[**Hint:** Consider a point $x_0 \in [0, 1]$ where $\tilde{u}$ achieves its maximum.]

8. Prove that the sequence $(u_n)_{n \geq 1}$ defined by (3), starting with $u_0 \equiv 1$, satisfies

$$\tilde{u} \leq u_n \quad \text{on } I,$$

and deduce that $\tilde{u} \leq u$ on I.
9. Show that

$$\int_0^1 (\tilde{u}^\alpha u - u^\alpha \tilde{u}) = 0.$$

10. Conclude that $\tilde{u} = u$ on I.

[**Hint:** Write $\tilde{u}^\alpha u - u^\alpha \tilde{u} = u\tilde{u}(\tilde{u}^{\alpha-1} - u^{\alpha-1})$ and note that $u^{\alpha-1} \leq \tilde{u}^{\alpha-1}$.]

We now present an alternative proof of existence. Set, for every $u \in H_0^1(I)$,

$$F(u) = \frac{1}{2}\int_0^1 (u'^2 + u^2) - \int_0^1 g(u),$$

where $g(t) = \frac{1}{\alpha+1}(t^+)^{\alpha+1}$, $0 < \alpha < 1$, and $t^+ = \max(t, 0)$.
11. Prove that there exists a constant C such that

$$F(u) \geq \frac{1}{2}\|u\|_{H^1}^2 - C\|u\|_{H^1}^{\alpha+1} \quad \forall u \in H_0^1(I).$$

12. Deduce that
$$m = \inf_{v \in H_0^1(I)} F(v) > -\infty,$$
and that the infimum is achieved.

[**Hint:** Let (u_n) be a minimizing sequence. Check that a subsequence (u_{n_k}) converges weakly in $H_0^1(I)$ to a limit u and that $\int_0^1 g(u_{n_k}) \to \int_0^1 g(u)$. The reader is warned that the functional F is *not* convex; why?]

13. Show that $m < 0$.

[**Hint:** Prove that $F(\varepsilon v) < 0$ for all $v \in H_0^1(I)$ such that $v^+ \not\equiv 0$ and for all ε sufficiently small.]

14. Check that
$$g(b) - g(a) \geq (a^+)^\alpha (b-a) \quad \forall a, b \in \mathbb{R}.$$

15. Let $u \in H_0^1(I)$ be a minimizer of F on $H_0^1(I)$. Prove that
$$\int_0^1 (u'v' + uv) = \int_0^1 (u^+)^\alpha v \quad \forall v \in H_0^1(I).$$

[**Hint:** Write that $F(u) \leq F(u + tv)$, apply question 14, and let $t \to 0$.]

16. Deduce that $u \in C^2([0,1])$ is a solution of
$$(5) \qquad \begin{cases} -u'' + u = (u^+)^\alpha & \text{on } I, \\ u(0) = u(1) = 0. \end{cases}$$
Prove that $u \geq 0$ on I and $u \not\equiv 0$.

17. Conclude that $u > 0$ on I using the strong maximum principle (see Problem 45).

PROBLEM 47 (8)

Poincaré–Wirtinger's inequalities

Let $I = (0, 1)$.

- A -

1. Prove that
$$(1) \qquad \|u - \overline{u}\|_{L^\infty(I)} \leq \|u'\|_{L^1(I)} \quad \forall u \in W^{1,1}(I), \text{ where } \overline{u} = \int_I u.$$

[**Hint:** Note that $\overline{u} = u(x_0)$ for some $x_0 \in [0,1]$.]

2. Show that the constant 1 in (1) is optimal, i.e.,
$$(2) \qquad \sup\{\|u - \overline{u}\|_{L^\infty(I)}; \ u \in W^{1,1}(I), \text{ and } \|u'\|_{L^1(I)} = 1\} = 1.$$

[**Hint:** Consider a sequence (u_n) of smooth functions on $[0, 1]$ such that $u'_n \geq 0$ on $(0, 1)$ $\forall n$, $u_n(1) = 1$ $\forall n$, $u_n(x) = 0$ $\forall x \in [0, 1 - \frac{1}{n}]$, $\forall n$.]

3. Prove that the sup in (2) is not achieved, i.e., there exists no function $u \in W^{1,1}(I)$ such that
$$\|u - \bar{u}\|_{L^\infty(I)} = 1 \quad \text{and} \quad \|u'\|_{L^1(I)} = 1.$$

4. Prove that

    (3) $$\|u\|_{L^\infty(I)} \leq \frac{1}{2}\|u'\|_{L^1(I)} \quad \forall u \in W_0^{1,1}(I).$$

    [**Hint:** Write that $|u(x) - u(0)| \leq \int_0^x |u'(t)|dt$ and $|u(x) - u(1)| \leq \int_x^1 |u'(t)|dt$.]

5. Show that $\frac{1}{2}$ is the best constant in (3). Is it achieved?

    [**Hint:** Fix $a \in (0, 1)$ and consider a function $u \in W_0^{1,1}(I)$ increasing on $(0, a)$, decreasing on $(a, 1)$, with $u(a) = 1$.]

6. Deduce that the following inequalities hold:

    (4) $$\|u - \bar{u}\|_{L^q(I)} \leq C\|u'\|_{L^p(I)} \quad \forall u \in W^{1,p}(I).$$

    and

    (5) $$\|u\|_{L^q(I)} \leq C\|u'\|_{L^p(I)} \quad \forall u \in W_0^{1,p}(I)$$

    with $1 \leq q \leq \infty$ and $1 \leq p \leq \infty$.
    Prove that the best constants in (4) and (5) are achieved when $1 \leq q \leq \infty$ and $1 < p \leq \infty$.

    [**Hint:** Minimize $\|u'\|_{L^p(I)}$ in the class $u \in W^{1,p}(I)$ such that $\|u - \bar{u}\|_{L^q(I)} = 1$, resp. $u \in W_0^{1,p}(I)$ and $\|u\|_{L^q(I)} = 1$.]

- **B** -

The next goal is to find the best constant in (4) when $p = q = 2$, i.e.,

(6) $$\|u - \bar{u}\|_{L^2(I)} \leq C\|u'\|_{L^2(I)} \quad \forall u \in H^1(I).$$

Set $H = \{f \in L^2(I); \int_I f = 0\}$ and $V = \{v \in H^1(I); \int_I v = 0\}$.

1. Check that for every $f \in H$ there exists a unique $u \in V$ such that
$$\int_I u'v' = \int_I fv \quad \forall v \in V.$$

2. Prove that $u \in H^2(I)$ and satisfies

$$\begin{cases} -u'' = f & \text{a.e. on } I, \\ u'(0) = u'(1) = 0. \end{cases}$$

3. Show that the operator $T : H \to H$ defined by $Tf = u$ is self-adjoint, compact, and that $\int_I fTf \geq 0 \ \forall f \in H$.

4. Let λ_1 be the largest eigenvalue of T. Prove that (6) holds with $C = \sqrt{\lambda_1}$ and that $\sqrt{\lambda_1}$ is the best constant in (6).

   [**Hint:** Use Exercise 6.24.]

5. Compute explicitly the best constant in (6).

- C -

1. Prove that

   (7) $$\|u - \bar{u}\|_{L^1(I)} \leq 2 \int_I |u'(t)| t(1 - t) dt \quad \forall u \in W^{1,1}(I).$$

2. Deduce that

   (8) $$\|u - \bar{u}\|_{L^1(I)} \leq \frac{1}{2}\|u'\|_{L^1(I)} \quad \forall u \in W^{1,1}(I).$$

3. Show that the constant $1/2$ in (8) is optimal, i.e.,

   (9) $$\sup\{\|u - \bar{u}\|_{L^1(I)}; \ u \in W^{1,1}(I), \ \text{and} \ \|u'\|_{L^1(I)} = 1\} = \frac{1}{2}.$$

4. Is the sup in (9) achieved?

PROBLEM 48 (8)

A nonlinear problem

Let $j : [-1, +1] \to [0, +\infty)$ be a continuous convex function such that $j \in C^2((-1, +1))$, $j(0) = 0$, $j'(0) = 0$, and

$$\lim_{t \uparrow +1} j'(t) = +\infty, \quad \lim_{t \downarrow -1} j'(t) = -\infty.$$

(A good example to keep in mind is $j(t) = 1 - \sqrt{1 - t^2}$, $t \in [-1, +1]$.) Given $f \in L^2(0, 1)$, define the function $\varphi : H_0^1(0, 1) \to (-\infty, +\infty]$ by

$$\varphi(v) = \begin{cases} \frac{1}{2} \int_0^1 v'^2 + \int_0^1 j(v) - \int_0^1 fv & \text{if } v \in H_0^1(0, 1) \text{ and } \|v\|_{L^\infty} \leq 1, \\ +\infty & \text{otherwise.} \end{cases}$$

1. Check that φ is convex l.s.c. on $H_0^1(0, 1)$ and that $\lim_{\|v\|_{H_0^1} \to +\infty} \varphi(v) = +\infty$.

2. Deduce that there exists a unique $u \in H_0^1(0, 1)$ such that

$$\varphi(u) = \min_{v \in H^1(0,1)} \varphi(v).$$

The goal is to prove that if $f \in L^\infty(0, 1)$ then $\|u\|_{L^\infty(0,1)} < 1$, $u \in H^2(0, 1)$, and u satisfies

(1)
$$\begin{cases} -u'' + j'(u) = f & \text{on } (0, 1), \\ u(0) = u(1) = 0. \end{cases}$$

3. Check that

$$j(t) - j(a) \geq j'(a)(t - a) \quad \forall t \in [-1, +1], \quad \forall a \in (-1, +1).$$

[**Hint:** Use the convexity of j.]

Fix $a \in [0, 1)$.

4. Set $v = \min(u, a)$. Prove that $v \in H_0^1(0, 1)$ and that

$$v' = \begin{cases} u' & \text{a.e. on } [u \leq a], \\ 0 & \text{a.e. on } [u > a]. \end{cases}$$

[**Hint:** Write $v = a - (a - u)^+$ and use Exercise 8.11.]

5. Prove that

$$\frac{1}{2} \int_{[u>a]} u'^2 \leq \int_{[u>a]} (f - j'(a))(u - a).$$

[**Hint:** Write that $\varphi(u) \leq \varphi(v)$, where v is defined in question 4. Then use question 3.]

6. Choose $a \in [0, 1)$ such that $f(x) \leq j'(a) \ \forall x \in [0, 1]$ and prove that $u(x) \leq a$ $\forall x \in [0, 1]$.

[**Hint:** Show that $\int_0^1 w'^2 = 0$, where $w = (u - a)^+$ belongs to $H_0^1(0, 1)$; why?]

7. Conclude that $\|u\|_{L^\infty(0,1)} < 1$.

[**Hint:** Apply the previous argument, replacing u by $-u$, $j(t)$ by $j(-t)$, and f by $-f$.]

8. Deduce that u belongs to $H^2(0, 1)$ and satisfies (1).

[**Hint:** Write that $\varphi(u) \leq \varphi(u + \varepsilon v)$ with $v \in H_0^1(0, 1)$ and ε small.]

9. Check that $u \in C^2([0, 1])$ if $f \in C([0, 1])$.

10. Conversely, show that any function $u \in C^2([0, 1])$ such that $\|u\|_{L^\infty(0,1)} < 1$, and satisfying (1), is a minimizer of φ on $H_0^1(0, 1)$.

    [**Hint:** Use question 3 with $t = v(x)$ and $a = u(x)$.]

    Assume now that $f \in L^2(0, 1)$. Set $f_n = T_n f$, where T_n is the truncation operation (defined in Chapter 4 after Theorem 4.12). Let u_n be the solution of (1) corresponding to f_n.

11. Prove that $\|j'(u_n)\|_{L^2(0,1)} \le C$ as $n \to \infty$.

    [**Hint:** Multiply (1) by $j'(u_n)$.]

12. Deduce that $\|u_n\|_{H^2(0,1)} \le C$.

13. Show that a subsequence (u_{n_k}) converges weakly in $H^2(0, 1)$ to a limit $u \in H^2(0, 1)$ with $u_{n_k} \to u$ in $C^1([0, 1])$. Prove that $|u(x)| < 1$ a.e. on $(0, 1)$, and $j'(u) \in L^2(0, 1)$.

    [**Hint:** Apply Fatou's lemma to the sequence by $j'(u_{n_k})^2$.]

14. Show that $j'(u_{n_k})$ converges weakly in $L^2(0, 1)$ to $j'(u)$ and deduce that (1) holds.

    [**Hint:** Apply Exercise 4.16.]

15. Deduce that $\|u\|_{L^\infty(0,1)} < 1$ if one assumes, in addition, that

$$\liminf_{t \uparrow 1} j'(t)(1 - t)^{1/3} > 0 \quad \text{and} \quad \limsup_{t \downarrow -1} j'(t)(1 + t)^{1/3} < 0.$$

    [**Hint:** Assume, by contradiction, that $u(x_0) = 1$ for some $x_0 \in (0, 1)$. Check that $|u'(x)| \le |x - x_0|^{1/2}\|u''\|_{L^2}$ $\forall x \in (0, 1)$ and $|u(x) - 1| \le \frac{2}{3}|x - x_0|^{3/2}\|u''\|_{L^2}$ $\forall x \in (0, 1)$. Deduce that $j'(u) \notin L^2(0, 1)$.]

PROBLEM 49 (8)

Min–max principles for the eigenvalues of Sturm–Liouville operators

Consider the Sturm–Liouville operator $Au = -(pu')' + qu$ on $(0, 1)$ with Dirichlet boundary condition $u(0) = u(1) = 0$. Assume that $p \in C^1([0, 1])$, $p(x) \ge \alpha > 0$ $\forall x \in [0, 1]$, and $q \in C([0, 1])$. Set

$$a(u, v) = \int_0^1 (pu'v' + quv) \quad \forall u, v \in H_0^1(0, 1).$$

Note that we make no further assumption on q, so that the bilinear form a need not be coercive. Fix M sufficiently large that $\tilde{a}(u, v) = a(u, v) + M \int_0^1 uv$ is coercive (e.g., $M > -\min_{x \in [0,1]} q(x)$). Let (λ_k) be the sequence of eigenvalues of A. The space $H = H_0^1(0, 1)$ is equipped with the scalar product $\tilde{a}(u, v)$, now denoted by $(u, v)_H$, and the corresponding norm $|u|_H = \tilde{a}(u, u)^{1/2}$. Given any $f \in L^2(0, 1)$, let $u \in H_0^1(0, 1)$ be the unique solution of the problem

$$\tilde{a}(u, v) = \int_0^1 fv \quad \forall v \in H_0^1(0, 1).$$

Set $u = Tf$ and consider T as an operator from H into itself.

1. Show that T is self-adjoint and compact.

   [**Hint:** Recall that the identity map from H into $L^2(0, 1)$ is compact.]

2. Let (λ_k) be the sequence of eigenvalues of A (in the sense of Theorem 8.22) with corresponding eigenfunctions (e_k), and let (μ_k) be the sequence of eigenvalues of T. Check that $\mu_k > 0 \ \forall k$ and show that

$$\lambda_k = \frac{1}{\mu_k} - M \quad \forall k \quad \text{and} \quad T(e_k) = \mu_k e_k \quad \forall k.$$

3. Prove that

$$(Tw, w)_H = \int_0^1 w^2 \quad \forall w \in H,$$

   and deduce that

$$\frac{1}{R(w)} = \frac{a(w, w)}{\int_0^1 w^2} + M \quad \forall w \in H, \ w \neq 0,$$

   where R is the Rayleigh quotient associated with T, i.e., $R(w) = \frac{(Tw, w)_H}{|w|_H^2}$ (see Problem 37).

4. Prove that

   (1) $$\lambda_1 = \min_{\substack{w \in H_0^1 \\ w \neq 0}} \left\{ \frac{a(w, w)}{\int_0^1 w^2} \right\},$$

   and $\forall k \geq 2$,

$$\lambda_k = \min \left\{ \frac{a(w, w)}{\int_0^1 w^2}; \ w \in H_0^1(0, 1), \ w \neq 0 \ \text{and} \ \int_0^1 w e_j = 0 \ \forall j = 1, 2, \ldots, k - 1 \right\}.$$

   [**Hint:** Apply question 2 in Problem 37 and show that $(w, e_j)_H = 0$ iff $\int_0^1 w e_j = 0$.]

5. Prove that $\forall k \geq 1$,

$$\lambda_k = \min_{\substack{\Sigma \subset H_0^1(0,1) \\ \dim \Sigma = k}} \max_{\substack{u \in \Sigma \\ u \neq 0}} \left\{ \frac{a(u, u)}{\int_0^1 u^2} \right\},$$

   and

$$\lambda_k = \max_{\substack{\Lambda \subset H_0^1(0,1) \\ \text{codim } \Lambda = k-1}} \min_{\substack{u \in \Lambda \\ u \neq 0}} \left\{ \frac{a(u,u)}{\int_0^1 u^2} \right\},$$

where Σ and Λ are closed subspaces of $H_0^1(0,1)$.

[**Hint:** Apply question 7 in Problem 37.]

6. Prove similar results for the Sturm–Liouville operator with Neumann boundary conditions.

   We now return to formula (1) and discuss further properties of the eigenfunctions corresponding to the first eigenvalue λ_1. In particular, we will see that there is a positive eigenfunction generating the eigenspace associated to λ_1.

7. Let $w_0 \in H_0^1(0,1)$ be a minimizer of (1) such that $\int_0^1 w_0^2 = 1$. Show that

$$Aw_0 = \lambda_1 w_0 \quad \text{on } (0,1).$$

8. Set $w_1 = |w_0|$. Check that w_1 is also a minimizer of (1) and deduce that

   (2) $\qquad\qquad\qquad Aw_1 = \lambda_1 w_1 \quad \text{on } (0,1).$

[**Hint:** Use Exercise 8.11.]

9. Prove that $w_1 > 0$ on $(0,1)$, $w_1'(0) > 0$, and $w_1'(1) < 0$.

[**Hint:** Apply the strong maximum principle to the operator $A + M$; see Problem 45.]

10. Assume that $w \in H_0^1(0,1)$ satisfies

$$Aw = \lambda_1 w \quad \text{on } (0,1).$$

Prove that w is a multiple of w_1.

[**Hint:** Recall that eigenvalues are simple; see Exercise 8.33. Find another proof that does not rely on the simplicity of eigenvalues; use w^2/w_1 as test function in (2).]

11. Show that any function $\psi \in H_0^1(0,1)$ satisfying

$$A\psi = \mu\psi \quad \text{on } (0,1), \quad \psi \geq 0 \quad \text{on } (0,1), \quad \text{and} \quad \int_0^1 \psi^2 = 1,$$

for some $\mu \in \mathbb{R}$, must coincide with w_1.

[**Hint:** If $\mu \neq \lambda_1$, check that $\int_0^1 \psi w_1 = 0$. Deduce that $\mu = \lambda_1$.]

PROBLEM 50 (8)

Another nonlinear problem

Let $q \in C([0, 1])$ and consider the bilinear form

$$a(u, v) = \int_0^1 (u'v' + quv), \quad u, v \in H_0^1(0, 1).$$

Assume that there exists $v_1 \in H_0^1(0, 1)$ such that

(1) $a(v_1, v_1) < 0.$

1. Check that assumption (1) is equivalent to

(2) $\lambda_1(A) < 0,$

where $\lambda_1(A)$ is the first eigenvalue of the operator $Au = -u'' + qu$ with zero Dirichlet condition.

2. Verify that

(3) $-\infty < m = \inf_{u \in H_0^1(0,1)} \left\{ \frac{1}{2} a(u, u) + \frac{1}{4} \int_0^1 |u|^4 \right\} < 0.$

[**Hint:** Use $u = \varepsilon v_1$ with $\varepsilon > 0$ sufficiently small.]

3. Prove that the inf in (3) is achieved by some u_0.

[*Warning:* The functional in (3) is *not* convex; why?]

    Our goal is to prove that (3) admits precisely two minimizers.

4. Prove that u_0 belongs to $C^2([0, 1])$ and satisfies

(4) $\begin{cases} -u'' + qu + u^3 = 0 & \text{on } (0, 1), \\ u(0) = u(1) = 0. \end{cases}$

5. Set $u_1 = |u_0|$. Show that u_1 is also a minimizer for (3). Deduce that u_1 satisfies (4).

[**Hint:** Apply Exercise 8.11.]

6. Prove that $u_1(x) > 0 \ \forall x \in (0, 1)$, $u_1'(0) > 0$, and $u_1'(1) < 0$.

[**Hint:** Choose a constant a so large that $-u_1'' + a^2 u_1 = f \geq 0$, $f \not\equiv 0$. Then use the strong maximum principle.]

7. Let $u_0 \in H_0^1(0, 1)$ be again any minimizer in (3). Prove that either $u_0(x) > 0$ $\forall x \in (0, 1)$, or $u_0(x) < 0 \ \forall x \in (0, 1)$.

[**Hint:** Check that $|u_0(x)| > 0 \ \forall x \in (0, 1)$.]

8. Let U_1 be any solution of (4) satisfying $U_1 \geq 0$ on $[0, 1]$, and $U_1 \not\equiv 0$. Set $\rho_1 = U_1^2$. Consider the functional

$$\Phi(\rho) = \int_0^1 \left(|(\sqrt{\rho})'|^2 + q\rho + \frac{1}{2}\rho^2 \right)$$

defined on the set

$$K = \left\{ \rho \in H_0^1(0, 1); \, \rho \geq 0 \text{ on } (0, 1) \text{ and } \sqrt{\rho} \in H_0^1(0, 1) \right\}.$$

Prove that

(5) $$\Phi(\rho) - \Phi(\rho_1) \geq \frac{1}{2} \int_0^1 (\rho - \rho_1)^2 \quad \forall \rho \in K.$$

[**Hint:** Let $u \in C_c^1((0, 1))$. Note that

$$2\frac{U_1' u u'}{U_1} \leq u'^2 + \frac{U_1'^2 u^2}{U_1^2} \quad \text{on } (0, 1),$$

and deduce (using integration by parts) that

$$\int_0^1 (u'^2 - U_1'^2) \geq -\int_0^1 \frac{U_1''}{U_1}(u^2 - U_1^2) \quad \forall u \in H_0^1(0, 1).$$

Then apply equation (4) to establish (5).]

9. Deduce that there exists exactly one nontrivial solution u of (4) such that $u \geq 0$ on $[0, 1]$. Denote it by U_0.

[*Comment:* There exist in general many sign-changing solutions of (4).]

10. Prove that there exist exactly two minimizers for (3): U_0 and $-U_0$.

PROBLEM 51 (8)

Harmonic oscillator. Hermite polynomials.

Let $p \in C(\mathbb{R})$ be such that $p \geq 0$ on $\mathbb{R}$. Consider the space

$$V = \left\{ v \in H^1(\mathbb{R}); \int_{-\infty}^{+\infty} pv^2 < \infty \right\}$$

equipped with the scalar product

$$(u, v)_V = \int_{-\infty}^{+\infty} (u'v' + uv + puv),$$

and the corresponding norm $|u|_V = (u, u)_V^{1/2}$.

1. Check that V is a separable Hilbert space.
2. Show that $C_c^\infty(\mathbb{R})$ is dense in V.

[Hint: Let ζ_n be a sequence of cut-off functions as in the proof of Theorem 8.7. Given $u \in V$, consider $\zeta_n u$ and then use convolution.**]**

Consider the bilinear form

$$a(u, v) = \int_{-\infty}^{+\infty} u'v' + puv, \quad u, v \in V.$$

In what follows we assume that there exist constants $\delta > 0$ and $A > 0$ such that

(1) $p(x) \geq \delta \quad \forall x \in \mathbb{R}$ with $|x| \geq A.$

3. Prove that a is coercive on V. Deduce that for every $f \in L^2(\mathbb{R})$ there exists a unique solution $u \in V$ of the problem

(2) $a(u, v) = \int_{-\infty}^{+\infty} fv \quad \forall v \in V.$

4. Assuming that $f \in L^2(\mathbb{R}) \cap C(\mathbb{R})$, show that u satisfies

(3) $\begin{cases} u \in C^2(\mathbb{R}), \\ -u'' + pu = f \quad \text{on } \mathbb{R}, \\ u(x) \to 0 \quad \text{as } |x| \to \infty. \end{cases}$

5. Conversely, prove that any solution u of (3) belongs to V and satisfies (2).

   [Hint: Multiply the equation $-u'' + pu = f$ by $\zeta_n^2 u$ and use the fact that a is coercive.**]**

   In what follows we assume that

(4) $\lim_{|x| \to \infty} p(x) = +\infty.$

6. Given $f \in L^2(\mathbb{R})$, set $u = Tf$, where u is the solution of (2). Prove that $T : L^2(\mathbb{R}) \to L^2(\mathbb{R})$ is self-adjoint and compact.

   [Hint: Using Corollary 4.27 check that $V \subset L^2(\mathbb{R})$ with compact injection.**]**

7. Deduce that there exist a sequence (λ_n) of positive numbers with $\lambda_n \to \infty$ as $n \to \infty$, and a Hilbert basis (e_n) of $L^2(\mathbb{R})$ satisfying

(5) $\begin{cases} e_n \in V \cap C^2(\mathbb{R}), \\ -e_n'' + pe_n = \lambda_n e_n \quad \text{on } \mathbb{R}. \end{cases}$

   In what follows we take $p(x) = x^2$.

8. Check that (5) admits a solution of the form $e_n(x) = e^{-x^2/2} P_n(x)$, where $\lambda_n = (2n + 1)$ and $P_n(x)$ is a polynomial of degree n.

Partial Solutions of the Problems

Problem 1

5. In view of Zorn's lemma (Lemma 1.1) it suffices to check that $\mathcal{F}$ is inductive. Let $(A_i)_{i \in I}$ be a totally ordered subset of $\mathcal{F}$. Set $A = \bigcap_{i \in I} A_i$ and check that A is nonempty, A is an extreme set of K, $A \in \mathcal{F}$, and A is an upper bound for $(A_i)_{i \in I}$.

6. Suppose not, that there are two distinct points $a, b \in M_0$. By Hahn–Banach (Theorem 1.7) there exists some $f \in E^\star$ such that $\langle f, a \rangle \neq \langle f, b \rangle$. Set

$$M_1 = \left\{ x \in M_0;\ \langle f, x \rangle = \max_{y \in M_0} \langle f, y \rangle \right\}.$$

   Clearly $M_1 \in \mathcal{F}$ and $M_0 \leq M_1$. Since M_0 is maximal, it follows that $M_1 = M_0$. This is absurd, since the points a and b cannot both belong to M_1.

8. Let K_1 be the closed convex hull of all the extreme points of K. Assume, by contradiction, that there exists some point $a \in K$ such that $a \notin K_1$. Then there exists some hyperplane strictly separating $\{a\}$ and K_1. Let $f \in E^\star$ be such that

$$\langle f, x \rangle < \langle f, a \rangle \quad \forall x \in K_1.$$

   Note that

$$B = \left\{ x \in K;\ \langle f, x \rangle = \max_{y \in K} \langle f, y \rangle \right\}$$

   is an extreme set of K such that $B \cap K_1 = \emptyset$. But B contains at least one extreme point of K; absurd.

9. (a) $\mathcal{E} = \{ x = (x_i);\ |x_i| = 1\ \forall i \}$,
   (b) $\mathcal{E} = \{ x = (x_i);\ |x_i| = 1\ \forall i,\ \text{and } x_i \text{ is stationary for large } i \}$,
   (c) $\mathcal{E} = \emptyset$,
   (d) $\mathcal{E} = \{ x = (x_i);\ \exists j \text{ such that } |x_j| = 1,\ \text{and } x_i = 0\ \forall i \neq j \}$,
   (e) $\mathcal{E} = \{ x = (x_i);\ \sum |x_i|^p = 1 \}$,
   (f) $\mathcal{E} = \emptyset$.
   To see that $\mathcal{E} = \emptyset$ in the case (f) let $f \in L^1(\mathbb{R})$ be any function such that $\int_{\mathbb{R}} |f| = 1$. By a translation we may always assume that $\int_{-\infty}^{0} |f| = \int_{0}^{\infty} |f| = 1/2$. Then

write $f = (g + h)/2$ with

$$g = \begin{cases} 2f & \text{on } (-\infty, 0), \\ 0 & \text{on } (0, +\infty), \end{cases} \quad \text{and} \quad h = \begin{cases} 0 & \text{on } (-\infty, 0), \\ 2f & \text{on } (0, +\infty). \end{cases}$$

Problem 2

Determine $\partial\varphi(x)$ for the function φ defined by $\varphi(x) = -\sqrt{x}$ for $x \geq 0$ and $\varphi(x) = +\infty$ for $x < 0$.

-A-

4. (a) $\partial\varphi(x) = F(x)$,
    (b) $\partial\varphi(x) = \frac{1}{\|x\|} F(x)$ if $x \neq 0$ and $\partial\varphi(0) = B_{E^\star}$,
    (c) $\partial\varphi(x) = \begin{cases} 0 & \text{if } x \in \text{Int } K, \\ \text{outward normal cone at } x & \text{if } x \in \text{Boundary of } K, \end{cases}$
    $\partial\varphi(x) = K^{\perp}$ if K is a linear subspace,
    (d) $\partial\varphi(x) = D\varphi(x) = $ differential of φ at x.
5. Study the following example: In $E = \mathbb{R}^2$ (equipped with the Euclidean norm),

$$\varphi = I_C \quad \text{with } C = \{[x_1, x_2]; (x_1 - 1)^2 + x_2^2 \leq 1\},$$

and

$$\psi = I_D \quad \text{with } D = \{[x_1, x_2]; x_1 = 0\}.$$

- B -

1. Let $C = \text{epi } \varphi$. Apply Hahn–Banach (first geometric form) with $A = \text{Int } C$ and $B = [x_0, \varphi(x_0)]$. Note that $A \neq \emptyset$ (why?). Hence there exist some $f \in E^\star$ and some constants k and a such that $\|f\| + |k| \neq 0$ and

$$\langle f, x \rangle + k\lambda \geq a \geq \langle f, x_0 \rangle + k\varphi(x_0) \quad \forall x \in D(\varphi), \forall \lambda \geq \varphi(x).$$

Check that $k > 0$ and deduce that $-\frac{1}{k} f \in \partial\varphi(x_0)$.

6. Note that $\inf_E (\tilde{\varphi} + \tilde{\psi}) = 0$, and so there exists some $g \in E^\star$ such that $\tilde{\varphi}^\star(-g) + \tilde{\psi}^\star(g) = 0$. Check that $f_0 - g \in \partial\varphi(x)$, and that $g \in \partial\psi(x)$; thus $f_0 \in \partial\varphi(x) + \partial\psi(x)$.

-C-

1. For every $R > 0$ and every $x_0 \in E$ we have

$$\varphi(x) \leq k(\|x_0\| + R) + C \equiv M(R) \quad \forall x \in E \text{ with } \|x - x_0\| \leq R.$$

Thus

$$\|f\| \leq \frac{1}{R} (k\|x_0\| + kR + C - \varphi(x_0)) \quad \forall f \in \partial\varphi(x_0).$$

Letting $R \to \infty$ we see that $\|f\| \le k \ \forall f \in \partial\varphi(x_0)$ and consequently

$$\varphi(x) - \varphi(x_0) \ge -k\|x - x_0\| \quad \forall x, x_0 \in E.$$

We have $D(\varphi^\star) \subset k B_{E^\star}$. Indeed, if $f \in D(\varphi^\star)$, write

$$\langle f, x \rangle \le \varphi(x) + \varphi^\star(f) \le k\|x\| + C + \varphi^\star(f).$$

Choosing $\|x\| = R$, we obtain

$$R\|f\| \le kR + C + \varphi^\star(f) \quad \forall R > 0$$

and the conclusion follows by letting $R \to \infty$.

2. Check, with the help of a basis of $\mathbb{R}^n$, that every point $x_0 \in A$ satisfies assumption (1).

-D-

The main difficulty is to show that if $f \in \partial I_C(x)$ with $\varphi(x) = 0$ and $f \ne 0$, then there exists some $\lambda > 0$ such that $f \in \lambda\partial\varphi(x)$. Apply Hahn–Banach (first geometric form) in $E \times \mathbb{R}$ to the convex sets $A = \mathrm{Int}(\mathrm{epi}\,\varphi)$ and $B = \{[y, 0] \in E \times \mathbb{R}; \langle f, y - x \rangle \ge 0\}$ (check that $A \cap B = \emptyset$). Thus, there exist some $g \in E^\star$ and some constant k such that $\|g\| + |k| \ne 0$ and

$$\langle g, y \rangle + k\mu \ge \langle g, z \rangle \quad \forall[y, \mu] \in \mathrm{epi}\,\varphi, \quad \forall[z, 0] \in B.$$

It follows, in particular, that $k \ge 0$ and that

$$\langle g, y \rangle + k\varphi(y) \ge \langle g, x \rangle \quad \forall y \in E.$$

In fact, $k \ne 0$ (since $k = 0$ would imply $g = 0$). Thus $-\frac{g}{k} \in \partial\varphi(x)$ (since $\varphi(x) = 0$). Moreover, $g \ne 0$ (why?). Finally, we have $\langle g, x \rangle \ge \langle g, z \rangle \ \forall[z, 0] \in B$ and consequently $\langle g, u \rangle \le 0 \ \forall u \in E$ such that $\langle f, u \rangle \ge 0$. It follows that $g = 0$ on the set $f^{-1}(\{0\})$. We conclude that there is a constant $\theta < 0$ such that $g = \theta f$ (see Lemma 3.2).

Problem 3

-A-

3. *Either* $x \in S(x_n) \ \forall n$ and then we have $\psi(x_{n+1}) \le \psi(x) + \varepsilon_{n+1} \ \forall n$. Passing to the limit one obtains $\psi(a) \le \psi(x)$ and a fortiori $\psi(x) - \psi(a) + d(x, a) \ge 0$. *Or* $\exists N$ such that $x \notin S(x_N)$ and then $x \notin S(x_n) \ \forall n \ge N$. It follows that

$$\psi(x) - \psi(x_n) + d(x, x_n) > 0 \quad \forall n \ge N.$$

Passing to the limit also yields $\psi(x) - \psi(a) + d(x, a) \ge 0$.

-B-

1. The set M equipped with the distance $d(x,y) = \lambda\|x-y\|$ is complete (since ψ is l.s.c.) and nonempty ($x_0 \in M$). Note that $\psi \geq 0$. By the result of part A there exists some $x_1 \in M$ such that

$$\psi(x) - \psi(x_1) + \lambda\|x-x_1\| \geq 0 \quad \forall x \in M.$$

If $x \notin M$ we have $\psi(x) > \psi(x_0) - \lambda\|x_0 - x\|$ (by definition of M), while $\psi(x_0) - \lambda\|x_0 - x\| \geq \psi(x_1) - \lambda\|x-x_1\|$ (since $x_1 \in M$).
Combining the two cases, we see that

$$\psi(x) - \psi(x_1) + \lambda\|x-x_1\| \geq 0 \quad \forall x \in E.$$

On the other hand, since $x_1 \in M$, we have $\psi(x_1) \leq \psi(x_0) - \lambda\|x_0 - x_1\|$. But $\psi(x_0) \leq \varepsilon$ and $\psi(x_1) \geq 0$. Consequently $\|x_0 - x_1\| \leq \varepsilon/\lambda$.
2. Consider the functions $\omega(x) = \psi(x) - \psi(x_1)$ and $\theta(x) = \lambda\|x-x_1\|$, so that $0 \in \partial(\omega + \theta)(x_1)$. We know that $\partial(\omega + \theta) = \partial\omega + \partial\theta$ and that $\partial\theta(x_1) = \lambda B_{E^*}$. It follows that $0 \in \partial\varphi(x_1) - f + \lambda B_{E^*}$.
3. Let us check that $D(\varphi) \subset D(\partial\varphi)$. Given any $x_0 \in D(\varphi)$, we know, from the previous questions, that $\forall\varepsilon > 0, \forall\lambda > 0, \exists x_1 \in D(\partial\varphi)$ such that $\|x_1 - x_0\| < \varepsilon/\lambda$. Clearly $R(\partial\varphi) \subset D(\varphi^*)$. Conversely, let us check that $D(\varphi^*) \subset \overline{R(\partial\varphi)}$. Given any $f_0 \in D(\varphi^*)$ we know that $\forall\varepsilon > 0, \exists x_0 \in D(\varphi)$ such that $f_0 \in \partial_\varepsilon\varphi(x_0)$, and thus $\forall\lambda > 0, \exists f_1 \in R(\partial\varphi)$ such that $\|f_1 - f_0\| < \lambda$.

-C-

1. Let $f_0 \in E^*$. Since $(I_C)^*(f_0) < \infty$, we know that $\forall\varepsilon > 0, \exists x_0 \in C$ such that $f_0 \in \partial_\varepsilon I_C(x_0)$. It follows that $\forall\lambda > 0, \exists x_1 \in C, \exists f_1 \in \partial I_C(x_1)$ with $\|f_1 - f_0\| \leq \lambda$. Clearly we have $\sup_{x \in C}\langle f_1, x\rangle = \langle f_1, x_1\rangle$.
2. Let x_0 be a boundary point of C. Then $\forall\varepsilon > 0, \exists a \in E, a \notin C$, such that $\|a - x_0\| < \varepsilon$. Separating C and $\{a\}$ by a closed hyperplane we obtain some $f_0 \in E^*$ such that $f_0 \neq 0$ and $\langle f_0, x - a\rangle \leq 0 \; \forall x \in C$. Of course, we may assume that $\|f_0\| = 1$. Thus, we have $\langle f_0, x - x_0\rangle \leq \varepsilon \; \forall x \in C$ and consequently $f_0 \in \partial_\varepsilon I_C(x_0)$. Applying the result of part B with $\lambda = \sqrt{\varepsilon}$ we find some $x_1 \in C$ and some $f_1 \in \partial I_C(x_1)$ such that $\|x_1 - x_0\| \leq \sqrt{\varepsilon}$ and $\|f_1 - f_0\| \leq \sqrt{\varepsilon}$. Since $f_1 \neq 0$ (provided $\varepsilon < 1$), we see that there exists a supporting hyperplane to C at x_1.

Problem 4

2. Argue by induction and apply question 7 of Exercise 1.23.
3. Note that $x = \frac{1}{2}[(x+y)+(x-y)]$, and so by convexity,

$$\psi_n(x) \leq \left[\frac{1}{2}\psi_n(x+y) + \psi_n(x-y)\right] \leq \frac{1}{2}[\varphi_n(x+y) + \psi_n(x-y)] \quad \forall x, y.$$

Thus $\psi_n(x) \le \psi_{n+1}(x)$. We have $\varphi_n \downarrow \theta$, $\psi_n \uparrow \tilde{\theta}$ and $\varphi_{n+1} = \frac{1}{2}(\varphi_n + \psi_n)$. Therefore $\theta = \tilde{\theta}$.

4. The sequence $(\varphi_n^\star)$ is nondecreasing and converges to a limit, denoted by ω. Since $\theta \le \varphi_n$, it follows that $\varphi_n^\star \le \theta^\star$ and $\omega \le \theta^\star$. On the other hand, we have $\langle f, x \rangle - \varphi_n(x) \le \varphi_n^\star(f) \; \forall x \in E, \forall f \in E^\star$. Thus $\langle f, x \rangle - \theta(x) \le \omega(f) \; \forall x \in E$, $\forall f \in E^\star$, that is, $\theta^\star \le \omega$. We conclude that $\omega = \theta^\star$.

5. Applying question 1 of Exercise 1.23, we see that $\psi_{n+1}^\star = \frac{1}{2}(\varphi_n^\star + \psi_n^\star)$. The sequence $(\psi_n^\star)$ is nonincreasing and thus it converges to a limit ζ such that $\zeta = \frac{1}{2}(\theta^\star + \zeta)$. It follows that $\zeta = \theta^\star$ (since $\zeta < \infty$).

-B-

From the convexity and the homogeneity of φ we obtain

$$\varphi(x + y) = \varphi\left(t\frac{x}{t} + (1 - t)\frac{y}{1 - t}\right)$$

$$\le t\varphi\left(\frac{x}{t}\right) + (1 - t)\varphi\left(\frac{y}{1 - t}\right) = \frac{1}{t}\varphi(x) + \frac{1}{1 - t}\varphi(y).$$

In order to establish (1) choose $x = \frac{1}{2}(X + Y)$ and $y = \frac{1}{2}(X - Y)$.

2. Using (1) we find that $\forall x, y \in E, \; \forall t \in (0, 1)$,

$$\varphi_{n+1}(x) = \frac{1}{2}\{\varphi_n(x) + \psi_n(x)\}$$

$$\le \frac{1}{2}\left\{\frac{1}{4t}\varphi_n(x + y) + \frac{1}{4(1 - t)}\varphi_n(x - y)\right.$$

$$\left. + \frac{1}{4t}\psi_n(x + y) + \frac{1}{4(1 - t)}\psi_n(x - y)\right\}.$$

Applying A1 and the induction assumption we have $\forall x, y \in E$ and $\forall t \in (0, 1)$,

$$\varphi_{n+1}(x) \le \frac{1}{2}\left\{\frac{2}{4t}\varphi_{n+1}(x + y) + \frac{1}{4(1 - t)}\left(2 + \frac{C}{4^n}\right)\psi_n(x - y)\right\}.$$

Choosing t such that $\frac{2}{4t} = \frac{1}{4(1-t)}(2 + \frac{C}{4^n})$, that is, $t = 1/2(1 + \frac{C}{4^{n+1}})$, we conclude that

$$\varphi_{n+1}(x) \le \frac{1}{2}\left(1 + \frac{C}{4^{n+1}}\right)\{\varphi_n(x + y) + \psi_n(x - y)\} \quad \forall x, y \in E.$$

It follows that $\varphi_{n+1}(x) \le \frac{1}{2}(1 + \frac{C}{4^{n+1}})\psi_{n+1}(x) \; \forall x \in E$.

3. With $x \ne y$ and $t \in (0, 1)$ write

$$\theta(tx + (1-t)y) \le \theta_n(tx + (1-t)y) + \frac{1}{2^n}\varphi_0(tx + (1-t)y)$$

$$\le t\theta_n(x) + (1-t)\theta_n(y) + \frac{1}{2^n}\varphi_0(tx + (1-t)y)$$

$$\le t\theta(x) + (1-t)\theta(y)$$

$$+ \frac{1}{2^n}\Big[\varphi_0(tx + (1-t)y) - t\varphi_0(x) - (1-t)\varphi_0(y)$$

$$+ \frac{C}{2^n}(t\varphi_0(x) + (1-t)\varphi_0(y))\Big]$$

$$< t\theta(x) + (1-t)\theta(y),$$

for n large enough, since φ_0 is strictly convex.

-C-

Take $\varphi_0(x) = \frac{1}{2}\|x\|_1^2$ and $\psi_0(x) = \frac{1}{2}\alpha^2\|x\|_2^2$, with $\alpha > 0$ sufficiently small. The norm $\|\ \|$ is defined through the relation $\theta(x) = \frac{1}{2}\|x\|^2$.

Problem 5

-B-

1. It suffices to prove that there is a constant $c > 0$ such that $B(0, c) \subset K$. By (iii) we have $\bigcup_{n=1}^{\infty} (nK) = E$ and thus $\bigcup_{n=1}^{\infty} (n\overline{K}) = E$. Applying Baire's theorem, one sees that $\text{Int}(\overline{K}) \ne \emptyset$, and hence there exist some $y_0 \in E$ and a constant $c > 0$ such that $B(y_0, 4c) \subset \overline{K}$. Since K is convex and symmetric it follows that $B(0, 2c) \subset \overline{K}$.

   We claim that $B(0, c) \subset K$. Fix $x \in E$ with $\|x\| < c$. There exist $y_1, z_1 \in P$ such that $\|y_1\| \le 1/2$, $\|z_1\| \le 1/2$ and $\|x - (y_1 - z_1)\| < c/2$. Next, there exist $y_2, z_2 \in P$ such that $\|y_2\| \le 1/4$, $\|z_2\| \le 1/4$, and

   $$\|x - (y_1 - z_1) - (y_2 - z_2)\| < c/4.$$

   Iterating this construction, one obtains sequences (y_n) and (z_n) in P such that $\|y_n\| \le 1/2^n$, $\|z_n\| \le 1/2^n$, and

   $$\left\|x - \sum_{i=1}^{n}(y_i - z_i)\right\| < c/2^n.$$

   Then write $x = y - z$ with $y = \sum_{i=1}^{\infty}y_i$ and $z = \sum_{i=1}^{\infty}z_1$, so that $x \in K$.
2. Write $x_n = y_n - z_n$ with $y_n, z_n \in P$, $\|y_n\| \le C/2^n$, and $\|z_n\| \le C/2^n$. Then $1 \le f(x_n) \le f(y_n)$. Set $u_n = \sum_{i=1}^{n}y_i$ and $u = \sum_{i=1}^{\infty}y_i$. On the one hand, $f(u_n) \ge n$, and on the other hand, $f(u - u_n) \ge 0$. It follows that $f(u) \ge n \ \forall n$; absurd.
3. Consider a complement of F (see Section 2.4).

-C-

(a) One has $F = P - P = E$; one can also check (i) directly: if $f \geq 0$ on P, then $|f(u)| \leq \|u\|_\infty f(1) \; \forall u \in E$.

(b) Here $F = \{u \in E; u(0) = u(1) = 0\}$ is a closed subspace of finite codimension.

(c) One has $F = E$. Indeed, if $u \in E$ there is a constant $c > 0$ such that $|u(t)| \leq ct(1-t) \; \forall t \in [0,1]$ and one can write $u = v - w$ with $w = ct(1-t)$ and $v = u + ct(1-t)$.

Problem 7

-A-

2. Fix $x \in M$ with $\|x\| \leq 1$. Let $\varepsilon > 0$. Since $\text{dist}(x, N) \leq a$, there exists some $y \in N$ such that $\|x - y\| \leq a + \varepsilon$, and thus $\|y\| \leq 1 + a + \varepsilon$. On the other hand, $\text{dist}(\frac{y}{\|y\|+\varepsilon}, L) \leq b$ and so $\text{dist}(y, L) \leq b(\|y\| + \varepsilon) \leq b(1 + a + 2\varepsilon)$. It follows that $\text{dist}(x, L) \leq a + \varepsilon + b(1 + a + 2\varepsilon) \; \forall \varepsilon > 0$.

-B-

In order to construct an example such that $A^\star + B^\star \neq (A + B)^\star$ it suffices to consider any unbounded operator $A : D(A) \subset E \to F$ that is densely defined, closed, and such that $D(A) \neq E$. Then take $B = -A$. We have $(A + B)^\star = 0$ with $D((A + B)^\star) = F^\star$, while $A^\star + B^\star = 0$ with $D(A^\star + B^\star) = D(A^\star)$. [Note that $D(A^\star) \neq F^\star$; why?].

3. $A + B$ is closed; indeed, let (u_n) be a sequence in E such that $u_n \to u$ in E and $(A + B)u_n \to f$ in F. Note that

$$\|Bu\| \leq k\|Au + Bu\| + k\|Bu\| + C\|u\| \quad \forall u \in D(A)$$

and thus

$$\|Bu\| \leq \frac{k}{1-k}\|Au + Bu\| + \frac{C}{1-k}\|u\| \quad \forall u \in D(A).$$

It follows that (Bu_n) is a Cauchy sequence. Let $Bu_n \to g$, and so $u \in D(B)$ with $Bu = g$. On the other hand, $Au_n \to f - Bu$, and so $u \in D(A)$ with $Au + Bu = f$. Clearly one has

$$\rho(A, A+B) = \sup_{\substack{u \in D(A) \\ \|u\|+\|Au\| \leq 1}} \inf_{v \in D(A)} \{\|u - v\| + \|Au - (Av + Bv)\|\}$$

$$\leq \sup_{\substack{u \in D(A) \\ \|u\|+\|Au\| \leq 1}} \|Bu\| \leq k + C.$$

4. The same argument shows that under assumption (H*), one has

$$\rho(A^\star, A^\star + B^\star) \leq k^\star + C^\star.$$

[There are some minor changes, since the dual norm on $E^\star \times F^\star$ is given by $\|[f, g]\|_{E^\star \times F^\star} = \max\{\|f\|_{E^\star}, \|g\|_{E^\star}\}$.]

5. Let $t \in [0, 1]$. For every $u \in D(A)$ one has

$$\|Bu\| \le k\|Au\| + C\|u\| \le k(\|Au + tBu\| + t\|Bu\|) + C\|u\|,$$

and thus

$$\|Bu\| \le \frac{k}{1-k}\|Au + tBu\| + \frac{C}{1-k}\|u\|.$$

Fix any $\varepsilon > 0$ such that $1/\varepsilon = n$ is an integer, $\frac{\varepsilon(k+C)}{1-k} \le \frac{1}{3}$, and $\frac{\varepsilon(k^\star + C^\star)}{1-k^\star} \le \frac{1}{3}$. Set $A_1 = A + \varepsilon B$, so that $A_1^\star = A^\star + \varepsilon B^\star$ and, moreover,

$$\|Bu\| \le \frac{k}{1-k}\|A_1 u\| + \frac{C}{1-k}\|u\| \quad \forall u \in D(A),$$

and also

$$\|B^\star v\| \le \frac{k^\star}{1-k^\star}\|A_1^\star v\| + \frac{C^\star}{1-k^\star}\|v\| \quad \forall v \in D(A^\star).$$

It follows that $(A_1 + \varepsilon B)^\star = A_1^\star + \varepsilon B^\star$, i.e., $(A + 2\varepsilon B)^\star = A^\star + 2\varepsilon B^\star$, and so on, step by step with $A_j = A + j\varepsilon B$ and $j \le n - 1$.

Problem 8

1. Let $\mathcal{T}$ be the topology corresponding to the metric d. Since $B_{E^\star}$ equipped with the topology $\sigma(E, E^\star)$ is compact, it suffices to check that the canonical injection $(B_{E^\star}, \sigma(E^\star, E)) \to (B_{E^\star}, \mathcal{T})$ is continuous. This amounts to proving that for every $f_0 \in B_{E^\star}$ and for every $\varepsilon > 0$ there exists a neighborhood $V(f^0)$ of f^0 for $\sigma(E^\star, E)$ such that

$$V(f^0) \cap B_{E^\star} \subset \{f \in B_{E^\star}; d(f, f^0) < \varepsilon\}.$$

Let (e^i) be the canonical basis of ℓ^1. Choose

$$V(f^0) = \{f \in E^\star; |\langle f - f^0, e^i \rangle| < \delta \ \forall i = 1, 2, \dots, n\}$$

with $\delta + (1/2^{n-1}) < \varepsilon$.

2. Note that $(B_{E^\star}, d)$ is a complete metric space (since it is compact). The sets F_k are closed for the topology $\mathcal{T}$, and, moreover, $\bigcup_{k=1}^\infty F_k = B_{E^\star}$ (since $\langle f, x^n \rangle \to 0$ for every $f \in E^\star$). Baire's theorem says that there exists some integer k_0 such that $\text{Int}(F_{k_0}) \ne \emptyset$.

3. Write $f^0 = (f_1^0, f_2^0, \dots, f_i^0, \dots)$ and consider the elements $f \in B_{E^\star}$ of the form

$$f = (f_1^0, f_2^0, \dots, f_N^0, \pm1, \pm1, \pm1, \dots),$$

so that

$$d(f, f^0) \leq \sum_{i=N+1}^{\infty} \frac{2}{2^i} < \rho.$$

Such f's belong to F_{k_0} and one has, for every $n \geq k_0$,

$$|\langle f, x^n \rangle| = \left| \sum_{i=1}^{\infty} f_i x_i^n \right| = \left| \sum_{i=1}^{N} f_i^0 x_i^n + \sum_{i=N+1}^{\infty} (\pm x_i^n) \right| \leq \varepsilon.$$

It follows that

$$\sum_{i=N+1}^{\infty} |x_i^n| \leq \varepsilon + \sum_{i=1}^{N} |f_i^0| \, |x_i^n| \leq \varepsilon + \sum_{i=1}^{N} |x_i^n|,$$

and thus

$$\sum_{i=1}^{\infty} |x_i^n| \leq \varepsilon + 2 \sum_{i=1}^{N} |x_i^n| \quad \forall n \geq k_0.$$

4. The conclusion is clear, since for each fixed i the sequence x_i^n tends to 0 as $n \to \infty$.
5. Given $\varepsilon > 0$, set

$$F_k = \{ f \in B_{E^\star}; \, |\langle f, x^n - x^m \rangle| \leq \varepsilon \; \forall m, n \geq k \}.$$

By the same method as above one finds integers k_0 and N such that

$$\|x^n - x^m\|_1 \leq \varepsilon + 2 \sum_{i=1}^{N} |x_i^n - x_i^m| \quad \forall m, n \geq k_0.$$

It follows that (x^n) is a Cauchy sequence in ℓ^1.
6. See Exercises 4.18 and 4.19.

Problem 9

-A-

1. A is open for the strong topology (since it is open for the topology $\sigma(E^\star, E)$). Thus (by Hahn–Banach applied in $E^\star$) there exist some $\xi \in E^{\star\star}, \xi \neq 0$, and a constant α such that

$$\langle \xi, f \rangle \leq a \leq \langle \xi, g \rangle \quad \forall f \in A, \quad \forall g \in B.$$

Fix $f_0 \in A$ and a neighborhood V of 0 for the topology $\sigma(E^\star, E)$ such that $f_0 + V \subset A$. We may always assume that V is symmetric; otherwise, consider $V \cap (-V)$. We have $\langle \xi, f_0 + g \rangle \leq \alpha \; \forall g \in V$, and hence there exists a constant C such that $|\langle \xi, g \rangle| \leq C \; \forall g \in V$. Therefore $\xi : E^\star \to \mathbb{R}$ is continuous for the topology $\sigma(E^\star, E)$. In view of Proposition 3.14 there exists some $x \in E$ such that $\langle \xi, f \rangle = \langle f, x \rangle \ \forall f \in E^\star$.

2. See the solution of Exercise 3.7.

3. Let V be an open set for the topology $\sigma(E^\star, E)$ that is convex, and such that $0 \in V$ and $V \cap (A - B) = \emptyset$. Separating V and $(A - B)$, we find some $x \in E$, $x \neq 0$, and a constant α such that

$$\langle f, x \rangle \leq \alpha \leq \langle g - h, x \rangle \quad \forall f \in V, \ \forall g \in A, \ \forall h \in B.$$

Since V is also a neighborhood of 0 for the strong topology, there exists some $r > 0$ such that $r B_{E^\star} \subset V$. Thus $\alpha \geq r\|x\| > 0$, which leads to a strict separation of A and B.

4. Let $f, g \in \overline{A}^{\sigma(E^\star, E)}$ and let V be a convex neighborhood of 0 for $\sigma(E^\star, E)$. Then $(f + V) \cap A \neq \emptyset$ and $(g + V) \cap A \neq \emptyset$. Thus $(tf + (1 - t)g + V) \cap A \neq \emptyset$ $\forall t \in [0, 1]$.

- B-

1. If E is reflexive, then $\overline{N}^{\sigma(E^\star, E)} = \overline{N} =$ the closure of N for the strong topology, since $\sigma(E^\star, E) = \sigma(E^\star, E^{\star\star})$ and N is convex. Let $E = \ell^1$, so that $E^\star = \ell^\infty$; taking $N = c_0$ we have $N^\perp = \{0\}$ and $N^{\perp\perp} = \ell^\infty$.

2. For every $x \in E$, set $\varphi(x) = \sup_{f \in E^\star}\{\langle f, x \rangle - \psi(f)\}$. Then $\varphi : E \to (-\infty, +\infty]$ is convex and l.s.c. In order to show that $\varphi \not\equiv +\infty$ and that $\varphi^\star = \psi$, one may follow the same arguments as in Proposition 1.10 and Theorem 1.11, except that here one uses question A3 instead of the usual Hahn–Banach theorem.

3. (i) One knows (Corollary 2.18) that $N(A) = R(A^\star)^\perp$ and thus $N(A)^\perp = R(A^\star)^{\perp\perp} = \overline{R(A^\star)}^{\sigma(E^\star, E)}$. If E is reflexive, then $N(A)^\perp = \overline{R(A^\star)}$.

(ii) Argue as in the proof of Theorem 3.24 and apply question A3.

4. Suppose, by contradiction, that there exists some $\xi \in B_{E^{\star\star}}$ such that $\xi \notin \overline{J(B_E)}^{\sigma(E^{\star\star}, E^\star)}$. Applying question A3 in $E^{\star\star}$, we may find some $f \in E^\star$ and a constant α such that

$$\langle f, x \rangle < \alpha < \langle \xi, f \rangle \quad \forall x \in B_E.$$

Thus $\|f\| \leq \alpha < \langle \xi, f \rangle \leq \|f\|$; absurd.

5. Assume, by contradiction, that there exists some $u_0 \in E$ with $\|u_0\| < 1$ and $Au_0 \notin \overline{\text{conv } A(S_E)}^{\sigma(E^\star, E)}$. Applying question A3, we may find some $x_0 \in E$ and a constant α such that

$$\langle Au, x_0 \rangle \leq \alpha < \langle Au_0, x_0 \rangle \ \forall u \in S_E;$$

thus $\langle Au - Au_0, x_0 \rangle < 0 \ \forall u \in S_E$. On the other hand, there is some $t > 0$ such that $\|u_0 + tx_0\| = 1$, and by monotonicity, we have $\langle A(u_0 + tx_0) - Au_0, x_0 \rangle \geq 0$; absurd.

Problem 10

- A -

1. $B_{E^\star}$ is compact and metrizable for the topology $\sigma(E^\star, E)$ (see Theorem 3.28). It follows, by a standard result in point-set topology, that there exists a subset in $B_{E^\star}$ that is countable and dense for $\sigma(E^\star, E)$. Let $\mathcal{T}$ denote the topology associated to the metric d. It is easy to see that the canonical injection $i: (B_E, \sigma(E, E^\star)) \to (B_E, \mathcal{T})$ is continuous (see part (b) in the proof of Theorem 3.28). [Note that in general, i^{-1} is not continuous; otherwise, B_E would be metrizable for the topology $\sigma(E, E^\star)$ and $E^\star$ would be separable (see Exercise 3.24). However, there are examples in which E is separable and $E^\star$ is not, for instance $E = L^1(\Omega)$ and $E^\star = L^\infty(\Omega)$.]

   Since B is compact for $\sigma(E, E^\star)$, it follows (by Corollary 2.4) that B is bounded. Thus B is a compact (metric) space for the topology $\mathcal{T}$ and, moreover, the topologies $\sigma(E, E^\star)$ and $\mathcal{T}$ coincide on B.

2. Consider the closed linear space spanned by the x_n's.

- B -

For each i choose $a_1 \in B_F$ such that $\langle g_i, a_i \rangle \geq 3/4$.

- C -

4. For each $\eta \in E^{\star\star}$ set $h(\eta) = \sup_{i \geq 1} \langle \eta, f_i \rangle$; the function $h: E^{\star\star} \to \mathbb{R}$ is continuous for the strong topology on $E^{\star\star}$, since we have $|h(\eta_1) - h(\eta_2)| \leq \|\eta_1 - \eta_2\| \; \forall \eta_1, \eta_2 \in E^{\star\star}$.

5. A subsequence of the sequence (x_n) converges to x for $\sigma(E, E^\star)$ (by assumption (Q) and we have $\langle \xi, f_i \rangle = \langle f_i, x \rangle \; \forall i \geq 1$.

   On the other hand, x belongs to the closure of $[x_1, x_2, \ldots, x_k, \ldots]$ for the topology $\sigma(E, E^\star)$ and thus also for the strong topology (by Theorem 3.7). In particular, $x \in \overline{M}$ and consequently $\xi - x \in \overline{M}$. It follows that $\xi = x$ since

$$0 = \sup_{i \geq 1} \langle \xi - x, f_i \rangle \geq \frac{1}{2} \|\xi - x\|.$$

-D-

1. A is bounded by assumption (Q) and Corollary 2.4. It follows that $\overline{A}^{\sigma(E^{\star\star}, E^\star)}$ is compact for the topology $\sigma(E^{\star\star}, E^\star)$ by Theorem 3.16. But the result of part C shows that $B = \overline{A}^{\sigma(E^{\star\star}, E^\star)}$, or more precisely that $J(B) = \overline{J(A)}^{\sigma(E^{\star\star}, E^\star)}$. Consequently $J(B)$ is compact for the topology $\sigma(E^{\star\star}, E^\star)$. Since the map $J^{-1}: J(E) \to E$ is continuous from $\sigma(E^{\star\star}, E^\star)$ to $\sigma(E, E^\star)$, it follows that B is compact for $\sigma(E, E^\star)$.

2. Already established in question C4.

Problem 11

-A-

2. Separating $\{0\}$ and C_1 we find some $x_1 \in E$ and a constant α such that $0 < \alpha < \langle f, x_1 \rangle \ \forall f \in C_1$. If needed, replace x_1 by a multiple of x_1.
3. One has to find a finite subset $A \subset E$ such that $A \subset (1/d_1) B_E$ and $Y_A = \emptyset$. We first claim that $\bigcap_{A \in \mathcal{F}} Y_A = \emptyset$, where $\mathcal{F}$ denotes the family of all finite subsets A in $(1/d_1) B_E$. Assume, by contradiction, that $f \in \bigcap_{A \in \mathcal{F}} Y_A$; we have

$$\langle f, x_1 \rangle \leq 1 \quad \text{and} \quad \langle f, x \rangle \leq 1 \ \forall x \in (1/d_1) B_E.$$

Thus $\|f\| \leq d_1$ and so $f \in C_1$; it follows that $\langle f, x_1 \rangle > 1$; absurd.
By compactness there is a finite sequence $A_1', A_2', \dots, A_j'$ such that $\bigcap_{i=1}^{j} Y_{A_i'} = \emptyset$. Set $A' = A_1' \cup A_2', \cdots \cup A_j'$. It is easy to check that $Y_{A'} = \emptyset$.
4. For every finite subset A in $(1/d_{k-1}) B_E$ consider the set

$$Y_A = \left\{ f \in C_k; \sup \left\{ \langle f, x \rangle; x \in \left(\bigcup_{i=1}^{k-1} A_i \right) \cup A \right\} \leq 1 \right\}.$$

One proves, as in question 3, that there is some A such that $Y_A = \emptyset$.
5. Write the set $\bigcup_{k=1}^{\infty} A_k$ as a sequence (x_n) that tends to 0.

- B -

1. Applying Hahn–Banach in c_0, there exist some $\theta \in \ell^1 (= (c_0)^\star$; see Chapter 11) with $\theta \neq 0$, and a constant α such that

$$\langle \theta, \xi \rangle \leq \alpha \leq \langle \theta, T(f) \rangle \quad \forall \xi \in c_0 \text{ with } \|\xi\| < 1, \quad \forall f \in C.$$

It follows that

$$0 < \|\theta\|_{\ell^1} \leq \alpha \leq \sum \theta_n \langle f, x_n \rangle \quad \forall f \in C.$$

Letting $x = \sum \theta_n x_n$, we obtain

$$\langle f, x \rangle \geq \alpha > 0 \quad \forall f \in C.$$

If needed, replace x by a multiple of x and conclude.
2. Fix any $f_0 \notin C$; set $\widetilde{C} = C - f_0$. Then $0 \notin \widetilde{C}$ and for each integer n the set $\widetilde{C} \cap (n B_{E^\star})$ is closed for $\sigma(E^\star, E)$. Hence, there is some $x \in E$ such that $\langle f, x \rangle \geq 1 \ \forall f \in \widetilde{C}$. The set $V = \{ f \in E^\star; \langle f - f_0, x \rangle < 1 \}$ is a neighborhood of f_0 for $\sigma(E^\star, E)$ and $V \cap C = \emptyset$.

Problem 12

- A -

2. Apply the results of questions 1, 7, and 4 in Exercise 1.23 to the functions $\varphi^\star$ and $\psi^\star$.

3. We have $\theta^{\star\star} = (\varphi + \psi)^\star$. Following the same argument as in the proof of Theorem 1.11, it is easy to see that epi $\theta^{\star\star} = \overline{\text{epi } \theta}$ (warning: in general, θ need not be l.s.c.).

Therefore we obtain $D(\theta^{\star\star}) \subset \overline{D(\theta)}$, i.e., $D((\varphi + \psi)^\star) \subset \overline{D(\varphi^\star) + D(\psi^\star)}$.

-B-

1. It suffices to check that for every fixed $x \in E$ the set $\langle M, x \rangle$ is bounded. In fact, it suffices to check that $\langle M, x \rangle$ is bounded below (choose $\pm x$). Given $x \in E$, $x \neq 0$, write $x = \lambda(a - b)$ with $\lambda > 0$, $a \in D(\varphi)$, and $b \in D(\psi)$. We have

$$\langle f - g, a \rangle \leq \varphi(a) + \varphi^\star(f - g),$$
$$\langle g, b \rangle \leq \psi(b) + \psi^\star(g),$$

and thus

$$-\left\langle g, \frac{x}{\lambda} \right\rangle \leq -\langle f, a \rangle + \varphi(a) + \psi(b) + \alpha \quad \forall g \in M.$$

Consequently $\langle M, x \rangle \geq C$, where C depends only on x, f, and α.

2. Use the same method as above.

3. Let $\alpha \in \mathbb{R}$ be fixed and let (f_n) be a sequence in $E^\star$ such that $\theta(f_n) \leq \alpha$ and $f_n \to f$. Thus, there is a sequence (g_n) in $E^\star$ such that $\varphi^\star(f_n - g_n) + \psi^\star(g_n) \leq \alpha + (1/n)$. Consequently, (g_n) is bounded and we may assume that $g_{n_k} \rightharpoonup g$ for $\sigma(E^\star, E)$. Since $\varphi^\star$ and $\psi^\star$ are l.s.c. for $\sigma(E^\star, E)$, it follows that $\varphi^\star(f - g) + \psi^\star(g) \leq \alpha$, and so $\theta(f) \leq \alpha$.

4. (i) We have $\theta = \theta^{\star\star} = (\varphi + \psi)^\star$.

(ii) Write that $(\varphi + \psi)^\star(0) = (\varphi^\star \nabla \psi^\star)(0)$ and note that

$$\inf_{g \in E^\star} \{\varphi^\star(-g) + \psi^\star(g)\}$$

is achieved by the result of question B1.

(iii) This is a direct consequence of (i).

Remark. Assumption (H) holds if there is some $x_0 \in D(\varphi) \cap D(\psi)$ such that φ is continuous at x_0.

Problem 13

- A -

1. By question 5 of Exercise 1.25 we know that

$$\lim_{\substack{\lambda \to 0 \\ \lambda > 0}} \frac{1}{2\lambda} \left(\|x + \lambda y\|^2 - \|x\|^2 \right) = \langle Fx, y \rangle.$$

If $\lambda < 0$ set $\mu = -\lambda$ and write

$$\frac{1}{2\lambda} \left(\|x + \lambda y\|^2 - \|x\|^2 \right) = -\frac{1}{2\mu} \left(\|x + \mu(-y)\|^2 - \|x\|^2 \right).$$

2. Let $t_n \to 0$ be such that $\langle F(x + t_n y), y \rangle \to \ell$. We have

$$\frac{1}{2} \left(\|x + \lambda y\|^2 - \|x + t_n y\|^2 \right) \geq \langle F(x + t_n y), (\lambda - t_n)y \rangle.$$

Passing to the limit (with $\lambda \in \mathbb{R}$ fixed) we obtain $\frac{1}{2} \left(\|x + \lambda y\|^2 - \|x\|^2 \right) \geq \lambda \ell$. Dividing by λ (distinguish the cases $\lambda > 0$ and $\lambda < 0$) and letting $\lambda \to 0$ leads to $\langle Fx, y \rangle = \ell$. The uniqueness of the limit allows us to conclude that

$$\lim_{t \to 0} \langle F(x + ty), y \rangle = \langle Fx, y \rangle$$

(check the details).

3. Recall that F is monotone by question 4 of Exercise 1.1.

   Alternative proof. It suffices to show that if $x_n \to x$ then $Fx_n \rightharpoonup Fx$ for $\sigma(E^\star, E)$. Assume $x_n \to x$. If E is reflexive or separable there is a subsequence such that $Fx_{n_k} \rightharpoonup f$ for $\sigma(E^\star, E)$. Recall that $\langle Fx_n, x_n \rangle = \|x_n\|^2$ and $\|Fx_n\| = \|x_n\|$. Passing to the limit we obtain $\langle f, x \rangle = \|x\|^2$ and $\|f\| \leq \|x\|$. Thus $f = Fx$; the uniqueness of the limit allows us to conclude that $Fx_n \rightharpoonup Fx$ for $\sigma(E^\star, E)$ (check the details).

- B -

1. If $x_n \to x$, then $Fx_n \rightharpoonup Fx$ for $\sigma(E^\star, E)$ and $\|Fx_n\| = \|x_n\| \to \|x\| = \|Fx\|$. It follows from Proposition 3.32 that $Fx_n \to Fx$.

2. Assume, by contradiction, that there are two sequences (x_n), and (y_n) such that $\|x_n\| \leq M$, $\|y_n\| \leq M$, $\|x_n - y_n\| \to 0$, and $\|Fx_n - Fy_n\| \geq \varepsilon > 0$. Passing to a subsequence we may assume that $\|x_n\| \to \ell$, and $\|y_n\| \to \ell$ with $\varepsilon \leq 2\ell$, so that $\ell \neq 0$. Set $a_n = x_n/\|x_n\|$ and $b_n = y_n/\|y_n\|$. We have $\|a_n\| = \|b_n\| = 1$, $\|a_n - b_n\| \to 0$, and $\|Fa_n - Fb_n\| \geq \varepsilon' > 0$ for n large enough. Since $E^\star$ is uniformly convex there exists $\delta > 0$ such that

$$\left\| \frac{Fa_n + Fb_n}{2} \right\| \leq 1 - \delta.$$

On the other hand, the inequality of question A4 leads to

$$2 \leq \|Fa_n + Fb_n\| + \|a_n - b_n\|;$$

this is impossible.

3. We have
$$\varphi(x) - \varphi(x_0) \geq \langle Fx_0, x - x_0 \rangle$$

and
$$\varphi(x_0) - \varphi(x) \geq \langle Fx, x_0 - x \rangle.$$

It follows that
$$0 \leq \varphi(x) - \varphi(x_0) - \langle Fx_0, x - x_0 \rangle \leq \langle Fx - Fx_0, x - x_0 \rangle$$

and therefore
$$|\varphi(x) - \varphi(x_0) - \langle Fx_0, x - x_0 \rangle| \leq \|Fx - Fx_0\| \, \|x - x_0\|.$$

The conclusion is derived easily with the help of question B1.

-C -

Write
$$\|f + g\| = \sup_{\substack{x \in E \\ \|x\| \leq 1}} \langle f + g, x \rangle$$
$$= \sup_{\substack{x \in E \\ \|x\| \leq 1}} \{\langle f, x + y \rangle + \langle g, x - y \rangle + \langle g, x - y \rangle - \langle f - g, y \rangle\}$$
$$\leq \frac{1}{2}\|f\|^2 + \frac{1}{2}\|g\|^2 - \langle f - g, y \rangle + \sup_{\substack{x \in E \\ \|x\| \leq 1}} \{\varphi(x + y) + \varphi(x - y)\}.$$

From the computation in question B3 we see that for every $x, y \in E$,
$$|\varphi(x + y) - \varphi(x) - \langle Fx, y \rangle| \leq \|F(x + y) - F(x)\| \, \|y\|$$

and
$$|\varphi(x - y) - \varphi(x) + \langle Fx, y \rangle| \leq \|F(x - y) - F(x)\| \, \|y\|.$$

It follows that for every $x, y \in E$,
$$\varphi(x + y) + \varphi(x - y) \leq 2\varphi(x) + \|y\|(\|F(x + y) - F(x)\| + \|F(x - y) - F(x)\|).$$

Therefore, if $\|f\| \leq 1$ and $\|g\| \leq 1$, we obtain for every $y \in E$,
$$\|f + g\| \leq 2 - \langle f - g, y \rangle + \|y\| \sup_{\substack{x \in E \\ \|x\| \leq 1}} \{\|F(x + y) - F(x)\| + \|F(x - y) - F(x)\|\}.$$

Fix $\varepsilon > 0$ and assume that $\|f - g\| > \varepsilon$. Since F is uniformly continuous, there exists some $\alpha > 0$ such that for $\|y\| \leq \alpha$ we have
$$\sup_{\substack{x \in E \\ \|x\| \leq 1}} \{\|F(x + y) - F(x)\| + \|F(x - y) - F(x)\|\} < \varepsilon/2.$$

On the other hand, there exists some $y_0 \in E$, $y_0 \neq 0$, such that $\langle f - g, y_0 \rangle \geq \varepsilon \|y_0\|$, and we may assume that $\|y_0\| = \alpha$. We conclude that

$$\|f + g\| \leq 2 - \varepsilon \|y_0\| + \frac{\varepsilon}{2} \|y_0\| = 2 - \frac{\varepsilon}{2} \alpha.$$

Problem 14

- A -

2. Assume that $x_n \to x$ in E and set $f_n = Sx_n$, so that $\forall f \in E^\star$,

(S1) $\qquad \frac{1}{2} \|f_n\|^2 + \varphi^\star(f_n) - \langle f_n, x_n \rangle \leq \frac{1}{2} \|f\|^2 + \varphi^\star(f) - \langle f, x_n \rangle.$

It follows that the sequence (f_n) is bounded (why?) and thus there is a subsequence such that $f_{n_k} \overset{\star}{\rightharpoonup} g$ for $\sigma(E^\star, E)$. Passing to the limit in (S1) (note that the function $f \mapsto \frac{1}{2} \|f\|^2 + \varphi^\star(f)$ is l.s.c. for $\sigma(E^\star, E)$), we find that

$$\frac{1}{2} \|g\|^2 + \varphi^\star(g) - \langle g, x \rangle \leq \frac{1}{2} \|f\|^2 + \varphi^\star(f) - \langle f, x \rangle \quad \forall f \in E^\star$$

(one uses also Proposition 3.13). Thus $g = Sx$; the uniqueness of the limit implies that $f_n \overset{\star}{\rightharpoonup} Sx$ (check the details). Returning to (S1) and choosing $f = Sx$, we obtain $\limsup \|f_n\|^2 \leq \|Sx\|^2$. We conclude with the help of Proposition 3.32 that $f_n \to Sx$.

- B -

1. The convexity of ψ follows from question 7 of Exercise 1.23. Equality (i) is a consequence of Theorem 1.12, and equality (ii) follows from question 1 of Exercise 1.24.
2. We have

$$\langle Sx, y \rangle \leq \psi(y) + \psi^\star(Sx) = \psi(y) + \langle Sx, x \rangle - \psi(x)$$

and thus

$$0 \leq \psi(y) - \psi(x) - \langle Sx, y - x \rangle \quad \forall x, y \in E.$$

Changing x into y and y into x, we obtain

$$0 \leq \psi(x) - \psi(y) - \langle Sy, x - y \rangle \quad \forall x, y \in E.$$

We conclude that

$$0 \leq \psi(y) - \psi(x) - \langle Sx, y - x \rangle \leq \langle Sy - Sx, y - x \rangle.$$

Problem 15

- A -

2. Note that $\psi(x) \geq \|x\| - \|a\|$ with $a \in A$ being fixed and thus $\psi(x) \to +\infty$ as $\|x\| \to \infty$; therefore c exists. In order to establish the uniqueness it suffices to check that

$$\varphi^2 \left(\frac{c_1 + c_2}{2} \right) < \frac{1}{2}\varphi^2(c_1) + \frac{1}{2}\varphi^2(c_2) \quad \forall c_1, c_2 \in E \text{ with } c_1 \neq c_2.$$

Let $c_1, c_2 \in E$ with $c_1 \neq c_2$. Fix some $0 < \varepsilon < \|c_1 - c_2\|$. In view of Exercise 3.29, and because A is bounded, there exists some $\delta > 0$ such that

$$\left\| \frac{(c_1 - y) + (c_2 - y)}{2} \right\| \leq \frac{1}{2}\|c_1 - y\|^2 + \frac{1}{2}\|c_2 - y\|^2 - \delta \quad \forall y \in A,$$

since $\|(c_1 - y) - (c_2 - y)\| > \varepsilon$. Taking $\sup_{y \in A}$ leads to

$$\varphi^2 \left(\frac{c_1 + c_2}{2} \right) \leq \frac{1}{2}\varphi^2(c_1) + \frac{1}{2}\varphi^2(c_2) - \delta.$$

3. We know that $\varphi(\sigma(A)) < \varphi(x) \ \forall x \in C, \ x \neq \sigma(A)$. If A is not reduced to a single point there exists some $x_0 \in A$, $x_0 \neq \sigma(A)$, and we have

$$\varphi(\sigma(A)) < \varphi(x_0) = \sup_{y \in A} \|x_0 - y\| \leq \text{diam } A.$$

- B -

1. Note that the sequence $(\varphi_n(x))$ is nonincreasing.
3. We have

$$\varphi(\sigma) \leq \varphi(\sigma_n) \leq \varphi_n(\sigma_n) \leq \varphi_n(x) \quad \forall x \in C.$$

Taking $x = \sigma$, we find that all the limits are equal. It is easy to see that the sequence (σ_n) is bounded, and thus for a subsequence, $\sigma_{n_k} \rightharpoonup \tilde{\sigma}$ weakly $\sigma(E, E^\star)$. Hence we have

$$\varphi(\tilde{\sigma}) \leq \liminf \varphi(\sigma_{n_k}) \leq \varphi(x) \quad \forall x \in C.$$

It follows that $\varphi(\tilde{\sigma}) = \inf_C \varphi$ and, by uniqueness, $\tilde{\sigma} = \sigma$. The uniqueness of the limit implies that $\sigma_n \rightharpoonup \sigma$ (check the details).

4. Assume, by contradiction, that there exist some $\varepsilon > 0$ and a subsequence (σ_{n_k}) such that $\|\sigma_{n_k} - \sigma\| > \varepsilon \ \forall k$. Using once more Exercise 3.29 we obtain some $\delta > 0$ such that

$$\varphi_{n_k}^2 \left(\frac{1}{2}(\sigma_{n_k} + \sigma) \right) \leq \frac{1}{2}\varphi_{n_k}^2(\sigma_{n_k}) + \frac{1}{2}\varphi_{n_k}^2(\sigma) - \delta \quad \forall k,$$

and since $\varphi \leq \varphi_{n_k}$, we deduce that

$$\varphi^2\left(\frac{1}{2}(\sigma_{n_k}+\sigma)\right) \le \frac{1}{2}\varphi^2_{n_k}(\sigma_{n_k}) + \frac{1}{2}\varphi^2_{n_k}(\sigma) - \delta \quad \forall k.$$

This leads to a contradiction, since φ is l.s.c.

5. Note that $\varphi(x) = \|x - a\|$ and thus $\sigma = a$.

6. Write

$$|x - a_n|^2 = |x - a + a - a_n|^2 = |x - a|^2 + 2(x - a, a - a_n) + |a - a_n|^2,$$

and thus

$$\varphi^2(x) = \limsup_{n\to\infty} |x - a_n|^2 = |x - a|^2 + \limsup_{n\to\infty} |a_n - a|^2 = |x - a|^2 + \varphi^2(a).$$

It follows that $\sigma = a$.

- C -

1. We have $\|a_{n+1} - Tx\| \le \|a_n - x\| \ \forall n, \forall x \in C$, and therefore $\varphi_{n+1}(Tx) \le \varphi_n(x)$ $\forall x \in C$. Passing to the limit leads to $\varphi(Tx) \le \varphi(x) \ \forall x \in C$. In particular $\varphi(T\sigma) \le \varphi(\sigma)$ and thus $T\sigma = \sigma$.

2. Let $x, y \in C$ be fixed points of T; set $z = tx + (1 - t)y$ with $t \in [0, 1]$. We have

$$\|Tz - x\| \le (1 - t)\|y - x\| \quad \text{and} \quad \|Tz - y\| \le t\|y - x\|$$

and therefore $\|Tz - x\| = (1 - t)\|y - x\|$, $\|Tz - y\| = t\|y - x\|$. The conclusion follows from the fact that E is strictly convex. (Recall that uniform convexity implies strict convexity; see Exercise 3.31).

Problem 16

- A -

1. We have $\langle Au - f, u \rangle \ge 0 \ \forall u \in D(A)$ and using (P) we see that $f = A0 = 0$.

2. Let (u_n) be a sequence in $D(A)$ such that $u_n \to x$ in E and $Au_n \to f$ in $E^\star$. We have $\langle Au - Au_n, u - u_n \rangle \ge 0 \ \forall u \in D(A)$. Passing to the limit we obtain $\langle Au - f, u - x \rangle \ge 0 \ \forall u \in D(A)$. From (P) we deduce that $x \in D(A)$ and $Ax = f$.

3. It is easy to check that if $t \in (0, 1)$, the convexity inequality

$$\langle A(tu + (1 - t)v), tu + (1 - t)v \rangle \le t\langle Au, u \rangle + (1 - t)\langle Av, v \rangle$$

is equivalent to $\langle Au - Av, u - v \rangle \ge 0$.

4. Let $u \in N(A)$; we have $\langle Av, v - u \rangle \ge 0, \ \forall v \in D(A)$. Replacing v by λv, we see that $\langle Av, u \rangle = 0 \ \forall v \in D(A)$; that is, $u \in R(A)^\perp$.

- B -

1. Note that $\langle A^\star v, v \rangle = \langle Av, v \rangle \ \forall v \in D(A) \cap D(A^\star)$.

2. The first claim is a direct consequence of (P) and the assumption that $v \notin D(A)$. Choosing $f = -A^\star v$, we have some $u \in D(A)$ such that $\langle Au + A^\star v, u - v \rangle < 0$ and consequently $\langle A^\star v, v \rangle > \langle Au, u \rangle \geq 0$.

3. Applying question A4 to $A^\star$ (this is permissible since $A^\star$ is monotone), we see that $N(A^\star) \subset N(A^{\star\star}) = N(A)$; therefore $N(A) = N(A^\star)$. We always have $\overline{R(A)} = N(A^\star)^\perp$ (see Corollary 2.18), and since E is reflexive, we also have $\overline{R(A^\star)} = N(A)^\perp$.

- C -

1. The map $u \in D(A) \mapsto [u, Au]$ is an isometry from $D(A)$, equipped with the graph norm, onto $G(A)$, which is a closed subspace of $E \times E^\star$.

2. Note that

$$\langle Au - f, u - x \rangle \geq -\|Au\| \, \|x\| - \|f\| \, \|u\| + \langle f, x \rangle.$$

3. Using the properties below (see Problem 13)

$$\lim_{t \to 0} \frac{1}{2t} (\|x + ty\|^2 - \|x\|^2) = \langle Fx, y \rangle \qquad \forall x, y \in E,$$

$$\lim_{t \to 0} \frac{1}{2t} (\|f + tg\|^2 - \|f\|^2) = \langle g, F^{-1}f \rangle \quad \forall f, g \in E^\star,$$

we find that for all $v \in D(A)$,

$$\langle Av, F^{-1}(Au_0 - f) \rangle + \langle F(u_0 - x), v \rangle + \langle Au_0 - f, v \rangle + \langle Av, u_0 - x \rangle = 0.$$

It follows that $F^{-1}(Au_0 - f) + u_0 - x \in D(A^\star)$ and

(S1) $A^\star[F^{-1}(Au_0 - f) + u_0 - x] + (Au_0 - f) + F(u_0 - x) = 0.$

4. Let $x \in E$ and $f \in E^\star$ be such that $\langle Au - f, u - x \rangle \geq 0 \; \forall u \in D(A)$. One has to prove that $x \in D(A)$ and $Ax = f$. We know that there exists some $u_0 \in D(A)$ satisfying (S1). Applying (M*) leads to

$$\langle Au_0 - f + F(u_0 - x), F^{-1}(Au_0 - f) + u_0 - x \rangle \leq 0,$$

that is,

$$\|Au_0 - f\|^2 + \|u_0 - x\|^2 + \langle Au_0 - f, u_0 - x \rangle + \langle F(u_0 - x), F^{-1}(Au_0 - f) \rangle \leq 0.$$

It follows that

$$\|Au_0 - f\|^2 + \|u_0 - x\|^2 \leq \|u_0 - x\| \, \|Au_0 - f\|;$$

therefore $u_0 = x$ and $Au_0 = f$.

5. Apply to the operator $A^\star$ the implication (M*) $\Rightarrow$ (P).

Problem 18

2. $[G(a) - G(b) - g(b)(a - b) = 0] \Leftrightarrow [g(a) = g(b)]$.
3. Passing to a subsequence we may assume that $a_{n_k} \to a$ (possibly $\pm\infty$). We have $\int_b^a (g(t) - g(b))dt = 0$ and therefore $g(a) = g(b)$. It follows that $g(a_{n_k}) \to g(b)$.
4. Note that

$$0 \leq \int |G(u_n) - G(u) - g(u)(u_n - u)| = \int G(u_n) - \int G(u) - \int g(u)(u_n - u)$$

and use assumption (ii). Then apply Theorem 4.9.
5. Since $g(u_n)$ is bounded in $L^{p'}$ (why?), we deduce (see Exercise 4.16) that $g(u_{n_k}) \to g(u)$ strongly in L^q for every $q \in [1, p')$. The uniqueness of the limit implies that $g(u_n) \to g(u)$.
6. If g is increasing then $u_{n_k} \to u$ a.e., and using once more Exercise 4.16 we see that $u_{n_k} \to u$ strongly in L^q for every $q \in [1, p)$.
7 and 9. Applying question 4 and Theorem 4.9, we know that there exists some function $f \in L^1$ such that

(S1) $|G(u_{n_k}) - G(u) - g(u)(u_{n_k} - u)| \leq f \quad \forall k.$

From (S1) and (3) we deduce that $|u_{n_k}|^p \leq \tilde{f}$ for some other function $\tilde{f} \in L^1$. The conclusion follows by dominated convergence.
8. I don't know.

Problem 19

4. Note that the set $\tilde{K} = \{u \in L^2(\mathbb{R}); u \geq 0 \text{ a.e.}\}$ is a closed convex subset of $L^2(\mathbb{R})$. Thus, it is also closed for the weak L^2 topology. It remains to check that $u \in L^1(\mathbb{R})$ and that $\int_\mathbb{R} u \leq 1$. Let $A \subset \mathbb{R}$ be any measurable set with finite measure. We have $\int_A u_n \to \int_A u$ since $\chi_A \in L^2(\mathbb{R})$ and thus $\int_A u \leq 1$. It follows that $u \in L^1(\mathbb{R})$ and that $\int_\mathbb{R} u \leq 1$. Next, write

$$\left| \int \frac{1}{|x|^\alpha}(u_n - u) \right| \leq \int_{[|x|>M]} \frac{1}{|x|^\alpha} |u_n - u| + \left| \int_{[|x|\leq M]} \frac{1}{|x|^\alpha}(u_n - u) \right|$$

$$\leq \frac{2}{M^\alpha} + \left| \int_{[|x|\leq M]} \frac{1}{|x|^\alpha}(u_n - u) \right|.$$

For each fixed M the last integral tends to 0 as $n \to \infty$ (since $u_n \rightharpoonup u$ weakly in $L^2(\mathbb{R})$). We deduce that

$$\limsup_{n\to\infty} \left| \int \frac{1}{|x|^\alpha}(u_n - u) \right| \leq \frac{2}{M^\alpha} \quad \forall M > 0.$$

5. Write

$$\int \frac{1}{|x|^\alpha} u(x)dx = \int_{[|x|>1]} \frac{1}{|x|^\alpha} u(x)dx + \int_{[|x|\le 1]} \frac{1}{|x|^\alpha} u(x)dx$$

$$\le \int u(x)dx + C\|u\|_2 \le 1 + C\|u\|_2 \quad \forall u \in K.$$

8. E is not reflexive. Assume, by contradiction, that E is reflexive and consider the sequence $u_n = \chi_{[n,n+1]}$. Since (u_n) is bounded in E, there is a subsequence u_{n_k} such that $u_{n_k} \rightharpoonup u$ weakly $\sigma(E, E^\star)$. In particular, $\int f u_{n_k} \to \int f u \, \forall f \in L^\infty(\mathbb{R})$ and therefore $\int f u = 0$ for every $f \in L^\infty(\mathbb{R})$ with compact support. It follows that $u = 0$ a.e. On the other hand, if we choose $f \equiv 1$ we see that $\int u = 1$; absurd.

Problem 20

- A -

2. Note that

$$f''(x) = \left(1 - \frac{1}{p}\right) x^{-2+(1/p)}[(1 - x^{1/p})^{p-2} - (1 + x^{1/p})^{p-2}] \le 0.$$

- B -

1. Replacing x by $f(x)$ and y by $g(x)$ in (1) and integrating over Ω, we obtain

$$\|f + g\|_p^p + \|f - g\|_p^p \le 2\int (|f(x)|^{p'} + |g(x)|^{p'})^{p/p'}.$$

On the other hand, letting $u(x) = |f(x)|^{p'}$ and $v(x) = |g(x)|^{p'}$ and using the fact that $p/p' \ge 1$, we obtain

$$\int (u + v)^{p/p'} = \|u + v\|_{p/p'}^{p/p'} \le (\|u\|_{p/p'} + \|v\|_{p/p'})^{p/p'}$$

$$= (\|f\|_p^{p'} + \|g\|_p^{p'})^{p/p'}.$$

Applying (2) with $x = \|f\|_p$ and $y = \|g\|_p$ leads to (5).

- C -

1. *Method* (i). By Hölder's inequality we have

$$\int u\varphi + v\psi \le \|u\|_p\|\varphi\|_{p'} + \|v\|_p\|\psi\|_{p'}$$

$$\le (\|u\|_p^{p'} + \|v\|_p^{p'})^{1/p'}(\|\varphi\|_{p'}^p + \|\psi\|_{p'}^p)^{1/p}.$$

Moreover, equality holds when $\varphi = |u|^{p-2}u\|u\|_p^\alpha$ and $\psi = |v|^{p-2}v\|v\|_p^\alpha$ with $\alpha = p' - p$. Applying the above inequality to $u = f + g$ and $v = f - g$, we obtain

$$(\|f+g\|_p^{p'} + \|f-g\|_p^{p'})^{1/p'} = \sup_{\varphi,\psi\in L^{p'}} \left\{ \frac{\int f(\varphi+\psi)+g(\varphi-\psi)}{[\|\varphi\|_{p'}^p + \|\psi\|_{p'}^p]^{1/p}} \right\}.$$

Using Hölder's inequality we obtain

$$\int f(\varphi+\psi)+g(\varphi-\psi) \le \|f\|_p \|\varphi+\psi\|_{p'} + \|g\|_p \|\varphi-\psi\|_{p'}$$

$$\le (\|f\|_p^p + \|g\|_p^p)^{1/p}(\|\varphi+\psi\|_{p'}^{p'} + \|\varphi-\psi\|_{p'}^{p'})^{1/p'}.$$

On the other hand, inequality (4) applied with p' in place of p says that

$$\|\varphi+\psi\|_{p'}^{p'} + \|\varphi-\psi\|_{p'}^{p'} \le 2(\|\varphi\|_{p'}^p + \|\psi\|_{p'}^p)^{p'/p},$$

and (6) follows.

Method (ii). Applying (1) with $x \to f(x)$, $y \to g(x)$ and $p \to p'$, we obtain

$$|f(x)+g(x)|^{p'} + |f(x)-g(x)|^{p'} \le 2(|f(x)|^p + |g(x)|^p)^{p'/p}$$

and thus

$$(|f(x)+g(x)|^{p'} + |f(x)-g(x)|^{p'})^{p/p'} \le 2^{p/p'}(|f(x)|^p + |g(x)|^p).$$

Integrating over Ω, we obtain, with the notation of Exercise 4.11,

$$[|f+g|^{p'} + |f-g|^{p'}]_{p/p'} \le 2(\|f\|_p^p + \|g\|_p^p)^{p'/p}.$$

The conclusion follows from the fact that $[u+v]_{p/p'} \ge [u]_{p/p'} + [v]_{p/p'}$ (since $p/p' \le 1$).

Problem 21

- A -

1. Use monotone convergence to prove that $\alpha(t+0) = \alpha(t)$. Note that if $f = \chi_\omega$ with $\omega \subset \Omega$ measurable, then $\alpha(1-0) = |\omega|$, while $\alpha(1) = 0$.
2. Given $t > 0$, let $\omega_n = [|f_n| > t]$, $\omega = [|f| > t]$, $\chi_n = \chi_{\omega_n}$, and $\chi = \chi_\omega$. It is easy to check that $\chi(x) \le \liminf \chi_n(x)$ for a.e. $x \in \Omega$ (distinguish the cases $x \in \omega$ and $x \notin \omega$). Applying Fatou's lemma, we see that

$$\alpha(t) = \int_\Omega \chi \le \liminf \int_\Omega \chi_n = \liminf \alpha_n(t).$$

On the other hand, let $\delta \in (0, t)$ and write

$$\int_{\Omega} \chi_n = \int_{[|f| \le t - \delta]} \chi_n + \int_{[|f| > t - \delta]} \chi_n \le \int_{[|f| \le t - \delta]} \chi_n + \alpha(t - \delta).$$

Since $\chi_n \to 0$ a.e. on the set $[|f| \le t - \delta]$, we have, by dominated convergence, $\int_{[|f| \le t - \delta]} \chi_n \to 0$. It follows that $\limsup \int_{\Omega} \chi_n \le \alpha(t - \delta) \, \forall \delta \in (0, t)$.

- B -

1. Consider the measurable function $H : \Omega \times (0, \infty) \to \mathbb{R}$ defined by

$$H(x, t) = \begin{cases} g(t) & \text{if } |f(x)| > t, \\ 0 & \text{if } |f(x)| \le t. \end{cases}$$

Note that

$$\int_{\Omega} H(x, t) d\mu = \alpha(t) g(t) \quad \text{for a.e. } t \in (0, \infty),$$

while

$$\int_0^{\infty} H(x, s) ds = \int_0^{|f(x)|} g(s) ds = G(|f(x)|) \quad \text{for a.e. } x \in \Omega.$$

Then use Fubini and Tonelli.

2. Given $\lambda > 0$ consider the function $\tilde{f} : \Omega \to \mathbb{R}$ defined by

$$\tilde{f}(x) = \begin{cases} f(x) & \text{on } [|f| > \lambda], \\ 0 & \text{on } [|f| \le \lambda], \end{cases}$$

so that its distribution function $\tilde{\alpha}$ is given by

$$\tilde{\alpha}(t) = \begin{cases} \alpha(\lambda) & \text{if } t \le \lambda, \\ \alpha(t) & \text{if } t > \lambda. \end{cases}$$

Apply to $\tilde{f}$ the result of question B1.

- C -

3. Use the inequality $\int_A |f| \le |A|^{1/p'} [f]_p$ with $A = [|f| > t]$ and note that $\int_A |f| \ge t\alpha(t)$.
4. Let $C = \sup_{t > 0} t^p \alpha(t)$. We have

$$\int_A |f| \le \alpha(\lambda) \lambda + \int_{\lambda}^{\infty} \alpha(t) dt + \lambda |A| \le C \left(1 + \frac{1}{p - 1}\right) \lambda^{1-p} + \lambda |A| \quad \forall \lambda > 0.$$

Choose $\lambda = |A|^{-1/p}$.

6. Write

$$\|f\|_p^p = p \int_0^\infty \alpha(t)t^{p-1}dt = p \int_0^\lambda \alpha(t)t^{p-1}dt + p \int_\lambda^\infty \alpha(t)t^{p-1}dt$$

$$\leq p[f]_q^q \int_0^\lambda \frac{t^{p-1}}{t^q}dt + p[f]_r^r \int_\lambda^\infty \frac{t^{p-1}}{t^r}dt$$

and choose λ appropriately.

Problem 22

1. Apply the closed graph theorem.
3. We know, by Problem 21, question B3, that $\|g_\lambda\|_1 = \int_0^\infty \gamma_\lambda(t)dt$. Applying question 2 and once more question B3 of Problem 21, we see that $\|g_\lambda\|_1 \leq N_1[\alpha(\lambda)\lambda + \int_\lambda^\infty \alpha(t)dt]$. On the other hand, since $\|f - g_\lambda\|_\infty \leq \lambda$, we have $[|f| > t] \subset [|g_\lambda| > t - \lambda]$.
4. By question 3 we know that

$$\int_\lambda^\infty \beta(s)ds \leq N_1 \left[\alpha(\lambda)\lambda + \int_\lambda^\infty \alpha(t)dt \right] \quad \forall \lambda > 0.$$

Multiplying this inequality by λ^{p-2} and integrating leads to

$$\int_0^\infty \lambda^{p-2}d\lambda \int_\lambda^\infty \beta(s)ds$$

$$\leq N_1 \left[\int_0^\infty \alpha(\lambda)\lambda^{p-1}d\lambda + \int_0^\infty \lambda^{p-2}d\lambda \int_\lambda^\infty \alpha(t)dt \right],$$

that is,

$$\frac{1}{p-1} \int_0^\infty \beta(s)s^{p-1}ds \leq N_1 \left(1 + \frac{1}{p-1} \right) \int_0^\infty \alpha(\lambda)\lambda^{p-1}d\lambda.$$

From question B3 of Problem 21 we deduce that $\|f\|_p^p \leq pN_1\|u\|_p^p$; finally, we note that $p^{1/p} \leq e^{1/e} \leq 2 \ \forall p \geq 1$.

Problem 23

- A -

1. The sets X_n are closed and $\bigcup_n X_n = X$. Hence, there is some integer n_0 such that $\text{Int}(X_{n_0}) \neq \emptyset$. Thus, there exists $A_0 \subset \Omega$ measurable with $|A_0| < \infty$, and there exists some $\rho > 0$ such that

$$\left[\chi_B \in X \text{ and } \int_\Omega |\chi_B - \chi_{A_0}| < \rho \right] \Rightarrow \left[\left| \int_B f_k \right| \leq \varepsilon \quad \forall k \geq n_0 \right].$$

We first claim that

(S1) $\int_A |f_k| \leq 4\varepsilon \quad \forall A \subset \Omega$ measurable with $|A| < \rho$, and $\forall k \geq n_0$.

Indeed, let $A \subset \Omega$ be measurable with $|A| < \rho$; consider the sets

$$B_1 = A_0 \cup A \quad \text{and} \quad B_2 = B_1 \backslash A.$$

We have

$$\int_\Omega |\chi_{B_1} - \chi_{A_0}| \leq |A| < \rho \quad \text{and} \quad \int_\Omega |\chi_{B_2} - \chi_{A_0}| \leq |A| < \rho,$$

and therefore

$$\left| \int_{B_1} f_k \right| \leq \varepsilon \quad \text{and} \quad \left| \int_{B_2} f_k \right| \leq \varepsilon \quad \forall k \geq n_0.$$

It follows that

$$\left| \int_A f_k \right| = \left| \int_{B_1} f_k - \int_{B_2} f_k \right| \leq 2\varepsilon \quad \forall k \geq n_0.$$

Applying the preceding inequality with A replaced by $A \cap [f_k > 0]$ and by $A \cap [f_k < 0]$, we are led to (S1). The conclusion of question 1 is obvious, since there exists some $\rho' > 0$ such that

$$\int_A |f_k| \leq 4\varepsilon \quad \forall A \subset \Omega \text{ measurable with } |A| < \rho', \quad \forall k = 1, 2, \ldots, n_0.$$

2. There is some integer n_0 such that $\text{Int}(Y_{n_0}) \neq \emptyset$. Thus, there exists $A_0 \subset \Omega$ measurable and there exists some $\rho > 0$ such that

$$[\chi_B \in Y \text{ and } d(\chi_B, \chi_A) < \rho] \Rightarrow \left[\left| \int_B f_k \right| \leq \varepsilon \, \forall k \geq n_0 \right].$$

Fix an integer j such that $2^{-j} < \rho$. We claim that

$$(S2) \qquad \int_A |f_k| \leq 4\varepsilon \; \forall A \subset \Omega \text{ measurable with } A \cap \Omega_j = \emptyset, \; \forall k \geq n_0.$$

Indeed, let $A \subset \Omega$ be measurable with $A \cap \Omega_j = \emptyset$; consider the sets

$$B_1 = A_0 \cup A \quad \text{and} \quad B_2 = B_1 \backslash A.$$

We have $d(\chi_{B_1}, \chi_{A_0}) \leq 2^{-j} < \rho$ and $d(\chi_{B_2}, \chi_{A_0}) \leq 2^{-j} < \rho$; therefore $|\int_{B_1} f_k| \leq \varepsilon$ and $|\int_{B_2} f_k| \leq \varepsilon$, $\forall k \geq n_0$. We then proceed as in question 1.

4. Let us prove, for example, that (i) $\Rightarrow$ (b). Suppose, by contradiction, that (b) fails. There exist some $\varepsilon_0 > 0$, a sequence (A_n) of measurable sets in Ω, and a sequence (f_n) in $\mathcal{F}$ such that $|A_n| \to 0$ and $\int_{A_n} |f_n| \geq \varepsilon_0 \, \forall n$. By the Eberlein–Šmulian theorem there exists a subsequence such that $f_{n_k} \rightharpoonup f$ weakly $\sigma(L^1, L^\infty)$. Thus (see question 3) (f_{n_k}) is equi-integrable and we obtain a contradiction.

5. Assume, for example, that $\int_A f_n \to \ell(A)$ for every $A \subset \Omega$ with A measurable and $|A| < \infty$. We claim that (b) holds.

Indeed, consider the sequence

$$X_n = \left\{ \chi_A \in X; \left| \int_A f_j - \int_A f_k \right| \le \varepsilon \ \forall j \ge n, \ \forall k \ge n \right\}.$$

In view of the Baire category theorem there exist n_0, $A_0 \subset \Omega$ measurable with $|A_0| < \infty$, and $\rho > 0$ such that

$$\left[\chi_B \in X \text{ and } \int_\Omega |\chi_B - \chi_{A_0}| < \rho \right] \Rightarrow \left[\left| \int_B f_k - \ell(B) \right| \le \varepsilon \ \forall k \ge n_0 \right].$$

Let $A \subset \Omega$ be measurable with $|A| < \rho$; with the same method as in question 1 one obtains

$$\left| \int_A f_k \right| \le 2\varepsilon + |\ell(B_2) - \ell(B_1)| \le 4\varepsilon + \left| \int_A f_{n_0} \right| \quad \forall k \ge n_0.$$

It follows that

$$\int_A |f_k| \le 8\varepsilon + 2 \int_A |f_{n_0}| \quad \forall A \text{ measurable with } |A| < \rho, \ \forall k \ge n_0,$$

and the conclusion is easy.

- **B** -

1. We have $\mathcal{F} \subset \mathcal{F}_\varepsilon + \varepsilon B_E \subset \mathcal{F}_\varepsilon + \varepsilon B_{E^{**}}$. But $\mathcal{F}_\varepsilon + \varepsilon B_{E^{**}}$ is compact for the topology $\sigma(E^{**}, E^\star)$ (since it is a sum of two compact sets). It follows that $\mathcal{G}$ is compact for $\sigma(E^{**}, E^\star)$. Also, since $\mathcal{G} \subset E + \varepsilon B_{E^{**}} \ \forall \varepsilon > 0$, we deduce that $\mathcal{G} \subset E$. These properties imply that $\mathcal{G}$ is compact for $\sigma(E, E^\star)$.

2. Given $\varepsilon > 0$ choose $\omega \subset \Omega$ measurable with $|\omega| < \infty$ such that $\int_{\Omega \setminus \omega} |f| \le \varepsilon/2$ $\forall f \in \mathcal{F}$, and choose n such that $\int_{[|f|>n]} |f| \le \varepsilon/2 \ \forall f \in \mathcal{F}$ (see Exercise 4.36). Set $\mathcal{F}_\varepsilon = (\chi_\omega T_n(f))_{f \in \mathcal{F}}$. Clearly, $\mathcal{F}_\varepsilon$ is bounded in $L^\infty(\omega)$ and thus it is contained in a compact subset of $L^1(\Omega)$ for $\sigma(L^1, L^\infty)$. On the other hand, for every $f \in \mathcal{F}$, we have

$$\int_\Omega |f - \chi_\omega T_n(f)| \le \int_\omega |f - T_n f| + \int_{\Omega \setminus \omega} |f| \le \int_{[|f|>n]} |f| + \int_{\Omega \setminus \omega} |f| \le \varepsilon.$$

Thus, $\mathcal{F} \subset \mathcal{F}_\varepsilon + \varepsilon B_E$ with $E = L^1(\Omega)$.

- **C** -

4. Applying A5 we know that (f_n) satisfies (b) and (c). In view of B2 the set (f_n) has a compact closure in the topology $\sigma(L^1, L^\infty)$. Thus (by Eberlein–Šmulian) there is a subsequence such that $f_{n_k} \rightharpoonup f$ weakly $\sigma(L^1, L^\infty)$. It follows that

$\ell(A) = \int_A f \; \forall A$ measurable. The uniqueness of the limit implies that $f_n \rightharpoonup f$ weakly $\sigma(L^1, L^\infty)$ (check the details).

- D -

1. Apply Exercise 4.36.
2. Set

$$\Phi(t) = \sup_{f \in \mathcal{F}} \int_{[|f| > t]} |f|,$$

so that $\Phi \geq 0$, Φ is nonincreasing, and $\lim_{t \to \infty} \Phi(t) = 0$. We may always assume that $\Phi(t) > 0 \; \forall t > 0$; otherwise, there exists some T such that $\|f\|_\infty \leq T$, for all $f \in \mathcal{F}$, and the conclusion is obvious. Consider a function $g : [0, \infty) \to (0, \infty)$ such that g is nondecreasing and $\lim_{t \to \infty} g(t) = \infty$. Set $G(t) = \int_0^t g(s) ds, t \geq 0$, so that G is increasing, convex, and $\lim_{t \to \infty} G(t)/t = +\infty$. We recall (see Problem 21) that for every f,

$$\int G(|f|) = \int_0^\infty \alpha(t) g(t) dt \quad \text{and} \quad \int_{[|f| > t]} |f| = \alpha(t)t + \int_t^\infty \alpha(s) ds.$$

Set $\beta(t) = \int_t^\infty \alpha(s) ds$, so that $\beta(t) \leq \Phi(t)$ and $\beta'(t) = -\alpha(t)$. We claim that if we choose $g(t) = [\Phi(t)]^{-1/2}$, then the corresponding function G has the required property. Indeed, for every $f \in \mathcal{F}$, we have

$$\int G(|f|) = \int_0^\infty \alpha(t) g(t) dt \leq \int_0^\infty -\beta'(t)[\beta(t)]^{-1/2} dt$$

$$= 2[\beta(0)]^{1/2} = 2 \left[\int_0^\infty \alpha(s) ds \right]^{1/2} = 2 \left[\int |f| \right]^{1/2} \leq C.$$

Problem 24

- B -

3. Clearly A is convex, and so is $\overline{A}^{\sigma(E^\star, E)}$ (see Problem 9, question A4). Suppose by contradiction that $\mu_0 \notin \overline{A}^{\sigma(E^\star, E)}$. By Hahn–Banach (applied in $E^\star$ with the weak* topology) there exist $f_0 \in C(\overline{\Omega})$ and $\beta \in \mathbb{R}$ such that

(S1) $$\int_\Omega u f_0 < \beta < \langle \mu_0, f_0 \rangle \quad \forall u \in A.$$

On the other hand, we have

(S2) $$\sup_{u \in A} \int_\Omega u f_0 = \|f_0\|_\infty;$$

indeed, A is dense in the unit ball of $L^1(\Omega)$ (by Corollary 4.23) and L^∞ is the dual of L^1 (see Theorem 4.14). Combining (S1) and (S2) yields

$$\|f_0\|_\infty \le \beta < \langle \mu_0, f_0 \rangle \le \|f_0\|_\infty,$$

since $\|\mu_0\| \le 1$. This is impossible.

4. B_{E^*} is metrizable because $E = C(\overline{\Omega})$ is separable (see Theorem 3.28). Since $\mu_0 \in \overline{A}^{\sigma(E^*,E)} \subset B_{E^*}$ there exists a sequence (v_n) in A such that $v_n \overset{*}{\rightharpoonup} \mu_0$. Then apply Proposition 3.13.

6. Clearly $\langle \mu, 1 \rangle \le \|\mu\| \; \forall \mu$. On the other hand, if $\|f\|_\infty \le 1$ and $\mu \ge 0$ we have $\langle \mu, f \rangle \le \langle \mu, 1 \rangle$ and thus $\|\mu\| = \sup_{\|f\|_\infty \le 1} \langle \mu, f \rangle \le \langle \mu, 1 \rangle$.

7. Set $A^+ = \{u \in A; u(x) \ge 0 \; \forall x \in \overline{\Omega}\}$. Repeat the same proof as in question 3 with A being replaced by A^+; check that

$$\sup_{u \in A^+} \int_\Omega u f_0 = \|f_0^+\|_\infty$$

and that

$$\langle \mu_0, f_0 \rangle \le \|f_0^+\|_\infty.$$

8. We claim that $\|u + \delta_a\|_{\mathcal{M}} = \|u\|_{L^1} + 1$. Clearly $\|u + \delta_a\|_{\mathcal{M}} \le \|u\|_{L^1} + 1$. To prove the reverse inequality, fix <u>any</u> $\varepsilon > 0$ and choose $r > 0$ sufficiently small that $\int_{B(a,r)} |u| < \varepsilon$. Let $\omega = \Omega \setminus \overline{B(a,r)}$ and pick $\varphi \in C_c(\omega)$ with $\|\varphi\|_{L^\infty(\omega)} \le 1$ and

$$\int_\omega u\varphi \ge \|u\|_{L^1(\omega)} - \varepsilon.$$

Then let $\theta \in C_c(B(a,r))$ be such that $\theta(a) = 1$ and $\|\theta\|_{L^\infty(\Omega)} \le 1$. Check that $\|\varphi + \theta\|_{L^\infty(\Omega)} \le 1$ and

$$\langle u + \delta_a, \varphi + \theta \rangle \ge \|u\|_{L^1(\Omega)} - 2\varepsilon + 1.$$

-D-

1. Clearly $L(f_1) + L(f_2) \le L(f_1 + f_2)$. For the reverse inequality, note that if $0 \le g \le f_1 + f_2$, then one can write $g = g_1 + g_2$ with $0 \le g_1 \le f_1$ and $0 \le g_2 \le f_2$; take, for example, $g_1 = \max\{g - f_2, 0\}$ and $g_2 = g - g_1$.

2. If $f = h + k$ with $h, k \in C(K)$, we have

$$f^+ - f^- = h^+ - h^- + k^+ - k^-,$$

so that

$$f^+ + h^- + k^- = h^+ + k^+ + f^-,$$

and thus

$$L(f^+) + L(h^-) + L(k^-) = L(h^+) + L(k^+) + L(f^-),$$

i.e.,

$$\mu_1(f) = \mu_1(h) + \mu_1(k).$$

Note that $L(f^+) \leq \|\mu\| \|f^+\|$ and $L(f^-) \leq \|\mu\| \|f^-\|$. Thus $|\mu_1(f)| \leq \|\mu\| \|f\|$. If $f \geq 0$ we have $\mu_1(f) = L(f) \geq 0$, so that $\mu_1 \geq 0$.

3. If $f \geq 0$, we have (taking $g = f$) $L(f) \geq \langle \mu, f \rangle$, so that $\langle \mu_1, f \rangle = L(f) \geq \langle \mu, f \rangle$, i.e., $\mu_2 = \mu_1 - \mu \geq 0$. Next, note that if $g \in C(K)$ and $0 \leq g \leq 1$, we have $-1 \leq 2g - 1 \leq 1$ and thus

$$\langle \mu, 2g - 1 \rangle \leq \|\mu\|.$$

Therefore

$$L(1) = \sup\{\langle \mu, g \rangle; 0 \leq g \leq 1\} \leq \frac{1}{2}(\langle \mu, 1 \rangle + \|\mu\|),$$

i.e.,

$$2\langle \mu_1, 1 \rangle = 2L(1) \leq \langle \mu_1, 1 \rangle - \langle \mu_2, 1 \rangle + \|\mu\|.$$

Thus

$$\|\mu_1\| + \|\mu_2\| = \langle \mu_1 + \mu_2, 1 \rangle \leq \|\mu\|$$

and consequently $\|\mu\| = \|\mu_1\| + \|\mu_2\|$.

-E-

One can repeat all the above proofs without modification. The only change occurs in question D3, where we have used the function 1, which is no longer admissible. We introduce, instead of 1, a sequence (θ_n) in E_0 such that $\theta_n \uparrow 1$ as $n \uparrow \infty$. Note that for every $v \in \mathcal{M}(\Omega)$, $v \geq 0$, we have $\langle v, \theta_n \rangle \uparrow \|v\|$.

If $g \in E_0$ and $0 \leq g \leq \theta_n$ we have $-\theta_n \leq 2g - \theta_n \leq \theta_n$ and thus $\langle \mu, 2g - \theta_n \rangle \leq \|\mu\|$. Hence

$$L(\theta_n) = \sup\{\langle \mu, g \rangle; 0 \leq g \leq \theta_n\} \leq \frac{1}{2}(\langle \mu, \theta_n \rangle + \|\mu\|)$$

i.e.,

$$2\langle \mu_1, \theta_n \rangle = 2L(\theta_n) \leq \langle \mu_1, \theta_n \rangle - \langle \mu_2, \theta_n \rangle + \|\mu\|.$$

Letting $n \to \infty$ yields $\|\mu_1\| + \|\mu_2\| \leq \|\mu\|$.

Problem 25

1. Let $v_0 \in C$ and let $u \in \Sigma \setminus \{0\}$; if $B(v_0, \rho) \subset C$ then $(u, v_0 + \rho z) \leq 0 \; \forall z \in H$ with $|z| < 1$. It follows that $(u, v_0) + \rho|u| \leq 0$. Conversely, let $v_0 \in H$ be such that $(u, v_0) < 0 \; \forall u \in \setminus \{0\}$. In order to prove that $v_0 \in C$, assume by contradiction that $v_0 \notin C$ and separate C and $\{v_0\}$.

2. If $u \in \Sigma$, then $(u, \omega) + \rho|u| \leq 0$; therefore $\rho|u| \leq 1$ for every $u \in K$.

4. If $(-C) \cap \Sigma = \emptyset$ separate $(-C)$ and Σ, and obtain a contradiction.

5. Since $a \in (-C) \cap \Sigma$ we may write $-a = \mu(w_0 - x_0)$ with $\mu > 0$ and $w_0 \in D$. On the other hand, since $a \in \Sigma \setminus \{0\}$ we have $(a, v) < 0 \; \forall v \in C$ and thus $(a, w - x_0) < 0 \; \forall w \in D$. It follows that $(x_0 - w_0, w - x_0) < 0 \; \forall w \in D$.

Problem 26

1. By Proposition 1.10 there exist some $g \in H$ and some constant C such that $\varphi(v) \geq (g, v) + C \ \forall v \in H$; therefore $I > -\infty$. Let (u_n) be a minimizing sequence, that is, $\frac{1}{2}|f - u_n|^2 + \varphi(u_n) = I_n \to I$. Using the parallelogram law we obtain

$$\left| f - \frac{u_n + u_m}{2} \right|^2 + \left| \frac{u_n - u_m}{2} \right|^2 = \frac{1}{2}\left(|f - u_n|^2 + |f - u_m|^2 \right)$$

$$= I_n + I_m - \varphi(u_n) - \varphi(u_m) \leq I_n + I_m - 2\varphi\left(\frac{u_n + u_m}{2} \right).$$

It follows that $\left| \frac{u_n - u_m}{2} \right|^2 \leq I_n + I_m - 2I$.

2. If u satisfies (Q) we have

$$\frac{1}{2}|f - v|^2 + \varphi(v) \geq \frac{1}{2}|f - u|^2 + \varphi(u) + \frac{1}{2}|u - v|^2 \quad \forall v \in H.$$

Conversely, if u satisfies (P) we have

$$\frac{1}{2}|f - u|^2 + \varphi(u) \leq \frac{1}{2}|f - v|^2 + \varphi(v) \quad \forall v \in H;$$

choose $v = (1 - t)u + tw$ with $t \in (0, 1)$ and note that

$$\frac{1}{2}|f - v|^2 = \frac{1}{2}|f - u|^2 + t(f - u, u - w) + \frac{t^2}{2}|u - w|^2,$$

and

$$\varphi(v) \leq (1 - t)\varphi(u) + t\varphi(w).$$

3. Choose $v = \bar{u}$ in (Q), $v = u$ in ($\bar{\text{Q}}$), and add.

5. By (Q) we have

$$(f - u, v) - \varphi(v) \leq (f - u, u) - \varphi(u) \quad \forall v \in H$$

and thus $\varphi^\star(f - u) = (f - u, u) - \varphi(u)$. It follows that

$$\frac{1}{2}|u|^2 + \varphi^\star(f - u) = -\frac{1}{2}|f - u|^2 - \varphi(u) + \frac{1}{2}|f|^2.$$

Letting $u^\star = f - u$, one obtains

$$\frac{1}{2}|f - u^\star|^2 + \varphi^\star(u^\star) = -\frac{1}{2}|f - u|^2 - \varphi(u) + \frac{1}{2}|f|^2,$$

and one checks easily that

$$-\frac{1}{2}|f - u|^2 - \varphi(u) + \frac{1}{2}|f|^2 \leq \frac{1}{2}|f - v|^2 + \varphi^\star(v) \quad \forall v \in H.$$

(Recall that $(u, v) \leq \varphi(u) + \varphi^\star(v)$.)

6. We have

$$(\text{P}_\lambda) \qquad \frac{1}{2}|f - u_\lambda|^2 + \lambda\varphi(u_\lambda) \leq \frac{1}{2}|f - v|^2 + \lambda\varphi(v) \quad \forall v \in H.$$

Using that fact that $\varphi(u_\lambda) \geq (g, u_\lambda) + C$, it is easy to see that $|u_\lambda|$ remains bounded as $\lambda \to 0$. We may therefore assume that $u_{\lambda_n} \rightharpoonup u_0$ weakly $(\lambda_n \to 0)$ with $u_0 \in \overline{D(\varphi)}$ (why?). Passing to the limit in (P_{λ_n}) (how?), we obtain

$$\frac{1}{2}|f - u_0|^2 \leq \frac{1}{2}|f - v|^2 \quad \forall v \in D(\varphi),$$

and we deduce that $u_0 = P_{\overline{D(\varphi)}} f$. The uniqueness of the limit implies that $u_\lambda \rightharpoonup u_0$ weakly as $\lambda \to 0$. To see that $u_\lambda \to u_0$ we note that

$$\frac{1}{2}\limsup_{\lambda \to 0}|f - u_\lambda|^2 \leq \frac{1}{2}|f - v|^2 \quad \forall v \in D(\varphi),$$

which implies that $\limsup_{\lambda \to 0}|f - u_\lambda| \leq |f - u_0|$ and the strong convergence follows.

Alternative proof. Combining (Q_λ) and (Q_μ) we obtain

$$\left(\frac{1}{\lambda}(u_\lambda - f) - \frac{1}{\mu}(u_\mu - f), u_\lambda - u_\mu\right) \leq 0 \quad \forall \lambda, \mu > 0.$$

We deduce from Exercise 5.3, question 1, that $(u_\lambda - f)$ converges strongly as $\lambda \to 0$ to some limit. In order to identify the limit one may proceed as above.

7. We have $\frac{1}{2}|f - u_\lambda|^2 + \lambda\varphi(u_\lambda) \leq \frac{1}{2}|f - v|^2 + \lambda\varphi(v) \; \forall v \in H$, and in particular, $|f - u_\lambda| \leq |f - v| \; \forall v \in K$. We may therefore assume that $u_{\lambda_n} \rightharpoonup u_\infty$ weakly $(\lambda_n \to +\infty)$ and we obtain $|f - u_\infty| \leq |f - v| \; \forall v \in K$. On the other hand, we have

$$\varphi(u_\lambda) \leq \frac{1}{2\lambda}|f - v|^2 + \varphi(v) \quad \forall v \in H,$$

and passing to the limit, we obtain $\varphi(u_\infty) \leq \varphi(v) \; \forall v \in H$. Thus, $u_\infty \in K$, $u_\infty = P_K f$ (why?), and $u_\lambda \rightharpoonup u_\infty$ weakly as $\lambda \to +\infty$ (why?). Finally, note that $\limsup_{\lambda \to +\infty}|f - u_\lambda| \leq |f - u_\infty|$.
If $K = \emptyset$, then $|u_\lambda| \to \infty$ as $\lambda \to +\infty$ (argue by contradiction).

8. If $f = 0$ check that $(1/\lambda)u_\lambda = -u^\star_{1/\lambda} \; \forall \lambda > 0$. In the general case (in which $f \neq 0$) denote by u_λ and $\overline{u_\lambda}$ the solutions of (P_λ) corresponding respectively to f and to 0. We know, by question 3, that $|u_\lambda - \overline{u_\lambda}| \leq |f|$ and thus $|\frac{1}{\lambda}u_\lambda - \frac{1}{\lambda}\overline{u_\lambda}| \to 0$ as $\lambda \to +\infty$.

Problem 27

- A -

3. By definition of the projection we have

$$|u_{2n+2} - u_{2n+1}| = |P_2 u_{2n+1} - u_{2n+1}| \leq |u_{2n} - u_{2n+1}|$$

(since $u_{2n} \in K_2$), and similarly

$$|u_{2n+1} - u_{2n}| = |P_1 u_{2n} - u_{2n}| \leq |u_{2n-1} - u_{2n}|.$$

It follows that

$$|u_{2n+2} - u_{2n+1}| \leq |u_{2n} - u_{2n-1}|.$$

- B -

To see that a_1 and a_2 may depend on u_0 take convex sets K_1 and K_2 as shown in Figure 9.

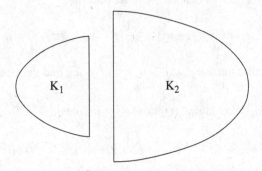

Fig. 9

Problem 28

-A -

1. (a) $\Rightarrow$ (b).   Note that $(u, Pv) = (Pu, Pv) = (Pu, v) \ \forall u, v \in H$.

   (b) $\Rightarrow$ (c).   We have $|Pu|^2 = (Pu, Pu) = (u, P^2 u) = (u, Pu) \ \forall u \in H$.

   (c) $\Rightarrow$ (d).   From (c) we have $((u - Pu) - (v - Pv), u - v) \geq 0 \ \forall u, v \in H$
   and therefore $(u, u - v) \geq 0 \ \forall u \in N(P)$ and $\forall v \in N(I - P)$.
   Replacing u by λu, we obtain (d).

   (d) $\Rightarrow$ (a).   Set $M = N(I - P)$ and check that $P = P_M$.

- B -

1. (b) $\Rightarrow$ (c).   Note that $(PQ)^2 = PQ$ and pass to the adjoints.
   (c) $\Rightarrow$ (a). QP is a projection operator and $\|QP\| \leq 1$. Thus QP is an orthogonal projection and therefore $(QP)^\star = QP$, that is, $PQ = QP$.

(i) Check that $N(I - PQ) = M \cap N$.

(ii) Applying (i) to $(I - P)$ and $(I - Q)$, we see that $(I - P)(I - Q)$ is the orthogonal projection onto $M^{\perp} \cap N^{\perp}$. Therefore $I - (I - P)(I - Q) = P + Q - PQ$ is the orthogonal projection onto $(M^{\perp} \cap N^{\perp})^{\perp} = \overline{M + N}$.

2. It is easy to check that

$$(a) \Rightarrow (b) \Leftrightarrow (c) \Rightarrow (d) \Leftrightarrow (e) \Leftrightarrow (f) \Rightarrow (a).$$

Clearly $(b) + (c) \Rightarrow (g)$. Conversely, we claim that $(g) \Rightarrow (b) + (c)$. Indeed, we have $PQ + QP = 0$. Multiplying this identity on the left and on the right by P, we obtain $PQ - QP = 0$; thus, $PQ = 0$. Finally, apply case (ii) of question B1.

3. Replace N by $N^{\perp}$ and apply question B2.

Problem 29

- A -

5. Note that $\sum_{i=0}^{n} |\mu_i - \mu_{i+1}|^2 \le |f - v|^2$ and that $|\mu_n - \mu_{n+1}| \le |\mu_i - \mu_{i+1}| \ \forall i = 0, 1, \ldots, n$.

- B -

2. Since $0 \in K$, the sequence $(|u_n|)$ is nonincreasing and thus it converges to some limit, say a. Applying the result of B1 with $u = u_n$ and $v = u_{n+i}$, we obtain

$$2|(u_n, u_{n+i}) - (u_{n+p}, u_{n+p+i})| \le 2(|u_n|^2 - |u_{n+p+i}|^2)$$
$$\le 2(|u_n|^2 - a^2).$$

Therefore $\ell(i) = \lim_{n \to \infty} (u_n, u_{u+i})$ exists and we have

$$|(u_n, u_{n+i}) - \ell(i)| \le |u_n|^2 - a^2 \equiv \varepsilon_n.$$

3. Applying to S the above result, we see that

$$|(\mu_n, \mu_{n+i}) - m(i)| \le \varepsilon'_n \quad \forall i, \ \forall n.$$

In particular, we have

$$||\mu_n|^2 - m(0)| \le \varepsilon'_n \quad \text{and} \quad |(\mu_n, \mu_{n+1}) - m(1)| \le \varepsilon'_n$$

and therefore

$$|m(0) - m(1)| \le 2\varepsilon'_n + |\mu_n||\mu_n - \mu_{n+1}| \to 0 \quad \text{as } n \to \infty.$$

It follows that $m(0) = m(1)$ and similarly, $m(1) = m(2)$, etc.

4. We have established that $|(\mu_n, \mu_{n+i}) - m(0)| \le \varepsilon'_n$ $\forall i, \forall n$. Passing to the limit
   as $i \to \infty$ we obtain $|(\mu_n, \mu) - m(0)| \le \varepsilon'_n$ and then, as $n \to \infty$, we obtain
   $|\mu|^2 = m(0)$. Thus, $|\mu_n| \to |\mu|$ and consequently $\mu_n \to \mu$ strongly.
5. Applying (1) and adding the corresponding inequalities for $i = 0, 1, \ldots, p - 1$,
   leads to

$$\left|\left(u_n, \frac{(n+p)}{p}\sigma_{n+p} - \frac{n}{p}\sigma_n\right) - X_p\right| \le \varepsilon_n.$$

We deduce that

$$\left|(u_n, \sigma_{n+p}) - X_p\right| \le \varepsilon_n + \frac{n}{p}|u_n|\left(|\sigma_{n+p}| + |\sigma_n|\right) \le \varepsilon_n + \frac{2n}{p}|f|^2$$

(since $|u_n| \le |f|$ $\forall n$).
6. We have

$$|X_p - X_q| \le 2\varepsilon_n + 2n\left(\frac{1}{p} + \frac{1}{q}\right)|f|^2 + |(u_n, \sigma_{n+p} - \sigma_{n+q})|$$

and thus $\limsup_{p,q\to\infty} |X_p - X_q| \le 2\varepsilon_n$ $\forall n$.
7. Write that

$$n^2|\sigma_n|^2 = \sum_{i=0}^{n-1} |u_i|^2 + 2\sum_{i=1}^{n-1}\sum_{j=0}^{n-i-1} (u_j, u_{j+i})$$

and apply (1).
8. Note that $\sum_{i=0}^{n-1}(n - i)\ell(i) = \sum_{j=1}^{n} jX_j$ and use the fact that $X_j \to X$ as
   $j \to \infty$.

Problem 30

- C -

3. Choose $\bar{\lambda} \in A$ and $\bar{\mu} \in B$ such that

$$\min_{\lambda\in A}\max_{\mu\in B} F(\lambda, \mu) = \max_{\mu\in B} F(\bar{\lambda}, \mu) \quad \text{and} \quad \max_{\mu\in B}\min_{\lambda\in A} F(\lambda, \mu) = \min_{\lambda\in A} F(\lambda, \bar{\mu}).$$

- D -

2. The sets B_u and A_v are compact for the weak topology. Applying the convexity
   of K in u and the concavity of K in v, we obtain

$$K\left(\sum_i \lambda_i u_i, v_j\right) \le \sum_i \lambda_i K(u_i, v_j)$$

and

$$K\left(u_i, \sum_j \mu_j v_j\right) \geq \sum_j \mu_j K(u_i, v_j).$$

It follows that

$$\sum_j \mu_j K\left(\sum_i \lambda_i u_i, v_j\right) \leq F(\lambda, \mu) \leq \sum_i \lambda_i K\left(u_i, \sum_j \mu_j v_j\right)$$

and in particular

$$\sum_j \mu_j K(\bar{u}, v_j) \leq F(\bar{\lambda}, \mu) \quad \forall \mu \in B',$$

$$\sum_i \lambda_i K(u_i, \bar{v}) \geq F(\lambda, \bar{\mu}) \quad \forall \lambda \in A'.$$

Applying (1), we see that

$$\sum_j \mu_j K(\bar{u}, v_j) \leq \sum_i \lambda_i K(u_i, \bar{v}) \quad \forall \lambda \in A', \ \forall \mu \in B'.$$

Finally, choose λ and μ to be the elements of the canonical basis.

Problem 31

- A -

3. Note that $\sum_{i,j} \lambda_i \lambda_j \langle Av_j, v_i - v_j \rangle = \frac{1}{2} \sum_{i,j} \lambda_i \lambda_j \langle Av_j - Av_i, v_i - v_j \rangle$.

- B -

2. For every $R > 0$ there exists some $u_R \in K_R$ such that $\langle Au_R, v - u_R \rangle \geq 0 \, \forall v \in K_R$. Choosing $v = 0$ we see that there exists a constant M (independent of R) such that $\|u_R\| \leq M \, \forall R$. Fix any $R > M$. Given $w \in K$, take $v = (1-t)u_R + tw$ with $t > 0$ sufficiently small (so that $v \in K_R$).
3. Take $K = E$. First, prove that there exists some $u \in E$ such that $Au = 0$. Then, replace A by the map $v \mapsto Av - f$ ($f \in E^*$ being fixed).

Problem 32

2. For $\varepsilon > 0$ small enough we have

$$\varphi(u_0) \leq \varphi(u_0 + \varepsilon v) = \max_{i \in I}\{|u_0 - y_i|^2 - c_i + 2\varepsilon(v, u_0 - u_i)\} + O(\varepsilon^2)$$

$$= \max_{i \in J(u_0)}\{|u_0 - y_i|^2 - c_i + 2\varepsilon(v, u_0 - y_i)\} + O(\varepsilon^2)$$

$$\leq \varphi(u_0) + 2\varepsilon \max_{i \in J(u_0)}\{(v, u_0 - y_i)\} + O(\varepsilon^2).$$

3. Argue by contradiction and apply Hahn–Banach.
4. Note that for every $u, v \in H$, we have

$$\varphi(v) - \varphi(u) \geq \max_{i \in J(u)} \{|v - y_i|^2 - |u - y_i|^2\} \geq 2 \max_{i \in J(u)} \{(u - y_i, v - u)\}.$$

5. Condition (1) is replaced by $0 \in \mathrm{conv}\,(\bigcup_{i \in J(u_0)}\{f'(u_0)\})$.

- B -

1. Letting $\sigma_x = \sum_{i \in I} \lambda_i x_i$ and $\sigma_y = \sum_{i \in I} \lambda_i y_i$, we obtain

$$\sum_{j \in I} \lambda_j |\sigma_y - y_j|^2 = -|\sigma_y|^2 + \sum_{j \in I} \lambda_j |y_j|^2 = \frac{1}{2} \sum_{i,j \in I} \lambda_i \lambda_j |y_i - y_j|^2$$

$$\leq \frac{1}{2} \sum_{i,j \in I} \lambda_i \lambda_j |x_i - x_j|^2 = -|\sigma_x|^2 + \sum_{j \in I} \lambda_j |x_j|^2$$

$$= -|\sigma_x - p|^2 + \sum_{j \in I} \lambda_j |x_j - p|^2.$$

2. Write $u_0 = \sum_{i \in J(u_0)} \lambda_j y_j$. By the result of B1 we have

$$\sum_{j \in J(u_0)} \lambda_j |u_0 - y_j|^2 \leq \sum_{j \in J(u_0)} \lambda_j |p - x_j|^2.$$

It follows that $\sum_{j \in J(u_0)} \lambda_j \varphi(u_0) \leq 0$ and thus $\varphi(u_0) \leq 0$.

Remark. One could also establish the existence of q by applying the von Neumann min–max theorem (see Problem 30, part D) to the function

$$K(\lambda, \mu) = \sum_{j \in I} \mu_j \left|\left(\sum_{i \in I} \lambda_i y_i\right) - y_j\right|^2 - \sum_{j \in I} \mu_j |p - x_j|^2.$$

- C -

1. Set $K_i = \{z \in H; |z - y_i| \leq |p - x_i|\}$ and $K = \overline{\mathrm{conv}\,(\bigcup_{i \in I}\{y_i\})}$. One has to show that $(\bigcap_{i \in I} K_i) \neq \emptyset$. This is done by contradiction and reduction to a finite set I.
2. Consider the ordered set of all contractions $T : D(T) \subset H \to H$ that extend S and such that $T(D(T)) \subset \overline{\mathrm{conv}\, S(D)}$. By Zorn's lemma it has a maximal element T_0 and $D(T_0) = H$ (why?).

Problem 33

1. Note that $|au_n| \le n|u|$, so that $u_n \in D(A)$. Moreover, $|u_n| \le |u|$ and $u_n \to u$ a.e.

2. Let $u_n \to u$ and $au_n \to f$ in L^p. Passing to a subsequence, we may assume that $u_n \to u$ a.e. and $au_n \to f$ a.e. Thus $au = f$.

3. If $D(A) = E$, the closed graph theorem (Theorem 2.9) implies the existence of a constant C such that

$$\int_\Omega |au|^p \le C \int_\Omega |u|^p \quad \forall u \in L^p.$$

Hence the mapping $v \mapsto \int_\Omega |a|^p v$ is a continuous linear functional on L^1. By Theorem 4.14 there exists $f \in L^\infty$ such that

$$\int_\Omega |a|^p v = \int_\Omega f v \quad \forall v \in L^1.$$

Thus $a \in L^\infty$.

4. $N(A) = \{u \in L^p; u = 0 \text{ a.e. on } [a \ne 0]\}$ and $N(A)^\perp = \{f \in L^{p'}; f = 0 \text{ a.e. on } [a = 0]\}$.

To verify the second assertion, let $f \in N(A)^\perp$. Then $\int_\Omega fu = 0 \ \forall u \in N(A)$. Taking $u = |f|^{p'-2} f \chi_{[a=0]}$, we see that $f = 0$ a.e. on $[a = 0]$.

5. $D(A^\star) = \{v \in L^{p'}; av \in L^{p'}\}$ and $A^\star v = av$.

Indeed, if $v \in D(A^\star)$, there exists a constant C such that

$$\left| \int_\Omega v(au) \right| \le C\|u\|_p \quad \forall u \in D(A).$$

The linear functional $u \in D(A) \mapsto \int_\Omega v(au)$ can be extended by Hahn–Banach (or by density) to a continuous linear functional on all of L^p. Hence, by Theorem 4.11, there exists some $f \in L^{p'}$ such that

$$\int_\Omega v(au) = \int_\Omega fu \quad \forall u \in D(A).$$

Given any $\varphi \in L^p$, take $u = (1 + |a|)^{-1} \varphi$, so that

$$\int_\Omega \frac{av}{1 + |a|} \varphi = \int_\Omega \frac{f}{1 + |a|} \varphi.$$

Thus $f = av \in L^{p'}$.

6. Assume that there exists $\alpha > 0$ such that $|a(x)| \ge \alpha$ a.e. Then A is surjective, since any $f \in L^p$ can be written as $au = f$, where $u = a^{-1} f \in D(A)$. Conversely, assume that A is surjective. Then $a \ne 0$ a.e. Moreover, $\forall f \in L^p$, $a^{-1} f \in L^p$. Applying question 3 to the function a^{-1}, we see that $a^{-1} \in L^\infty$.

7. $EV(A) = \{\lambda \in \mathbb{R}; \ |[a = \lambda]| > 0\},$

             $\rho(A) = \{\lambda \in \mathbb{R}; \ \exists \varepsilon > 0 \text{ such that } |a(x) - \lambda| \geq \varepsilon \text{ a.e. on } \Omega\},$

and

$$\sigma(A) = \{\lambda \in \mathbb{R}; \quad \forall \varepsilon > 0, |[|a - \lambda| < \varepsilon]| > 0\}.$$

Set $M = \sup_\Omega a$ and let us show that $M \in \sigma(A)$. By definition of M we know that $a \leq M$ a.e. on Ω and $\forall \varepsilon > 0 \ |[a > M - \varepsilon]| > 0$. Thus, $\forall \varepsilon > 0$ $|[|a - M| < \varepsilon]| > 0$ and therefore $M \in \sigma(A)$.

Note that $\sigma(A)$ coincides with the smallest closed set $F \subset \mathbb{R}$ such that $a(x) \in F$ a.e. in Ω. (The existence of a smallest such set can be established as in Proposition 4.17.)

10. Let us show that $\sigma(A) = \{0\}$. Let $\lambda \in \sigma(A)$ with $\lambda \neq 0$. Then $\lambda \in EV(A)$ (by Theorem 6.8) and thus $|[a = \lambda]| > 0$. Set $\omega = [a = \lambda]$. Then $N(A - \lambda I)$ is a finite-dimensional space not reduced to $\{0\}$. On the other hand, $N(A - \lambda I)$ is clearly isomorphic to $L^p(\omega)$. Then ω consists of a finite number of atoms (see Remark 6 in Chapter 4) and it has at least one atom, since $L^p(\omega)$ is not reduced to $\{0\}$. Impossible.

Problem 34

- A -

1. Clearly $0 \notin EV(T)$. Assume that $\lambda \in EV(T)$ and $\lambda \neq 0$. Let u be the corresponding eigenfunction, so that

$$\frac{1}{x} \int_0^x u(t)dt = \lambda u(x).$$

Thus $u \in C^1((0, 1])$ and satisfies

$$u = \lambda u + \lambda x u'.$$

Integrating this ODE, we see that $u(x) = Cx^{-1+1/\lambda}$, for some constant C. Since $u \in C([0, 1])$, we must have $0 < \lambda \leq 1$. Conversely, any $\lambda \in (0, 1]$ is an eigenvalue with corresponding eigenspace $Cx^{-1+1/\lambda}$.

3. We already know that $[0, 1] \subset \sigma(T) \subset [-1, +1]$. We will now prove that for any $\lambda \in \mathbb{R}, \lambda \notin \{0, 1\}$, the equation

(S1) $Tu - \lambda u = f \in E$

admits at least one solution $u \in E$.

Assuming that we have a solution u, set $\varphi(x) = \int_0^x u(t)dt$. Then

$$\varphi - \lambda x \varphi' = xf,$$

and hence we must have

$$\varphi(x) = \frac{1}{\lambda} x^{1/\lambda} \int_x^1 t^{-1/\lambda} f(t) dt + C x^{1/\lambda},$$

for some constant C. Therefore

(S2) $\quad u(x) = \varphi'(x) = \frac{1}{\lambda^2} x^{-1+1/\lambda} \int_x^1 t^{-1/\lambda} f(t) dt - \frac{1}{\lambda} f(x) + \frac{C}{\lambda} x^{-1+1/\lambda}.$

If $\lambda < 0$ or if $\lambda > 1$ we must choose

(S3) $$C = -\frac{1}{\lambda} \int_0^1 t^{-1/\lambda} f(t) dt$$

in order to make u continuous at $x = 0$, and then the unique solution u of (S1) is given by

(S4) $\quad u(x) = (T - \lambda I)^{-1} f = -\frac{1}{\lambda^2} x^{-1+1/\lambda} \int_0^x t^{-1/\lambda} f(t) dt - \frac{1}{\lambda} f(x),$

with

$$u(0) = \frac{1}{1 - \lambda} f(0).$$

It follows that $\sigma(T) = [0, 1]$ and $\rho(T) = (-\infty, 0) \cup (1, \infty)$.

When $0 < \lambda < 1$, the function u given by

(S5) $$u(x) = \frac{1}{\lambda^2} x^{-1+1/\lambda} \int_x^1 t^{-1/\lambda} f(t) dt - \frac{1}{\lambda} f(x),$$

with

$$u(0) = \frac{1}{1 - \lambda} f(0),$$

is still a solution of (S1). But the solution of (S1) is not unique, since we can add to u any multiple of $x^{-1+1/\lambda}$. Hence, for $\lambda \in (0, 1)$, the operator $(T - \lambda I)$ is surjective but not injective.

When $\lambda = 0$, the operator T is injective but not surjective. Indeed for every $u \in E$, $Tu \in C^1((0, 1])$.

When $\lambda = 1$, $(T - I)$ is not injective and is not surjective. We already know that $N(T - I)$ consists of constant functions. Suppose now that u is a solution of $Tu - u = f$. Then $f(0) = u(0) - u(0) = 0$ and therefore $(T - I)$ is not surjective.

4. A direct computation gives

$$\|T_\varepsilon u - Tu\|_{L^q(0,1)} \le \frac{\varepsilon^{1/q}}{(q-1)^{1/q}} \|u\|_{L^\infty(0,1)} \quad \text{if } q > 1,$$

and

$$\|T_\varepsilon u - Tu\|_{L^1(0,1)} \le \varepsilon \log(1 + 1/\varepsilon)\|u\|_{L^\infty(0,1)}.$$

Thus $\|T_\varepsilon - T\|_{\mathcal{L}(E,F)} \to 0$. Clearly $T_\varepsilon \in \mathcal{K}(E, F)$ (why?), and we may then apply Theorem 6.1 to conclude that $T \in \mathcal{K}(E, F)$.

- B -

1. It is convenient to write

$$Tu(x) = \frac{1}{x}\int_0^x u(t)dt = \int_0^1 u(xs)ds$$

and therefore

$$(Tu)'(x) = \int_0^1 u'(xs)sds.$$

2. Assume that $\lambda \in EV(T)$. By question A1 the corresponding eigenfunction must be $u(x) = Cx^{-1+1/\lambda}$. This function belongs to $C^1([0, 1])$ only when $0 < \lambda \le 1/2$ or $\lambda = 1$.

3. We will show that if $\lambda \notin [0, \frac{1}{2}] \cup \{1\}$, then $(T - \lambda I)$ is bijective. Consider the equation

$$Tu - \lambda u = f \in C^1([0, 1]).$$

When $\lambda < 0$ or $\lambda > 1$ we know, by part A, that if a solution exists, it must be given by (S4). Rewrite it as

$$u(x) = -\frac{1}{\lambda^2}\int_0^1 s^{-1/\lambda} f(xs)ds - \frac{1}{\lambda}f(x),$$

and thus $u \in C^1([0, 1])$.

When $1 > \lambda > 1/2$, we know from part A that (S1) admits solutions $u \in C([0, 1])$. Moreover, all solutions u are given by (S2). We will see that there is a (unique) choice of the constant C in (S2) such that $u \in C^1([0, 1])$. Write

$$u(x) = x^{-1+1/\lambda}\left[\frac{1}{\lambda^2}\int_x^1 t^{-1/\lambda}(f(t) - f(0))dt + \frac{f(0)}{\lambda^2 - \lambda} + \frac{C}{\lambda}\right]$$
$$- \frac{f(x)}{\lambda} - \frac{f(0)}{\lambda^2 - \lambda}.$$

A natural choice for C is such that

$$\frac{1}{\lambda^2}\int_0^1 t^{-1/\lambda}(f(t) - f(0))dt + \frac{f(0)}{\lambda^2 - \lambda} + \frac{C}{\lambda} = 0,$$

and then u becomes

$$u(x) = -\frac{1}{\lambda^2}x^{-1+1/\lambda}\int_0^x t^{-1/\lambda}(f(t) - f(0))dt - \frac{f(x)}{\lambda} - \frac{f(0)}{\lambda^2 - \lambda}.$$

Changing variables yields

$$u(x) = -\frac{1}{\lambda^2}\int_0^1 s^{-1/\lambda}(f(xs) - f(0))ds - \frac{f(x)}{\lambda} - \frac{f(0)}{\lambda^2 - \lambda}.$$

Direct inspection shows that indeed $u \in C^1([0, 1])$ with

$$u'(x) = -\frac{1}{\lambda^2}\int_0^1 s^{1-1/\lambda}f'(xs)ds - \frac{f'(x)}{\lambda}.$$

- C -

1. We have

$$\int_0^1 |Tu(x)|^p dx = -\frac{1}{p-1}|\varphi(1)|^p + \frac{p}{p-1}\int_0^1 |\varphi(x)|^{p-1}(\text{sign }\varphi(x))\varphi'(x)\frac{dx}{x^{p-1}},$$

and therefore, by Hölder,

$$\int_0^1 |Tu(x)|^p dx \le \frac{p}{p-1}\left[\int_0^1 \left|\frac{\varphi(x)}{x}\right|^p dx\right]^{\frac{p-1}{p}}\left[\int_0^1 |\varphi'(x)|^p dx\right]^{\frac{1}{p}},$$

i.e.,

$$\|Tu\|_p^p \le \frac{p}{p-1}\|Tu\|_p^{p-1}\|u\|_p.$$

3. Clearly $0 \notin EV(T)$. Suppose that $\lambda \in EV(T)$ and $\lambda \ne 0$. As in part A we see that the corresponding eigenfunction is $u = Cx^{-1+1/\lambda}$. This function belongs to $L^p(0, 1)$ iff $0 < \lambda < p/(p-1)$.

5. Assume that $\lambda < 0$. Let us prove that $\lambda \in \rho(T)$. For $f \in C([0, 1])$, let Sf be the right-hand side in (S4). Clearly

$$|Sf(x)| \le \frac{1}{\lambda^2}\frac{1}{x}\int_0^x |f(t)|dt + \frac{1}{\lambda}|f(x)|.$$

Therefore S can be extended as a bounded operator from $L^p(0, 1)$ into itself. Since we have

$$(T - \lambda I)S = S(T - \lambda I) = I \quad \text{on } C([0, 1]),$$

the same holds on $L^p(0, 1)$. Consequently $\lambda \in \rho(T)$.

Suggestion for further investigation: prove that for $\lambda \in (0, \frac{p}{p-1})$ the operator $(T - \lambda I)$ is surjective from $L^p(0, 1)$ onto itself. *Hint*: start with formula (S5) and show that $\|u\|_p \le C\|f\|_p$ using the same method as in questions C1 and C2.

6. $(T^\star v)(x) = \int_x^1 \frac{v(t)}{t} dt$.
7. Check that T_ε is a compact operator from $L^p(0, 1)$ into $C([0, 1])$ with the help of Ascoli's theorem. Then prove that

$$\|T_\varepsilon - T\|_{\mathcal{L}(L^p, L^q)} \leq C\varepsilon^{\frac{1}{q} - \frac{1}{p}}.$$

Problem 35

- A -

1. Clearly $\|T^\star T\| \leq \|T\|^2$. On the other hand,

$$|Tx|^2 = (Tx, Tx) = (T^\star Tx, x) \leq \|T^\star T\| \, |x|^2.$$

Thus $\|T\|^2 \leq \|T^\star T\|$.
2. By induction we have

$$\|T^{2^k}\| = \|T\|^{2^k} \quad \forall \text{ integer } k.$$

Given any integer N, fix k such that $N \leq 2^k$.
Then

$$\|T\|^{2^k} = \|T^{2^k}\| = \|T^N T^{2^k - N}\| \leq \|T^N\| \, \|T\|^{2^k - N},$$

and thus

$$\|T\|^N \leq \|T^N\|.$$

- B -

Set

$$X = \|T_{j_1}^\star T_{k_1} T_{j_2}^\star T_{k_2} \cdots T_{j_N}^\star T_{k_N}\|.$$

By assumption (1) we have

$$X \leq \omega^2(j_1 - k_1)\omega^2(j_2 - k_2) \cdots \omega^2(j_N - k_N),$$

and by assumption (2),

$$X \leq \|T_{j_1}^\star\|\omega^2(k_1 - j_2)\omega^2(k_2 - j_3) \cdots \omega^2(k_{N-1} - j_N)\|T_{k_N}\|$$
$$\leq \omega^2(0)\omega^2(k_1 - j_2)\omega^2(k_2 - j_3) \cdots \omega^2(k_{N-1} - j_N),$$

since $\|T_i\| = \|T_i^\star T_i\|^{1/2} \leq \omega(0)$.

Multiplying the above estimates, we obtain

$$X \leq \omega(0)\omega(j_1 - k_1)\omega(k_1 - j_2) \cdots \omega(j_{N-1} - k_{N-1})\omega(k_{N-1} - j_N)\omega(j_N - k_N).$$

Summing over k_N, then over j_N, then over k_{N-1}, then over $j_{N-1}, \ldots$, then over k_2, then over j_2, then over k_1, yields a bound by σ^{2N}. Finally, summing over j_1 gives the bound $m\sigma^{2N}$.

3. We have

$$\|(U^*U)^N\| \leq \sum_{j_1} \sum_{k_1} \sum_{j_2} \sum_{k_2} \cdots \sum_{j_N} \sum_{k_N} \|T_{j_1}^* T_{k_1} T_{j_2}^* T_{k_2} \cdots T_{j_N}^* T_{k_N}\| \leq m\sigma^{2N}.$$

Therefore

$$\|U\| \leq m^{1/2N}\sigma,$$

and the desired conclusion follows by letting $N \to \infty$.

Problem 36

1. To see that R_k is closed, note that $(I - T)^k = I - S$ for some $S \in \mathcal{K}(E)$ and apply Theorem 6.6.
2. Suppose $R_{q+1} = R_q$ for some $q \geq 1$. Then $R_{k+1} = R_k \ \forall k \geq q$. On the other hand, we cannot have $R_{k+1} \neq R_k \ \forall k \geq 1$ (see part (c) in the proof of Theorem 6.6).
4. From Theorem 6.6(b) and (d) we have

$$R_k = N((I - T^*)^k)^\perp$$

and thus

$$\text{codim } R_k = \dim N((I - T^*)^k) = \dim N((I - T)^k) = \dim N_k.$$

5. Let $x \in R_p \cap N_p$. Then $x = (I - T)^p \xi$ for some $\xi \in E$ and $(I - T)^p x = 0$. It follows that $\xi \in N_{2p} = N_p$ and thus $x = 0$. On the other hand,

$$\text{codim } R_p = \dim N_p;$$

combining this with the fact that $R_p \cap N_p = \{0\}$, we conclude that $E = R_p + N_p$.
6. $(I - T)R_p = R_{p+1} = R_p$. Theorem 6.6(c) applied in the space R_p allows us to conclude that $(I - T)$ is also injective on R_p.
7. It suffices to show that $N_2 = N_1$. Let $x \in N_2$. Then $(I - T)^2 x = 0$ and thus $|(I - T)x|^2 = ((I - T)x, (I - T)x) = ((I - T)^2 x, x) = 0$.

Problem 37

10. From question 9 we know that $v_k^{(n)}$ is nondecreasing in n and $v_k^{(n)} \leq \mu_k \ \forall k \geq 1$ and $\forall n$. Thus it suffices to prove that

(S1)
$$\liminf_{n\to\infty} v_k^{(n)} \geq \mu_k.$$

In fact, using question 9, one has, $\forall k < n$,

$$\max_{\substack{\Sigma \subset V^{(n)} \\ \dim \Sigma = k}} \min_{\substack{x \in \Sigma \\ x \neq 0}} R(x) = v_k^{(n)}.$$

Note that from the assumption on $V^{(n)}$,

(S2) $$\lim_{n\to\infty} P_{V^{(n)}}(x) = x \quad \forall x \in H.$$

Thus $P_{V^{(n)}}(e_1), \ldots, P_{V^{(n)}}(e_k)$ are linearly independent for $n \geq N_k$ sufficiently large (depending on k, but recall that k is fixed); this implies

(S3) $$v_k^{(n)} \geq \min_{\substack{x\in E(n,k)\\x\neq 0}} R(x),$$

where $E(n, k)$ is the space spanned by $\{P_{V^{(n)}}(e_1), \ldots, P_{V^{(n)}}(e_k)\}$. However, it is clear from (S2) that

(S4) $$\lim_{n\to\infty} \min_{\substack{x\in E(n,k)\\x\neq 0}} R(x) = \min_{\substack{x\in E_k\\x\neq 0}} R(x).$$

Inequality (S1) follows from (S3), (S4), and question 1.

Problem 38

- A -

2. Use Exercise 6.25 or apply question 1 to the operator $(I_F + K)$, that satisfies (1) (why?). Then write
$$T \circ (S \circ M) = I_F - P.$$

3. Clearly $R(I_F - P)$ is closed and codim $R(I_F - P)$ is finite. By Proposition 11.5 we know that any space $X \supset R(I_F - P)$ is also closed and has finite codimension. In particular, (1)(a) holds.

Next, we have
$$U^\star \circ T^\star = I_{F^\star} - P^\star,$$

where $P^\star$ is a compact operator (since P is). Thus we may argue as above and conclude that $R(U^\star)$ is closed. From Theorem 2.19 we infer that $R(U)$ is also closed.

We now prove that $N(T)+R(U)+\Sigma_1 = E$ for some finite-dimensional space Σ_1. Given any $x \in E$, write $x = x_1+x_2$ with $x_1 = x-U(Tx)$ and $x_2 = U(Tx)$. Note that $Tx_1 = Tx - (T \circ U)(Tx) = P(Tx)$ by (3). Therefore $x_1 \in T^{-1}(R(P)) = N(T) + \Sigma_1$, where Σ_1 is finite-dimensional, since $R(P)$ is. Consequently, any $x \in E$ belongs to $N(T) + R(U) + \Sigma_1$.

Finally, we prove that $N(T) \cap R(U) \subset \Sigma_2$ with Σ_2 finite-dimensional. Indeed, let $x \in N(T)\cap R(U)$. Then $x = Uy$ for some $y \in F$ and $Tx = (T \circ U)(y) = 0$. Thus, by (3), $y-Py = 0$ and therefore $y \in R(P)$. Consequently $x \in U(R(P)) = \Sigma_2$, which is finite-dimensional, since $R(P)$ is. Applying Proposition 11.7, we conclude that $N(T)$ admits a complement in E.

- B -

(4) $\Rightarrow$ (6). Let U_0 be as in question 1 of part A. Then $U_0 \circ T = I$ on X. Given any $x \in E$ write $x = x_1 + x_2$ with $x_1 \in X$ and $x_2 \in N(T)$. Then

$$(U_0 \circ T)(x) = (U_0 \circ T)(x_1) = x_1 = x - x_2 = x - \widetilde{P}x,$$

where $\widetilde{P}$ is a finite-rank projection onto $N(T)$.

(5) $\Rightarrow$ (6). Use Exercise 2.26.

(6) $\Rightarrow$ (4). From (6) it is clear that $\dim N(T) < \infty$. Also, since $T^\star \circ (\widetilde{U})^\star = I_{E^\star} - (\widetilde{P})^\star$, we may apply part A ((2) $\Rightarrow$ (1)) to $T^\star$ in $E^\star$ and deduce that $R(T^\star)$ is closed in $E^\star$. Therefore $R(T)$ is closed in F.

As in question 3 of part A, we construct finite-dimensional spaces Σ_3 and Σ_4 in F such that

$$N(\widetilde{U}) + R(T) + \Sigma_3 = F,$$
$$N(\widetilde{U}) \cap R(T) \subset \Sigma_4,$$

and we conclude (using Proposition 11.7) that $R(T)$ admits a complement.

- C -

1. Note that $Q \circ T = T$ and thus $U \circ T = U_0 \circ Q \circ T = U_0 \circ T = I - \widetilde{P}$.
2. Use (2) $\Rightarrow$ (1) and (5) $\Rightarrow$ (4).
3. Let $Z \subset F$ be a closed subspace. From Proposition 11.13 we know that Z has finite codimension iff $Z^\perp$ is finite-dimensional, and then $\operatorname{codim} Z = \dim Z^\perp$. Apply this to $Z = R(T)$, with $Z^\perp = N(T^\star)$ (by Proposition 2.18).
4. We already know that $\dim N(T^\star) = \operatorname{codim} R(T) < \infty$. Next, we have $\dim N(T) < \infty$, and thus $\operatorname{codim} N(T)^\perp < \infty$ (by Proposition 11.13). But $N(T)^\perp = R(T^\star)$ (by Theorem 2.19). Therefore $\operatorname{codim} R(T^\star) < \infty$ and, moreover, $\operatorname{codim} R(T^\star) = \dim N(T)$.
5. From Theorem 2.19 we know that $R(T)$ is closed. Since $N(T^\star) = R(T)^\perp$ is finite-dimensional, Proposition 11.11 yields that $\operatorname{codim} R(T) < \infty$. Since $R(T^\star) = N(T)^\perp$ and $\operatorname{codim} R(T^\star) < \infty$, we deduce from Proposition 11.11 that $\dim N(T) < \infty$.
6. Write $T = J(I_E + J^{-1} \circ K)$. By Theorem 6.6 we know that $(I_E + J^{-1} \circ K) \in \Phi(E, E)$ and $\operatorname{ind}(I_E + J^{-1} \circ K) = 0$. Thus $T \in \Phi(E, F)$ and $\operatorname{ind} T = 0$, since J is an isomorphism.

   Conversely, assume that $T \in \Phi(E, F)$ and $\operatorname{ind} T = 0$. Let X be a complement of $N(T)$ in E and let Y be a complement of $R(T)$ in F. Since $\operatorname{ind} T = 0$, we have $\dim N(T) = \dim Y$. Let Λ be an isomorphism from $N(T)$ onto Y. Given $x \in E$, write $x = x_1 + x_2$ with $x_1 \in X$ and $x_2 \in N(T)$. Set $Jx = Tx_1 + \Lambda x_2$. Clearly J is bijective and $Tx = Tx_1 = Jx - \Lambda x_2$ is a desired decomposition.
7. Use a pseudoinverse.
8. Let X and Y be as in question 6. Set $\widetilde{E} = E \times Y$ and $\widetilde{F} = F \times N(T)$. Consider the operator $\widetilde{T} : \widetilde{E} \to \widetilde{F}$ defined by

$$\widetilde{T}(x, y) = (Tx + Kx, 0).$$

Clearly

$$R(\widetilde{T}) = R(T + K) \times \{0\},$$
$$N(\widetilde{T}) = N(T + K) \times Y.$$

Thus $\widetilde{T} \in \Phi(\widetilde{E}, \widetilde{F})$ and

$$\text{codim } R(\widetilde{T}) = \text{codim } R(T + K) + \dim N(T),$$
$$\dim N(\widetilde{T}) = \dim N(T + K) + \dim Y = \dim N(T + K) + \text{codim } R(T).$$

We claim that $\widetilde{T} = \widetilde{J} + \widetilde{K}$, where $\widetilde{J}$ is bijective from $\widetilde{E}$ onto $\widetilde{F}$, and $\widetilde{K} \in \mathcal{K}(\widetilde{E}, \widetilde{F})$. Indeed, writing $x = x_1 + x_2$, with $x_1 \in X$ and $x_2 \in N(T)$, we have

$$\widetilde{T}(x, y) = (Tx_1 + Kx, 0) = \widetilde{J}(x, y) + \widetilde{K}(x, y),$$

where

$$\widetilde{J}(x, y) = (Tx_1 + y, x_2)$$

and

$$\widetilde{K}(x, y) = (Kx, 0) - (y, x_2).$$

Clearly $\widetilde{J}$ is bijective and $\widetilde{K}$ is compact (since y and x_2 are finite-dimensional variables). Applying question 6, we see that

$$\text{ind } \widetilde{T} = 0 = \dim N(\widetilde{T}) - \text{codim } R(\widetilde{T}).$$

It follows that

$$\text{ind}(T + K) = \dim N(T + K) - \text{codim } R(T + K) = \dim N(T) - \text{codim } R(T).$$

9. Let V be a pseudoinverse of T and set $\varepsilon = \|V\|^{-1}$ (any $\varepsilon > 0$ if $V = 0$). From (8)(b) we have

$$V \circ (T + M) = I_E + (V \circ M) + \widetilde{K}.$$

If $\|M\| < \varepsilon$ we see that $\|V \circ M\| < 1$, and thus $W = I_E + (V \circ M)$ is bijective from E onto E (see Proposition 6.7). Multiplying the equation

$$V \circ (T + M) = W + \widetilde{K}$$

on the left by T and using (8)(a) yields

$$T + M = (T \circ W) + (T \circ \widetilde{K}) - K \circ (T + M).$$

Since W is bijective, it is clear (from the definition of $\Phi(E, F)$) that $T \circ W \in \Phi(E, F)$ and $\text{ind}(T \circ W) = \text{ind } T$. Applying the previous question, we conclude that $T + M \in \Phi(E, F)$ and $\text{ind}(T + M) = \text{ind}(T \circ W) = \text{ind } T$.

11. Check that $V_1 \circ V_2$ is a pseudoinverse for $T_2 \circ T_1$.

12. Note that $H_0(x_2, x_2) = (T_1 x_1, T_2 x_2)$, so that ind $H_0 =$ ind $T_1 +$ ind T_2. On the other hand, $H_1(x_2, x_2) = (x_2, -T_2(T_1 x_1))$, so that ind $H_1 =$ ind$(T_2 \circ T_1)$.

13. ind $V = -$ ind T by (8), questions 6 and 12.

- D -

1. ind $T = \dim E - \dim F$, since $\dim R(T) = \dim E - \dim N(T)$ and codim $R(T) = \dim F - \dim R(T)$.

2. When $|\lambda| < 1$, ind$(S_r - \lambda I) = -1$ and ind$(S_\ell - \lambda I) = +1$. When $|\lambda| > 1$, $(S_r - \lambda I)$ and $(S_\ell - \lambda I)$ are bijective; thus ind$(S_r - \lambda I) = 0$ and ind$(S_\ell - \lambda I) = 0$.

Problem 41

- A -

1. Assume by contradiction that $a \in (\text{Int } P) \cap (-P)$. From Exercise 1.7 we have $0 = \frac{1}{2}a + \frac{1}{2}(-a) \in \text{Int } P$ and this implies $P = E$.

2. Suppose not; then there exists a sequence (x_n) in P such that $x_n + u \to 0$. Since $(x_n + u) - u = x_n \in P$, we obtain at the limit $-u \in P$. This contradicts question 1.

3. Clearly $u \neq 0$ (since $0 \notin \text{Int } P$ by (2)). From (3) we have $Tu \in \text{Int } C$ and thus $B(Tu, \rho) \subset C$ for some $\rho > 0$. Then choose $0 < r < \rho/\|u\|$.

4. Since $\lambda x = T(x + u) \geq Tu \geq ru$, we have $\frac{\lambda}{r}x \geq u$. Assuming $(\frac{\lambda}{r})^n x \geq u$, we obtain $(\frac{\lambda}{r})^n Tx \geq Tu$ and thus $(\frac{\lambda}{r})^n(\lambda x - Tu) \geq Tu \geq ru$. Hence $(\frac{\lambda}{r})^n \lambda x \geq ru$, i.e., $(\frac{\lambda}{r})^{n+1}x \geq u$. On the other hand, $\lambda x = T(x + u) \in \text{Int } P$, which implies that $\lambda > 0$ (by question 1). If we had $0 < \lambda < r$ we could pass to the limit as $n \to \infty$ and obtain $-u \in P$, which is impossible (again by question 1).

5. The map $x \mapsto (x + u)/\|x + u\|$ is clearly continuous on P (by question 2). $F(P) \subset T(B_E) \subset K$ since $T \in \mathcal{K}(E)$.

6. When replacing u by εu, the constant α in question 2 may change, but the constant r in question 3 remains unchanged.

7. We have $\lambda_\varepsilon \|x_\varepsilon\| = \|T(x_\varepsilon + \varepsilon u)\| \leq \|T\| \|x_\varepsilon + \varepsilon u\|$ and therefore $\|x_\varepsilon\| \leq \|T\|$. Hence $\lambda_\varepsilon \leq \|T\| + \varepsilon\|u\|$. Passing to a subsequence $\varepsilon_n \to 0$, we may assume that $\lambda_{\varepsilon_n} \to \mu_0$ and $Tx_{\varepsilon_n} \to \ell$ (since $T \in \mathcal{K}(E)$). Hence $x_{\varepsilon_n} \to x_0$ with $x_0 \in P$, $\mu_0 = \|x_0\| \geq r$ and $Tx_0 = \mu_0 x_0$, so that $x_0 \in \text{Int } P$ by (3).

- B -

1. The set $\Sigma = \{s \in [0, 1]; (1 - s)a + sb \in P\}$ is a closed interval (since P is convex and closed). Then $\sigma = \max \{s; s \in \Sigma\}$ has the required properties by Exercise 1.7.

2. We cannot have $\mu = 0$ (otherwise, $0 \in \text{Int } P$) and we cannot have $\mu < 0$ (otherwise, $-x \in (\text{Int } P) \cap (-P)$). Thus $\mu > 0$, and then $x \in \text{Int } P$, which implies $-x \notin P$. Note that x_0 and x play symmetric roles: $x_0, x \in \text{Int } P$, $-x_0 \notin P$, $-x \notin P$, $Tx_0 = \mu_0 x_0$ with $\mu_0 > 0$, and $Tx = \mu x$ with $\mu > 0$. Set $y = x_0 - \tau_0 x$, where $\tau_0 = \tau(x_0, -x)$. Then $y \in P$ (from the definition of σ and τ). Moreover, $y \neq 0$ (otherwise $x = mx_0$ with $m = 1/\tau_0$). Thus $Ty \in \text{Int } P$. But

$$Ty = Tx_0 - \tau_0 Tx = \mu_0 x_0 - \tau_0 \mu x.$$

Hence $x_0 + \frac{\tau_0 \mu}{\mu_0}(-x) \in \text{Int } P$. From the definition of τ_0 we deduce that $\frac{\tau_0 \mu}{\mu_0} < \tau_0$ and therefore $\mu < \mu_0$. Reversing the roles of x_0 and x yields $\mu_0 < \mu$. Hence we obtain a contradiction. Therefore $x = mx_0$ for some $m > 0$ and then $\mu = \mu_0$.

3. If $x \in P$ or $-x \in P$ we deduce the first part of the alternative from question 2. We may thus assume that $x \notin P$ and $-x \notin P$. We will then show that $|\mu| < \mu_0$. If $\mu = 0$ we are done. Suppose that $\mu > 0$ and let $\tau_0 = \tau(x_0, x)$. Set $y = x_0 + \tau_0 x$, so that $y \in P$. We have $y \neq 0$ (otherwise $-x \in P$) and thus

$$Ty = \mu_0 x_0 + \tau_0 \mu x \in \text{Int } P.$$

Hence $x_0 + \frac{\tau_0 \mu}{\mu_0} x \in \text{Int } P$. From the definition of τ_0 we deduce that $\frac{\tau_0 \mu}{\mu_0} < \tau_0$, and thus $\mu < \mu_0$.

Suppose now that $\mu < 0$. Let $\tau_0 = \tau(x_0, x)$ and $\tilde{\tau}_0 = \tau(x_0, -x)$. Set $y = x_0 + \tau_0 x$ and $\tilde{y} = x_0 - \tilde{\tau}_0 x$, so that $y, \tilde{y} \in P$ and $y \neq 0$, $\tilde{y} \neq 0$. As above, we obtain

$$x_0 + \frac{\tau_0 \mu}{\mu_0} x \in \text{Int } P \quad \text{and} \quad x_0 - \frac{\tilde{\tau}_0 \mu}{\mu_0} x \in \text{Int } P.$$

Thus

$$x_0 + \frac{\tau_0 |\mu|}{\mu_0}(-x) \in \text{Int } P \quad \text{and} \quad x_0 + \frac{\tilde{\tau}_0 |\mu|}{\mu_0} x \in \text{Int } P.$$

From the definition of τ_0 and $\tilde{\tau}_0$ we deduce that

$$\frac{\tau_0 |\mu|}{\mu_0} < \tilde{\tau}_0 \quad \text{and} \quad \frac{\tilde{\tau}_0 |\mu|}{\mu_0} < \tau_0.$$

Therefore

$$\frac{|\mu|}{\mu_0} < \min\left\{\frac{\tilde{\tau}_0}{\tau_0}, \frac{\tau_0}{\tilde{\tau}_0}\right\} \leq 1.$$

4. Using question 3 with $\mu = \mu_0$ yields $N(T - \mu_0 I) \subset \mathbb{R}x_0$.

5. In view of the results in Problem 36 it suffices to show that $N((T - \mu_0 I)^2) = N(T - \mu_0 I)$. Let $x \in E$ be such that $(T - \mu_0 I)^2 x = 0$. Using question 4 we may write $Tx - \mu_0 x = \alpha x_0$ for some $\alpha \in \mathbb{R}$. We need to prove that $\alpha = 0$. Suppose not, that $\alpha \neq 0$. Set $y = \frac{x}{\alpha}$, so that $Ty - \mu_0 y = x_0$. Then $T^2 y = \mu_0 Ty + Tx_0 = \mu_0^2 y + 2\mu_0 x_0$. By induction we obtain $T^n y = \mu_0^n y + n\mu_0^{n-1} x_0$ for all $n \geq 1$, which we may write as

$$T^n\left(x_0 - \frac{\mu_0 y}{n}\right) = -\frac{\mu_0^{n+1}}{n} y.$$

Since $x_0 \in \text{Int } P$, we may choose n sufficiently large that $x_0 - \frac{\mu_0 y}{n} \in P$. Since $T^n(P) \subset P$, we conclude that $-y \in P$. Thus $T^n(-y) \in P$. Returning to the equation $T^n y = \mu_0^n y + n\mu_0^{n-1} x_0 \; \forall n \geq 1$, we obtain $-y - \frac{n}{\mu_0} x_0 \in P$, i.e., $-x_0 - \frac{\mu_0}{n} y \in P$. As $n \to +\infty$ we obtain $-x_0 \in P$. Impossible.

Thus we have established that the geometric multiplicity of μ_0 (i.e., $\dim(N - \mu_0 I)$) is one, but also that the algebraic multiplicity is one.

Problem 42

1. We have, $\forall x \in C$,

$$\|Tx - Tx_0\| \leq \|T\| \, \|x - x_0\| \leq \|T\| r \leq \frac{1}{2} \|Tx_0\|,$$

and by the triangle inequality,

$$\|Tx_0\| - \|Tx\| \leq \|Tx - Tx_0\| \leq \frac{1}{2} \|Tx_0\|.$$

Thus $\|y\| \geq \frac{1}{2} \|Tx_0\| \; \forall y \in T(C)$, and therefore also $\forall y \in \overline{T(C)}$. Since $Tx_0 \neq 0$, we see that $0 \notin \overline{T(C)}$.

2. By assumption (1), $\mathcal{A}_y$ is dense in E, and consequently $\mathcal{A}_y \cap B(x_0, r/2) \neq \emptyset$, i.e, there exists $S \in \mathcal{A}$ such that $\|Sy - x_0\| < r/2$.

3. We have, $\forall z \in B(y, \varepsilon)$,

$$\|Sz - x_0\| \leq \|S(z - y)\| + \|Sy - x_0\| \leq \|S\|\varepsilon + \frac{r}{2}.$$

Then choose $\varepsilon = \frac{r}{2\|S\|}$.

4. If $x \in C$, then $Tx \in B(y_j, \frac{1}{2} \varepsilon_{y_j})$ for some $j \in J$. Therefore $q_j(x) \geq \frac{1}{2} \varepsilon_{y_j}$ and thus $q(x) \geq \min_{j \in J}\{\frac{1}{2} \varepsilon_{y_j}\}$.

5. The functions q_j are continuous on E and the function $1/q$ is continuous on C. Thus F is continuous on C. Write

$$F(x) - x_0 = \frac{1}{q(x)} \sum_{j \in J} q_j(x) \left[S_{y_j}(Tx) - x_0 \right].$$

Note that $q_j(x) \geq 0 \; \forall x \in E$ and $q_j(x) > 0$ implies $\|Tx - y_j\| < \varepsilon_{y_j}$. Using the result of question 2 with $z = Tx$ and $y = y_j$ yields

$$\|S_{y_j}(Tx) - x_0\| \leq r.$$

Therefore

$$q_j(x)\|S_{y_j}(Tx) - x_0\| \leq q_j(x)r \quad \forall x \in E, \quad \forall j \in J,$$

and thus

$$\|F(x) - x_0\| \leq r.$$

6. Let $Q = \overline{T(C)}$, so that Q is compact. Thus $R_j = S_{y_j}(Q)$ is compact, and so is $[0, 1]R_j$ (since it is the image of $[0, 1] \times R_j$ under the continuous map

$(t, x) \mapsto tx$). Finally, $\sum_{j \in J}[0, 1]R_j$ is also compact (being the image under the map $(x_1, x_2, \ldots) \mapsto \sum_{j \in J} x_j$ of a product of compact sets).

7. Each operator $S_{y_j} \circ T$ is compact by Proposition 6.3. Since $\mathcal{K}(E)$ is a subspace (see Theorem 6.1), we see that $U \in \mathcal{K}(E)$ ($q_j(\xi)$ is a constant). From Theorem 6.6 we know that $F = N(I - U)$ is finite-dimensional.

Writing that $F(\xi) = \xi$ gives

$$\frac{1}{q(\xi)} \sum_{j \in J} q_j(\xi) S_{y_j}(T\xi) = \xi,$$

and by definition of U,

$$U(\xi) = \frac{1}{q(\xi)} \sum_{j \in J} q_j(\xi) S_{y_j}(T\xi) = \xi.$$

8. We need to show that $Ta = U(Ta)$ $\forall a \in Z$. Note that $S_{y_j} \in \mathcal{A}$ (by the construction of question 2; thus $S_{y_j} \circ T \in \mathcal{A}$ and $U \in \mathcal{A}$. From the definition of $\mathcal{A}$ it is clear that $U \circ T = T \circ U$. Let $a \in Z$, so that $a = Ua$. Then $Ta = T(Ua) = U(Ta)$.

The space Z is finite-dimensional and thus $Z \neq E$ (this is the only place where we use the fact that E is infinite-dimensional). Clearly Z is closed and $Z \neq \{0\}$, since $\xi \in Z$ (and $\xi \in C$ implies $\xi \neq 0$ by question 1). Thus Z is a nontrivial closed invariant subspace of T.

9. Nontrivial subspaces have dimension one. Thus the only nontrivial invariant subspaces are of the form $\mathbb{R}x_0$ with $x_0 \neq 0$ and $Tx_0 = \alpha x_0$ for some $\alpha \in \mathbb{R}$. Therefore it suffices to choose any T with no real eigenvalue, for example a rotation by $\pi/2$.

Problem 43

1. $|T(u+v)|^2 = |Tu|^2 + |Tv|^2 + 2(T^\star Tu, v)$ and $|T^\star(u+v)|^2 = |T^\star u|^2 + |T^\star v|^2 + 2(TT^\star u, v)$.

3. By Corollary 2.18 (and since H is reflexive) we always have $\overline{R(T)} = N(T^\star)^\perp$ and $\overline{R(T^\star)} = N(T)^\perp$.

4. Since $f \in R(T)$, we have $f = Tv$ for some $v \in H$. Using question 3 we may decompose $v = v_1 + v_2$ with $v_1 \in \overline{R(T)}$ and $v_2 \in N(T)$. Then $f = Tv = Tv_1$ and we choose $u = v_1$.

5. We have by question 1 $|u_n - u_m| = |T^\star(y_n - y_m)| = |T(y_n - y_m)| \to 0$ as $m, n \to \infty$. Thus Ty_n is a Cauchy sequence; let $z = \lim_{n \to \infty} Ty_n$. Then $T^\star Ty_n = TT^\star y_n$ with $TT^\star y_n = Tu_n \to Tu = f$ and $T^\star Ty_n \to T^\star z$. Thus $T^\star z = f$.

6. In question 5 we have proved that $R(T) \subset R(T^\star)$. Applying this inclusion to $T^\star$ (which is also normal) gives $R(T^\star) \subset R(T)$.

7. Clearly $\|T^2\| \leq \|T^2\|$. For the reverse inequality write $|Tu|^2 = (T^\star Tu, u) \leq |T^\star Tu| \, |u|$. Since T is normal, we have $|T^\star Tu| = |TTu| \leq \|T^2\| \, |u|$. Therefore $\|T\|^2 = \sup_{u \neq 0} \frac{|Tu|^2}{|u|^2} \leq \|T^2\|$.

8. When $p = 2^k$ we argue by induction on k. Indeed, $\|T^{2^{k+1}}\| = \|S^2\|$, where $S = T^{2^k}$. Since S is normal, we have $\|S^2\| = \|S\|^2$. But $\|S\| = \|T\|^{2^k}$ from the induction assumption. Therefore $\|T^{2^{k+1}}\| = \|T\|^{2^{k+1}}$.

For a general integer p, choose any k such that $2^k \geq p$. We have

$$\|T\|^{2^k} = \|T^{2^k}\| = \|T^{2^k-p}T^p\| \leq \|T^{2^k-p}\| \|T^p\| \leq \|T\|^{2^k-p}\|T^p\|.$$

Thus $\|T\|^p \leq \|T^p\|$, and since $\|T^p\| \leq \|T\|^p$, we obtain $\|T^p\| = \|T\|^p$.

9. Let $u \in N(T^2)$. Then $Tu \in N(T) \cap R(T) \subset N(T) \cap N(T)^\perp$ by question 2. Therefore $Tu = 0$ and $u \in N(T)$. The same argument shows that $N(T^p) \subset N(T^{p-1})$ for $p \geq 2$, and thus $N(T^p) \subset N(T)$. Clearly $N(T) \subset N(T^p)$ and therefore $N(T^p) = N(T)$.

Problem 44

- A -

2. Clearly $T^* \circ T = I$ implies $|Tu| = |u| \; \forall u \in H$. Conversely, write $|T(u+v)|^2 = |u + v|^2$ and deduce that $(Tu, Tv) = (u, v) \; \forall u, v \in H$, so that $T^* \circ T = I$.

3. (a) $\Rightarrow$ (b). $T^* \circ T = I$ and T bijective imply that $T^* = T^{-1}$, so that T^* is also bijective.

   (b) $\Rightarrow$ (c). $T^* \circ T = I$ implies that T^* is surjective. If T^* is also injective, then T^* is bijective and $T = (T^*)^{-1}$. Hence $T \circ T^* = I$.

   (c) $\Rightarrow$ (d).   Obvious.

   (d) $\Rightarrow$ (e). $T^* \circ T = I$ implies that T^* is surjective. If T^* is an isometry, it must be a unitary operator.

   (e) $\Rightarrow$ (a).   Apply (a) $\Rightarrow$ (e) to T^*.

4. In $H = \ell^2$ the right shift S_r defined by $S_r(x_1, x_2, x_3, \dots) = (0, x_1, x_2, \dots)$ is an isometry that is not surjective.

5. Let $f_n \in R(T)$ with $f_n \to f$. Write $f_n = Tu_n$ and $|u_n - u_m| = |f_n - f_m|$, so that (u_n) is Cauchy sequence and $u_n \to u$ with $f = Tu$. Given $v \in H$, set $g = TT^*v$. Then $g \in R(T)$ and we have $\forall x \in H$,

$$(v - g, Tx) = (v, Tx) - (TT^*v, Tx) = (v, Tx) - (v, TT^*Tx) = 0,$$

since $T^* \circ T = I$. Thus $v - g \in R(T)^\perp$ and consequently $g = P_{R(T)}v$.

6. Assume that T is an isometry. Write $(T - \lambda I) = (I - \lambda T^*) \circ T$. Assume $|\lambda| < 1$. Then $\|\lambda T^*\| < 1$ and thus $(I - \lambda T^*)$ is bijective. When T is a unitary operator we deduce that $(T - \lambda I)$ is bijective; therefore $(-1, +1) \subset \rho(T)$ and hence $\sigma(T) \subset (-\infty, -1] \cup [+1, +\infty)$, so that $\sigma(T) \subset \{-1, +1\}$ (since $\sigma(T) \subset [-1, +1]$). On the other hand, if T is *not* a unitary operator and $|\lambda| < 1$, we see that $(T - \lambda I)$ cannot be bijective; therefore $(-1, +1) \subset \sigma(T)$, so that $\sigma(T) = [-1, +1]$ (since $\sigma(T)$ is closed and $\sigma(T) \subset [-1, +1]$).

7. T is an isometry from H onto $T(H)$. If $T \in \mathcal{K}(H)$ then $T(B_H) = B_{T(H)}$ is compact. Hence $\dim T(H) < \infty$ by Theorem 6.5. Since T is bijective from H onto $T(H)$, it follows that $\dim H < \infty$.

8. If T is skew-adjoint then $(Tu, u) = (u, T^\star u) = -(u, Tu)$ and thus $(Tu, u) = 0$. Conversely, write $0 = (T(u + v), (u + v)) = (Tu, v) + (Tv, u)$ $\forall u, v \in H$, so that $Tu + T^\star u = 0$ $\forall u \in H$.

9. Assume $\lambda \neq 0$. Then $(T - \lambda I) = -\lambda(I - \frac{1}{\lambda}T)$, and the operator $(I - \frac{1}{\lambda}T)$ satisfies the conditions of the Lax–Milgram theorem (Corollary 5.8). Thus $(T - \lambda I)$ is bijective.

10. From question 9 we know that $1 \notin \sigma(T)$, and thus $(T - I)^{-1}$ is well defined. From the relation $(T - I) \circ (T + I) = (T + I) \circ (T - I)$ we deduce that $U = (T - I)^{-1} \circ (T + I)$. Similarly $U \circ T = (T - I)^{-1} \circ (T + I) \circ T = T \circ (T+I) \circ (T-I)^{-1} = T \circ U$ because $(T+I) \circ T \circ (T-I) = (T-I) \circ T \circ (T+I)$. Next, we have $U^\star = (T^\star - I)^{-1} \circ (T^\star + I)$ and thus $U^\star \circ U = (T^\star - I)^{-1} \circ (T^\star + I) \circ (T + I) \circ (T - I)^{-1} = I$, since $(T^\star + I) \circ (T + I) = (T^\star - I) \circ (T - I)$ because $T^\star + T = 0$.

    Thus U is an isometry. On the other hand, $U = (T + I) \circ (T - I)^{-1}$ is bijective since $-1 \in \rho(T)$ by question 9.

11. By assumption we have $U^\star \circ U = I$. Thus $(T^\star - I)^{-1} \circ (T^\star + I) \circ (T + I) \circ (T - I)^{-1} = I$. This implies $(T^\star + I) \circ (T + I) = (T^\star - I) \circ (T - I)$, i.e., $T^\star + T = 0$.

- B -

1. (i) Trivial.
   (ii) If $\dim H < \infty$, standard linear algebra gives $\dim N(T) = \dim N(T^\star)$.
   (iii) If T is normal, then $N(T) = N(T^\star)$.
   (iv) $\dim N(T) = \dim N(T^\star) < \infty$ by Theorem 6.6.

2. If $T = S_\ell$, a left shift, then $\dim N(T) = 1$ and $T^\star = S_r$ satisfies $N(T^\star) = \{0\}$.

3. We have $T^\star = P \circ U^\star$ and thus $T^\star \circ T = P \circ U^\star \circ U \circ P = P^2$ by question A.2.

4. From the results of Problem 39 we know that P must be a square root of $T^\star \circ T$, and that P is unique.

5. Suppose that $T = U \circ P = V \circ P$ are two polar decompositions. Then $U = V$ on $R(P)$ and by continuity $U = V$ on $\overline{R(P)}$. But $P^2 = T^\star \circ T$ implies $N(P) = N(T)$. Thus $\overline{R(P)} = N(P^\star)^\perp = N(P)^\perp = N(T)^\perp$ (since $P^\star = P$).

6. From the relation $T = U \circ P$ we see that $U(R(P)) \subset R(T)$. In fact, we have $U(R(P)) = R(T)$; indeed, given $f \in R(T)$ write $f = Tx$ for some $x \in H$, and then $U(Px) = f$, so that $f \in U(R(P))$.

    By continuity U maps $\overline{R(P)} = N(T)^\perp$ into $\overline{R(T)} = N(T^\star)^\perp$. Since U is an isometry, the space $U(N(T)^\perp)$ is closed (by the standard Cauchy sequence argument). But $U(N(T)^\perp) \supset R(T)$ and therefore $U(N(T)^\perp) = \overline{R(T)} = N(T^\star)^\perp$. Using the property $(Ux, Uy) = (x, y)$ $\forall x, y \in H$ we find that $(Ux, Uy) = 0$ $\forall x \in N(T)^\perp$, $\forall y \in N(T)$. Thus $U(y) \in N(T^\star)^{\perp\perp} = N(T^\star)$ $\forall y \in N(T)$. Consequently $J = U_{|N(T)}$ is an isometry from $N(T)$ into $N(T^\star)$.

7. Let P be the square root of $T^\star \circ T$. We now construct the isometry U. First define $U_0 : R(P) \to R(T)$ as follows. Given $f \in R(P)$, there exists some $u \in H$ (not necessarily unique) such that $f = Pu$. We set

$$U_0 f = Tu.$$

This definition makes sense; indeed, if $f = Pu = Pu'$, then $u - u' \in N(P) = N(T)$, so that $Tu = Tu'$. Moreover,

$$|U_0 f| = |Tu| = |Pu| = |f| \quad \forall f \in R(P).$$

In addition we have $U_0(R(P)) = R(T)$. Indeed, we already know that $U_0(R(P)) \subset R(T)$. The reverse inclusion follows from the identity $U_0(Pu) = Tu \; \forall u \in H$.

Let $\widetilde{U}_0$ be the extension by continuity of U_0 to $\overline{R(P)}$. Then $\widetilde{U}_0$ is an isometry from $\overline{R(P)} = N(T)^\perp$ into $\overline{R(T)} = N(T^\star)^\perp$. But $R(\widetilde{U}_0) \supset R(U_0) = R(T)$ and therefore (as above) $R(\widetilde{U}_0) \supset \overline{R(T)} = N(T^\star)^\perp$. Hence $\widetilde{U}_0$ is an isometry from $N(T)^\perp$ onto $N(T^\star)^\perp$.

Finally, we extend $\widetilde{U}_0$ to all of H as follows. Given $x \in H$, write

$$x = x_1 + x_2$$

with $x_1 \in N(T)^\perp$ and $x_2 \in N(T)$. Set

$$Ux = \widetilde{U}_0 x_1 + J x_2.$$

Then

$$|Ux|^2 = |\widetilde{U}_0 x_1|^2 + 2(\widetilde{U}_0 x_1, J x_2) + |J x_2|^2 = |x_1|^2 + |x_2|^2 = |x|^2,$$

since $\widetilde{U}_0 x_1 \in N(T^\star)^\perp$ and $x_2 \in N(T^\star)$ (by (1)).

Clearly $U(Pu) = U_0(Pu) = Tu \; \forall u \in H$, and therefore we have constructed a polar decomposition of T.

8. The construction of question 7 shows that $R(U) = N(T^\star)^\perp \oplus R(J)$. Thus $R(U) = H$ if $R(J) = N(T^\star)$, and then U is a unitary operator.

9. If T is a normal operator then $N(T) = N(T^\star)$ (see Problem 43). Thus (2) is satisfied and we may apply question 8. Next, we have $T^\star = P \circ U^\star$, and since T is normal we can write

$$(P \circ U^\star) \circ (U \circ P) = T^\star \circ T = T \circ T^\star = (U \circ P) \circ (P \circ U^\star),$$

which implies that

$$P^2 = U \circ P^2 \circ U^\star,$$

and thus

$$P^2 \circ U = U \circ P^2.$$

Applying the result of question C2 in Problem 39 we deduce that $P \circ U = U \circ P$.

10. We have $P^2 = T^\star \circ T \in \mathcal{K}(H)$. This implies that $P \in \mathcal{K}(H)$. Indeed, let (u_n) be a sequence in H with $|u_n| \le 1$. Passing to a subsequence (still denoted by u_n), we may assume that $u_n \rightharpoonup u$ and $P^2 u_n \to P^2 u$. Then $|P(u_n - u)|^2 = (P^2(u_n - u), u_n - u) \to 0$, so that $Pu_n \to Pu$. Hence $P \in \mathcal{K}(H)$.

11. We have $T^\star \circ T \in \mathcal{K}(H)$, since $T \in \mathcal{K}(H)$ and its square root P is compact (see part D in Problem 39).

12. Let (e_n) be an orthonormal basis of H consisting of eigenvectors of T^*T, with corresponding eigenvalues (λ_n), so that $\lambda_n \ge 0 \ \forall n$ and $\lambda_n \to 0$ as $n \to \infty$. Let $I = \{n \in \mathbb{N}; \lambda_n > 0\}$. Consider the isometry U_0 defined on $R(P)$ with values in $R(T)$ constructed in question 7; we have $U_0 \circ P = T$ on H.

Set $f_n = U_0(e_n)$ for $n \in I$; this is well defined, since $Pe_n = \sqrt{\lambda_n} e_n$, so that $e_n \in R(P)$ when $n \in I$. Then $(f_n)_{n \in I}$ is an orthonormal system in H (but it is not a basis of H, since $f_n \in R(U_0) \subset \overline{R(T)} \neq H$ in general). Choose any basis of H, still denoted by $(f_n)_{n \in \mathbb{N}}$, containing the system $(f_n)_{n \in I}$. For $u \in H$, write

$$u = \sum_{n \in \mathbb{N}} (u, e_n) e_n,$$

so that

$$Pu = \sum_{n \in \mathbb{N}} \sqrt{\lambda_n}(u, e_n) e_n = \sum_{n \in I} \sqrt{\lambda_n}(u, e_n) e_n,$$

and then

$$Tu = U_0(Pu) = \sum_{n \in I} \sqrt{\lambda_n}(u, e_n) f_n = \sum_{n \in \mathbb{N}} \sqrt{\lambda_n}(u, e_n) f_n.$$

Clearly

$$T^\star v = \sum_{n \in \mathbb{N}} \sqrt{\lambda_n}(v, f_n) e_n.$$

Set

$$T_N u = \sum_{n=1}^{N} \alpha_n (u, e_n) f_n,$$

so that $T_N \in \mathcal{K}(H)$ (since it is a finite-rank operator). Then $\|T_N - T\| \le \max_{n \ge N+1} |\alpha_n|$, so that $\|T_N - T\| \to 0$ as $N \to \infty$, provided $\alpha_n \to 0$ as $n \to \infty$; thus $T \in \mathcal{K}(H)$ by Corollary 6.2.

Problem 45

12. Consider the equation

$$\begin{cases} -u'' + k^2 u = f & \text{on } (0, 1), \\ u(0) = u(1) = 0. \end{cases}$$

The solution is given by

$$u(x) = \frac{\sinh(kx)}{k \sinh k} \int_0^1 f(s) \sinh(k(1-s))ds - \frac{1}{k} \int_0^x f(s) \sinh(k(x-s))ds.$$

A tedious computation shows that

$$u(x) \geq \frac{k}{\sinh k} x(1-x) \int_0^1 f(s)s(1-s)ds.$$

Next, suppose that $p \equiv 1$ and u satisfies

$$\begin{cases} -u'' + qu = f & \text{on } (0, 1), \\ u(0) = u(1) = 0. \end{cases}$$

Write

$$-u'' + k^2 u = f + (k^2 - q)u.$$

We already know that $u \geq 0$. Choosing the constant k sufficiently large we have $f + (k^2 - q)u \geq f$, and we are reduced to the previous case.

In the general case, consider the new variable

$$y = \frac{1}{L} \int_0^x \frac{1}{p(t)} dt, \quad \text{where } L = \int_0^1 \frac{1}{p(t)} dt.$$

Set $v(y) = u(x)$. Then

$$u_x(x) = v_y(y) \frac{1}{Lp(x)}$$

and

$$(p(x)u_x)_x = v_{yy}(y) \frac{1}{L^2 p(x)}.$$

Therefore the problem

$$\begin{cases} -(pu')' + qu = f & \text{on } (0, 1), \\ u(0) = u(1) = 0 \end{cases}$$

becomes

$$\begin{cases} -v_{yy}(y) + L^2 p(x)q(x)v(y) = L^2 p(x)f(x) & \text{on } (0, 1), \\ v(0) = v(1) = 0, \end{cases}$$

and we are reduced to the previous case, noting that $x(1-x) \sim y(1-y)$.

Problem 46

12. Let (u_n) be a minimizing sequence i.e., $F(u_n) \to m$. We have

$$F(u_n) = \frac{1}{2} \int_0^1 (u_n'^2 + u_n^2) - \int_0^1 g(u_n) \leq C.$$

On the other hand we may use Young's inequality (see (2) in Chapter 4, and the corresponding footnote) with $a = (t^+)^{\alpha+1}$ and $p = 2/(\alpha+1)$, so that $p > 1$ since $\alpha < 1$. We obtain

$$g(u_n) \leq \varepsilon u_n^2 + C_\varepsilon \quad \forall \varepsilon > 0.$$

Choosing, e.g., $\varepsilon = 1/4$ we see that (u_n) is bounded in $H_0^1(I)$. Therefore we may extract a subsequence (u_{n_k}) converging weakly in $H_0^1(I)$, and strongly in $C(\bar{I})$ (by Theorem 8.8), to some limit $u \in H_0^1(I)$. Therefore

$$\liminf_{k \to \infty} \int_0^1 (u_{n_k}'^2 + u_{n_k}^2) \geq \int_0^1 (u'^2 + u^2)$$

and

$$\lim_{k \to \infty} \int_0^1 g(u_{n_k}) = \int_0^1 g(u).$$

Consequently $F(u) \leq m$, and thus $F(u) = m$.

Problem 47

- A -

2. Choose a sequence (u_n) proposed in the hint. We have

$$\bar{u}_n = \int_{1-\frac{1}{n}}^1 u_n(x)dx$$

and thus $|\bar{u}_n| \leq 1/n$. On the other hand

$$\|u_n - \bar{u}_n\|_{L^\infty(I)} \geq u_n(1) - \bar{u}_n \geq 1 - \frac{1}{n},$$

and

$$\|u_n'\|_{L^1(I)} = \int_0^1 u_n'(x)dx = u_n(1) - u_n(0) = 1.$$

3. Suppose, by contradiction, that the sup is achieved by some function $u \in W^{1,1}(I)$, i.e.,

$$\|u - \bar{u}\|_{L^\infty(I)} = 1 \text{ and } \|u'\|_{L^1(I)} = 1.$$

We may assume, e.g., that there exists some $x_0 \in [0, 1]$ such that

(S1) $$u(x_0) - \bar{u} = +1.$$

On the other hand,

(S2) $\qquad \bar{u} = \int_0^1 \left(u(x) - \min_{[0,1]} u \right) dx + \min_{[0,1]} u \geq \min_{[0,1]} u = u(y_0),$

for some $y_0 \in [0, 1]$. Combining (S1) and (S2) we obtain

$$u(x_0) - u(y_0) \geq 1.$$

But

$$u(x_0) - u(y_0) \leq \int_0^1 |u'(x)| dx = 1.$$

Therefore all the inequalities become equalities, and in particular $u \equiv \min_{[0,1]} u$. This contradicts (S1).

6. Set

$$m = \inf\{\|u'\|_{L^p(I)}; u \in W^{1,p}(I) \text{ and } \|u - \bar{u}\|_{L^q(I)} = 1\},$$

and let (u_n) be a minimizing sequence, i.e., $\|u_n'\|_{L^p(I)} \to m$ and $\|u_n - \bar{u}_n\|_{L^q(I)} = 1$. Without loss of generality we may assume that $\bar{u}_n = 0$. Therefore (u_n) is bounded in $W^{1,p}(I)$. We may extract a subsequence (u_{n_k}) converging weakly in $W^{1,p}(I)$ when $p < \infty$ (and (u_{n_k}') converges weak* in $L^\infty(I)$ when $p = \infty$) to some limit $u \in W^{1,p}(I)$. By Theorem 8.8 we may also assume that $u_{n_k} \to u$ in $C(\bar{I})$ (since $p > 1$). Clearly we have

$$\|u'\|_{L^p(I)} \leq m, \quad \bar{u} = 0, \text{ and } \|u\|_{L^q(I)} = 1.$$

- B -

1. Apply Lax–Milgram in V equipped with the H^1-norm, to the bilinear form $a(u, v) = \int_I u'v'$. Note that a is coercive (e.g., by question A6).
2. Let $w \in C_c^1(I)$. Choosing $v = (w - \bar{w})$ we obtain

$$\int_I u'w' = \int_I f(w - \bar{w}) = \int_I fw \quad \forall w \in C_c^1(I).$$

We deduce that $u \in H^2(I)$ and $-u'' = f$. Similarly we have

$$\int_I u'w' = \int_I fw \quad \forall w \in H^1(I)$$

and thus $u'(0) = u'(1) = 0$ (since $w(0)$ and $w(1)$ are arbitrary).
4. We have $\sigma(T/\lambda_1) \subset [0, 1]$. Applying Exercise 6.24 ((v) $\Rightarrow$ (vi)) we know that

$$\lambda_1(Tf, f) \geq |Tf|^2 \quad \forall f \in H$$

and we deduce that

$$\lambda_1 \int_0^1 u'^2 \geq \int_0^1 u^2 \quad \forall u \in W,$$

where

$$W = \left\{ u \in H^2(0,1); \, u'(0) = u'(1) = 0 \text{ and } \int_0^1 u = 0 \right\}.$$

On the other hand, given any $u \in V$, there exists a sequence $u_n \in W$ such that $u_n \to u$ in H^1. (Indeed let $\varphi_n \in C_c^1(I)$ be a sequence such that $\varphi_n \to u'$ in $L^2(I)$ and set $u_n(x) = \int_0^x \varphi_n(t)dt + c_n$, where the constant c_n is adjusted so that $\int_0^1 u_n = 0$.) Therefore we obtain

$$\|u\|_{L^2(I)} \le \sqrt{\lambda_1} \|u'\|_{L^2(I)} \quad \forall u \in V.$$

Choosing an eigenfunction e_1 of T corresponding to λ_1, and letting $u_1 = Te_1$ we obtain

$$\|u_1\|_{L^2(I)} = \sqrt{\lambda_1} \|u_1'\|_{L^2(I)}.$$

The eigenvalues of T are given by $\lambda_k = \frac{1}{k^2\pi^2}, k = 1, 2, \ldots$. Therefore the best constant in (6) is $1/\pi$.

- C -

1. Write, for $u \in W^{1,1}(I)$,

$$\int_0^1 |u(x) - \bar{u}|dx = \int_0^1 \left| u(x) - \int_0^1 u(y)dy \right| dx \le \int_0^1 \int_0^1 |u(x) - u(y)|dxdy$$

$$\le \int_0^1 dx \int_0^x dy \int_y^x |u'(t)|dt + \int_0^1 dx \int_x^1 dy \int_x^y |u'(t)|dt$$

$$= 2 \int_0^1 |u'(t)|t(1-t)dt$$

by Fubini.

3. Choose a function $u \in W^{1,1}(I)$ such that $u(x) = -\frac{1}{2} \, \forall x \in \left(0, \frac{1}{2} - \varepsilon\right), u(x) = +\frac{1}{2} \, \forall x \in \left(\frac{1}{2} + \varepsilon, 1\right), \bar{u} = 0$ and $u' \ge 0$, where $\varepsilon \in \left(0, \frac{1}{2}\right)$ is arbitrary. Then $\|u'\|_{L^1} = 1$ and $\|u\|_{L^1} \ge \frac{1}{2} - \varepsilon$.

4. There is no function $u \in W^{1,1}(I)$ such that $\|u - \bar{u}\|_{L^1(I)} = \frac{1}{2}$ and $\|u'\|_{L^1(I)} = 1$. Suppose, by contradiction, that such a function exists. Then

$$\frac{1}{2} = \|u - \bar{u}\|_{L^1(I)} \le 2 \int_I |u'(t)|t(1-t)dt \le \frac{1}{2} \int_I |u'(t)|dt = \frac{1}{2},$$

since $2t(1-t) \le \frac{1}{2} \, \forall t \in (0,1)$. All the inequalities become equalities and therefore $\left(\frac{1}{4} - t(1-t)\right)|u'(t)| = 0$ a.e. Hence $u' = 0$ a.e. Impossible.

Problem 49

7. We have

$$\lambda_1 = a(w_0, w_0) \le \frac{a(w_0 + tv, w_0 + tv)}{\|w_0 + tv\|_{L^2}^2} \quad \forall v \in H_0^1(0, 1), \quad \forall t \text{ sufficiently small.}$$

Therefore we obtain

$$\lambda_1 \left(1 + 2t \int_0^1 w_0 v + t^2 \int_0^1 v^2 \right) \le \lambda_1 + 2ta(w_0, v) + t^2 a(v, v),$$

and consequently

$$\lambda_1 \int_0^1 w_0 v = a(w_0, v) \quad \forall v \in H_0^1(0, 1),$$

i.e., $A w_0 = \lambda_1 w_0$ on $(0, 1)$.

8. We know from Exercise 8.11 that $w_1 = |w_0| \in H_0^1(0, 1)$ and $|w_1'| = |w_0'|$ a.e. Therefore $a(w_1, w_1) = a(w_0, w_0)$, and thus w_1 is also a minimizer for (1). We may then apply question 7.

10. Here is another proof which does not rely on the fact that all eigenvalues are simple. (This proof can be adapted to elliptic PDE's in dimension > 1.) It is easy to see (using question 9) that w^2/w_1 belongs to $H_0^1(0, 1)$. Therefore we have

$$\int_0^1 (Aw_1) \frac{w^2}{w_1} = \lambda_1 \int_0^1 w^2 = \int_0^1 (Aw)w.$$

Integrating by parts we obtain

$$\int_0^1 pw_1' \left(\frac{2ww'}{w_1} - \frac{w^2}{w_1^2} w_1' \right) + qw^2 = \int_0^1 pw'^2 + qw^2,$$

and therefore

$$\int_0^1 p \left(w' - \frac{w_1' w}{w_1} \right)^2 = 0.$$

Consequently $(\frac{w}{w_1})' = \frac{1}{w_1}(w' - \frac{w_1' w}{w_1}) = 0$, and therefore w is a multiple of w_1.

Problem 50

2. Note that

$$\int_0^1 |q|u^2 \le \left(\int_0^1 q^2 \right)^{1/2} \left(\int_0^1 u^4 \right)^{1/2} \le \varepsilon \int_0^1 u^4 + C_\varepsilon \int_0^1 q^2.$$

Choosing $\varepsilon = 1/8$ we deduce that, $\forall u \in H_0^1(0, 1)$,

$$\frac{1}{2} a(u, u) + \frac{1}{4} \int_0^1 u^4 \ge \frac{1}{2} \int_0^1 u'^2 + \frac{1}{8} \int_0^1 u^4 - C.$$

Partial Solutions of the Problems

3. Let (u_n) be a minimizing sequence, i.e., $\frac{1}{2}a(u_n, u_n) + \frac{1}{4}\int_0^1 u_n^4 \to m$. Clearly (u_n) is bounded in $H_0^1(0, 1)$. Passing to a subsequence, still denoted by u_n, we may assume that $u_n \rightharpoonup u_0$ weakly in $H_0^1(0, 1)$ and $u_n \to u_0$ in $C([0, 1])$. Therefore $\liminf_{n\to\infty} \int_0^1 u_n'^2 \geq \int_0^1 u_0'^2$, $\int_0^1 qu_n^2 \to \int_0^1 qu_0^2$ and $\int_0^1 u_n^4 \to \int_0^1 u_0^4$. Consequently $a(u_0, u_0) + \frac{1}{4}\int |u_0|^4 \leq m$, and thus u_0 is a minimizer.

4. We have

$$\frac{1}{2}a(u_0, u_0) + \frac{1}{4}\int_0^1 u_0^4 \leq \frac{1}{2}a(u_0 + tv, u_0 + tv) + \frac{1}{4}\int_0^1 (u_0 + tv)^4$$

$$= \frac{1}{2}a(u_0, u_0) + ta(u_0, v) + \frac{1}{4}\int_0^1 (u_0^4 + 4u_0^3 tv) + O(t^2).$$

Taking $t > 0$ we obtain

$$a(u_0, v) + \int_0^1 u_0^3 v \geq O(t).$$

Letting $t \to 0$ and choosing $\pm v$ we are led to

$$a(u_0, v) + \int_0^1 u_0^3 v = 0 \quad \forall v \in H_0^1(0, 1).$$

6. Recall that $u_1 \not\equiv 0$ since $\frac{1}{2}a(u_1, u_1) + \frac{1}{4}\int_0^1 u_1^4 = m < 0$. On the other hand we have

$$-u_1'' + a^2 u_1 = (a^2 - q - u_1^2)u_1 = f \geq 0$$

and $f \not\equiv 0$ (provided $a^2 - q - u_1^2 > 0$). We deduce from the strong maximum principle (see Problem 45) that $u_1 > 0$ on $(0, 1)$, $u_1'(0) > 0$, and $u_1'(1) < 0$.

8. Let $u \in C_c^1((0, 1))$; we have, using integration by parts,

$$-\int_0^1 U_1'' \frac{u^2}{U_1} = \int_0^1 U_1' \left(\frac{2uu'}{U_1} - \frac{u^2 U_1'}{U_1^2} \right) \leq \int_0^1 u'^2,$$

and therefore

(S1) $\quad \int_0^1 u'^2 - U_1'^2 \geq -\int_0^1 \frac{U_1''}{U_1}(u^2 - U_1^2) = -\int_0^1 (q + U_1^2)(u^2 - U_1^2).$

By density, inequality (S1) holds for every $u \in H_0^1(0, 1)$. Assume $\rho \in K$ and set $u = \sqrt{\rho}$. Then $u \in H_0^1(0, 1)$, and we have

$$\Phi(\rho) - \Phi(\rho_1) = \int_0^1 u'^2 + qu^2 + \frac{1}{2}u^4 - U_1'^2 - qU_1^2 - \frac{1}{2}U_1^4.$$

Using (S1) we see that

$$\Phi(\rho) - \Phi(\rho_1) \geq \int_0^1 -\left(q + U_1^2\right)\left(u^2 - U_1^2\right) + qu^2 + \frac{1}{2}u^4 - qU_1^2 - \frac{1}{2}U_1^4$$

$$= \int_0^1 \frac{1}{2}u^4 + \frac{1}{2}U_1^4 - U_1^2 u^2 = \frac{1}{2}\int_0^1 \left(u^2 - U_1^2\right)^2.$$

Problem 51

1. The mapping $v \mapsto Tv = (v', v, \sqrt{p}v)$ is an isometry from V into $L^2(\mathbb{R})^3$. It is easy to check that $T(V)$ is a closed subspace of $L^2(\mathbb{R})^3$, and therefore V is complete. V is separable since $L^2(\mathbb{R})^3$ is separable.

3. Let $u \in C_c^\infty(\mathbb{R})$. We have $\forall x \in [-A, +A]$,

(S1) $\qquad |u(x) - u(-A)| \leq \int_{-A}^{+A} |u'(t)|dt \leq \sqrt{2A}\|u'\|_{L^2(\mathbb{R})}.$

On the other hand $u^2(-A) = 2\int_{-\infty}^{-A} uu'$ and therefore
(S2)
$$|u(-A)|^2 \leq \int_{-\infty}^{-A} |u|^2 + \int_{-\infty}^{+\infty} |u'|^2 \leq \frac{1}{\delta}\int_{-\infty}^{+\infty} p|u|^2 + \int_{-\infty}^{+\infty} |u'|^2 \leq Ca(u, u).$$

Combining (S1) and (S2) we obtain

(S3) $\qquad |u(x)| \leq Ca(u, u)^{1/2} \quad \forall x \in [-A, +A],$

and consequently

$$\int_{-A}^{+A} |u|^2 \leq Ca(u, u).$$

Next, write that

$$\int_{-\infty}^{+\infty} |u|^2 \leq \int_{|x| \leq A} |u|^2 + \int_{|x| \geq A} |u|^2 dx \leq Ca(u, u).$$

Since

$$\int_{-\infty}^{+\infty} |u'|^2 \leq a(u, u) \quad \text{and} \quad \int_{-\infty}^{+\infty} p|u|^2 \leq a(u, u),$$

we conclude that $a(u, u) \geq \alpha\|u\|_V^2 \ \forall u \in C_c^\infty(\mathbb{R})$, for some $\alpha > 0$.

4. It is clear that $u \in H^2(I)$ for every bounded open interval I and u satisfies $-u'' + pu = f$ a.e. on I. Since p, u and f are continuous on I we deduce that $u \in C^2(I)$. On the other hand, $u(x) \to 0$ as $|x| \to \infty$ by Corollary 8.9 (recall that $V \subset H^1(\mathbb{R})$).

5. We have

(S4) $\qquad \int_{\mathbb{R}} u'(2\zeta_n'\zeta_n u + \zeta_n^2 u') + \int_{\mathbb{R}} p\zeta_n^2 u^2 = \int_{\mathbb{R}} f\zeta_n^2 u.$

But

$$a(\zeta_n u, \zeta_n u) = \int_{\mathbb{R}} (\zeta_n u' + \zeta'_n u)^2 + \int_{\mathbb{R}} p \zeta_n^2 u^2 = \int_{\mathbb{R}} f \zeta_n^2 u + \int_{\mathbb{R}} \zeta'^2_n u^2 \quad \text{by (S4)}.$$

Thus (since $|\zeta_n| \le 1$)

$$a(\zeta_n u, \zeta_n u) \le \|f\|_{L^2(\mathbb{R})} \|\zeta_n u\|_{L^2(\mathbb{R})} + \frac{C}{n^2} \int_{n \le |x| \le 2n} u^2.$$

Since $u(x) \to 0$ as $|x| \to \infty$ we see that $\frac{1}{n^2} \int_{n \le |x| \le 2n} u^2 \to 0$ as $n \to \infty$. Using the fact that a is coercive on V we conclude that $\|\zeta_n u\|_V \le C$. It follows easily that $u \in V$. Returning to (S3) we obtain

$$a(u, v) = \int_{-\infty}^{+\infty} fv \quad \forall v \in C_c^\infty(\mathbb{R}),$$

and by density the same relation holds $\forall v \in V$.

6. Let $\mathcal{F} = \{u \in V; \|u\|_V \le 1\}$. We need to show that $\mathcal{F}$ has compact closure in $L^2(\mathbb{R})$. For this purpose we apply Corollary 4.27. Recall (see Proposition 8.5) that

$$\|\tau_h u - u\|_{L^2(\mathbb{R})} \le |h| \|u'\|_{L^2(\mathbb{R})}$$

and therefore

$$\lim_{|h| \to 0} \|\tau_h u - u\|_{L^2(\mathbb{R})} = 0 \text{ uniformly in } u \in \mathcal{F}.$$

On the other hand, given any $\varepsilon > 0$ we may fix a bounded interval I such that $|p(x)| > \frac{1}{\varepsilon^2} \; \forall x \in \mathbb{R} \setminus I$. Therefore

$$\int_{\mathbb{R} \setminus I} |u|^2 \le \varepsilon^2 \int_{\mathbb{R}} p|u|^2 \le \varepsilon^2 \|u\|_V^2 \le \varepsilon^2 \quad \forall u \in \mathcal{F}.$$

Notation

General notations

A^c	complement of the set A
$E^\star$	dual space
$\langle\,,\,\rangle$	scalar product in the duality $E^\star$, E
$[f = \alpha] = \{x;\, f(x) = \alpha\}$	
$B(x_0, r)$	open ball of radius r centered at x_0
$B_E = \{x \in E;\, \|x\| \leq 1\}$	
epi $\varphi = \{[x, \lambda];\, \varphi(x) \leq \lambda\}$	
$\varphi^\star$	conjugate function
$\mathcal{L}(E, F)$	space of bounded linear operators from E into F
$M^\perp$	orthogonal of M
$D(A)$	domain of the operator A
$G(A)$	graph of the operator A
$N(A)$	kernel (= null space) of the operator A
$R(A)$	range of the operator A
$\sigma(E, E^\star)$	weak topology on E
$\sigma(E^\star, E)$	weak* topology on $E^\star$
$\rightharpoonup$	weak convergence
J	canonical injection from E into $E^{\star\star}$
p'	conjugate exponent of p, i.e., $\frac{1}{p} + \frac{1}{p'} = 1$
a.e.	almost everywhere
$|A|$	measure of the set A
supp f	support of the function f
$f \star g$	convolution product of f with g
ρ_n	sequence of mollifiers
$(\tau_h f)(x) = f(x + h)$	shift of the function f
$\omega \subset\subset \Omega$	ω strongly included in Ω, i.e., $\overline{\omega}$ is compact and $\overline{\omega} \subset \Omega$
P_K	projection onto the closed convex set K

| | | Hilbert norm

$\rho(T)$ resolvent set of the operator T

$\sigma(T)$ spectrum of the operator T

$EV(T)$ the set of eigenvalues of the operator T

$J_\lambda = (I + \lambda A)^{-1}$ resolvent of the operator A

$A_\lambda = A J_\lambda$ Yosida approximation of the operator A

$\nabla u = \left(\frac{\partial u}{\partial x_1}, \frac{\partial u}{\partial x_2}, \ldots, \frac{\partial u}{\partial x_N} \right)$ gradient of the function u

$D^\alpha u = \dfrac{\partial^{|\alpha|} u}{\partial x_1^{\alpha_1} \partial x_2^{\alpha_2} \partial x_N^{\alpha_N}}, \; \alpha = (\alpha_1, \alpha_2, \ldots, \alpha_N), |\alpha| = \sum_{i=1}^{N} \alpha_i$

$\Delta u = \displaystyle\sum_{i=1}^{N} \frac{\partial^2 u}{\partial x_i^2}$ Laplacian of u

$\mathbb{R}_+^N = \{x = (x', x_N) \in \mathbb{R}^{N-1} \times \mathbb{R}; \, x_N > 0\}$

$Q = \{x = (x', x_N) \in \mathbb{R}^N \times \mathbb{R}; \, |x'| < 1 \text{ and } |x_N| < 1\}$

$Q_+ = Q \cap \mathbb{R}_+^N$

$Q_0 = \{x \in Q; \, x_N = 0\}$

$(D_h u)(x) = \dfrac{1}{|h|} (u(x + h) - u(x))$

$\dfrac{\partial u}{\partial n}$ outward normal derivative

Function spaces

$\Omega \subset \mathbb{R}^N$ open set in $\mathbb{R}^N$

$\partial \Omega = \Gamma$ boundary of Ω

$L^p(\Omega) = \{u : \Omega \to \mathbb{R}: u \text{ is measurable and } \int_\Omega |u|^p < \infty\}, 1 \le p < \infty$

$L^\infty(\Omega) = \{u : \Omega \to \mathbb{R}: u \text{ is measurable and } |u(x)| \le C \text{ a.e. in } \Omega \text{ for some}$
constant $C\}$

$C_c(\Omega)$ space of continuous functions with compact
support in Ω

$C^k(\Omega)$ space of k times continuously differentiable
functions on $\Omega, k \ge 0$

$C^\infty(\Omega) = \underset{k \ge 0}{\cap} C^k(\Omega)$

$C^k(\overline{\Omega})$ functions in $C^k(\Omega)$ such that
for every multi-index α with $|\alpha| \le k$,
the function $x \mapsto D^\alpha u(x)$ admits a continuous
extension to $\overline{\Omega}$

$C^\infty(\overline{\Omega}) = \underset{k \ge 0}{\cap} C^k(\overline{\Omega})$

$C^{0,\alpha}(\overline{\Omega}) = \left\{ u \in C(\overline{\Omega}); \; \underset{\substack{x,y \in \Omega \\ x \ne y}}{\sup} \dfrac{|u(x) - u(y)|}{|x - y|^\alpha} < \infty \right\}$ with $0 < \alpha < 1$

$C^{k,\alpha}(\overline{\Omega}) = \{u \in C^k(\Omega); \, D^j u \in C^{0,\alpha}(\overline{\Omega}) \quad \forall j, |j| \le k\}$

$W^{1,p}(\Omega), W_0^{1,p}(\Omega), W^{m,p}(\Omega), H^1(\Omega), H_0^1(\Omega), H^m(\Omega)$ Sobolev spaces

References

Adams, R. A., [1] *Sobolev spaces*, Academic Press, 1975.

Agmon, S., [1] *Lectures on Elliptic Boundary Value Problems*, Van Nostrand, 1965, [2] *On positive solutions of elliptic equations with periodic coefficients in $\mathbb{R}^n$, spectral results and extensions operators on Riemannian manifolds* in Differential Equations (Knowles, I. W. and Lewis, R. T., eds.), North-Holland, 1984, pp. 7–17.

Agmon, S., Douglis, A. and Nirenberg, L., [1] *Estimates near the boundary for solutions of elliptic partial differential equations satisfying general boundary value conditions* I, Comm. Pure Appl. Math. **12** (1959), pp. 623–727.

Akhiezer, N. and Glazman, I., [1] *Theory of Linear Operators in Hilbert Space*, Pitman, 1980.

Albiac, F. and Kalton, N., [1] *Topics in Banach Space Theory*, Springer, 2006.

Alexits, G., [1] *Convergence Problems of Orthogonal Series*, Pergamon Press, 1961.

Ambrosetti, A. and Prodi, G., [1] *A Primer of Nonlinear Analysis*, Cambridge University Press, 1993.

Ambrosio, L., Fusco, N., and Pallara, D., [1] *Functions of Bounded Variation and Free Discontinuity Problems*, Oxford University Press, 2000.

Apostol, T. M., [1] *Mathematical Analysis* (2nd ed.), Addison–Wesley, 1974.

Ash, J. M. (ed.), [1] *Studies in Harmonic Analysis*, Mathematical Association of America, Washington D.C., 1976.

Aubin, J. P., [1] *Mathematical Methods of Game and Economic Theory*, North-Holland, 1979, [2] *Applied Functional Analysis*, Wiley, 1999, [3] *Optima and Equilibria*, Springer, 1993.

Aubin, J. P. and Ekeland, I., [1] *Applied Nonlinear Analysis*, Wiley, 1984.
Aubin, Th., [1] *Problèmes isopérimétriques et espaces de Sobolev*, C. R. Acad. Sci. Paris **280** (1975), pp. 279–281, and J. Diff. Geom. **11** (1976), pp. 573–598, [2] *Nonlinear Analysis on Manifolds, Monge–Ampère equations*, Springer, 1982, [3] *Some Nonlinear Problems in Riemannian Geometry*, Springer, 1998.

Auchmuty, G., [1] *Duality for non-convex variational principles*, J. Diff. Eq **50** (1983), pp. 80–145.

Azencott, R., Guivarc'h, Y., and Gundy, R. F. (Hennequin, P. L., ed.), [1] *Ecole d'été de probabilités de Saint-Flour* VIII-1978, Springer, 1980.

Bachman, G., Narici, L., and Beckenstein, E., [1] *Fourier and Wavelet Analysis*, Springer, 2000.

Baiocchi, C. and Capelo, A., [1] *Variational and Quasivariational Inequalities. Applications to Free Boundary Problems*, Wiley, 1984.

Balakrishnan, A., [1] *Applied Functional Analysis*, Springer, 1976.

Baouendi, M. S. and Goulaouic, C., [1] *Régularité et théorie spectrale pour une classe d'opérateurs elliptiques dégénérés*, Archive Rat. Mech. Anal. **34** (1969), pp. 361–379.

Barbu, V., [1] *Nonlinear Semigroups and Differential Equations in Banach Spaces*, Noordhoff, 1976, [2] *Optimal Control of Variational Inequalities*, Pitman, 1984.

Barbu, V. and Precúpanu, I., [1] *Convexity and Optimization in Banach Spaces*, Noordhoff, 1978.

Beauzamy, B., [1] *Introduction to Banach Spaces and Their Geometry*, North-Holland, 1983.

Benedetto, J. J. and Frazier, M. W. (eds.), [1] *Wavelets: Mathematics and Applications*, CRC Press, 1994.

Bensoussan, A. and Lions, J. L., [1] *Applications of Variational Inequalities in Stochastic Control*, North-Holland, Elsevier, 1982.

Benyamini, Y. and Lindenstrauss, J., [1] *Geometric Functional Analysis*, American Mathematical Society, 2000.

Bérard, P., [1] *Spectral Geometry: Direct and Inverse Problems*, Lecture Notes in Math., 1207, Springer, 1986.

Berger, Marcel, [1] *Geometry of the spectrum*, in Differential Geometry (Chern, S. S. and Osserman, R., eds.), Proc. Sympos. Pure Math., Vol. 27, Part 2;, American Mathematical Society, 1975, pp. 129–152.

Berger, Melvyn, [1] *Nonlinearity and Functional Analysis*, Academic Press, 1977.

Bergh, J. and Löfstróm J., [1] *Interpolation Spaces: An Introduction*, Springer, 1976.

Bers, L., John, F. and Schechter, M., [1] *Partial Differential Equations*, American Mathematical Society, 1979.

Bombieri, E., [1] *Variational problems and elliptic equations*, in Mathematical Developments Arising from Hilbert Problems (Browder, F., ed.), Proc. Sympos. Pure Math., Vol. 28, Part 2;, American Mathematical Society, 1977, pp. 525–536.

Bonsall, F.F., [1] *Linear operators in complete positive cones*, Proc. London Math. Soc. **8** (1958), pp. 53–75.

Bourbaki, N., [1] *Topological Vector Spaces*, Springer, 1987.

Bourgain, J. and Brezis, H., [1] *On the equation* div$Y = f$ *and application to control of phases*, J. Amer. Math. Soc. **16** (2003), pp. 393–426, [2] *New estimates for elliptic equations and Hodge type systems*, J. European Math. Soc. **9** (2007), pp. 277–315.

Brandt, A., [1] *Interior estimates for second order elliptic differential (or finite-difference) equations via the maximum principle*, Israel J. Math. **76** (1969), pp. 95–121, [2] *Interior Schauder estimates for parabolic differential (or difference) equations*, Israel J. Math. **7**, (1969), pp. 254–262.

Bressan, A., [1] *BV solutions to systems of conservation laws*, in Hyperbolic Systems of Balance Laws, Lectures Given at the C.I.M.E. Summer School Held in Cetraro, Italy, July 14-21, 2003 (Marcati, P., ed.), Lecture Notes in Math., 1911, Springer, 2007, pp. 1–78.

Brezis, H., [1] *Opérateurs maximaux monotones et semi-groupes de*, contractions dans les espaces de Hilbert North-Holland, 1973, [2] *Periodic solutions of nonlinear vibrating strings and duality principles*, Bull. Amer. Math. Soc. **8** (1983), 409–426.

Brezis, H. and Browder, F., [1] *Partial differential equations in the 20th century*, Advances in Math. **135** (1998), pp. 76–144.

Brezis, H., Coron, J. M. and Nirenberg, L., [1] *Free vibrations for a nonlinear wave equation and a theorem of P. Rabinowitz*, Comm. Pure Appl. Math. **33** (1980), pp. 667–689.

Brezis, H. and Lieb E. H., [1] *A relation between pointwise convergence of functions and convergence of functionals*, Proc. Amer. Math. Soc. **88** (1983), pp. 486–490.

Browder, F., [1] *Nonlinear operators and nonlinear equations of evolution*, Proc. Sympos. Pure Math., Vol. 18, Part 2, American Mathematical Society, 1976.

Bui, H. D., [1] *Duality and symmetry lost in solid mechanics*, C. R. Mécanique **336** (2008), pp. 12–23.

Buttazzo, G., Giaquinta, M. and Hildebrandt S., [1] *One-Dimensional Variational Problems. An Introduction*, Oxford University Press, 1998.

Cabré X. and Caffarelli L., [1] *Fully Nonlinear Elliptic Equations*, American Mathematical Society, 1995.

Caffarelli, L., Nirenberg, L. and Spruck, J., [1] *The Dirichlet problem for nonlinear second-order elliptic equations I: Monge-Ampère equations*, Comm. Pure Appl. Math. **37** (1984), pp. 339–402.

Caffarelli L. and Salsa S., [1] *A Geometric Approach to Free Boundary Problems*, American Mathematical Society, 2005.

Carleson, L., [1] *On convergence and growth of partial sums of Fourier series*, Acta Math. **116** (1966), pp. 135–157.

Cazenave, T. and Haraux, A., [1] *An Introduction to Semilinear Evolution Equations*, Oxford University Press, 1998.

Chae, S. B., [1] *Lebesgue Integration*, Dekker, 1980.

Chan, T. F. and Shen, J., [1] *Image Processing and Analysis - Variational, PDE, Wavelet, and Stochastic Methods*, SIAM Publisher, 2005.

Chavel, I., [1] *Eigenvalues in Riemannian Geometry*, Academic Press, 1994.

Chen, Ya-Zhe and Wu, Lan-Cheng, [1] *Second Order Elliptic Equations and Elliptic Systems*, American Mathematical Society, 1998.

Choquet, G, [1] *Topology*, Academic Press, 1966, [2] *Lectures on Analysis* (3 volumes), Benjamin, 1969.

Choquet-Bruhat, Y., Dewitt-Morette, C., and Dillard-Bleick, M., [1] *Analysis, Manifolds and Physics*, North-Holland, 1977.

Chui, C. K., [1] *An Introduction to Wavelets*, Academic Press, 1992.

Ciarlet, Ph., [1] *The Finite Element Method for Elliptic Problems* (2nd ed.), North-Holland, 1979.

Clarke, F., [1] *Optimization and Nonsmooth Analysis*, Wiley, 1983.

Clarke, F. and Ekeland, I., [1] *Hamiltonian trajectories having prescribed minimal period*, Comm. Pure Appl. Math. **33** (1980), pp. 103–116.

Coddington, E. and Levinson, N., [1] *Theory of Ordinary Differential Equations*, McGraw Hill, 1955.

Cohn, D. L., [1] *Measure Theory*, Birkhäuser, 1980.

Coifman, R. and Meyer, Y., [1] *Wavelets: Calderón–Zygmund and Multilinear Operators*, Cambridge University Press, 1997.

Conway, J. B., [1] *A Course in Functional Analysis*, Springer, 1990.

Courant, R. and Hilbert, D., [1] *Methods of Mathematical Physics* (2 volumes), Interscience, 1962.

Crank J., [1] *Free and Moving Boundary Problems*, Oxford University Press, 1987.

Croke C.B., Lasiecka I., Uhlmann G., and Vogelius M., eds, [1] *Geometric Methods in Inverse Problems and PDE Control*, Springer, 2004.

Damlamian, A., [1] *Application de la dualité non convexe à un problème non linéaire à frontière libre (équilibre d'un plasma confiné)*, C. R. Acad. Sci. Paris **286** (1978), pp. 153–155.

Daubechies, I., [1] *Ten Lectures on Wavelets*, CBMS-NSF, Reg. Conf. Ser. Appl. Math., Vol. 61, SIAM, 1992.

Dautray, R. and Lions, J.-L., [1] *Mathematical Analysis and Numerical Methods for Science and Technology* (6 volumes), Springer, 1988.

David, G., [1] *Wavelets and Singular Integrals on Curves and Surfaces*, Lecture Notes in Math., 1465, Springer, 1992.

Davies, E., [1] *One parameter semigroups*, Academic Press, 1980, [2] *Linear operators and their spectra*, Cambridge University Press, 2007, [3] *Heat kernels and Spectral Theory*, Cambridge University Press, 1989.

de Figueiredo, D. G. and Karlovitz, L., [1] *On the radial projection in normed spaces*, Bull. Amer. Math. Soc. **73** (1967), pp. 364–368.

Deimling, K., [1] *Nonlinear Functional Analysis*, Springer, 1985.

Dellacherie, C. and Meyer, P.-A., [1] *Probabilités et potentiel*, Hermann, Paris, 1983.

Deville, R., Godefroy, G., and Zizler, V., [1] *Smoothness and Renormings in Banach Spaces*, Longman, 1993.

DeVito, C., [1] *Functional Analysis*, Academic Press, 1978.

DiBenedetto, E., [1] *Partial Differential Equations* (2nd ed.), Birkhäuser, 2009.

Diestel, J., [1] *Geometry of Banach spaces: Selected Topics*, Springer, 1975, [2] *Sequences and Series in Banach Spaces*, Springer, 1984.

Dieudonné, J., [1] *Treatise on Analysis* (8 volumes), Academic Press, 1969, [2] *History of Functional Analysis*, North-Holland, 1981.

Dixmier, J., [1] *General Topology*, Springer, 1984.

Dugundji, J., [1] *Topology*, Allyn and Bacon, 1965.

Dunford, N. and Schwartz, J. T, [1] *Linear Operators* (3 volumes), Interscience (1972).

Duvaut, G. and Lions, J. L., [1] *Inequalities in Mechanics and Physics*, Springer, 1976.

Dym, H. and McKean, H. P., [1] *Fourier Series and Integrals*, Academic Press, 1972,.

Edwards, R., [1] *Functional Analysis*, Holt–Rinehart–Winston, 1965.

Ekeland, I., [1] *Convexity Methods in Hamiltonian mechanics*, Springer, 1990.

Ekeland, I. and Temam, R., [1] *Convex Analysis and Variational Problems*, North-Holland, 1976.

Ekeland, I. and Turnbull, T., [1] *Infinite-Dimensional Optimization and Convexity*, The University of Chicago Press, 1983.

Enflo, P., [1] *A counterexample to the approximation property in Banach spaces*, Acta Math. **130** (1973), pp. 309–317.

Evans, L. C., [1] *Partial Differential Equations*, American Mathematical Society, 1998.

Evans, L. C. and Gariepy, R. F., [1] *Measure Theory and Fine Properties of Functions*, CRC Press, 1992.

Fife, P., [1] *Mathematical aspects of reacting and diffusing systems*, Lecture Notes in Biomathematics, 28, Springer, 1979.

Folland, G.B., [1] *Introduction to Partial Differential Equations*, Princeton University Press, 1976, [2] *Real Analysis*, Wiley, 1984.

Fonseca, I., Leoni, G., [1] *Modern Methods in the Calculus of Variations: L^p Spaces*, Springer, 2007.

Franklin, J., [1] *Methods of Mathematical Economics: linear and Nonlinear Programming, Fixed Point Theorems*, SIAM, 2002.

Friedman, A., [1] *Partial Differential Equations of Parabolic Type*, Prentice Hall, 1964, [2] *Partial, differential Equations*, Holt–Rinehart–Winston, 1969, [3] *Foundations of Modern Analysis*, Holt–Rinehart–Winston, 1970, [4] *Variational Principles and Free Boundary Problems*, Wiley, 1982.

Garabedian, P., [1] *Partial Differential equations*, Wiley, 1964.

Germain, P., [1] *Duality and convection in continuum mechanics*, in Trends in Applications of Pure Mathematics to Mechanics (Fichera, G., ed.), Monographs and Studies in Math., Vol. 2, Pitman, 1976, pp. 107–128, [2] *Functional concepts in continuum mechanics*, Meccanica **33** (1998), pp. 433–444.

Giaquinta, M., [1] *Multiple Integrals in the Calculus of Variations and Nonlinear Elliptic Systems*, Princeton University Press, 1983.

Gilbarg, D. and Trudinger, N., [1] *Elliptic Partial Differential Equations of Second Order*, Springer, 1977.

Gilkey, P., [1] *The Index theorem and the Heat Equation*, Publish or Perish, 1974.

Giusti, E., [1] *Minimal Surfaces and Functions of Bounded Variation*, Birkhäuser, 1984, [2] *Direct Methods in the Calculus of Variations*, World Scientific, 2003.

Goldstein, J., [1] *Semigroups of Operators and Applications*, Oxford University Press, 1985.

Gordon, C., Webb, L., and Wolpert, S., [1] *One cannot hear the shape of a drum*, Bull. Amer. Math. Soc. **27** (1992), pp. 134–138.

Grafakos, L., [1] *Classical and Modern Fourier Analysis* (2nd ed.), Springer, 2008.

Granas, A. and Dugundji, J., [1] *Fixed Point Theory*, Springer, 2003.

Grisvard, P., [1] *Équations différentielles abstraites*, Ann. Sci. ENS **2** (1969), pp. 311–395.

Gurtin, M., [1] *An Introduction to Continuum Mechanics*, Academic Press, 1981.

Halmos, P., [1] *Measure Theory*, Van Nostrand, 1950, [2] *A Hilbert Space Problem Book* (2nd ed.), Springer, 1982.

Han, Q. and Lin, F. H., [1] *Elliptic Partial Differential Equations*, American Mathematical Society, 1997.

Harmuth, H. F., [1] *Transmission of Information by Orthogonal Functions*, Springer, 1972.

Hartman, Ph., [1] *Ordinary Differential Equations*, Wiley, 1964.

Henry, D., [1] *Geometric Theory of Semilinear Parabolic Equations*, Springer, 1981.

Hernandez, E. and Weiss, G., [1] *A First Course on Wavelets*, CRC Press, 1996.

Hewitt, E. and Stromberg, K., [1] *Real and Abstract Analysis*, Springer, 1965.

Hille, E., [1] *Methods in Classical and Functional Analysis*, Addison–Wesley, 1972.

Hiriart-Urruty, J. B. and Lemaréchal, C., [1] *Convex Analysis and Minimization Algorithms* (2 volumes), Springer, 1993.

Holmes, R., [1] *Geometric Functional Analysis and Its Applications*, Springer, 1975.

Hörmander, L., [1] *Linear Partial Differential Operators*, Springer, 1963, [2] *The Analysis of Linear Partial Differenetial Operators*, (4 volumes), Springer, 1983–1985, [3] *Notions of Convexity*, Birkhäuser, 1994.

Horvath, J., [1] *Topological Vector Spaces and Distributions*, Addison–Wesley, 1966.

Ince, E. L., [1] *Ordinary Differential Equations*, Dover Publications, 1944.

James, R. C., [1] *A non-reflexive Banach space isometric with its second conjugate space*, Proc. Nat. Acad. Sci. USA **37** (1951), pp. 174–177.

Jost, J., [1] *Partial Differential Equations*, Springer, 2002.

Kac, M., [1] *Can one hear the shape of a drum?* Amer. Math. Monthly **73** (1966), pp. 1–23.

Kahane, J. P. and Lemarié-Rieusset, P. G., [1] *Fourier Series and Wavelets*, Gordon Breach, 1996.

Kaiser, G., [1] *A Friendly Guide to Wavelets*, Birkhäuser, 1994.

Kakutani, S., [1] *Some characterizations of Euclidean spaces*, Jap. J. Math. **16** (1939), pp. 93–97.

Karlin, S., [1] *Mathematical Methods and Theory in Games, Programming, and Economics* (2 volumes), Addison–Wesley, 1959.

Kato, T., [1] *Perturbation Theory for Linear Operators*, Springer, 1976.

Katznelson, Y., [1] *An Introduction to Harmonic Analysis*, Dover Publications, 1976.

Kelley, J. L., [1] *Banach spaces with the extension property*, Trans. Amer. Math. Soc. **72** (1952), pp. 323–326.

Kelley, J. and Namioka, I., [1] *Linear Topological Spaces* (2nd ed.), Springer, 1976.

Kinderlehrer, D. and Stampacchia, G., [1] *An Introduction to Variational Inequalities and Their Applications*, Academic Press, 1980.

Klainerman, S., [1] *Introduction to Analysis*, Lecture Notes, Princeton University, 2008.

Knapp, A., [1] *Basic Real Analysis*, Birkhäuser, 2005, [2] *Advanced Real Analysis*, Birkhäuser, 2005.

Knerr, B., [1] *Parabolic interior Schauder estimates by the maximum principle*, Archive Rat. Mech. Anal. **75** (1980), pp. 51–58.

Kohn, J.J. and Nirenberg, L., [1] *Degenerate elliptic parabolic equations of second order*, Comm. Pure Appl. Math. **20** (1967), pp. 797–872.

Kolmogorov, A. and Fomin, S., [1] *Introductory Real Analysis*, Prentice-Hall, 1970.

Köthe, G., [1] *Topological Vector Spaces* (2 volumes), Springer, 1969–1979.

Krasnoselskii, M., [1] *Topological Methods in the Theory of Nonlinear Integral Equations*, Mcmillan, 1964.

Kreyszig, E., [1] *Introductory Functional Analysis with Applications*, Wiley, 1978.

Krylov, N. V., [1] *Lectures on Elliptic and Parabolic Equations in Hölder Spaces*, American Mathematical Society, 1996, [2] *Lectures on Elliptic and Parabolic Equations in Sobolev Spaces*, American Mathematical Society, 2008.

Ladyzhenskaya, O. and Uraltseva, N., [1] *Linear and quasilinear elliptic equations*, Academic Press, 1968.

Ladyzhenskaya, O., Solonnikov, V., and Uraltseva, N., [1] *Linear and Quasilinear Equations of Parabolic Type*, American Mathematical Society, 1968.

Lang, S., [1] *Real and Functional Analysis*, Springer, 1993.

Larsen, R., [1] *Functional Analysis: An Introduction*, Dekker, 1973.

Lax, P., [1] *Functional Analysis*, Wiley, 2002.

Levitan, B. M., [1] *Inverse Sturm-Liouville problems*, VNU Science Press, 1987.

Lieb, E., [1] *Sharp constants in the Hardy-Littlewood-Sobolev and related inequalities*, Ann. Math. **118** (1983), pp. 349–374.

Lieb, E. H. and Loss, M., [1] *Analysis*, American Mathematical Society, 1997.

Lin, F. H. and Yang, X. P., [1] *Geometric Measure Theory – An Introduction*, International Press, 2002.

Lindenstrauss, J. and Tzafriri, L., [1] *On the complemented subspaces problem*, Israel J. Math. **9** (1971), pp. 263–269, [2] *Classical Banach Spaces*, (2 volumes), Springer, 1973–1979.

Lions, J. L., [1] *Problèmes aux limites dans les équations aux dérivées partielles*, Presses de l'Université de Montreal, 1965, [2] *Optimal Control of Systems Governed by Partial Differential Equations*, Springer, 1971, [3] *Quelques méthodes de résolution des problèmes aux limites non linéaires*, Dunod-Gauthier Villars, 1969.

Lions, J. L. and Magenes, E., [1] *Non-homogeneous Boundary Value Problems and Applications* (3 volumes), Springer, 1972.

Magenes, E., [1] *Topics in parabolic equations: some typical free boundary problems*, in Boundary Value Problems for Linear Evolution Partial Differential Equations (Garnir, H. G., ed.), Reidel, 1977.

Mallat, S., [1] *A Wavelet Tour of Signal Processing*, Academic Press, 1999.

Malliavin, P., [1] *Integration and Probabilities*, Springer, 1995.

Martin, R. H., [1] *Nonlinear Operators and Differential Equations in Banach Spaces*, Wiley, 1976.

Mawhin, J., R. Ortega, R., and Robles-Pérez, A. M., [1] *Maximum principles for bounded solutions of the telegraph equation in space dimensions two and three and applications*, J. Differential Equations **208** (2005), pp. 42–63.

Mawhin, J. and Willem, M., [1] *Critical Point Theory and Hamiltonian Systems*, Springer, 1989.

Maz'ja, V. G., [1] *Sobolev Spaces*, Springer, 1985.

Meyer, Y., [1] *Wavelets and Operators*, Cambridge University Press, 1992. [2] *Wavelets, Vibrations and Scaling*, American Mathematical Society, 1998, [3] *Oscillating Patterns in Image Processing and Nonlinear Evolution Equations*, American Mathematical Society, 2001.

Mikhlin, S., [1] *An Advanced Course of Mathematical Physics*, North-Holland, 1970.

Miranda, C., [1] *Partial Differential Equations of Elliptic Type*, Springer, 1970.

Mizohata, S., [1] *The Theory of Partial Differential Equations*, Cambridge University Press, 1973.

Morawetz, C., [1] L^p inequalities, Bull. Amer. Math. Soc. **75** (1969), pp. 1299–1302.

Moreau, J. J, [1] Fonctionnelles convexes, Séminaire Leray, Collège de France, 1966. [2] Applications of convex analysis to the treatment of elastoplastic systems, in Applications of methods of Functional Analysis to Problems in Mechanics, Sympos. IUTAM/IMU (Germain, P. and Nayroles, B., eds.), Springer, 1976.

Morrey, C., [1] Multiple Integrals in the Calculus of Variations, Springer, 1966.

Morton K.W. and Mayers D.F., [1] Numerical Solutions of Partial Differential Equations (2nd ed.), Cambridge University Press, 2005.

Munkres, J. R., [1] Topology, A First Course, Prentice Hall, 1975.

Nachbin, L., [1] A theorem of the Hahn-Banach type for linear transformations, Trans. Amer. Math. Soc. **68** (1950), pp. 28–46.

Nečas, J. and Hlaváček, L., [1] Mathematical Theory of Elastic and Elastoplastic Bodies. An Introduction, Elsevier, 1981.

Neveu, J., [1] , Mathematical Foundations of the Calculus of Probability, Holden-Day, 1965.

Nirenberg, L., [1] On elliptic partial differential equations, Ann. Sc. Norm. Sup. Pisa **13** (1959), pp. 116–162, [2] Topics in Nonlinear Functional Analysis, New York University Lecture Notes, 1974, reprinted by the American Mathematical Society, 2001, [3] Variational and topological methods in nonlinear problems, Bull. Amer. Math. Soc. **4**, (1981), pp. 267–302.

Nussbaum, R., [1] Eigenvectors of nonlinear positive operators and the linear Krein-Rutman theorem, in Fixed Point Theory (Fadell, E. and Fournier, G., eds.), Lecture Notes in Math., 886, Springer, 1981, pp. 309–330.

Oleinik, O. and Radkevitch, E., [1] Second Order Equations with Non-negative Characteristic Form, Plenum, 1973.

Osserman, R., [1] Isoperimetric inequalities and eigenvalues of the Laplacian, in Proc. Int. Congress of Math. Helsinski, 1978, and Bull. Amer. Math. Soc. **84** (1978), pp. 1182–1238.

Pazy, A., [1] Semigroups of Linear Operators and Applications to Partial Differential Equations, Springer, 1983.

Pearcy, C. M., [1] Some Recent Developments in Operator Theory, CBMS Reg. Conf. Series, American Mathematical Society, 1978.

Phelps, R., [1] Lectures on Choquet's Theorem, Van Nostrand, 1966, [2] Convex Functions, Monotone Operators and Differentiability, Lecture Notes in Math., 1364, Springer, 1989.

Pietsch, A., [1] History of Banach Spaces and Linear Operators, Birkhäuser, 2007.

Pinchover, Y. and Rubinstein, J., [1] An Introduction to Partial Differential Equations, Cambridge University Press, 2005.

Protter, M. and Weinberger, H., [1] Maximum Principles in Differential Equations, Prentice-Hall, 1967.

Pucci, P. and Serrin, J., [1] The Maximum Principle, Birkhäuser, 2007.

Rabinowitz, P., [1] Variational methods for nonlinear eigenvalue problems, in Eigenvalues of Nonlinear Problems CIME, Cremonese, 1974, [2] Théorie du degré topologique et applications à des problèmes aux limites non linéaires, Lecture Notes written by H. Berestycki, Laboratoire d'Analyse Numérique, Université Paris VI, 1975.

Reed, M. and Simon, B., [1] Methods of Modern Mathematical Physics (4 volumes), Academic Press, 1972–1979.

Rees, C., Shah, S., and Stanojevic, C., [1] *Theory and Applications of Fourier Series*, Dekker, 1981.

Rockafellar, R. T., [1] *Convex Analysis*, Princeton University Press, 1970, [2] *The Theory of Subgradients and Its Applications to Problems of Optimization: Convex and Nonconvex Functions*, Helderman Verlag, Berlin, 1981.

Royden, H. L., [1] *Real Analysis*, Macmillan, 1963.

Rudin, W., [1] *Functional Analysis*, McGraw Hill, 1973, [2] *Real and Complex Analysis*, McGraw Hill, 1987.

Ruskai, M. B. et al. (eds.), [1] *Wavelets and Their Applications*, Jones and Bartlett, 1992.

Sadosky, C., [1] *Interpolation of Operators and Singular Integrals*, Dekker, 1979.

Schaefer, H., [1] *Topological Vector Spaces*, Springer, 1971.

Schechter, M., [1] *Principles of Functional Analysis*, Academic Press, 1971, [2] *Operator Methods in Quantum Mechanics*, North-Holland, 1981.

Schwartz, J. T., [1] *Nonlinear Functional Analysis*, Gordon Breach, 1969.

Schwartz, L., [1] *Théorie des distributions*, Hermann, 1973, [2] *Geometry and Probability in Banach spaces*, Bull. Amer. Math. Soc., **4** (1981), 135–141, and Lecture Notes in Math., 852, Springer, 1981, [3] *Fonctions mesurables et ⋆-scalairement mesurables, propriété de Radon-Nikodym*, Exposés 4, 5 et 6, Séminaire Maurey-Schwartz, École Polytechnique (1974–1975).

Serrin, J., [1] *The solvability of boundary value problems*, in Mathematical Developments Arising from Hilbert Problems (Browder, F., ed.), Proc. Sympos. Pure Math., Vol. 28, Part 2, American Mathematical Society, 1977, pp. 507–525.

Simon, L., [1] *Lectures on Geometric Measure Theory*, Australian National University, Center for Mathematical Analysis, Canberra, 1983, [2] *Schauder estimates by scaling*, Calc. Var. Partial Differential Equations **5** (1997), pp. 391–407.

Singer, I., [1] *Bases in Banach Spaces* (2 volumes) Springer, 1970.

Singer, I. M., [1] *Eigenvalues of the Laplacian and invariants*, of manifolds, in Proc. Int. Congress of Math. Vancouver, 1974.

Sperb, R., [1] *Maximum Principles and Their Applications*, Academic Press, 1981.

Stampacchia, G., [1] *Equations elliptiques du second ordre à coefficients discontinus*, Presses de l'Université de Montreal, 1966.

Stein, E., [1] *Singular Integrals and Differentiability Properties of Functions*, Princeton University Press, 1970.

Stein, E. and Weiss, G., [1] *Introduction to Fourier Analysis on Euclidean Spaces*, Princeton University Press, 1971.

Stoer, J. and Witzgall, C., [1] *Convexity and Optimization in Finite Dimensions*, Springer, 1970.

Strauss, W., [1] *Partial Differential Equations: An Introduction*, Wiley, 1992.

Stroock, D., [1] *An Introduction to the Theory of Large Deviations*, Springer, 1984.

Stroock, D. and Varadhan, S., [1] *Multidimensional Diffusion Processes*, Springer, 1979.

Struwe, M., [1] *Variational Methods: Applications to Nonlinear Partial Differential Equations and Hamiltonian Systems*, Springer, 1990.

Szankowski, A., [1] $B(H)$ *does not have the approximation property*, Acta Math. **147** (1981), pp. 89–108.

Talenti, G., [1] *Best constants in Sobolev inequality*, Ann. Mat. Pura Appl. **110** (1976), pp. 353–372.

Tanabe, H., [1] *Equations of Evolution*, Pitman, 1979.

Taylor, A. and Lay, D., [1] *Introduction to Functional Analysis*, Wiley, 1980.

Taylor, M., [1] *Partial Differential Equations*, vols. I–III, Springer, 1996.

Temam, T, [1] *Navier–Stokes Equations*, North-Holland, 1979.

Temam, R. and Strang, G., [1] *Duality and relaxation in the variational problems of plasticity*, J. de Mécanique **19** (1980), pp. 493–528, [2] *Functions of bounded deformation*, Archive Rat. Mech. Anal. **75** (1980), pp. 7–21.

Toland, J. F., [1] *Duality in nonconvex optimization*, J. Math. Anal. Appl. **66** (1978), pp. 399–415, [2] *A duality principle for nonconvex optimization and the calculus of variations*, Arch. Rat. Mech. Anal. **71**, (1979), 41–61, [3] *Stability of heavy rotating chains*, J. Diff. Eq. **32** (1979), pp. 15–31, [4] *Self-adjoint operators and cones*, J. London Math. Soc. **53** (1996), pp. 167–183.

Treves, F., [1] *Topological Vector Spaces, Distributions and Kernels*, Academic Press, 1967, [2] *Linear Partial Differential Equations with Constant Coefficients*, Gordon Breach, 1967, [3] *Locally Convex Spaces and Linear Partial Differential Equations*, Springer, 1967, [4] *Basic Linear Partial Differential Equations*, Academic Press, 1975.

Triebel, H., [1] *Theory of function spaces*, (3 volumes), Birkhäuser, 1983–2006.

Uhlmann, G. (ed.), [1] *Inside Out: Inverse Problems and Applications*, MSRI Publications, Volume 47, Cambridge University Press, 2003.

Volpert, A. I., [1] *The spaces BV and quasilinear equations*, Mat. USSR-Sbornik **2** (1967), pp. 225–267.

Weinberger, H., [1] *A First Course in Partial Differential Equations*, Blaisdell, 1965, [2] *Variational Methods for Eigenvalue Approximation*, Reg. Conf. Appl. Math., SIAM, 1974.

Weidmann, J., [1] *Linear Operators and Hilbert Spaces*, Springer, 1980.

Weir, A. J., [1] *General Integration and Measure*, Cambridge University Press, 1974.

Wheeden, R. and Zygmund, A., [1] *Measure and Integral*, Dekker, 1977.

Willem, M., [1] *Minimax Theorems*, Birkhäuser, 1996.

Wojtaszczyk P., [1] *A Mathematical Introduction to Wavelets*, Cambridge University Press, 1997.

Yau, S. T., [1] *The role of partial differential equations in differential geometry*, in Proc. Int. Congress of Math. Helsinki, 1978.

Yosida, K, [1] *Functional Analysis*, Springer, 1965.

Zeidler, E., [1] *Nonlinear Functional Analysis and Its Applications*, Springer, 1988.

Zettl, A., [1] *Sturm–Liouville Theory*, American Mathematical Society, 2005.

Ziemer, W., [1] *Weakly Differentiable Functions*, Springer, 1989.

Index

Printed in the United States
by Baker & Taylor Publisher Services